2014 IEEE Workshop on Wide Bandgap Power Devices and Applications

(WiPDA 2014)

Knoxville, Tennessee, USA
13-15 October 2014

IEEE Catalog Number: CFP14WBP-POD
ISBN: 978-1-4799-5494-0

Copyright © 2014 by the Institute of Electrical and Electronic Engineers, Inc
All Rights Reserved

Copyright and Reprint Permissions: Abstracting is permitted with credit to the source. Libraries are permitted to photocopy beyond the limit of U.S. copyright law for private use of patrons those articles in this volume that carry a code at the bottom of the first page, provided the per-copy fee indicated in the code is paid through Copyright Clearance Center, 222 Rosewood Drive, Danvers, MA 01923.

For other copying, reprint or republication permission, write to IEEE Copyrights Manager, IEEE Service Center, 445 Hoes Lane, Piscataway, NJ 08854. All rights reserved.

***This publication is a representation of what appears in the IEEE Digital Libraries. Some format issues inherent in the e-media version may also appear in this print version.**

IEEE Catalog Number: CFP14WBP-POD
ISBN 13:   978-1-4799-5494-0

Additional Copies of This Publication Are Available From:

Curran Associates, Inc
57 Morehouse Lane
Red Hook, NY 12571 USA
Phone: (845) 758-0400
Fax: (845) 758-2633
E-mail: curran@proceedings.com
Web: www.proceedings.com

2014 IEEE Workshop on Wide Bandgap Power Devices and Applications (WiPDA 2014)

Knoxville, Tennessee, USA
13-15 October 2014

IEEE Catalog Number: CFP14WBP-POD
ISBN: 978-1-47995-494-0

Table of Contents

Technical Sessions A1L-A: WBG Characterization

Oct. 14, 2014 10:30 am – 11:45 am

Session chair: *John Hostetler*

4500 Volt Si/SiC Hybrid Module Qualification for Modern Megawatt Scale Wind Energy Inverters..*1*
William Erdman, *Cinch LLC*
David Grider, *CREE Inc.*
Edward VanBrunt, *CREE Inc.*

Understanding the Limitations and Impact Factors of Wide Bandgap Devices' High Switching-Speed Capability in a Voltage Source Converter......................................*7*
Zheyu Zhang, *University of Tennessee*
Fred Wang, *University of Tennessee*
Leon M. Tolbert, *University of Tennessee*
Benjamin J. Blalock, *University of Tennessee*
Daniel Costinett, *University of Tennessee*

The Development of a High-Voltage Power Device Evaluation Platform...................*13*
Lixing Fu, *Ohio State University*
Xuan Zhang, *Ohio State University*
He Li, *Ohio State University*
Xintong Lu, *Ohio State University*
Jin Wang, *Ohio State University*

Technical Sessions A1L-B: WBG Application Challenges

Oct. 14, 2014 10:30 am – 11:45 am

Session chairs: *Chingchi Chen and Jürgen Schuderer*

Packaging SiC Power Semiconductors - Challenges, Technologies and Strategies....*18*
Jürgen Schuderer, *ABB Corporate Research*
Umamaheswara Vemulapati, *ABB Corporate Research*
Felix Traub, *ABB Corporate Research*

The Opportunities and Challenges of Wide-Band-Gap Technologies for Automotive Applications (Presentation only)

Chingchi Chen, *Ford Motor Company*
Ming Su, *Ford Motor Company*

Application-Based Review of GaN HFETs..24

Edward Jones, *University of Tennessee*
Fred Wang, *University of Tennessee*
Burak Ozpineci, *Oak Ridge National Laboratory*

Technical Sessions A2L-A: GaN Devices I

Oct. 14, 2014 1:30 pm – 3:10 pm

Session chair: *Gaudenzio Meneghesso*

Vertical GaN Electronic Devices on Bulk-GaN Substrates (Presentation only)

David Bour, *Avogy*
Hui Nie, *Avogy*
Quentin Diduck, *Avogy*
Ozgur Aktas, *Avogy*
Tom Prunty, *Avogy*
Andrew Edwards, *Avogy*
Gangfeng Ye, *Avogy*
Ming Zhang, *Avogy*
Isik Kizilyalli, *Avogy*

Advances in Reliability and Operation Space of High-voltage GaN Power Devices Grown on Si Substrates..30

Yifeng Wu, *Transphorm Incorporated*
Jose Guerrero, *Transphorm Incorporated*
J. McKay, *Transphorm Incorporated*
Kurt Smith, *Transphorm Incorporated*

Application Specific Device Characterization and Datasheet Parameters for Commercial (600V) GaN-on-Si-Based Conversion Switches (Presentation only)

Tim McDonald, *International Rectifier Corporation*
Deepak Veereddy, *International Rectifier Corporation*
Mohamed Imam, *International Rectifier Corporation*

Degradation Mechanisms of AlGaN/GaN HEMTs on Sapphire, Si, and SiC Substrates under Proton Irradiation..*33*

Andrew Koehler, *U.S. Naval Research Laboratory*
Travis Anderson, *U.S. Naval Research Laboratory*
Jennifer Hite, *U.S. Naval Research Laboratory*
Bradley Weaver, *U.S. Naval Research Laboratory*
Marko Tadjer, *U.S. Naval Research Laboratory*
Micheal Mastro, *U.S. Naval Research Laboratory*
Jordan Greenlee, *U.S. Naval Research Laboratory*
Petra Specht, *University of California Berkeley*
Matthew Porter, *Naval Postgraduate School*
Todd Weatherford, *Naval Postgraduate School*
Karl Hobart, *Naval Research Laboratory*
Fritz Kub, *Naval Research Laboratory*

Technical Sessions A2L-B: Power Module

Oct. 14, 2014 1:30 pm – 3:10 pm

Session chairs: *Laura Marlino and Alan Mantooth*

High-Temperature SiC Power Module with Integrated SiC Gate Drivers for Future High-Density Power Electronics Applications..*36*

Bret Whitaker, *Arkansas Power International, Inc.*
Zach Cole, *Arkansas Power International, Inc.*
Brandon Passmore, *Arkansas Power International, Inc.*
Daniel Martin, *Arkansas Power International, Inc.*
Ty McNutt, *Arkansas Power International, Inc.*
Alex Lostetter, *Arkansas Power International, Inc.*
M. Nance Ericson, *Oak Ridge National Laboratory*
S. Shane Frank, *Oak Ridge National Laboratory*
Charles L. Britton, *Oak Ridge National Laboratory*
Laura D. Marlino, *Oak Ridge National Laboratory*
Alan Mantooth, *University of Arkansas, Fayetteville*
Matt Francis, *University of Arkansas, Fayetteville*
Ranjan Lamichhane, *University of Arkansas, Fayetteville*
Paul Shepherd, *University of Arkansas, Fayetteville*
Michael Glover, *University of Arkansas, Fayetteville*

Development of Packaging Technologies for Advanced SiC Power Modules............*42*

Zhenxian Liang, *Oak Ridge National Laboratory*
Fred Wang, *University of Tennessee*
Leon M. Tolbert, *University of Tennessee*

A 10-kW SiC Inverter with A Novel Printed Metal Power Module With Integrated Cooling Using Additive Manufacturing..*48*

Madhu Chinthavali, *Oak Ridge National Laboratory*
Curt Ayers, *Oak Ridge National Laboratory*
Steven Campbell, *Oak Ridge National Laboratory*
Randy Wiles, *Oak Ridge National Laboratory*
Burak Ozpineci, *Oak Ridge National Laboratory*

10 kV, 120 A SiC MOSFET Modules for a Power Electronics Building Block (PEBB)
....*55*

Christina DiMarino, *Virginia Polytechnic Institute and State University*
Igor Cvetkovic, *Virginia Polytechnic Institute and State University*
Zhiyu Shen, *Virginia Polytechnic Institute and State University*
Rolando Burgos, *Virginia Polytechnic Institute and State University*
Dushan Boroyevich, *Virginia Polytechnic Institute and State University*

Technical Sessions A3L-A: GaN Devices II

Oct. 14, 2014 3:30 pm – 4:45 pm

Session chair: *Tim McDonald*

Process Optimization of Multicycle Rapid Thermal Annealing of Mg-implanted GaN.....*59*

Jordan D. Greenlee, *National Research Council*
Boris N. Feigelson, *Naval Research Laboratory*
Travis J. Anderson, *Naval Research Laboratory*
Marko J. Tadjer, *American Society for Engineering Education*
Jennifer K. Hite, *Naval Research Laboratory*
Charles R. Eddy Jr., *Naval Research Laboratory*
Karl D. Hobart, *Naval Research Laboratory*
Francis J. Kub, *Naval Research Laboratory*

Design and Fabrication of High Current AlGaN/GaN HFET for Gen III Solid State Transformer..*63*

In-Hwan Ji, *NSF FREEDM Center at North Carolina State University*
Sizhen Wang, *NSF FREEDM Center at North Carolina State University*
Bongmook Lee, *NSF FREEDM Center at North Carolina State University*
Haotao Ke, *NSF FREEDM Center at North Carolina State University*
Veena Misra, *NSF FREEDM Center at North Carolina State University*
Alex Q. Huang, *NSF FREEDM Center at North Carolina State University*

Low ON-state Resistance of GaN PiN Rectifiers grown on FS-GaN Substrate (Presentation only)

Jeomoh Kim, *Georgia Institute of Technology*
Tsung-Ting Kao, *Georgia Institute of Technology*
Mi-hee Ji, *Georgia Institute of Technology*
Yi-Che Lee, *Georgia Institute of Technology*
Teeradetch *Georgia Institute of Technology*
Detchprohm, *Georgia Institute of Technology*
Russell Dupis, *Georgia Institute of Technology*
Shyh-Chiang Shen, *Georgia Institute of Technology*

Technical Sessions A3L-B: Gate Drive and Isolation Circuits

Oct. 14, 2014 3:30 pm – 4:45 pm

Session chairs: *Jin Wang and Bulent Sarlioglu*

An Integrated Gate Driver in 4H-SiC for Power Converter Applications....................66

M. Nance Ericson, *Oak Ridge National Laboratory*
S. Shane Frank, *Oak Ridge National Laboratory*
Charles L. Britton, *Oak Ridge National Laboratory*
Laura D. Marlino, *Oak Ridge National Laboratory*
Devon D. Janke, *Oak Ridge National Laboratory*
Dianne B. Ezell, *Oak Ridge National Laboratory*
Sei-Hyung Ryu, *Cree Inc.*
Ranjan Lamichhane, *University of Arkansas*
A. Matt Francis, *University of Arkansas*
Paul D. Shepherd, *University of Arkansas*
Michael D. Glover, *University of Arkansas*
H. Alan Mantooth, *University of Arkansas*
Bret Whitaker, *Arkansas Power International, Inc.*
Zach Cole, *Arkansas Power International, Inc.*
Brandon Passmore, *Arkansas Power International, Inc.*
Ty McNutt, *Arkansas Power International, Inc.*

Understanding the Influence of Dead-time on GaN Based Synchronous Boost Converter
....75

Di Han, *University of Wisconsin-Madison*
Bulent Sarlioglu, *University of Wisconsin-Madison*

Discussions on the Semiconductor-based Galvanic Isolation................................75

Xuan Zhang, *Ohio State University*
Lixing Fu, *Ohio State University*
Mingzhi Leng, *Ohio State University*
Jin Wang, *Ohio State University*

Poster Sessions

Oct. 14, 2014 5:15 pm – 8:00 pm

Investigation of Driver Circuits for GaN HEMTs in Leaded Packages......................81

Zhan Wang, *Transphorm, Incorporated*
Jim Honea, *Transphorm, Incorporated*
Yuxiang Shi, *Florida State University*
Hui Li, *Florida State University*

An Isolated Bi-directional Soft-Switched High-Frequency-AC Link DC-AC Converter Using SiC MOSFETs...88

Mengqi Wang, *North Carolina State University*
Suxuan Guo, *North Carolina State University*
Qingyun Huang, *North Carolina State University*
Wensong Yu, *North Carolina State University*
Alex Q.Huang, *North Carolina State University*

Dependence of Ti/C Ratio on Ohmic Contact with TiC Electrode in AlGaN/GaN Structure...94

Mari Okamoto, *Tokyo Institute of Technology*
Kuniyuki Kakushima, *Tokyo Institute of Technology*
Yoshinori Kataoka, *Tokyo Institute of Technology*
A. *Nishiyama,Tokyo Institute of Technology*
Nobuyuki Sugii, *Tokyo Institute of Technology*
Hitosh Wakabayashi, *Tokyo Institute of Technology*
Kazuo Tsutsui, *Tokyo Institute of Technology*
Hiroshi Iwai, *Tokyo Institute of Technology*
Wataru Saito, *Toshiba Corporation*

Impact of Current Measurement on Switching Characterization of GaN Transistors....98

Jennifer Lautner, *University of Erlangen-Nuremberg*
Bernhard Piepenbreier, *University of Erlangen-Nuremberg*

The Influence of SiC/SiO$_2$ Interface Morphology on the Electrical Characteristics of SiC MOS Structure...103

Li Liu, *XiDian University*
Chunkun Jiao, *Auburn University*
Yi Xu, *Rutgers University*
Gang Liu, *Rutgers University*
Leonard Feldman, *Rutgers University*
Sarit Dhar, *Auburn University*

Optically-Switched High-Voltage Bipolar SiC Device..*107*

Supid K. Mazumber, *University of Illinois-Chicago*
Alireza Mojab, *University of Illinois-Chicago*
Lin Cheng, CREE Inc.
Anant Agarwal, *Department of Energy*
Charles Scozzie, U.S. Army Research Laboratory

Technical Sessions B1L-A: Gate Dielectrics

Oct. 15, 2014 10:30 am – 11:45 am

Session chair: *Sarit Dhar*

Enhanced Oxidation of SiC Substrates using La_2O_3 Capped Annealing and a Proposal for Uniform LaSiON Gate Dielectric Formation...*110*

Yiming Lei, *Tokyo Institute of Technology*
Shu Munekiyo, *Tokyo Institute of Technology*
Takamasa Kawanago, *Tokyo Institute of Technology*
Kuniyuki Kakushima, *Tokyo Institute of Technology*
Hitoshi Wakabayashi, *Tokyo Institute of Technology*
Kazuo Tsutsui, *Tokyo Institute of Technology*
Hiroshi Iwai, *Tokyo Institute of Technology*
Masayuki Furuhashi, *Mitsubishi Electric Corporation*
Naruhisa Miura, *Mitsubishi Electric Corporation*

Passivation of SiO_2/SiC Interface with La_2O_3 Capped Oxidation........................*114*

Shu Munekiyo, *Tokyo Institute of Technology*
Yiming Lei, *Tokyo Institute of Technology*
Kenji Natori, *Tokyo Institute of Technology*
Hiroshi Iwai, *Tokyo Institute of Technology*
Takamasa Kawanago, *Tokyo Institute of Technology*
Kuniyuki Kakushima, *Tokyo Institute of Technology*
K. Kataoka, *Tokyo Institute of Technology*
A. Nishiyama, *Tokyo Institute of Technology*
Nobuyuki Sugii, *Tokyo Institute of Technology*
Hitoshi Wakabayashi, *Tokyo Institute of Technology*
Kazuo Tsutsui, *Tokyo Institute of Technology*
Masayuki Furuhashi, *Mitsubishi Electric Corporation*
Naruhisa Miura, *Mitsubishi Electric Corporation*
S. Yamakawa, *Mitsubishi Electric Corporation*

Effect of Post Deposition Annealing for High Mobility 4H-SiC MOSFET Utilizing Lanthanum Silicate and Atomic Layer Deposited SiO$_2$..........................117

Xiangyu Yang, *North Carolina State University*
Bongmook Lee, *North Carolina State University*
Veena Misra, *North Carolina State University*

Technical Sessions B1L-B: High Efficiency Power Converters

Oct. 15, 2014 10:30 am – 11:45 am

Session chairs: *Robert Dean and Benjamin Blalock*

Wide Band Gap Power Devices Based High Efficiency Power Converters for Data Center Application..........................121

Weimin Zhang, *University of Tennessee*
Ben Guo, *University of Tennessee*
Fan Xu, *University of Tennessee*
Yutian Cui, *University of Tennessee*
Yu Long, *University of Tennessee*
Fred Wang, *University of Tennessee*
Leon M. Tolbert, *University of Tennessee*
Benjamin J. Blalock, *University of Tennessee*
Daniel J. Costinett, *University of Tennessee*

125 W Multiphase GaN/Si Hybrid Point of Load Converter for Improved High Load Efficiency..........................127

Luke L. Jenkins, *Auburn University*
Benjamin K. Rhea, *Auburn University*
William Abell, *Auburn University*
Frank T. Werner, *Auburn University*
Christopher G. Wilson, *Auburn University*
Robert N. Dean, *Auburn University*
Daniel K. Harris, *Auburn University*

A Full-Bridge Current-Source Isolated DC/DC Converter with Reduced Number of Switches and Voltage Stresses for Photovoltaic Applications..........................133

Feng Guo, *Ohio State University*
Lixing Fu, *Ohio State University*
He Li, *Ohio State University*
Mohammed Alsolami, *Ohio State University*
Xuan Zhang, *Ohio State University*
Jie Zhang, *Hubei University of Technology*
Jin Wang, *Ohio State University*

Technical Sessions B2L-A: SiC II

Oct. 15, 2014 3:20 pm – 5:00 pm

Session chair: *Avinash Kashyap*

Reliability and Stability of SiC power MOSFETs and Next-Generation SiC MOSFETs...*139*

Brett Hull, *Cree Incorporated*
Scott Allen, *Cree Incorporated*
Jon Zhang, *Cree Incorporated*
Don Gajewski, *Cree Incorporated*
Vipindas Pala, *Cree Incorporated*
Jim Richmond, *Cree Incorporated*
Sei-Hyung Ryu, *Cree Incorporated*
Michael O'Loughlin, *Cree Incorporated*
Edward VanBrunt, *Cree Incorporated*
Lin Cheng, *Cree Incorporated*
Al Burk, *Cree Incorporated*
Jeff Casady, *Cree Incorporated*
David Grider, *Cree Incorporated*
John Palmour, *Cree Incorporated*

6.5 kV Normally-off JFETs Technology Status...*143*

John L. Hostetler, *United Silicon Carbide, Incorporated*
Petre Alexandrov, *United Silicon Carbide, Incorporated*
Xueqing Li, *United Silicon Carbide, Incorporated*
Leonid Fursin, *United Silicon Carbide, Incorporated*
Anup Bhalla, *United Silicon Carbide, Incorporated*

Silicon Carbide Transient Voltage Suppressor for Next Generation Lightning Protection...*147*

Avinash Kashyap, *GE Global Research Center*
Peter Sandvik, *GE Global Research Center*
James McMahon, *GE Global Research Center*
Alexander Bolotnikov, *GE Global Research Center*
Jeffrey Erlbaum, *GE Global Research Center*
Emad Andarawis, *GE Global Research Center*

Temperature Dependent Design of Silicon Carbide Schottky Diodes.................*151*

Rahul Radhakrishnan, *Global Power Technologies Group*
Tony Witt, *Global Power Technologies Group*
Richard Woodin, *Global Power Technologies Group*

Technical Sessions B2L-B: GaN Power Converters

Oct. 15, 2014 3:20 pm – 5:00 pm

Session chairs: *Eric Persson and Kevin Bai*

Matching GaN characteristics to power circuit topology maximizes performance benefit (presentation only)
Eric Persson, *International Rectifier Corporation*

A 12 to 1 V Five Phase Interleaving GaN POL Converter for High Current Low Voltage Applications...*155*
Benjamin Rhea, *Auburn University*
Luke Jenkins, *Auburn University*
William Abell, *Auburn University*
Frank Werner, *Auburn University*
Christopher Wilson, *Auburn University*
Robert Dean, *Auburn University*
Daniel Harris, *Auburn University*

High-Frequency Wireless Charging System Study Based on Normally-off GaN HEMTs...*159*
Yongsheng Fu, *Kettering University*
Lei Shi, *Kettering University*
Hua Bai, *Kettering University*

Reflected Wave Phenomenon in Motor Drive Systems Using Wide Bandgap Devices...*164*
Mark Scott, *Ohio State University,*
Jared Brockman, *Ohio State University,*
Boxue Hu, *Ohio State University,*
Lixing Fu, *Ohio State University,*
Longya Xu, *Ohio State University,*
Jin Wang, *Ohio State University,*
Rachid Darbali Zamora, *University of Puerto Rico*

2014 WiPDA Tutorials

Oct. 13, 2014, 1:00 pm- 3:00 pm

Chair: Madhu Chinthavali

Driving and Characterization of Wide Bandgap Semiconductors for Voltage Source Converter Applications...*169*

Fred Wang and **Zheyu Zhang**, University of Tennessee

Chair: Daniel Costinett

Reliability and High Field Related Issues in GaN-HEMT Devices..............................211

Gaudenzio Meneghesso, University of Padova

Oct. 13, 2014, 3:30 pm- 5:30 pm

Chair: Madhu Chinthavali

WBG Devices Enabled MV Power Converters for Utility Applications – Opportunities and Challenges...297

Ram Adapa, Electric Power Research Institute;

Subhashish Bhattacharya, North Carolina State University

Chair: Daniel Costinett

Packaging Technologies to Exploit the Attributes of WBG Power Electronics............360

Zhenxian Liang

The 2nd IEEE Workshop on Wide Bandgap
Power Devices and Application

WiPDA 2014 Leadship

Organization Committee

General Chair:
Burak Ozpineci, *Oak Ridge National Laboratory*

Vice Chairs:
Siddharth Rajan, *The Ohio State University*
Madhu Chinthavali, *Oak Ridge National Laboratory*

Technical Program Chairs:
Aminul Huque, *Electric Power Research Institute*
Robert Kaplar, *Sandia National Laboratories*

Local Organizing Committee Chairs:
Daniel Costinett, *The University of Tennessee*
Karen Nolen, *Oak Ridge National Laboratory*
Tom King, *Oak Ridge National Laboratory*

Treasurer:
Fred Wang, *The University of Tennessee*

Publication Chair:
Liming Liu, *ABB Inc.*

Tutorial Chairs:
Madhu Chinthavali, *Oak Ridge National Laboratory*
Siddharth Rajan, *The Ohio State University*

Publicity Chair:

Bulent Sarlioglu, *University of Wisconsin*

Website:

Leon Tolbert, *The University of Tennessee*
Yutian Cui, *The University of Tennessee*

Past Chair and Panel Organizer

Jin Wang, *The Ohio State University*

Steering Committee

Vladimir Blasko, *United Technology Research Center*
Dushan Boroyevich, *Virginia Tech*
Paul Chow, *Rensselaer Polytechnic Institute*
Henry Chung, *City University of Hong Kong*
Daniel Costinett, *The University of Tennessee,* **Chair**
Don Disney, *Avogy, Inc.*
David Grider, *Cree, Inc.*
Robert Kaplar, *Sandia National Laboratories*
Alex Huang, *North Carolina State University*
Liming Liu, *ABB Inc.*
Alan Mantooth, *University of Arkansas*
Umesh Mishra, *University of California, Santa Barbara*
Burak Ozpineci, *Oak Ridge National Laboratory*
Eric Persson, *International Rectifier*
Siddharth Rajan, *The Ohio State University*
Leon Tolbert, *The University of Tennessee*
Jin Wang, *The Ohio State University,* **Chair**
Joe Weimer, *Air Force Research Laboratory*
Longya Xu, *The Ohio State University*

www.wipda2014.org

THE 2ND IEEE WORKSHOP on WIDE BANDGAP POWER DEVICES AND APPLICATIONS

WiPDA

2014 EVENT AGENDA
Knoxville, Tennessee

MONDAY, OCTOBER 13TH		
10:00 a.m. – 5:30 p.m.	Registration and Exhibit Setup	Park Level Atrium
	Tutorial Session I – Crystal Room *Chair: Madhu Chinthavali*	*Tutorial Session II – Medallion & Carriage Rooms* *Chair: Daniel Costinett*
Time:	Title, Author, and Affiliation	Title, Author, and Affiliation
1:00 p.m. – 3:00 p.m.	*"Driving and Characterization of Wide Bandgap Semiconductors for Voltage Source Converter Applications,"* **Fred Wang** and **Zheyu Zhang**, University of Tennessee	*"Reliability and High Field Related Issues in GaN-HEMT Devices,"* **Gaudenzio Meneghesso**, University of Padova
3:00 p.m. – 3:30 p.m.	**BREAK – Pre-Function Hallway Meeting Space Hallway**	
3:30 p.m. – 5:30 p.m.	*"WBG Devices Enabled MV Power Converters for Utility Applications – Opportunities and Challenges,"* **Ram Adapa**, Electric Power Research Institute; **Subhashish Bhattacharya**, North Carolina State University	*"Packaging Technologies to Exploit the Attributes of WBG Power Electronics,"* **Zhenxian Liang**, Oak Ridge National Laboratory; **Fred Wang** and **Leon Tolbert**, University of Tennessee
6:00 p.m. – 8:00 p.m.	Welcome Reception with Exhibits & Displays **Johney Green**, Oak Ridge National Laboratory	Medallion & Carriage Rooms

TUESDAY, OCTOBER 14TH		
TIME:	EVENT:	LOCATION:
7:00 a.m. – 8:00 a.m.	Author's Breakfast	Cumberland Ballroom
7:30 a.m. – 4:00 p.m.	Registration	Park Level Atrium
8:00 a.m. – 8:10 a.m.	Welcome **Burak Ozpineci**, Oak Ridge National Laboratory	Medallion & Carriage Rooms
	Keynote Sessions – Medallion & Carriage Rooms *Chair: Leon Tolbert*	
8:10 a.m. – 8:40 a.m.	*"A Vision for U.S. Leadership in Wide Bandgap Power Electronics,"* **Anant Agarwal**, Department of Energy	
8:40 a.m. – 9:10 a.m.	*"Is GaN a Game Changing Device?,"* **Fred Lee**, Virginia Tech	
9:10 a.m. – 9:40 a.m.	*"Ultra High Voltage SiC Power Devices and Their Applications in Solid State Transformer and Circuit Breaker for Future Electric Power Grid,"* **Alex Huang**, North Carolina State University	
9:40 a.m. – 10:10 a.m.	*"Next-Generation Silicon Carbide Power Modules Enable New Levels of Performance,"* **Ty McNutt**, Arkansas Power Electronics International	
10:10 a.m. – 10:30 a.m.	**BREAK – Pre-Function Hallway Meeting Space Hallway**	
	WBG Characterization Technical Session – Crystal Room *Chair: John Hostetler*	*WBG Application Challenges Technical Session - Medallion & Carriage Rooms* *Chair: Chingchi Chen and Jürgen Schuderer*
10:30 a.m. – 10:55 a.m.	*"4500 Volt Si/SiC Hybrid Module Qualification for Modern MegaWatt Scale Wind Energy Inverters,"* **William Erdman**, Cinch LLC	*"Packaging SiC Power Semiconductors – Challenges, Technologies and Strategies,"* **Jürgen Schuderer, Umamaheswara Vemulapati**, and **Felix Traub**, ABB Corporate Research
10:55 a.m. – 11:20 a.m.	*"Understanding the Limitations and Impact Factors of Wide Band-gap Devices' High Switching-Speed Capability in Voltage Source Converter,"* **Zheyu Zhang, Fred Wang, Leon M. Tolbert, Benjamin J. Blalock**, and **Daniel Costinett**, University of Tennessee	*"The Opportunities and Challenges of Wide-Band-Gap Technologies for Automotive Applications,"* **Chingchi Chen**, and **Ming Su**, Ford Motor Company
11:20 a.m. – 11:45 a.m.	*"The Development of a High-Voltage Power Device Evaluation Platform,"* **Lixing Fu, Xuan Zhang, He Li, Xintong Lu**, and **Jin Wang**, Ohio State University	*"Application-Based Review of GaN HFETs,"* **Edward Jones** and **Fred Wang**, University of Tennessee; **Burak Ozpineci**, Oak Ridge National Laboratory
11:45 a.m. – 1:30 p.m.	**Attendees Lunch – Tennessee Ballroom** **Steering Committee Lunch – Private Dining Area** *(Conference Room at the entrance to onsite Windows on the Park Restaurant)*	

TUESDAY, OCTOBER 14TH (CONTINUED)

TIME:	EVENT:	LOCATION:
	GaN Devices Technical Session I – Crystal Room **Chair: Gaudenzio Meneghesso**	**Power Module Technical Session –** **Medallion & Carriage Rooms** **Chair: Laura Marlino and Alan Mantooth**
1:30 p.m. – 1:55 p.m.	"Vertical GaN Electronic Devices on Bulk-GaN Substrates," **David Bour, Hui Nie, Quentin Diduck, Ozgur Aktas, Tom Prunty, Andrew Edwards, Gangfeng Ye, Ming Zhang, and Isik Kizilyalli,** Avogy	"High-Temperature SiC Power Module with Integrated SiC Gate Drivers for Future High-Density Power Electronics Applications," **Bret Whitaker, Zach Cole, Brandon Passmore, Daniel Martin, Ty McNutt, and Alex Lostetter,** Arkansas Power International, Inc.; **Nance Ericson, Shane Frank, and Charles Britton,** Oak Ridge National Laboratory
1:55 p.m. – 2:20 p.m.	"Advances in Reliability and Operation Space of High-voltage GaN Power Devices Grown on Si Substrates," **Yifeng Wu, Jose Guerrero, and Kurt Smith,** Transphorm Incorporated	"Development of Packaging Technologies for Advanced SiC Power Modules," **Zhenxian Liang,** Oak Ridge National Laboratory
2:20 p.m. – 2:45 p.m.	"Application Specific Device Characterization and Datasheet Parameters for Commercial (600V) GaN-on-Si-Based Conversion Switches," **Tim McDonald, Deepak Veereddy, and Mohamed Imam,** International Rectifier Corporation	"A 10-kW SiC Inverter with A Novel Printed Metal Power Module With Integrated Cooling Using Additive Manufacturing," **Madhu Chinthavali, Curt Ayers, Steven Campbell, Randy Wiles, and Burak Ozpineci,** Oak Ridge National Laboratory
2:45 p.m. – 3:10 p.m.	"Degradation Mechanisms of AlGaN/GaN HEMTs on Sapphire, Si, and SiC Substrates under Proton Irradiation," **Andrew Koehler, Travis Anderson, Jennifer Hite, Bradley Weaver, Marko Tadjer, and Jordan Greenlee,** U.S. Naval Research Laboratory; **Petra Specht,** University of California Berkeley; **Matthew Porter** and **Todd Weatherford,** Naval Postgraduate School	"10 kV, 120 A SiC MOSFET Modules for a Power Electronics Building Block (PEBB)," **Christina DiMarino, Igor Cvetkovic, Rolando Burgos, and Dushan Boroyevich, Zhiyu Shen,** Virginia Tech
3:10 p.m. – 3:30 p.m.	**BREAK – Pre-Function Hallway Meeting Space Hallway**	
	GaN Devices Technical Session II – Crystal Room **Chair: Tim McDonald**	**Gate Drive and Isolation Circuits Technical Session –** **Medallion & Carriage Rooms** **Chair: Jin Wang and Bulent Sarlioglu**
3:30 p.m. – 3:55 p.m.	"Process Optimization of Multicycle Rapid Thermal Annealing of Mg-implanted GaN," **Jordan Greenlee,** National Research Council; **Boris Feigelson, Travis Anderson, Jennifer Hite, Charles Eddy, Jr., Karl Hobart, and Francis Kub,** Naval Research Laboratory; **Marko Tadjer,** American Society for Engineering Education	"An Integrated Gate Driver in 4H-SiC for Power Converter Applications," **Nance Ericson,** Oak Ridge National Laboratory
3:55 p.m. – 4:20 p.m.	"Design and Fabrication of High Current AlGaN/GaN HFET for Gen III Solid State Transformers," **In-Hwan Ji, Sizhen Wang, Bongmook Lee, Haotao Ke, Veena Misra, and Alex Q. Huang,** NSF FREEDM Center at North Carolina State University	"Understanding the Influence of Dead-time on GaN Based Synchronous Boost Converter," **Di Han, and Bulent Sarlioglu,** University of Wisconsin-Madison

TUESDAY, OCTOBER 14TH (CONTINUED)

TIME:	EVENT:	LOCATION:
4:20 p.m. – 4:45 p.m.	"Low ON-state Resistance of GaN PiN Rectifiers grown on FS-GaN Substrate," **Jeomoh Kim, Tsung-Ting Kao, Mi-hee Ji, Yi-Che Lee, Teeradetch Detchprohm, Russell Dupis, and Shyh-Chiang Shen,** Georgia Tech	"Discussions on the Semiconductor-based Galvanic Isolation," **Xuan Zhang, Lixing Fu, Mingzhi Leng, and Jin Wang,** Ohio State University
4:45 p.m. – 5:15 p.m.	**Travel to University of Tennessee, West Club Neyland Stadium**	
	Poster Session & Banquet – University of Tennessee, West Club at Neyland Stadium	
5:15 p.m. – 8:30 p.m.	❖ "Investigation of Drive Circuits for GaN HEMTs in Leaded Packages," **Zhan Wang** and **Jim Honea,** Transphorm, Incorporated; **Yuxiang Shi** and **Hui Li,** Florida State University ❖ "An Isolated Bi-directional Soft-Switched High-Frequency-AC Link DC-AC Converter Using SiC MOSFETs," **Mengqi Wang, Qingyun Huang, Wensong Yu, and Alex Huang,** North Carolina State University ❖ "Dependence of Ti/C Ratio on Ohmic Contact with TiC Electrode in AlGaN/GaN Structure," **Wataru Saito,** Toshiba Corporation **Mari Okamoto, Kuniyuki Kakushima, Yoshinori Kataoka, Kenji Natori, Hitoshi Wakaabayashi, and Hiroshi Iwai,** Tokyo Institute of Technology; ❖ "Impact of Current Measurement of Switching Characterization of GaN Transistors," **Jennifer Lautner** and **Bernhard Piepenbreier,** University of Erlangen-Nuremberg ❖ "The Influence of SiC/SiO2 Interface Morphology on the Electrical Characteristics of 4H-SiC MOS Structure," **Li Liu,** XiDian University; **Chunkun Jiao** and **Sarit Dhar,** Auburn University; **Yi Xu, Gang Liu, and Leonard Feldman,** Rutgers University ❖ "Optically-Switched High-Voltage Bipolar SiC Device," **Supid K. Mazumder,** University of Illinois-Chicago	
8:00 p.m. – 10:00 p.m.	Tour of the University of Tennessee-Knoxville's Laboratories	1st floor of the Min Kao Building

WEDNESDAY, OCTOBER 15TH

TIME:	EVENT:	LOCATION:
7:00 a.m. – 8:00 a.m.	Author's Breakfast	Cumberland Ballroom
7:30 a.m. – 11:30 a.m.	Registration	Park Level Atrium

WEDNESDAY, OCTOBER 15TH *(CONTINUED)*

TIME:	EVENT:	LOCATION:
keynote	**Keynote Sessions – Medallion & Carriage Rooms** **Chair: Fred Wang**	
8:00 a.m. – 8:10 a.m.	Welcome **Burak Ozpineci**, Oak Ridge National Laboratory	
8:10 a.m. – 8:40 a.m.	*"The New York Power Electronics Manufacturing Consortium at CNSE-SUNY Poly,"* **Michael Liehr**, State University of New York System	
8:40 a.m. – 9:10 a.m.	*"Industrial Readiness of SiC Power Devices,"* **Ljubisa Stevanovic**, General Electric	
9:10 a.m. – 9:40 a.m.	*"Merits of Epitaxial Re-growth Technologies for Advanced SiC Device Concepts,"* **Adolf Schoner**, Ascatron AB	
9:40 a.m. – 10:10 a.m.	*"Power America: Next Generation Power Electronics Manufacturing Innovation Institute,"* **Dennis Kekas**, North Carolina State University	
10:10 a.m. – 10:30 a.m.	**BREAK – Pre-Function Hallway Meeting Space Hallway**	

	Gate Dielectrics – Crystal Room **Chair: Sarit Dhar**	**High Efficiency Power Converters –** **Medallion & Carriage Rooms** **Chair: Robert Dean and Benjamin Blalock**
10:30 a.m. – 10:55 a.m.	*"Enhanced Oxidation of SiC Substrates using La2O3 Capped Annealing and a Proposal for Uniform LaSiON Gate Dielectric Formation,"* **Yiming Lei, Shu Munekiyo, Takamasa Kawanago, Kuniyuki Kakushima, Hitoshi Wakabayashi, Kazuo Tsutsui, Hiroshi Iwai**, Tokyo Institute of Technology; **Masayuki Furuhashi**, and **Naruhisa Miura**, Mitsubishi Electric Corporation	*"Wide Band Gap Power Devices Based High Efficiency Power Converters for Data Center Application,"* **Weimin Zhang, Ben Guo, Fan Xu, Yutian Cui, Yu Long, Fred Wang, Leon Tolbert, Benjamin J. Blalock**, and **Daniel Costinett**, University of Tennessee
10:55 a.m. – 11:20 a.m.	*"Passivation of SiO2/SiC Interface with La2O3 Capped Oxidation,"* **Shu Munekiyo, Yiming Lei, Takamasa Kawanago, Kuniyuki Kakushima, Hitoshi Wakabayashi, Kazuo Tsutsui**, and **Hiroshi Iwai**, Tokyo Institute of Technology; **Masayuki Furuhashi** and **Naruhisa Miura**, Mitsubishi Electric Corporation	*"125 W Multiphase GaN/Si Hybrid Point of Load Converter for Improved High Load Efficiency,"* **Luke L. Jenkins, Benjamin K. Rhea, William E. Abell, Frank T. Werner, Christopher G. Wilson, Robert N. Dean**, and **Daniel K. Harris**, Auburn University
11:20 a.m. – 11:45 a.m.	*"Effect of Post Deposition Annealing for High Mobility 4H-SiC MOSFET Utilizing Lanthanum Silicate and Atomic Layer Deposited SiO2,"* **Xiangyu Yang, Bongmook Lee**, and **Veena Misra**, North Carolina State University	*"A Full-Bridge Current-Source Isolated DC/DC Converter with Reduced Number of Switches and Voltage Stresses for PV Applications,"* **Feng Guo, Lixing Fu, He Li, Mohammed Alsolami, Xuan Zhang, Jie Zhang**, and **Jin Wang**, Ohio State University
11:45 a.m. – 1:30 p.m.	**LUNCH – Tennessee Ballroom**	
1:30 p.m. – 3:10 p.m.	*Panel Discussion* **Jin Wang**, Ohio State University, Moderator **Anant Agarwal**, Department of Energy; **Eric Persson**, International Rectifier Corporation; **Lin Cheng**, CREE, Incorporated; **Tom Baker**, Catapillar; **Hao Huang**, General Electric Aviation	Medallion & Carriage Rooms

WEDNESDAY, OCTOBER 15TH *(CONTINUED)*

TIME:	EVENT:	LOCATION:
3:10 p.m. – 3:30 p.m.	**BREAK- Pre-Function Hallway Meeting Space Hallway**	

	SiC Devices Technical Session II – Crystal Room **Chair: Avinash Kashyap**	**GaN Power Converters Technical Session –** **Medallion & Carriage Rooms** **Chair: Eric Persson and Kevin Bai**
3:30 p.m. – 3:45 p.m.	*"Reliability and Stability of SiC power MOSFETs and Next-Generation SiC MOSFETs"* **Scott Allen, Brett Hull, Jon Zhang, Don Gajewski, Vipindas Pala, Jim Richmond, Sei-Hyung Ryu, Michael O'Loughlin**, and **Edward VanBrunt**, Cree Incorporated	*"Matching GaN Characteristics to Power Circuit Topology Maximizes Performance Benefit"* **Eric Persson**, Internation Rectifier Corporation
3:45 p.m. – 4:10 p.m.	*"6.5 kV Normally-off JFETs Technology Status,"* **John L. Hostetler, Anup Bhalla, Petre Alexandrov**, and **Xueqing Li**, United Silicon Carbide, Incorporated	*"A 12 to 1 V Five Phase Interleaving GaN POL Converter for High Current Low Voltage Applications,"* **Benjamin Rhea, Luke Jenkins, William Abell, Frank Werner, Christopher Wilson, Robert Dean**, and **Daniel Harris**, Auburn University
4:10 p.m. – 4:35 p.m.	*"Silicon Carbide Transient Voltage Suppressor for Next Generation Lightning Protection,"* **Avinash Kashyap, Peter Sandvik, James McMamahon, Alexander Bolotnikov, Jeffrey Erlbaum**, and **Emad Andarawis**, GE Global Research Center	*"High-Frequency Wireless Charging System Study Based on Normally-off GaN HEMTs,"* **Hua Bai**, Kettering University
4:35 p.m. – 5:00 p.m.	*"Temperature Dependence Design of Silicon Carbide Schottky Diodes,"* **Rahul Radhakrishnan, Tony Witt**, and **Richard Woodin**, Global Power Technologies Group	*"Reflected Wave Phenomenon in Motor Drive Systems Using Wide Bandgap Devices,"* **Mark Scott, Jared Brockman, Boxue Hu, Lixing Fu**, and **Longya Xu**, Ohio State University; **Rachid Darbali Zamora**, University of Puerto Rico
	**END OF 2014 IEEE WiPDA WORKSHOP**	

4500 Volt Si/SiC Hybrid Module Qualification for Modern Megawatt Scale Wind Energy Inverters

William L. Erdman
Cinch LLC
Moraga, Ca

David Grider and Edward VanBrunt
Cree Inc.
4600 Silicon Dr.
Raleigh, NC 27703
David_grider@cree.com

Abstract-This paper discusses recent trends in wind turbine electrical architecture resulting from utility interconnect and public siting requirements. These trends together with constant need to reduce the cost-of-energy place a premium on wind turbine inverter efficiency. This need is addressed by the use of a medium voltage, three-level inverter with 4HN structure silicon carbide barrier diode located in the clamping diode location. A comparison with a conventional silicon PIN clamping diode is covered in detail.

Keywords– Wind turbine; wind energy; medium voltage; three-level inverter; silicon carbide; SiC; barrier diode

I. INTRODUCTION

Windturbine electrical design approaches have changed drastically in recent years to address a series of issues including turbine reliability, utility interconnect requirements, public acceptance in the form of aesthetics and noise abatement and of course, a continued reduction in turbine Cost-of-Energy (COE). In Section II of this paper, a review of recent turbine electrical architectures is presented along with a new proposed architecture by the National Renewable Energy Laboratory (NREL) team as part of the Department of Energy (DOE) "Advanced Drive Train Project." The proposed architecture places the converter and padmount transformer within the turbine tower which then places a difficult thermal management problem on the converter design along with a desire to increases energy capture, both being addressed by an improvement in converter efficiency. The efficiency improvement is accomplished by applying a medium voltage, three-level inverter utilizing silicon carbide (SiC) barrier diodes in the clamping section of the inverter. In Section III a simplified presentation of the three-level inverter is presented which highlights the importance of SiC clamping diode. The SiC barrier diode characteristics are presented and summarized in switching energy improvements and a rigorous power semiconductor loss model completes the paper.

II. EFFICIENCY MOTIVATIONS IN MW SCALE TURBINES

A. Recent History of Wind Turbine Electrical Architectures

Fig. 1 depicts the evolution of turbine electrical architecture over time and as a function of ever increasing turbine ratings.

In this figure, a multi-megawatt rated turbine is shown along with principal drivetrain components including generator, power converter, and padmount transformer. From the mid-1980s thru 2000, turbines were sub-megawatt rated and often the tower was of truss, or lattice construction (not shown). In these early turbines, the power converter and padmount transformer were necessarily ground-mounted and stood separate from the turbine tower. Converter and transformer cooling in these early turbines was not an issue, although proper outdoor environmental rating of the converter was important. Beginning in the early 2000s, turbines had grown to the megawatt scale and tube towers became the accepted norm. While the tube tower suffered from a higher capital cost, there were many advantages to this structure, including allowance of higher cutout wind speeds, improved tower resonance characteristics, aesthetics and lower installation and maintenance costs. In moving to the tube tower, the configuration of Fig. 1(a) remained popular and converter and transformer requirements did not change. As turbine ratings grew further, towards 1.5MW and 2.0MW, the concept of incorporating the converter and padmount transformer uptower, or in the nacelle, gained some popularity. The advantage of this approach was that much of the turbine wiring and construction which had taken place in the field could be accomplished on the manufacturing floor. Additionally, pendant cables which run the length of the tower (80–100 meters today) would be at collection system voltages of 20-34.5 kV, significantly reducing cable copper weights and cost. The disadvantages of the approach were that substantial weight was added to the nacelle, influencing tower dynamics and costs, the converter was difficult to maintain, and the uptower location subjected the transformer and converter to non-stationary shock and vibration loads with the converter subjected to high nacelle temperatures. This turbine electrical architecture began to point to the premium that could be placed on reducing converter losses in an effort to keep surrounding ambient temperatures at a reasonable level. Fig. 1(c) shows a competing architecture in response to some of the disadvantages of 1(b). This configuration shows the converter at ground level and moved to the inside of the tower. This new locating of the converter was viewed as "natural" and an environmental (from the converter perspective) and aesthetic improvement. The tube tower was viewed as a natural thermal conducting "chimney" for power losses created by the power converter and internal, surrounding ambient temperature concerns were glossed over. Thousands of turbines were fielded in the US using this configuration and it was not until

978-1-4799-5494-0/14 $31.00 © 2014 IEEE

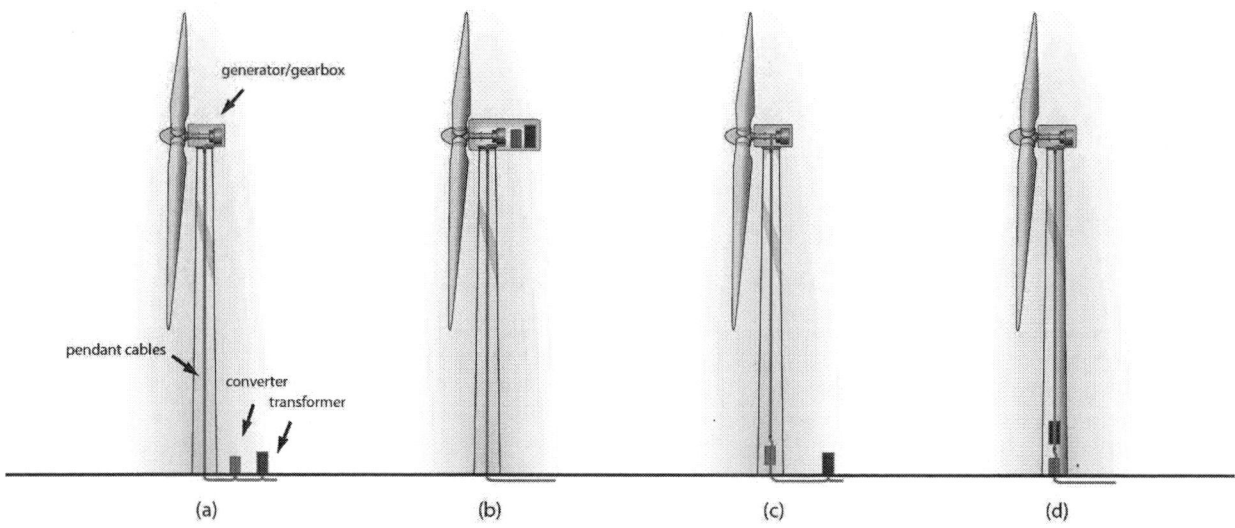

Fig. 1(a) – (d) Evolution of wind turbine electrical architectures over time and as a function of MW rating.

2–3 years of field data became available that tower-converter thermal problems came clearly into focus. Just as with the up-tower configuration of Fig. 1(b), a focus on converter losses became apparent to turbine manufacturers and operators. Positive features of configuration 1(c) include ground level converter maintenance, improved audible performance (reduction in switching frequency audible noise), and improved acceptance of turbine aesthetics. A negative aspect of the configuration in Fig. 1(c), relative to 1(b), is the use of low voltage (690 V typical), large copper cross section, pendant cables.

B. The NREL/DOE Advanced Drivetrain Architecture

Fig. 1(d) depicts the proposed drivetrain electrical architecture being applied by the NREL team in the DOE's "Advanced Drivetrain Project." This architecture addresses new and anticipated utility interconnect, audible noise, aesthetic, and offshore/onshore unifying requirements. The architecture has been in use for the last 3-5 years; principally in offshore windturbine designs and often with a rotor-fed, partially rated 0.33 PU converter (discussed later). The NREL version of the Fig. 1(d) architecture addresses the large pendant cable issue by moving the generator and converter to medium voltage (3.3 kV). It also adopts a full, 1.0 PU stator-fed conversion system with the padmount transformer being moved into the tower; this configuration amplifies the need to reduce power losses at the converter level. It will be shown that the use of 4.5 kV Si/SiC hybrid power semiconductor modules used in a three-level medium voltage, voltage-source inverter will play an important role in reducing the converter power losses.

C. Full Conversion - Partial Conversion Tradeoffs

Another factor that has an important thermal influence on the configurations of Fig. 1 is whether the turbine applies a doubly-fed, wound rotor generator or a stator-fed conventional squirrel-cage or synchronous generator. The former uses a partially rated, bi-directional, back-to-back converter in the generator's rotor circuit. The rating of this converter is proportional to the slip of the generator at rated power which is usually restricted to 33%. In the stator fed case, the converter must carry 100% of the turbine power. With converters of the same efficiency, the stator fed approach will dissipate three-times as much power as the doubly-fed system. Despite this, the stator fed approach has many advantages over the doubly-fed approach including the elimination of slip-ring and brush assembly, and most importantly, the complete decoupling of the generator from the utility during utility fault conditions. These severe utility fault events can cause substantial torque transients on the drivetrain and the complete decoupling which exists in the stator case allows for greater control during these events. The collection of new and emerging utility-interconnect requirements favors the completely decoupled, fully rated stator-fed systems, and this is the basis for its selection in the Advanced Drive Train. This topic is widely discussed in the research literature; [1] is particularly concise in its comparison of the two approaches, giving the advantage to full conversion systems.

Placing a combined 100% turbine rated converter and padmount transformer within the turbine tower, puts a significant premium on the reduction of converter losses and on the adoption of a wide band gap barrier diode.

D. Incremental Value of Converter Losses

In addition to the thermal management issues discussed above, turbine Annual Energy Production (AEP) is another motivation for reducing converter losses. Fig. 2 shows a simple study of the incremental value of converter (drivetrain) efficiency. A 95% efficiency curve serves as the starting point and a second 94% efficiency curve is generated according to $\eta_{94\%} = \eta_{95\%} - 1\%$, maintaining a constant 1%

TABLE I

Turbine Architecture	Fig. 1 (a) [1]	Fig. 1 (b) [1]	Fig. 1 (c) [1]	Fig. 1 (d) [1]	Fig. 1 (d) [2]
Uptower Losses	180 kW	354 kW	180 kW	180 kW	112 kW
Downtower Losses (In-Tower)	-	-	150 kW	174 kW	70 kW
Downtower Losses (Out-of-Tower)	174 kW	-	24 kW	-	-
Total losses	354 kW	354 kW	254 kW	354 kW	182 kW
Drivetrain Efficiency	88.20%				94.00%

Notes: 1) 3 MW turbine at rated power, $\eta_{generator}$ = 0.97, $\eta_{gearbox}$ = 0.97, $\eta_{converter}$ = 0.95 (back-to-back 0.975), $\eta_{transformer}$ = 0.992. These are rounded approximate values to represent the magnitude of losses; 2) 3 MW turbine at rated power, $\eta_{generator}$ = 0.9775, $\eta_{gearbox}$ = 0.985, $\eta_{converter}$ = 0.985 (single, passive rectifier uptower), $\eta_{transformer}$ = 0.992.

difference over the entire load range. The two efficiencies result in two (non-discernable) power curves for the turbine as shown in Fig. 2(b). These two power curves are processed through an IEC Class II wind regime with 8.5 m/s mean wind speed; this provides the AEP pertaining to each efficiency curve. The table in Fig. 2(c) shows that the 95% efficiency converter produces an additional 118,948 kWh of annual energy. If this efficiency improvement can be captured at no additional capital cost, then the efficiency improvement reduces the COE of the turbine; an important improvement in and of itself. This COE model was manipulated differently however. The Turbine Capital Cost (TCC) was increased by 1.3% to offset the increased AEP. This then gives the initial capital cost value of $103,000 for the 1% improvement in converter efficiency. This example is meant to bound the question of initial capital value, specific values are highly dependent on wind regime and financing variables.

III. NEUTRAL POINT CLAMP (NPC) INVERTER AND THE SIC BARRIER DIODE

The Neutral Point Clamp (NPC) three-level inverter shown in Fig. 3 (a) was first introduced by Nabae, et al. in 1981, [2], and its operation is covered in detail in [3]. Advantages of this topology over the more traditional two-level inverter include the addition of a third, zero-voltage output state, and reduced switching losses which occur at ½ of the DC bus voltage due to the neutral point clamping diode. A comparison between two-level and three-level in wind energy applications is provided in [4]-[5], where efficiency advantages of this architecture are highlighted. This topology has become popular due to its many advantages and has been extended in recent years from its high(er) voltage origins to low voltage systems as well. Originally, the topology exploited low voltage semiconductors in high voltage applications by requiring only ½ of the DC bus voltage to appear across the various semiconductors. In a presentation by Schweizer, et al. in [6], the use of a barrier diode in the clamping position of each inverter phase provides a significant efficiency improvement when the converter is operating at (or near) unity power factor. Wind turbine power factor requirements typically include a +/- 0.95 range, which is consistent with Schweizer's power factor requirement. In non-unity power factor applications, there is further benefit by the substitution of IGBT flyback barrier diodes, as well. To achieve the target efficiencies of the intower wind tur-

bine converter, the two-step approach of using a three-level inverter (over a two-level) together with the application of a SiC barrier clamping diode provides a realistic opportunity of addressing the issues raised in Section II.

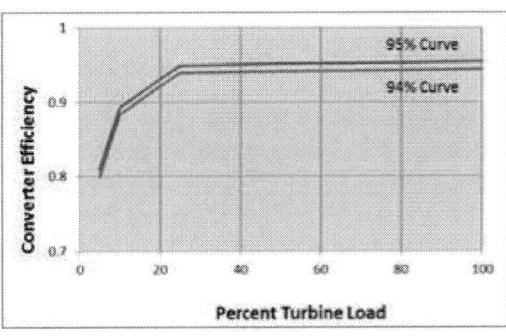

Fig. 2 (a) 94% and 95% converter efficiency profiles as a function of load, (b) 5 MW power curves resulting from the two efficiencies and (c) COE model used in determining incremental efficiency value; arrows identify increases AEP and allowable increase in TCC.

A. Circuital Reduction of the NPC Inverter

Operation of the NPC inverter has been covered in the identified references. However, it is worthwhile to simplify a single phase of the NPC circuit where barrier diode clamping operation can be focused upon. Development of this simplified circuit begins with Fig. 3 (a), which schematically represents an NPC single phase. Fig. 3 (b) shows inverter operation including the fundamental component of output voltage and the PWM, or instantaneous output voltage. Since the converter is operating at unity power factor, the output current (not shown) is in phase with the output voltage. The circuit is symmetrical and the solid black top portion is operated during the positive half cycle of operation. The solid grey bottom portion operates during negative half cycle operation. The load is shown in dotted black to represent its connection over the entire sinusoidal cycle. Due to positive/negative half cycle symmetry, only positive half cycle operation requires development.

During the positive half cycle, auxiliary IGBT Q2 remains in the on-state and is not switched allowing it to be modeled as a simple conductor; its forward voltage drop of 3-4 volts can be neglected relative to the +/-2.5 kV power supply voltage. The main IGBT, Q1, is switched at the modulating frequency, thereby connecting the load to the positive power supply when Q1 is gated to the on-state. When Q1 is gated to the off-state, load current is commutated to the clamping diode D5, resulting in the zero-voltage output state as shown in 3(b). Repetitively, load current is commutated back-and-forth between Q1 and D5. This half cycle operation results in the simplified circuit of Fig. 3 (c), which is identical to the simple buck chopper often used in DC power supplies. The only difference here is that the modulation pattern provides a sinusoidal weighted output from the inverter while a constant DC weighted output is typically provided in the DC power supply application. The same barrier diode benefits that have been identified in DC power supplies accrue to the unity power factor NPC inverter. These benefits result from the significant reduction in reverse recovery losses associated with barrier diodes relative to a PIN diode. Fig. 3 (c) also shows the Q1, D5 commutation path which is the critical circuital path and sensitive to physical layout of the devices. A description of the 4500 V SiC barrier diode and comparison of static and dynamic characteristics against a 4500 V Si PIN diode helps to explain advantages.

B. Development and Characteristics of the SiC Barrier Diode

For the Advanced Drive Train project, a 4500 V Si IGBT module was developed with a substituted SiC barrier diode. The IGBT module current rating is 1200 A, provided in a 140x190 mm isolated baseplate package. The module contains a quantity of 36, 40 amp rated, 8 mm x 10 mm 4HN-SiC diode chips. The clamping diode is rated at 800 amps and is provided in a 130 x 140 mm isolated baseplate package. The clamping diode utilizes a quantity of twenty, 40 amp SiC diodes of the same design. The target three-level inverter will run at 3.3 kV AC with a 2.3 MW rating leading to a line current of a 400 A RMS. The target switching frequency is 1.0 kHz., with split +/-2.5 kV DC bus voltage.

1) Static Characteristics

Forward conduction characteristics for the Si PIN and SiC barrier diode are shown in Fig. 4 (next page). It can be seen that the SiC diode has higher conduction losses as a result of the higher voltage drop relative to the PIN diode. Furthermore, the SiC diode has a definite positive temperature coefficient associated with it. In the three-level inverter, conduction losses are dominated by the IGBT conduction losses, so this penalty is small and substantially offset by advantages in switching losses as will be quantified later. The results in Fig. 4 were measured at the 40 amp, chip level and scaled to the module level at the time of paper submission. Module static testing will be performed when module assembly is complete.

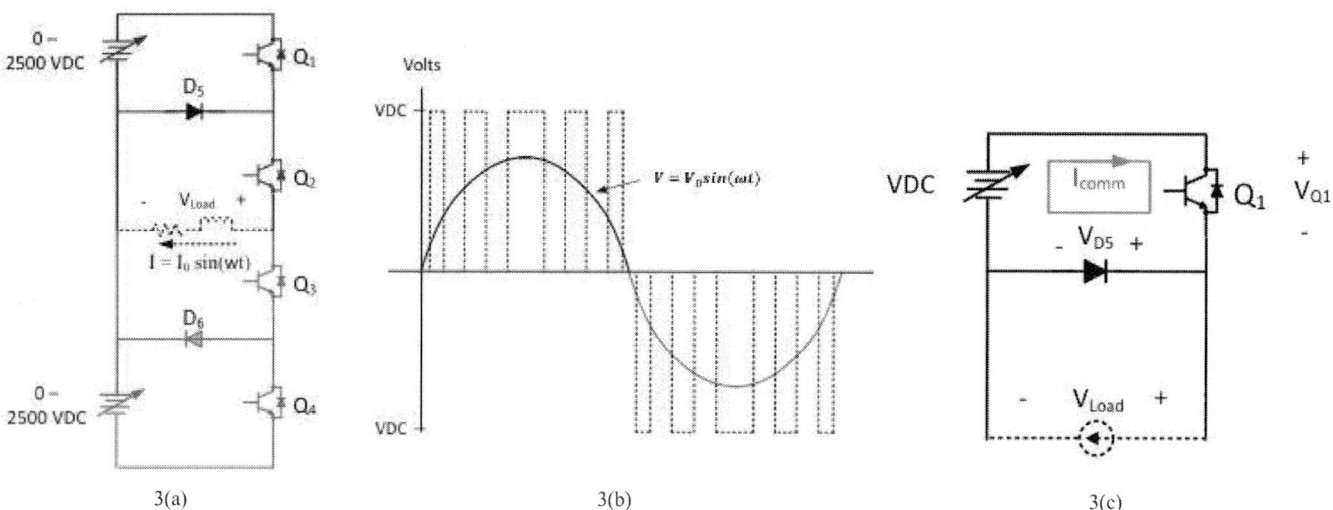

Fig. 3(a) Single phase NPC inverter, (b) desired fundamental output voltage and actual PWM output; the black positive fundamental corresponds to the black positive portion in (a), likewise with grey circuit and fundamental, (c) simplified reduction to the classical buck converter in which the diode recovery characteristics are dominant in switching losses.

978-1-4799-5494-0/14 $31.00 © 2014 IEEE

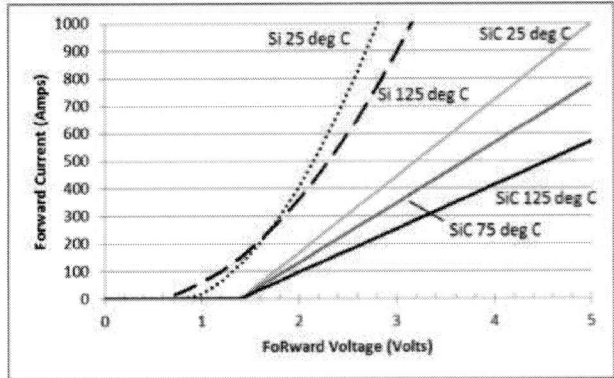

Fig. 4 Forward conduction characteristics if Si and SiC diode.

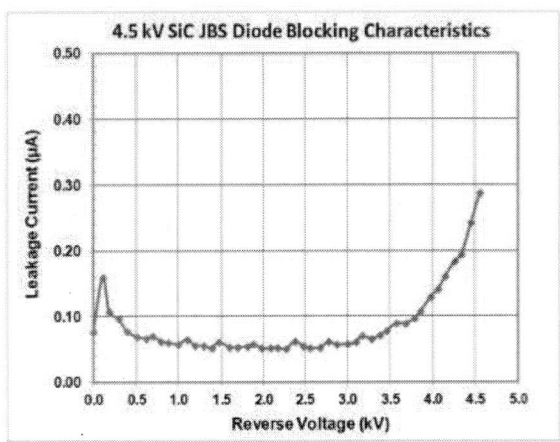

Fig. 5 Reverse leakage current for the SiC diode at 25° C.

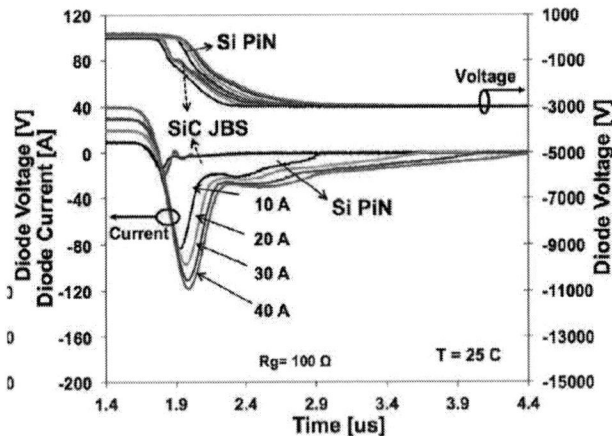

Fig. 6 Barrier diode recovery characteristics compared to Si PIN characteristics at 25° C.Fig. 5 Reverse leakage current for the SiC diode at 25° C.

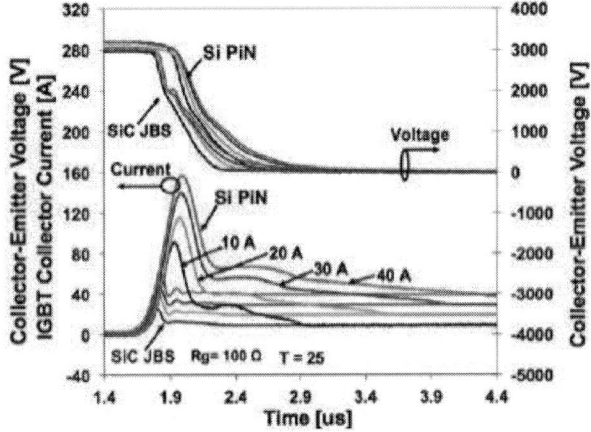

Fig. 7 Effect of diode recovery on IGBT turn-on losses at 25° C.

Fig. 5 shows the reverse blocking capabilities of the SiC diode as recently tested. It can be seen that the reverse leakage current of the SiC diode is approximately 570 µ amps at 4.5 kV, approximately 1/1000 of the data sheet rating reported for the Si PIN diode when scaled to an equivalent 40 amp level.

2) Dynamic Characteristics

The dynamic characteristics of the SiC barrier diode reported on here are at the 40 amp chip level and are from an earlier process (described in [7]). Dynamic characteristics

for the module discussed herein are not available at the time of paper submission. Fig. 6 shows the reverse recovery characteristics of the SiC and equivalent PIN diode at a DC bus voltage of 3.0 kV taken at 25° C. Module data for this project will be taken at 125° C where further benefits of the SiC diode will be realized. The SiC diode recovery is essentially independent of forward current, and also independent of temperature (not shown). The impact of reverse recovery characteristics on IGBT turn-on losses is captured in Fig. 7.

Table II summarizes dynamic characteristics into switching energies for the IGBT and diode. All values are scaled for

TABLE II Summary of Switching Energies

Current	Si IGBT with Si PIN Diode		Si IGBT with SiC Barrier Diode		Si PIN Diode	SiC Barrier Diode
	E_{SW-ON} [1]	E_{SW-OFF} [1]	E_{SW-ON} [1]	E_{SW-OFF} [1]	E_{REC} [2]	E_{REC} [1]
(Amps)	(Joules)	(Joules)	(Joules)	(Joules)	(Joules)	(Joules)
200	1.11	0.88	0.23	0.88	0.71	0.023
400	1.65	1.53	0.46	1.53	1.12	0.023
600	2.39	2.51	0.65	2.51	1.38	0.023
800	3.00	3.35	0.99	3.35	1.52	0.023
1200	4.65	4.81	1.42	4.81	1.64	0.023

[1] Data Source is [7], [2] Data Source is [8]. All measurements will be confirmed in full inverter phase test fixture.

978-1-4799-5494-0/14 $31.00 © 2014 IEEE

voltage and temperature of 5 kV and 25° C respectively. Measurements at 125° C will be taken upon module completion. The combination of static and dynamic characteristics serve as inputs into the inverter efficiency model in the nextsection. At 125° C, the reverse recovery charge of the PIN diode goes up substantially, while the barrier diode characteristics are essentially unchanged further enhancing the benefits of the barrier diode.

C. Inverter Efficiency Model

To summarize the advantages for the SiC barrier diode in a three-level inverter, the efficiency model from [9] was used. Fig. 8 is a plot of losses versus percent power where 100% represents 2.3 MW. The loss values presented are for the upper (or lower) portion of a single inverter phase and therefore represent 1/6 of the total losses. Switching losses are seen to increase linearly with power and dominate conduction losses for both inverter cases at the 1.0 kHz switching frequency. Conduction losses overlay one-another and are dominated by the (same) two IGBT's conducting at any given time.

Nontrivial line-side reactor losses are added to the semiconductor losses to determine the overall inverter efficiency as shown in Fig. 9. An improvement of 1% is observed at the 10% power level and an efficiency improvement of 0.6% (14 kW total) is observed at the 100% power level. The economic value of this improvement could be determined using the methods mentioned in Table I and by using a specific wind regime. Initial capital value would be in excess of $50k in most Class II wind sites. These results are intentionally presented for the line side inverter only. In the case where a passive rectifier is used on the generator side, the line-side inverter efficiency dominates the system. When an active rectifier is used, a similar analysis to that performed here (described in [9]) can be applied to the generator-side active rectifier. In this later case, the losses are approximately double those shown in Fig. 8 resulting in an approximate mathematical squaring of the efficiencies for the entire converter. The advantages of the Si/SiC approach are amplified in the active rectifier case.

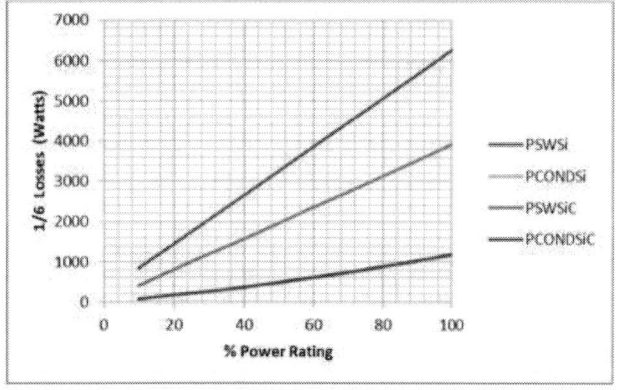

Fig. 8 Semiconductor loses for Si and SiC based inverters.

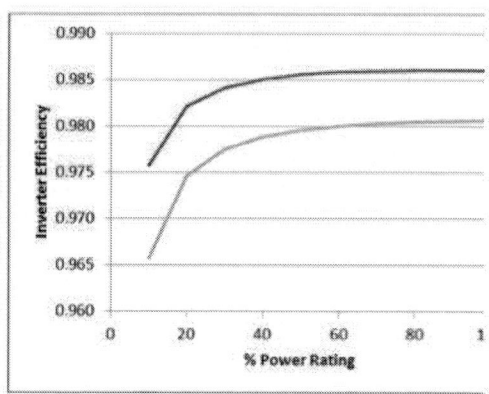

Fig. 9 Efficiency plots for the two inverters with magnetics included.

ACKNOWLEDGMENT

The authors would like to acknowledge the financial support of US Department of Energy and the leadership provided by the National Renewable Energy Laboratory.

REFERENCES

[1] R. J. Nelson, H. Ma, and N. M. Goldenbaum, "Fault Ride-Through Capabilities of Siemens Full-Converter Wind Turbines," Power and Energy Society General Meeting, 2011 IEEE, 24-29, July, 2011.

[2] A. Nabae, I. Takahashi and H. Akagi, "A New Neutral-Point-Clamped PWM Inverter," IEEE Transactions on Industry Applications, Vol IA-17, NO. 5, Spetember/October, 1981.

[3] B. Wu, "High Power Converters and AC Drives," IEEE Press, Piscataway, NJ, nad John Wiley, Hoboken, NJ, 2006, pp 149 - 173.

[4] L. Kumar, M.A. Hasnat, "Multilevel Inverter in Grid Connected PMSG Based Wind Power Generation Systems," International Journal of Electronic and Electrical Engineering, ISSN 0974-2174 Volume 7, Number 8, 2014, pp 867 -872.

[5] B. Backlund, M. Rahimo, S. Klaka, J. Siefken, "Topologies, Voltage Ratings and State of the Art Power Semiconductor Devices for Medium Voltage Wind Energy Conversion," IEEE, 2009 Power Electronics and Machines in Wind Applications Conference, PEMWA, June, 2009, Lincoln, USA.

[6] M. Schweizer, T. Friedli, and J.W. Kolar, "Comparative Evaluation of Advanced 3-Level Inverter/Converter Topologies Against 2-Level Systems," Swiss Federal Institute of Tchnology, Zurich, Power Electronic Systems Laboratory, European Center for Power Electronics, Nuremberg, DE, 2010.

[7] T. Duong, A. Hefner, K. Hobart, S.-H. Ryu, D. Grider, D. Berning, J. M. Ortiz-Rodriguez, E. Imhoff, J. Sherbondy, "Performance of Hybrid 4.5 kV JBS Freewheeling Diode and Si IGBT," Proc. of 2011 IEEE Applied Power Electronics Conference, p. 1057.

[8] "RM800DG-90F High Voltage Diode Module Data Sheet," Available from Powerex and Mitsubishi websites.

[9] I. Staudt, A. Wintrich, K. Haddad, V. Cardi, "Numerical Loss Calculation and Simulation Tool for 3L NPC Converter Design," Proceeding of the 2011 PCIM Conference, Nuremberg, DE.

Understanding the Limitations and Impact Factors of Wide Bandgap Devices' High Switching-Speed Capability in a Voltage Source Converter

Zheyu Zhang, Fred Wang, Leon M. Tolbert, Benjamin J. Blalock, and Daniel Costinett

Center for Ultra-wide-Area Resilient Electric Energy Transmission Networks (CURENT)
Department of Electrical Engineering and Computer Science,
The University of Tennessee, Knoxville, TN 37996-2250, USA
zzhang31@utk.edu

Abstract— This paper focuses on understanding the key impacting factors for switching speed of wide bandgap (WBG) devices in a voltage source converter. First, the constraints and challenges of WBG devices during fast switching transients are summarized. Special attention is given to the transient gate-source and drain-source voltages. Second, the impacts of major components in voltage source converter, including gate drivers, parasitics, inductive loads, and cooling systems, on the switching performance of power devices are systematically investigated. The critical parameters for each component are highlighted. Finally, design criteria are suggested to maximize switching speed of WBG devices.

Keywords—Wide Bandgap devices, Switching performance, Voltage source converter

I. INTRODUCTION

With the development of power semiconductor device technologies, especially wide bandgap (WBG) device technologies, state-of-the-art devices, such as Gallium Nitride (GaN) and Silicon Carbide (SiC) transistors, allow faster switching speed, leading to lower switching loss, shorter dead time for a phase-leg, and higher switching frequency. All of these are beneficial to high power efficiency, quality and density [1-3].

However, high switching-speed of WBG devices makes their switching behavior more susceptible to the cirucit parasitics and noise. In the end, unlike the excellent switching performance of WBG devices tested for manufacturer's datasheets using the double pulse test (DPT) that is a well-accepted method for dynamic characterization of power devices in both industry and academia, the observed switching performance in actual power converters is almost always worse. Previously reported work observed that in a SiC based current source rectifier the turn-off energy was increased by a factor of 5, with 3 times longer current fall time compared with the test result of the DPT under the same operating conditions [4]. Similarly, in a voltage source inverter, due to the increased overshoot current and

slower turn-off time, the total switching energies were increased by a factor of 1.5 to 1.8 [5].

To fully utilize the potential advantages of wide bandgap device technologies, the key factors which impact the high switching speed capability must be understood. This paper focuses on the voltage source converter (VSC) because it is the widely applied topology in many popular applications. Also, the three-phase voltage source inverter fed motor drives are selected as the representative VSC under evaluation since it covers almost all the components in a voltage source converter. Thus, the generality of the following analysis is not lost. Fig. 1 displays the configuration comparison between the double pulse tester and three-phase voltage source inverter. As can be observed, in addition to the interaction between gate drivers and devices in the DPT, the switching performance of the device under test (i.e., lower switch in phase A) is affected by the interference between upper and lower switches in a phase-leg. Also, in a three-phase voltage source inverter, the other phase-legs (i.e., phase B and C) and heat sink may become the possible impacting factors. Moreover, in the motor drive system, the motor/cable load will impact the switching performance as well. Consequently, a systematical investigation of the limitations and impact factors of WBG devices' high switching-speed capability is necessary.

This paper first summarizes the constraints and challenges of WBG semiconductors during fast switching transition. Special attention is given on the transient gate-source and drain-source voltages. Second, the impacts of major components in three-phase voltage source inverter, including gate drivers, parasitics, inductive loads and heat sink on the switching performance are investigated. The critical parameters for each component are highlighted and categorized according to their effects on the high switching-speed performance of WBG devices. Finally, design criteria of predefined critical parameters of components are suggested to maximize switching speed of WBG devices as well as to limit the switching voltage spike within a safe operation range.

978-1-4799-5494-0/14 $31.00 © 2014 IEEE

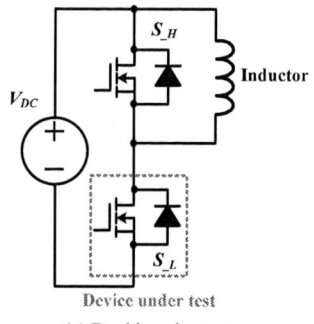

(a) Double pulse tester.

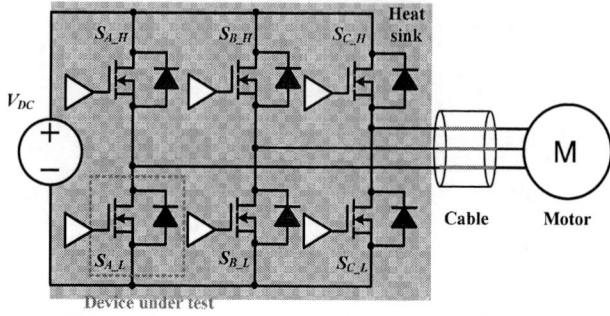

(b) Three-phase voltage source inverter fed motor/cable load.

Fig.1. Configuration comparison between double pulse tester and three-phase voltage source inverter fed motor drives.

II. CONSTRAINTS AND CHALLENGES OF HIGH SWITCHING-SPEED CAPABILITY OF WBG DEVICES

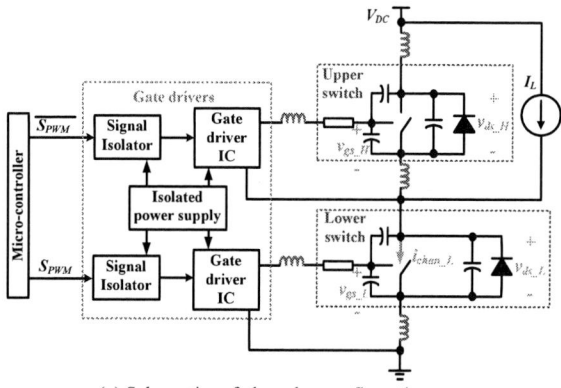

(a) Schematics of phase-leg configuration.

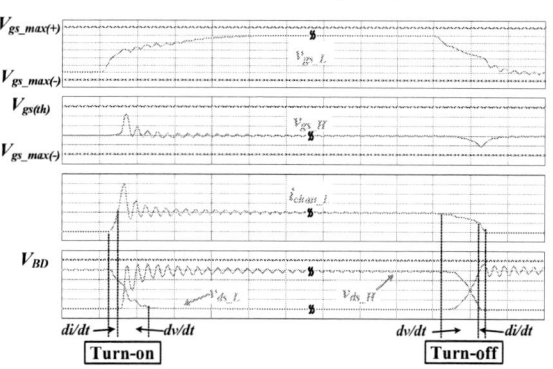

(b) Transient voltage waveforms.

Fig. 2. Constraints of voltage spike.

During switching transients, voltage spikes across drain-source terminals and gate-source terminals of power devices must be limited within the safe operating ranges; otherwise, the devices' reliability is affected. Specifically, as shown in Fig. 2, during the switching transient of the lower switch in a phase-leg, which is the basic cell of the VSC, the following requirements must be satisfied:

$$v_{ds\_L} < V_{BD}, \ v_{ds\_H} < V_{BD} \tag{1}$$

$$V_{gs\_\max(-)} < v_{gs\_L} < V_{gs\_\max(+)} \tag{2}$$

$$V_{gs\_\max(-)} < v_{gs\_H} < V_{gs(th)} \tag{3}$$

where v_{ds_L}, v_{ds_H} refer to drain-source voltages of the lower and upper switches, respectively; v_{gs_L}, v_{gs_H} represent to gate-source voltages of the lower and upper switches; V_{BD} is the breakdown voltage; $V_{gs\_max(+)}$ and $V_{gs\_max(-)}$ are the positive and negative maximum allowable gate voltages, respectively; and $V_{gs(th)}$ is the threshold voltage. Since transient voltage overshoots become worse as the switching-speed increases, the maximum switching-speed represents the upper limit.

A. Drain-source Voltage Overshoot

As can be observed in Fig. 2(b), there are two voltage spikes across the drain-source terminals of power devices: v_{ds_H} during the turn-on transient and v_{ds_L} during the turn-off transient. The mechanisms causing turn-on and turn-off over-voltage are different.

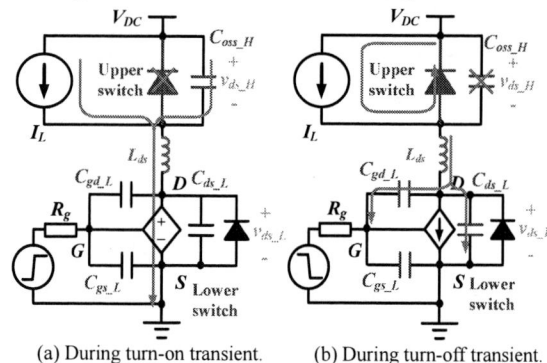

(a) During turn-on transient. (b) During turn-off transient.

Fig. 3. Mechanisms causing voltage overshoot.

During the turn-on transient of the lower switch, the over-voltage occurs during the dv/dt subinterval (see in Fig. 2(b)). As shown in Fig. 3(a), the lower switch can be represented as a voltage source with dv/dt; and the upper switch is modeled as an output capacitance since the body diode has started to block the dc bus voltage. Thus, the over-voltage across the drain-source terminals of the upper switch is originated by a dv/dt voltage source across the resonant network formed by the power loop parasitic inductance L_{ds} in series with the output capacitance of the upper switch C_{oss_H}. Fig. 4(a) displays the simplified model for the turn-on over-voltage, where R_{stray} is the stray resistance in the power loop. Fast switching speed leads to high dv/dt, resulting in serious resonance and high over-voltage. In addition to the excitation, L_{ds} is a critical parameter that

978-1-4799-5494-0/14 $31.00 © 2014 IEEE

significantly affects the resonant period and damping factor.

During the turn-off transient of the lower switch, the over-voltage will be observed during the *di/dt* subinterval (see in Fig. 2(b)). As illustrated in Fig. 3(b), the low switch is modeled as a current source with *di/dt*; and the upper switch is modeled as a body diode operating in conduction mode and its output capacitance C_{oss_H} is bypassed. Therefore, a *di/dt* current source is the excitation added across the parallel resonant network consisting of L_{ds} and the output capacitance of the lower switch C_{oss_L} that induces the turn-off over-voltage, as shown in Fig. 4(b). Similar to the turn-on over-voltage, the turn-off switching speed determines *di/dt*, and then influences the turn-off over-voltage. Also, L_{ds} is the main contributor to the resonance.

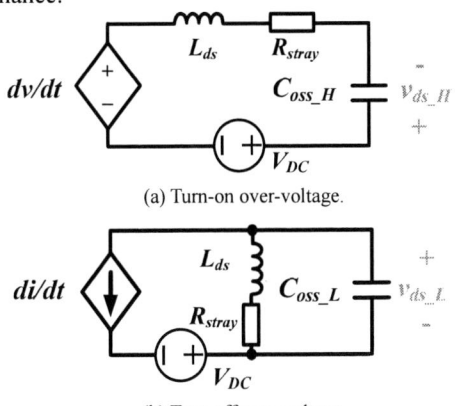

(a) Turn-on over-voltage.

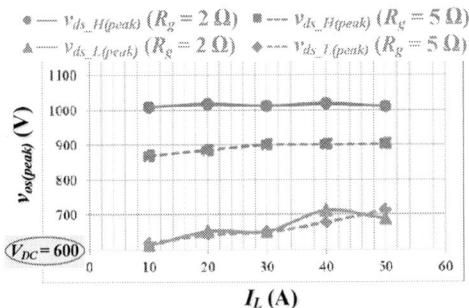

(b) Turn-off over-voltage.

Fig. 4. Model of drain-source over-voltage.

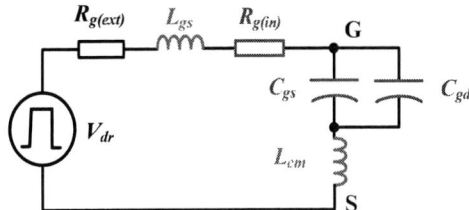

Fig. 5. Comparison between turn-off and turn-on over-voltage.

In the end, although the mechanisms causing turn-on and turn-off over-voltages are different, they both become worse as switching speed increases and damping factors in resonant networks decrease. WBG devices with high switching-speed capability and small on-state resistance make this issue more serious. Also note that compared to the turn-on over-voltage, the turn-off over-voltage has been thoroughly investigated [6, 7]. However, based on the phase-leg power module built using Cree's CPM2-1200-0025B SiC MOSFETs and CPW5-1200-Z050B SiC Schottky diode, the test data shows that the turn-on over-voltage is higher than the turn-off over-voltage (see Fig. 5). It is therefore necessary to pay special attention to the turn-on over-voltage when using WBG devices in a VSC.

B. Gate-source voltage spikes

In the phase-leg configuration, gate-source voltages of both the lower and upper switches should be taken into account. For the lower switch, which is the operating device in a phase-leg, the gate-source voltage spikes are induced by the resonance between the parasitic inductance and capacitance associated in the gate loop, as shown in Fig. 6, where L_{gs} and L_{cm} refer to the gate loop inductance and common source inductance, respetively; C_{gs} and C_{gd} represent gate-source capacitance and Miller capacitance, respetively; $R_{g(in)}$ and $R_{g(ext)}$ are the internal gate resistance of the power device and external gate resistance. Usually, for SiC devices, this gate voltage spike is not severe due to the sufficiently safe margin between the output voltage of the gate driver and gate voltage ratings of power devices, as well as relatively large $R_{g(in)}$. However, for certain GaN devices, such as EPC transistors, because of the low $V_{gs\_max(+)}$, this gate-source over-voltage may overstress the gate terminals of power devices [8]. In general, considering the optimized layout design and adjustable damping factor (i.e., $R_{g(ext)}$) in the gate loop, gate-source voltage spikes of the lower switch are not critical [9].

Fig. 6. Equivalent circuit of gate loop with parasitics.

Additionally, in the phase-leg configuration, the switching process of the lower switch will affect gate-source voltage of the upper one, which in turn might worsen the switching performance of the lower switch and the reliability of upper switch. This interaction between the two switches is called cross-talk and is illustrated in Fig. 7 [10].

During the turn-on transient of the lower switch, the drain-source voltage rise on the upper switch induces a current that is coupled through its Miller capacitance C_{gd_H}. This induced current generates a voltage drop across the gate resistance R_{g_H}, and then a positive spurious v_{gs_H}. If v_{gs_H} exceeds $v_{gs(th)}$, a shoot-through problem occurs. During the turn-off transient of the lower switch, similarly, a negative spurious v_{gs_H} would be induced. Once v_{gs_H} is less than $V_{gs\_max(-)}$, a device's reliability is degraded. The peak value of this spurious gate voltage is shown as [10]

$$v_{gs\_H(max)} = \frac{dv}{dt} \times R_{g\_H} C_{gd\_H} \times (1 - e^{-\frac{V_{DC}}{\frac{dv}{dt} \times C_{iss\_H} R_{g\_H}}})$$

(4)

where *dv/dt* is the slew rate of drain-to-source voltage of the lower switch, partially representing the switching speed. R_{g_H}, C_{gd_H} and C_{iss_H} refer to the resistance in the gate loop, Miller capacitance and input capacitance of the upper switch. It is obvious that in addition to the parameters related to the power device itself, the

978-1-4799-5494-0/14 $31.00 © 2014 IEEE

switching speed will significantly affect gate-source voltage spikes. For WBG devices, fast switching capability worsens the peak value of v_{gs_H}. In the meantime, the low $V_{gs(th)}$ and $V_{gs\_max(-)}$ shrink the safe margin of the gate voltage, enabling v_{gs_H} to easily exceed the required reliable range as in (3) [11].

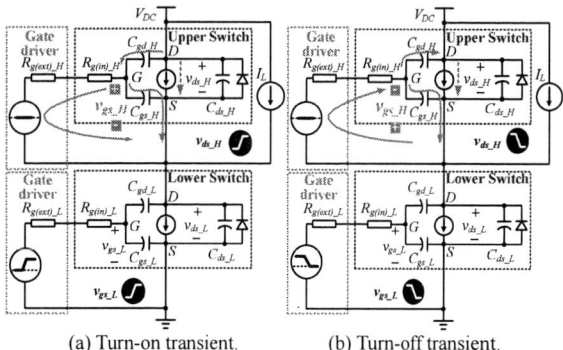

(a) Turn-on transient. (b) Turn-off transient.

Fig. 7. Mechanisms causing cross-talk.

In total, voltage spikes in the power loop and gate loop, especially v_{ds_H} and v_{gs_H}, are the constraints for fast switching. According to the inherent properties of WBG devices, such as small on-state resistance, low $v_{gs\_max(+)}$, $v_{gs\_max(-)}$, and $v_{gs(th)}$, this issue becomes more challenging.

III. IMPACTING FACTORS IN THREE-PHASE VOLTAGE SOURCE INVERTER FOR MOTOR DRIVES

The impact of each component in a three-phase voltage source inverter on the switching performance is systematically investigated, including gate driver and parasitics in a phase-leg, inductive load, and heat sink in three-phase motor drive systems.

A. Gate Driver

As can be observed in Fig. 2(a), the gate driver primarily consists of the gate driver integrated circuit (IC), signal isolator, and isolated power supply. Among them, the gate driver IC directly hooks up with the gate terminals of power devices and is one of the dominant components to determine the switching performance of power devices [12]. Fig. 8 depicts the schematics of phase-leg configuration with gate driver IC. According to its functionality, the gate driver IC can be simplified as a PWM voltage source in series with internal resistance, where the voltage source can be analytically expressed by rise (fall) time and amplitude of gate driver output voltage (t_r, t_f and V_p in Fig. 8), the internal resistance represents the pull-up (-down) resistance of gate driver induced by transistors S_1 and S_2. It is the pull-up (-down) resistance of gate driver, rise (fall) time, and amplitude of gate driver output voltage that determine the gate driver capability, and then affect the power device switching speed. Meanwhile, under different operating conditions, the roles of these parameters on switching speed are different: if the rise (fall) time of gate driver output voltage is smaller than the delay interval of switching time, which is a reasonable assumption for high voltage SiC devices, the

dominant factors are pull-up (-down) resistance and voltage amplitude. Otherwise, the switching speed depends largely on the rise (fall) time, such as low voltage GaN transistors [10].

Also, the gate driver output voltage and external gate resistance impact the spurious gate voltage of the upper switch induced by cross-talk, and then influence the switching speed.

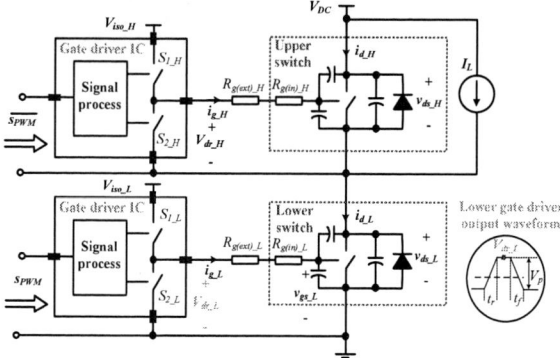

Fig. 8. Impact of gate driver IC.

Additionally, signal isolator and isolated power supply, as the auxiliary components to transmit PWM signals and power gate driver IC, would also limit the switching speed if their dv/dt immunity capability is less than the dv/dt during switching transients of the power devices [13]. For example, generally, the maximum dv/dt immunity capability of the commercially available signal isolator is around 35 kV/μs to 50 kV/μs. However, the dv/dt of SiC devices has already approached up to 80 kV/μs. Even worse, for GaN transistors, dv/dt of 150 kV/μs has been reported [13].

B. Parasitics

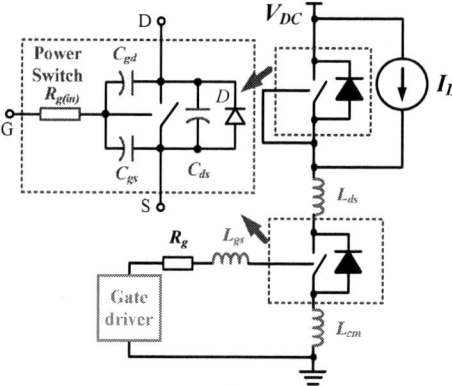

Fig. 9. Parasitics involving in the switching commutation loop.

Parasitics are another critical impacting factor to limit switching speed [9]. As shown in Fig. 9, there are mainly six parasitics involved in switching commutation loop, including parasitics in the gate loop, L_{gs} and C_{gs}, parasitics in the power loop L_{ds} and C_{ds}, as well as mutual parasitics shared in both the gate and power loop, L_{cm} and C_{gd}. Among them, power loop inductance L_{ds}, common source inductance L_{cm} and Miller capacitance C_{gd} significantly impact the switching performance of power devices, especially for

978-1-4799-5494-0/14 $31.00 © 2014 IEEE

fast switching WBG devices.

As mentioned in Section II, power loop inductance L_{ds} is the crucial contributor to the parasitic ringing and overshoot voltage. In addition, the mutual parasitics L_{cm} and C_{gd} provide negative feedback paths from the power loop to the gate loop due to di/dt and dv/dt, leading to slower switching speed and more switching losses [9].

C. Inductive Loads

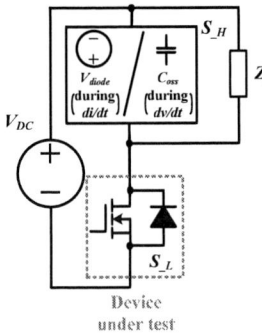

Fig. 10. Simplified DPT circuit.

In the DPT circuit, as shown in Fig. 1(a), the load inductor is paralleled with the upper switch, assuming the lower switch is selected as the device under test (DUT). During the switching transient, the upper switch remains off, and can be modeled as a conducting diode when the switching current is commutating or an output capacitance while the switching voltage is changing. Thus, the DPT circuit can be redrawn as in Fig. 10, where Z_L refers to the inductive load impedance. If Z_L is always much greater than the equivalent impedance of the upper switch during the switching transient, the impact of inductive load on the switching performance can be ignored; otherwise, Z_L must be taken into consideration.

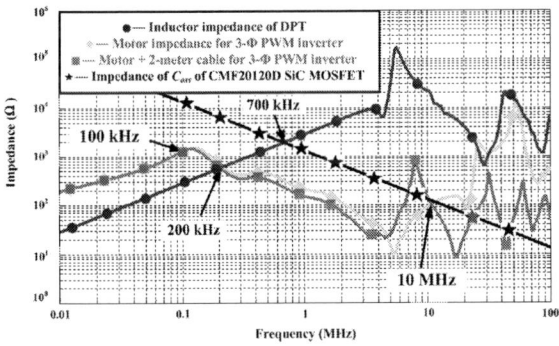

Fig. 11. Impedance comparison among different inductive loads.

In the typical DPT, the load inductor of DPT is optimally-designed so that it induced parasitics such as equivalent parallel capacitance (EPC) are fairly small, enabling the impedance of the load inductor dependence on the switching related frequency spectrum to be much larger than the impedance of the upper switch. Therefore, the load inductor negligibly affects the switching performance in the DPT.

However, in adjustable speed drive systems, the parasitics of the induction motor significantly decreases its high frequency impedance, as shown in Fig. 11. Measured using the Agilent 4294A impedance analyzer, the impedance of the load inductor for the DPT is greater than that of the 7.5 kW induction motor beyond

200 kHz, although the inductance of the DPT based inductor is smaller than that of the induction motor. In addition, the impedance of the induction motor is even smaller than that of the output capacitance of Cree's CMF20120D SiC MOSFET (120 pF in datasheet) up to 10 MHz. On the other hand, above 700 kHz, the impedance of the DPT based inductor is greater than that of the device's output capacitance. Consequently, due to the small impedance at high frequency, the inductor motor more significantly impacts the switching behavior of DUT.

Fig. 11 also provides the impedance comparison between the induction motor and the induction motor plus 2-meter power cable. One can readily observe that the impedances of the induction motor and the induction motor plus power cable are identical below 100 kHz, but significantly different at high frequency. Especially above 10 MHz, the impedance of the induction motor with power cable is always smaller than that of the induction motor, and is even comparable to the C_{oss} impedance of the SiC MOSFET under evaluation. According to the aforementioned analysis, the more serious impact of the induction motor with power cable on the switching performance can be predicted. Also, for the higher power rating induction motor with longer power cable in many practical applications, the associated impedance Z_L at high frequency becomes lower [14]. Thus, the motor load selected in our study case indicates a conservative impact of the induction motor on the switching performance.

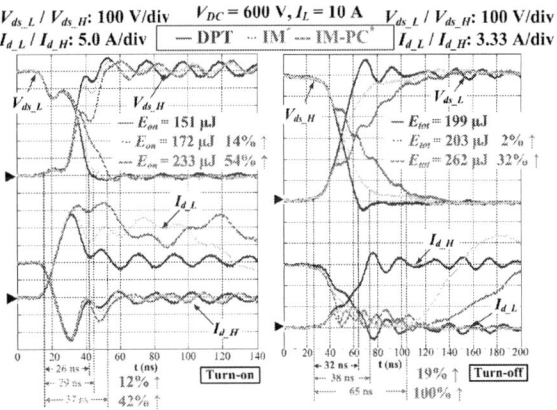

* where IM and PC refer to induction motor, power cable.

Fig. 12. Switching waveform comparison of different inductive loads.

Fig. 12 shows the switching waveform comparison among different inductive loads. As compared to the switching behavior with the DPT load inductor, the parasitics of the induction motor cause the turn-on and turn-off switching time to increase from 26 ns to 29 ns and 32 ns to 38 ns, respectively. The total energy loss slightly increases as well. After inserting a 2-meter power cable between the inverter and motor, the switching performance becomes even worse: switching time increases up to 42% during turn-on, and doubles during turn-off; an additional 32% of energy loss is dissipated during the switching transient.

D. Heat Sink

In a three-phase voltage source inverter, SiC devices are attached to a heat sink for thermal management. Usually a thin layer of insulating material is used to separate the SiC devices from the electrically conductive heat sink. Thus, a parasitic capacitance is formed between the drain base plate of the SiC devices and the common heat sink plate, as shown in Fig. 13 [5]. In the end, this capacitance is paralleled with devices, which increases their effective output capacitance, and then negatively affects the switching speed.

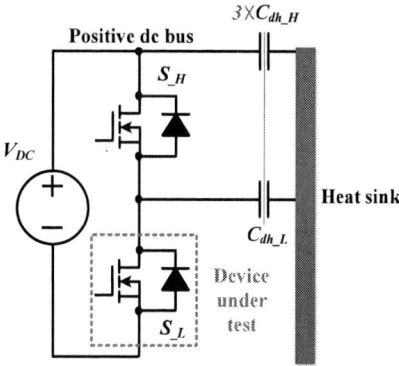

Fig. 13. Capacitive coupling between devices and heat sink.

V. CONCLUSIONS

According to the inherent properties of WBG devices, the impacting factors of high switching-speed capability of WBG devices in a voltage source converter can be categorized into three groups according to their effects on the switching performance. Parameters in the first group directly limit the switching speed, such as dv/dt immunity capability of the gate driver isolation, common source inductance, parasitics of inductive loads, and capacitance induced by the heat sink. The second group of parameters increases the spikes of gate-source and drain-source voltages, requiring the switching speed be reduced to guarantee the reliability of devices. These include the gate/power loop parasitic inductances. The third parameters' group increases the switching speed with the penalty of deteriorating the over-voltage, such as gate driver capability, external gate resistance.

To maximize switching speed of WBG devices as well as to limit the switching voltage spike within safe operational range, parameters in the first and second groups should be optimized as much as possible, such as using two heat sinks to mitigate the capacitive coupling effect [5], adopting advanced packaging techniques to minimize parasitics in the switching loop [15, 16], and employing dedicated auxiliary networks to compensate for the inductive load parasitics. For the gate driver in the third group, considering the tradeoff between the fast switching and transient voltage spike, the active gate driver scheme is preferred [11].

ACKNOWLEDGEMENT

The authors would like to thank the II-VI Foundation for supporting this research. This work made use of the Engineering Research Center Shared Facilities supported by the Engineering Research Center Program of the National Science Foundation and DOE under NSF Award Number EEC-1041877 and the CURENT Industry Partnership Program.

The author would also like to thank Dr. Zhenxian Liang, Dr. Puqi Ning, Ben Guo, Zhiqiang Wang, and Weimin Zhang for their contributions to this paper.

REFERENCES

[1] B. Callanan. CREE CPWR AN-08. Application considerations for silicon carbide MOSFETs [Online]. Available: http://www.cree.com

[2] A. Lidow, J. Strydom, M. d. Rooij, and D. Reusch, *GaN Transistors for Efficient Power Conversion.* El Segundo, CA: Power Conversion Publications, 2014.

[3] Z. Weimin, X. Zhuxian, Z. Zheyu, F. Wang, L. M. Tolbert, and B. J. Blalock, "Evaluation of 600 V cascode GaN HEMT in device characterization and all-GaN-based LLC resonant converter," in *Energy Conversion Congress and Exposition (ECCE), 2013 IEEE,* 2013, pp. 3571-3578.

[4] C. J. Cass, W. Yi, R. Burgos, T. P. Chow, F. Wang, and D. Boroyevich, "Evaluation of SiC JFETs for a Three-Phase Current-Source Rectifier with High Switching Frequency," in *Applied Power Electronics Conference, APEC 2007 - Twenty Second Annual IEEE,* 2007, pp. 345-351.

[5] I. Josifovic, J. Popovic-Gerber, and J. A. Ferreira, "Improving SiC JFET Switching Behavior Under Influence of Circuit Parasitics," *Power Electronics, IEEE Transactions on,* vol. 27, pp. 3843-3854, 2012.

[6] L. Qian, W. Shuo, A. C. Baisden, W. Fei, and D. Boroyevich, "EMI Suppression in Voltage Source Converters by Utilizing dc-link Decoupling Capacitors," *Power Electronics, IEEE Transactions on,* vol. 22, pp. 1417-1428, 2007.

[7] C. Zheng, D. Boroyevich, P. Mattavelli, and K. Ngo, "A frequency-domain study on the effect of DC-link decoupling capacitors," in *Energy Conversion Congress and Exposition (ECCE), 2013 IEEE,* 2013, pp. 1886-1893.

[8] http://epc-co.com/epc/.

[9] C. Zheng, D. Boroyevich, and R. Burgos, "Experimental parametric study of the parasitic inductance influence on MOSFET switching characteristics," in *Power Electronics Conference (IPEC), 2010 International,* 2010, pp. 164-169.

[10] Z. Zheyu, Z. Weimin, F. Wang, L. M. Tolbert, and B. J. Blalock, "Analysis of the switching speed limitation of wide band-gap devices in a phase-leg configuration," in *Energy Conversion Congress and Exposition (ECCE), 2012 IEEE,* 2012, pp. 3950-3955.

[11] Z. Zheyu, F. Wang, L. M. Tolbert, and B. J. Blalock, "Active Gate Driver for Crosstalk Suppression of SiC Devices in a Phase-Leg Configuration," *Power Electronics, IEEE Transactions on,* vol. 29, pp. 1986-1997, 2014.

[12] L. Balogh, "Design and application guide for high speed MOSFET gate drive circuits," in *Power Supply Design Seminar,* 2001.

[13] Z. Wei, H. Xiucheng, F. C. Lee, and L. Qiang, "Gate drive design considerations for high voltage cascode GaN HEMT," in *Applied Power Electronics Conference and Exposition (APEC), 2014 Twenty-Ninth Annual IEEE,* 2014, pp. 1484-1489.

[14] B. Mirafzal, G. L. Skibinski, R. M. Tallam, D. W. Schlegel, and R. A. Lukaszewski, "Universal Induction Motor Model With Low-to-High Frequency-Response Characteristics," *Industry Applications, IEEE Transactions on,* vol. 43, pp. 1233-1246, 2007.

[15] L. Zhengyang, H. Xiucheng, Z. Wenli, F. C. Lee, and L. Qiang, "Evaluation of high-voltage cascode GaN HEMT in different packages," in *Applied Power Electronics Conference and Exposition (APEC), 2014 Twenty-Ninth Annual IEEE,* 2014, pp. 168-173.

[16] L. Zhenxian, N. Puqi, F. Wang, and L. Marlino, "Planar bond all: A new packaging technology for advanced automotive power modules," in *Energy Conversion Congress and Exposition (ECCE), 2012 IEEE,* 2012, pp. 438-443.

The Development of A High-Voltage Power Device Evaluation Platform

Lixing Fu, Xuan Zhang, He Li, Xintong Lu and Jin Wang

Department of Electrical and Computer Engineering
The Ohio State University
Columbus, OH
fu.141@osu.edu

Abstract—This paper presents the development of a high voltage evaluation platform to investigate both the static and switching characteristics of silicon (Si) and silicon carbide (SiC) based high voltage (HV) power devices. The test methodologies, design, implementations, and safety guidelines for the HV device evaluation are elaborated. A 15 kV rated double pulse tester is built to study the dynamic behaviors of HV power devices. The gate drive circuit and main circuit are designed to achieve clean waveforms and safe operation. Meanwhile, a high voltage high frequency load inductor is designed with finite element analysis (FEA) tools, taking into consideration the non-uniform voltage distribution on the windings under sharp voltage impulse. Both the design details and test results are presented.

Keywords — high voltage; SiC; wide bandgap; MOSFET; characterization

I. INTRODUCTION

Power electronics has been playing an essential role for the purpose of realizing power conversion and power conditioning in the modern power systems. Power semiconductor device is one of the key factors that decide its development [1]–[3]. The functionality of power devices can directly limit the capability of the design, realization and control of power electronics. From electric tubes, thyristors to IGBTs, power semiconductors are developing towards the aim of easy control, high reliability, high switching speed and low losses. All these aims can lead power electronic systems to be more robust and more efficient with higher power density. For a lot of utility applications, high voltage (HV) power devices, especially HV IGBTs, have been widely used. Large scale long distance high voltage dc (HVDC) transmission, unified power factor correction and solid state transformers are all realized and improved with HV power devices for high efficiency and high reliability of power delivery [3].

Currently, the highest voltage rating of Si based IGBTs is 6.5 kV. In order to achieve higher blocking voltage, a large number of power devices need to be connected in series, which increases the total cost and downgrades the system reliability. With the development of wide bandgap (WBG) materials, SiC based devices have been developed with higher breakdown voltage, lower conduction loss and faster switching speed compared to Si based devices. The higher breakdown voltage can significantly reduce the number of required series-connected devices or modules for HV applications. The lower conduction loss helps shrink the heat sink and improves

the system efficiency. Moreover, the higher switching frequency enables better pulse modulation to guarantee high power quality, and the use of smaller energy storage components (capacitors and inductors) and filters. As results, SiC power devices are considered as promising candidates of the next-generation HV devices.

For the purpose of better utilization of the HV power devices in practical applications, this paper presents the development of a HV device evaluation platform for comprehensive static and dynamic characterizations. First, the static characterizations aim to provide accurate information for the converter loss estimation and device modeling. A curve tracer extracts the I/V plots at elevated gate biases and temperatures, while some self-designed peripheral circuits can apply HV bias on the device under test (DUT) for HV related measurement. Second, the switching characterizations aim to deliver inductive hard switching waveforms, which help with the switching behavior interpretation and switching loss estimation. A double pulse tester is built to study the inductive switching behaviors of HV power devices. For the purpose of realizing clean switching waveforms with minimized voltage and current oscillation, both the gate drive loop and main circuit loop require stringent layout. Meanwhile, an HV load inductor has been designed with finite element analysis (FEA) tools. Considering the non-uniform voltage distribution on windings under sharp voltage impulse, parasitics based circuit simulations help understand the possible voltage distribution of the inductor windings at high dv/dt and provide enough guidance for insulation design. For HV tests, a series of safety design and test considerations are presented, regarding the grounding, clearance and protection issues for HV-device testing which require extra care. Test setup and results of a 4.5 kV rated Si MOSFET and a 15 kV rated SiC MOSFET are presented.

II. STATIC CHARACTERIZATION OF HV POWER DEVICE

The static characteristics are the parameters that decides the switching performance and efficiency of a power semiconductor device. They usually include: 1) device I/V family curves; 2) threshold voltage; 3) device transconductance; 4) device leakage current at different voltage biases; 5) device third quadrant characteristics (or parallel diode characteristics); and 6) device parasitic capacitances. When the device static characteristics are extracted properly, they can be utilized for appropriately circuit design, and furthermore, for building device behavior based model. The simulation results of device switching based on device modeling can be used to validate the switching test.

A power device analyzer (curve tracer) is usually used for the parameter extraction of DUT. At relatively low voltage, most of device static characteristics can be obtained with guaranteed accuracy. However, due to the high blocking voltage of HV power semiconductor devices, the leakage current and parasitic capacitances at HV condition cannot be easily measured with currently available equipment. Therefore, external circuitries are need to apply HV on the device for measurement [8].

Fig. 1 presents the output capacitance measurement of a 15 kV rated SiC MOSFET up to 3 kV. The C_{OSS} of DUT reaches around 4 nF at low voltage and drops to 30 pF at HV due to the increase of depletion region. Fig. 2 shows the IV family curves of the same device. It can be observed that the $R_{ds(on)}$ becomes stable after the gate voltage rises above 13 V. The measured $R_{ds(on)}$ is 1.48 Ohm at the Vgs = 15 V. This is much smaller than the $R_{ds(on)}$ of highest voltage rated Si MOSFET (4.5 kV). The HV SiC MOSFET also shows good third quadrant characteristics, as shown in Fig. 3. When the gate voltage is above its threshold voltage, the $R_{ds(on)}$ of DUT drops

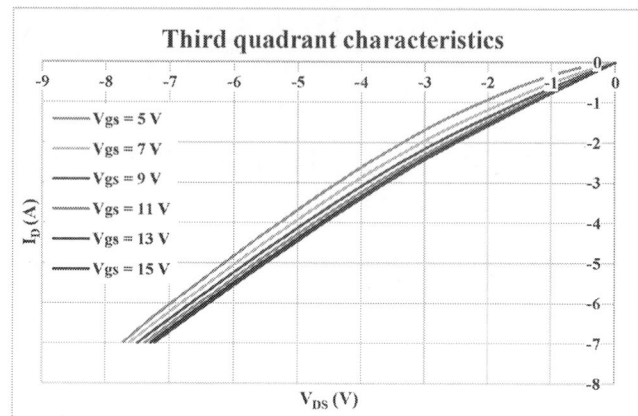

Fig. 3 Third quadrant characteristics of 15 kV rated SiC MOSFET

to around 1 Ohm, which helps reduce the conduction loss in synchronized rectification significantly.

III. SWITCHING CHARACTERIZATION OF HV POWER DEVICE

The switching behavior of a power semiconductor device usually involves several different stages [6]. During switching transients, the overlaps of device voltage and current cause switching losses. High dv/dt and di/dt can lead to voltage and current overshoot, while their induced electromagnetic interference can affect the whole circuit operation. To better study the switching characteristics of HV device, a clamped inductive switching circuit needs to be built, and associated measurements need to have high accuracy [9].

A. Circuit design and hardware realization

For inductive switching, a typical double pulse test (DPT) circuit is employed in this test. The circuit diagram is shown in Fig. 4. After the load inductor is charged to aimed current value, the first turn-off and second turn-on will deliver the information of hard switching. In practice, due to the parasitic components of the circuit such as stray inductance and capacitance, there could be certain ringings in the measured switching waveform. To mitigate this problem, stringent circuit layout is needed to minimize loop stray inductance. In Fig. 5, a test setup for a 4.5 kV MOSFET is shown. In this circuit, copper sheets are used to connect the DUT and diode with the dc capacitor. Minimum length of copper is utilized in the layout, and the circuit components are placed as close as possible to the HV dc capacitor. Therefore, the circuit stray

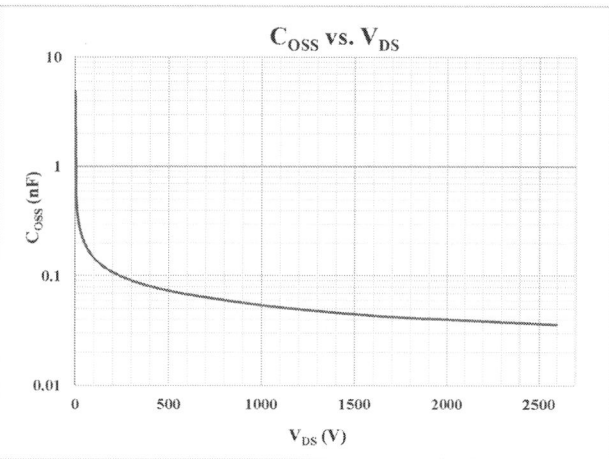

Fig. 1 Output capacitance of 15 kV SiC MOSFET

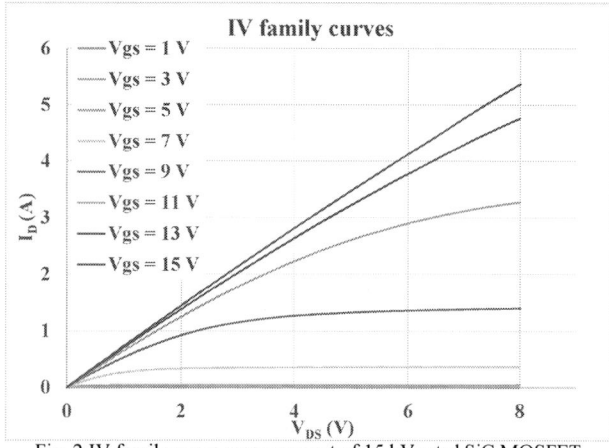

Fig. 2 IV family curve measurement of 15 kV rated SiC MOSFET

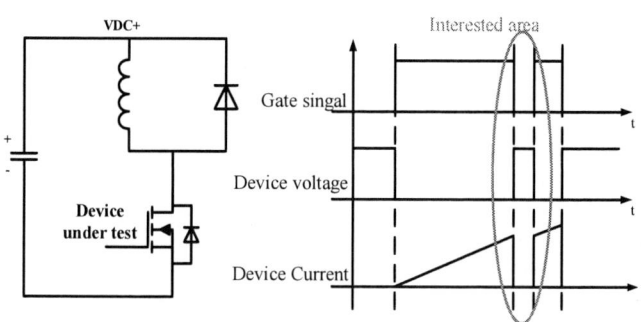

Fig. 4 Double pulse test circuit and control scheme

978-1-4799-5494-0/14 $31.00 © 2014 IEEE

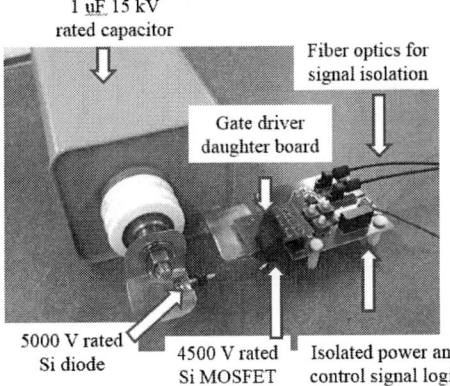

Fig.5 Circuit prototype for HV device switching characterization

inductance can be reduced into an acceptable range which enables small ringing in the switching waveforms.

For HV applications, circuit insulation has been considered as a major issue. When small sizing is needed to reduce circuit parasitics, the clearances among points with large potential difference need to be considered with care. Meanwhile, sharp corners on the circuit connections should be avoided severe corona issue.

B. Gate drive design considerations

Gate drive circuits for high voltage power devices mainly requires more considerations regarding isolation and safety. In a switching mode power supply, the ground of a gate drive circuit usually shifts with high dv/dt. And the floating ground can have high voltage bias up to tens of kilovolts. In this case, high isolation voltage needs to be achieved by the gate drive to guarantee circuit safe operation. Therefore, in the HV gate drive design, the gating logic signal is realized with optic fiber, and the isolated power supply needs to be carefully selected or

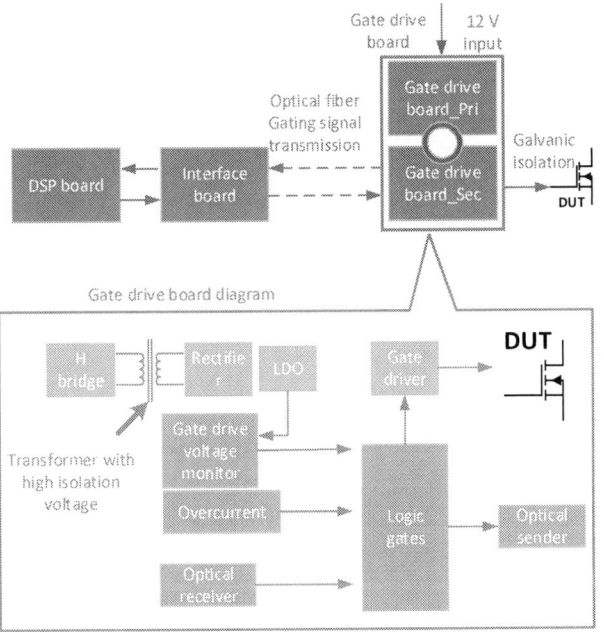

Fig. 6 HV device gate drive system configuration

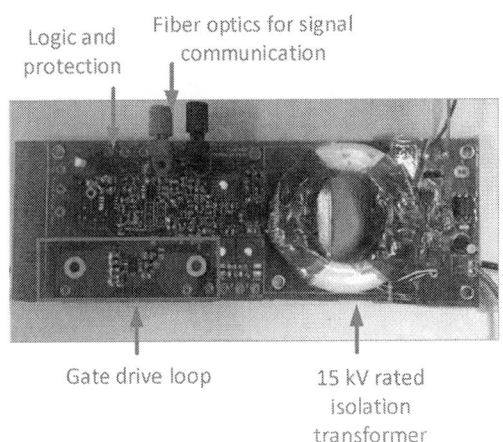

Fig. 7 High voltage device gate drive board

designed. With transformer based isolations, enough creepage distance between primary and secondary side is essential, and high design margin also needs to be considered. Meanwhile, the coupling capacitance between primary and secondary side needs to be as small as possible to minimize common mode noises [10].

Fig. 5 shows the control, isolation and protection scheme of the gate drive circuit for the 15 kV HV power device. And Fig. 6 shows the circuit prototype. It can be seen that a large isolation transformer with windings distributed along the two sides of the core guarantees enough creepage distance, and Kapton tape is used to realize winding to core insulation. Meanwhile, hysteresis gate voltage control and overcurrent protection is realized on the circuit, with hardware based protection and fault signal feedback to the control system through optical fiber.

Gate drive circuitries are always low voltage circuits compared to the HV main loop. Therefore, in the whole system layout, it is critical to make sure that the main circuit loop does not have much overlap with the gate drive loop. The clearance among different gate drive circuits also needs to be carefully considered since their potentials can have huge difference.

C. Measurement approaches

The switching transients of a HV power device involves fast rising and falling edge. The detection of these sharp edges requires high bandwidth measurement. As an approximation, the bandwidth of rising/falling edge is usually estimated according to equation

$$f = \frac{0.35}{t_r},$$

and the bandwidth of probes requires $3-5$ times of margin. At the same time, due to the short duration of switching transient, deskew is essential since even nanoseconds of delay would significantly affect the results of switching loss.

D. Experimental results

In the switching test, the 4.5 kV rated power MOSFET has been characterized with dc voltage up to 4 kV. A Tektronix

978-1-4799-5494-0/14 $31.00 © 2014 IEEE

P6015A HV single-ended probe is used for drain to source voltage measurement. A passive probe with high bandwidth is inserted between gate and source. Meanwhile, a 30 MHz rated rogowski coil is used to measure the drain to source current. The switching waveforms are shown as in Fig. 8. It can be observed that the waveforms are obtained with ringings in the

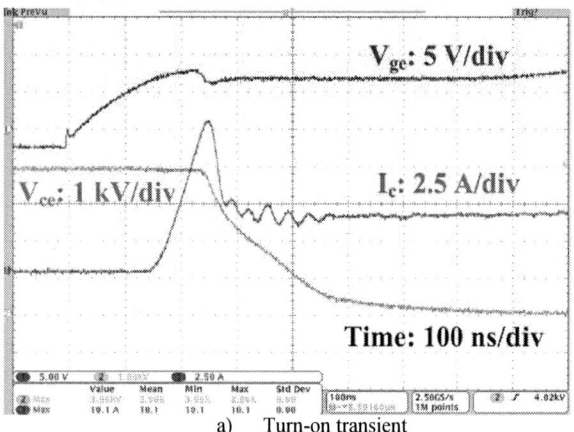

a) Turn-on transient

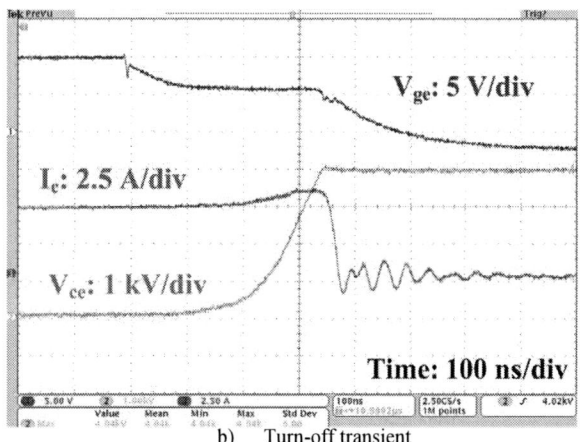

b) Turn-off transient

Fig. 8 Switching transients in inductive switching of a 4.5 kV HV device

acceptable range. According to the switching waveform, the rise time for turn-on and fall time for turn-off are 229 ns and 138 ns, respectively. The turn-on switching loss is 2.929 mJ and the turn-off loss is 1.868 mJ.

IV. HIGH VOLTAGE HIGH FREQUENCY INDUCTOR DESIGN

For switching characterization, an inductive load is needed to sustain both high voltage and high dv/dt. Due to the unavoidable inductor parasitic capacitances, high dv/dt will lead to severe voltage non-uniform distribution along the windings, which means one winding needs to withstand much higher voltage than the ideal case [7]. To further explore the voltage non-uniform distribution and apply it to the winding insulation design, FEA and associated tools are utilized to conduct this study.

In this paper, the inductor is designed with 20 kV voltage rating, and 20 A load current. The design of this inductor include both parasitics extraction and voltage distribution simulation for insulation design.

A. Parasitic capacitance extraction

The load inductor is aimed to be charged to 10 A in 2.5 us. To realize higher inductance, Metglas amorphous core with high saturation flux density is adopted in this design. The detailed design parameters are listed in the Table 1.

TABLE 1. INDUCTOR DESIGN SPECIFICATIONS

Voltage rating	20 kV
Current rating	30 A
inductance	5 mH
Air gap	1.27 mm
Turns	47
Core	Metglas AMCC-1000

There are mainly two kinds of parasitic capacitances included in an inductor: turn-to-turn capacitance and turn-to-core capacitance. With the detailed design parameter, a FEA model is constructed in Maxwell for 3-dimentional simulation. To guarantee the accuracy of the simulation, the proposed insulation layers, core shape, bobbin, air gap and winding distances are all carefully defined considering the real case. The designed model is shown as Fig. 9.

According to the simulation, the turn-to-core capacitances keep consistent and have small variations. However, the turn-to-turn capacitances keeps consistent for most turns in the middle but become large at two sides. This is because of the fringing effect of the windings at two sides, where the electric

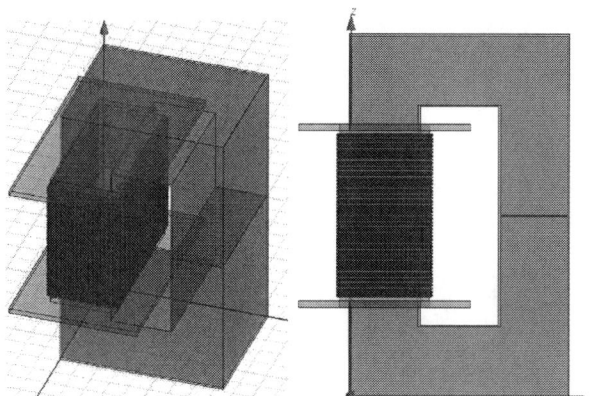

Fig. 9 Maxwell 3D model for evaluating inductor parasitic capacitance

magnetic field lines are not in parallel.

B. Simulation of winding voltage distribution

With the extracted parasitic capacitances from FEA, a circuit based simulation model, which is shown is Fig. 10, is

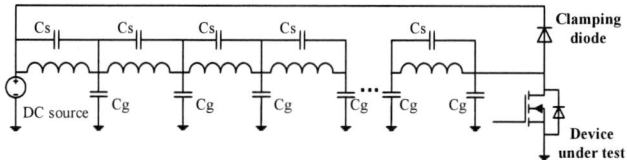

Fig. 10 Equivalent switching circuit considering inductor parasitic capacitance

978-1-4799-5494-0/14 $31.00 © 2014 IEEE

conducted to observe the voltage non-uniform distribution along the windings. The two kinds of parasitic capacitances are modeled for each turn. A fast-ramped voltage change is applied to the inductor through power device switching, and the turn-to-turn voltage is observed in the simulation. The simulations results in Fig. 11 show that with higher voltage slew rate, the peak voltage stress on the winding will be higher. With a 90 ns of voltage rise time and 15 kV dc voltage, less than 2 kV of turn-to-turn voltage pulse can be applied to a single turn when

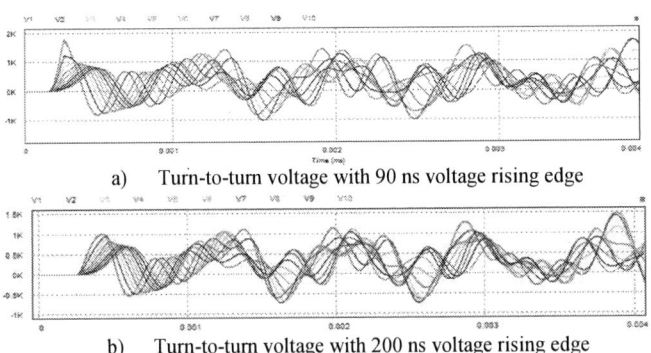

a)   Turn-to-turn voltage with 90 ns voltage rising edge

b)   Turn-to-turn voltage with 200 ns voltage rising edge
Fig. 11 Turn-to-turn voltage with varied dv/dt

Fig. 12 High voltage high frequency inductor prototype

this high dv/dt happens across the inductor. Based on this data, 8 kV rated Kapton polyimide based insulation is utilized on inductor windings. On the other hand, fiberglass is used as bobbin and 30 kV of insulation has been experimentally demonstrated between winding and core.

V.   CONCLUSION AND FUTURE WORK

This paper presents the detailed development of an HV power device evaluation platform. The methods of switching and static characterizations are introduced. An HV inductor is built for high voltage and high dv/dt applications based on FEA. The gate drive design, main circuit layout considerations are presented. The test results of a 4.5 kV Si MOSFET and a 15 kV SiC MOSFET are shown. Future work will be focused on the switching tests and applications of the 15 kV SiC MOSFET, as well as its switching behavior based modeling.

REFERENCE

[1] Adamczyk, A.; Teodorescu, R.; Mukerjee, R.N.; Rodriguez, P.; "Overview of FACTS devices for wind power plants directly connected to the transmission network", *International Symposium on Industrial Electronics*, pp. 3742 – 3748, 2010.

[2] H. Abu-Rub, J. Holtz, J. Rodriguez, and G. Baoming, "Medium voltage multilevel converters – state of the art, challenges and requirements in industrial applications", *IEEE Trans. Ind. Electron.*, Vol. 57, no. 88, pp. 2581 – 2596, Aug. 2010.

[3] N. Flourentzou, V.G. Agelidis, and G.D. Demetriades, "VSC-Based HVDC Power Transmission Systems: An Overview," *IEEE Trans. Power Electron.*, 2009. 24(3): pp. 592-602.

[4] S. Bhattacharya, T.F.Zhao, G.Y.Wang, S.Dutta, S.Baek, Y.Du, B. Parkhideh, X.H.Zhou, and A.Q.Huang, "Design and development of Generation-I silicon based solid state transformer," *in Proc. IEEE Applied Power Electronics Conf. and Expo., 2010*, pp.1666-1673

[5] X. She, A. Q. Huang, and R. Burgos, "Review of solid state transformer technologies and their applications in power distribution system," *IEEE J. Emerg. Sel. Topics Power Electron.*, vol. 1, no. 3, pp. 186–198, Sep.2013.

[6] Z. Chen, "Characterization and modeling of high-switching-speed behavior of SiC active devices", *MS thesis, Virginia Polytechnic Institute and State University*, 2009.

[7] H. Masdi, N. Mariun, A. Mohamed, "Study of Impulse Voltage Distribution in Transformer Windings", *Power and Energy, IEEE Intl. Conf.* on Power and Energy, 2010, pp.379-383.

[8] P. Ralston, T. H. Duong, Y. Nanying, D. W. Berning, C. Hood, A. R. Hefner, and K. Meehan, "High-Voltage Capacitance Measurement System for SiC Power MOSFETs, " *in Proc. IEEE Energy Conversion Congress and Exposition (ECCE)*, 2009, pp. 1472 – 1479.

[9] F. Filsecker, R. Alvarez, S. Bernet, "Characterization of a new 6.5 kV 1000 A SiC diode for medium voltage converters," *in Proc. IEEE Energy Conversion Congress and Exposition (ECCE)*, 2012, pp 2253 – 2260.

[10] A. Kadavelugu, S. Bhattacharya, "Design considerations and development of gate driver for 15 kV SiC IGBT," *in Proc. IEEE Applied Power Electronics Conf. and Expo., 2014, pp. 1494 – 1501.*

978-1-4799-5494-0/14 $31.00 © 2014 IEEE

Packaging SiC Power Semiconductors - Challenges, Technologies and Strategies

Jürgen Schuderer, Umamaheswara Vemulapati, Felix Traub

ABB Corporate Research
Segelhofstrasse 1K
5405, Baden-Dättwil, SWITZERLAND
Email: juergen.schuderer@ch.abb.com

Abstract—**In this paper a comprehensive review on SiC power module challenges and technology approaches is given. These challenges originate from SiC-specific reduced chip area, from high power and high loss density operation, from high temperature capability, from fast switching transients and from high electric field operation. New approaches based on advanced integrated cooling, low-profile integrated topside bonding and low-inductance architectures are outlined and discussed.**

Keywords—SiC packaging, SiC challenges, SiC power module design

I. INTRODUCTION

Silicon-based power electronics is approaching its power density limits. The use of WBG semiconductors such as SiC promises huge potential to increase power density, to reduce losses, to reduce size / weight / cost of magnetics, and to allow for ultra-high voltage (UHV) devices. These benefits however come at the cost of significant challenges for semiconductor packaging. It is the goal of this study to summarize these challenges and key requirements, to review SiC power module technology status, and to highlight design strategies.

II. REDUCED CHIP AREA AND NEED FOR IMPROVED POWER HANDLING CAPABILITY

Lower specific on-resistance and switching energy of SiC semiconductors can allow operation at significantly improved efficiency. However, due to high specific costs and possible defects in SiC substrates, chips are typically much smaller than Si devices in the same voltage class. Therefore, to allow for a cost and performance optimized application, power module design must be capable of handling high levels of local current and loss densities. A first issue of small chip area is that the thermal resistance approximately decreases quadratically with the chip edge length L

$$R_{th} \approx d \cdot \lambda^{-1} \cdot (L + d \cdot \tan\alpha)^{-2} \qquad (1)$$

where d is the thickness of a spreader, λ is the heat conductivity of the spreader and α its thermal spreading angle [1]. This means that these smaller chips need more emphasis on heat spreading and cooling. Since chips need to be paralleled in multichip modules, an optimized chip distance

must be applied in order to have sufficient heat spreading at minimum chip distances (to keep stray inductance and package footprint small). Another aspect that should be mentioned is that current and loss density will increase even more by heading for a bi-mode operation of the MOSFET when utilizing its internal body diode. In this case switch and freewheeling diode are in the same chip and will lead to twice the current load and respective losses that need to be cooled away.

To summarize, due to the shrinked die size and general miniaturization, improvement of heat spreading and reduction of thermal resistances is a key aspect for WBG power modules. Strategies rely on 1) elimination of the number of interfaces in the thermal path, e.g., by integrated heat sinks, 2) double-side cooling for topside integrated package designs, 3) advanced liquid cooling approaches such as indirect microchannel coolers (e.g., including two-phase flow boiling), and direct immersion, jet impingement and spray cooling, 4) advanced heat spreader technologies, e.g., based on carbon composites or planar vapor chambers / heat pipes, and finally 5) thinner bond lines and increased flatness on heat spreaders [2].

Another issue of small chip-level active area is the difficulty to place the required number of wirebonds to achieve sufficient current-carrying capability, in particular for over-currents [3]. For example, a 15mil (380μm) diameter Al wire has a typical fusing current (current at which the wire melts) of 25A and requires a 1.6mm² bond pad area. State-of-the-art 1.2kV, 60A rated MOSFETs (Cree CPM2-1200-0040B) have an active source area of about 8.3mm². For the given wire, a maximum of 5 wire bonds can be placed on the available area providing a fusing current limit of 125A = 2 x I_{nom} only! In addition, also for short time-duration surge currents it is a disadvantage to have small chips with few bond wires representing a small thermal capacitance. It is therefore important to head for new topside interconnect approaches to increase topside bonding cross section and thermal mass for improved surge current capability.

Finally, it has been mentioned that current density in shrinked chips might become so high that issues of electromigration might occur [3]. For Al, electromigration becomes relevant for current densities above $\sim 10^5 A/cm^2$ [4]. Such high current densities can probably only be achieved in

978-1-4799-5494-0/14 $31.00 © 2014 IEEE

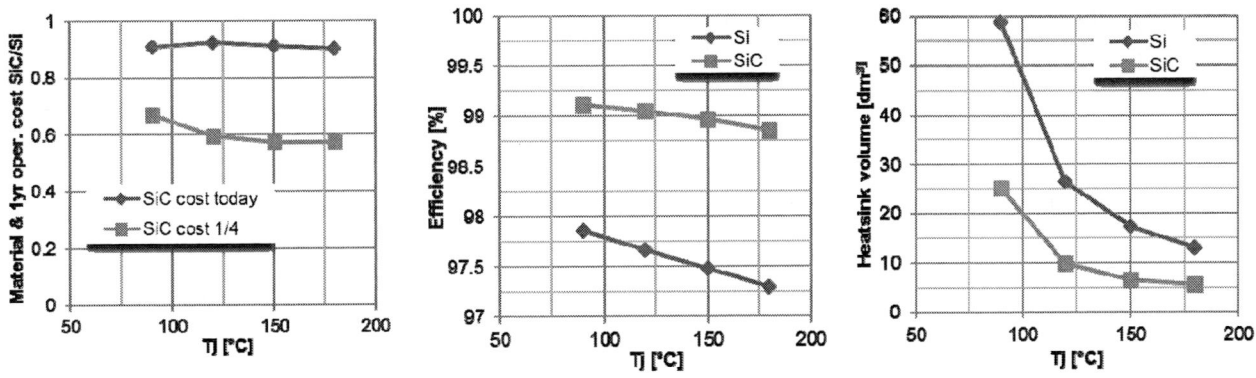

Fig. 1. Optimized SiC vs. Si package performance in 3p-2L converter at switching frequency of 16kHz [6].

very much localized contact areas of device cell structures [3]. On package level it is rather unlikely that SiC devices are operated in an area where electromigration becomes relevant.

III. HIGH-TEMPERATURE OPERATION AND NEED FOR NEW PACKAGE MATERIALS

High-T electronics has attracted lots of R&D activities and related funding over many years with a new enabler being the availability of WBG power semiconductors. It is possible to distinguish three different motivations to operate power electronics at high temperatures $\geq$ 200°C: First, there are certain applications where the ambient temperature or coolant temperature require operation above today's limits of ~150°C. These are for example automotive motor drives when using engine coolant, oil & gas drilling and extraction, geothermal drilling and extraction, avionics power supplies, space power supplies and military applications [5]. The second motivation and the biggest driving force for higher junction temperature is to increase the power handling capability in mainstream applications. Cost is intended to be saved 1) by reduction of expensive chip area and 2) by reduction of cooling cost when applying natural vs. forced convection, air vs. liquid cooling, and by general reduced heat sink size. Thirdly, yet another motivation is to use short-term high-T capability in applications with respective over-load mission profiles that lead either to an overload exposure smaller than the thermal time constant of the power module, or that result into an acceptable lifetime reduction since occurring only rarely.

For mainstream applications, the option of SiC high-T operation > 200°C must be considered carefully. The most important effect to consider for the SiC MOSFET is its strongly positive temperature coefficient of resistance that leads to an increase of equivalent resistance with temperature. Some system aspects of this effect have been recently investigated [6]. In this study a 1kV, 200kW, 3p-2L inverter topology for motor drives has been optimized for cost and efficiency as function of switching frequency and junction temperature by systematic variation of various package design parameters for SiC vs. Si devices (SiC MOSFET, Si IGBT). Results on the ratio of SiC vs. Si material & 1-yr operational costs, converter efficiency, and heats sink volume are plotted in Fig. 1. We can see that SiC operational cost and volume benefit saturate at $T_j \sim$ 150-180°C due to the increasing losses of the SiC MOSFET with temperature. Note that this situation looks

different for bipolar devices like the SiC IGBT that have a very small positive temperature coefficient of resistance and in this sense are better suited for high-T / high current density operation. Additionally to what was said, the following issues of high-T operation are significant:

1) Based on available power module fatigue / creep degradation models, lifetime under thermal cycles will reduce by a factor of ~50 when changing from Tj = 150 to 200°C. This is a challenge, but novel advanced bonding methods will probably allow to cope with these demands. Examples are die attach by nano-particles or transient liquid phase bonding [7], and topside contact by Cu wire, Cu foil or buffers [7-9]. In addition, new failure modes like substrate brittle crack and Al metallization reconstruction pose issues [10]. A powerful approach to overcome the limitations of package materials and interconnects at high temperature is to simply leave them out as much as possible, a strategy that is sometimes called "Un-packaging" [11].

2) Temperature limitation of other components at system level like capacitors and gate units can become critical. In addition, the heat sinks, operated at temperatures beyond 100°C, need to be protected from external access and thermally isolated from the rest of the enclosure.

3) State-of-the-art encapsulation materials (standard Silicone gels) are only recommended up to ~150-175°C. Polymers in the focus for high-T operation are i) spin-coated films based on silicones, benzocyclobutenes and polyimides, ii) conformal coatings based on deposited parylenes, iii) underfills and mold compounds based on silica-filled epoxy, and iv) potting compounds based on new high-T silicone gels & elastomers [12]. In addition, hydro-set ceramics based on SiC and Al_2O_3 have been studied [13].

4) Finally, device reliability concerns such as MOSFET threshold voltage drift and any other diffusion driven degradation mechanism are expected to be more pronounced at high temperatures.

IV. FAST SWITCHING TRANSIENTS AND NEED FOR ADVANCED POWER MODULE ARCHITECTURES

One of the most important drivers to implement SiC power semiconductors is the possible reduction of switching energy attained by very fast switching. A switching loss reduction of

978-1-4799-5494-0/14 $31.00 © 2014 IEEE

more than a factor of 6.6 against state-of-the-art Si IGBTs with similar ratings has been recently attained in power modules [14]. The fast switching transients however lead to issues with device and package-internal electromagnetic parasitics. Among them are voltage over-shoots, parasitic turn-on in half-bridge configurations, ringing, asynchronous switching of parallel devices and differential and common mode emissions.

A. Parasitics

When considering a half-bridge module with paralleled switches on high-side and low-side, the most important parasitic inductances, their unintended effects and some mitigation strategies are summarized below:

1) Commutation loop stray inductance L_σ: L_σ is formed by the equivalent series inductance (ESL) of the DC-link capacitor, the inductances of the DC link busbars and/or PCBs and the inductance of the package between the DC+ and DC- terminals. It is mainly responsible for i) voltage overshoots during turn-off caused by current fall, ii) slower commutation during switching and respective higher switching losses, and iii) increased EMI emissions caused by ringing of the resonant circuit formed by the stray inductance and the drain-source capacitances of the chips. Design strategies for mitigation rely for example on i) planar topside interconnects replacing bond wires [15-17], ii) low inductive circuit connections based on striplines [18], flex-foils [19], PCBs [20] and laminated busbars, and iii) parallel commutation cells with on-board de-coupling capacitors [21, 22]. In addition, a symmetrical layout (static & dynamic) of paralleled switches is needed to avoid current imbalance during conduction and switching.

2) Gate loop inductance L_G: L_G is formed by the circuitry from the driver board to the gate contact pad of the semiconductors. It is responsible i) for delaying the gate-source voltage build-up and thereby reduces maximum achievable switching frequency, ii) for transient current imbalance during switching of parallel chips for asymmetric L_G, and iii) for ringing in the gate signals by forming a resonance circuit with the gate-source capacitance. Design strategies for mitigation rely for example on low inductive gate circuit layout, on gate inductance balancing for all switches and by usage of damping resistors.

3) Gate circuit coupling inductance $M_{\sigma-G}$: $M_{\sigma-G}$ describes the coupling between the power loop stray inductance and the gate loop inductance. It acts as an additional (typically opposite) voltage source in the gate circuit. Due to the high load current transients, even small couplings can significantly distort gate signals and limit switching speed. Design strategies for mitigation rely for example on low-inductance gate circuit layout by striplines, by Kelvin type auxiliary source connections, and by using advanced control schemes.

In addition to inductances, the physical wiring of the package induces also capacitive parasitics. Consideration of parasitic capacitances will become increasingly important for fast switching chips because they are conduction paths for high frequent currents. Parasitic capacitors can 1) slow down voltage transients, 2) produce overcurrent spikes during switching, 3) increase EMI emissions by forming resonant circuits with stray inductances and, 4) provide EMC coupling

paths, e.g., via the heat sink system [23]. Mitigation strategies rely on reducing the size of AC connectors and avoiding geometrical overlaps with DC+, DC-, and GND connectors. For example, the flip-chip in polymer approach [22] minimizes parasitic capacitive coupling to the heat sink by avoiding metallized areas on AC potential.

One major parasitic capacitor is the gate-drain (Miller) capacitance of a semiconductor switch, which is involved in the issue of parasitic turn-on. This effect has been reported as a major limiting factor for the switching speed in advanced power modules [14, 22]: When turning on a switch in a half-bridge, e.g., on the high-side, a voltage change dV/dt occurs across the low-side switch and a current flows through its Miller capacitance, the gate resistor and internal driver gate resistors. If the respective voltage drop across the gate resistor exceeds the low-side gate threshold voltage, a parasitic turn-on will happen. Different mitigation solutions have been proposed [25]: 1) Use separate gate resistor for turn-on and turn-off, 2) use a negative power supply to increase margin to threshold voltage, 3) apply an active clamp technique (shorten the gate to source path by an additional switch), 4) use ferrite beads on the gate path in order to suppress high-frequent currents [26] and 5) reduce switching speed by active voltage control on the cost of increasing switching losses [14].

In conclusion, SiC fast switching power modules require significantly reduced parasitics. This requires low-inductance topside interconnects, the development of compact integrated packages, and the implementation of low-inductance stripline terminals, control circuits and vertical interconnects. In addition, as outlined in [24] and [21], a "paradigm shift" might be required towards the packaging of full commutation cells including passives and drivers rather than semiconductors only. The traditional approach of optimizing individual converter elements might not be sufficient and neglects the strong interactions when realizing a converter-wide electromagnetic, thermal and insulation design.

B. Interconnect Technologies for Power Module Integration

When heading for highly integrated power module architectures, bond wire topside interconnects are an issue for topside integration. Therefore, a summary of alternative bonding technology for power module topside integration shall be given here. These technologies are particularly interesting for WBG package design due to the following potential benefits over conventional wire-bonded packages: 1) Short interconnects reduce parasitics and improve EMC, 2) increased topside electric contact area for current capability and surge current robustness, 3) enabler for topside heat transfer (double-side cooling), and 4) enabler for topside integration of gate drive and protection. Technologies can be categorized into planar area bonding, solder bumps, Cu posts, flex foils, as well as by metallization approaches:

Planar connects realize an area solder or sinter contact making use of the maximum available topside contact area which allows best conditions for double side cooling, e.g., [27], [28]. In solder bumping an array of solder balls is used to interconnect the chip topside to a substrate. Specific solder bumping technologies have been developed to fulfill power electronics requirements of topside insulation and high-current capability. These are for example the "triple-stacked solder bumps" [29] or the "power bumps" with Cu inlets [30]. An advantage of solder bumps is that different bump diameter can be flexibly processed to adjust for different chip thicknesses and insulation demands. In metal post interconnects an array of Cu pieces (pins, posts) are connecting the topside of the dies to a substrate such as a DBC [31] or PCB [20]. Good current capability and flexible height adjustment for insulation are advantages of the post technology. In flex foil interconnects a flexible Cu sheet [32] or Cu/Polyimide/Cu sheet [19] is bonded to the topside of the dies. Metallization based approaches are realized by sputtering and electroplating of metallization layers and vias with chips being priory embedded in polyimide carriers [33], dielectric conformal coatings [34] or epoxy prepreg layers [17]. Both, flex foil and metallization approaches, have the advantage that topside multilayer circuit structures can be realized that can provide very low-inductance power and control circuit lines.

C. Advanced Power Module Architectures

This section reviews some advanced power module concepts and demonstrators that are grouped in the following three categories: 2D conventional module architecture, 2.5D module architecture, and 3D module architecture:

1) "2D conventional packages" are based on wire bonds (topside contact), soldering / sintering (die attach) and screw, plug, press-in & spring contacts (terminals). Significant progress has been achieved on advanced wire-bonded 2D packages. Infineon's 1.7kV, 480A SiC JFET stripline demonstrator with a reported stray inductance of only 5nH is certainly an advanced 2D architecture [35]. Another interesting concept is GE's blade connector [36], in which a 5nH switch module is put into a laminated busbar socket with distributed DC-link capacitors realizing an overall commutation loop stray inductance of 8nH only. Finally, first 2D high-T SiC packages are now commercialized such as the APEI HT-3000 series reaching ratings as high as 1.2kV, 550A, Tj = 200°C [37].

2) "2.5D packages" are wirebond-less modules with integrated topside for low-inductance interconnect. A huge number of 2.5D planar area bonded demonstrators in DBC sandwich can be found in literature with many of them heading for automotive applications. An interesting version of this technology is the Nottingham / Denso DBC sandwich which foresees double-side jet impingement cooling on the DBC backsides [38]. Mitsubishi should be highlighted since they have announced a product line-up for EV based on their planar area bonded direct beam lead epoxy-mold packages [39]. Finally, University of Arkansas has demonstrated a planar bonded DBC sandwich for blocking voltages as high as 6.5kV [40], and ABB Corporate Research has developed a high-power 1.2kV, 600A sandwich based on metal matrix composite substrates [28]. The Alstom power bump technology, announced for commercialization by the company aPSI3D, is an example of a 2.5D solder bump sandwich concept [41]. The technology has been demonstrated for a blocking voltage of 3.3kV. Fuji has developed a new 2.5D power module [20] utilizing Cu posts. An interesting aspect of the module is that instead of DBC substrates, a combination of a thick insulated

Cu substrate and PCB has been chosen (with the Cu acting as an excellent heat spreader). Finally, Semikron's all-sintered SKiN® package [19] should be highlighted as the first flex foil type of package close to commercialization. From the metallization-based packages it seems that the Chip-in-Polymer approach [17] represents a successful concept with commercialized LV products available such as Infineon's Dr. Blade® [42] and Schweizer Electronics p^2 Pack® [43].

3) "3D packages" are based on stacked power semiconductors either by chip-on-chip, chip-on-cooler or chip-on-substrate stacking. This transition from side-by-side towards stacked die aims at obtaining smaller package size, improved electrical parasitics, improved thermal performance and reduced warpage stresses due to larger system stiffness. In contrast to microelectronics, there is only little literature available on attempts for 3D stacking in power electronics. Some advanced concepts in this direction are the Nottingham chip-on-chip approach [44] and the ORNL microchannel cooled cold plate stack [45].

V. HV DEVICES AND NEED FOR ADVANCED INSULATION COORDINATION

The higher field strengths permitted with SiC aid in two ways to provide novel device designs not possible to achieve with Si: 1) Unipolar devices like MOSFETs can be designed for MV levels up to several kV and 2) bipolar devices like IGBTs can be designed for ultra-high blocking voltages >> 10kV with a record being a 27kV pin diode demonstrated lately [46]. This can allow 1) the use of simple two-level voltage source converter topologies at MV level, e.g. for distribution grid applications like solid state transformers, or 2) the reduction of the number of levels in UHV stacks, e.g., as used for HVDC and FACTS (less devices in series → compact converter design, less control boards & control efforts, less coolers, etc.).

One issue that both device types share is that due to the higher device internal electric field, also the field stress in the passivation stack and at the chip surfaces is higher. Fig. 2 shows a comparison between a Cree 10kV SiC diode [47] vs. an ABB 6.5kV Si diode. The average electric field for the chip / gel interface at the edge termination is around 3 times higher for the SiC diode. Any contamination in the form of particles and mobile ions, any material defect like pinholes / cracks in the passivation layers, any possible electrochemical driven corrosion processes due to moisture, and any delamination / insufficient adhesion of encapsulation becomes extremely critical at such high surface field strengths.

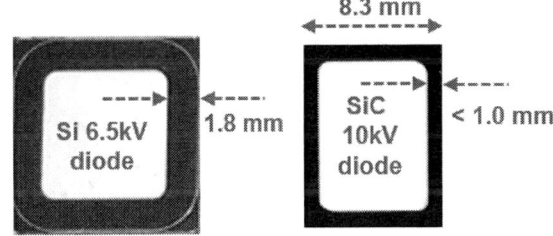

Fig. 2.   3 x higher electric field at edge termination of a SiC [47] vs. Si diode.

Another aspect that needs insulation consideration is the presence of nearby conducting surfaces for topside integrated package designs. In order to assess the effects, edge termination field simulations have been conducted for a 1.7kV rated Si IGBTs, see Fig. 3. Three configurations are compared: A conventional design with free-space Silicone gel around the chip (Fig. 3a) vs. a topside integrated design with a planar metal trace in 120μm (Fig. 3b) and 360μm (Fig. 3c) distance to the topside of the chip. In addition, the dielectric permittivity has been varied from 3.6 to 15 for the topside integrated simulation models in order to asses a refractive field grading approach.

Simulations revealed first of all that even for the very compact design (1360V / 120μm / ε_r = 15), neither breakdown

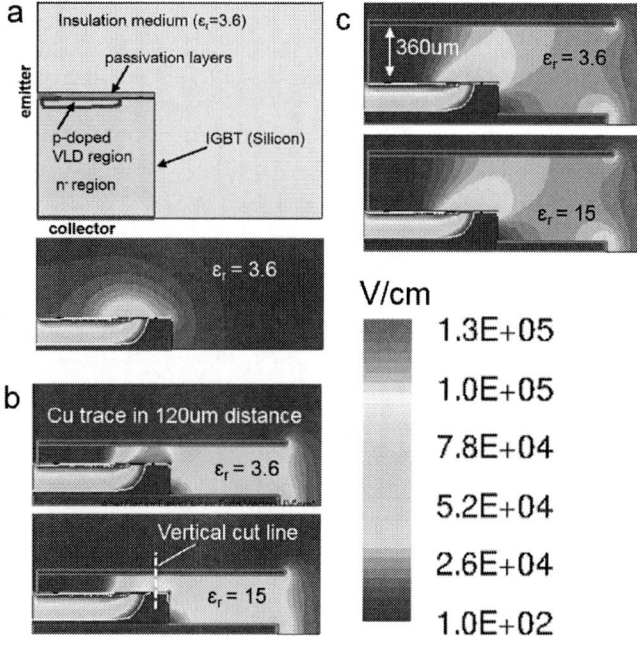

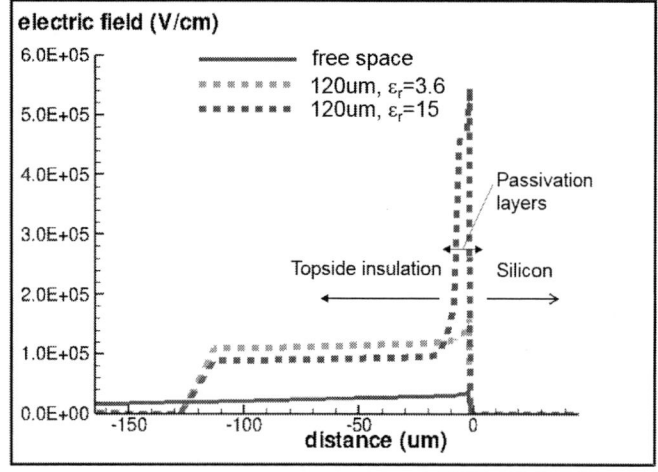

Fig. 3. a) 1.7kV IGBT model and field distribution in conventional design for a blocking voltage of 1360V; Field distribution for topside integrated design b) for Cu trace in 120um distance; c) for Cu trace in 360um distance.

voltage nor device internal field distribution are affected. What has changed however, is the field in the topside insulation: Whereas the maximum electric field is 9.0 / 3.3 kV/mm for the conventional design for a distance of 5 / 100 μm from the passivation layers, it is 13.0 / 11.0 kV/mm for the 120μm-integrated design, and 9.8 / 4.3 kV/mm for the 360μm-integrated layout at ε_r = 3.6. In addition, the high-permittivity encapsulation helps to reduce the E-field in the topside insulation (peak field reduced by 25%). However, as shown in Fig. 4, this comes at the cost of significantly increased field stresses in the lower-permittivity device passivation layers.

In conclusion, high strength topside encapsulation materials (coatings, glob tops, underfills and molds) are needed to cope with increased insulation demands resulting from SiC-specific higher device internal and external fields.

When considering UHV >10kV bipolar devices, a list of additional packaging issues require careful analysis: 1) Due to higher insulation demands, a thicker ceramic insulator will be needed leading to increased thermal resistance. 2) The effects of partial discharges, insulation breakdown, dendritic growth under humidity, etc. triggered by field enhancements of metallic parts at high voltage potential are already an issue of today's MV-rated Si IGBT modules and become more critical at higher voltages. Thereby, a main concern are the "sharp" Cu metallization edges of DBC substrates [48]. 3) Another aspect for consideration of bipolar devices is their small or even negative temperature coefficient of resistance [49]. In order to achieve safe paralleling of these devices, careful binning and very symmetric package R_{th} is mandatory. 4) Finally, there is a design conflict when on the one hand heading for fast switching requiring compact package architectures, and heading for large clearance and creepage distances on the other hand in order to insulate UHV devices and terminals. Solid insulation concepts that avoid clearance distances should be implemented.

VI. CONCLUSIONS

In this study, requirements for SiC power modules have been reviewed. The challengers involved are considered as significant especially when heading for high-T, hard-switching and UHV operation. Nevertheless, it was also shown that advanced packaging technology based on load cycle reliable low temperature bonding, advanced cooling, bonding methods for topside integration, and low-inductance module architectures are available and have been successfully demonstrated.

ACKNOWLEDGMENT

The authors want to thank Fabian Mohn, Daniel Kearney, Andrei Mihaila and Uwe Drofenik for very valuable discussions and input to this study.

REFERENCES

[1] A. Volke, M Hornkamp, "IGBT modules, technologies, driver and application", Infineon Technologies AG, 2011.

[2] X. Tong, "Advanced materials for thermal management of electronic packaging", Springer Series in Advanced Microelectronics 30, 2011.

Fig. 4. Vertical cut (see cut line in Fig. 3b) of E-field distribution for conventional vs. topside-integrated design (120μm, ε_r = 3.6 and 15).

[3] P. Friedrichs, "SiC power devices as enbaler for high power density – spects and prospect", Materials Science Forum, Vols. 778-780, pp. 1104-09, 2014.

[4] M. Ohring "Reliability and Failure of Electronic Materials and Devices", Academic Press, 1998.

[5] Yole development report, "SiC market 2013", 2013.

[6] J. Schuderer, U. Drofenik, B. Agostini, F. Brem, F. Mohn, F. Canales., "MV power module reliability and robustness challenges", ECPE workshop "Power electronics packaging", Switzerland, Nov. 2013.

[7] K. Guth, N Oeschler, L. Böwer, R. Speckels, G. Strotmann, N. Heuck, S. Krasel, A. Ciliox, "New assembly and interconnect technologies for power modules", 7th Intern. Conf. on Integ.Power Systems (CIPS), 2012.

[8] U. Scheuermann, "Reliability of planar SKiN interconnect technology", 7th Intern. Conf. on Integrated Power Systems (CIPS), 2012.

[9] J. Rudzki. M. Becker, R. Eisele, M. Poech, F. Osterwald, "Power modules wirth increased power density and reliability using Cu wire bonds on sintered metal buffer layers", 8th Intern. Conf. on Integrated Power Systems (CIPS), 2014.

[10] S. Kicin, J. Hamers "Assessment of selected materials and assembly technologies for power electronics modules with the capability to operate at high temperatures", 15th European Conf. on Power Electronics and Applications (EPE), 2013.

[11] T. Stockmeier, "From packaging to "un"-packaging – trends in power semiconductor modules", 20th Intern. Symposium on Power Semiconductor Devices and IC's (ISPSD), 2008.

[12] Y. Yao, G. Lu, D. Boroyevich, K. Ngo, "Survey of high-temperature polymeric encapsulants for power electronics packaging", CPES research paper 2013.

[13] B. Grummel, R. McClure, Z. Lei, A. Gordon, L. Chow, Z. Shen, "Design considerations for high temperature SiC power modules", 34th Annual Conf. of IEEE Industrial Electronics (IECON), 2008.

[14] D. Heer, R Beyerer, D. Domes, "SiC JFET in half-bridge configuration – parasitic turn-on at current commutation", Intern. Exhibition and Conf. for Power Electronics, Intelligent Motion (PCIM Europe), 2014.

[15] A. Gowda et al., "Power overlay packaging platform for high performance electronics", Chip Scale Review, Sep/Oct 2012.

[16] Z. Liang, F. Lee, G. Lu, "Embedded power – an integration packaging technology for IPEMs", Intern. Journal of Microcircuits and Electronic Packaging, 23(4), 2000.

[17] L. Boettcher S. Karaszkiewicz, D. Manessis, A. Ostmann, "Development of embedded power electronics modules", 4th Electronic System-Integration Technology Conf. (ESTC), 2012.

[18] R. Beyerer, D. Domes, "Power circuit design for clean switching", 6th Intern. Conf. on Integrated Power Systems (CIPS), 2010.

[19] P. Beckedahl, M. Spang, O. Tamm, "Breakthrough into the third dimension – Sintered multi layer flex for ultra low inductance power modules", 8th Intern. Conf. on Integrated Power Systems (CIPS), 2014.

[20] Y. Hinata, M. Horio, Y. Ikeda, R.Yamada, Y. Takahashi, "Full SiC Power Module with Advanced Structure and its Solar Inverter Application", 28th Annual IEEE Applied Power Electronics Conf. and Exposition (APEC), 2013.

[21] M. Johnson, A. Castellazzi, R. Skuriat, P. Evans, J. Li, P. Agyakwa, "Integrated high power modules", 7th Intern. Conf. on Integrated Power Systems (CIPS), 2012.

[22] E. Hoene, A. Ostmann, C. Marczok, "Packaging very fast switching semiconductors, 8th Intern. Conf. on Integ. Power Systems (CIPS), 2014.

[23] A. Domurat-Linde, E. Hoene, "Analysis and reduction of radiated EMI of power modules", 7th Int. Conf. on Integ. Power Systems (CIPS), 2012.

[24] J. Popovic, J. Ferreira, J. van Dyk, F. Pansier "System integration of GaN converters – paradigm shift", 8th Intern. Conf. on Integrated Power Systems (CIPS), 2014.

[25] "Mitigation Methods for Parasitic Turn-on Effect due to Miller Capacitor", White Paper, Avago Technologies, http://www.avagotech.com/docs/AV02-0599EN

[26] F. Bjoerk, J. Hancock, G. Deboy, "CoolMOS Cp – How to make most beneficial use of the latest generation of super junction technology devices", Infineon Technologies AG, Application Note, Feb. 2007.

[27] A. Narazaki et al, "Direct beam lead bonding for trench MOSFET & CSTBT", 17th Intern. Symposium on Power Semiconductor Devices and ICs (ISPSD), 2005.

[28] S. Kicin, M. Laitinen, C. Haederli, J. Sikanen, R. Grinberg, J. Fabian, A. Hamidi, "Concept of low-voltage AC drive based on double-side cooled IGBT press-pack modules", 14th European Conf. on Power Electronics and Applications (EPE), 2011.

[29] J. Bai, G. Lu, X. Liu, "Flip-chip on flex integrated power electronic modules for high density power integration", IEEE Trans on Advanced Packaging, 26(1), 2003.

[30] M. Memet-Guynnet, "New structure of power integrated modules", 4th Intern. Conf. on Integrated Power Systems (CIPS), 2006.

[31] S. Haque et al, "An innovative technique for packaging power electronic building blocks using metal posts interconnected parallel plate structures", IEEE Trans on advanced packaging, 22(2), 1999.

[32] S. Wen, D. Huff, G. Lu, "Dimple array interconnect technique for packaging power electronic devices and modules", Intern. Symposium on Power Semiconductor Devices and ICs, Osaka, 2001.

[33] R. Fillion, E. Delgado, R. Beaupre, P. McConnelee, "High power planar interconnect for high frequency converters", 6th Electronics Packaging Technology Conf. (EPTC), 2004.

[34] K. Weidner, M. Kaspar, N. Seliger, "Planar Interconnect Technology for Power Module System Integration", 7th Intern. Conf. on Integrated Power Systems (CIPS), 2012.

[35] R. Bayerer, "1700V SiC-JFET half-bridge module", International SiC Power Electronics Application Workshop (ISiCPEAW), 2014.

[36] L. Stevanovic, R. Beaupre, E. Delgado, A. Gowda "Low-inductance power module with blade connector", 25th Applied Power Electronics Conf. and Exposition (APEC), 2010.

[37] B. Passmore et al, "Three-dimensional packaging for wide bandgap based discrete and multichip packages", 25th Applied PE Conf. and Exposition (APEC), 2014, Industry session 3D power packaging.

[38] M. Johnson, C. Buttay, S. Rashid, F. Udrea, G. Amaratunga, P. Ireland, R. Malhan, "Compact Double-Side Liquid-Impingement-Cooled Integrated Power Electronic Module", 19th Int. Symp. on Power Semiconductor Devices and IC's (ISPSD), 2007.

[39] M. Ishihara, Inokuchi, M. Honsberg, E. Thal, "Power Modules for Electric and Hybrid Vehicles", in Bodo's Power Systems, June 2014.

[40] H. Zhang, S. Ang, A. Mantooth, J. Balda, "A 6.5kV wire-bondless. Double-sided cooling power electronic module", IEEE Energy Conversion Congress and Exposition (ECCE), 2012.

[41] M. Mermet-Guyennet, A. Castellazzi, P. Lasserre, J. Saiz, "3D integration of power semiconductor devices based on surface bump technology", 5th Intern. Conf. on Integ. Power Systems (CIPS), 2008.

[42] H. Timmer, K. Pressel, G. Beer, R. Bergmann, "Interconnect Technologies for System-In-Package Integration", 15th IEEE Electronics Packaging Technology Conference (EPTC), 2013.

[43] T. Gottwald, et al, "Increasing packaging density and thermal performance with minimized parasitic for high power inverters ", Intern. Exhibition and Conf. for PE, Intelligent Motion (PCIM Europe), 2012.

[44] J. Li, A. Castellazzi, A. Solomon, C. Johnson, "Reliable Integration of Double-Sided Cooled Stacked Power Switches based on 70 um Thin IGBTs and Diodes", 7th Int. Conf. on Int. Power Systems (CIPS), 2012.

[45] Z. Liang "Integration of cooling function into 3-D power module packaging", 25th Applied Power Electronics Conf. and Exposition (APEC), 2014, Industry session 3D power packaging.

[46] T. Kimoto, "Ultra-high voltage devices for future power infrastructure", www.compundsemiconductor.net, March 2014.

[47] D. Grider et al, "10kV/120A SiC DMOSFET half H-bridge power modules for 1 MVA solid state power substation", IEEE Electric Ship Technologies Symposium, 2011.

[48] J. Fabian, S. Hartmann, A Hamidi, "Analysis of Insulation Failure Modes in High Power IGBT Modules", 40th IAS Annual Meeting Conference Record, Vol. 2, pp. 799 – 805, 2005.

[49] E. Johnson, O. Saadeh, H. Mantooth, J. Balda, S. Ang, A. Agarwal, "An Analysis of Paralleled SiC Bipolar Devices", IEEE Power Electronics Specialists Conference (PESC), 2008.

978-1-4799-5494-0/14 $31.00 © 2014 IEEE

Application-Based Review of GaN HFETs

Edward A. Jones and Fred Wang
Dept. of Electrical Engineering and Computer Science
University of Tennessee
Knoxville, TN
E-mail: edward.jones@utk.edu, fred.wang@utk.edu

Burak Ozpineci
Power Electronics and Electric Machinery Group
Oak Ridge National Lab
Oak Ridge, TN
E-mail: burak@ornl.gov

Abstract—Normally-off GaN-on-Si heterojunction field-effect transistors (HFETs) have been developed with up to 650 V blocking capability, fast switching, and low conduction losses in commercial devices. The natively depletion-mode device can be modified to be normally-off using a variety of techniques. For a power electronics engineer accustomed to Si-based converter design, there is inherent benefit to understanding the unique characteristics and challenges that distinguish GaN HFETs from Si MOSFETs. Dynamic R_{ds-on}, self-commutated reverse conduction, gate voltage and current requirements, and the effects of very fast switching are explained from an applications perspective. This paper reviews available literature on commercial and near-commercial GaN HFETs, to prepare engineers with Si-based power electronics experience to effectively design GaN-based converters.

Keywords—*GaN, Gallium Nitride, HFET, HEMT*

I. INTRODUCTION

Wide bandgap (WBG) semiconductors are a game-changing technology for power electronics, enabling high-voltage and high-frequency converters with improved efficiency, smaller size, and lower cost. The higher breakdown field and drift velocity of WBG devices produce lower losses than comparable silicon (Si) devices at high voltage, typically in the hundreds of volts. The two most promising WBG semiconductors are silicon carbide (SiC) and gallium nitride (GaN). GaN shows more benefits than SiC in several areas, especially carrier mobility. Although GaN does not offer the high thermal conductivity of SiC, GaN has a very stable R_{ds-on} over temperature, making it suitable for high temperature applications [1].

According to [2], 11 of the major 23 Si power switch manufacturers have GaN research programs. Ongoing research involves fabricating normally-off devices with higher breakdown voltage, lower R_{ds-on}, and current collapse suppression, as well as designing fast gate drive circuits that can handle the unique requirements of GaN transistors.

Vertical GaN devices are a popular research area, but GaN wafer fabrication technology is still a major challenge. Instead, lateral devices called heterojunction field-effect transistors (HFETs), also known as high electron mobility transistors (HEMTs), offer an effective alternative that take advantage of the high-mobility layer formed at the heterojunction between GaN and AlGaN. HFETs have been fabricated on substrates of Si, SiC, and sapphire. Si is

considerably cheaper than the alternatives, and its conductive properties offer unique advantages as a substrate [1]-[3].

This paper will review normally-off GaN-on-Si power HFETs rated in the hundreds of volts, as well as device characteristics that are relevant to the converter designer. Although there are noteworthy achievements in the areas of vertical device fabrication, normally-on devices, and other substrate materials, this paper will focus on transistors that are commercially available, or that have been manufactured by industry research teams. For engineers with experience only in Si-based converter design, there is inherent usefulness in understanding the difference between GaN HFETs and Si MOSFETs to effectively design a GaN-based converter and understand experimental results.

II. HFET DEVICE STRUCTURE

Unlike a MOSFET, there is no intentional doping or p-n junction in an HFET, due to the two-dimensional electron gas (2DEG) generated by the piezoelectric polarization at the heterojunction between GaN and AlGaN. Fig. 1 shows the basic structure. A thick GaN layer is epitaxially deposited on the substrate wafer, then a multi-layer buffer that allows for crystal lattice strain relief before the AlGaN cap layer is deposited on top. The 2DEG is a dense native layer of electrons that creates a normally-on low-resistance current path between the drain and source. To turn the device off, a Schottky gate is deposited on the AlGaN cap layer that depletes the 2DEG when a negative voltage is applied. Therefore, the HFET is a fundamentally depletion-mode (normally-on) device.

Because there is doping and no source-connected body region, there is no body diode as in a Si MOSFET. Instead, the symmetry of the drain and source allows for self-commutated reverse conduction as described later in Section IV.

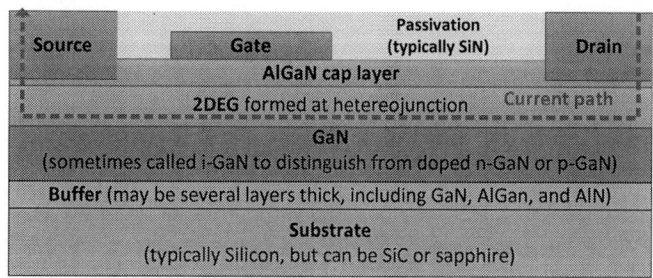

Fig. 1. GaN HFET basic device structure.

978-1-4799-5494-0/14 $31.00 © 2014 IEEE

Additionally, the HFET has no avalanche breakdown, so irreversible device breakdown occurs with overvoltage [1]-[3].

Depletion-mode devices are less desirable in power converters due to reliability concerns. Normally-off HFETs can be fabricated through three distinct methods: non-insulated gate enhancement mode, insulated gate enhancement mode, and cascode.

III. TYPES OF NORMALLY-OFF DEVICES

Fig. 2 shows a simplified gate structure of each of the popular normally-off device designs described below, and Table 1 summarizes the latest published data from several leading companies developing GaN HFETs.

A. Enhancement-mode, non-insulated gate

The first commercially available GaN FET was the EPC eGaN FET, shown in Fig. 2 (a). A p-doped GaN cap is deposited beneath the gate, lifting the threshold to a positive voltage. The result is a diode-like gate characteristic, with a positive gate current when V_{gs} is sufficiently above the threshold voltage [1],[4].

The gate injection transistor (GIT) offered by Panasonic uses a similar deposition of p-doped AlGaN beneath the gate, shown in Fig. 2 (b). The gate current in a GIT is on the order of mA. The injected holes benefit device performance by reducing R_{ds-on} as the gate current increases [5].

Enhancement-mode behavior can also be achieved by etching the AlGaN layer beneath the gate, resulting in a recessed gate as shown in Fig. 2 (c). The recess must be etched to a very precise depth, in order to leave the 2DEG layer intact and accurately control the threshold voltage [2]. The Sanken device employs the recessed gate architecture along with a NiOx gate electrode to produce a normally-off device [6].

B. Enhancement-model, insulated gate

Several normally-off HFET architectures have been developed with insulated gates to reduce gate current and subsequent gate driver losses. These devices are sometimes referred to as MOSHFETs or MISHFETs because of the insulating/oxide layer between the AlGaN layer and gate [1].

HRL has used a CF$_4$ plasma treatment to inject negative charges into the AlGaN beneath the gate, effectively depleting the 2DEG under the gate when V_{gs} is below a positive threshold. A gate insulator is then deposited between the plasma-treated region and the gate as shown in Fig. 2 (d), preventing significant gate current from flowing [7].

The latest HRL devices have been redesigned to use an insulated recessed gate architecture with no plasma treatment, as shown in Fig. 2 (e). HRL uses AlN as the insulator in their recessed gate HFET, which is crucial to this device's enhancement-mode performance [8]. NEC has also developed an insulated recessed gate method, using a multi-layer AlGaN structure called a piezo-neutralization technique (PNT) layer [9].

If the recess is deep enough that it penetrates the AlGaN layer completely, the 2DEG layer is permanently removed beneath the gate. The resulting structure is called a hybrid

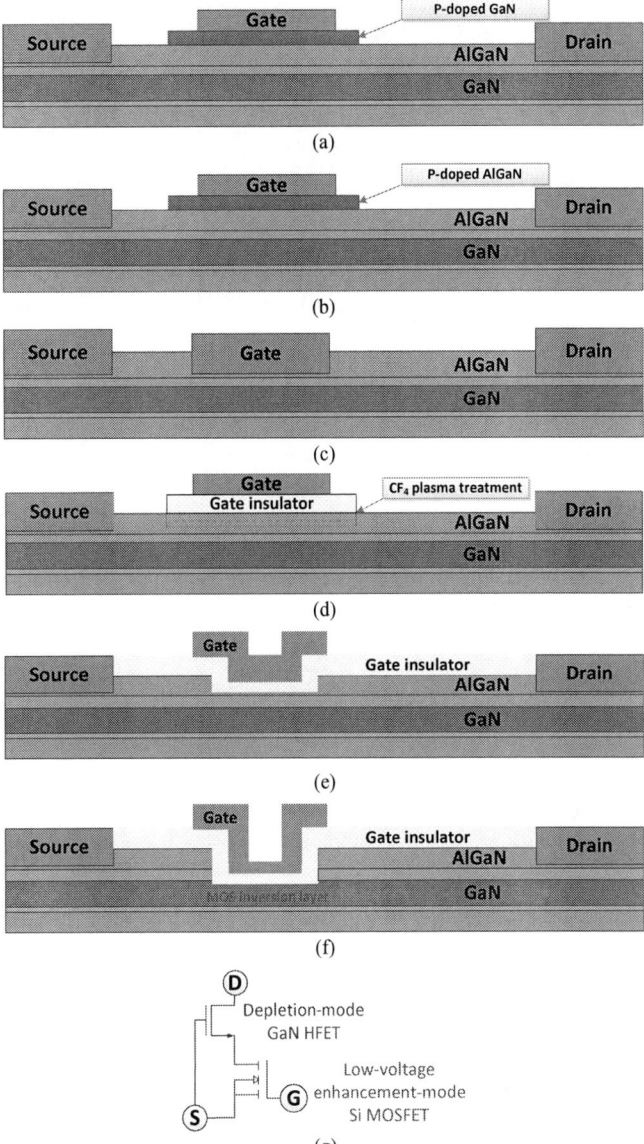

Fig. 2. Methods to build a normally-off GaN HFET (a) P-doped GaN under gate; (b) P-doped AlGaN under gate; (c) Recessed gate; (d) Plasma treatment under gate; (e) Recessed AlN gate insulator; (f) Hybrid MOS-HFET; (g) cascode structure.

MOS-HFET, because a typical MOSFET inversion layer must be formed in the GaN beneath the gate insulator to connect the 2DEG on either side of the recess. The hybrid MOS-HFET is shown in Fig. 2 (f), and has been fabricated by both Sharp and Furukawa [10],[11].

GaN Systems, Inc. has recently begun selling engineering samples of their 650 V insulated-gate enhancement-mode HFETs on the open market, and released a preliminary datasheet [12]. However, there have been no publications to date that describe their technique to achieve enhancement-mode operation. One interesting property of the GaN Systems devices is that its gate is safe up to 10 V, with a recommended v_{gs} of 7 V. This could allow for easier implementation with a standard MOSFET gate driver, whereas many other enhancement-mode GaN gates will rupture above 5 or 6 V.

978-1-4799-5494-0/14 $31.00 © 2014 IEEE

TABLE 1: PUBLISHED SPECIFICATIONS FOR NORMALLY-OFF GaN-ON-Si HFETs

Manufacturer	Gate structure	V_{rated}	I_{max}	R_{ds-on}	$V_{gs,th}$	Q_g
EPC (*eGaN-FET*) [4]	P-doped GaN cap	200 V	12 A	25 mΩ	1.4 V	7.5 nC
Sharp [10]	Hybrid MOS-HFET, insulated	400 V	200 mA/mm *	--	5.2 V	--
Fujitsu [15]	Cascode	600 V	20 A	100 mΩ	2.3 V	14 nC
Panasonic (*GIT*) [5]	P-doped AlGaN cap	600 V	15 A	65 mΩ	1.2 V	11 nC
Transphorm (*EZ-GaN*) [13]	Cascode	600 V	17 A	150 Ω	1.8 V	6.2 nC
IR (*GaNpowIR*) [14]	Cascode	600 V	10 A	160 mΩ	3 V	--
HRL [8]	Recessed gate, insulated	600 V	10.5 A	350 mΩ	1.3 V	--
MicroGaN [16]	Cascode	600 V	--	320 mΩ	2.0 V	--
Gan Systems [12]	Unknown enhancement-mode, insulated	650 V	30 A	52 mΩ	1.6 V	6.5 nC
Sanken [6]	Recessed gate w/ NiOx gate electrode (non-insulated)	800 V	20 A	72 mΩ	0.8 V	--
NEC [9]	Recessed gate, insulated	1000 V	240 mA/mm *	5.0 mΩ-cm2 *	1.5 V	--
HRL [7]	F-plasma treatment, insulated	1200 V	5.5 A	500 mΩ	0.6 V	--
Furukawa [11]	Hybrid MOS-HFET, insulated	1700 V	--	7.1 mΩ-cm2 *	2.0 V	--

* In some cases, only specific on-resistance and current per gate-length were available rather than packaged device parameters

C. Cascode

The third method to achieve normally-off behavior is to incorporate a low-voltage enhancement-mode Si MOSFET in a cascode structure, as shown in Fig. 2 (g). The MOSFET can be controlled directly by a traditional gate driver, and its output clamps the depletion-mode GaN HFET's gate input to behave as a normally-off device. The Si gate input is more rugged and easier to drive, but cascode package parasitics tend to be higher. The gate input to the depletion-mode GaN device is typically not accessible in the external cascode contacts, so the GaN switching speed cannot be controlled directly with gate drive design as in enhancement-mode HFETs.

Transphorm, IR, Fujitsu, and MicroGaN have all produced cascode-type GaN transistors rated at 600 V, and the Transphorm device has already seen a limited commercial release [13]-[16].

IV. GaN-BASED CONVERTER DESIGN CHALLENGES

The HFET is fundamentally a different device from the Si MOSFET, and it is important to recognize the specific challenges of designing a GaN-based converter. Several of these challenges are explained in this section.

A. Dynamic R_{ds-on}

The polarization of the heterojunction make HFETs susceptible to an undesirable memory effect called current collapse. This phenomenon produces a dynamic R_{ds-on} that increases proportionally to the blocking voltage seen during each switching period. Fig. 3 shows the mechanisms of current collapse as well as long-term R_{ds-on} degradation, which is a similar but distinct concern.

When the device is off, the gate-drain voltage is reverse biased with the majority of the output voltage according to (1).

$$v_{dg} = v_{ds} - v_{gs} \qquad (1)$$

The unevenly distributed electric field causes electrons to be trapped at surface states in the passivation layer near the gate edge. This gate-edge effect produces a "virtual gate" that

continues to reduce 2DEG density after the device turns on. The captured surface electrons have a longer time constant than those intentionally injected directly beneath the gate during the off-state, so high off-state blocking voltage results in higher R_{ds-on} during the following on-state. Additionally, the "hot electron effect" from switching causes more electron trapping, penetrating deeper traps at impurities in lower layers and causing long-term degradation [17]-[21]. Dynamic R_{ds-on} is difficult to capture accurately even with state-of-the-art oscilloscopes, due to the difference in scale between on-state and off-state voltage, so [21] proposes a voltage clamping circuit to verify current collapse suppression.

Current collapse can be suppressed using field plates to reduce the electric field strength around the gate edge and distribute it more evenly between the source and drain. Field plates can be added to the gate or source terminals, as shown in Fig. 3 with dashed lines. The semiconductive substrate in GaN-on-Si HFETs also acts as a lower field plate to better distribute the electric field. Field plates have the added benefit of increasing overall blocking voltage, so most GaN HFETs

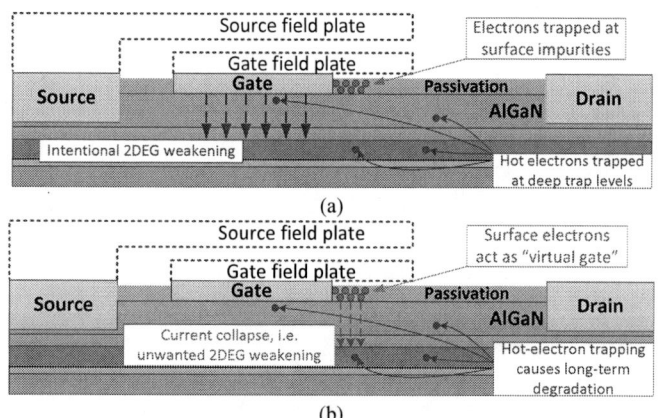

Fig. 3. Dynamic R_{ds-on} phenomenon. Field plates shown with dashed lines reduce trapped electrons and lessen the dynamic R_{ds-on} phenomenon. (a) During off-state, 2DEG is intentionally weakened to turn off the device. (b) During the next on-state, unwanted 2DEG weakening occurs due to current collapse.

978-1-4799-5494-0/14 $31.00 © 2014 IEEE

now incorporate multiple field plates [22]. Improved SiN passivation techniques with fewer impurities have also helped to suppress current collapse in the latest devices.

In fact, most of the leading GaN device manufacturers such as EPC, Panasonic, HRL, Transphorm, and IR have stated that dynamic R_{ds-on} in their devices is no longer a substantial issue [5],[7],[14],[23],[24]. Although the current collapse problem has been significantly reduced, it is important to be aware that the phenomenon still exists, as it may have some effect on modeling and loss analysis.

B. Self-commutated reverse conduction

As mentioned previously, the HFET does not demonstrate a body diode effect in reverse conduction. However, a negative V_{ds} will turn the device on in reverse, with the drain behaving as a source and the source as a drain. This self-commutating reverse conduction (SCRC) appears similar to a diode characteristic. The crucial difference is that there is no reverse recovery in enhancement-mode HFETs. On the other hand, the SCRC conduction loss can be higher than in a conventional Si MOSFET's body diode. The voltage drop between the source and drain can only be as low as the V_{gd} threshold voltage of the device, according to (2).

$$v_{sd} = V_{gd,th} - v_{gs} + i_d R_{ds-on} \qquad (2)$$

Because the gate is fabricated closer to the source than the drain, this $V_{gd,th}$ tends to be slightly higher than the specified $V_{gs,th}$. The total voltage drop from source to drain will also increase proportionally as the off-state gate drive voltage is reduced below 0 V. If negative voltage must be applied to prevent spurious turn-on, the SCRC losses will suffer. In addition to the diode-like voltage drop caused by the threshold, there is a further voltage drop as current flows from source to drain, with an R_{ds-on} that is approximately the same as the forward R_{ds-on}. Fig. 4 shows a simplified output characteristic of a typical enhancement-mode HFET in first and third quadrants.

Similar to a MOSFET, the device can conduct in reverse with the same voltage drop as seen in forward conduction, when v_{gs} is above its positive threshold voltage. This is the preferred method of HFET reverse conduction, useful in synchronous rectification topologies [1],[4],[5],[25].

The high SCRC loss can be a major factor in GaN-based voltage source converter design, because a period of "body diode" conduction occurs during dead time. Minimizing dead time helps to mitigate these losses, but that is difficult to ensure across the full load range. Adding an anti-parallel diode is often suggested, but this would increase cost and junction capacitance, and possibly introduce reverse recovery losses. A three-level gate driver was proposed by [26] to reduce losses during dead time, by holding the gate voltage near the threshold (but still below it) during part of the switching transition. A capacitor-less gate drive was proposed by [27] to minimize SCRC loss as well as switching losses.

The method of reverse conduction in cascode-type devices is completely different from enhancement-mode devices. The low-voltage Si MOSFET does have a body diode, which conducts when the cascode device is reverse-biased, thereby

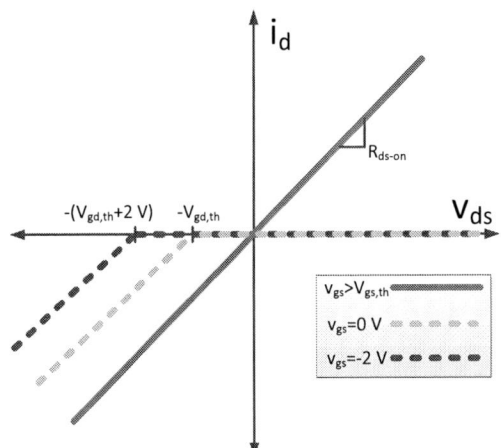

Fig. 4. Typical GaN HFET output characteristic in 1st and 3rd quadrant, showing reverse conduction diode-like voltage drop with different design choices for off-state v_{gs}.

clamping the GaN HFET gate in on-state. Conduction losses are therefore much lower in cascode reverse conduction, however there is a small amount of reverse recovery associated with the Si MOSFET body diode [1].

C. Gate current and voltage

In non-insulated gate enhancement-mode HFETs such as the EPC and Panasonic devices, the p-doped layer beneath the gate causes a diode-like current/voltage relationship, with current flowing at some voltage above the turn-on threshold as shown in Fig. 5. Higher gate current decreases R_{ds-on}, so there is a tradeoff between driving losses and conduction losses.

For either insulated or non-insulated gate devices, the gate capacitance must be charged and discharged very rapidly to achieve a very high switching frequency. The TI LM5114 provides separate turn-on and turn-off paths so that source and sink gate resistors can be chosen independently without the need for a diode [28]. For devices with a high gate current, an additional voltage drop will occur across the gate resistor, so the gate drive IC may need a higher allowable output voltage supply. In that case, the Analog Devices ADP36xx family is designed for very fast switching FETs with up to 18 V [29]. Another option is the Silicon Laboratories Si823x series, which includes both low-side and high-side drive and incorporates digital signal isolation. However, this IC cannot be used with most insulated-gate enhancement-mode devices (which typically have recommended v_{gs} between 4 and 5 V), because the minimum output supply voltage is 6.5 V [30].

As shown in Table 1, most enhancement-mode GaN HFETs have a threshold voltage between 1 and 2 V. However, extra care must be taken to maintain the gate voltage below the maximum at all times to prevent gate rupture (typically around 6 V). The TI 5113 and 5114 incorporate voltage-limiting protection to prevent accidental gate voltage overshoot [28].

D. High di/dt and dv/dt effects

In addition to static requirements for gate current and voltage, the very short switching time for WBG devices

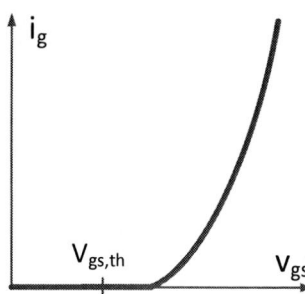

Fig. 5. Typical gate characteristic for GaN HFET with non-insulated gate.

results in a high di/dt and dv/dt during the switching period. These two phenomena have serious consequences for the gate drive circuit and board design.

The very short current rise time and fall time results in a high di/dt in both the gate loop and the power loop, inducing voltage across any parasitic inductances. Exceeding the rated output voltage on a GaN HFET will result in catastrophic dielectric breakdown, because there is no avalanche breakdown [31]. Also, as mentioned previously, the enhancement-mode gate is subject to rupture above its rated voltage. To prevent any overvoltage, the power loop and gate loop must employ a very dense board design to reduce stray inductance.

High di/dt in the power loop will also affect the switching time in the gate loop through the common-source inductance. Excessive common-source inductance can come from the device packaging or board design, and it will slow down the turn-on transition and increase switching losses. Decoupling the gate and power loops with a Kelvin source helps to mitigate this effect, along with low-inductance package and careful board design. A Kelvin source connection is already offered in some HFETs, such as the Transphorm and GaN Systems devices.

The high dv/dt in the power loop can induce current in the device junction capacitances (C_{ds}, C_{gs}, and C_{gd}), leading to spurious Miller turn-on and subsequent shoot-through conditions. This cross-talk can cause excessive switching losses and limit switching speed. Devices with high Miller charge Q_{gd} relative to Q_{gs} tend to be more susceptible to cross-talk issues. Cross-talk can be mitigated with a very low gate resistance, but this also increases the risk of gate overvoltage by underdamping the gate loop resonance. Typically, a gate resistance can be chosen that prevents both conditions from occurring, but this becomes more difficult when the device Q_{gd} is high and its threshold voltage is low.

The parasitic inductance and junction capacitances also lead to high-frequency ringing, which can generate radiative EMI and interfere with converter operation [32]-[34].

Inserting a capacitor in parallel with C_{gs} intentionally slows down the switching transition, thereby mitigating many of these dv/dt and di/dt issues, but it increases driving loss and limits switching frequency as a consequence.

To minimize gate-loop parasitics, several options have been proposed to integrate the gate drive circuit into the module with one or more GaN devices. Panasonic has developed a monolithic inverter IC with the gate drive fabricated on the same Si substrate as six GaN HFETs. They have also used this method to fabricate a monolithic low-voltage DC-DC converter IC [35],[36]. The HRL multi-chip module (MCM) packages the gate drive PCB inside a half bridge module, with a published turn-on time of 1.4 ns at 400 V [37],[38]. Sanken has recently demonstrated a 600 V single-device module with integrated gate drive, and they have announced an initial offering in 10 A and 20 A QFN packages [39],[40].

V. Conclusion

GaN-on-Si HFETs have matured in recent years, so that power electronics engineers can design commercial converters using GaN transistors. Normally-off devices rated at 200, 600, and 650 V are already on the market, and there are more coming soon. It would be easy to treat the HFET as a standard Si MOSFET, and simply adjust the converter design slightly based on the device datasheet. However, it requires a more thorough understanding to take full advantage of GaN transistors and design a rugged, high switching frequency, high efficiency converter.

After choosing a particular GaN device and understanding its unique characteristics and requirements, the gate drive and PCB can be designed to minimize power losses. Experimental results can then be more meaningfully understood, accounting for dynamic R_{ds-on} that varies with blocking voltage, higher dead time power losses, cross-talk and ringing.

Acknowledgment

This work made use of facilities supported by the Engineering Research Center Program of the National Science Foundation and the Department of Energy under NSF Award Number EEC-1041877 and the CURENT Industry Partnership Program, as well as facilities provided by Oak Ridge National Laboratory. Funding was provided in part by the Bredesen Center for Interdisciplinary Research and Graduate Education at the University of Tennessee, Knoxville.

The author would also like to thank Dr. Daniel Costinett, Dr. Syed K. Islam, Weimin Zhang, and Zheyu Zhang for their contributions to this review.

References

[1] A. Lidow, J. Strydom, M. de Rooij, D. Reusch, *GaN Transistors for Efficiency Power Conversion, 2nd Edition.* El Segundo, CA: Power Conversion Publications, Sept. 2014.

[2] J. Roberts, H. Lafontaine, C. McKnight-Macneil, "Advanced SPICE models applied to high power GaN devices and integrated GaN drive circuits," presented at IEEE Applied Power Electronics Conference and Exposition, Fort Worth, TX, Mar. 2014, pp. 493-496.

[3] M. Su, C. Chen, S. Rajan, "Prospects for the application of GaN power devices in hybrid electric vehicle drive systems," *Semiconductors Science and Technology*, vol. 28, no. 7, Jul. 2013.

[4] Efficient Power Conversion, "EPC-2010 enhancement mode power transistor," device datasheet, Feb. 2013.

[5] Panasonic, "600V GaN power transistor," presented at IEEE Applied Power Electronics Conference and Exposition, Long Beach, CA, Mar. 2013.

[6] N. Kaneko, O. Machida, M. Yanagihara, S. Iwakami, R. Baba, H. Goto, A. Iwabuchi, "Normally-off AlGaN/GaN HFETs using NiOx gate with

recess," *Proceedings of the 21st International Symposium on Power Semiconductor Devices & IC's*, Barcelona, Spain, 14-18 June 2009, pp.25-28.

[7] R. Chu, A. Corrion, M. Chen, R. Li, D. Wong, D. Zehnder, B. Hughes, K. Boutros, "1200-V normally off GaN-on-Si field-effect transistors with low dynamic on-resistance," *IEEE Electron Device Letters*, vol. 32, no. 5, May 2011, pp. 632-634.

[8] R. Chu, B. Hughes, M. Chen, D. Brown, R. Li, S. Khalil, D. Zehnder, S. Chen, A. Williams, A. Garrido, M. Musni, K. Boutros, "Normally-Off GaN-on-Si transistors enabling nanosecond power switching at one kilowatt," presented at *71st Annual Device Research Conference*, Notre Dame, Indiana, 23-26 June 2013, pp.199-200.

[9] K. Ota, K. Endo, Y. Okamoto, Y. Ando, H. Miyamoto, H. Shimawaki, "A normally-off GaN FET with high threshold voltage uniformity using a novel piezo neutralization technique," presented at International Electron Devices Meeting, Baltimore, MD, 7-9 Dec. 2009, pp. 1-4.

[10] T. Oka, T. Nozawa, "AlGaN/GaN Recessed MIS-Gate HFET With High-Threshold-Voltage Normally-Off Operation for Power Electronics Applications," *IEEE Electron Device Letters*, July 2008, vol.29, no.7, pp.668-670.

[11] N. Ikeda, R. Tamura, T. Kokawa, H. Kambayashi, Y. Sato, T. Nomura, S. Kato, "Over 1.7 kV normally-off GaN hybrid MOS-HFETs with a lower on-resistance on a Si substrate," *Proceedings of the 23rd International Symposium on Power Semiconductor Devices & ICs*, San Diego, CA, 23-26 May 2011, pp. 284-287.

[12] GaN Systems, "GS66508P-E03, 650V enhancement mode GaN transistor," preliminary datasheet, 29 Aug. 2014.

[13] Transphorm, "TPH3006LD GaN Power Low-loss Switch," device datasheet, 9 Apr. 2014.

[14] T. McDonald, "GaN Based Power Conversion: Moving On!" presented at IEEE Applied Power Electronics Convention, Long Beach, CA, Mar. 2013.

[15] T. Hirose, M. Imai, K. Joshin, K. Watanabe, "Dynamic performances of GaN-HEMT on Si in cascode configuration," presented at IEEE Applied Power Electronics Conference and Exposition, Fort Worth, TX, Mar. 2014, pp. 174-181.

[16] MicroGaN Gmbh, "MicroGaN 600V N-OFF Switch," Preliminary device datasheet for MGG1T0617D-CO, Accessed 3 Sept. 2014, Available: http://www.microgan.com/includes/products/MGG1T0617D-CO.pdf

[17] D. Jin, J. del Alamo, "Mechanisms responsible for dynamic on-resistance in GaN high-voltage HEMTs," *Proceedings of the 24th International Symposium on Power Semiconductor Devices and ICs*, Bruges, Belgium, 3-7 Jun. 2012, pp. 333-336.

[18] W. Saito, T. Nitta, Y. Kakiuchi, Y. Saito, K. Tsuda, I. Omura, M. Yamaguchi, "On-resistance modulation of high voltage GaN HEMT on sapphire substrate under high applied voltage," *IEEE Electron Device Letters*, vol. 28, no. 8, Aug. 2007, pp. 676-678.

[19] T. Mizutani, Y. Ohno, M. Akita, S. Kishimoto, K. Maezawa, "A study on current collapse in AlGaN/GaN HEMTs induced by bias stress," *IEEE Transactions on Electron Devices*, vol. 50, no. 10, Oct. 2003, pp. 2015-2020.

[20] G. Meneghesso, G. Verzellesi, F. Danesin, F. Rampazzo, F. Zanon, A. Tazzoli, M. Meneghini, E. Zanoni, "Reliability of GaN high-electron-mobility transistors: state of the art and perspectives," in *IEEE Transactions on Device and Materials Reliability*, vol. 8, no. 2, Jun. 2008, pp. 332-343.

[21] B. Lu, T. Palacios, D. Risbud, S. Bal, D. Anderson, "Extraction of dynamic on-resistance in GaN transistors under soft- and hard-switching conditions," presented at IEEE Compound Semiconductor Integrated Circuit Symposium, Waikoloa, HI, Oct. 2011.

[22] N. Ikeda, Y. Niiyama, H. Kambayashi, Y. Sato, T. Nomura, S. Kato, S. Yoshida, "GaN power transistors on Si substrates for switching applications," in *Proceedings of the IEEE*, vol. 98, iss. 7, Jul. 2010, pp. 1151-1161.

[23] Y. Ma, "EPC GaN transistor application readiness: phase three testing," 2011. Available: http://epc-co.com/epc/documents/product-training/EPC_Phase_Three_Rel_Report.pdf

[24] C. Blake, "GaN technology: what the power designer needs to know," presented in Power Sources Manufacturer Association Power Technology Roadmap Forum, Jul. 12, 2012.

[25] H. Umegami, M. Yamamoto, Y. Nozaki, O. Machida, "A Novel High Efficiency Gate Drive Circuit for Normally Off Type GaN-FET," presented at IEEE Energy Conversion Congress and Exposition, Raleigh, NC, 15-20 Sept. 2012, pp.2954-2960.

[26] X. Ren, D. Reusch, S. Ji, Z. Zhang, M. Mu, F. Lee, "Three-level driving method for GaN power transistor in synchronous buck converter," presented at IEEE Applied Power Electronics Conference and Exposition, 15-20 Sept. 2012, Raleigh, NC, pp. 2949-2953.

[27] F. Hattori, H. Umegami, M. Yamamoto, "Capacitor-less gate drive circuit capable of high-efficiency operation for non-insulating-gate GaN FETs," *IEEE Transactions on Electron Devices*, vol. 60, no. 10, Oct. 2013, pp. 3249-3255.

[28] Texas Instruments, "Gate drivers for enhancement mode GaN power FETs," 2012. Available: http://www.ti.com/lit/ml/slyb189/slyb189.pdf.

[29] Analog Devices, "ADP3623/ ADP3624/ ADP3625/ ADP3633/ ADP3634/ ADP3635 High Speed, Dual, 4A MOSFET Driver with Thermal Protection," device datasheet, 2009.

[30] Silicon Laboratories, "Si823x," Apr. 2014. Available: https://www.silabs.com/Support%20Documents/TechnicalDocs/Si823x.pdf.

[31] M. Briere, "Understanding the breakdown characteristics of lateral GaN-based HEMTs," *How2Power Today*, 16 Sept. 2013. Available: http://www.how2power.com/newsletters/

[32] Z. Zhang, B. Guo, F. Wang, L. Tolbert, B. Blalock, Z. Liang, P. Ning, "Methodology for switching characterization evaluation of wide band-gap devices in a phase-leg configuration," presented at IEEE Applied Power Electronics Conference and Exposition, Fort Worth, TX, Mar. 2014, pp. 2534-2541.

[33] Z. Zhang, W. Zhang, F. Wang, L. Tolbert, B Blalock, "Analysis of the switching speed limitation of wide band-gap devices in a phase-leg configuration," presented at IEEE Energy Conversion Congress and Exposition, Raleigh, NC, Sept. 2012, pp. 3950-3955.

[34] D. Reusch, J. Strydom, "Understanding the effect of PCB layout on circuit performance in a high frequency gallium nitride based point of load converter," presented at IEEE Applied Power Electronics Conference and Exposition, Long Beach, CA, Mar. 2013, pp. 649-655.

[35] S. Tamura, Y. Anda, M. Ishida, Y. Uemoto, T. Ueda, T. Tanaka, D. Ueda, "Recent advances in GaN power switching devices," presented at IEEE Compounds Semiconductor Integrated Circuit Symposium, Monterey, CA, Oct. 2010.

[36] S. Ujita, Y. Kinoshita, H. Umeda, T. Morita, S. Tamura, M. Ishida, T. Ueda, "A Compact GaN-based DC-DC Converter IC with High-Speed Gate Drivers Enabling High Efficiencies," *Proceedings of the 26th International Symposium on Power Semiconductor Devices & ICs*, Waikoloa, Hawaii, 15-19 Jun. 2014, pp. 51-54.

[37] F. Luo, Z. Chen, L. Xue, P. Mattavelli, D. Boroyevich, B. Hughes, "Design considerations for GaN HEMT multichip half-bridge module for high-frequency power converters," presented at IEEE Applied Power Electronics Conference and Exposition, Fort Worth, TX, Mar. 2014, pp. 537-544.

[38] B. Hughes, R. Chu, J. Lazar, S. Hulsey, A. Garrido, D. Zehnder, M. Musni, K. Boutros, "Normally-off GaN switching 400V in 1.4ns using an ultra-low resistance and inductance gate drive," presented at IEEE Workshop on Wide Bandgap Power Devices and Applications, Columbus, OH, Oct. 2013, pp. 76-79.

[39] Sanken, "Next Generation Power Semiconductors - Commitment to GaN / SiC Development," Accessed 3 Sept. 2014. Available: http://www.semicon.sanken-ele.co.jp/en/guide/GaNSiC.html/

[40] PowerPulse, "Driver Integrated GaN Transistors Announced by Sanken, Techno-Frontier 2014," 25 Jul. 2014, Available: http://www.powerpulse.net/story.php?storyID=30441;s=07252014

978-1-4799-5494-0/14 $31.00 © 2014 IEEE

Advances in Reliability and Operation Space of High-voltage GaN Power Devices on Si Substrates

Y.-F. Wu, J. Guerrero, J. McKay and K. Smith

Transphorm Inc., 115 Castilian Drive, Goleta, CA 93117, USA

Email: ywu@transphormusa.com

Abstract—GaN-on-Si power devices have advantages in both performance and the potential for low cost and high volume production. Advances have been made in extended reliability tests exhibiting an intrinsic life time >10^6 hours for qualified 600-V GaN HEMT products. Experimental kV-class devices have shown CW operation at 800V with >99% efficiency at 100 kHz and a 2:1 boost ratio. Both are exciting developments reaffirming the validity of the new power semiconductor technology.

I. INTRODUCTION

The need for post-Si power devices has led to the development of both SiC [1] and GaN [2] products. While both shown performance breakthroughs, the GaN-on-Si platform has the potential for lower cost and higher throughput due to the availability of inexpensive large-diameter Si wafers and associated fabrication infrastructures. The 1st generation 600-V GaN/Si HEMTs passed JEDEC qualification in 2013 and entered the power electronics market [2]. The stringent sampling scheme and zero-failure requirement for qualification ensure a substantial level of technology maturity. However, the qualification test time is typically limited to 1000 hours, insufficient to predict the true lifetime of these devices. Therefore extended reliability tests well exceeding 1000 hours and above specification limits are necessary to confirm the robustness and to predict intrinsic lifetime of these devices. Furthermore, there is great interest and debate on whether GaN/Si is limited to under 600-V applications. A demonstration of true kv-class GaN/Si devices operating at near kV with high performance will shed light to the real application potential of this semiconductor platform.

II. 600-V GAN HEMTs & EXTENDED LIFE TESTS

The devices in this study are production parts TPH3006PS from Transphorm based on a cascode structure presented before [2]. It incorporates a normally-off low-voltage Si MOSFET at the input and a normally-on high-voltage GaN HEMT at the output. The combined device is normally-off having a gate threshold of the Si MOSFET and a blocking voltage of the GaN HEMT. The GaN HEMT is built on Si substrate with an insulating GaN buffer for voltage blocking, a AlGaN barrier layer, source-drain ohmic contacts to 2-demential electron gas at the AlGaN/GaN interface, a metal gate & field plate structure surrounded by dielectrics. The finished hybrid device is in a Quiet-Tab^{TM} TO-220 package with patented Gate-Source-Drain (G-S-D) pin arrangement and the tab as a low-inductance source terminal. These 600V-rated devices have passed full product qualification including 1000-hour of high temperature reverse bias (HTRB) at 80% rated voltage and 150 °C for 3 fabrication lots of 77 pcs each with zero failure.

While device qualification consisted of dc and mechanical stress tests is important, equally relevant is the high temperature operation life test (HTOL) that mimics the actual hard switching conditions in applications. This was done using 7 GaN HEMTs in their TO-220 packages, each operating as the main switch in a boost converter at a 300-kHz PWM frequency and a 200V:400V boost ratio. The device under test (DUT) was bolted on a hot plate set at 170 °C resulting in a junction temperature (Tj) of 175 °C calculated by the thermal resistance and device dissipation. Other circuit components were arranged on a printed circuit board (PCB) suspended by an air gap to avoid overheating. Conversion loss is plotted against time in Fig.1. The conversion losses show small fluctuations mostly due to the imperfect room temperature control. It appears that there is a slight but persistence loss increase from 2000 to 3000 hours. However, after the devices were removed from the circuit, parametric test showed no

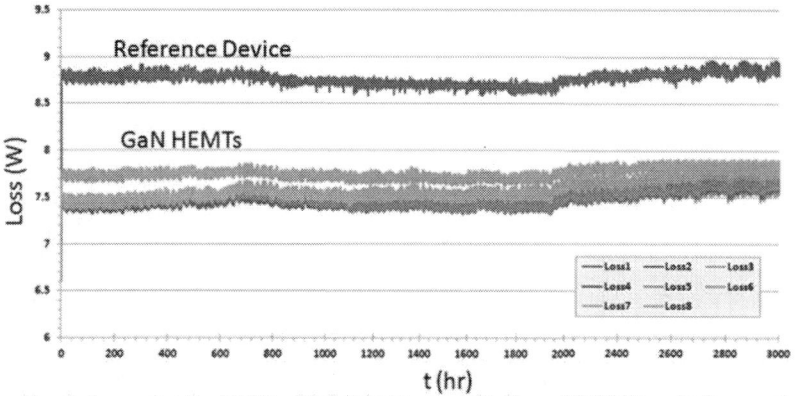

Fig. 1. Loss plot for HTOL of (7) 600-V-rated GaN-on-Si HEMT switches and a reference device to 3000 hours at Tj=175°C. Each device operated in a boost converter at 300kHz with 200V:400V boost ratio & 410W output power.

Table 1. Leakage at 600V and dynamic on-resistance of DUTs before and after 3000hr HTOL, exhibiting no parametric degradation.

	Leakage current @ 600V (uA)		Ron_dynamic @ 480V, 4A, 60s (mΩ)	
Dev	0hrs	3000hrs	0hrs	3000hrs
8	2.291	1.713	155.3	154.1
9	2.387	1.735	151.9	148.0
10	2.354	1.764	150.6	149.8
11	2.403	1.791	148.0	144.7
12	2.515	1.843	145.2	144.0
13	2.488	1.843	151.9	148.0
14	2.469	1.812	145.9	146.0

978-1-4799-5494-0/14 $31.00 © 2014 IEEE

discernible degradation as summarized in Table 1. Inspection of the PCB circuit revealed severe oxidation due to the long term exposure to the high temperature environment, which was responsible for the increase of conversion loss. The TO220 packages were also significantly tarnished, potentially increasing the thermal resistance and adding to the conversion loss. As a result, this HTOL test shows that the actual GaN device robustness is superior to the typical package and surrounding circuits.

HTRB test is less affected by package oxidation and circuit conductivity degradations. It is a key test for GaN power devices, therefore data with multiple fabrication lots totaled more than 300 devices up to 8000 hours were collected as shown in Fig. 2. There was no device failure until 3000 hours and the failure rate was only 2% at 8000 hours. Compared to typical Si designs with a significant failure rate at 5000 hours, this 1st generation GaN HEMTs show a robustness design and reproducibility. The failure mode is low-level random defects which is not intrinsic hence can be minimized when scaled from the previous small fabrication facilities to larger production lines.

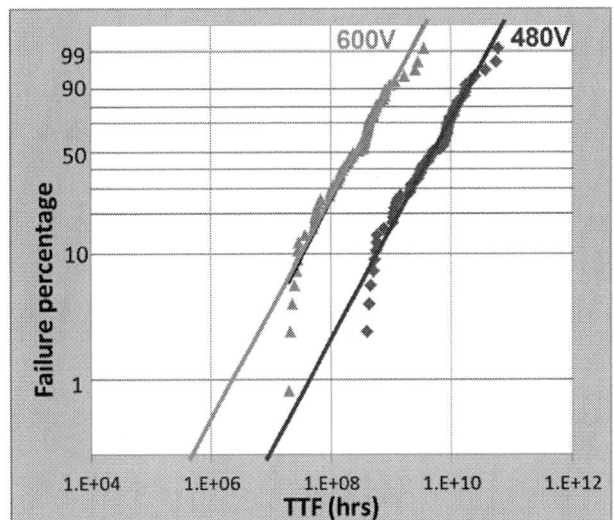

Fig. 3. Over-voltage life-time test result of 600V-rated GaN HEMTs based on failure rate from 900V to 1200V extrapolated to 600V with acceleration factors using the TDDB model. With a 90% confidence at 1% failure rate, the life time at 600V is >10^6 hours.

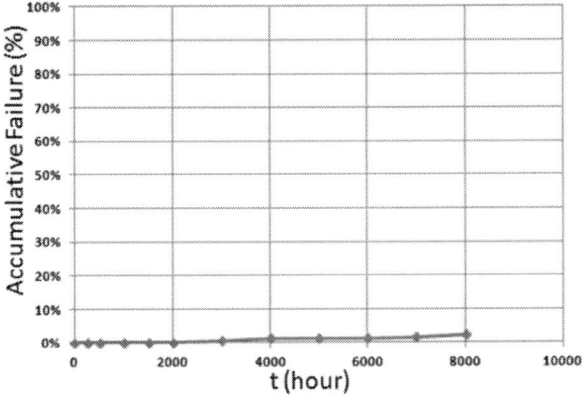

Fig. 2. High temperature reverse bias (HTRB) test result of more than 300 units of 600V-rated GaN-on-Si HEMTs up to 8000 hours, showing no failure until 3000 hr and only 2% failure at 8000 hr.

To assess the intrinsic life time of these 600V-rated GaN switches, high voltage off state (HVOS) tests has been carried out at three stress levels of 1050V, 1100V and 1150V respectively. The intention is to obtain a life time distribution and associated voltage acceleration factors for each stress level, which then can be converted to an equivalent life time plot at the rated voltage of 600V. Up to now, 2 sets of stresses at 1100V and 1150V at 85°C have completed. The data fits the time dependent dielectric breakdown model (TDDB). The percent of failure (or unreliability) is plotted against the equivalent life time at 480V and 600V as shown in Fig.3. It is seen that with 90% confidence at 1% failure rate, the life times at 480V and 600V are >10^7 and > 10^6 hours respectively. This shows outstanding intrinsic reliability, conveying a high confidence to application users.

III.   KV-CLASS GaN HEMTs & PERFORMANCE TEST

Earlier data indicates immediate potential for the GaN/Si HEMT to be engineered to operate at above 1kV. Such experimental devices have been made with enhanced buffer voltage blocking and electric field management targeting a 1.2-kV rating. These devices show maximum pulsed current of 50A at a drain bias of 10 V and a dynamic on-resistance about 180 mΩ when measured within 1 μs after reverse blocking 800 V for 10 minutes. The input gate charge is 10 nC for a V_{GS} swing of 0-8V and the output charging energy is 13.8 μJ for a V_{DS} rise from 0 to 800V.

Inductive switching waveforms at a bus voltage of 800V and inductor current of 8A are shown in Fig. 4. It is seen that

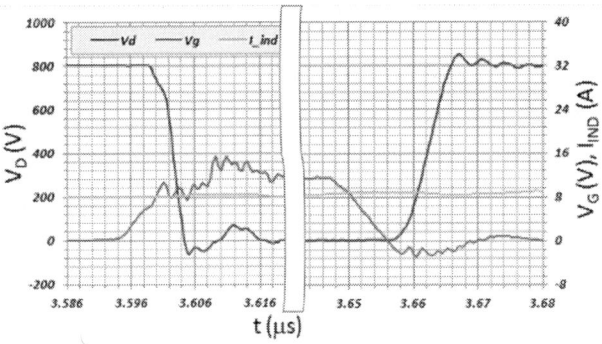

Fig. 4. Hard switching waveform of an 1.2kV experimental GaN HEMT operating at 800V dc bus with 8A of inductive current.

a conventional gate driver with 0 to 12 V output is well suited to driving the GaN Transistor. The output turn-on and turn-off transients are healthy with minor overshoots under such high voltage and high current conditions. Detailed measurements yielded turn-on propagation delay time and fall time of 4.3 and 4 ns, as well as turn-off delay time and rise time of 10 and 6 ns, respectively. The voltage transient at turn-on is as high as 210 V/ns, which is important to minimize current-voltage cross-over losses. These rise and fall times are about a factor of 2-5 less than that of the competing SiC MOSFET [4],

978-1-4799-5494-0/14 $31.00 © 2014 IEEE   31

which are direct indicators of lower switching losses and are credited to the high mobility and high electron velocity of the 2-dimensional-electron-gas in the GaN HEMT structure.

A boost converter was constructed to evaluate device performance in actual applications. Important components include a 1200 V/5A SiC Schottky diode as the boost diode, 5-uF film capacitors for input and output charge storage, as well as a 1-mH high-Q inductor for energy transfer. The boost inductance was calculated to ensure hard-switched operation in the power and frequency of interest. A 4700-pF surface-mount ceramic capacitor at the diode cathode was added for high-frequency decoupling and four sets of high-frequency L-C filters were employed to reduce noise at the dc power input/output nodes as well as the voltage sense nodes. The converter design strives for text-book simplicity with a tight layout, without any gate drive resistor and snubbing components. The Quiet-Tab^{TM} GaN HEMT package allows the GaN device to be mounted on a grounded heat-sink without insulation, which facilitates heat dissipation and minimizes source inductance.

The converter was tested with 400V:800V boost ratio at 100-kHz Pulse-Width-Modulation (PWM). The gate driver consumption was 20 mW which is negligible for system efficiency and confirms the benefit of a low device gate charge. Efficiency measurement was done using precision dc meters with accuracy better than 0.05% and the system error was verified to be within 0.02% using zero-loss by-pass. Each data point was recorded after thermal equilibrium which took about 300 to 450 seconds. The performance as a function of output power up to 3.2 kW is plotted in Fig. 5 with all component losses included. The converter achieves greater than 98% efficiency from 400W and above with a peak value >99% located at 1.5-2.2 kW. This very-high efficiency

obtained at 100 kHz matches that of SiC MOSFET at 30 kHz with the same conversion ratio and rail voltages [3]. The ability of GaN to achieve 99% efficiency at 3x higher frequency allows shrinkage of energy storage components, enabling more compact and lower-cost power systems.

Loss analysis was performed with measurements at varying PWM frequencies with each fixed load [4]. It reveals that at 2 kW and 100 kHz, the switching and conduction losses of the GaN HEMT are 4.3 and 3.0 W respectively, which totals to 7.3 W or 37% of the total converter loss. Although there is much room to improve the GaN HEMT, passive component upgrades are necessary to match the new wide-band-gap power device.

IV. CONCLUSION

Recent advances in extended reliability tests of industry's 1st generation 600-V GaN-on-Si switching devices show outstanding robustness including no measurable degradation after 3000 hours of HTOL test at 175°C and an estimated intrinsic life time $>10^6$ hours based on electric field acceleration.

Development effort to further extend the voltage operation space resulted in true kV class GaN HEMTs with low dynamic on-resistance of 180 mΩ and superior switching characteristics at an 800V bus. A 400V/800V boost converter with the GaN HEMT as the switching device achieved >99% efficiency at 100 kHz, setting a new state-of-the-art for power semiconductors.

REFERENCES

[1] J.W. Palmour et al, "Silicon carbide power MOSFETs: Breakthrough performance from 900 V up to 15 kV", 2014 *IEEE 26th International Symposium on Power Semiconductor Devices & IC's (ISPSD)*, p79-82, June 15-19, 2014.

[2] Y.-F. Wu, J. Gritters, L. Shen, R.P. Smith, J. McKay, R. Barr and R. Birkhahn, "Performance and Robustness of First Generation 600-V GaN-on-Si Power Transistors", Proceeding, *1st IEEE Workshop on Wide Bandgap Devices and Applications*, S1-002, Ohio, Oct 27-19, 2013.

[3] B. Callanan & J. Rice, "Comparative High Frequency Performance of SiC MOSFETs Under Hard-switched Conditions", *PCIM Europe 2012*, Conference Proceeding, p318-324, May 8-10, 2012, Nuremberg. Also see: http://www.power-mag.com/pdf/feature_pdf/1342108658_Cree_feature_Layout_1.pdf

[4] Y.-F. Wu, J. Gritters, L. Shen, R. P. Smith, and B. Swenson, "kV-Class GaN-on-Si HEMTs Enabling 99% Efficiency Converter at 800 V and 100 kHz", *IEEE TRANSACTIONS ON POWER ELECTRONICS*, VOL. 29, NO. 6, p2634-2637, JUNE 2014.

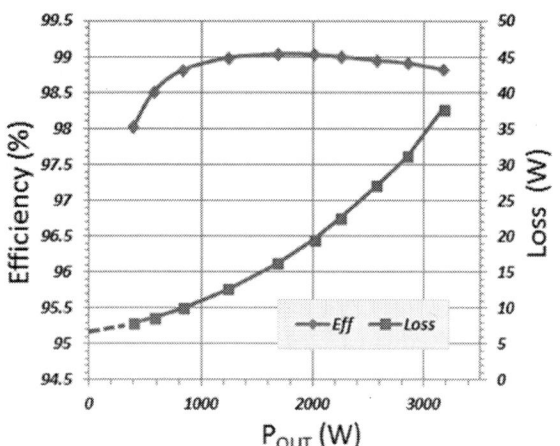

Fig. 5. Performance of a 1.2-kV GaN HEMT in a 400V:800V boost converter showing 99% peak efficiency at 100 kHz.

Degradation Mechanisms of AlGaN/GaN HEMTs on Sapphire, Si, and SiC Substrates under Proton Irradiation

Andrew D. Koehler[1], Travis J. Anderson[1], Jennifer K. Hite[1], Bradley D. Weaver[1], Marko J. Tadjer[2], Michael A. Mastro[1], Jordan D. Greenlee[3], Petra Specht[4], Matthew Porter[5], Todd R. Weatherford[5], Karl D. Hobart[1], Fritz J. Kub[1]

[1]Power Electronics, Naval Research Laboratory, Washington, DC, USA, andrew.koehler@nrl.navy.mil
[2]American Society of Engineering Education Postdoctoral Fellow Residing at NRL
[3]National Research Council Postdoctoral Fellow Residing at NRL
[4]University of California Berkeley, CA 94720
[5]Naval Postgraduate School, Monterey, CA 93943

Abstract—The degradation mechanisms of AlGaN/GaN high electron mobility transistors (HEMTs) grown on sapphire, Si, and SiC substrates, under 2 MeV proton irradiation are investigated. It was determined by electron channeling contrast imaging that the threading dislocation density of the AlGaN/GaN epitaxial layers is highest on sapphire substrates and lowest on SiC substrates. Photoluminescence spectroscopy confirmed the material quality order from worst to best to be AlGaN/GaN grown on sapphire, Si, and SiC substrates, respectively. The radiation response of sheet carrier density was not statistically different for HEMTs on each substrate, however the mobility degraded more for HEMTs with less initial dislocations (on SiC) than more defective HEMTs (on sapphire).

Keywords—GaN, HEMT; proton radiation

I. INTRODUCTION

Gallium nitride technology is appealing for power amplifiers and converters in radiation intensive environments [1]. AlGaN/GaN high electron mobility transistors (HEMTs) have been shown to exhibit strong radiation hardness to protons [2-7]. However, the specific mechanisms accountable the resilience of AlGaN/GaN HEMTs to displacement damage is currently undetermined. A likely explanation is that GaN has a large number of pre-existing defects; therefore the addition of more damage will have little impact. Since GaN technology is rapidly improving and the material is becoming less defective, it is possible that with material improvements that the radiation response is becoming more sensitive. A systematic study is performed, where AlGaN/GaN HEMT structures are grown on sapphire (Al_2O_3), Si, and SiC substrates, in order to vary the defect densities to investigate the role of pre-existing defects on the radiation hardness of the HEMT electrical performance.

II. EXPERIMENTAL

AlGaN/GaN epitaxial layers were grown on sapphire, Si, and SiC substrates by metal organic chemical vapor deposition (MOCVD). The epitaxial layers were nominally identical with a >1 μm thick unintentionaly doped GaN buffer layer topped with a 17-25 nm AlGaN barrier layer (comprised 27-30% aluminium concentration). The AlGaN/GaN hereostructures on each substrate were processed in a single fabrication lot. First, HEMT mesa isolation was perfomed by a Cl_2-based inductively coupled plasma (ICP) etch. Then, ohmic metal was formed by electron-beam deposition of Ti/Al/Ni/Au, lift-off, and rapid thermal anneal. Then, the Schottky Ni/Au gate metal was deposited and patterned by lift-off. Finally, all samples received 100 nm PECVD SiN_x passivation, followed by a SF_6 reactive ion etch (RIE), contact window etch, and a thick overlay metal deposition of Ti/Au. HEMTs, transmission line measurement (TLM) structures, capacitance-voltage (C-V) dots, and van der Pauw structures were fabricated on each HEMT sample.

The fabricated HEMT wafer samples were mounted on a copper plate and subjected to 2 MeV protons at room temperature, using a tandem Van de Graaff particle accelerator. The device terminals were left floating and the sample was incrementally exposed to seven fluences, beginning at 1×10^{12} H^+/cm^2, up to a final fluence of 6×10^{14} H^+/cm^2. Electrical characterization of static (DC) current-voltage (*I-V*), pulsed *I-V*, and Hall parameters was performed iteratively at each fluence of 2 MeV proton irradiation.

III. RESULTS AND DISCUSSION

Prior to device fabrication and irradiation, the epitaxial AlGaN/GaN layers were imaged by scanning electron microscopy (SEM). The threading dislocation density (TD) of the epitaxial layers were determined by electron channeling contrast imaging (ECCI) [8] to be 8.6×10^{-8}, 8.2×10^{-8}, and 1.0×10^{-8} cm^{-2} on sapphire, Si, and SiC, substrates, respectively (Figure 1). Lattice and thermal mismatch between the substrate and epitaixal films is responsible for the difference observed in the initial material defect density.

978-1-4799-5494-0/14 $31.00 © 2014 IEEE

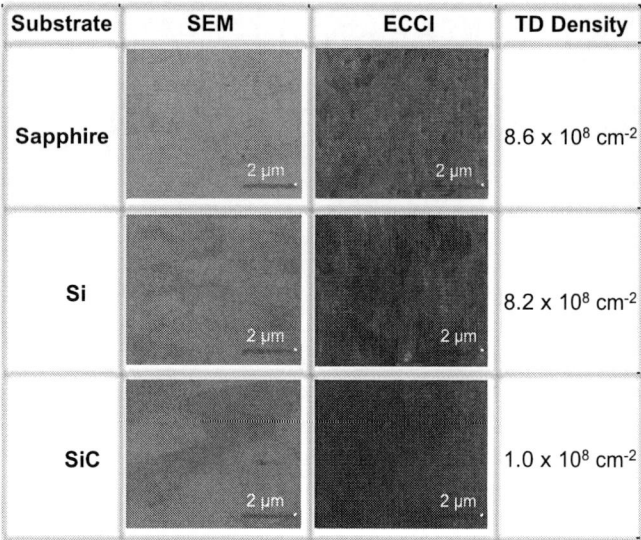

Figure 1. SEM and ECCI images of the epitaxial AlGaN/GaN layers, before irradiation, on sapphoire, Si, and SiC substrates with the coooresponding threading dislocation density (TD).

Substrate	SEM	ECCI	TD Density
Sapphire			8.6×10^8 cm^{-2}
Si			8.2×10^8 cm^{-2}
SiC			1.0×10^8 cm^{-2}

Photoluminescence (PL) spectroscopy was performed to characterize the initial GaN material defect quality. As seen in Figure 2, HEMTs on SiC show the largest band edge peak and HEMTs on sapphire show the smallest GaN band edge peak at 365 nm (3.4 eV). Conversely, the yellow defect band (~500 nm to ~700 nm) signal is largest for the HEMTs on sapphire than the HEMTs on SiC substrates. This is consistent with the ECCI results, the order of material quality from most to least defective is GaN on sapphire, Si, and SiC, respectively.

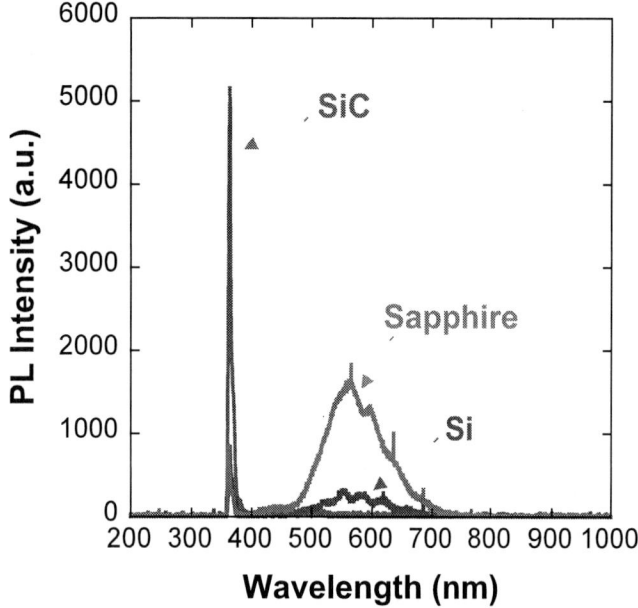

Figure 2. PL spectrum from HEMTs on sapphire, Si, and SiC substrates, before irradiation.

Representative static and pulsed I-V characteristics are shown for HEMTs on sapphire, Si, and SiC substrates

irradation in Figure 3 before and after 2 MeV proton irradiation (6×10^{14} H$^+$/cm^2). Each device demonstrated degradation caused by the 2 MeV protons in such parameters as on-resistance ($\Delta R_{ON,Sapphire} = 48.2\%$, $\Delta R_{ON,Si} = 86.0\%$ and $\Delta R_{ON,SiC} = 51.7\%$), maximum drain current ($\Delta I_{Dmax,Sapphire} = -36.6\%$, $\Delta I_{Dmax,Si} = 51.8\%$, $\Delta I_{Dmax,SiC} = 46.2\%$). Before irradiation, the HEMTs on sapphire demonstrate the largest dispersion between 20 ns pulsed (with the device grounded during the quiescent period) and static I-V, indicative of the most charge trapping.

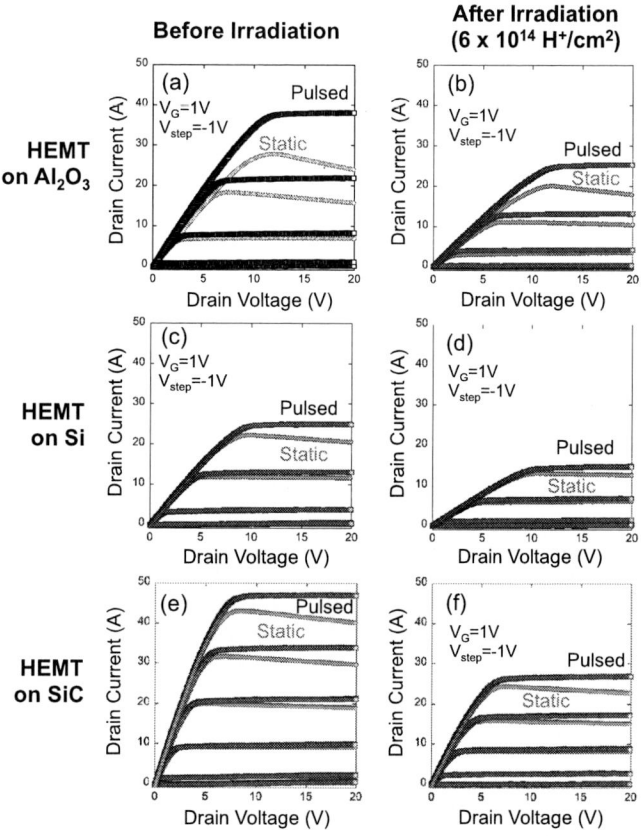

Figure 3. Pulsed and static I_D-V_{DS} curves for HEMTs on sapphire, Si, and SiC substrates before and after irradiation of 6×10^{14} H$^+$/cm^2.

To further investigate the observed radiation-induced degradation, Hall measurements were performed on van der Pauw structures. First, the two-dimensional electron gas (2DEG) sheet carrier density (n_s) was measured before radiation, as well as after each fluence of 2 MeV proton irradiation. The sheet carrier density for the HEMT structures on each substrate began to reduce around 1×10^{14} H$^+$/cm^2, as shown in Figure 4. The material quality did appear to impact the magnitude reduction in sheet carrier density, indicating that a comparable amount of radiation-induced charge trapping occurred regardless of the initial defect density. Neither the heterostructure nor the Al composition was fundamentally altered by irradiation, since the maximum fluence of 2 MeV protons in this experiment is orders of magnitude less than the amount required to amorphize the material [9]. The capacitance of large area (diameter = 150 μm) C-V dots was unchanged by the irradiations. Therefore, the polarization was also unaffected by the irradiation, and the sheet carrier density

978-1-4799-5494-0/14 $31.00 © 2014 IEEE

degradation is likely due to charge trapped at radiation-induced defects [9,10] screening the 2DEG. This is also reflected in the threshold voltage (V_T) shift, which is proportional to the sheet carrier density, as shown prevously [2]. However, the percent change in 2DEG mobility (μ_{2DEG}) caused by irradiation trended with the epitaixial material quality, as shown in Figure 5. A second-order, substrate-dependent effect was observed which degrades the 2DEG mobility more for HEMTs on higher quality substrates (e.g. SiC) than HEMTS on lower quality substrates (e.g. sapphire). Scattering of 2DEG electrons by trapped charge in radiation-induced defects near the 2DEG is more pronounced for higher quality material, with higher mobility and less defects, while the mobility of more defective material is more resistant to radiation-induced damage.

IV. SUMMARY

The combination of scattering and screening of 2DEG charge by radiation-induced trapped charge is responsible for degradation of HEMTs on sapphire, Si, and SiC substrates. The basic response for HEMTs on each substrate was similar, however HEMTs on less defective epitaxial layers with high mobility exhibit larger mobiltiy degradation, whereas HEMTs on more defective epitaxial layers were more resistant to radiation damage. To probe this substrate effect further, additional experiments on bulk GaN substrates, with significantly fewer dislocations would provide additional insight.

REFERENCES

[1] B.D. Weaver, P.A. Martin, J.B. Boos and C.D. Cress, "Displacement damage effects in AlxGa1-xN/GaN high electron mobility transistors, IEEE Trans. Nucl Sci., Vol. 59, No. 6, Dec. 2012, 3077.

[2] T.J. Anderson, A.D. Koehler, J.D. Greenelee, B.D. Wweaver, M.A. Mastro, J.K. Hite, C.R. Eddy Jr., F.J. Kub, "Substrate-Dependent Effects on the Response of AlGaN/GaN HEMTs to 2-MeV Proton Irradiation" IEEE Electron Device Letters Vol. 35, No 8, pp. 826-828, 2014.

[3] S.M. Khanna, J. Webb, H. Tang, A.J. Houdayer and C. Carlone, "2 MeV proton radiation damage studies of gallium nitride films through low temperature photoluminescence spectroscopy measurements," IEEE Trans. Nucl. Sci. 47 (2000) 2322.

[4] A. Ionascut-Nedelcescu, C. Carlone, A. Houdayer, H.J. Bardeleben, J.-L. Cantin and S. Raymond, "Radiation Hardness of Gallium Nitride," IEEE Trans. Nucl. Sci. 49 (2002) 2733.

[5] S.J. Pearton, R. Deist, F. Ren, L. Liu, A.Y. Polyakov and J. Kim, "Review of raadiation damage in GaN-based materials and devices," J. Vac. Sci. Technol. 31 No. 5, (2013) 05081-1. .

[6] T. Roy, E.-X. Zhang, Y.S. Puzyrev, D.M. Fleetwood,, R.D. Scrimpf, B.K. Choi, A.B. Hmelo and S.T. Pantelides, "Process dependence of proton-induced degradation in GaN HEMTs," IEEE Trans. Nucl. Sci. 57 (2010) 3060.

[7] A. D. Koheler, T.J. Anderson, B.D. Weaver, M.J. Tadjer, K.D. Hobart, F.J. Kub. "Degradation of Dynamic ON-Resistance of AlGaN/GaN HEMTs Under Proton Irradiation" Proceedings of the 2013 IEEE Workshop on Wide Bandgap Power Devices and Applications (WiPDA), 112-114 (2013)

[8] Y. N. Picard, J. D. Caldwell, M. E. Twigg, C. R. Eddy Jr., M. A. Mastro, R. L. Henry, R. T. Holm, P. G. Neudeck, A. J. Trunek and J. A. Powell "Nondestructive analysis of threading dislocations in GaN by electron channeling contrast imaging" Appl. Phys. Lett. 91, 094106 (2007).

[9] J.F. Ziegler, M.D. Ziegler, J.P. Biersack. "SRIM – The stopping and range of ions in matter" Nucl. Instrum. Methods Phys. Res., Sect. B. 258, 1818-1823 (2010)

[10] A.P. Karmarkar, B. Jun, D.M. Fleetwood, R.D. Schrimpf, R.A. Weller, B.D. White, L.J. Brillsonb and U.K. Mishra, "Proton irradiation effects on GaN-based high electron mobility transistors with Si-doped AlxGa1-xN and thick GaN cap layers," IEEE Trans. Nucl. Sci. 51 (2004) 293.

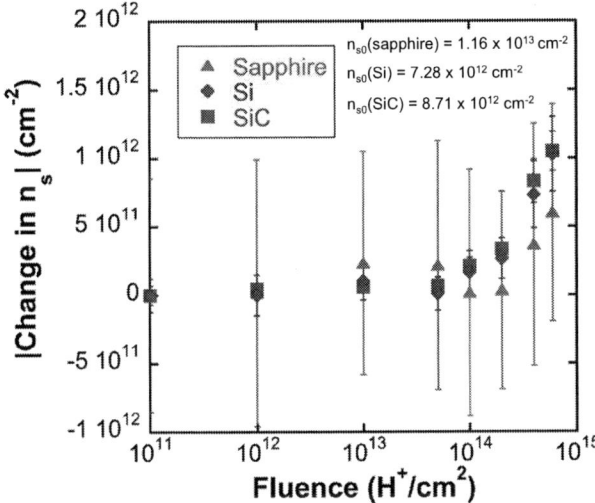

Figure 4. Magnitude reduction in sheet carrier density for HEMTs fabricated on different substrates. Uncertainty shows no statistical difference in the reduction in sheet carrier density.

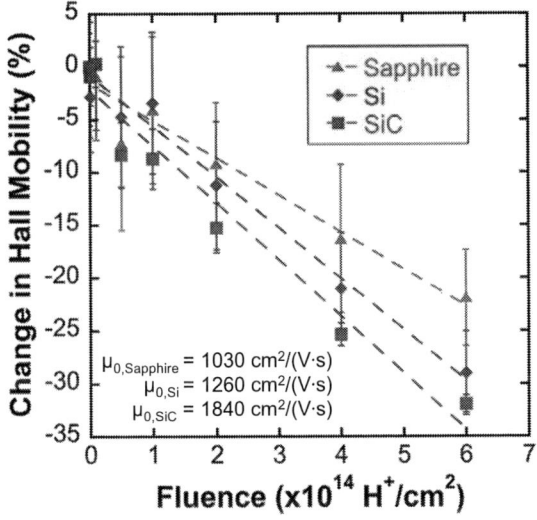

Figure 5. The percent change in Hall mobility shown for HEMTs fabricated on different substrates. Different degradation level occurs for HEMTs on each substrate.

High-Temperature SiC Power Module with Integrated SiC Gate Drivers for Future High-Density Power Electronics Applications

Bret Whitaker, Zach Cole,
Brandon Passmore, Daniel
Martin, Ty McNutt, and
Alex Lostetter
Arkansas Power Electronics Intl. Inc.
Fayetteville, AR

M. Nance Ericson, S. Shane
Frank, Charles L. Britton, and
Laura D. Marlino
Oak Ridge National Laboratory
Oak Ridge, TN

Alan Mantooth, Matt Francis,
Ranjan Lamichhane, Paul
Shepherd, and Michael Glover
University of Arkansas
Fayetteville, AR

Abstract—This paper presents the testing results of an all-silicon carbide (SiC) intelligent power module (IPM) for use in future high-density power electronics applications. The IPM has high-temperature capability and contains both SiC power devices and SiC gate driver integrated circuits (ICs). The high-temperature capability of the SiC gate driver ICs allows for them to be packaged into the power module and be located physically close to the power devices. This provides a distinct advantage by reducing the gate driver loop inductance, which promotes high-frequency operation, while also reducing the overall volume of the system through higher levels of integration. The power module was tested in a bridgeless-boost converter to showcase the performance of the module in a system level application. The converter was initially operated with a switching frequency of 200 kHz with a peak output power of approximately 5 kW. The efficiency of the converter was then evaluated experimentally and optimized by increasing the overdrive voltage on the SiC gate driver ICs. Overall a peak efficiency of 97.7% was measured at 3.0 kW output. The converter's switching frequency was then increased to 500 kHz to prove the high-frequency capability of the power module. With no further optimization of components, the converter was able to operate under these conditions and showed a peak efficiency of 95.0% at an output power of 2.1 kW.

Keywords—Silicon Carbide, Power Module, Power Electronics

I. INTRODUCTION

The growing number of electronic devices in mobile systems, such as electric vehicles (EV) and plug-in hybrid electric vehicles (PHEV), increases the demand for power electronics on-board the vehicles. High-density power electronics are required to accommodate the electrical demand while minimizing the impact on the vehicle. One method for achieving high-density power electronics is through high-frequency operation. This allows for a reduction in the size of magnetic components, such as inductors and transformers, as well as passive filtering components, such as capacitors. A second method for densification is by operating the system at elevated temperatures. The size and or complexity of the thermal management system, in most cases a heat sink, can be minimized if the circuitry can operate at higher temperatures. Also, the need for water chiller plate cooling can be eliminated. A third method for volume reduction can be found through

high levels of integration for integrated circuits (ICs), power modules, and system level packaging. In order to maximize these effects and minimize converter volume, high-frequency and high-temperature capable components with high levels of integration are required.

Silicon carbide (SiC) is a semiconductor material that can operate efficiently in high-temperature environments [1] and at high switching frequencies [2]. These properties make SiC devices ideal for high-density applications. SiC power devices are widely commercially available and are being researched for use in high-performance power electronic systems [2–6]. An example of the volume reduction that can be achieved by using SiC power devices is shown in Fig. 1 where a prototype on-board battery charger [6] is compared to a commercial on-board charger that is installed in the 2010 model Toyota Prius Plug-In Hybrid. The prototype on-board battery charger, which was designed under the same project as the work presented here, showed very high-density, however, it utilized silicon gate driver ICs. Unlike the SiC power devices which are capable of 225 °C, the commercial silicon gate driver ICs can

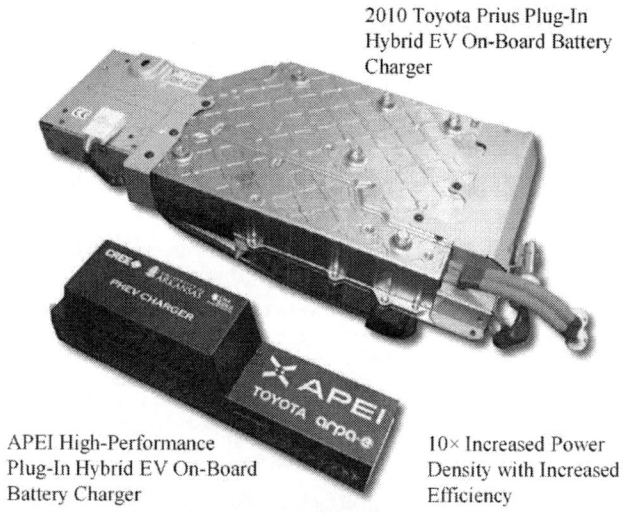

2010 Toyota Prius Plug-In
Hybrid EV On-Board Battery
Charger

APEI High-Performance
Plug-In Hybrid EV On-Board
Battery Charger

10× Increased Power
Density with Increased
Efficiency

Fig. 1. Size comparison of 6 kW APEI, Inc. prototype charger to 2.8 kW 2010 model Toyota Prius Plug-In Hybrid battery charger.

only operate at 125 °C. The lower operating temperature of the gate driver ICs required them to be located on a separate board above the power module. Ultimately this reduced the performance of the gate driver circuit by adding inductance to the gate loop while also increasing the volume of the system.

A high-temperature capable, high-performance gate driver will allow for further densification of power electronic systems. These devices would enable a new breed of intelligent power module (IPM) that would contain both the power devices as well as the gate driver ICs on the same substrate in close physical proximity. Such an IPM would reduce the size of the overall system by increasing the level of integration while also providing critical electrical advantages. The close proximity of the gate driver ICs to the power devices would provide the lowest gate loop inductance and maximize the switching performance of the power devices. To realize an IPM such as this requires a high-temperature capable gate driver with the voltage and current capability required to efficiently drive SiC power devices.

This paper presents an IPM that utilizes a SiC gate driver to achieve a highly integrated, all SiC solution. The IPM is developed and then tested in a bridgeless-boost converter. The operation of the IPM is verified and an optimized efficiency of 97.7% was measured at a switching frequency of 200 kHz and an output power of 3.0 kW. The limits of the IPM were then pushed and the converter was operated at a switching frequency of 500 kHz where a peak efficiency of 95.0% was measured.

II. SiC Gate Driver Integrated Power Module

A. SiC Gate Driver

The SiC gate driver utilized in this work was fabricated in a 2-μm 4H-SiC process selected to enable future die level integration with a SiC power MOSFET. Limitations in this power device optimized process restricted the devices available for use in the gate driver circuit to n-channel MOSFETs and resistors. The simplified driver architecture is shown in Fig. 2 and is the second generation of a previously reported gate driver [7]. The input uses four inverter stages (X_1-X_4) to amplify the 5V logic-level input signal to sufficient levels for the output stage buffers (X_5 and X_6). The totem pole output stage is composed of an n-channel pull-down device (M_{PD}) and a source follower pull-up device (M_{PU}), sized to sink and source the high-currents required for high-frequency switching of a power SiC MOSFET. The use of multiple supplies enabled the optimization of each of the primary functional blocks. The

gate driver output rise time was improved by overdriving the gate of the output source follower stage, facilitated by a separate connection to of the power supply to X_6 (V_{dd-OD}). The output voltage level of the circuit is defined by the voltage rail of the output pull-up device (V_{dd-OS}). A significant improvement in the output rise time was obtained by connecting the source and body of the output stage pull-up device, thus minimizing the associated body effect. Since an n-well process was used for this work, isolation of the pull-up device body from the negative rail required use of a separate body-isolated MPU device die to implement the necessary source-body connection.

B. Power Module

The SiC gate drivers were packaged into a full bridge power module designed specifically for high-frequency and high-temperature operation [8]. The IPM uses a high-temperature PCB that is attached directly to the power substrate. This PCB contains high-frequency capacitors as well as power and signal connection points for the IPM. The location of the capacitors minimizes the loop inductance in the power path which in turn maximizes the switching performance of the module. By placing the capacitors directly on the module almost all voltage overshoot associated with switching events is eliminated. The original prototype, dubbed the X-5 which did not contain gate drivers, was modified slightly to accommodate the additional die area needed for the SiC gate drivers on the substrate. It was also further optimized to maximize the thermal and electrical performance of the integrated system. The two high-side quadrants (Q_1 and Q_2) are populated with a single 20 A SiC Schottky diode per position. The two low-side quadrants (Q_3 and Q_4) contain two parallel SiC MOSFETs and one antiparallel 20 A SiC Schottky diode per position. The IPM is shown in Fig. 3.

III. Boost Converter

The X-5 power module with integrated SiC gate drivers was designed specifically for use in a bridgeless-boost converter. This topology is popular for grid-connected active

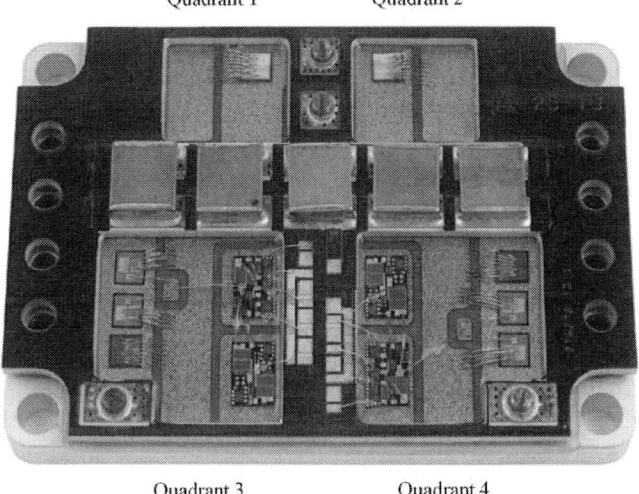

Fig. 3. IPM with integrated SiC gate drivers and local high-frequency bypass capacitors.

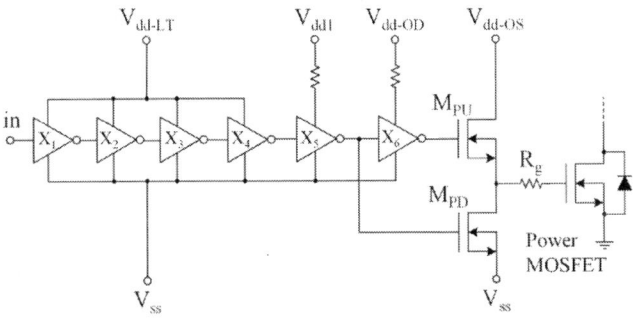

Fig. 2. Simplified gate driver architecture and connection to power MOSFET.

978-1-4799-5494-0/14 $31.00 © 2014 IEEE   37

rectifiers where high power quality is required. This topology can implement power factor correction (PFC) to minimize grid current harmonic content and make the load appear resistive. The circuit schematic for a bridgeless-boost converter is shown in Fig. 4. Through the addition of MOSFETs to Q_1 and Q_2, the topology is capable of bi-directional power conversion which can be utilized in future charger applications.

The selection of the bridgeless-boost topology was based on the previously designed prototype on-board battery charger that utilized silicon gate drivers [6]. The bridgeless-boost converter in the prototype charger represented the first power stage of a two stage system. This topology was selected to demonstrate the SiC gate drivers because it was deemed the lowest risk of the two stages used by the prototype charger because it requires half the number of active switch positions. Additionally this topology is hard-switched and the performance of the gate drivers was thought to provide a larger impact on the converter's switching loss and therefore overall efficiency.

The power stage was designed to replicate that in the prototype on-board charger. The same power devices, in both part number and quantity, were used and the power module packaging was very similar. The unipolar control scheme was utilized along with the same coupled inductors that share a common core. One slight difference is this hardware utilized a smaller DC-link capacitor because testing was only performed with DC operating points. Because it was only tested at DC the high capacitance required on the DC-link of the previous converter was not necessary. A summary of the power stage design components and design parameters is given in Table 1. The bridgeless-boost converter using the modified X-5 power module with integrated SiC gate drivers was fabricated on a heat sink and is shown in Fig 5. The circuit board that is placed on top of the modified X-5 power module is responsible for galvanic signal isolation and provides convenient pin headers to receive logic level gate driver commands.

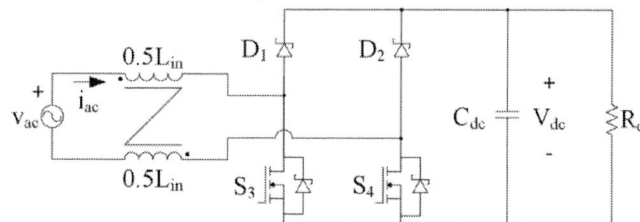

Fig. 4. Bridgeless-boost converter power factor correction circuit schematic.

Fig. 5. Bridgeless-boost converter utilizing X-5 power module with integrated SiC gate drivers.

Table 1
Summary of Power Stage Design

Parameter	Value
Switching frequency (f_{sw})	200 kHz
Input inductance (L_{in})	80 µH
DC-link capacitance (C_{dc})	10 µF
SiC Schottky diode (D_1–D_4)	CPW4-1200-S020B
SiC MOSFET (S_3–S_4)	CPMF-1200-S080B

IV. EXPERIMENTAL RESULTS

A. Basic Operation

The converter was tested by applying a DC input voltage and resistively loading the output. Device S_3 was modulated and S_4 was held on to conduct current. This emulates operation under a positive half cycle of an AC input voltage. The duty cycle was selected to boost 240 V up to 350 V. The converter was first operated under these conditions at a switching frequency of 200 kHz. The initial testing was performed with all power rails of the gate driver (V_{dd-LT}, V_{dd1}, V_{dd-OD}, and V_{dd-OS}) connected to a common 20 V from an external power supply. The ground for the gate driver was connected to a second external power supply that biased V_{ss} to -4 V relative to the common sources of S_3 and S_4. The gate drivers apply 20 V to

the gate of the MOSFET to turn on and -4 V to the gate to turn the MOSFET off. Waveforms highlighting the operation of the converter are shown in Fig. 6.

B. Operation with Various V_{dd-OD} Voltages

The performance of the gate driver with increased V_{dd-OD} voltages was then tested to optimize the overall system performance. The converter was operated with an input voltage of 240 V, a fixed switching frequency of 200 kHz, a fixed duty cycle, and a load of 24 Ω. For this test V_{dd-LT}, V_{dd1} and V_{dd-OS} were connected to a common 20 V supply and the ground of the gate driver was again biased to 4 V below the common sources of S_3 and S_4. V_{dd-OD} was initially set to 20 V and then linearly increased to 30 V. It is important to note that changing V_{dd-OD} does not change the voltage applied to the power MOSFET gate, it only affects the gate driver buffer stage. During this sweep of V_{dd-OD}, the fall time of V_{ds3} was measured from 90% to 10% of the steady state voltages. These turn-on times for S_3 are plotted in Fig. 7. Since turn-on loss is a function of the fall time, it is shown that higher V_{dd-OD} decreases the turn on loss in the power MOSFET.

To illustrate the effect of V_{dd-OD}, the output power level of the converter was swept by adjusting the load resistance and the efficiency of the converter was measured from approximately 1 kW up to nearly 5 kW output power. The sweeps were performed under the same previously used conditions for V_{dd-LT}, V_{dd1}, V_{dd-OS}, and V_{ss}. Full power sweeps were conducted for V_{dd-OD} equal to 20 V, 24 V, and 28 V. Overall a peak efficiency of 97.7% was found at 3.0 kW with V_{dd-OD} equal to 28 V and is shown in Fig. 8.

C. Higher Frequency Operation

The converter was then operated at a higher switching frequency to push the limits of the SiC gate drivers. This testing was performed in the same way as the previous testing

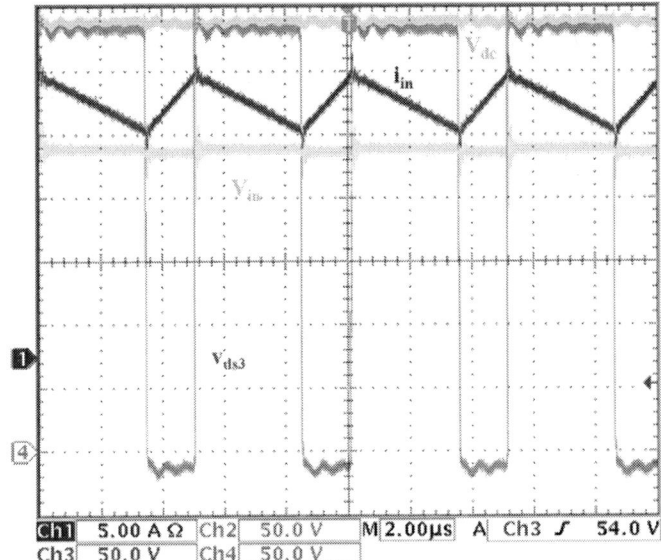

Fig. 6. Boost converter operational waveforms with V_{in} = 240 V, V_{dc} = 350 V, R_o = 24 Ω, $V_{dd\text{-}LT}$ = V_{dd1} = $V_{dd\text{-}OS}$ = $V_{dd\text{-}OD}$ = 20 V, V_{ss} = -4 V, and f_{sw} = 200 kHz.

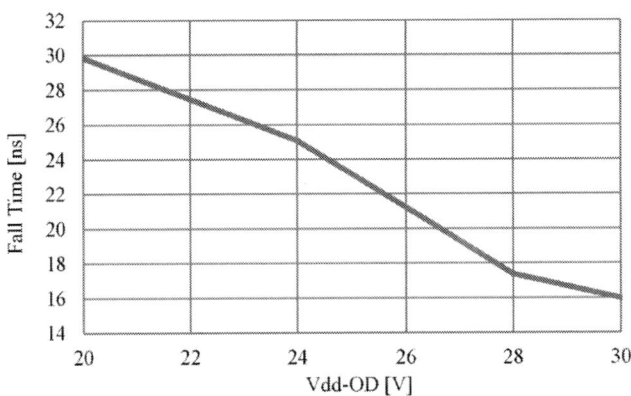

Fig. 7. Drain-to-source fall time (90% to 10%) vs. $V_{dd\text{-}OD}$ with $V_{dd\text{-}LT}$ = V_{dd1} = $V_{dd\text{-}OS}$ = 20 V, V_{ss} = -4 V, V_{in} = 240 V, V_{dc} = 350 V, R_o = 24 Ω, and f_{sw} = 200 kHz.

at lower switching frequencies. The SiC gate driver power rails were again connected to a common 20 V power supply and V_{ss} was again biased to -4 V. The operation of the converter was first verified with $V_{dd\text{-}OD}$ set to 20 V. Waveforms showing converter operation at 500 kHz are shown in Fig. 9.

The resistive load was then adjusted to perform power sweeps at the increased frequency. $V_{dd\text{-}OD}$ was initially set to 20 V but then was increased to a maximum of 25 V. Power loss in the gate driver prevented testing at higher voltages. Full power sweeps were completed at both the 20 V and 25 V conditions. A peak efficiency under the 500 kHz condition was found to be 95.0% at a power level of 2.1 kW. These efficiency results are shown in Fig. 10.

V. CONCLUSION

In this work an all SiC intelligent power module was presented as a solution for future high-density power electronics applications. The IPM contained both SiC power devices and SiC gate driver ICs and was designed to optimize the performance of both devices. The IPM was tested in a

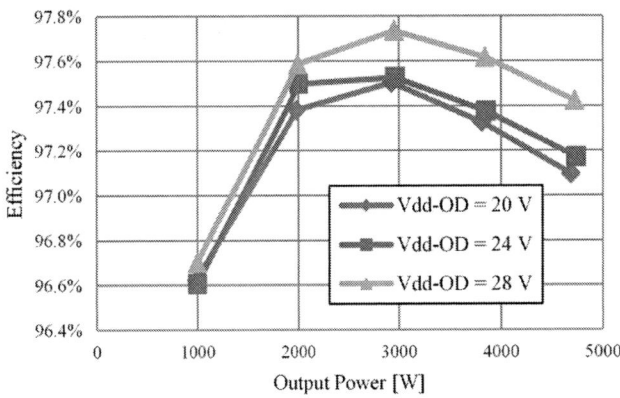

Fig. 8. Boost converter efficiency vs. output power for various values of $V_{dd\text{-}OD}$ with $V_{dd\text{-}LT}$ = V_{dd1} = $V_{dd\text{-}OS}$ = 20 V, V_{ss} = -4 V, V_{in} = 240 V, V_{dc} = 350 V, and f_{sw} = 200 kHz.

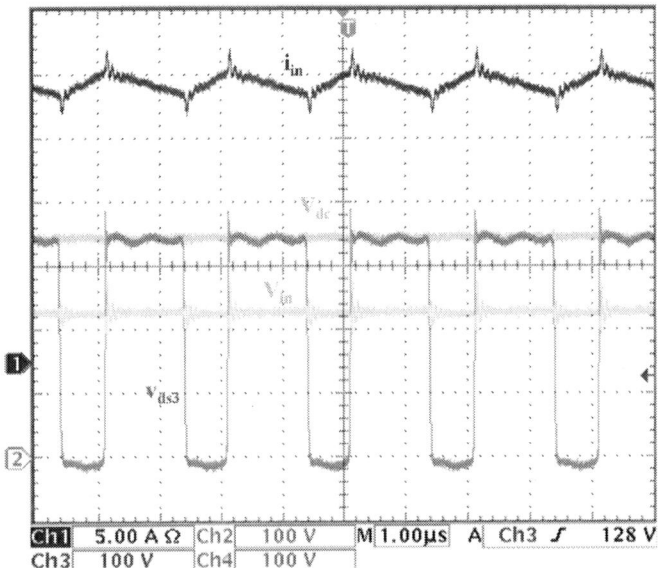

Fig. 9. Boost converter operational waveforms with V_{in} = 240 V, V_{dc} = 350 V, R_o = 24 Ω, $V_{dd\text{-}LT}$ = V_{dd1} = $V_{dd\text{-}OS}$ = $V_{dd\text{-}OD}$ = 20 V, V_{ss} = -4 V, and f_{sw} = 500 kHz.

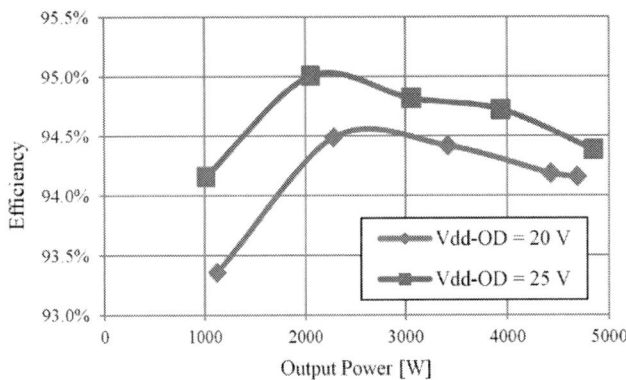

Fig. 10. Boost converter efficiency vs. output power for various values of $V_{dd\text{-}OD}$ with $V_{dd\text{-}LD}$ = V_{dd1} = $V_{dd\text{-}OS}$ = 20 V, V_{ss} = -4 V, V_{in} = 240 V, V_{dc} = 350 V, and f_{sw} = 500 kHz.

978-1-4799-5494-0/14 $31.00 © 2014 IEEE

power electronics converter to demonstration the functionality of the module and to measure the performance of the full system. A bridgeless-boost converter was built and shown to operate as expected with DC input and output voltages with a switching frequency of 200 kHz. The efficiency of converter was measured and improved through an optimization of the overdrive voltage of the gate driver ICs. It was found that increasing this voltage improved the turn-on time of the MOSFETs and reduced the switching losses of the converter. An overall peak efficiency of 97.7% was measured at an output power level of 2.9 kW when switching at 200 kHz.

The performance of the IPM under increased stress was then evaluated by testing the converter at an increased frequency. The switching frequency was increased to 500 kHz and the converter was again found to operate properly. For this condition a peak efficiency of 95.0% was found at an output power level of 2.1 kW. The overdrive voltage corresponding to the peak efficiency was found to be 25 V. The overall efficiency for this case was found to be much lower than the 200 kHz case due to increased switching losses for the converter.

Overall this work proved the feasibility of an all SiC IPM and demonstrated its operation under realistic conditions in a power converter. Future improvements for the IPM could be made with an improved SiC gate driver manufacturing process that would allow for a CMOS output buffer stage. Additional future work could see the integration of the SiC gate driver onto the same die as the power device.

DISCLAIMER

The information, data, or work presented herein was funded in part by an agency of the United States Government. Neither the United States Government nor any agency thereof, nor any of their employees, makes any warranty, express or implied, or assumes any legal liability or responsibility for the accuracy, completeness, or usefulness of any information, apparatus, product, or process disclosed, or represents that its use would not infringe privately owned rights. Reference herein to any specific commercial product, process, or service by trade name, trademark, manufacturer, or otherwise does not necessarily constitute or imply its endorsement, recommendation, or favoring by the United States Government or any agency thereof. The views and opinions of authors expressed herein do not necessarily state or reflect those of the United States Government or any agency thereof.

REFERENCES

[1] T. Funaki, J. C. Balda, J. Junghans, A. S. Kashyap, H. A. Mantooth, F. Barlow, T. Kimoto and T. Hikihara, "Power conversion with SiC devices at extremely high ambient temperatures," *IEEE Trans. Power Electron.*, vol. 22, no. 4, p. 1321–1329, Jul. 2007.

[2] A. M. Abou-Alfotouh, A. V. Radun, H. Chang, and C. Winterhalter, "A 1-MHz hard-switched silicon carbide DC–DC converter," *IEEE Trans. Power. Electron.*, vol. 21, no. 4, pp. 880–889, Jul. 2006.

[3] Wrzecionko, B.; Bortis, D.; Kolar J. W., "A 120°C ambient temperature forced air-cooled normally-off SiC JFET automotive inverter system," *IEEE Trans. Power Electron.*, vol. 29, no. 5, p. 2345–2358, May 2014.

[4] Rabkowski, J.; Peftitsis, D.; Nee, H. P., "Parallel-operation of discrete SiC BJTs in a 6-kW/250-kHz DC/DC boost converter," *IEEE Trans. Power Electron.*, vol. 29, no. 5, p. 2482–2491, May 2014.

[5] Wang, R.; Boroyevich, D.; Ning, P.; Wang, Z; Wang, F.; Mattavelli, P.; Ngo, K. T. D.; Rajashekara, K., "A high-temperature SiC three-phase AC-DC converter design for >100°C ambient temperature," IEEE Trans. Power Electron., vol. 28, no. 1, p. 555–572, Jan. 2013.

[6] Whitaker, B.; Barkley, A.; Cole, Z.; Passmore, B.; Martin, D.; McNutt, T.; Lostetter, A.; Lee, J.; Shiozaki, K., "A high-density, high-efficiency, isolated on-board vehicle battery charger utilizing silicon carbide power devices," *IEEE Trans. Power Electron.*, vol. 29, no. 5, p. 2606–2617, May 2014.

[7] Ericson, N.; Frank, S.; Britton, C.; Marlino, L.; Sei-Hyung Ryu; Grider, D.; Mantooth, A.; Francis, M.; Lamichhane, R.; Mudholkar, M.; Shepherd, P.; Glover, M.; Valle-Mayorga, J.; McNutt, T.; Barkley, A.; Whitaker, B.; Cole, Z.; Passmore, B.; Lostetter, A., "A 4H silicon carbide gate buffer for integrated power systems," *IEEE Trans. Power Electron.*, vol. 29, no. 2, p. 539–542, Feb. 2014.

[8] Z. Cole, B. Passmore, B. Whitaker, A. Barkley, T. McNutt and A. B. Lostetter, "Packaging of high frequency, high temperature silicon carbide (SiC) multichip power module (MCPM) bi-directional battery chargers for next generation hybrid electric vehicles," in *International Symposium on Microelectronics (IMAPS 2012)*, San Diego, CA, Sep. 2012.

Gap in pagination due to formatting issues.

Page 41

Development of Packaging Technologies for Advanced SiC Power Modules

Zhenxian Liang
Power Electronics and Electric Machinery Group
OAK RIDGE NATIONAL LABORATORY
Oak Ridge, TN, USA
liangz@ornl.gov

Fred Wang and Leon Tolbert
Department of EECS
The University of Tennessee
Knoxville, TN, USA
fred.wang@utk.edu, and tolbert@utk.edu

Abstract—**A set of packaging technologies has been developed for promotion of SiC power devices in automotive applications. The technical advances include integrating single side cooling, three-dimensional (3-D) planar electrical interconnection, and integrated double sided direct cooling. The further integration of these features into one packaging process has been demonstrated with highly integrated SiC phase leg power module prototypes. The comprehensive improvements in module's electrical, thermal performance and manufacturability help exploit fully the attributes provided exclusively by the wide bandgap (WBG) power semiconductors. The technical advancements lead to cost-effectiveness, high efficiency, high power density power conversion in electric drive system in modern vehicles.**

Keywords—SiC, power module; integrated packaging; high density; double sided cooling; direct cooling

I. INTRODUCTION

It has been well known that the wide bandgap (WBG) power semiconductors, such as silicon carbide (SiC) and gallium nitride (GaN), provide superior power switching features: high current density, low loss high switching speed and high operation temperature over its counterpart silicon (Si). The state of the art (SOA) power electronics module and inverter/converter packaging technologies based on the wire-bond interconnection scheme have been successfully developed by WBG power semiconductor manufactures [1-5]. Conventional processes have been used for packaging multiple SiC bare chips (or dies) in these modules. The packaging elements include die attach and top interconnection, power substrate, power and signal terminals, baseplate, encapsulate, etc. Specifically, the interconnection and the die attach are accomplished by multiple bond wires (from top of the dies) and reflowed solder areas (from bottom of the dies) onto the power substrate, respectively. In converters/inverters, the modules need to be attached to a heat sink or cold plate using forced air or liquids for cooling. For good thermal transfer the thermal interface material such as thermal grease is usually applied between them. In general, while they fulfill the designed electrical, thermal and mechanical functions, these hybrid package elements will bring

parasitic effects on power semiconductors' operation, furthermore, define the capabilities of semiconductors. Specially, they limit the full exploitation of WBGs' attributes, and in turn, limit the benefits of WBG power electronics in cost, efficiency and density. These parasitic effects can be characterized by a set of technical metrics such as thermal impedance (resistance and capacitance), electrical impedance (resistance and inductance), thermo-mechanical properties (power, thermal cycling numbers and vibration ruggedness), etc. Looking at things in terms of these parameters and their relationships to power module's performance criteria (efficiency, reliability, cost, density, etc.) makes evaluation of power module technologies, to identify the shortcomings of existing technology and develop new concepts, relatively easier. In past years, many techniques in SiC power module packaging, focused on improvements of these technical parameters, have been developed to promote successfully the application of WBG power semiconductors.

II. SiC POWER MODULE WITH INTEGRATED COOLING

The target all-SiC power module is a one-phase-leg configuration composed of SiC metal-oxide-semiconductor field effect transistors (MOSFETs) and SiC Schottky barrier diodes (SBDs), as illustrated in Fig. 1, which is a basic

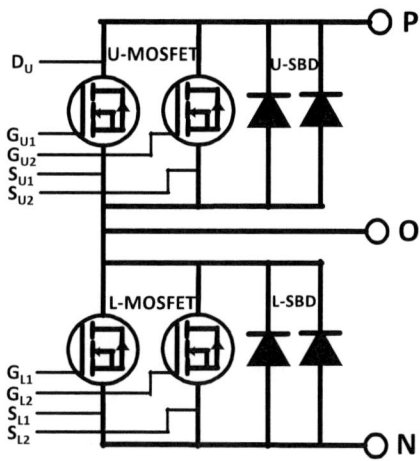

Fig. 1. Electrical diagram of an 100A/1200V SiC phase leg module (U-upper unit, L=lower unit), consisting of two paralleled 50A/1200V dies.

Research sponsored by the Advanced Power Electronics and Electric Motors Program, DOE Office of Vehicle Technologies, under contract DE-AC05-00OR22725 with UT-Battelle, LLC.

building block for various automotive power converters and inverters. The power MOSFETs and diodes are commercially available, in the form of bare SiC die, both with rating of 50 A and 1,200 V. The current rating of the power module can be multiplied by paralleling more dies. Fig. 1 shows the topology for a 100 A, 1,200 V phase leg all-SiC power module. The switch units consist of two paralleled bare dies for each switch.

Fig. 2(a) presents photo of a prototype of this module fabricated in house. The uniqueness is that it features an integrated direct cooling configuration [6], as schematically depicted in Fig. 2 (b), where a cold-base plate replaces the pure base plate in conventional package. The employed cold-base plate is made of a flat copper tube with criss-crossed fins inside, as shown in the X-ray image in Fig. 2(c). In this module package the power stage on the direct bond copper (DBC) ceramic power substrate is directly bonded onto this cold-base plate by solder. Thus, it not only eliminates the thermal grease and base plate, compared to conventional power module

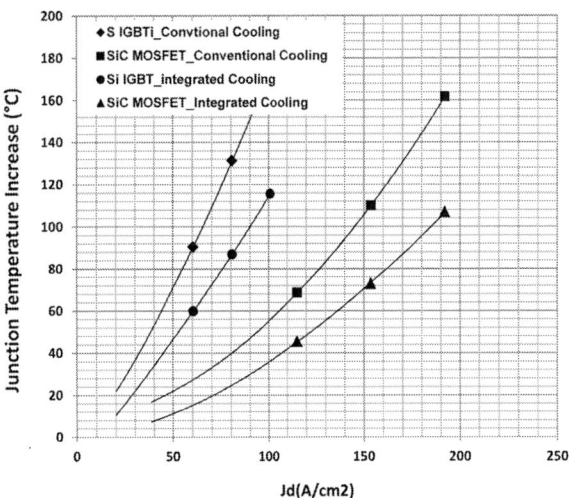

Fig. 3　Graph of power device temperature increase vs. current density with different device/package combinations.

thermal package scheme, but also realizes integrated direct cooling.

This cooling integration results in that specific thermal resistivity of the integrated cooling package is more than 33% lower than that of a conventional (first-generation) package, as presented in [6]. For assessment of the effect of this improvement and advantages of SiC devices over its silicon (Si) counterparts, four samples with different package/device combination have been prototyped. The Si insulated gate bipolar transistor (IGBT)/PiN diode dies are with the same rating of SiCs'. It is experimentally characterized that the SiC MOSFET allowed reductions of 45% in die area, 25% in voltage drop (at 40 A), and 78% in switching power losses. The improvement combining high efficient integrated cooling can be observed in increase of allowed current density in the die. For this measurement, a junction temperature increase vs. current density relationships for four different module packages have been calculated, as shown in Fig. 3. It can be seen that cooling configuration is greatly significant on the junction temperature of power devices. With a fixed current density, for example 100 A/cm², the increase in the junction temperature ranges from ~36 to 170°C between SiC/integrated cooling and Si/conventional cooling combinations. The higher temperatures lead to a significant reduction of the power module lifetime. In other words, if they keep the same current density in the die, this integrated SiC power module will serve much longer time.

On the other hand, if the maximum operational boundary is set based on temperature limits (determined by the package reliability and the semiconductor limitations), the maximum handling current density of a power switch can be determined accordingly. For instance, the allowed current density of a power device for four combinations increases from 66 A/cm² to 185 A/cm² for a 100°C temperature increase, which can be a case for a coolant temperature of 25°C and a junction temperature of 125°C. These maximum current density values define the minimum die area of the power semiconductors for a designed system current (or power) level. It is well known that die size is a dominant factor in power module cost. The 3× die

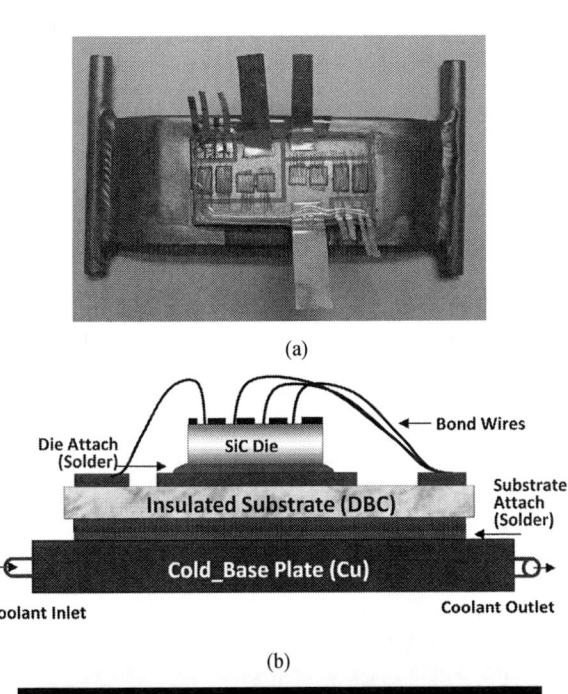

(a)

(b)

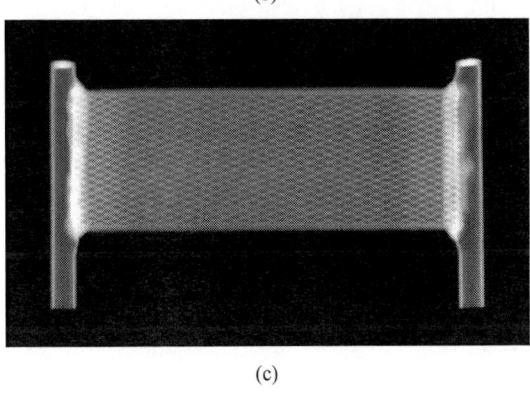

(c)

Fig. 2　100 A, 1,200 V SiC phase leg power modules with integrated cooling: (a) Top view photo of the prototype; (b) Cross-sectional view schematics of the module package; (c) X-ray view of the cold-baseplate used in the module.

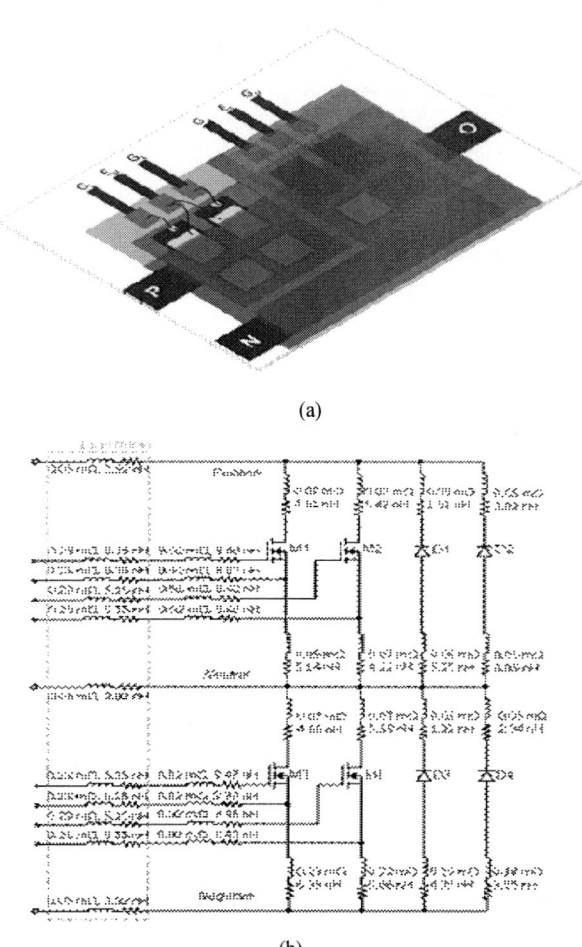

(a)

(b)

Fig.4 (a) three-dimensional (3-D) planar power interconnection configuration; (b) lumped element model of packaging interconnection.

substrates. The electrical interconnection is achieved by conductively bonding these dies from both top and bottom sides to the copper layers on the two substrates, which are patterned to form circuitry corresponding with the electrode pad layout on the dies. The DBC substrates provide electrical insulation via the internal ceramics (such as Al_2O_3) sandwiched between the two copper layers. This symmetric planar bonded package offers flexibility in the arrangement of the switch dies. As shown in Fig. 4(a) the upper switch pair and lower switch pair in the phase leg are oriented in a face-up/face down configuration. This makes optimal use of the die's vertical semiconductor (electrode) structure, in which the electrodes are arranged on both the top and bottom surfaces of the dies. Compared to traditional wire bond layout, where all dies are placed in a face up orientation on the X-Y plane of the substrate, the die layout in Fig. 4(a) makes the main current flow loops in the X-Z plane of the package. The thickness (in Z direction) of the switch die is typically 0.1mm, while its length (X direction) or width (Y direction) is in the range of 10mm. Thus, the enclosed area of the main current loop in this new interconnection configuration is reduced dramatically, leading to a significant reduction in electrically parasitic inductance and resistance.

Fig. 4(b) presents the circuit model with lumped packaging parameters obtained by employing an electromagnetic simulation tool (MAXWELL Q3D Extractor) which calculates the electromagnetic field and extracts the parasitic components through current conduction paths. The sums of lumped elements along a main power path from the positive terminal through the upper MOSFET and then the lower diode to the negative terminal are the representative parameters of the packaging structure. In this planar bonded power module, the total inductance L is 13.58nH, while the total resistance R equals 0.25mΩ. For comparison, the parasitic electrical parameters in a conventional wire bonded power module using the same dies were also obtained through simulations. The analogous values for the major parameters defined above are L=53.3nH, and R=2.45mΩ, respectively.

B. Integrated Manufacture Process

The module structure described above reduces the number of tiny wire bonds used in conventional packaging to a few identical planar area bonds. It also makes the batch processing possible. To do so, a special process method has been developed in which, as shown in Fig. 5(a), all packaging components are assembled into a fixture first. The components include the patterned DBC substrates, multiple bare dies and shims, preformed bonding material (such as solder prform), power terminals and signal pins, etc. The second processing step involves heating the assemblies simultaneously to form the bonds and creating the final package. The simplicity of the process helps reduce costs and improve the manufacturability of the modules. This technology is called 'Planar_Bond_All' (PBA), representing the features inherent in the packaging structure as well as the fabrication process.

Fig. 5(b) is a photograph of the prototype power module employing the PBA packaging technology. It is a single phase leg (half bridge) electrical configuration, comprised of two pairs of single SiC MOSFET and diode dies in parallel, rated at

area reduction from Si/conventional cooling to SiC/integrated cooling combination is significant for reducing the system cost.

The analysis and data discussed are closely related to a defined operation condition (here, for continuous switching at duty cycle, D=0.5, and frequency, f=5 kHz). The contributions of the packaging technology and the power semiconductor technology to the improvements in power module efficiency, cost, and reliability can be further analyzed using these performance parameters for any specific application cases.

III. PLANAR-BOND-ALL PACKAGING OF SiC MODULE

Planar-bond-all is a power module packaging scheme, significantly different from the conventional wire-bond packaging scheme. The key element is that the top interconnections of the dies utilize planar, larger area bond to the electrodes, instead of multiple, tiny wire bonds. This change results in possibility to integrate many superior packaging concepts into SiC power modules.

A. Three-Dimensional (3-D) Planar Power Interconnection

Fig. 4(a) is a schematic of the planar interconnection configuration for the phase leg power module. The four MOSFET and four diode dies are mounted between two DBC

978-1-4799-5494-0/14 $31.00 © 2014 IEEE

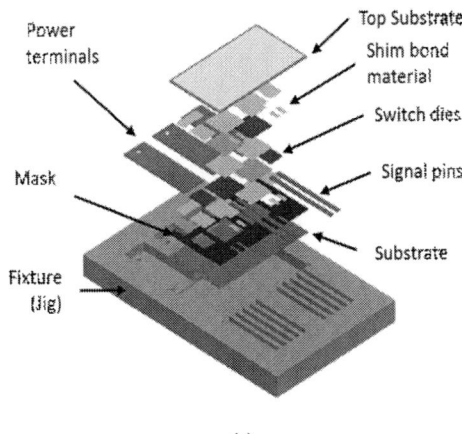

(a)

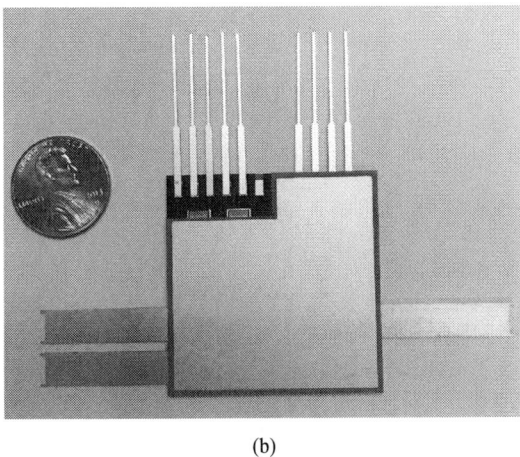

(b)

Fig. 5   (a) schematic of components assembing step in Planar-Bond-All packaging manufacture process; (b) top view photo of a SiC PBA 100A/1200V phase leg power package.

100A/1200V. The module body measures 40 mm×40 mm×2 mm, excluding the terminals and pins. A soldering process is employed to form the planar bonding, due to its high electrical conductivity and ease of processing. The special top metallization of the power semiconductor dies was produced during the wafer level processing. This enables reliable soldering bonds. The power terminals [positive, neutral, and negative (P, O, and N)] and signal pins (G_U, S_U, and G_L, S_L) are also mounted between the two substrates to form compact input/output connections.

C. Double Sided Integrated Direct Cooling

The dual side planar bond structure allows both top and bottom surfaces of the power package to be thermal dissipation paths. The heat generated in the dies can be more efficiently removed by applying double sided cooling, through forced air or liquid heat transfer. Additional, direct attachment (bond) of heat sink or cold plate leads to a further reduction in the thermal resistance, as discussed above.

Fig. 6(a) shows an example of integrated double sided cooling configuration, where two pin-fin base plates made of copper are directly bond to power package (Fig. 5(b)) from

(a)

(b)

Fig. 6   Double sided integrated direct cooling of PBA module: (a) photo of power module with integrated doubl sided pin fin heat sinks; (b) graph of power module in a coolant manifold.

both sides by another soldering process. The effective pin fin array area is 38mm x 38mm, the fin height is 7mm.

This module can be directly cooled by air. For higher efficiency, forced liquid cooling is usually employed in most applications. Fig. 6(b) is a photo of a prototype in which the power module is assembled into a special coolant manifold. The manifold provides the liquid passageways for coolant passing through the pin fins.

IV.   Performance Evaluation

The benefits gained through the packaging, as described above, can be demonstrated by the system performance improvement of the PBA module in power converters and inverters.

Various experimental measurements and multi-physics simulation tools have been used to intensively characterize the electrical and thermal performance of the module packaging and are summarized in the following sections.

A. Electrical Characterization

The electrical performance of power modules can be drastically influenced by their "parasitic" electric parameters (resistance, inductance and capacitance) adding to the semiconductors intrinsic electric parameter network.

The parasitic inductances in the current loop will cause the voltage overshoot superimposed on the blocking voltage of the MOSFET while it was turned on and off, as shown in Fig. 7(a). From the current and voltage waveforms, it can be seen that there is only 30 V of overshoot adding to 600V bus voltage. This is attributed to its smaller parasitic inductance (13.58nH), compared to 53.3nH in the conventional module. On the other side, the voltage overshoot is dependent the switching speed (current decrease slope). The smaller inductance allows the MOSFETs in PBA module operate at higher frequency, which is a critical way to increase the power density of SiC inverters.

The parasitic inductance and resistance also increase the power loss of the SiC switches. It has been demonstrated that the PBA module reduces this parasitics related losses by 75%, compared to conventional one [7].

B. Thermal Characterization

The thermal performance of a power module can be modeled as multiple orders of thermal resistance and capacitance networks. The thermal capacitance determines the transient response of the power module to power dissipation in the switches; while the thermal resistance determines the cooling efficiency.

The thermal parameters can be obtained by experimentally thermal tests through emulating operation of the power modules. A method for determining junction temperatures involves measuring the voltage drop of the body diode in SiC MOSFET has been employed [6]. By analyzing the cool down temperature curves and input electric power data, the thermal parameters (resistance and capacitance) of the module can be extracted. For a fair evaluation of the thermal performance of a packaging assembly, a specific thermal resistance, $\theta_{ja,sp}$ was taken as a figure of merit, which is a normalized parameter by die area (i. e., die area*thermal resistance). Here, the thermal path includes all thermal stacking elements from coolant inlet to the junction of switch die. Fig. 7(b) presents the measured specific thermal resistance value of the planar bonded module with double sided cooling. For comparison, the thermal resistance of the wire-bond package and single sided cooling package were also measured using the same method. The wire bonded module was mounted on a cold plate with thermal grease, and has a specific thermal resistivity of 0.54 cm$^2 \cdot$°C/W; while the integrated single side cooled module, as shown in Fig. 2(a), features a thermal resistivity of 0.47 cm$^2 \cdot$°C/W. The double sided cooling of the PBA module assembly reduces the specific thermal resistivity to 0.33 cm$^2 \cdot$°C/W, which is 38% lower than that of the first generation (wire bond) module assembly.

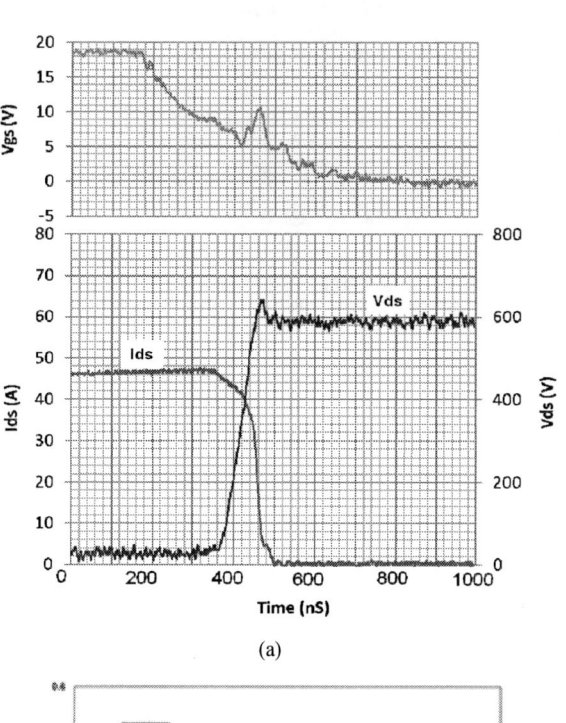

(a)

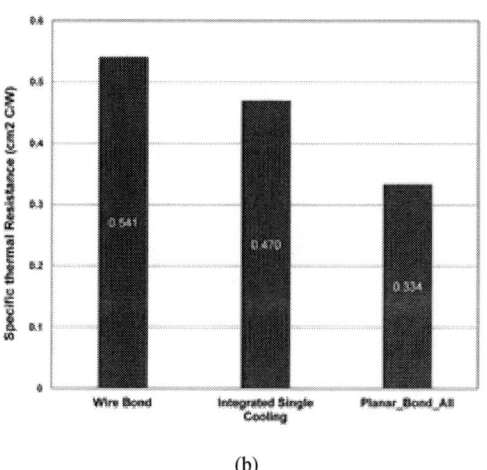

(b)

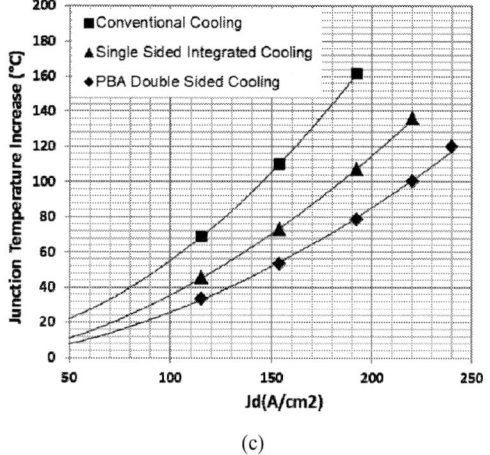

(c)

Fig. 7 Performance of PBA SiC power module: (a) turn-off switching wavefoms; (b) specific thermal resistance and comparison with conventional and single sided cooling modules; and (c) comparison of junction temperature vs current density for different packages.

C. Effects on Power Electronics Systems

As described in section II, the reduction in both power losses and thermal resistance help increase the operational current (power) density in the SiC die. This can be seen in the graph of junction temperature versus current density, as shown in Fig. 7(c). For 100°C temperature increase, the characteristic current density of the MOSFET in the module is 145 A/cm^2, 184 A/cm^2 and 220 A/cm^2, respectively for convention cooling, single sided integrated cooling and double sided direct cooling. It is almost double of the power processing capability of the SiC power devices.

V.   CONCLUSION

The packaging technologies developed to exploit the superior attributes of SiC power devices integrate low parasitics power interconnection and double sided cooling capability, as well as simplified manufacture processes. The benefits gained form these innovations have been demonstrated by comprehensive improvement in power conversion efficiency (low losses), and cost-effectiveness through reducing power semiconductor content and high productivity manufacturing. The future research will focus on the reliability evaluation and technical development for packaging reliability enhancement, enabling SiC switches to operate at higher temperature, for further advance of power electronics systems in in modern electric drive vehicles.

ACKNOWLEDGMENT

Research was sponsored by the Advanced Power Electronics and Electric Motors Program, DOE Vehicle Technologies Office, under contract DE-AC05-00OR22725 with UT-Battelle, LLC. The U.S. government retains and the publisher, by accepting the article for publication, acknowledges that the U.S. government retains a nonexclusive, paid-up, irrevocable, worldwide license to publish or reproduce the published form of this manuscript, or allow others to do so, for U.S. government purposes.

The authors also would like to thank Dr. Burak Ozpineci, Mr. Madhu S. Chinthavali for fruitful discussion and Mr. Randy H Wiles, Mr. Curtis W. Ayers, for their help in fabrication of packaging parts, and Mr. Zhiqiang (Jack) Wang for electrical stimulation.

REFERENCES

[1] Masafumi Horio, Yuji Iizuka, Yoshinari Ikeda, "Packaging Technologies for SiC Power Modules," FUJI ELECTRIC REVIEW, vol. 58 no. 2, pp. 75–80, November 2011.

[2] http://www.cree.com/power/products/sic-power-modules/sic-modules/cas100h12aml.

[3] Daniel Domes, Christoph Messelke, and Peter Kanschat, "1st Industrialized 1200 V SiC JFET Module for High Energy Efficiency Applications," PCIM Europe2011, 2011.

[4] Shuhei Nakata, Naruhisa Miura, and Yoshiyuki Nakaki, "SiC Power Device Technology," Mitsubishi Electric ADVANCE, June 2011.

[5] T. Nakamura, M. Aketa, Y. Nakano, M. Sasagawa, and T.Otsuka, "Novel Developments Towards Increased SiC Power Device and Module Efficiency," in Proceedings of the 2012 IEEE Energytech, Cleveland, Ohio, USA, May 29–31, 2012, p. 126. Also, https://www.rohm.com/web/global/full-sic-power-modules.

[6] Zhenxian Liang, Puqi Ning, Fred (Fei) Wang, "Development of Advanced All-SiC Power Modules," IEEE Transaction on Power Electronics, Vol. 29, No. 5, May 2014, pp.2289-2294.

[7] Zhenxian Liang, Puqi Ning, Fred Wang, and Laura Marlino, "A Phase – Leg Power Module Packaging with Optimized Planar Interconnection and Integrated Double Sided Cooling," IEEE Journal of Emerging and Selected Topics in Power Electronics, Vol. 2, No. 3, Sept., 2014, pp.487-495. Also presented at the Fourth IEEE Energy Conversion Congress and Exposition, Raleigh, NC, September 16-20, 2012, entitled "Planar Bond All: A New Packaging Technology for Advanced Automotive Power Modules".

A 10-kW SiC Inverter with A Novel Printed Metal Power Module With Integrated Cooling Using Additive Manufacturing

Madhu Chinthavali, Curt Ayers, Steven Campbell, Randy Wiles, Burak Ozpineci
Power Electronics and Electric Machinery Group, Oak Ridge National Laboratory
Oak Ridge, TN 37831 USA
E-mails: chinthavalim@ornl.gov, ayersca@ornl.gov, campbellsl@ornl.gov,rwiles@ornl.gov, ozpinecib@ornl.gov

Abstract— With efforts to reduce the cost, size, and thermal management systems for the power electronics drivetrain in hybrid electric vehicles (HEVs) and plug-in hybrid electric vehicles (PHEVs), wide band gap semiconductors including silicon carbide (SiC) have been identified as possibly being a partial solution. This paper focuses on the development of a 10-kW all SiC inverter using a high power density, integrated printed metal power module with integrated cooling using additive manufacturing techniques. This is the first ever heat sink printed for a power electronics application. About 50% of the inverter was built using additive manufacturing techniques.

I. INTRODUCTION

In order to take advantage of the high temperature operation capability of Wide Band Gap (WBG) devices, device packages that can withstand high temperatures are required. Various organizations are working on high temperature packaging for high temperature devices. Several high temperature packages which include discrete device packages to power modules have been reported in the past several years [1-4]. The novel packaging concepts primarily focus on improving the existing packages or design new packages utilizing new materials and or processing techniques for better reliability and performance. Even though the novel packages enable the devices to work at higher temperatures the theoretical advantages like the current density of WBG devices are not realized because of the interconnects needed for the power module to access the device terminals. Also the novel designed packages need further development to be used in a full system.

The other factors that limit the system designers from reaping the benefits of the WBG technology are the low voltage electronics and the passives. This is because even though the packages and the power devices can handle high temperatures, the low power electronics which drive the power devices are limited to a maximum temperature of 200 °C (Silicon on insulator (SOI) based technology). The Si based electronics are limited to 125 °C. The SOI based electronics can work up to 200 °C however, they are expensive. The high temperature (more than 200 °C)

electronics have been reported as being feasible [5, 6]; however, they have not been built to show their performance. It could be many years even before a logic level high temperature transistor could be built. This creates a void in the power module industry especially the intelligent power module (IPM) products, which include the electronics inside the module. Similarly, the passive components in an inverter have a low operating temperature and cannot be operated in close proximity to the high temperature WBG devices. This leads to increase in volume and reduction in power density of the system. The high temperature passives are currently being developed to address the high temperature operation requirement. However, similar to the electronic components the cost will be much higher compared to the low temperature components.

To address the problems mentioned above, a system level approach for packaging designs needs to be developed. Complex 3D packaging structures with integrated interconnects can reduce the steps in assembly and increase the power densities of the power electronic systems. The recent advancements in additive manufacturing technology promise an exciting future trend for this technology to make inroads to the power electronics industry. *Additive manufacturing techniques will enable the development of complex 3D geometries, which will result in size and volume reductions at the system level by integrating the low temperature components with high temperature active devices and reducing the material used for building the heat exchangers in inverters.* Oak Ridge National Laboratory (ORNL) is one of the leading research organizations in the world that has developed expertise in additive manufacturing in the last few years. ORNL's Power Electronics and Electric Machines (PEEM) team recognized the potential of this technology for power electronics system packaging for all the above-mentioned reasons and took the first step towards achieving a completely printed inverter concept. A 10-kW all SiC inverter with aluminum based printed power module with integrated cooling system and a printed plastic lead frame was built using additive manufacturing techniques. This is the first inverter prototype built using additive manufacturing techniques. This paper presents the design and development of the inverter and characterization of a high temperature 1200 V, 100 A all-SiC module.

This manuscript has been authored by UT-Battelle, LLC, under Contract DE-AC05-00OR22725 with the U.S. Department of Energy. The United States Government retains and the published, by accepting this article for publication, acknowledges that the United States Government retains a nonexclusive, paid, irrevocable, world-wide license to publish or reproduce the published form of this manuscript, or allows others to do so, for the Unites States Government purposes.

978-1-4799-5494-0/14 $31.00 © 2014 IEEE

II. POWER MODULE DEVELOPMENT

Two modules were developed for this project in order to systematically evaluate the discrete devices and then build the phase leg required for the inverter. 1200 V, 50 A discrete SiC MOSFET and diodes were used in this project. The SiC MOSFET-based phase-leg module designed and developed in the internal packaging facility at ORNL is shown in Fig. 1.

Fig. 1: ORNL high-temperature 1200V, 50 A SiC phase-leg module.

A. Characterization of the SiC module

Figure 2 shows forward characteristics of the (1200 V, 100 A) SiC MOSFET for different operating temperatures from 25 to 150°C in 25°C increments. SiC MOSFETs exhibit a linear relationship between the voltage and current and can be modeled as resistors. The diode forward characteristics are shown in Fig. 3.

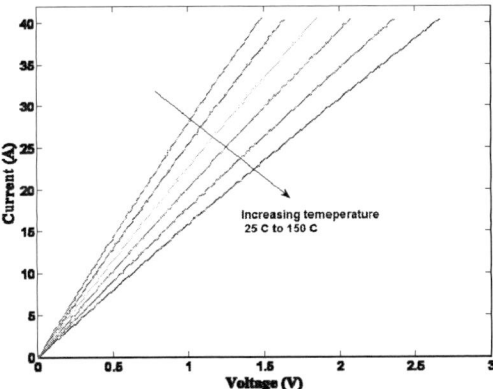

Fig. 2: Forward characteristics of 1200 V, 50 A SiC MOSFET.

A standard commercial driver IC from IXYS, IXDN509, was used in the drive circuit for both devices. The driver can provide a peak output current of 9 A with a maximum output resistance of 1 Ω. The output stage of the drive includes resistor R2 and capacitor C1 for transient current and parallel resistor R1 for static current. The device turn-on and turn-off times are controlled by selecting the values of capacitor C1, and the resistor dampens the oscillation caused by C1 and the parasitic inductance of the circuit. A negative gate voltage of −5 V was chosen for the SiC MOSFET. The resistor R1 in the range of 9–15 Ω for Vcc of 20 V, the capacitor C2 in the range of 10–100 nF, and the resistor R2 in the range of 1–7 Ω were tried to achieve a high dynamic gate current during switching. A schematic of the gate drive circuit topology is shown in Fig. 4.

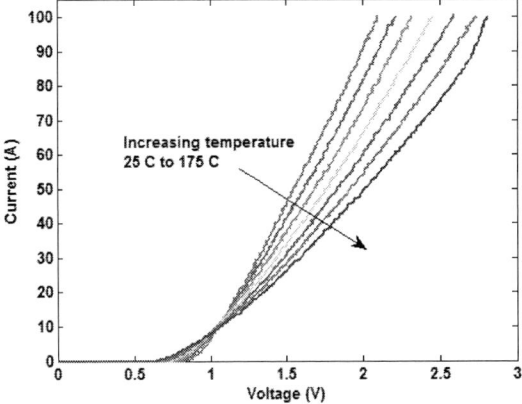

Fig. 3: Forward characteristics of 1200 V, 50 A SiC diode.

Switching measurements (double pulse tests) were performed to characterize the SiC MOSFET dynamically. The test setup is shown in Fig. 5. A load inductance of 120 uH and a 1200 V, 30 A SiC JBS freewheeling diode were used for all the tests. The experimental switching waveforms of the SiC MOSFET are shown in Fig.6 The total energy losses of the devices were obtained at 800 and 600 V and at different currents up to 35 A (Figs. 7 and 8). It should be noted that the diode reverse recovery losses are included in the MOSFET energy losses.

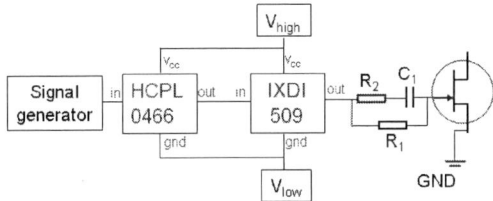

Fig. 4: Schematic of the gate drive circuit.

Fig. 5: Experimental test setup for dynamic characterization of the 1200 V, 50 A SiC MOSFET and diode.

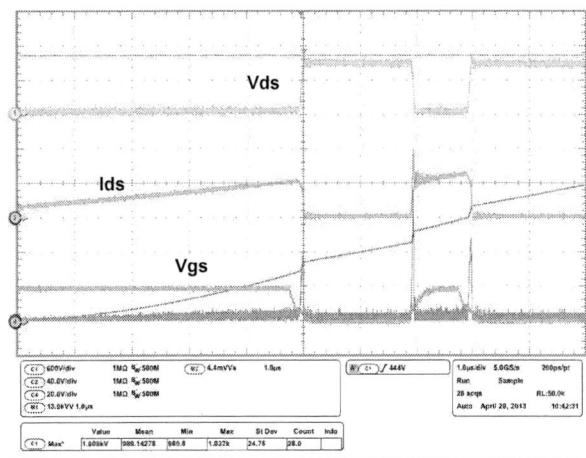

Fig. 6: Experimental switching waveforms of SiC MOSFET.

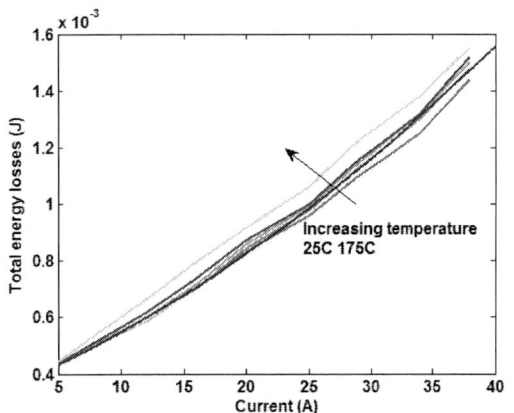

Fig. 7: Total energy losses of the SiC MOSFET at 600 V.

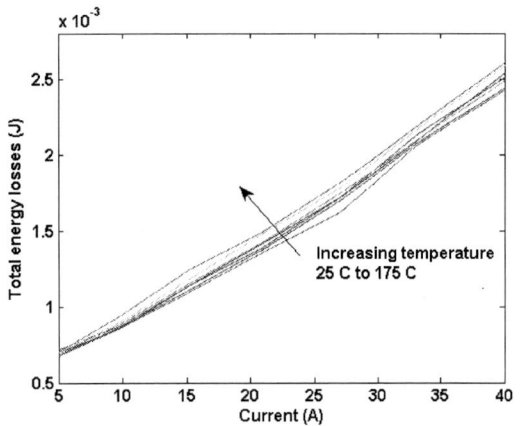

Fig. 8: Total energy losses of the SiC MOSFET at 800 V.

Fig. 9 shows the layout of the die and interconnection pattern. The module includes four 1200 V, 50 A SiC MOSFETs and two 1200 V, 50 A SiC diodes, which form a totem pole or phase leg (half-bridge) configuration, a building block for an inverter. The total module power ratings were designed to handle the currents required for a 10 kW inverter. The package is designed to work at a temperature of at least 200°C ambient. The electrical interconnection is achieved by bonding aluminum (Al) wires on top of the dies and soldering

dies on the Cu traces of DBC substrates, which offer electrical insulation with their ceramic slice inside. The layout of multiple SiC dies on the DBC substrate and their interconnections was optimized to reduce parasitic electric parameters.

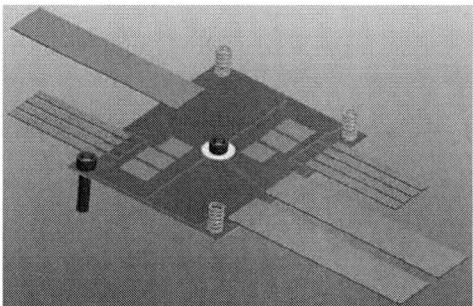

Fig. 9: Power module layout design.

III. INVERTER DESIGN

A. Overall design

The current required for the 10 kW inverter can be calculated to be approximately 37 A peak current per phase assuming the battery voltage to be 350 V dc with square mode of operation and the power factor (p. f.) is 0.6. This operating condition represents the worst-case scenario for the peak power, through the inverter, required to drive the motor. The dc bus capacitors was designed to handle a maximum rms ripple current (~90 A). The capacitors used in this design are not brick type and they are smaller individual capacitors connected in parallel to ensure better cooling and also for cost reduction. The cost reduction is simply because these are off the shelf components as opposed to a custom design brick capacitor with integrated bus bars. A total capacitance of 200 uF was obtained with five 900 V, 40 uF capacitors. The dc bus bars for this inverter were laminated and designed with inserts for each individual capacitor.

The power loss of the 1200V, 100 A (two 50 A SiC MOSFETs and two 50 A diodes) devices was calculated using the test data obtained through device characterization as shown in Fig. 10 , and 11. The loss per device is calculated as a function of power factor, operating voltage, peak current through each device (37/2=~16.5 A per switch), switching frequency, and modulation index. It can be clearly seen that for M=1 the maximum loss, for a 175C maximum junction temperature design, is ~30 W. The loss through the diode for M=1 is approximately 16 W. The total loss per switch is ~ 46 W. This number sets the design specifications for the heat removal system for the entire inverter for an operating junction temperature of 175°C maximum. The total loss for the inverter is simply six times the loss per module, which is approximately 276 W. It should be noted that for different values of M and power factors the losses would vary between the MOSFET and diodes. Based on the simulated values the average efficiency of this inverter is estimated to be around

98-99 % over a wide range of operating conditions. A CAD model of the overall design is shown in Fig.12.

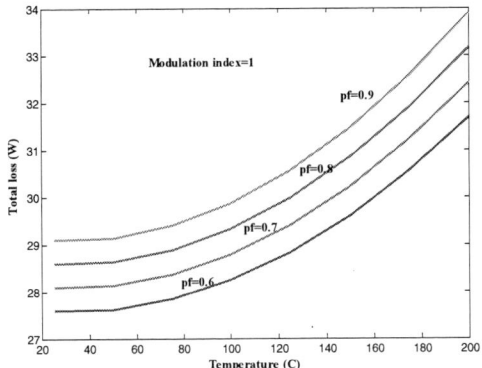

Fig. 10: Loss per switch in a 1200 V, 100 A SiC MOSFET module at 350 Vdc, 37 A,10 kHz, and M=1.

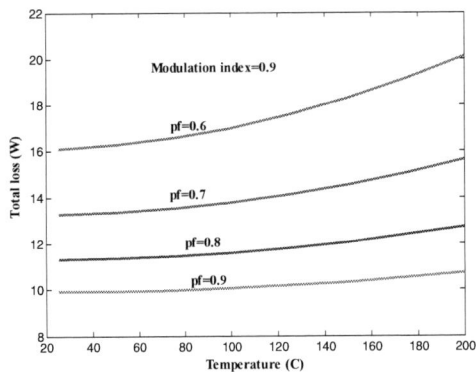

Fig. 11: Loss per switch in a 1200 V, 100 A SiC diode module at 350 Vdc, 37 A,10 kHz, and M=0.9.

Fig. 10 12: CAD model of the inverter design.

B. Gate drive design for the module

The gate drive boards were individually designed for each phase leg with separate gate drivers for upper and lower switches (Fig. 13). Commercially available gate driver chips from Rohm were used in this inverter. The gate driver has galvanic isolation up to 3000 Vrms and integrated overcurrent protection, under voltage lockout, and temperature feedback features. One important feature of this gate drive is that it has miller clamp protection, which is very important for fast switching devices like SiC MOSFETs. This feature prevents the upper and lower switch from faulting through switching transients.

The driver can provide a peak output current of 5 A with a maximum output resistance of 1 Ω. The maximum rise time is 45 ns for a capacitive load of 10 nF with VCC=24 V, according to the data sheet specifications. Each switch in the phase leg was designed for a gate voltage of 25 V (+20 V to -5 V). The devices were switched with an external gate resistance of 4 ohms to ensure fast turn-on and turn off switching times. The peak current required was slightly higher than 6 A, and a buffer stage was added to the output of the gate driver to handle the peak current. The over current protection circuit was designed using the desat detection feature of the gate driver chip. The gate driver protection features were designed and tested using a single phase test bed.

Fig. 13: Gate driver board for a single phase power module.

C. Cooling system design

The cooling system for this prototype is single sided cooling for the power module substrates. The heat sink is not a simple conventional structure, but capitalizes on additive manufacturing (AM) capabilities to build complex internal structures for better heat transfer capability throughout the unit. The reason for doing this is to allow lower-temperature components to be placed as close to the high temperature devices as physically possible to achieve two goals: (1)- reduce electrical parasitics, and (2)- allow reduction of volume and mass of the overall system package. Additive manufacturing (printed metals) will allow optimized design for heat transfer in the smallest possible package.

With AM, complexity is basically free, so any shape or grouping of shapes can be imagined and modeled for performance. The heat sink used for this inverter is a complex unit made using direct melted laser sintered (DMLS) process, which has flow paths internal to the heat sink that allow the complex geometry around the heat generating paths. Fig.14 shows an X-ray of internal structure of the printed power module. The x-ray confirms that there are no clogged channels inside the module with unremoved powder or bridged metal structures. Since this is still a maturing technology it is very important to understand the material properties and compare them to conventional aluminum alloys to further improve or optimize the design. A material analysis was performed on the AM power module and compared to regular 6061 aluminum alloy. Figs. 15 and 16 show materials analysis results on AM aluminum as compared to 6061 aluminum. The results show that the strength of AM alloy is very similar to the 6061 alloy.

978-1-4799-5494-0/14 $31.00 © 2014 IEEE   51

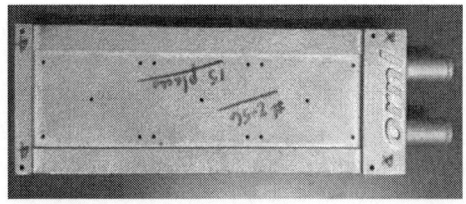

Fig.14 – crossectional view of AM model heatsink and X-ray of that view.

become close to equal around 150°C which is near the center of this inverter's operating range. It would be beneficial to improve the performance in the 100-150°C range for this application.

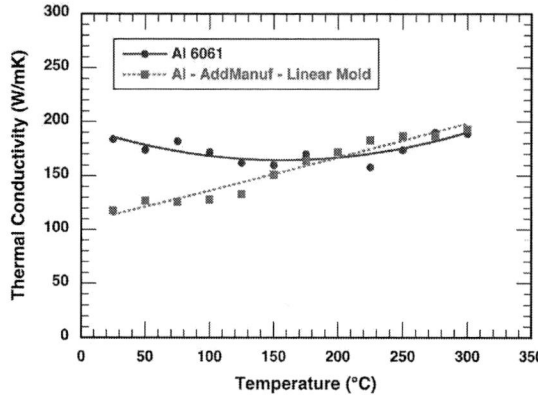

Fig. 17 – ORNL test results for thermal conductivity for AM aluminum sample (AlSi10Mg) vs conventional 6061 aluminum

IV. INVERTER ASSEMBLY AND TESTING

The final power module assembly is shown in Fig. 18. The DBC substrates were directly mounted to the printed module enclosure using spring pressure and thermal grease as the heat transfer medium from the lower side of the DBC substrates. The lead frame was built using a plastic printer at the ORNLs MDF. The boltholes in the lead frame served as the strain relief structure for the power leads from the substrates. The plastic lead frame also provided the support for the gate driver pc boards and the control board through integrated plastic stand-offs. The material for this first prototype was made using fused deposition melting (FDM) and is common ABS plastic – this can be replaced on the next prototype with an FDM plastic that has a temperature rating near 200C. The gate driver leads were soldered to flexible wires and were connected to the gate driver boards. The total volume of the inverter is ~1.5 L. 2.8 L (92 cm x91cmx178cm).

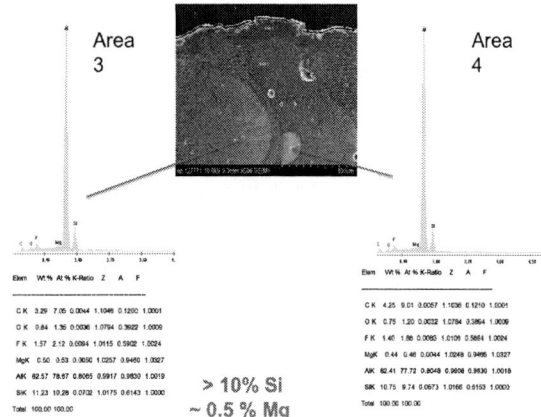

Fig.15: Material composition of AM Aluminum (AlSi10Mg)

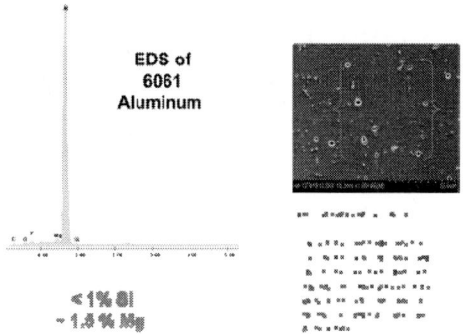

Fig. 16: Material composition of 6061 Aluminum

Below in Fig. 17 are additional tests showing thermal conductivity comparisons between conventional 6061 aluminum, which is used on many ORNL prototype heat sinks, and the typical AM aluminum alloy used in DMLS, AlSi10Mg. ORNL will be pursuing development of an aluminum alloy material that works well in the DMLS process that has improved thermal performance. At the lower temperatures, 6061 aluminum has a significantly better thermal conductivity than the AM aluminum, but they

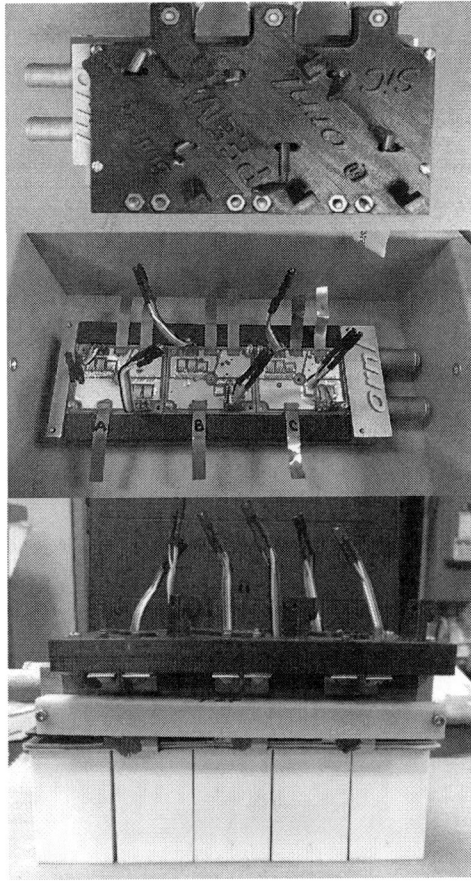

Fig. 18: Final assembled inverter and the power module prototype.

Fig. 19: Experimental setup for evaluating inverters performance.

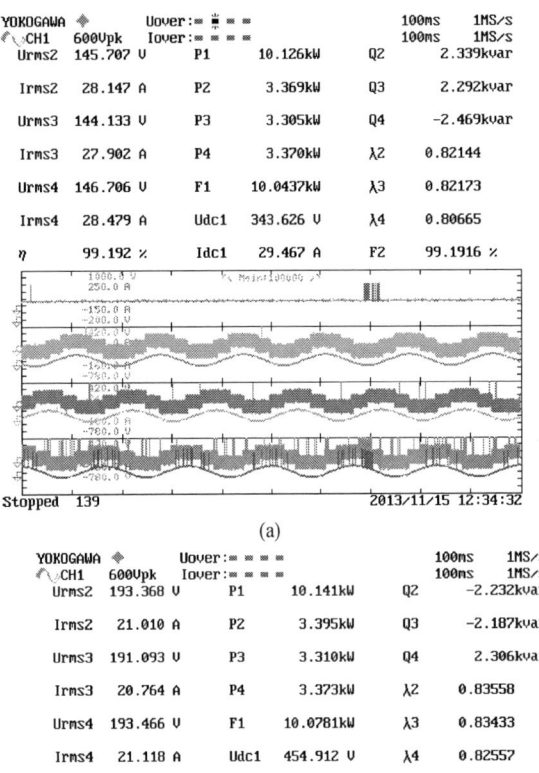

Fig.20: Experimental waveforms of 10 kW SiC inverter. (a) Screen shot of 350 V dc-link operation (b) screen shot of 450 V dc-link operation.

The experimental test setup is shown in Fig. 19. The test equipment that was used for this experiment includes Tektronix DPO 7104 1GHz, a TEK differential probe P5205 100 MHz bandwidth, and a Tektronix 404 XL current probe. PZ4000 power analyzer from Yokogawa was power measurements. The neutral point connection from the three phase resistive load was used for the power measurement setup in the power analyzer. For this test, the dc-link voltage was fixed at nominal operating voltage (350 V) to the maximum bus voltage (450 V). The load resistance was set to the minimum value, and changing the modulation index controlled the current. The coolant was set at 20°C at a flow rate of 1.5 gpm. The open-loop frequency of operation and the PWM frequency were fixed and the current command was varied for a particular dc-link voltage. The command current was increased in steps without exceeding the power rating of the inverter or of the load. The coolant temperature was changed to 60°C and data were recorded for a wide range of current and switching frequencies. The experimental waveforms for 350 V and 450 V operation are shown in Fig. 20.

The efficiency versus output-power plot for several operating conditions is shown in Fig. 21. Inverter efficiencies are higher at the 450 V than at the 325 V operating condition, as expected. Fig. 22 shows that the inverter efficiency does not change much as the switching frequency increases. The overall inverter efficiency is ~99% for different operating conditions. The total operating power density based on the test conditions of the inverter is ~7 kW/L and the designed power density is much higher. This inverter can be further pushed to higher operating power by increasing the dc-link voltage and current, and a power density of at least four times more can be realized.

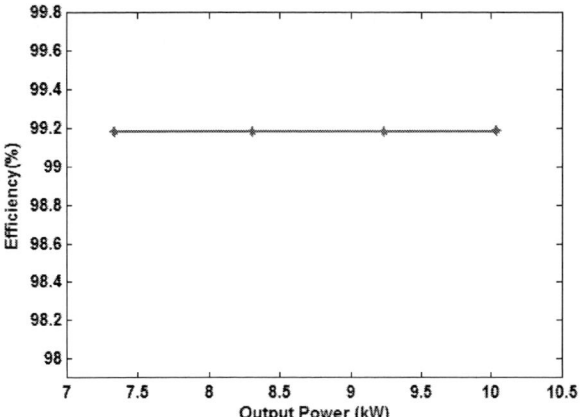

Fig. 21: Inverter efficiency vs. output power.

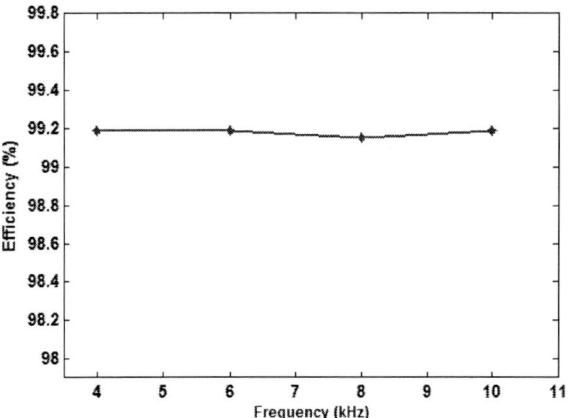

Fig. 22: Inverter efficiency vs. switching frequency.

VI. CONCLUSIONS AND RECOMMENDATIONS

In this paper, design, development and testing of a SiC inverter with printed power module AM heat sink was presented. The 1200 V, 100 A SiC module characterizations with detailed gate drive design was also presented. The total operating power density of the laboratory prototype inverter is ~7 kW/L and the average operating efficiency of the inverter for a wide range operating conditions is around 99%. However, based on the design the power density can potentially be four times more for higher power with the same power module. This prototype is the first inverter built using additive manufacturing techniques. The material analysis showed that there is no problem with the strength of material when it compares to Al-6061. It was also shown that there is no compromise in the performance of the inverter because of different material used. From ORNL's perspective

this is the first step toward realizing a full inverter built using additive manufacturing techniques.

ACKNOWLEDGEMENT

The authors would like to thank Andrew Wereszczak at ORNL for his support in material characterization.

REFERENCES

[1] S. Kubota, T. Sakurai, and H. Okada, "Size and Weight Reduction Technology for a Hybrid System," *SAE World Congress & Exhibition*, 2009.

[2] H. Ohtsuka and F. Anraku, "Development of inverter for 2006 model year Civic hybrid," *IEEE Power Conversion Conference*, 2007, pp. 1596–1600.

[3] D. Bortis, B. Wrzecionko, and J. W. Kolar, "A 120C ambient temperature forced air-cooled normally-off SiC JFET automotive inverter system," *IEEE Applied Power Electronics Conference and Exposition (APEC)*, 6-11 March 2011, pp. 1282 – 1289.

[4] S. Sato, k. Matsui, Y. Zushi, Y. Murakami, S. Tanimoto, H. Sato, and H. Yamaguchi, "Forced-Air-Cooled 10 kW Three-Phase SiC Inverter with Output Power Density of more than 20 kW/L, " *Materials Science Forum*, 2011, Vols. 679-680, pp. 738-741.

[5] Kashyap, A.S., Cheng-Po Chen ; Tilak, V. Compact modeling of silicon carbide lateral MOSFETs for extreme environment integrated circuits

[6] www.grc.nasa.gov/WWW/RT/2006/RI/RIS-neudeck.html.

10 kV, 120 A SiC MOSFET Modules for a Power Electronics Building Block (PEBB)

Christina DiMarino, Igor Cvetkovic, Zhiyu Shen, Rolando Burgos, Dushan Boroyevich
The Bradley Department of Electrical and Computer Engineering
Virginia Tech
Blacksburg, VA
dimaricm@vt.edu

Abstract—**10 kV, 120 A all-SiC half-bridge modules were tested and simulated for use in a 4 kV, 100 A power electronics building block (PEBB), which are standard, multifunctional units that can replace specialized devices in order to simplify the design and reduce the cost of power electronic systems. This PEBB design consists of four 10 kV SiC MOSFET modules in an H-bridge configuration. Each switch in the module contains twelve parallel 10 kV SiC DMOSFETs, and six parallel 10 kV SiC junction barrier Schottky (JBS) diodes. Double-pulse tests (DPTs) were conducted on the 10 kV modules up to 4.7 kV and 100 A in order to evaluate their hard switching capabilities. The DPT results revealed fast switching, with rise and fall times of 170 ns and 460 ns, respectively, and minimal ringing and overshoot. A Saber model of the module was then compared to the experimental results, which showed sufficient agreement with the testing waveforms. This model was then used to simulate the PEBB operation.**

Keywords—*10 kV SiC MOSFET; SiC module; power electronics building block; PEBB; characterization; simulation*

I. INTRODUCTION

The concept of a power electronics build block (PEBB) is the integration of fundamental components, such as power devices, gate drives, and control schemes, which can be used in a variety of applications [1]-[2]. The versatility of the PEBB allows for high return on investment, and also has the potential to reduce the cost, size, weight, loss, design complexity, installation, and maintenance of power electronics systems [1]-[2]. In the design of a PEBB, stress on the power devices, parasitic inductances, switching speed, losses, thermal management, and protection must all be taken into consideration [1]-[2]. Accordingly, wide bandgap semiconductors are of great interest due to their superior electrical and thermal properties. In particular, the high critical electric field and thermal conductivity of silicon carbide (SiC) allow for the conception of high power, efficient, and compact PEBBs.

The development of 10 kV SiC MOSFETs [3], and 10 kV SiC junction barrier Schottky (JBS) diodes [4] gave rise to the fabrication of high voltage, high speed power modules. These modules were critical components in the development of a high density 1 MVA solid-state power substation, which employed soft switching techniques and was capable of switching up to

20 kHz [6]. When compared with 6.5 kV Si IGBT modules, these 10 kV SiC modules demonstrated higher efficiency and switching frequency potential [5].

The modules are in a phase-leg configuration, with each switch (top and bottom) comprising twelve parallel 10 kV, 10 A SiC DMOSFETs, and six parallel 10 kV, 10 A SiC JBS diodes [6]. Each module has two gate drives, one for the top switch and another for the bottom switch, with desaturation protection. Two of these half-bridge modules can be assembled into an H-bridge, to which decoupling capacitors can be added to minimize the impact of parasitic inductance on the switching performance (Fig. 1). Furthermore, bulk capacitors with discharge resistors, IGBT short circuit protection, and inductors can be included to create a high power PEBB (Fig. 2). The PEBB design shown in Fig. 2 will be implemented into an impedance measurement unit (IMU). This IMU will perform series and shunt injection in order to identify the impedance of a source or load in an electric power system. These impedances are then used to determine the small-signal stability of the system.

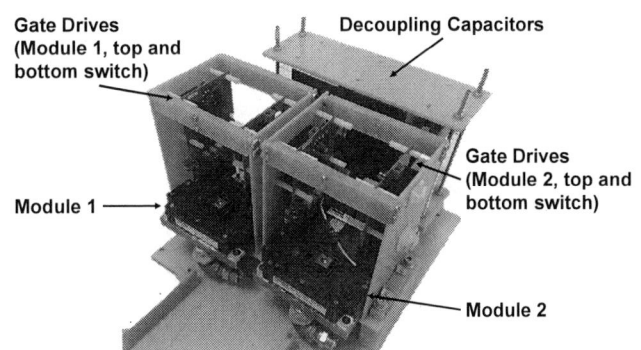

Fig. 1. SiC H-bridge.

Prior to integrating the SiC half-bridge modules into the PEBB, double-pulse tests (DPTs) were performed in order to evaluate their hard switching characteristics. Simulations were also carried out and compared to the experimental results in order to verify the model. This detailed switching model can then be used to simulate the full PEBB integration such that the control and power stage designs can be verified. The PEBB model will further allow designers to evaluate the PEBB

This research was sponsored by the Office of Naval Research.

978-1-4799-5494-0/14 $31.00 © 2014 IEEE

performance in other applications. After completing DPTs on each module, DPTs were then carried out on the fully integrated PEBB to determine its switching capabilities. Accordingly, this paper details the methods used for conducting this high power testing, and presents the results of DPTs on the modules and PEBB. A Saber simulation model for the modules, and the resulting simulation waveforms will also be presented.

and drain current values, the width of the first pulse needed to achieve the desired current level can be determined. At the end of this first pulse, the turn off switching waveforms for the device under test can be captured. In order to obtain the turn on switching waveforms under the same conditions, a short second pulse is applied. The rising edge of this second pulse gives the turn on switching waveforms for the device under test.

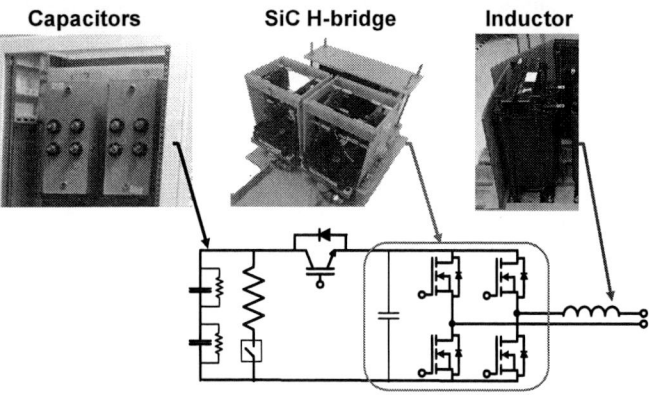

Fig. 2. SiC PEBB.

II. METHODS

A. SiC Module Double-Pulse Tests

Each module and gate drive was subjected to DPTs up to 4.7 kV and 100 A in order to evaluate their hard switching performance. The voltage was limited to 4.7 kV due to the voltage rating of the bulk capacitors, and because the peak voltage seen by the PEBB in the IMU is 4.6 kV. The module DPT hardware setup is shown in Fig. 3. A 1-mH inductor with high voltage wire for the winding was connected across the top switch of the module. The top switch was kept off with a gate-source voltage of -8 V supplied by its gate drive such that the antiparallel JBS diode would freewheel the current in the inductor when the bottom switch is off.

The bottom switch received the double-pulse signal through a fiber optic cable. The gate-source votlage of the bottom switch was monitored using a low-voltage differential probe (Tektronix TDP1000), the drain-source voltage of the bottom switch was measured with a high-voltage passive probe from Tektronix (P6105A), and the drain current was sensed with a Rogowski coil (PEM CWT Ultra Mini). DPTs were initially conducted using a low-inductance current shunt (T&M Research SSDN). The Rogowski coil and current shunt waveforms proved to be in good agreement, and thus the remaining tests were conducted with the Rogowski coil since the connections necessary for inserting the shunt into the testing setup add additional parasitics.

A high voltage (15 kV) power supply from TDK-Lambda (203/303L) was used to charge two series 850 μF, 2.5 kVDC bulk capacitors. Once the capacitors are charged, the power supply is disconnected. A double-pulse signal is then sent from the function generator, to the gate drive of the bottom switch using a fiber optic cable. From the dc bus voltage, inductance,

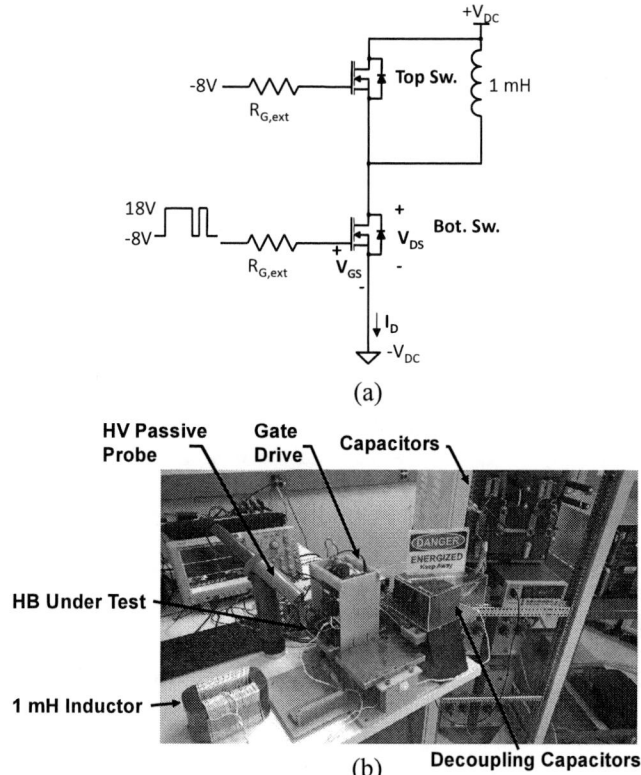

Fig. 3. Module DPT (a) schematic and (b) hardware setup.

B. PEBB Double-Pulse Tests

After the hard switching capability of the modules was evaluated, DPTs were then carried out on the PEBBs. It is crucial to perform DPTs on the PEBB in order to determine the impact of added parasitic elements on its hard switching behavior. The parasitics introduced during the PEBB integration include inductance from the busbar and series IGBT short-circuit protection, and the winding capacitance of the PEBB inductor. For this testing, the PEBB inductor was connected across the switch not under test. For instance, if the bottom switch of one of the two modules in the H-bridge were to be evaluated in the DPT, then the inductor would be connected across the top switch. The top switch would then be gated off such that the antiparallel JBS diode would freewheel the current (Fig. 3a). The top and bottom switches of the second module in the H-bridge would be kept off at a gate-source voltage of -8 V. For the PEBB DPTs, the switching of both the top and bottom switches of each module in the H-bridge was assessed. This results in a total of four DPTs conducted per PEBB.

978-1-4799-5494-0/14 $31.00 © 2014 IEEE   56

III. RESULTS

A. SiC Module Double-Pulse Test Results

The module DPTs revealed fast switching, high slew rates, and limited ringing and overshoot. Waveforms from one of the DPTs conducted on the bottom switch at 4.7 kV and 100 A is shown in Fig. 4. Table I lists the overshoot, undershoot, rise and fall times, and slew rates from the waveform. Compared to medium-voltage Si IGBTs, these SiC modules demonstrate fast switching. This is evident when comparing the turn off times in particular. A faster turn off transient will also correspond to a lower turn off switching loss.

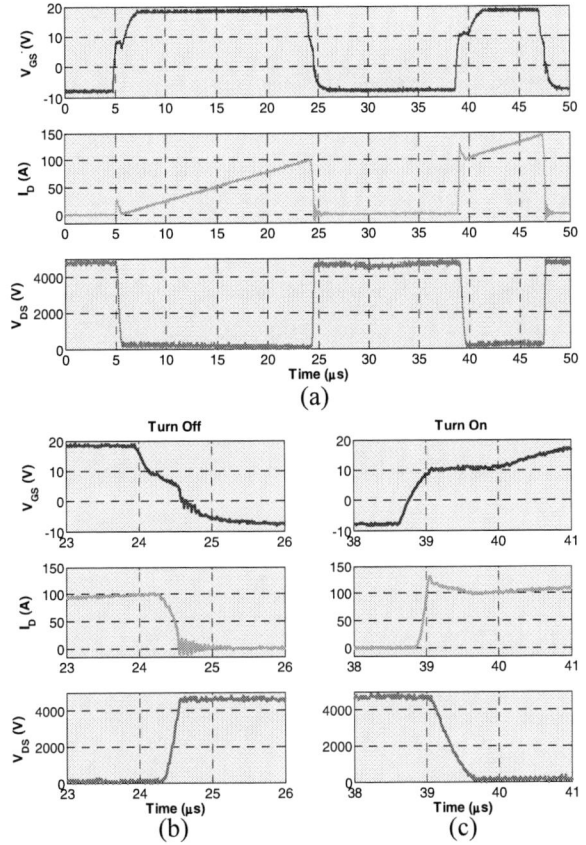

Fig. 4. Module DPT waveforms (a) full double-pulse, (b) turn off, and (c) turn on.

TABLE I. SiC MODULE DPT RESULTS

Parameter	Value		
	V_{GS}	I_D	V_{DS}
Overshoot	None	30 A	None
Undershoot	None	10 A	None
Rise time[a]	2020 ns	110 ns	170 ns
Rise slew rate	0.01 V/ns	0.73 A/ns	22.12 V/ns
Fall time[a]	980 ns	240 ns	460 ns
Fall slew rate	0.02 V/ns	0.33 A/ns	8.17 V/ns

[a.] Parameter measured from 10 % to 90 % of the steady-state value.

B. PEBB Double-Pulse Test Results

Fig. 5 shows the turn off and turn on waveforms from one of the PEBB DPTs conducted on the bottom switch at 4.7 kV and 80 A. A lower peak drain current was used for the PEBB DPTs in order to prevent the drain current from surpassing the IGBT overcurrent protection limit at the end of the second pulse, since, as shown by Fig. 4, the drain current can reach up to 150 A. The PEBB DPT waveforms were compared to the results of the module DPTs for the same switch (shown as the black curves). From the turn off waveforms shown in Fig. 5a, it can be observed that less ringing was experienced in the drain current turn off waveform for the PEBB DPT compared to that of the module DPTs. The greater ringing in the module DPTs could be due to the added source inductance from the current shunt connection. This figure also shows a lower drain-source voltage slew rate for the PEBB DPT. This could be partially due to the lower peak drain current; the peak drain current of the module DPTs was 100 A, whereas the peak drain current of the PEBB DPTs was 80 A. The turn off waveforms, shown in Fig. 5b, reveal that the drain current overshoot increased by approximately 30 A for the PEBB DPT. This larger overshoot could be attributed to the larger winding capacitance of the PEBB inductor compared to that of the inductor used for the module DPTs.

The top switches of each module in the PEBB were also tested. Due to the high voltage across the top switch, the drain-source voltage could not be measured, and a high voltage differential probe was used for the gate-source voltage measurement. The Rogowski coil was connected around the drain terminal of the top switch to sense the drain current. The waveforms matched well with those from the bottom switch DPTs.

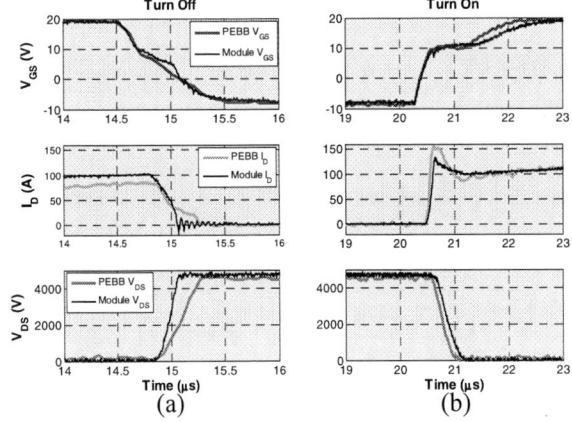

Fig. 5. PEBB DPT waveforms (a) turn off, and (b) turn on.

IV. SIMULATION

Another objective of this work was to develop a detailed switching model of the PEBB. This model would allow for verification of the control and power stage designs during full integration of the IMU. Additionally, the model can be used to examine the versatility of the PEBB by simulating different applications. A Saber model of the 10 kV, 120 A SiC half-bridge was developed by the National Institute of Standards and Technology (NIST). The model proved to have excellent agreement with the measured results as reported in [7]. In this work, a circuit that resembles the module DPT setup was generated in Saber using these half-bridge models (Fig. 6), and the results were compared to the experimental waveforms (Fig. 7). As shown by Fig. 7, the simulation and experimental waveforms are in good agreement. However, there is some discrepancy during the turn on process; the experimental results revealed faster gate-source voltage and drain current rise, and lower drain-source voltage slew rate than the simulation waveforms. Though, for this application, these differences are minor.

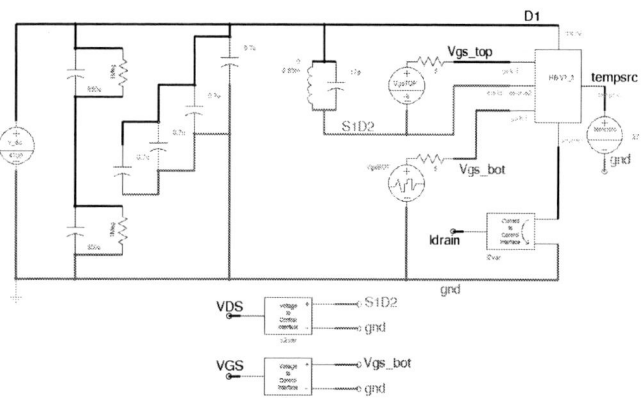

Fig. 6. Saber schematic for the module DPT simulations.

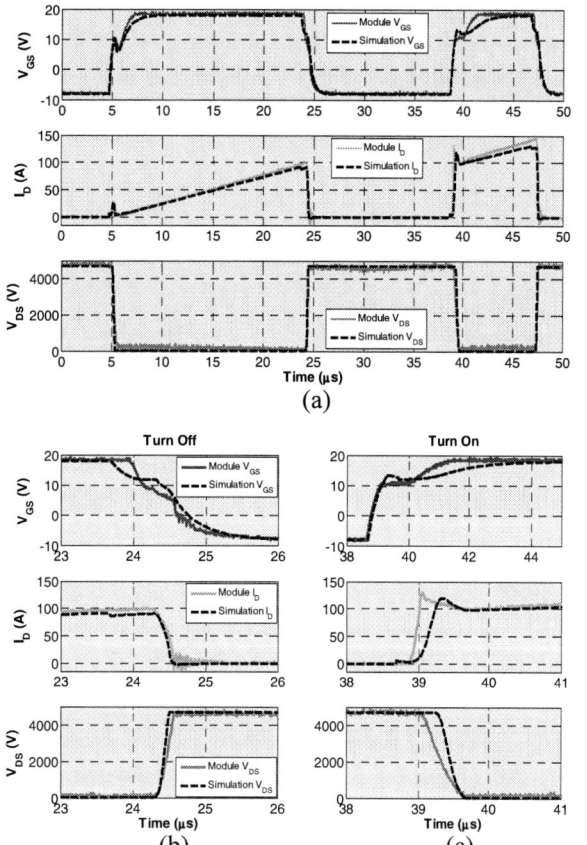

Fig. 7. Comparison of experimental and simulation (dashed black curves) DPT waveforms (a) full double-pulse, (b) turn off, and (c) turn on.

V. CONCLUSION

The dynamic characterization of 10 kV SiC MOSFET/JBS diode power modules were presented. The experimental results revealed fast switching times, with rise and fall times of 170 ns and 460 ns, respectively, and a drain-source voltage slew rate of 22 V/ns. Minimal ringing and marginal current overshoot were also observed. DPTs were also performed after the PEBB was fully integrated. Compared to the module DPTs, the PEBB DPTs showed reduced ringing, and a larger current overshoot (approximately 60 A). These phenomena demonstrate the impact of the other circuit elements on the switching behavior of the modules. The winding capacitance of the inductor will influence the drain current overshoot seen by the module during the turn on process, and the parasitic inductance will contribute to ringing.

The module DPT circuit was also simulated, and compared to the experimental results. Although the model showed good agreement with the experimental waveforms for the turn off process, some discrepancies existed for the turn on process. However, for the purposes of this work, this level of agreement is sufficient.

VI. FUTURE WORK

The future work entails running the PEBBs under continuous operation. The PEBBs will first be operated in a buck configuration, followed by a boost. During this continuous testing, the modules will be liquid cooled and their temperatures will be monitored. Further work also involves finalizing and verifying the PEBB simulation model.

REFERENCES

[1] T. S Ericsen, "Physics based design, the future of modeling and simulation," Acta Polytechnica, vol. 45, no. 4, pp. 59–64, 2005.

[2] T. Ericsen, Y. Khersonsky, P. Schugart, P. Steimer, "PEBB- Power electronics building blocks, from concept to reality," in *Proc. IEEE PEMD*, pp. 12-16, 2006.

[3] S. H. Ryu, S. Krishnaswami, B. Hull, J. Richmond, A. Agarwal, A. Hefner, "10 kV, 5 A 4H-SiC power DMOSFET," in *Proc. IEEE ISPSD*, pp. 1-4, 2006.

[4] B. A. Hull, J. J. Sumakeris, M. J. O'Loughlin, Q. Zhang, J. Richmond, A. R. Powell, E. A. Imhoff, K. D. Hobart, A. Rivera-Lopez, A. R. Hefner, "Performance and stability of large-area 4H-SiC 10-kV junction barrier Schottky rectifiers," *IEEE Trans. Electron. Devices*, vol. 55, no. 8, pp. 1864-1870, 2008.

[5] H. Mirzaee, A. De, A. Tripathi, S. Bhattacharya, "Design comparison of high-power medium-volage converters based on a 6.5-kV Si-IGBT/Si-PiN diode, a 6.5-kV Si-IGBT/SiC-JBS diode, and a 10-kV SiC-MOSFET/SiC-JBS diode," *IEEE Trans. Ind. Appl.*, vol. 50, no. 4, July/Aug. 2014.

[6] D. Grider, M. Das, R. Raju, M. Schutten, S. Leslie, J. Ostop, A. Hefner, "10 kV/120 A SiC DMOSFET half H-bridge power modules for 1 MVA solid state power substation," in *IEEE ESTS*, pp. 131-134, 2011.

[7] T. H. Duong, A. Rivera-Lopez, A. R. Hefner, "Circuit simulation model for a 100 A, 10 kV half-bridge SiC MOSFET/JBS power module," in *Proc. IEEE APEC*, pp. 913-917, 2008.

Process Optimization of Multicycle Rapid Thermal Annealing of Mg-implanted GaN

Jordan D. Greenlee*[†], Boris N. Feigelson[†], Travis J. Anderson[†], Marko J. Tadjer[‡†], Jennifer K. Hite[†], Charles R. Eddy Jr.[†], Karl D. Hobart[†], Francis J. Kub[†]

Email: jordan.greenlee.ctr@nrl.navy.mil

*National Research Council 500 Fifth St. NW Washington, DC 20001, USA
[†]Naval Research Laboratory 4555 Overlook Ave. SW Washington, DC 20375, USA
[‡]American Society for Engineering Education 1818 N St. NW Washington, DC 20036, USA

Abstract— To understand the capabilities and limitations of the multicycle rapid thermal annealing (MRTA) process, the key features of the process must be systematically explored. This work is dedicated to a study of the first step in the MRTA process – conventional annealing. A fundamental study was performed to determine the optimal conventional annealing temperature for the MRTA process by monitoring crystal quality and surface morphology. It was determined that both surface damage and crystal quality recovery must be considered when optimizing the MRTA process. Furthermore, the effect of performing the MRTA process with and without a preliminary conventional annealing step was investigated. It was determined that the conventional annealing step is a crucial part of the MRTA process.

Keywords—GaN; Implantation; p-type GaN;

I. INTRODUCTION

GaN and related III-nitride alloys possess several properties that make them attractive for a wide range of applications. The wide bandgap coupled with decent mobility make GaN suitable for high voltage vertical devices. The radiation hardness of III-nitride materials makes them attractive for space and military applications [1]. The inherent piezoelectric effects in AlGaN and GaN have enabled HEMTs for high power and high frequency applications [2]. Furthermore, the tunable direct bandgap is attractive for photovoltaic and optoelectronic applications [3].

For many III-nitride based devices, the formation of selective area p- and n-type regions by ion implantation is needed. The post-implantation activation of dopants by annealing requires implantation damage repair and temperatures high enough to enable the implanted dopants to substitute atoms in their proper lattice position. The activation of n-type dopants in GaN is achievable with relatively low temperatures (1250 °C) [4]. Activation of Mg via annealing requires higher temperatures (>1300 °C). Annealing and activating Mg is complicated by the additional difficulty that any N vacancies created in the annealing procedure will compensate the activated Mg. Co-implantation of Mg with another species such has P or N has been used to avoid compensation [5]. At these high annealing temperatures, in standard experimental setups, GaN decomposes. This decomposition leaves behind a roughened surface and a defective crystalline structure, both of which are unfavorable for III-nitride based device applications.

It is because of this difficulty in avoiding GaN decomposition that the activation of implanted Mg in GaN has been approached many times in the past with limited success. More recently, the highest activation efficiency of implanted Mg has been demonstrated using the MRTA process [6, 7]. GaN decomposition is limited in several ways during the MRTA process. First, the MRTA process is performed at a moderate nitrogen overpressure of 24 bar. Second, the samples are capped with a dual-layer AlN cap consisting of a thin high quality crystalline layer grown at high temperature followed by a thicker AlN layer grown at a lower temperature for mechanical support. AlN caps have been shown to mitigate GaN decomposition during high temperature annealing. Finally, the MRTA process is comprised of two thermal steps – a conventional anneal on the order of tens of minutes to prepare the sample for higher temperatures by removing some of the crystalline damage introduced by Mg implantation, followed by rapid heating/cooling cycles. In this work, a systematic study is commenced to explore the effects of the separate steps in the MRTA process. Specifically, the conventional annealing step will be investigated in this work.

II. EXPERIMENTAL SETUP

A. Sample Preparation

The samples used in this study were grown simultaneously on a 2 inch sapphire substrate. The layers grown starting from the substrate are of a 2 um thick undoped GaN layer followed by a ~4 nm high temperature AlN layer and a 25 nm low temperature AlN layer. The wafer was subsequently implanted with a three energy implant profile, which was simulated using the SRIM simulation package. The implantation profile and energies are shown in Figure 1. After implantation, the samples were annealed under a nitrogen pressure of 24 bar at temperatures ranging from 1000 °C to 1200 °C for 30 minutes.

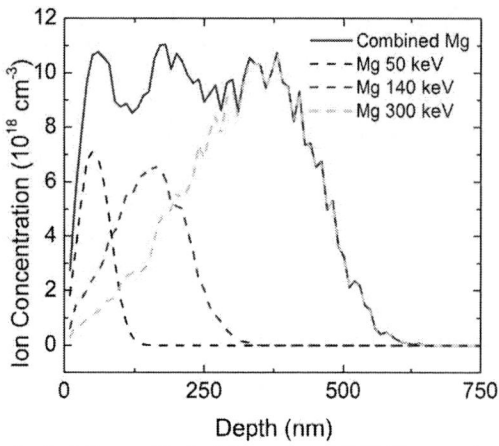

Figure 1. Implantation profile for Mg implanted GaN. The energies used are 50, 140, and 300 keV to achieve a roughly uniform concentration profile of 1 x 10^{19} to a depth of around 500 nm.

B. Characterization

The conventional annealing step for the MRTA process was investigated by monitoring crystal quality and surface morphology. Crystal quality before and after annealing were characterized using Raman spectroscopy while the surface morphology of the Mg-implanted films was characterized using atomic force and Nomarski microscopies.

III. RESULTS AND DISCUSSION

The implantation of GaN and post implantation annealing presents a steep challenge. The implantation of dopants in any crystal creates atomic displacements. Upon implantation of Mg into GaN, as shown in Figure 2, peaks appear in the Raman spectrum not associated with unimplanted GaN, AlN, or sapphire substrate.

A. Raman Spectroscopy

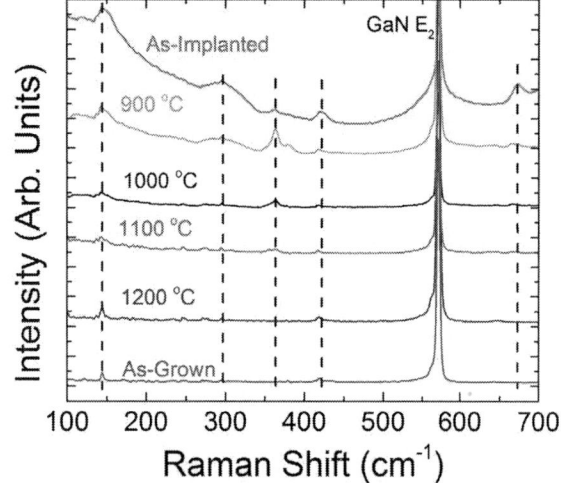

Figure 2. Raman spectra of implanted and annealed Mg-implanted GaN. As the conventional annealing temperature is increased, the peaks associated with implantation damage decrease

These peaks associated with implantation damage arise after Mg implantation, and the majority of the peaks decrease for increasing conventional annealing temperature. Specifically, five separate peaks arise from implantation damage. The peaks are located at approximately at 145, 300, 360, 420, and 670 cm^{-1}, and have been associated with implantation damage in previous Raman spectroscopic investigations of implanted GaN regardless of implantation species [8]. Interestingly, the peak at 360 cm^{-1} increases after annealing at 900 C. This is consistent with previous reports, and has been attributed to the evolution of simple defects into more complex defects [8].

B. Atomic Force Microscopy

AFM was used to examine the surface morphology of the samples both before and after annealing. Samples annealed below 1200 C did not show a change in surface morphology. As shown in Figures 3 and 4, the surface morphology drastically changes after a 1200 °C anneal for 30 minutes. There are portions of the sample where the GaN has decomposed under the surface, resulting in extremely rough areas. Before annealing, the sample has an RMS surface roughness of 2.62 nm RMS. After annealing, the conventionally annealed sample's RMS roughness has increased nearly ten times to 21.9 nm. This increased roughness and the associated crystal quality degradation is detrimental for devices, and sets an upper limit to the conventional annealing temperature in the MRTA process.

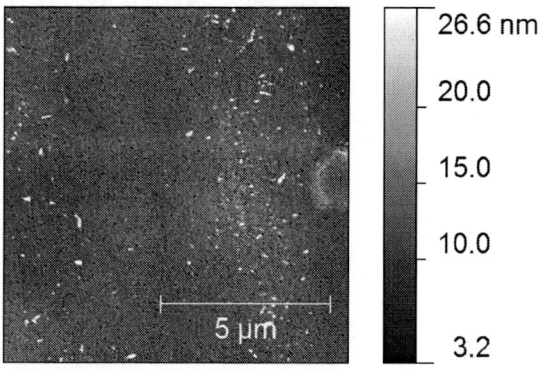

Figure 3. AFM image of the as grown surface of the GaN sample. The RMS surface roughness is 2.62 nm. The surface morphology is unchanged for samples annealed up to 1200 °C.

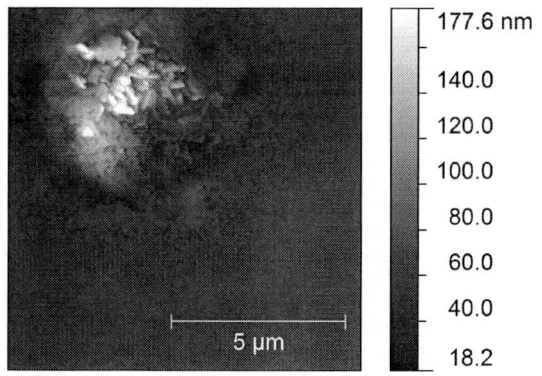

Figure 4. AFM image of the 1200 °C conventional annealed sample. After a 1200 °C thermal treatment, the GaN underneath the AlN begins to decompose and the surface morphology roughens significantly.

C. Necessity of Conventional Annealing in the MRTA Process

To understand the necessity of the conventional annealing step in the MRTA process, two samples grown at the same time were annealed using the same MRTA processing conditions with the exception of the conventional annealing step. The temperature pulses for both samples include 40 pulses up to 1300 °C. The first sample did not receive a conventional anneal prior to the MRTA process. A Nomarski micrograph of the first sample is shown in Fig. 5. Without a conventional annealing step before the MRTA process, the sample was badly damaged. In comparison, as shown in Fig. 6, the second sample, which did receive a preliminary conventional anneal showed little to no surface damage on the sample after the MRTA process, proving that the conventional anneal is a necessary step in the MRTA process.

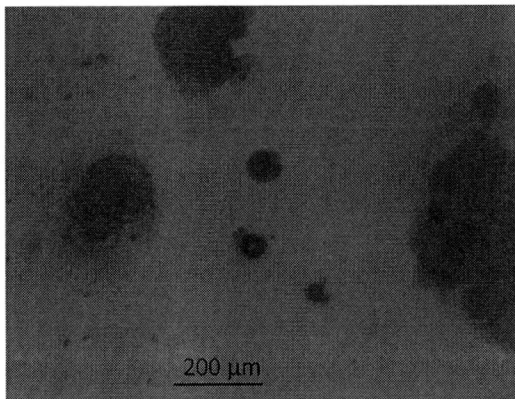

Figure 5. Nomarski micrograph of a half-implanted GaN sample that underwent the second half of the MRTA process – rapid heating and cooling pulses. The surface is badly damaged, with areas exceeding 200 μm laterally demonstrating decomposition of the underlying GaN.

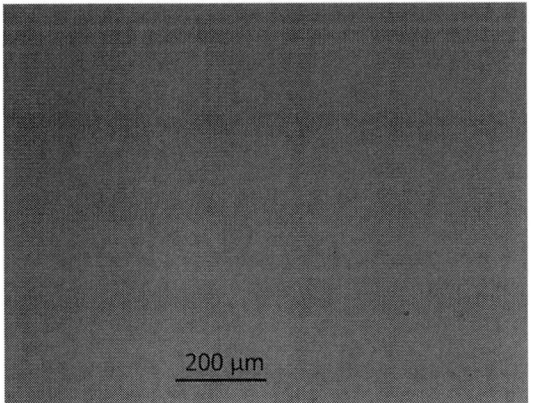

Figure 6. Nomarski micrograph of a half-implanted sample that underwent the full MRTA process including the conventional anneal and the rapid heating and cooling pulses. The sample does not exhibit surface degradation compared to the sample that did not receive the conventional annealing step.

As shown in the Raman spectroscopy data in Fig. 2, the crystalline quality of the implanted GaN improves dramatically after a conventional anneal. This repair from the conventional anneal is necessary to properly prepare the crystalline lattice for the >1300 °C temperatures involved with the second step in the MRTA process.

D. MRTA and Electrical Activation

The MRTA process is able to achieve high annealing temperatures while simultaneously avoiding GaN decomposition. This combination allows for the highest activation percentages of Mg-implanted GaN demonstrated by any annealing process [6]. The optimization of the rapid heating and cooling cycles to achieve the maximum activation ratio is the subject of ongoing research, but a preliminary study of the electrical properties of Mg implanted in GaN and annealed under MRTA conditions has been performed.

Samples have been annealed at peak temperatures ranging from 1341.5 to 1421.8 °C. It was determined that the sample annealed at the lowest temperature studied and had the shortest total time of 7.2 seconds above 1340 °C had the highest mobility of 40 cm^2/Vs. This sample also exhibited the lowest sheet resistance where R_{sh}= 2.69 kΩ/□ and an activation efficiency of 8.2% [6]. The sample annealed with the highest amount of time (125.8 s) above 1340 °C resulted in a lower mobility of 5 cm^2/Vs and a higher R_{sh} of 42.6 kΩ/□. The activation efficiency of the sample annealed at higher temperatures for longer time was 1.4 %. The higher sheet resistance and mobility of the sample annealed at higher temperatures for longer time is consistent with the introduction of additional defects, such as nitrogen vacancies, in the material. Electrically, nitrogen vacancies are donors and thus these additional defects will compensate the activated Mg and lower mobility due to increased scattering.

IV. CONCLUSIONS

In conclusion, it was determined that a preliminary conventional annealing step is necessary for the MRTA process. A preliminary conventional step on the order of tens of minutes prepares the sample structurally for the rapid heating and cooling pulses. This structural evolution was investigated using Raman spectroscopy. It was further determined that conventional anneals performed above 1200 °C resulted in surface damage due to the decomposition of GaN into Ga and N. Thus, the conventional annealing step should be kept below 1200 °C.

ACKNOWLEDGMENT

This research was performed while J. D. Greenlee held a National Research Council Research Associateship Award at the Naval Research Laboratory. M. J. Tadjer acknowledges support from the American Society for Engineering Education. Research at the Naval Research Laboratory was supported by the Office of Naval Research.

REFERENCES

[1] T. J. Anderson, A. D. Koehler, J. D. Greenlee, B. D. Weaver, M. A. Mastro, J. K. Hite, C. R. Eddy, K. J.

Kub, and K. D. Hobart, "Substrate-Dependent Effects on the Response of AlGaN/GaN HEMTs to 2-MeV Proton Irradiation," *IEEE Electr. Device Lett.,* vol. 35, pp. 826-828, 2014.

[2] T. J. Anderson, A. D. Koehler, K. D. Hobart, M. J. Tadjer, T. I. Feygelson, J. K. Hite, B. B. Pate, F. J. Kub, and C. R. Eddy, "Nanocrystalline Diamond-Gated AlGaN/GaN HEMT," *IEEE Electr. Device Lett.,* vol. 34, pp. 1382-1384, 2013.

[3] M. Asif Khan, J. N. Kuznia, D. T. Olson, M. Blasingame, and A. R. Bhattarai, "Schottky barrier photodetector based on Mg‐doped p‐type GaN films," *Appl. Phys. Lett.,* vol. 63, pp. 2455-2456, 1993.

[4] C. E. Hager, K. A. Jones, M. A. Derenge, and T. S. Zheleva, "Activation of ion implanted Si in GaN using a dual AlN annealing cap," *J. Appl. Phys.,* vol. 105, p. 033713, 2009.

[5] K. T. Liu, Y. K. Su, S. J. Chang, and Y. Horikoshi, "Magnesium/nitrogen and beryllium/nitrogen

coimplantation into GaN," *J. Appl. Phys.,* vol. 98, p. 073702, 2005.

[6] T. J. Anderson, B. N. Feigelson, F. J. Kub, M. J. Tadjer, K. D. Hobart, M. A. Mastro, J. K. Hite, and C. R. E. Jr., "Activation of Mg implanted in GaN by multicycle rapid thermal annealing," *Electronics Lett.,* vol. 50, pp. 197-198, 2014.

[7] B. N. Feigelson, T. J. Anderson, M. Abraham, J. A. Freitas, J. K. Hite, C. R. Eddy, and F. J. Kub, "Multicycle rapid thermal annealing technique and its application for the electrical activation of Mg implanted in GaN," *J. Cryst. Growth,* vol. 350, pp. 21-26, 2012.

[8] W. Limmer, W. Ritter, R. Sauer, B. Mensching, C. Liu, and B. Rauschenbach, "Raman scattering in ion-implanted GaN," *Appl. Phys. Lett.,* vol. 72, pp. 2589-2591, 1998.

Design and Fabrication of High Current AlGaN/GaN HFET for Gen III Solid State Transformer

In-Hwan Ji, Sizhen Wang, Bongmook Lee, Haotao Ke, Veena Misra, and Alex Q. Huang

Future Renewable Electric Energy Delivery and Management (FREEDM) Systems Center

North Carolina State University

Raleigh, NC, U.S.A

iji@ncsu.edu

Abstract— We have successfully designed and fabricated the high current AlGaN/GaN HFET using 4 masks process. We have achieved the maximum on-current of 7.3A at 1.5V with TO-3 power package. Multi-finger cell-design was implemented to achieve high current extraction from specific active area. In addition, the mesa-free design was adopted to reduce the process steps by employing the self-isolation with surrounded Schottky gate. Length of unit finger is 1000um. Total width of multi-finger is 80 mm. The measured Ronsp and breakdown voltage are 3.4mΩ-cm2 and 450V, respectively. Although the designed breakdown voltage (BV) was 600V, BV reduction has been observed after fabrication due to the process variation. However, overall figure of merit of this device is 2 times superior to that of Super Junction Power MOSFET.

Keywords—HFET, High current, AlGaN/GaN, Power GaN, SST, FREEDM

I.   INTRODUCTION

Future Renewable Electric Energy Delivery and Management (FREEDM) system requires high performance AlGaN/GaN power HFET devices suitable to next generation Solid State Transformer (SST) which enables high efficiency energy conversion with low cost.[1] AlGaN/GaN HFET device is essential switch for low voltage stage in SST circuit as shown in Fig. 1. Especially, high current capability of the HFET is necessary functionality for high power conversion as well as high speed operation.[2][3] To improve the current conduction capability, layout structure of array and multi-metal layer have been reported[4][5]. However, these approach requires high design complexity and additional process steps which increase device fabrication cost and reduce the yield. This paper presents the cost-effective design of large size AlGaN/GaN HFET device and successful fabrication of the high current HFET power device for Gen III SST application.

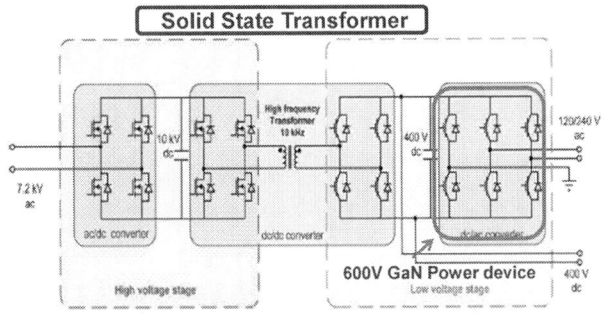

Fig. 1.Single-phase SST circuit topology

This work is supported by the U.S. National Science Foundation under award EEC-0812121.

II.   DEVICE STRUCTURE AND FABRICATION

The AlGaN/GaN heterojuction was grown on Si (111) substrate by metal-organic chemical vapor deposition (MOCVD) and consists of 4 nm GaN, 25 nm $Al_{0.23}Ga_{0.77}N$, 1 nm AlN, 1.5 μm GaN, 4.5 μm GaN/ AlGaN/ AlN layer. In order to minimize process steps, conventional HFET structure has been adopted as shown in Fig. 2.

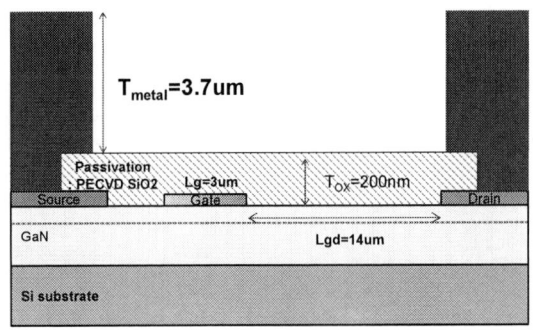

Fig. 2. Cross-sectional view of proposed device

Total number of mask layers is 4 as summarized in Table I. For ohmic contact formation, a Ti/Al/Ni/Au (30/ 200/ 30/100 nm) metal stack was deposited by the e-beam evaporation. A rapid thermal annealing (RTA) process was performed at 850 °C for 30 sec in an N_2 ambient. The gate metal, Ni/Au (50/ 100 nm), was deposited by the same e-beam evaporation. After gate definition, no additional RTA was not applied. The 200 nm SiO_2 as a passivation layer was deposited by the plasma-enhanced chemical vapor deposition (PECVD) and then the contact etching process was performed.

Table I. Process Summary

Mask	Process step	Film	Thickness
1	S/D Ohmic metallization	Ti/Al/Ni/Au	360nm
2	Gate metallization	Ni/Au	150nm
3	Contact etch	PECVD SiO2	200nm
4	Top metallization	Ti/Al	3.7um

For high current GaN power device, highly conductive top metallization is very critical process to minimize the contact resistance. In our device architecture, the metal thickness should be more than 3.0μm to conduct more than 10A. 3.7um Aluminum was deposited by the same e-beam evaporation. The measured sheet resistance of Aluminum metal film is

13.87 mΩ/□. Fig. 3 shows an SEM cross section image of the resulting metal pattern. To define thick Aluminum metal finger, Chlorine based ICP RIE etching has been applied, which imply CMOS friendly metallization process without electro-plating.

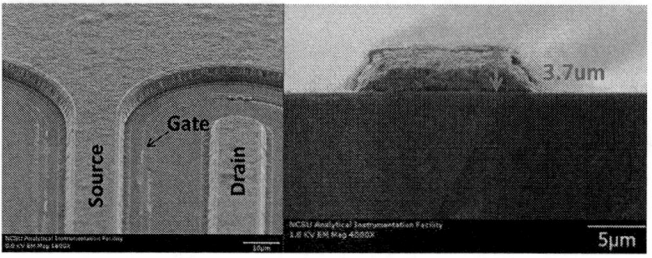

Fig. 3. FE-SEM image taken after CMOS compatible top metal process(left), 3.7 um thick metallization(Ti/Al) without electro-plating(right)

III. RESULTS AND DISCUSSIONS

A. Layout Design

Design of long finger shaped unit cell has been widely accepted for high current lateral GaN devices.[4][6] However, these comb shape design requires mesa isolation process to avoid the electrical short between source and drain by removing the 2DEG region out of device active region. To eliminate mesa isolation process, we employed mesa-free self-isolated design by surrounded Schottky gate as shown in Fig. 4.

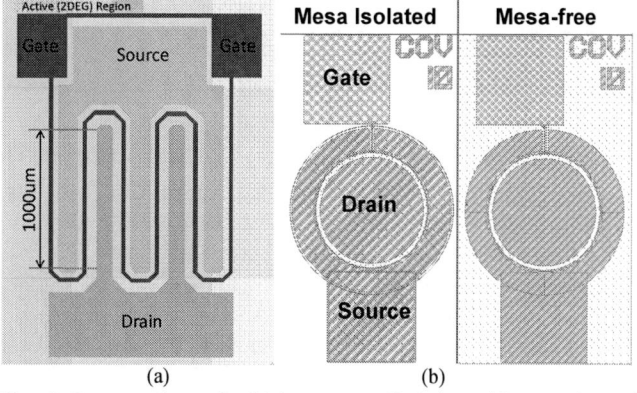

Fig. 4, Layout scheme for high current cell design with mesa-free self-isolation design by surrounded Schottky gate (a) Test structure for mesa-free design (b)

Mesa-free self-isolation design has been verified by using test structure as shown in Fig. 4.(b). Mesa isolated device has no 2DEG region out of source region by RIE etching. On the other hand, Mesa-free device has no definition of active region by removing of 2DEG. In Fig. 5, transfer, forward I-V and leakage characteristics show that mesa-free one is 96% identical to mesa-isolated one in terms of Ron. Both devices have identical threshold voltage of -3.5V and leakage current level at the same Vds of 100V.

Surrounded Schottky gate isolates the drain region from source region successfully without sacrificing trans-conductance and blocking capability, which enables several process steps to be removed such as photolithography, dry etching, ashing and strip.

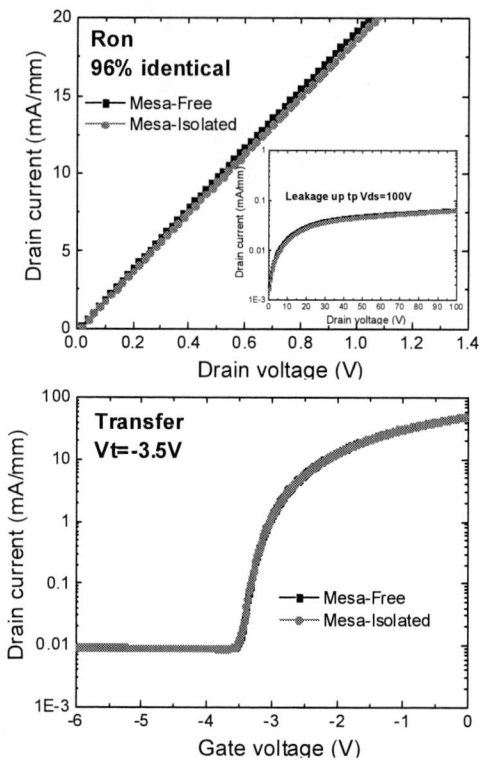

Fig. 5. Output and leakage (left) and transfer (right) characteristics of test structures

B. High Current Devices

High current device was designed with total gate width of 80mm and active area of 2.041×10^{-2} cm². For high current measurement, diced device has been mounted to TO-3 power package with the wedge wire-bonding and sealed with Silicone paste. The packaged device shows the high maximum on-current of 7.3A at 1.5V and low Ronsp of 3.4mΩ-cm2 at Ids of 1A. However, it is difficult to grab breakdown voltage of the high current device once the breakdown occurs because catastrophic current filaments damage Schottky gate in a second. The maximum voltage of 450V was observed before the device was destroyed.

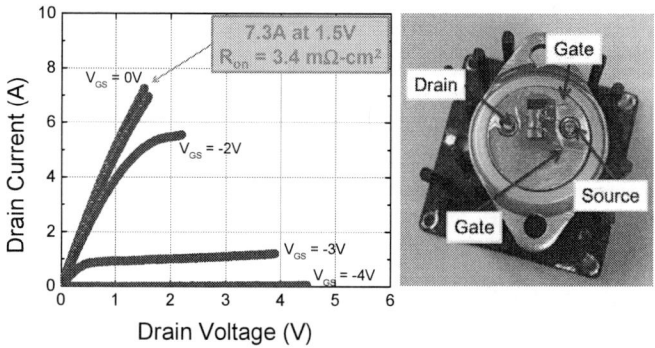

Fig. 6. Id-Vds curve of the fabricated high current AlGaN/GaN HFET (Left). TO-3 Packaged Device (Right).

Although the high current device showed the breakdown voltage of 450V, unit test device with the same design parameters exhibits the stable blocking behavior up to 600V as shown in Fig. 7. Because these test element unit cells has been fabricated at the same time with high current devices, our design scheme and architecture will support 600V blocking capability if process variation issues are eliminated.

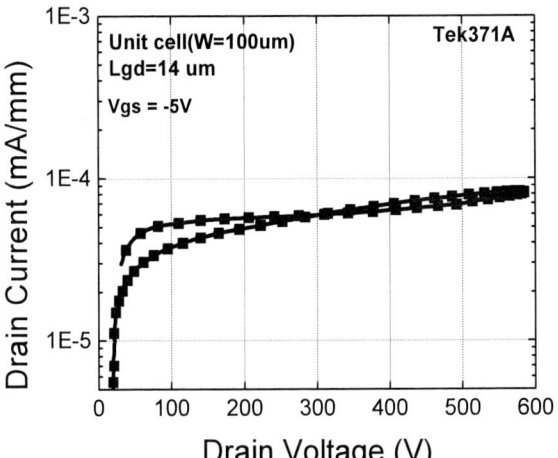

Fig. 7. Forward blocking capabilitiy of Test unit cell device

Test unit cell device showed an on-resistance value of 1.7 $m\Omega$-cm2 as shown in Fig. 8, which is half of high current device. This on-resistance increase with large device is believed to be due to the additional wiring and cabling added in the packaged device. The suppression of parasitic resistance of high current device will enable lower on-resistance.

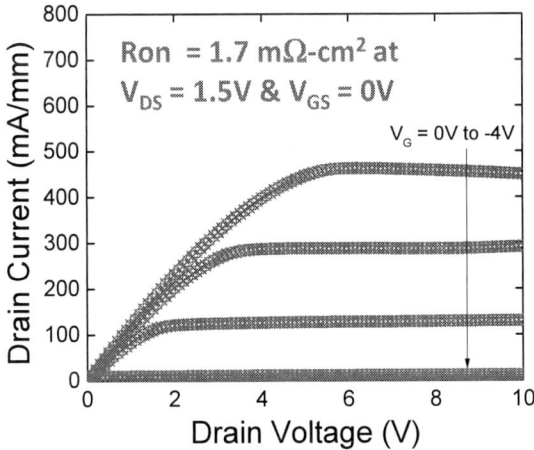

Fig. 8. Forward I-V characteristics of Test unit cell device

In spite of parasitic resistance of packaged device, the high current device have been demonstrated to have 30% the specific Ron(Ronsp) of Super-junction(SJ) Power MOSFET. It is well-known that Ronsp of commercialized 600V Power MOSFET has 10 ~11mΩ-cm2 as shown in Fig. 9. This allows operation of the SST circuit in higher efficiency.

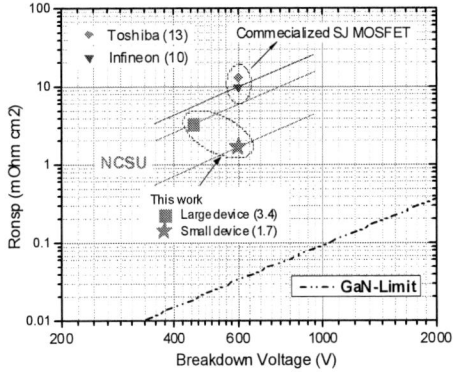

Fig. 9. Benchmark of Figure of Merit (BFOM)

IV. CONCLUSION

High current AlGaN/GaN HFET using 4 masks process has been successfully designed and verified by demonstrating the maximum on-current of 7.3A at 1.5V with TO-3 power package. Mesa-free self-isolation layout enables cost-effective fabrication of high power GaN device by reducing the process steps. Although the breakdown voltage of 450V and Ronsp of 3.4 mΩ-cm2 were achieved due to process variation and parasitic resistance, the overall figure of merit of this device is 2 times superior to that of Super Junction Power MOSFET. Low voltage stage of SST circuit in FREEDM system will be operated in higher efficiency thanks to this achievement.

ACKNOWLEDGMENT

This work was supported by ERC Program of the National Science Foundation under Award Number EEC-0812121.

REFERENCES

[1] Huang, Alex Q., and Jay Baliga. "FREEDM System: Role of power electronics and power semiconductors in developing an energy internet.", ISPSD 2009, pp. 9-12, IEEE, 2009

[2] Yang, Liyu, Tiefu Zhao, Jun Wang, and Alex Q. Huang. "Design and analysis of a 270kW five-level DC/DC converter for solid state transformer using 10kV SiC power devices." In Power Electronics Specialists Conference 2007, pp. 245-251. IEEE, 2007

[3] Bhattacharya, Subhashish, Tiefu Zhao, Gangyao Wang, Sumit Dutta, Seunghun Baek, Yu Du, Babak Parkhideh, Xiaohu Zhou, and Alex Q. Huang. "Design and development of generation-I silicon based solid state transformer." In Applied Power Electronics Conference and Exposition (APEC), 2010, pp. 1666-1673. IEEE, 2010.

[4] Reiner, R., P. Waltereit, F. Benkhelifa, H. Walcher, R. Quay, M. Schlechtweg, and O. Ambacher. "Novel Layout and Packaging for Lateral, Low-Resistance GaN-on-Si Power Transistors." In Integrated Power Systems (CIPS), 2014 8th International Conference on, pp. 1-5. VDE, 2014.

[5] Hwa-Young Ko, Jinhong Park, Youngje Jo, Hojung Lee, Won-Seok Choi, Seung Yup Jang, Eu Jin Hwang, Kwang-Choong Kim, and T. Jang. "Development of Over 100A, 600V Normally-Off GaN HFETs using Active Isolation and Multi-Level Metallization" In ISPSD 2014, pp. 59, IEEE, 2014

[6] Ikeda, Nariaki, Syuusuke Kaya, Jiang Li, Yoshihiro Sato, Sadahiro Kato, and Seikoh Yoshida. "High power AlGaN/GaN HFET with a high breakdown voltage of over 1.8 kV on 4 inch Si substrates and the suppression of current collapse." In Power Semiconductor Devices and IC's, 2008. ISPSD'08. 20th International Symposium on, pp. 287-290. IEEE, 2008.

An Integrated Gate Driver in 4H-SiC for Power Converter Applications

M. Nance Ericson, S. Shane Frank,
Charles L. Britton, Laura D. Marlino,
Devon D. Janke, Dianne B. Ezell
Oak Ridge National Laboratory, Oak Ridge, TN
ericsonmn@ornl.gov

Ranjan Lamichhane, A. Matt Francis, Paul D. Shepherd,
Michael D. Glover, H. Alan Mantooth
University of Arkansas, Fayetteville, AR

Sei-Hyung Ryu
CREE Semiconductor, Durham, NC

Bret Whitaker, Zach Cole,
Brandon Passmore, Ty McNutt
Arkansas Power Electronics Intl. Inc.
Fayetteville, AR

Abstract—**A gate driver fabricated in a 2-μm 4H silicon carbide (SiC) process is presented. This process was optimized for vertical power MOSFET fabrication but accommodated integration of a few low-voltage device types including N-channel MOSFETs, resistors, and capacitors. The gate driver topology employed incorporates an input level translator, variable power connections, and separate power supply connectivity allowing selection of the output signal drive amplitude. The output stage utilizes a source follower pull-up device that is both overdriven and body source connected to improve rise time behavior. Full characterization of this design driving a SiC power MOSFET is presented including rise and fall times, propagation delays, and power consumption. All parameters were measured to elevated temperatures exceeding 300°C. Details of the custom test system hardware and software utilized for gate driver testing are also provided.**

Keywords—*Gate Driver, SiC, Silicon Carbide, Wide Bandgap, Power Electronics*

I. INTRODUCTION

The recent commercialization of wide bandgap (WBG) power devices fabricated in silicon carbide (SiC) and Gallium Nitride (GaN) has provided new opportunities for electronic applications to operate at higher frequencies, voltages, and temperatures exceeding the inherent capabilities of silicon. When integrated into power conversion modules and traction drives, these WBG-based devices provide the potential for higher efficiency operation, lower system mass and size, and higher power density over silicon-based circuits due to the inherent properties of the WBG materials [1]. Industry is now commercializing limited WBG devices, namely power switches and diodes. However, recent literature has reported on the ability to fabricate WBG support electronics, including gate drivers and other low-voltage monitoring circuits in SiC, establishing the benefits of higher degrees of WBG electronics systems integration [2,3]. This work presents the furthering of these efforts by producing a second generation gate driver based on a previously published design [4].

Oak Ridge National Laboratory (ORNL) is managed by UT Battelle, LLC for the U. S. Department of Energy under Contract No. DE-AC05-000R22725.

II. SiC INTEGRATED GATE DRIVER

A. Gate Driver Design

The gate driver topology described in this paper is shown in Figure 1. Improvements of this design over a previously published version include the addition of an input level translator, multiple power connectivity options, a substrate isolated output pull-up device, and general device resizing to improve the gate driver performance for moderate levels of gate overdrive. Referring to Fig. 1, the input level translator is composed of 4 inverters (X_1-X_4) each having a source follower output stage which ensures that the output feeding the following stage has a sufficiently low logic level to correctly drive following stages. One risk with having multiple cascaded, resistor-transistor-logic (RTL) – based inverter stages is that the pull-up resistor value and associated pull-down device must be sized appropriately to minimize the associated RC-controlled rise time while ensuring a suitable logic '0' is obtained at each inverter output. Failure to reach this condition satisfactorily may result in inverter output logic low levels moving closer to undefined levels as signals propagate resulting in the eventual inability to produce valid logic levels late in the inverter or logic chain. Use of this inverter topology mitigated this risk at the cost of reduced logic '1' output levels. Note that high output levels are limited to VDD – V_{GS} for a source follower output stage. Buffer stages X_5 and X_6 were built as RTL inverters and provide the gain necessary to translate the logic '1' output to full positive rail values. Connectivity options to VDD-OD provide the means to 'program' the amount of current available for charging the capacitive gate of the output stage pull-up device (M_{PU}). These

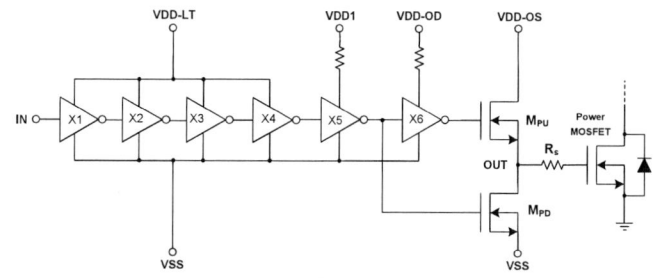

Fig. 1. Simplified gate driver architecture and connection to power MOSFET.

connections are best optimized during testing and are made at the packaging level. The combination of connections allows the selection of pull-up resistor values ranging from 94Ω–210Ω.

The process used for fabricating the design utilized a p-substrate and did not have an n-well option. This prevented the output pull-up device (M_{PU}) from having a source-body connection, resulting in a significant increase in rise time due to the body effect. To mitigate this undesired effect, the M_{PU} device was fabricated as a second die and packaged such that the isolated substrate could be source connected to minimize the body effect.

The final significant design change from the original gate driver involved complete resizing of the circuit to maximize the dynamic performance (limited by the output rise time) at low to moderate levels of gate overdrive (VDD-OD). The target overdrive level was selected as 32V rather than the 40-V overdrive used in the initial design, as lower overdrive levels provide improved reliability, and HSPICE simulations indicated 32V was sufficient.. To resize the design for improved performance, the circuit was first simulated with the expected output device capacitance (power MOSFET gate model) and the upper and lower output device sizes were determined for the case where ideal output stage drive sources were used. The sizing of the lower and upper output stage drive inverters, X_5 and X_6 respectively, were estimated by calculating the associated RC time constant desired, determining the associated pull-up resistor R, and then solving for the size of the pull-down device. This ratio of pull-up resistor to pull-down MOSFET size was set to produce a 'low' voltage output level of <1V. Note that with CMOS design the ratio of N-type to P-type device width/length is generally approximated by the ratio of the device's transconductance parameters (KP) provided in the SPICE model. The KP of a device is commonly defined as $\mu_0 C_{ox}$, where μ_0 is the carrier mobility and C_{ox} is the oxide capacitance per unit area which often produces a ratio of ~2.5 for standard CMOS processes. However, this design does not have PMOS devices, so the alternate pull-up resistor to pull-down MOSFET W/L ratio was calculated and then validated via HSPICE simulation using a custom model generated specifically for use with the devices available in this process [5].

B. Layout Considerations

The SiC process used for fabrication of the design was a P-Epi, 2-μm, single metal, NMOS-only process, optimized for fabrication of vertical SiC powerMOS switching devices. Having only one metal layer for signal and power routing required the occasional use of polysilicon as a routing layer. Efforts were made to only use polysilicon routing for traces having very low current requirements. A goal of limiting current densities to $\leq 1mA/\mu m^2$ for all conductive traces was followed during layout. In addition, diffusion resistors were used rather than polysilicon resistors to improve thermal coupling to the SiC substrate, and resistors were separated where possible to minimize concentrated areas of heating. Following layout and extraction, the circuit was simulated using HSPICE.

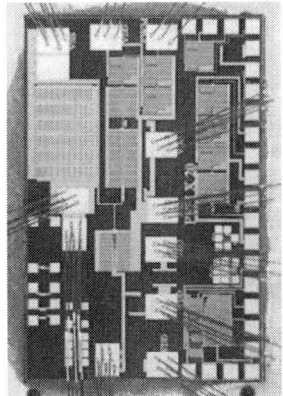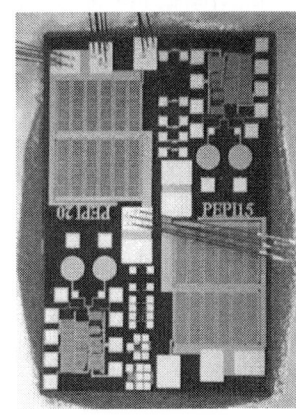

Fig. 2. SiC gate driver die (left) and separately fabricated pull-up device die (right).

III. EXPERIMENTAL RESULTS

A. Gate Driver Test System

The gate driver was fabricated as a chip set along with a number of other test cells and circuits. The chip set (gate driver die and pull-up device die) are shown in Fig. 2. A test setup was also designed to facilitate the evaluation and characterization of the SiC gate buffer. The die were bonded to a daughterboard carrier PCB (Fig. 3) which was mounted on a motherboard having the SiC powerMOS load device, supply filtering, input control signal isolation, and power supply connections. A mechanical stand was designed and constructed to fixture the motherboard and provide temperature control of a cold finger that thermally contacted the backside of the daughterboard. The opposite side of the cold finger was heated using a commercial hotplate and a thermocouple was placed between the daughterboard and the cold finger for monitoring the temperature. Initially, an IR imaging camera was used to determine the temperature difference between the thermocouple and the gate driver chip as a function of the gate driver power consumption. These tests indicated that the gate driver die temperature was approximately equal to the measured thermocouple temperature plus 7.7°C/Watt of gate driver power consumption. A custom LabVIEW-based software program was developed to enable automated testing of gate driver units. The program included the capability to

Fig. 3. Daughterboard with bonded SiC gate driver die set for testing.

perform voltage sweeps as well as data collection, display, and storage.

B. Measurement Results

Following fabrication of the SiC gate driver die, 27 gate buffer modules were assembled using the daughterboard PCBs as shown in Fig. 3, and were tested using the developed test system. Initial tests were performed at room temperature to verify basic functionality and to investigate the effect of using different supply voltage configurations. These tests showed that increasing VDD-LT beyond 20V had no significant effect on the overall dynamic performance of the gate driver except that larger supply voltages tended to increase the delay through the unit. Similarly, increasing VDD1 beyond the nominal 20V level did not significantly improve the output fall time, which was within expectations. The output voltage drive level (VDD-OS) was not a significant factor in performance optimization as the voltage is fixed by the drive requirements for the switching device used. However, increasing VDD-OS does increase the measured rise time since there is a slew limit imposed on the output drive signal due to limited source and sink currents in the output stage.

Tests were performed on all 27 of the assembled gate buffer modules to determine the approximate yield and room temperature performance. For these tests the rise time, fall time, delays, and power consumption were measured for VDD-OD swept from 20V-40V, with all power supply voltages at 20V, and with a single powerMOS load. These tests indicated a yield of ~70% for the gate driver die pair.

The measured rise and fall times are plotted for 10 SiC driver units as a function of VDD-OD (see Fig. 4). As expected, the VDD-OD had a very significant impact on the rise time of the gate driver output as the affected buffer stage directly drives the source follower output stage pull-up device (M_{PU}). However, the measured fall times are inversely proportional to the value of VDD-OD. This behavior is due to a voltage division between the pull-up resistor tied to VDD-OD and the X_6 output NMOS on-resistance which acts to keep M_{PU} slightly on during M_{PD} pull-down. Thus, a matching rise and fall time of ~50ns can be observed for VDD-OD ~30V-31V.

Subsequent tests were performed on two of the units over a die temperature range of approximately 90°C to 340°C. The rise times, fall times, propagation delays, and power dissipations were measured for each of the test conditions. This data is plotted in Figures 5-9. In these figures each of the curves represents the measured parameter (rise time, fall time, delay1, delay2, power) at a different VDD-OD value, plotted as a function of the calculated die temperature. Note that the first data point of each curve is the die temperature with the hot plate at room temperature. Here the die temperature is well above room temperature due to self-heating of the die. Consequently, for matched rise and fall times of ~45ns, VDD-OD=31V is the preferred case. Under this condition the rise time and fall time variation over the measured temperature range was 6.3% and 2.7%, respectively. The propagation delays of Figures 7 and 8 are very similar for all VDD-OD values, and match well over the temperature range tested. Both show a similar temperature dependence with a ~22% variation over the measured temperature range. The measured power consumption is shown in Fig. 9 and varied from 8.2W - 6.8W over the temperature range tested for VDD-OD=31V. The power decrease with temperature is due to the positive temperature coefficient of the n-diffusion resistors used in the design. This data shows that the gate buffer operates very well over the temperature range tested.

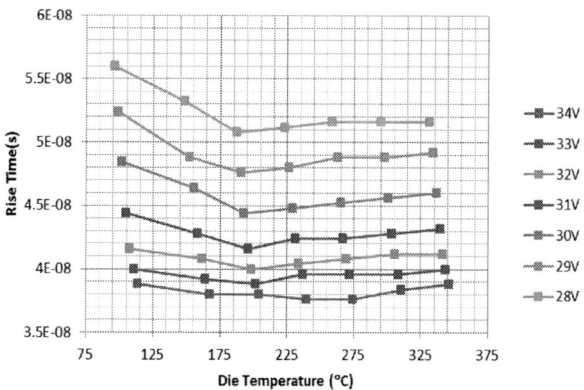

Fig. 5. Measured rise time over temperature (VDD-OD = 28V-34V, other supplies voltages = 20V, single powerMOS load).

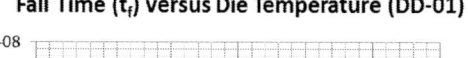

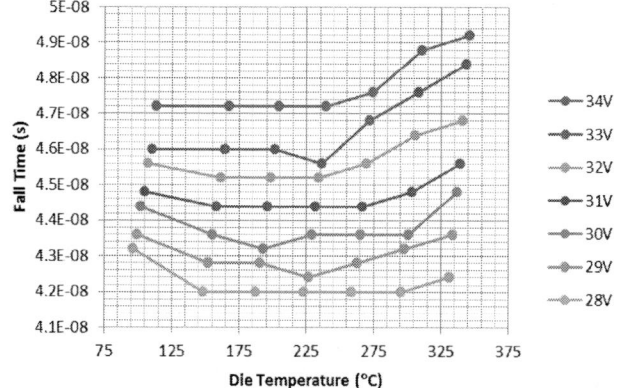

Fig. 6. Measured fall time over temperature (VDD-OD = 28V-34V, other supplies voltages = 20V, single powerMOS load).

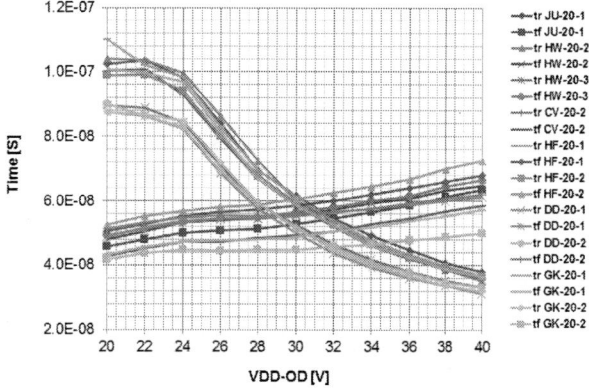

Fig. 4. Measured rise and fall times vs. VDD-OD for 10 SiC gate driver units (other supply voltages = 20V, single powerMOS load).

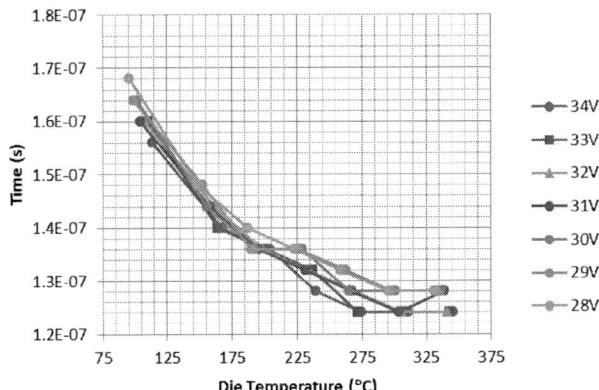

Fig. 7. Measured delay1 (low to high transition) over temperature (VDD-OD = 28V-34V, all other supplies voltages = 20V, single powerMOS load).

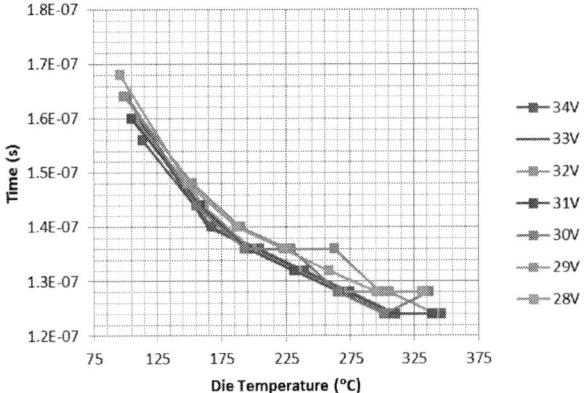

Fig. 8. Measured delay2 (high to low transition) over temperature (VDD-OD = 28V-34V, other supplies voltages = 20V, single powerMOS load).

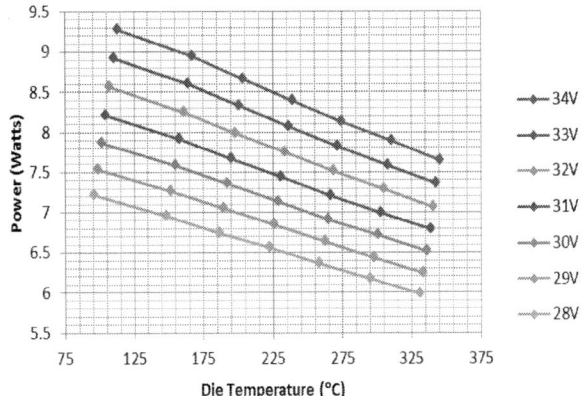

Fig. 9. Measured power consumption over temperature (VDD-OD = 28V-34V, all other supplies voltages = 20V, single powerMOS load).

IV. CONCLUSIONS

In this work an integrated SiC gate driver designed to operate at moderate overdrive voltages was presented. Specifics of the design, layout, and testing were summarized

both at room temperature and over a die temperature range of 90°C to 340°C. An automated test system was developed and implemented allowing for voltage sweeps to be performed to evaluate the effect of supply voltages on dynamic performance. Overdriving of the output stage pull-up device through the use of higher VDD-OD values produced a significant decrease in rise time, as expected. Conversely, overdriving was shown to produce an undesired increase in the fall time due to reduced turn-off ability of the output stage pull-up device at high levels of overdrive. At room temperature, a VDD-OD of ~31V produced similar rise and fall times of ~50ns and nearly matched propagation delays of ~165ns, while consuming ~8.2W.

Two gate driver units were fully characterized over a die temperature range of 90°C to 340°C while driving a powerMOS load. With VDD-OD=31V, the units produced rise and fall times of ~45ns, delays ranging from ~165ns to ~125ns, and power consumption of ~8.2W to ~6.8W. The associated variation in the rise times, fall times, and delays were 6.3%, 2.7%, and 22%, respectively.

This design performed significantly better than its predecessor due to improvements in the overall topology, improved device sizing, and separate packaging of the output stage pull-up device enabling elimination of the body effect. The results of this work support the feasibility of a single die-integrated gate driver and powerMOS device.

ACKNOWLEDGMENTS

This project was supported as part of ARPAe Grant No. DE-AR0000111, focused on a high power density bidirectional charger for plug-in electric vehicle applications.

REFERENCES

[1] J. B. Casady, R. W. Johnson, "Status of silicon carbide (SiC) as a wide-bandgap semiconductor for high-temperature applications: A review," *Solid-State Electronics*, vol. 39, issue 10, pp. 1409-1422, October 1996.

[2] J. A. Valle-Mayorga, A. Rahman, and H. A. Mantooth, "A SiC NMOS Linear Voltage Regulator for High-Temperature Applications," *IEEE Trans. Power Electron.*, vol. 29, no. 5, pp. 2321–2328, 2014.

[3] M. Glover, P. Shepherd, M. Francis, M. Mudholkar, H. A. Mantooth, M. Ericson, S. Frank, C. Britton, L. Marlino, T. McNutt, A. Barkley, B. Whitaker, and A. Lostetter, "A UVLO circuit in SiC compatible with power MOSFET integration," *IEEE Journal of Emerging and Selected Topics in Power Electronics*, vol. 2, no. 3, pp. 425–433, Sep. 2014.

[4] Ericson, N.; Frank, S.; Britton, C.; Marlino, L.; Sei-Hyung Ryu; Grider, D.; Mantooth, A.; Francis, M.; Lamichhane, R.; Mudholkar, M.; Shepherd, P.; Glover, M.; Valle-Mayorga, J.; McNutt, T.; Barkley, A.; Whitaker, B.; Cole, Z.; Passmore, B.; Lostetter, A., "A 4H silicon carbide gate buffer for integrated power systems," *IEEE Trans. Power Electron.*, vol. 29, no. 2, p. 539–542, Feb. 2014.

[5] M. Mudholkar, S. Ahmed, M. N. Ericson, S. S. Frank, C. L. Britton, and H. A. Mantooth, "Datasheet Driven Silicon Carbide Power MOSFET Model," *IEEE Trans. Power Electron.*, vol. 29, no. 5, pp. 2220–2228, 2014.

Understanding the Influence of Dead-time on GaN Based Synchronous Boost Converter

Di Han, *Student Member, IEEE*, Bulent Sarlioglu[1], *Senior Member, IEEE*
Wisconsin Electric Machines and Power Electronics Consortium (WEMPEC)
University of Wisconsin–Madison
Madison, WI, USA
[1]bulent@engr.wisc.edu

Abstract—Gallium Nitride (GaN) based power switching devices are known to be superior to conventional Si devices due to properties such as low loss and fast switching speed. However, as a new technology, some special characteristics of GaN-based power devices still remain unknown to the public. This paper tries to investigate the effect of dead-time on a GaN-based synchronous boost converter, as compared to its Si counterpart. It is found out that GaN-based converter is more sensitive to dead-time related loss as a result of fast switching transition and high reverse voltage drop. Improper selection of dead-time value can offset its advantage over Si converter. Analyses also show that GaN HEMTs have different switching characteristics compared to Si MOSFET due to low C_{gd} to C_{ds} ratio and lack of reverse recovery. These critical findings will help power electronic engineers take better advantage of GaN technology in synchronous rectification and inverter applications.

Keywords—dead-time; gallium nitride power switching devices; synchronous boost converter

I. INTRODUCTION

Despite the tremendous development in power electronics technology in the past several decades, the trends towards higher power, more efficient, lighter weight, and smaller size power electronic converters in various applications have never stopped. The improvements in semiconductor switching devices have always been one of the primary sources of momentum to push the technology forward. Due to the fact that the state-of-the-art Si based power devices are approaching their performance limits set by the physical properties of Si material, a lot of attention is paid to wide band gap (WBG) materials, such as silicon carbide (SiC) and gallium nitride (GaN), in recent years [1]-[3]. With their superior properties, WBG material-based power switching devices are predicted as the enabling technology for next generation power electronics [4]-[6]. As the WBG devices are maturing more than ever before, there is an urgent need for understanding the special characteristics of WBG devices in switching converter applications [7]-[11]. This paper will be devoted to studying the dead-time effect in a GaN-based synchronous boost converter (see Fig. 1) and comparing it with the conventional Si converter to show different characteristics of GaN devices.

Superior physical properties of GaN materials include lower intrinsic carrier concentration, higher electric breakdown field, and larger saturated electron drift velocity when compared to silicon. These properties lead to larger voltage

TABLE I. CHARACTERISTIC PARAMETERS OF GAN FET AND SI MOSFET USED IN THE STUDY

Part number	EPC2001	RJK1053DPB
Voltage rating V_{DS}	100 V	100 V
Current rating I_D (Continuous)	25 A	25 A
On-state resistance R_{DSon} (Max)	7 mΩ	13 mΩ
Total gate charge Q_g (Max)	8 nC	43 nC
Gate to source charge Q_{gs}	2.3 nC	19 nC
Gate to drain charge Q_{gd}	2.2 nC	12.5 nC
Input capacitance C_{iss} (@ V_{DS}=50V)	850 pF	6050 pF
Output capacitance C_{oss} (@ V_{DS}=50V)	450 pF	220 pF
Reverse transfer capacitance C_{rss} (@ V_{DS}=50V)	20 pF	93 pF
Source-drain forward voltage (@ I_D= −10A, V_{gs}=0V)	2.2 V	0.75 V

ratings, higher temperature capability, faster switching frequency and transients, and lower semiconductor losses of GaN-based power switching devices.

As an example, Table I compares some characteristics of a GaN HEMT (EPC2001) and a Si MOSFET (RJK1053DPB). The two devices have the same voltage and current rating, but the GaN HEMT has only half the on-state resistance and one-fifth the gate charge compared to the Si MOSFET. The low on-state resistance guarantees the low conduction loss, and the small gate charge results in the fast switching transition and low switching loss of the GaN device. Hence, GaN-based power converters feature low loss and high efficiency, and they can be operated at very high switching frequencies to increase the power density, as has been reported in a lot of relevant literature [12]-[14]. However, as a new technology, many special characteristics of GaN devices still remain unknown to the public and need to be studied and understood in a timely fashion. With this purpose, this paper will compare and analyze the influences of dead-times on GaN-based and Si-based synchronous boost converters.

The dead-time effects on GaN converters and Si converters are different due to different reverse conduction capabilities and parasitic capacitance values of GaN HEMTs and Si MOSFETs, as will be analyzed in detail later in this paper. The write-up is organized in the following way. Section II will provide thorough theoretical analysis of the dead-time effects on GaN devices based on device data and the switching model. In Section III, parametric study will be carried out in circuit

978-1-4799-5494-0/14 $31.00 © 2014 IEEE

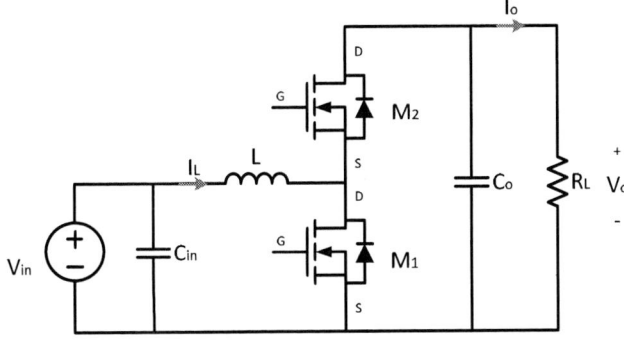

Fig. 1. Synchronous boost converter topology under study

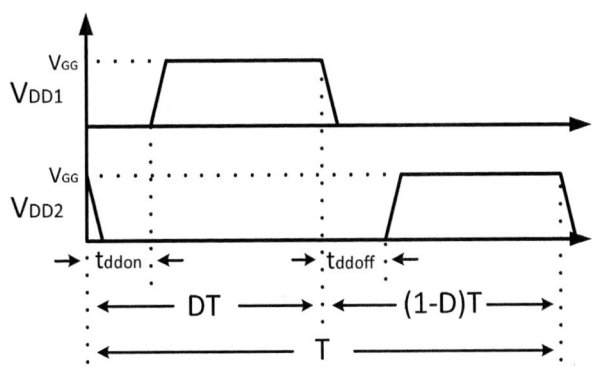

Fig. 2. Timing sequence of gate driver voltages

simulation to compare the semiconductor loss and switching behavior of GaN and Si devices with respect to different dead-times used. A brief conclusion will be drawn in Section IV.

II. THEORETICAL ANALYSIS

A. Dead-Time Definition

The converter topology used in this study is a synchronous boost converter as shown in Fig. 1. The lower switch M_1 is the main switch, also called the control switch. The upper switch M_2 turns on complementarily with M_1 to provide a low resistance path for the inductor current to flow to load when M_1 is off. Hence, M_2 is called a synchronous switch. To prevent the two switches from turning on simultaneously, which will short-circuit the power source V_{in}, safety margins are needed between the on-state gate voltages of the two switches, known as dead-times.

The timing sequence of gate drive output voltages of M_1 and M_2 (V_{DD1} and V_{DD2}) are illustrated in Fig. 2. As can be seen, the rise and fall times of gate driver voltages are also taken into consideration. Two dead-times are inserted into one switching cycle T. The dead-time right before the turn-on of control switch M_1 is called turn-on dead-time (t_{ddon}) and defined as the time period from when V_{DD2} starts to fall to when V_{DD1} starts to rise. Similarly, the dead-time right after the turn-off of M_1 is called turn-off dead-time (t_{ddoff}) and defined as the period from when V_{DD1} starts to fall to when V_{DD2} starts to rise.

Special attention should be paid to the fact that the definitions of dead-times used here are based on the output voltages of gate drivers, which are distinguished from the dead-times based on digital gating signals. The driver voltage based definitions are preferred since they are directly related to the actual switching behaviors of devices, and thus are more useful than the signal based definitions.

Based on the definition above, dead-time values can also become negative numbers, which means that the two gate driver output voltages overlap with each other. Although negative dead-times are not desired, it can happen due to the mismatch of propagation delay or even noises on the two gating channels, especially when dead-time is selected

aggressively. Hence, the effect of negative dead-times will also be included in the later sections.

B. Reverse Conduction Capability

In a synchronous boost converter, the synchronous switch M_2 only conducts current in the reverse direction. When gate voltage V_{DD2} is applied, M_2 is turned on and can conduct reverse current just like in the positive direction, presenting itself as an on-state resistance. When no gate voltages are applied, i.e. during dead-times, both M_1 and M_2 are turned off, but current still has to flow through M_2 in the reverse direction. During this time, the behaviors of Si MOSFET and GaN HEMT are completely different.

As is well known, MOSFET devices have a parasitic diode, which consists of a p-n junction that resides in the device structure and appears to be anti-parallel connected to the MOSFET. This parasitic diode is often referred as a "body diode," which is responsible for the reverse current conduction when MOSFET is off. Since the body diode is p-n junction-based, i.e. a minority carrier diode, reverse recovery effect can be observed during diode turn-off at the end of a dead-time period.

In contrast, GaN HEMTs do not possess a parasitic p-n junction structure; hence they have no physical diode inside the device. However, the transistor channel is able to conduct reverse current at no gate voltage when the drain-to-source is reverse-biased. This reverse conduction capability is called "majority carrier body diode" by the GaN device manufacturer EPC. Hence, the term "body diode" will also be used for GaN HEMT in this paper to simplify the terminology.

The body diode of GaN HEMT is different from a Si MOSFET body diode in two aspects. First, since it is a majority carrier-based conduction channel, the GaN diode does not have any reverse recovery effect. Hence, it will induce less switching loss during M_1 turn-on compared to the Si counterpart. Second, the GaN diode has considerably larger voltage drop than the Si diode as shown in Table I. The source-to-drain voltage drop of GaN device at 10 A reverse current is more than three times that of the Si device. This indicates higher conduction loss of GaN HEMT than Si MOSFET during dead-time periods. As a result, using too long dead-

TABLE II. SPECIFICATIONS OF SYNCHRONOUS BOOST CONVERTER UNDER STUDY

Specifications	Value
Input voltage V_{in} (V)	24
Output voltage V_o (V)	80
Power rating P (W)	160
Switching frequency f (kHz)	400
Choke inductor L (μH)	20
Input capacitor C_{in} (μF)	10
Output capacitor C_o (μF)	60

times will significantly increase the loss on GaN converter, and counteract the low loss advantage of GaN devices.

C. Switching Characteristics

A common solution to reduce the excessive diode conduction losses is to reduce the length of dead-times. However, if too short dead-times are used, one switch will start to turn-on before the other one is fully turned off. That is, the switching transitions of the two switches will interact with each other, leading to not only loss but also current spikes that are harmful to the device. In the worst case, this interaction will result in shoot-through problems.

As can be observed in Table I, the gate-to-source charge Q_{gs} and input capacitance C_{iss} of GaN HEMT is one-seventh to one-eighth of those of Si MOSFET. This is an advantage of GaN device in that its switching transition is shorter and generates less switching loss than Si device. However, the fast switching speed also makes GaN device more vulnerable to short dead-time problems. Several nanoseconds' difference in dead-time can fully turn on one GaN HEMT to create a destructive shoot-through, but can only partially turn on a Si MOSFET, whose high resistance will limit the short circuit current and help limit the resulted loss and damage.

III. CIRCUIT SIMULATION AND RESULTS

In order to validate the comparative analysis given in the last section, circuit simulation is carried out in LTSpice. By parametrically varying the dead-times values, the behavior of a GaN based converter and a Si based converter is compared.

A. Simulation Considerations

In the simulation, the same two devices shown in Table I, EPC 2001 and RJK1053DPB, are used under comparison. The SPICE models of the two devices are directly obtained from their manufacturers. The specifications of the synchronous boost converter under study are listed in Table II. The gate drive circuitry is realized by two programmable independent voltage sources whose output voltage rise time of 7 ns and fall time of 1.5 ns are also incorporated. The external gate resistances applied to the two devices are both 1 ohm. In addition, losses on passive components in the converter are considered by adding equivalent series resistances (ESR) to enhance fidelity of the simulation.

To evaluate the effects of turn-on dead-time t_{ddon} and turn-off dead-time t_{ddoff} separately, two sets of simulation are carried out. First, t_{ddon} is fixed at 0 ns while t_{ddoff} is increased from 0 ns to 30 ns with 5 ns increments. Then, t_{ddoff} is fixed at 10 ns

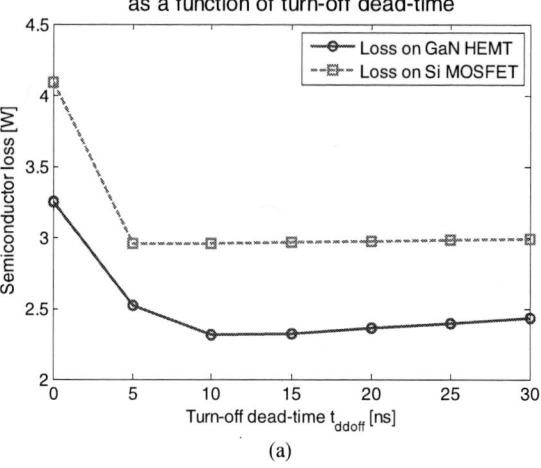

(a)

(b)

Fig. 3. Loss comparison on power devices as a function of (a) turn-off dead-time t_{ddoff} (b) turn-on dead-time t_{ddon}

while t_{ddoff} is varied from -10 ns to 20 ns with 5 ns increments. Again, the negative dead-time values indicate that on-state gate voltages of two switches overlap with each other. These values are chosen based on the observation that the optimal values t_{opt} for t_{ddon} and t_{ddoff} are around 0 ns and 10 ns, respectively, for the given converter operation conditions. Hence, the simulated ranges of dead-times are determined to be from ($t_{opt}-10$) ns to ($t_{opt}+20$) ns.

For each simulation case, the total semiconductor losses on two switches (M_1 and M_2) as well as some switching waveforms are recorded for both the GaN converter and Si converter. Simulation results and comparison will be given in the following parts of the paper.

B. Total Semiconductor Losses Versus Dead-Times

The simulated total semiconductor losses in GaN HEMT- and Si MOSFET-based converters are plotted in Fig. 3 as a function of different dead-times. This total semiconductor loss includes all the losses on two power switches in a converter, i.e. conduction loss, switching loss, and dead-time related loss.

As can be observed from Fig. 3(a), for all the turn-off dead-time values simulated, the loss on GaN devices is always smaller than that on Si devices, thanks to the low on-state resistance and fast switching transition of GaN HEMTs. However, more importantly, the loss on GaN HEMTs increases faster than that on Si MOSFET as the turn-off dead-time increases above the optimal value of 10 ns. For example, as t_{ddoff} is increased from 10 ns to 30 ns, the loss on GaN device increases by 0.117 W while the loss on Si device only increases by 0.027 W. This is due to the high reverse conduction loss of GaN HEMT as explained earlier. It can be expected from the trend that as turn-off dead-time further increases, GaN devices loss will be equal or even larger than that of Si devices, and completely compromises its advantages on the loss aspect.

To analyze Fig. 3(b), the simulated range of turn-on dead-time (-10 ns to 20 ns) can be divided into three segments: -10 ns to 0 ns, 0 ns to 10 ns, and 10 ns to 20 ns. The third segment (10 ns to 20 ns) shows the same phenomena described above: loss on GaN increase faster than on Si as turn-on dead-time increases due to the high voltage drop of GaN body diode.

The second segment shows very different phenomena as compared to the third segment because of the reverse recovery of Si body diode. When turn-on dead-time is less than 10 ns, the Si body diode of upper switch M_2 cannot fully turn on during the dead-time. The minority charge is not well established, and the reverse recovery charge and related loss during diode turn-off is reduced. The longer the dead-time, the more minority charge is accumulated, and the larger loss will be resulted as diode turns off. Hence, a fast increase in Si loss is observed when turn-on dead-time increases from 0 to 10 ns. But the GaN diode is majority carrier based channel and does not have reverse recovery effect; hence, the loss increase on GaN in this segment is still caused by diode conduction.

The first segment contains all negative dead-time values, which lead to the shoot-through in the converters. As can be seen, loss increases dramatically as the dead-time value goes negative and is too large to be shown in the plot when dead-time further decreases. Also noticeable is that the loss on GaN increases much faster than that of Si when dead-time decreases. For example, when t_{ddon} is decreased from 0 to -5 ns, the loss on Si devices will increase by 0.83 W, and the GaN loss will increase by 8.15 W. This is due to the fast turn-on capability of GaN device, which is able to create a low resistance short circuit path within a few nanoseconds. This observation also correlates well with the prediction made in the last section.

C. Switching Waveforms Versus Different Dead-Times

Some of the phenomena mentioned above can also be observed in switching waveforms of the power devices. The gate to source voltages of two switches v_{gs1}, v_{gs2} and the drain current of the control switch I_{D1} in the GaN converter and the Si converter are plotted in Fig. 4 and Fig. 5 with respect to different dead-time values.

Fig. 4 shows the changing waveforms during M_1 turn-off as turn-off dead-time t_{ddoff} varies. For small t_{ddoff} values, current spikes can be observed on I_{D1}, which indicates the interaction

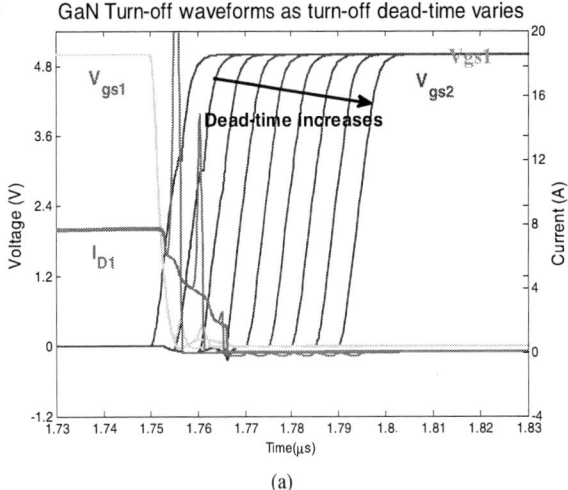

(a)

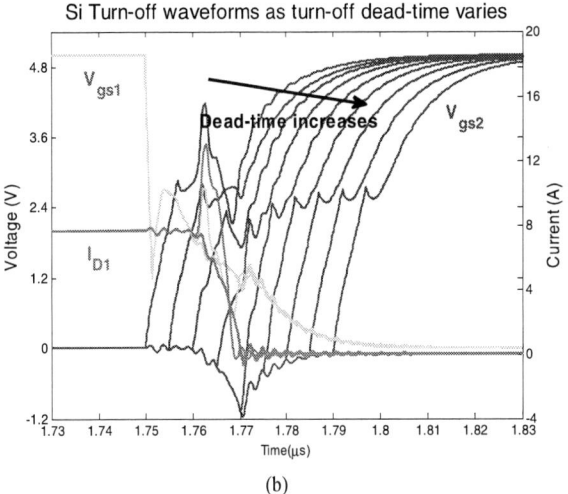

(b)

Fig. 4. Turn-off waveforms when turn-off dead-time varies from 0 ns to 30 ns with 5 ns increments (a) GaN converter (b) Si converter. Green: V_{gs} for control switch M_1. Blue: V_{gs} for synchronous switch M_2. Red: I_D for control switch M_1.

between the upper and lower switches as well as high semiconductor loss. By comparing Fig. 4(a) and 4(b), one can see that current spikes on the GaN device are much higher than those on the Si device. Again, this is because the GaN HEMT can be turned on faster than Si MOSFET. In addition, obvious Miller plateau can be observed on V_{gs1} for Si MOSFET, but not on GaN HEMT. This is due to the low C_{gd}/C_{ds} ratio of the GaN device.

The changing waveforms during M_1 turn-on with respect to different turn-on dead-times t_{ddon} are presented in Fig. 5. Similar to what is shown in Fig. 4, the GaN device has higher current spikes than the Si device due to its fast turn-on speed. However, different from the turn-off waveforms, the spikes always exist on I_{D1} during turn-on despite of the t_{ddon} used, and the causes of large current spikes are different for different segments of t_{ddon} values. The same method of segmentation used for Fig. 3(b) can be applied here again. In the first segment, shoot-through results in large currents.

In the case of Si MOSFET, the spikes are mainly caused by the reverse recovery of body diode in segment 2 and 3. Hence,

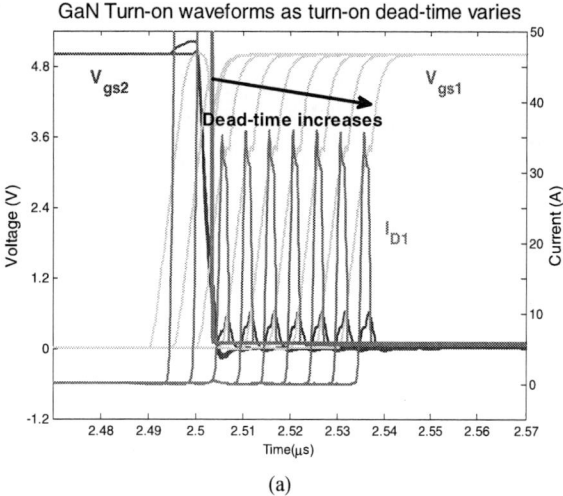

(a)

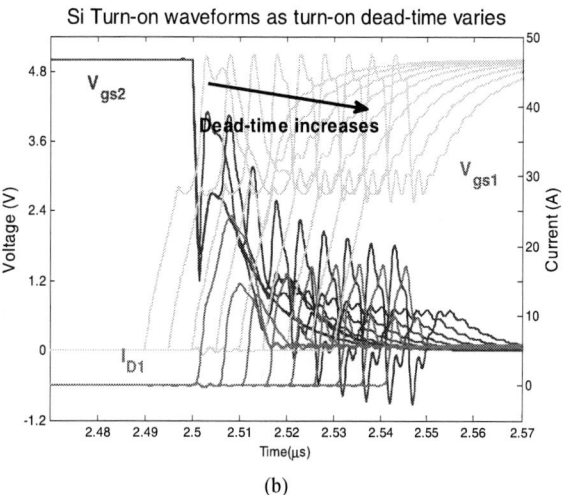

(b)

Fig. 5. Turn-on waveforms when turn-on dead-time varies from -10 ns to 20 ns with 5 ns increments (a) GaN converter (b) Si converter. Green: V_{gs} for control switch M_1. Blue: V_{gs} for synchronous switch M_2. Red: I_D for control switch M_1.

the amplitude of the spikes shows an increasing trend when t_{ddon} increases from 0 ns to 10 ns, which also correlate with the observations made earlier in Fig. 3(b).

In contrast, for GaN HEMT, discharging of the output capacitance of M_2 is the cause of the current spike when t_{ddon} is in second and third segment. As a result, the amplitude of the current peak is almost constant for these t_{ddon} values.

IV. CONCLUSION

In this paper, the different effects of dead-time on GaN-based and Si-based synchronous boost converters are comparatively analyzed from the loss aspect. In addition, as dead-time varies, the trends of change in switching waveforms of GaN and Si devices are also compared.

The study has found that the high reverse voltage drop of GaN HEMT causes high device losses during dead-times. For example, 20 ns increase in dead-time only leads to 0.027 W loss increase in Si devices, but results in 0.117 W loss increase

in GaN devices, which is 4 times larger. This high dead-time related loss of GaN devices can offset its advantage over Si MOSFET if too long dead-time values are used. In addition, GaN device is also more sensitive to shoot-through problems because of its fast turn-on speed. For example, 5 ns overlap in gating voltages of two switches only leads to 0.83 W shoot-through losses on Si devices, but will result in 8.15 W shoot-through losses on GaN devices, which is 10 times higher and destructive.

In order to fully utilize the potential benefits of GaN devices in power electronics applications, optimization of the dead-time values is an important topic for future research.

REFERENCES

[1] A. Elasser and T. P. Chow, "Silicon carbide benefits and advantages for power electronics circuits and systems," *Proc. IEEE*, vol. 90, no. 6, pp. 969- 986, June 2002.

[2] J. Millan, P. Godignon, X. Perpina, A. Perez-Tomas, and J. Rebollo, "A survey of wide bandgap power semiconductor devices," *IEEE Trans. Power Electron.*, vol. 29, no. 5, p. 2155-2163, May 2014.

[3] N. Kaminski, "State of the art and the future of wide band-gap devices," in *Proc. IEEE Power Electron. Appl.*, Sept. 2009, pp. 1–9.

[4] D. Han, J. Noppakunkajorn, and B. Sarlioglu, "Comprehensive efficiency, weight, and volume comparison of SiC and Si-based bidirectional DC-DC converters for hybrid electric vehicles," *IEEE Trans. Veh. Technol.*, no. 99, pp. 1-1, 2014.

[5] D. Han, A. Ogale, S. Li, Y. Li, and B. Sarlioglu, "Efficiency characterization and thermal study of GaN based 1 kW inverter," in *Proc. IEEE Appl. Power Electron. Conf.*, March 2014, pp. 2344-2350.

[6] D. Han, J. Noppakunkajorn, B. Sarlioglu, "Analysis of a SiC three-phase voltage source inverter under various current and power factor operations," in *Proc. Ind. Electron. Soc. Annu. Conf.*, Nov. 2013, pp. 447-452.

[7] M. Esposto, A. Chini, and S. Rajan, "Analytical Model for Power Switching GaN-Based HEMT Design," *IEEE Trans. Electron Devices*, vol. 58, no. 5, pp. 1456-1461, May 2011.

[8] X. Huang, Z. Liu, Q. Li, and F. C. Lee, "Evaluation and application of 600 V GaN HEMT in cascode structure," *IEEE Trans. Power Electron.*, vol. 29, no. 5, pp. 2453-2461, May 2014.

[9] D. Reusch, and J. Strydom, "Understanding the effect of PCB layout on circuit performance in a high-frequency gallium-nitride-based point of load converter," *IEEE Trans. Power Electron.*, vol. 29, no. 4, pp. 2008-2015, Apr. 2014.

[10] J. Noppakunkajorn, Di Han, and B. Sarlioglu, "Analysis of high-speed PCB with SiC devices by investigating turn-off overvoltage and interconnection inductances influence," in *Proc. IEEE Appl. Power Electron. Conf. Expo.*, March 2014, pp. 2628-2631.

[11] D. Han, W. Lee, J. Noppakunkajorn, and B. Sarlioglu, "Investigating the influence of interconnection parasitic inductance on the performance of SiC based DC-DC converters in hybrid vehicles," in *Proc. IEEE Transp. Elect. Conf. Expo.*, June 2014, pp. 1-7.

[12] Y.-F. Wu, M. Jacob-Mitos, M. L. Moore, and S. Heikman, "A 97.8% dfficient GaN HEMT boost converter with 300-W output power at 1 MHz," *IEEE Electron Device Lett.*, vol. 29, no. 8, pp. 824-826, Aug. 2008.

[13] M. Rodriguez, Y. Zhang, and D. Maksimovic, "High-frequency PWM buck converters using GaN-on-SiC HEMTs," *IEEE Trans. Power Electron.*, vol. 29, no. 5, pp. 2462-2473, May 2014.

[14] T. LaBella, J.-S. J. Lai, "A hybrid resonant converter utilizing a bidirectional GaN AC switch for high-efficiency PV applications," *IEEE Trans. Ind. Appl.*, no. 99, pp. 1-1, 2014.

Discussions on the Semiconductor-based Galvanic Isolation

Xuan Zhang, Lixing Fu, Mingzhi Leng, and Jin Wang

Department of Electrical and Computer Engineering
The Ohio State University
Columbus, Ohio – United States of America
wang@ece.osu.edu

Abstract— This paper presents the discussions on a semiconductor-based solution for galvanic isolation. It is a switched-capacitor (SC) isolated dc/dc converter based on the SC isolation cell structure, which aims to utilize the semiconductor switches to sustain isolation voltage and block common-mode (CM) leakage current. The converter delivers an active power with ZCS operation. The circuit structure and operation principle are introduced. To achieve high efficiency, the converter requires high voltage rating, low output capacitance, and low on-resistance of the switches, which makes the SiC devices the best choice, as shown in a comparison between the state-of-the-art Si and SiC devices. To validate the galvanic isolation, the isolation voltage and touch current (TC) must be evaluated according to the industrial safety standards such as IEC60950-1. Series connection of the SC isolation cells contributes to higher isolation voltage of the. It is also expected to provide smaller equivalent series capacitance which has higher impedance to block the line-frequency CM leakage current. However, study shows that the charging transients of the switch output capacitance results in a CM leakage current. This issue is verified by the test results from a prototype made of 2 SC isolation cells in series. This CM leakage current remains as a challenge for this isolation solution.

I. INTRODUCTION

In electric systems where two sections of the system have different ground potentials, galvanic isolation is required in between to provide high impedance and block common-mode leakage current. Traditionally, this is done by using a transformer. However, the magnetic core and multiple windings of the transformer contributes to manufacture complexity and cost. Moreover, for isolated dc/dc conversion, inverting circuit and rectifying circuit are necessary on the two sides of the transformer.

As alternative approaches to provide galvanic isolation, inductive coupling, capacitive coupling, and microwave coupling have been introduced for different applications and power levels. Inductive coupling is essentially using an air-core transformer to transfer power wirelessly, and it is used in isolated power conversion [1][2], inductive heating [3][4],

and wireless power transfer applications [5][6]. In capacitive coupling, a LC resonant tank replaces the transformer, utilizing the Y-class capacitor to provide high impedance at low frequency, but deliver energy at the high resonant frequency [7]-[9]. This approach has been applied in LED driver application [8] and communication applications [9]. In microwave coupling, wireless power transmission is realized using an electromagnetic resonant coupler, for isolated gate-drive application [10].

For isolated dc/dc power conversion, the switched-capacitor (SC) isolation cell was introduced in [11][12], which utilizes semiconductor switches to provide galvanic isolation. This solution aims to eliminate magnetic components, and provide simple structure, easy control, and high efficiency. However, in the literatures, the isolation capability of the SC isolation cell is not evaluated. To validate the galvanic isolation, the cell must meet the industrial safety standards such as IEC60950-1 which specifies the isolation voltage rating and touch current (TC). For example, the isolation voltage should be no less than 3-kV rms ac for applications with 600-V working voltage. These standard requirements call for high voltage rating, low output capacitance, and low on-resistance of the switches, which is beyond what the state-of-the-art Silicon (Si) devices can provide.

In this paper: 1) an isolated dc/dc converter based on SC isolation cell is introduced. The circuit descriptions and operation principles are presented; 2) a comparison between the state-of-the-art Si and SiC devices shows that the SiC device is the best choice to achieve good performance; 3) based on the industrial safety standard IEC60950-1, the evaluation methods for the converter's isolation voltage and TC are discussed; and 5) an isolated dc/dc converter is built using 1.7-kV SiC MOSFETs. The converter is made of 2 SC isolation cells connected in series. The preliminary test results of the TC is presented, and the current loop is analyzed. The CM leakage current remains as a challenge for this isolation solution.

II. CIRCUIT DESCRIPTIONS AND POWER DELIVERY PRINCIPLES

A. SC Isolation Cell Descriptions and Operation Principles

The circuit structure of the SC isolation cell is shown in Fig. 1 (a), and the operation waveforms are shown in Fig. 1 (b). The SC isolation cell has ZCS on and off as a result of the resonance between L_{s1}~L_{s4} and C_1~C_3, so the switching loss can be reduced. The two resonant loops of the two switching modes are signified in Fig. 1 (a). Because there exists the on-resistance of the switches, the parasitic resistance of the PCB traces and capacitors, the resonant circuit is in fact a series RLC circuit. The switch currents under different damping factors are shown in Fig. 1 (c).

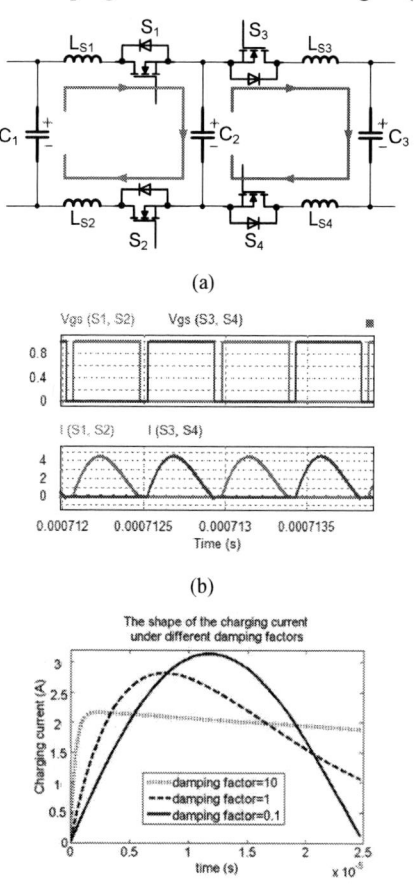

(a)

(b)

(c)

Figure 1. (a) SC isolation cell topology; (b) operation waveform of the SC isolation cell: (c) switch currents under different damping factors.

The resonance frequency (f_0) and the damping factor (δ) of the circuit are

$$\begin{cases} f_0 = \dfrac{1}{2\pi\sqrt{LC}} \\ \delta = \dfrac{R}{2}\sqrt{\dfrac{C}{L}} \end{cases}, \quad (1)$$

where R is the total series parasitic resistance in the loop; L is the total series parasitic inductance in the loop; and C is the equivalent series capacitance in the loop.

The conduction loss is caused by the switch on-resistance and the parasitic resistance of the PCB trace and the capacitors. The switching loss is the loss of the switch output capacitance during turn-on transients and the gate-drive loss.

B. The SC Isolation Converter

The proposed SC isolated dc/dc converter is shown in Fig. 2 (a). The converter is composed of several SC isolation cells connected in series. These cells take turns to operate and deliver energy to the next one. The enable signals of the cells are shown in Fig. 2 (b). This structure and control strategy enables a higher isolation voltage. The SC isolation cell topology and its operation waveforms are shown in Fig. 2. In the cell, S_1~S_4 are the switches; L_{s1}~L_{s4} are the parasitic inductance; and C_1~C_3 are the switched capacitors. When a cell is not enabled, it has the same isolation voltage rating as the switch's voltage rating. When it is enabled and starts to operate, it loses its isolation voltage blocking capability because its switch body diodes would conduct. However, out of the n series-connected cells, there is only one cell operating at a time, so the other (n-1) cells connected will continue to provide the isolation voltage blocking capability.

The number n needed in the converter depends on the voltage rating of the switches. For example, to meet the 3-kV$_{rms}$ ac isolation voltage requirement from the standard IEC60950-1, if 1.2-kV rated switches are used, the converter needs 5 cells in series, because the 4 cells which are not enabled can provide a 4.8-kV isolation voltage rating; If 4.5-kV rated switches are used, the converter needs 2 cells in series, since the 1 cell which is not enabled can provide a 4.5-kV isolation voltage rating.

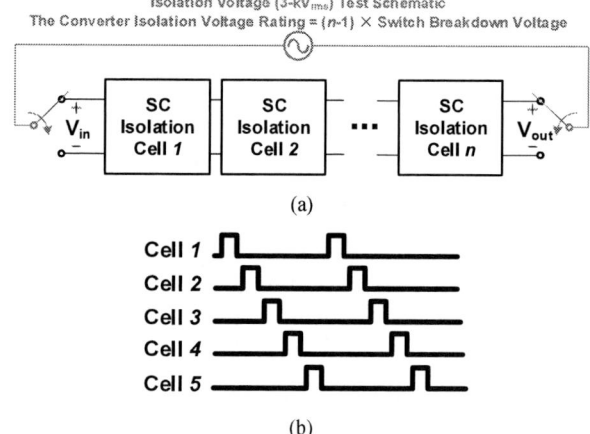

(a)

(b)

Figure 2. (a) Proposed SC isolated dc/dc converter and the insulation test schematic; (b) enable signals of the SC isolation cells (n=5 for example).

C. Design Considerations

The SC isolation cell must perform good efficiency. In order to reduce the switching loss, lower switching frequency and lower switch output capacitance are preferred. In order to reduce the conduction loss, the RMS current of the switches must be minimized, so a smaller damping factor is preferred. As shown in (1), this requires low R and an optimized combination of C and L. To achieve a compact design, it is preferred to eliminate extra inductors, so the stray inductance of the PCB trace could be used as L. Meanwhile, ceramic

Part Number	Material	Continuous Drain Current @ 25 C	Drain-Source Breakdown Voltage	Drain-Source On-State Resistance (R_{ds_on})	Gate Capacitance (C_{gs}) @ V_{ds}=200 V	Output Capacitance (C_{oss}) @ V_{ds}=200 V
C2M0025120D	SiC	90 A	1200 V	0.025 Ω	2770 pF	350 pF
C2M1000170D	SiC	4.9 A	1700 V	1.1 Ω	191 pF	20 pF
IXFX26N120P	Si	20 A	1200 V	0.46 Ω	16000 pF	<735 pF
WPH4003	Si	2.5 A	1700 V	8.2 Ω	850 pF	<70 pF
IXTK5N250	Si	5 A	2500 V	8.8 Ω	8560 pF	<250 pF
IXTL2N450	Si	2 A	4500 V	20 Ω	6860 pF	<200 pF

TABLE II. COMPARISON BETWEEN THE STATE-OF-THE-ART SI AND SIC DEVICES FOR THE SC ISOLATION CONVERTER

capacitors are selected for C_1~C_3 because of their low ESR and low profile. Ceramic capacitors rated at 450 V or higher generally has capacitance no more than 2.2 µF, so there is not much freedom in the selection of C. As results, low on-resistance of the switches is the most critical parameter, because it enables the use of the stray inductance of PCB trace, and also limits the switching frequency.

A comparison between the state-of-the-art Si and SiC MOSFETs is shown in Table I. It shows that: 1) at the same voltage rating, SiC MOSFETs have much smaller on-resistance and therefore have a lower conduction loss; 2) at the same voltage rating, SiC MOSFETs have smaller output capacitance and gate capacitance which leads to lower switching loss. As results, SiC MOSFETs are the best choices for the proposed solution.

III. TOUCH CURRENT EVALUATION

A. Industrial safety standards requirements

The TC path and measurement network is shown in Fig. 3. The TC limitation is specified by many standards for different equipment, such as the IEC 60950-1. If the output GND is not earthed (such as Class II equipment) and the dc input of the SC isolated converter is the output of the front-end power factor correction (PFC) stage, when the diode of the input bridge is on, the ac input is applied to the body-impedance network (Z_{TC}) via the SC isolation converter. Thus, the impedance of the SC isolation converter is the dominant one in the TC path.

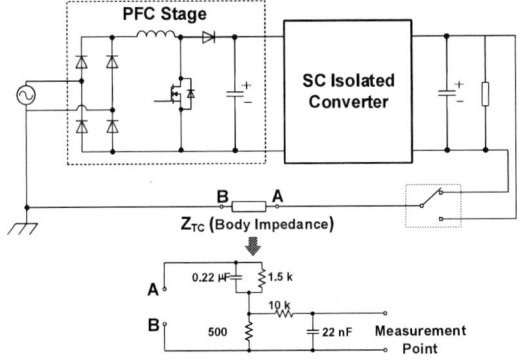

Figure 3. TC path scenario and the measurement network.

The equivalent series capacitance of the SC isolated converter should be specially designed to pass the required TC limitation. As shown in Fig. 1 and Fig. 2, it is expected that the output capacitance of the switches in the SC cells which

are not enabled serve as the series capacitance in the circuit. With the SiC MOSFETs and the series connection of the SC isolation cells applied, the equivalent series capacitance is expected to be very small, which results in a high line-frequency impedance. However, further study shows this is not the case.

B. TC analysis

A case study of the SC isolation converter is conducted to evaluate the TC. The test setup schematic is shown in Fig. 4. Two SC isolation cells, a body-impedance network Z_{TC}, and an external ac voltage source (V_s) are connected in series. The inductance L_{11}~L_{42} are the parasitic inductance, and C_1~C_6 are the dc-link capacitors. The 2 SC cells take turns to operate, which means that only one cell out of the two is operating at a time. When one cell is operating, only one pair of switches on the same half side of the cell is turned on at a time. In sum, the sequence of the switches to be turned on is as follows:

$S_{11}\&S_{12}$ → $S_{21}\&S_{22}$ → $S_{31}\&S_{32}$ → $S_{41}\&S_{42}$ → $S_{11}\&S_{12}$ …

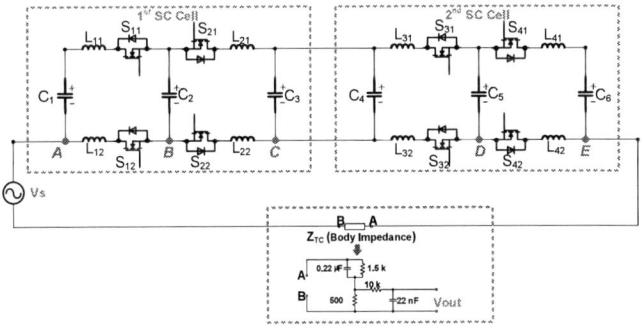

Figure 4. TC measurement setup.

Within a switching cycle, there are 4 switching states. Assume V_s>0, at the turn-on transients, the applied potential on the 5 labeled points A, B, C, D, and E are in Table II.

TABLE II. SWITCH STATUS AND JUNCTION POINT POTENTIALS

State#	Turned-On Switches	Potential of point A	Potential of point B	Potential of point C	Potential of point D	Potential of point E
1	S_{11}, S_{12}	V_s	V_s	V_s	0	0
2	S_{21}, S_{22}	V_s	V_s	V_s	0	0
3	S_{31}, S_{32}	V_s	0	0	0	0
4	S_{41}, S_{42}	V_s	0	0	0	0

At turning-on transients of state 1 and 4, a step voltage is applied on the switch output capacitance (C_{oss}), resulting in a

978-1-4799-5494-0/14 $31.00 © 2014 IEEE

pulse TC (I_{TC}) flowing through the C_{oss} and Z_{TC}. The equivalent circuits for each state at turning-on transients are shown in Fig. 5. The converter's conduction paths for the transients can be divided into 2 parallel branches, which are connected by the dc-link capacitors C_1 and C_6. Since the C_1 and C_6 has much larger capacitance than the switch output capacitance, they are considered shorted during the transients.

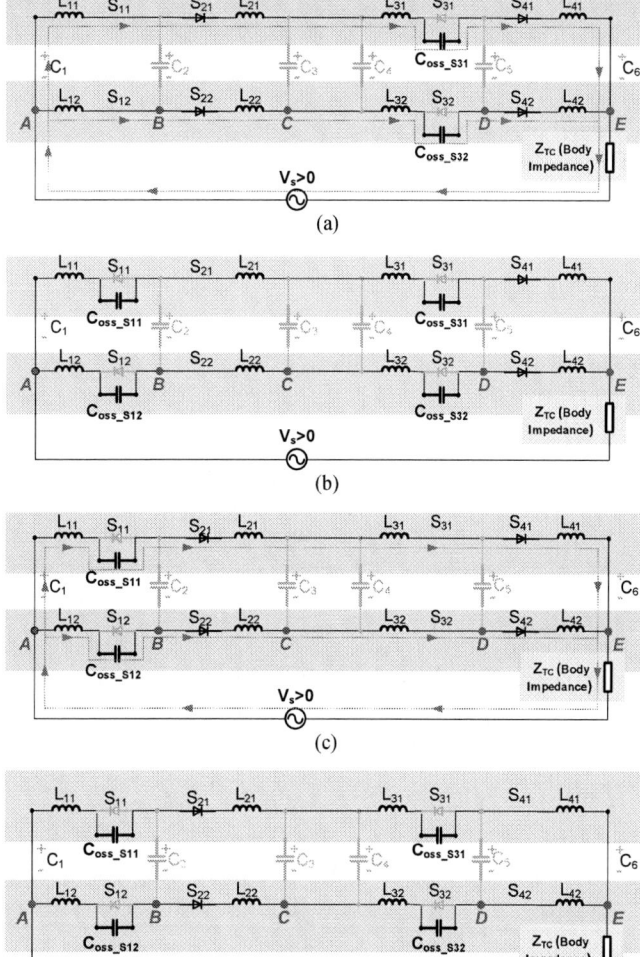

Figure 5. Equivalent circuits during swithcing transients in a swithcing cycle when: (a) S_{11}&S_{12} are turned on; (b) S_{21}&S_{22} are turned on; (c) S_{31}&S_{32} are turned on; (d) S_{41}&S_{42} are turned on.

The equivalent TC loop for each branch of the converter is shown in Fig. 6 (a). It is assumed the total stray inductance (L_{stray}) for each branch is 60 nH, and the 1.7-kV SiC MOSFET (C2M1000170D) is selected as the switch, of which the C_{oss} is about 30 pF. As results, the V_s and TC waveforms are shown in Fig. 6 (b).

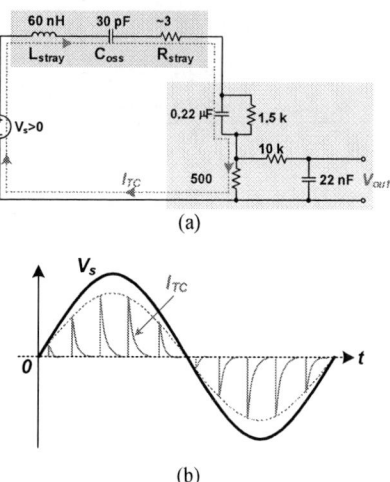

Figure 6. (a) equivalent circuit of one branch of the SC isolation converter and the TC loop; (b) the waveforms of V_s and I_{leak}.

It can be seen from Fig. 6 that even though the C_{oss} is very small, the loop impedance of the equivalent circuit in Fig. 6 (a) for a voltage step change is still small, so a pulse I_{TC} flows through the Z_{TC} repetitively at the switching frequency. This results in an output voltage (V_{out}) at the measurement port of the Z_{TC}. Due to the low-pass filter which is composed of the 10-kΩ resistor and the 22-nF capacitor, the V_{out} waveform is continuous and it has the same frequency as V_s.

IV. PRELIMINARY EXPERIMENTAL RESULTS

A preliminary 400-V/400-V SC isolated converter is built to verify the power-delivery operation and evaluate the TC. Fig. 7 shows a SC isolation cell prototype with the prototype parameters in Table III. The schematic of the cell is shown in Fig. 1 (a).

TABLE III. PROTOTYPE PARAMETERS OF THE SC ISOLATION CELL

V_{in}, V_o	Input and output voltage rating	400 V
C_1, C_3	Input and output capacitors	6.6 μF
C_2	Intermediate capacitor	150 nF
L_{11}~L_{22}	Resonance inductance (PCB stray inductance plus inserted extra inductance)	100 nH
S_{11}~S_{22}	1.7-kV SiC MOSFET	C2M1000170D
f_s	Switching frequency	1 MHz

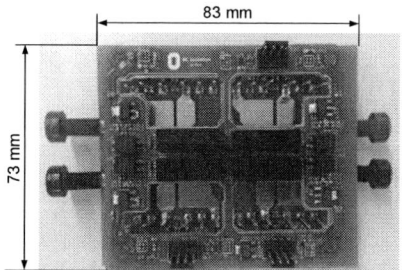

Figure 7. a 400-V/400-V SC isolation cell prototype.

The preliminary experimental waveforms to verify the power-delivery of the SC isolation cell is shown in Fig. 8. It can be seen that the resonant current is achieved, which results in ZCS on and off. More power-delivery experimental results for this SC cell structure were presented in [12].

The TC test setup schematic is shown in Fig. 4. The control scheme is described in the section III. *B*. The TC is supposed to be tested with an applied external ac voltage at line frequency. However, to evaluate the converter's CM impedance at different frequency, a signal generator is used as the external ac voltage source, and 4 frequencies including 60 Hz, 600 Hz, 6 kHz and 60 kHz are tested. The experimental waveforms are presented in Fig. 9.

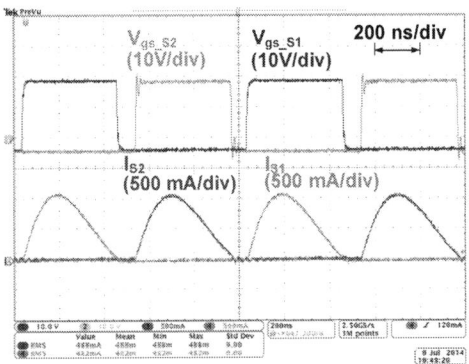

Figure 8. Experimental waveforms of the isolation cell to verify the power delivery.

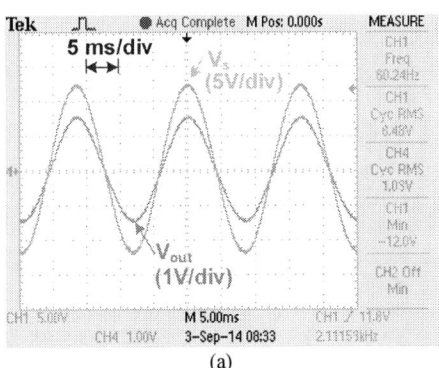

(a)

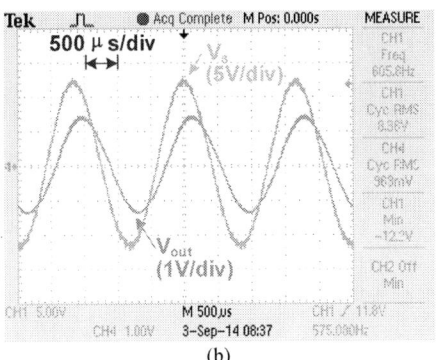

(b)

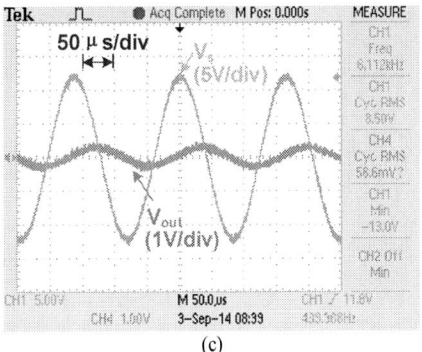

(c)

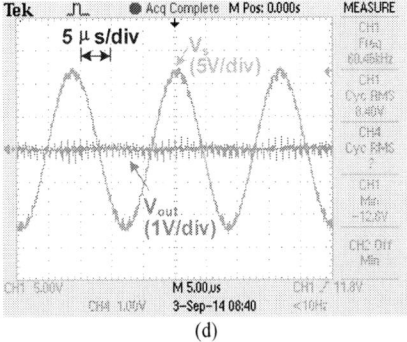

(d)

Figure 9. Experimental waveforms of the isolation cell to evaluate the TC, with an external ac voltage source operating at the frequency of (a) 60 Hz, (b) 600 Hz, (c) 6 kHz, and (d) 60 kHz.

It can be seen from Fig. 9 that: (1) at line frequency (60 Hz), the Z_{TC} has a high V_{out}, which means the SC isolation converter has a low CM impedance at line frequency. This V_{out} is caused by the pulse TC flowing through the Z_{TC} at the switching frequency and accumulating charge in the output low-pass filter, which is composed of the 10-kΩ resistor and the 22-nF capacitor; (2) there exists a phase shift between the V_s and V_{out}, and the phase shift gets larger at higher frequency of V_s. Also, the magnitude of V_{out} gets smaller at higher frequency of V_s. These phase angle and magnitude changes are caused by the output low-pass filter.

V. CONCLUSIONS

This paper presents the discussions on a semiconductor-based solution for galvanic isolation. The circuit can deliver an active power with ZCS on and off. To validate the galvanic isolation, the isolation voltage and touch current (TC) need to be evaluated according to the safety standard IEC60950-1. Study shows that the charging transients of the switch output capacitance results in a CM TC. This issue is verified by the test results from a prototype made of 2 SC isolation cells in series. This CM TC remains as a challenge for this isolation solution.

REFERENCES

[1] S. C. Tang, S. Y. Hui, and H. S.-H. Chung, "A low-profile low-power converter with coreless PCB isolation transformer," *IEEE Trans. Power Electron.*, vol. 16, no. 3, pp. 311-315, May 2001.

[2] H. B. Kotte, R. Ambatipudi, and K. Bertilsson, "High-Speed (MHz) Series Resonant Converter (SRC) Using Multilayered Coreless Printed

Circuit Board (PCB) Step-Down Power Transformer," *IEEE Trans. Power Electron.*, vol. 28, no. 3, pp. 1253-1264, Mar. 2013.

[3] Fairchild, "Induction Heating System Topology Review," [Online]. Available: https://www.fairchildsemi.com/application-notes/AN/AN-9012.pdf

[4] U. Schwarzer and R. W. De Doncker, "Power losses of IGBTs in an inverter prototype for high frequency inductive heating applications," *The 27th Annual Conference of the IEEE Industrial Electronics Society, 2001. (IECON '01).*, vol. 2, pp. 793-798, 2001.

[5] A. P. Sample, D. A. Meyer, and J. R. Smith, "Analysis, Experimental Results, and Range Adaptation of Magnetically Coupled Resonators for Wireless Power Transfer," *IEEE Trans. Ind. Electron.*, vol. 58, no. 2, pp. 544-554, Feb. 2011.

[6] B. L. Cannon, J. F. Hoburg, D. D. Stancil, and S. C. Goldstein, "Magnetic Resonant Coupling As a Potential Means for Wireless Power Transfer to Multiple Small Receivers," *IEEE Trans. Power Electron.*, vol. 24, no. 7, pp. 1819-1825, July 2009.

[7] J. Zhu, M. Xu, J. Sun, and C. Wang, "Novel capacitor-isolated power converter," *Proc. IEEE Energy Convers. Congr. Expo.*, pp. 1824–1828, 2010,.

[8] J. Zhang, J. Wang, and X. Wu, "A Capacitor-Isolated LED Driver With Inherent Current Balance Capability," *IEEE Trans. Ind. Electron.*, vol. 59, no. 4, pp. 1708-1716, Apr. 2012 .

[9] W. Kun and S. Sanders, "Contactless USB — A capacitive power and bidirectional data transfer system," *IEEE Twenty-Ninth Annual Applied Power Electronics Conference and Exposition (APEC)*, pp. 1342-1347, Mar. 2014.

[10]S. Nagai, N. Negoro, T. Fukuda, N. Otsuka, T. Ueda, T. Tanaka, and D. Ueda, "A one-chip isolated gate driver with Drive-by-Microwave technologies," *2012 IEEE International Symposium on Radio-Frequency Integration Technology (RFIT)*, pp. 165-167, Nov. 2012.

[11]P.K. Peter, V. Agarwal, "Analysis and Design Of a Ground Isolated Switched Capacitor DC-DC Converter", *IEEE International symposium on industrial electronics, Bari, Italy*, July 2010, pp. 632-637.

[12]M. J. Scott, K. Zou, E. Inoa, R. Duarte, Y. Huang, and J. Wang, "A Gallium Nitride switched-capacitor power inverter for photovoltaic applications," *IEEE Twenty-Seventh Annual Applied Power Electronics Conference and Exposition (APEC)*, pp. 46-52, Feb. 2012.

Investigation of Driver Circuits for GaN HEMTs in Leaded Packages

Zhan Wang, Jim Honea

Power Electronics Circuit Group

Transphorm Inc.

Goleta, CA, USA
zhanwang@transphormusa.com,
jhonea@transphormusa.com

Yuxiang Shi, Hui Li

Center for Advance Power System

Florida State University

Tallahassee, FL, USA
shi@caps.fsu.edu,
hli@caps.fsu.edu

Abstract— **GaN devices have superior performance over Si-based devices, but even a very small parasitic inductance makes GaN HEMT hard to drive. Many advanced leadless packages are proposed for lower parasitic inductance to fully realize advantages of GaN devices. However, leaded packages are still dominant in industrial applications because of their simplicity for PCB assembly and capability for a wide variety of heat-sinking techniques. In this paper, the performance of a basic half bridge power circuit based on GaN HEMTs with TO package (TO-220) is analyzed, and GaN device driver design is also discussed. Simulation and experimental results are provided for validation.**

Keywords—GaN HEMT; leaded package; driver; ferrite bead

I. INTRODUCTION

Currently widely used in LEDs, Gallium Nitride (GaN) is the next generation in power electronics. The GaN transistor combines low switching and conduction losses, offering reduced energy loss of more than 50 percent compared to conventional silicon-based power conversion designs. Transphorm Inc. has established the industry's first qualified 600-V GaN device platform with its TPH3006PS GaN HEMT. The TO-220-packaged device features R_{ds}(on) of 150 mΩ, Q_{rr} of 56 nC and high-frequency switching capability that enables compact lower cost systems [1].

Two types of TO-220 600-V GaN HEMTs are provided which have Source Kelvin Tab and Drain Tab, respectively. Soldering the tab directly to the PCB is a good way for eliminating the lead parasitic inductance so as to reduce ringing on gate and switching node, but it is not widely used in industry due to the issues of reflow soldering and heat dissipation limitation [2]. For the high power application, low thermal resistance heat sinks are necessary to mount on the tabs of GaN HEMTs, and PCB layout with leads connection should be studied.

II. GATE DRIVER CIRCUIT FOR GAN HEMTS

The GaN device manufactured by Transphorm, Inc. is the cascode structure which incorporates a normally-off low-voltage (LV) Si MOSFET at the input and a normally-on high-

voltage (HV) GaN HEMT at the output, as shown in Fig. 1. The switch is turned on and off by controlling the low voltage MOSFET, so the gate driver circuit is same as high voltage Si MOSFET. Fig. 2 shows a half bridge circuit schematic with non-isolated half-bridge boot-strap driver. By adjusting the gate resistor, the switching speed of LV MOSFET can be controlled, but it has little impact on GaN HEMT switching speed. Slowing down the switching speed helps to reduce the oscillation on the gate and makes the circuit stable. However the switching loss will increase due to the increasing switching time.

The TO-220 package unavoidably adds inductance in the source lead [3], as shown in Fig. 3. This inductance cannot be further reduced, and so its impact must be recognized and mitigated. The normal expectation when a voltage signal develops across a source impedance is that it will subtract from

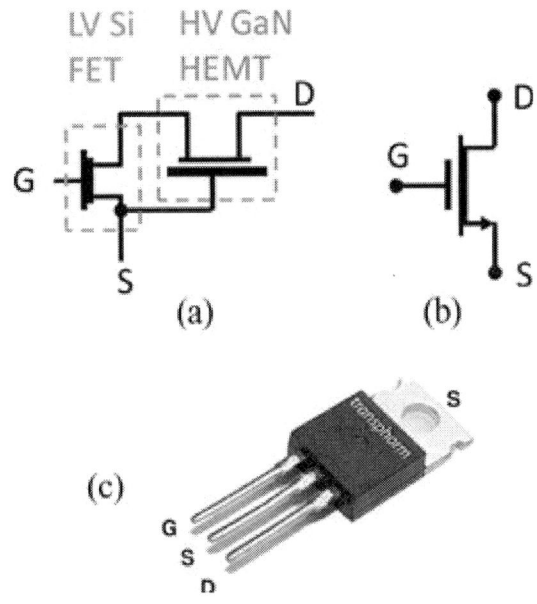

Fig. 1: GaN hybrid HEMT incorporating a LV normally-off Si FET and an HV normally-on GaN HEMT (a) to achieve a combined, normally-off device (b) in a Quiet-Tab TO-220 package (c).

978-1-4799-5494-0/14 $31.00 © 2014 IEEE

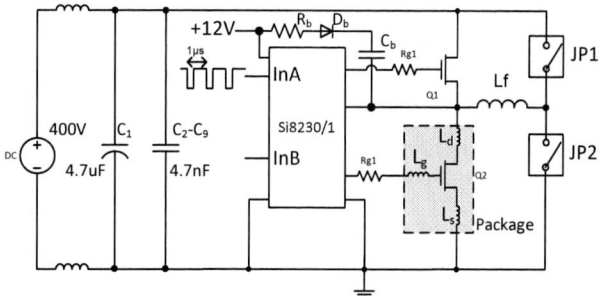

Fig. 2: Half bridge circuit schematic, JP1 is for low side device testing, JP2 is for high side testing.

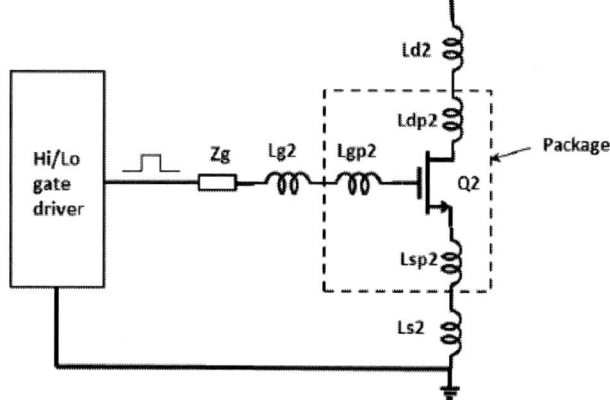

Fig. 3: input loop showing parasitic inductances internal and external to the transistor package

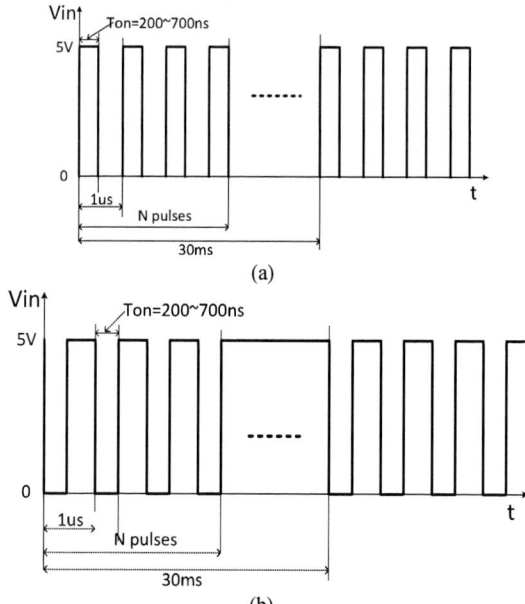

Fig. 4: Driver input signals for multi-pulse testing. (a) High side switching test when JP2 shorted; (b) Low side switching test when JP1 shorted.

the gate-drive voltage and slow the turn on. Slowing the di/dt transition by reducing i_g will indeed reduce the corresponding source voltage during that part of the turn-on transient. Using a gate resistor in the place of Z_g in Fig. 3 would accomplish this. A gate resistor will not, however, provide the additional function of limiting slew rate at the switching node: $d(V_{ds})/dt$. This is because the feedback capacitance, C_{rss}, of the cascode combination is so low. Simply choosing a gate driver with a lower output current is a better way to limit i_g and di/dt. The voltage oscillation on the gate is mainly caused by the common source inductance. Since the threshold gate voltage of MOSFET is only 2.1 V, the oscillation voltage spike may make the device fault turn on. Under this consideration, negative voltage bias on the gate driver is proposed in the previous research [4], and boot strap half bridge driver circuit is not appropriate.

III. SIMULATION AND SWITCHING TEST RESULTS

A. Boot strap driver circuit

Using traditional boot strap driving method without negative voltage bias, the oscillation reduction on the gate signal is necessary. Adding a gate resistor and RC snubber to damp the ringing is a good practice in the application. In order

for fast turning off the device, a reversed diode paralleled with R_g is applied, while due to the junction capacitance of the diode, it has less effect to turn off the GaN device fast. It is also found that a small SMD ferrite bead effectively opposes transient high di/dt current and inhibits coupling of the signal, although the voltage waveform on the gate pin is not obviously damped.

Due to the very high di/dt rating, Si8230/1 driver from Silicon Labs Inc. is selected as it consists of a half bridge MOSFETs driver and digital isolator, which provides best-in-class noise immunity due to its less than 2 pF coupling capacitance [5]. The IC can be configured as non-isolated boot strap driver with up to 1.5 kV driver to driver differential voltage, and also as an isolated driver with up to 5 kV isolation rating.

As shown in Fig. 2, for the half bridge circuit testing, the DC bus voltage is 400 V, several 4.7 nF X7R ceramic capacitors are placed between the drain of high side device and source of low side device as decoupling capacitors. 60 µH small inductor is connected to the Ground for high side device hard switching testing, and it also can be connected to the DC positive rail for low side device hard switching testing. Si8231 single input driver is selected for obtaining complementary output driving signals with 100 ns dead time. Multi-pulse driving waveforms from signal generator are shown in Fig. 4. N pulses every 30 ms interval are generated, and the pulse period is 1 µs. By adjusting pulse width and number of pulses, the switching current can be controlled. A 300 Ω at 100 MHz SMD ferrite bead from Bourns Inc. is chosen as Z_g. The same circuit model is set up in the LTspice for simulation verification.

B. Simulation and experimental results using ferrite beads

Fig. 5 shows the simulated and measured V_{gs} waveforms during turn-on time; voltage oscillation can be observed in both figures. This oscillation is caused by the source inductance and couples to the gate pin. The real internal gate voltage is clean, indicating that the device didn't turn on or off incorrectly, as shown in Fig. 6. Fig. 7 shows the high side device can hard switch at 41.6 A with V_{ds} undershoot no more than -100 V; Fig. 8 shows the low side device can hard switch at 27 A with V_{ds} spikes no more than 600 V. The device can switch in such high current rating that the circuit is robust to survive at high current stress transient conditions. Due to the undershoot voltage applying on the boot strap circuit during charging the boot strap capacitor, a 10 ~ 15 Ω, 0.5 W resistor R_b is connected to the circuit to limit the inrush current and an ultrafast 600 V 1 A diode is chosen as D_b.

C. Effect of ferrite beads

Different value of ferrite beads are applied and compared. Low side hard switching waveforms are shown in Fig. 9. It can be seen that, large ferrite bead helps to reduce the voltage spikes, but it brings longer turn on / off time delay. As shown in Fig. 10, in the condition of 15 A switching current, the turn

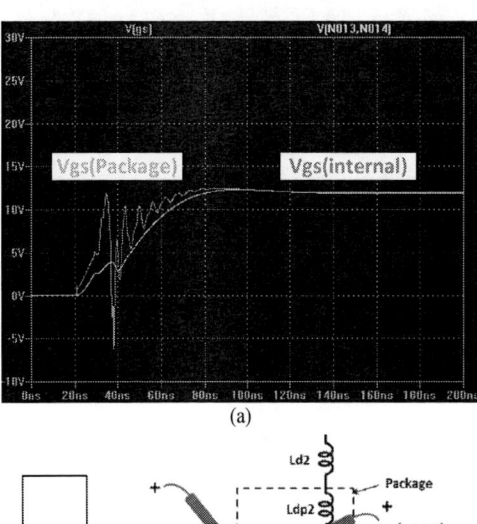

(a)

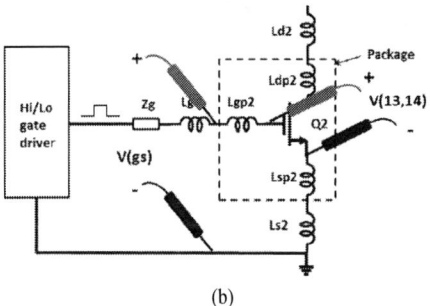

(b)

Fig. 6: Simulation showing the measured V_{gs} and internal V_{gs}. (a) Simulation waveforms; (b) Voltage probing point.

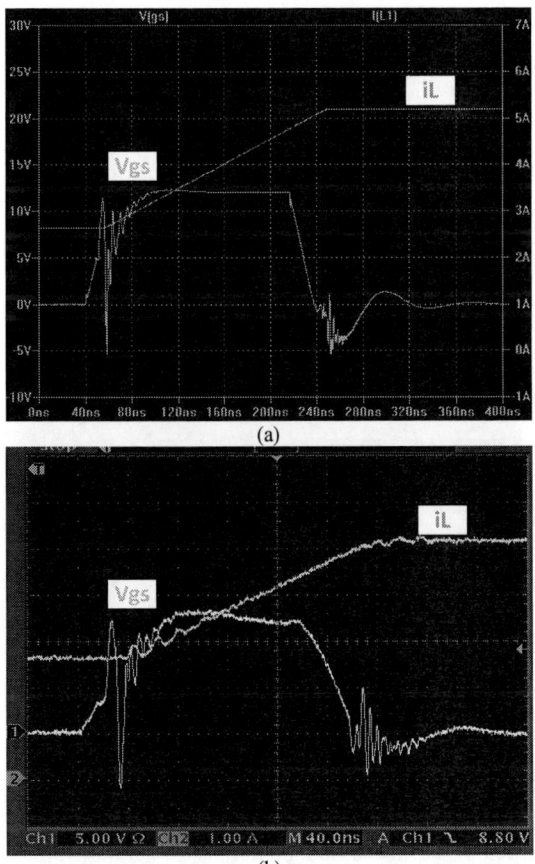

(a)

(b)

Fig. 5: Comparison of simulated and measured Vgs waveforms and output current. (a) Simulated V_{gs} (on package pins); (b) Actual Vgs in the experiment (on package pins).

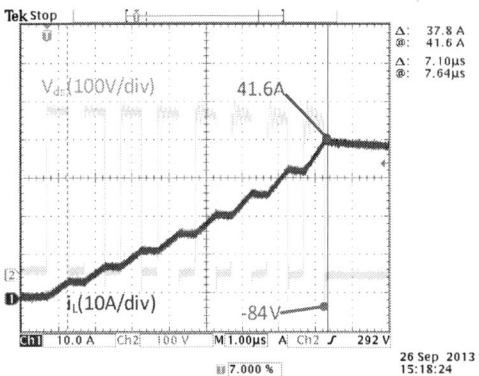

Fig. 7: High side switch multi-pulses test.

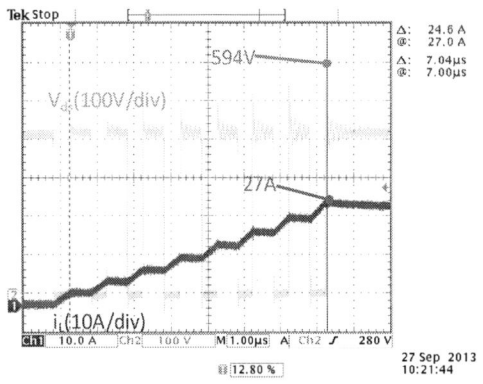

Fig. 8: Low side switch multi-pulses test.

978-1-4799-5494-0/14 $31.00 © 2014 IEEE

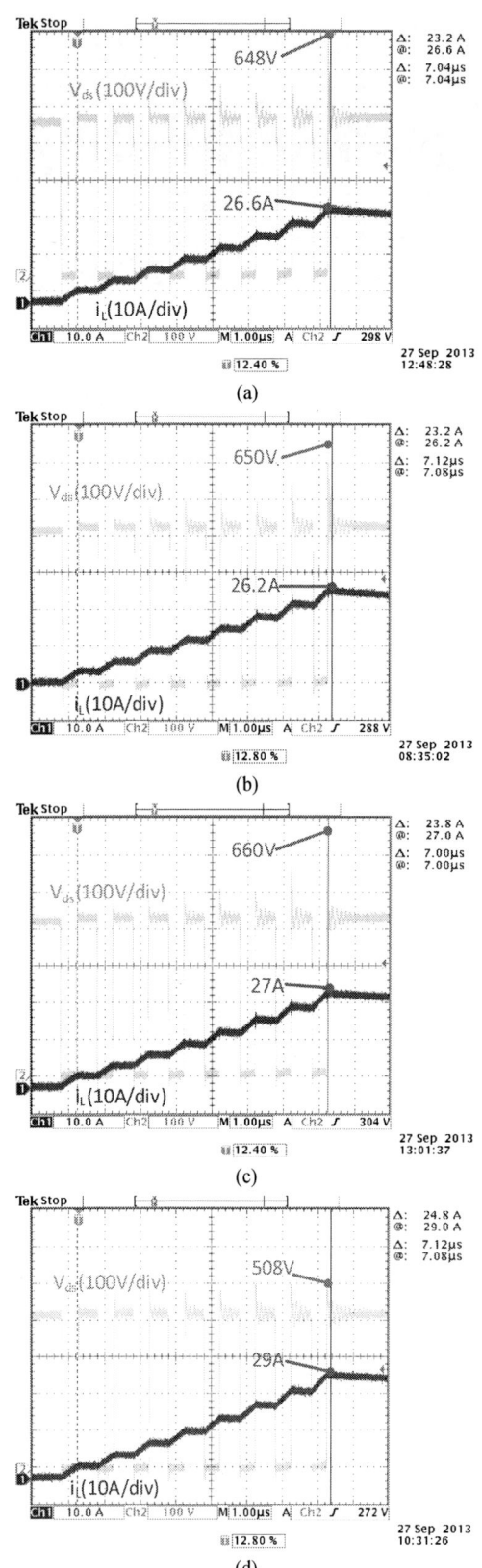

Fig. 9: Low side switching waveforms with different ferrite beads. (a) TDK 80 Ω MMZ2012D800B; (b) TDK 120 Ω MMZ2012D121B; (c) Wurth 120 Ω 74279202; (d) Taiyo 430 Ω BK2125HS431-T.

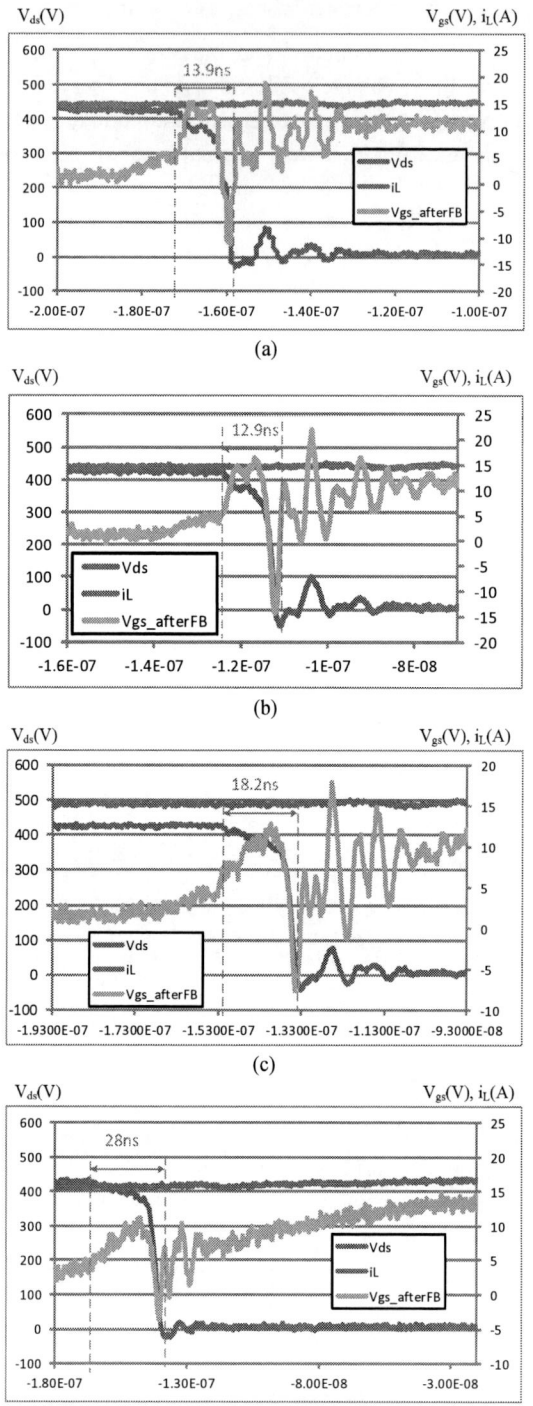

Fig. 10: Turn-on crossover time with different ferrite beads. (a) TDK 80 Ω MMZ2012D800B; (b) TDK 120 Ω MMZ2012D121B; (c) Bourns 300 Ω MU2029-301Y; (d) Taiyo 430 Ω BK2125HS431-T.

on crossover time of V_{ds} and I_d increases from 12 to 28 ns with the ferrite beads' value increasing. Compared to $Z_g = 0$ at 10A in [6], there is only additional 2 ns longer for 120 Ω ferrite bead even at 15 A. In the high switching frequency application, 120 Ω or less value of ferrite bead is proposed. Fig. 11 shows the ferrite bead solution does not reduce the switching speed.

978-1-4799-5494-0/14 $31.00 © 2014 IEEE

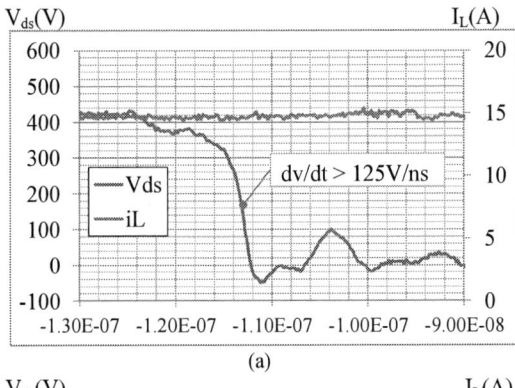

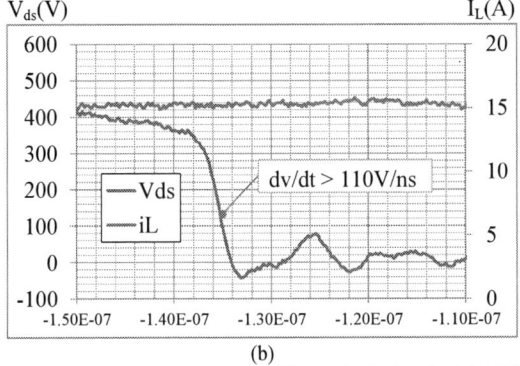

Fig. 11: Turn on dv/dt at 15 A with different ferrite beads. (a) TDK 120 Ω ferrite bead; (b) Bourns 300 Ω ferrite bead.

Even with 300 Ω ferrite beads, the dv/dt is still over 110 V/ns, indicating the efficiency will not be impacted.

D. Off-state margin and negative gate voltage

The previous analysis shows the internal gate voltage of the actively switched transistor. Also of interest is the behavior of the opposite transistor which should be held off during the transient. Fig. 12 shows a half bridge where the high-side transistor is switched and the low-side transistor carries the free-wheeling current. When Q_H turns on, the high dv/dt at the switching node (S) will induce current I_{GD} which charges C_{GD} of Q_L. Some of this current will flow in C_{GS}, raising the gate voltage with a high enough ratio of C_{GD}/C_{GS}, V_{gs} could potentially increase enough to turn on the low-side switch. However, in the cascode device this ratio is extremely low. For the TPH3006, for example, C_{GD} is 4.5 pF while C_{GS} is 720 pF at $V_{gs} = 0$ and over 2000 pF at the onset of turn-on.

A more valid concern than capacitive coupling through C_{GD} of the cascode pair is coupling through C_{GD_Si} of the low-voltage silicon FET. The drain-source voltage of the silicon FET, V_{ds_Si} in Fig. 12, only rises to $-V_{th}$ of the GaN HEMT, but will do so with high dv/dt. The ratio of C_{GD_Si}/C_{GS} in this case is significant. Fig. 13 compares external gate waveforms for measured and simulated cases. The simulation also includes the internal V_{gs} of the silicon FET. It is seen that V_{gs} does rise quickly toward the threshold voltage when V_{ds_Si} changes. However, because the rise in V_{ds_Si} corresponds to turn-off of

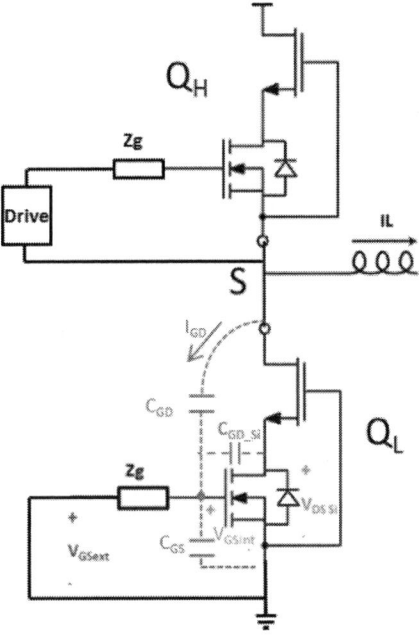

Fig. 12: Capacitances of the off-state transistor: Q_H is actively switched, Q_L is off.

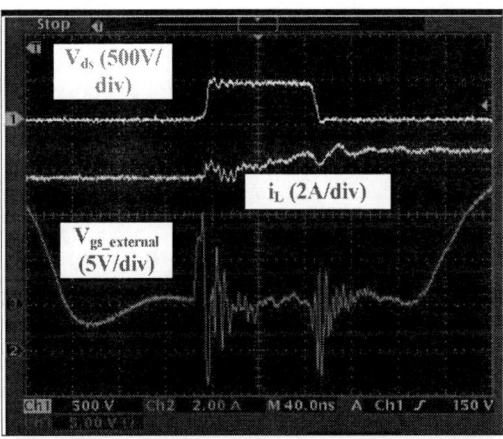

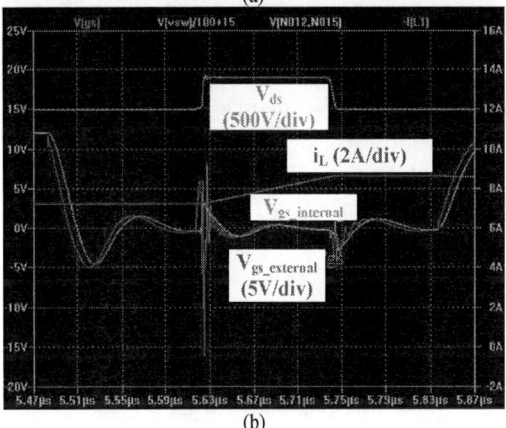

Fig. 13: Capacitances of the off-state transistor: Q_H is actively switched, Q_L is off.switched, Q_L is off.

978-1-4799-5494-0/14 $31.00 © 2014 IEEE   85

Fig. 14: Photo of half bridge converter with TPH3006 GaN HEMTs

the GaN HEMT, any drain current induced in the silicon FET due to the rise in V_{gs} must first discharge the GaN gate before external current flows.

IV. EFFICIENCY TEST

As shown in Fig. 14, a half bridge boost converter operating in synchronized mode was designed for 200 V input and 400 V output. 15 μF/630 V and 9.4 μF/630 V film capacitors are applied in the input and DC bus output. A 1 mH/76 mΩ boost inductor is employed, which is made up of two layers of AWG18 windings and a MPP core C055439A2 from Magnetics Inc. The back-to-back TPH3006PS GaN HEMTs as high side and low side devices are mounted on a standard heat sink with 2.6 °C/W at natural convection condition. Both tabs of GaN HEMTs are insulated from the heat sink as it can be connected to the power ground or the earth. The capacitance between the Tab of low side HEMT and surface of heat sink will not bring switching loss as the low side has the quiet source tab. If high side device is TPH3006PS, thick ceramic thermal pad or low permittivity insulator should be used to reduce the capacitance between TO-220 tab and heat sink so as to eliminate the extra switching loss. A fan placed towards to the heat sink providing 80 CFM air flow can keep the maximum temperature of heat sink below 95 °C at 1650 W output power. It is impossible to achieve such high output power using surface mount package device such as PQFN device without using aggressive active cooling method. In Fig. 15, it can be seen that the efficiency is over 99% over the wide load range for 50 kHz switching frequency. And the peak efficiency is 99.2% at 530 W output power, while for 100 kHz switching frequency it is still over 98.9% at 680 W output power.

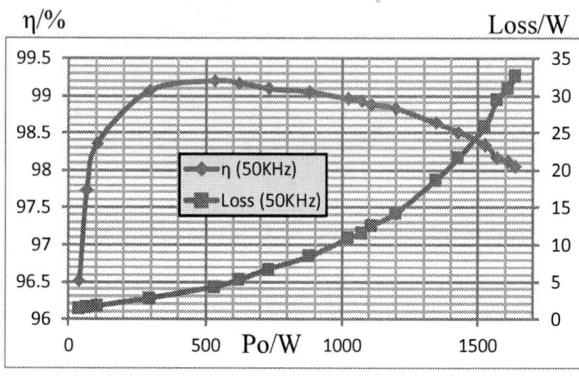

(a)

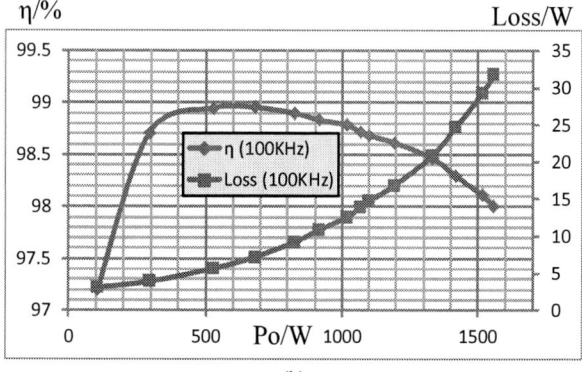

(b)

Fig. 15: Efficiency and power loss vs. output power at switching frequency: (a) 50 kHz; (b) 100 kHz

V. CONCLUSION AND FUTURE WORK

In this paper, the driver circuit of half bridge for leaded package GaN HEMTs is analyzed, and the effect of different ferrite beads is discussed in detail. A simple and robust boot strap half bridge driver circuit is proposed for high power half bridge converter applications. The simulation and experimental results shows the GaN HEMTs can safely operate at very high current rating in hard switching condition, while the switching speed does not slow down thus helping maintain the high efficiency.

In the future work, in order to reduce the high voltage spike due to the PCB layout limitation in application, RC or RCD snubber circuit will be explored although it will bring additional 0.1 ~ 0.2% efficiency drop. Isolated power supply circuit for providing negative bias voltage and insulation will also be analyzed, which is also very popular in high power applications. However, due to lack of high speed common mode transient immunity (CMTI) ICs and power supplies, high speed switching noise may result in severe interference between power circuit and signal circuit.

REFERENCES

[1] Transphorm Inc., Datasheet of TPH3006PS [Online]. Available: www.transphormusa.com

[2] Wu, Y.-F.; Gritters, J.; Shen, L.; Smith, R.P.; McKay, J.; Barr, R.; Birkhahn, R., "Performance and robustness of first generation 600-V GaN-on-Si power transistors," *Wide Bandgap Power Devices and Applications (WiPDA), 2013 IEEE Workshop on* , vol., no., pp.6,10, 27-29 Oct. 2013.

[3] Zhengyang Liu; Xiucheng Huang; Lee, F.C.; Qiang Li, "Package Parasitic Inductance Extraction and Simulation Model Development for the High-Voltage Cascode GaN HEMT," *Power Electronics, IEEE Transactions on* , vol.29, no.4, pp.1977,1985, April 2014.

[4] Wei Zhang; Xiucheng Huang; Lee, F.C.; Qiang Li, "Gate drive design considerations for high voltage cascode GaN HEMT," *Applied Power Electronics Conference and Exposition (APEC), 2014 Twenty-Ninth Annual IEEE* , vol., no., pp.1484,1489, 16-20 March 2014.

[5] Silicon Labs, Datasheet of Si823x [Online]. Available: www.silabs.com

[6] Huang, Xiucheng; Liu, Zhengyang; Li, Qiang; Lee, Fred.C., "Evaluation and application of 600V GaN HEMT in cascode structure," *Applied Power Electronics Conference and Exposition (APEC), 2013 Twenty-Eighth Annual IEEE* , vol., no., pp.1279,1286, 17-21 March 2013.

An Isolated Bi-directional Soft-switched High-frequency-AC Link DC-AC Converter Using SiC MOSFETs

Mengqi Wang, Suxuan Guo, Qingyun Huang, Wensong Yu and Alex Q. Huang

Department of Electrical and Computer Engineering
North Carolina State University
Raleigh, USA
mwang7@ncsu.edu

Abstract—An isolated bi-directional soft-switched DC-AC converter with high-frequency-AC (HFAC) link using Silicon Carbide (SiC) MOSFETs is presented in this paper. A unipolar-SPWM oriented modulation technique is proposed to enable the full-bridge (FB) stage to realize zero-voltage-switching (ZVS) and the cycloconverter stage to realize zero-current-switching (ZCS). Furthermore, the proposed modulation technique allows half of the switches in cycloconverter to work at line frequency (LF) instead of switching frequency, which significantly reduces switching loss. Because of SiC MOSFET's low on-resistance and great switching performance under high frequency conditions, they are used for all the switches such to further reduce the switching loss and conduction loss. Thus the switching frequency can be pushed to a much higher level, i.e. 50-100 kHz, which largely reduces the profile of the transformer and inductor. Therefore, the power-density of the converter is highly improved. The advantages of utilizing SiC MOSFETs are validated by simulation and experimental results.

Keywords—HFAC link, DC-AC converter, soft-switching, SiC MOSFET

I. INTRODUCTION

Wide-Bandgap (WBG) devices, such as Silicon Carbide (SiC) MOSFETs, have emerged as a promising alternative that pushes the limits of the power semiconductors [1]. The SiC material has superior performances in electrical breakdown field, thermal conductivity, electron saturated drift velocity, and irradiation tolerance [2]. Those remarkable advantages enable the SiC devices to work at higher voltage, frequency and temperature. Therefore, SiC device-based converters is expected to achieve high efficiency and high power-density as well [3], [4].

In present isolated AC grid-fed UPS, distributed renewable energy systems and propulsion systems, when low profile and weight are required, the power conversion is usually realized by a two-stage DC-AC converter: full-bridge

(FB) stage convert DC to HFAC power, and cycloconverter enables the HFAC power to fed to utility grid through a HFAC transformer link [5], [6]. The power conversion is usually realized by full-bridge (FB) stage cascaded by cycloconverter stage through a HFAC transformer link [7]. Most common devices used for the AC switches in cycloconverter are Silicon-controlled rectifier (SCR) and Insulated-gate bipolar transistor (IGBT). SCR and IGBT switches are only suitable for medium frequency, i.e., 5 kHz, because of high switching losses and conduction losses.

Si MOSFETs should be a better option to operate under high frequency, but for high voltage applications, the reverse recovery issue is still severe thus the switching frequency is quite limited [8]. In order to mitigate the reverse recovery loss of the Si MOSFET body diode, people suggest paralleling Schottky diode to the device. However, problem still exists even with Schottky diode. Section II will address this issue in detail. Thus we proposed to use SiC MOSFETs for the DC-AC converter.

Voltage spikes introduced by the HF transformer leakage inductance is another important issue. Adding an RC snubber branch to the HFAC side of cycloconverter is a straightforward and widely used approach, which is easy to carry out. But its effectiveness depends on different conditions. Moreover, the power dissipation on the snubber resistor lowers the efficiency. Sree and Mohan proposed adding an energy recovery circuit to clamp the voltage spikes [9]. The energy is recovered to the DC source by a transformer plus a diode rectifier bridge. This approach comprises additional four diodes and a transformer which leads to power loss due to the energy circulating in the system. In reference [10], the authors proposed an active clamp approach. A pair of bidirectional AC switch with a snubber capacitor are adopted to absorb the energy stored in the leakage inductance. However, additional AC switch and driver circuit result in circulating energy and increase the

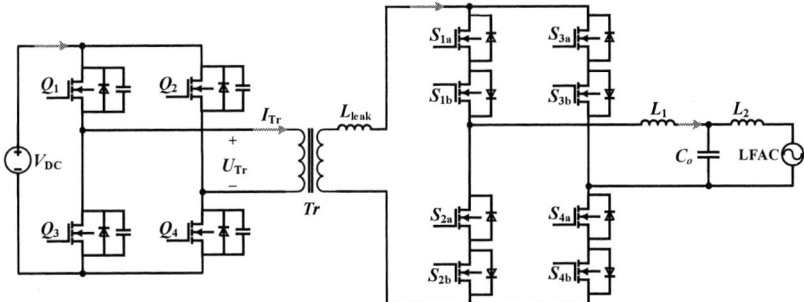

Figure 1. ZVZCS HFAC link DC-AC converter schematic

control complexity as well. Norrga presented a soft-switched AC-DC converter without auxiliary circuit in [11]. Power dissipation and circulation is largely reduced but complicated modulation and capacitor voltage detection are needed. In addition, all switches in the literature reviewed above are working at high switching frequency. We proposed a novel unipolar-SPWM-oriented modulation technique to not only enable the FB to realize ZVS, and cycloconverter to realize ZCS without any additional circuit nor components, but also allows half of the switches in cycloconverter to operate at 60 Hz instead of switching frequency. Thus switching loss is significantly reduced. The converter schematic is shown in Fig. 1. We also utilize SiC MOSFETs for all the switches to push the frequency to 54 kHz to increase the converter power-density while maintaining a high efficiency.

The rest of the paper are organized as follows: Section II performs a detailed analysis on why SiC devices are chosen for the specific application. Section III presents the proposed HFAC link DC-AC converter and explains how the ZVS for full-bridge and ZCS for the cycloconverter are realized. Section III provides simulation and experimental results to illustration the proposed structure and modulation technique. Finally, Section IV concludes the innovation and contribution of this paper.

II. ADVANTAGES ANALYSIS ON UTILIZING SiC DEVICES FOR HFAC LINK DC-AC CONVERTER

SCR and IGBT are commonly used in existing high power cycloconverter which is composed of bi-directional AC switches. Fig. 2 demonstrates the different structure of SCR and non-SCR AC switches. For the SCRs, two of them are parallel connected to allow current flow in both directions or to block current from both directions. For the other non-SCR devices, such as IGBT, Si or SiC MOSFETs, two of them are anti-series connected to form one pair of AC switch. Therefore the two anti-series connected body diodes are able to block current from both directions.

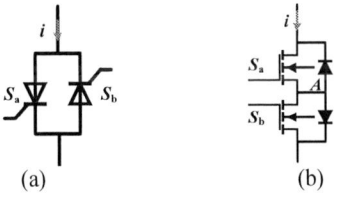

(a)   (b)

Figure 2. AC switches. (a) SCR AC switch; (b) non-SCR AC switch

Generally speaking, SCR and IGBT have higher power capacity than the newly developed semiconductor devices and are proven to be reliable and economical [12]. However, SCR will only turn off when the current drops below the hold-current, thus it is not suitable to work as a "current breaker". Moreover, SCR switches much slower than Si MOSFETs, which limits its working frequency to be at low to medium range. IGBT switches produce high switching losses and conduction losses, so they are only suitable for medium frequency (i.e., 5 kHz) applications as well.

Si MOSFET is able to work at a much higher frequency, such as 100-500 kHz, but high voltage MOSFET has severe reverse recovery issue, thus people suggest paralleling Schottky diode to the device. For high voltage applications, such as higher than 400 V, SiC Schottky is a more suitable choice for its low forward voltage drop, which is around 1.5 V. It is worth mentioning that connection as shown in Fig. 2 (a) does not work well since the forward voltage drop of the Si MOSFET body diode is usually around 1.5 V as well. There will still be current flow through the body diode that can cause reverse recovery losses. Thus the connection shown as Fig. 2 (b) is recommended. The source of both MOSFETs are disconnected to prevent current flow through the body diode. However, the problem within this structure is that the conduction loss will be very high since there are always diodes conducting.

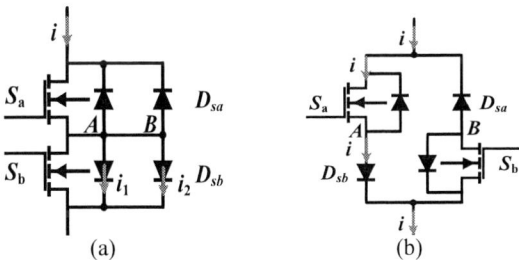

(a)   (b)

Figure 3. Current flow in Si AC switches with paralleling Schottky diodes. (a) direct connection; (b) source-disconnected connection

Compared to Si MOSFET, SiC MOSFET has a much better switching performance. The reverse recovery of the SiC MOSFET body diode is small enough to enable the device to work above 100 kHz. Even if paralleling Schottky diodes is needed, the connection does not need to be like Fig. 3 (b). The reason is that the forward voltage drop of the SiC

978-1-4799-5494-0/14 $31.00 © 2014 IEEE   89

MOSFET body diode is around 3.5 V, which is much higher than that of the SiC Schottky diode. The current will flow through the Schottky diode automatically as shown in Fig. 4.

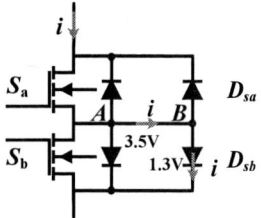

Figure 4. Current flow in SiC AC switches with paralleling Schottky diodes.

In order to perform a further quantitative analysis, we set the system specifications as shown in Table I.

Table I. System specifications

Rated power	5 kW
DC input voltage	400 V
AC output voltage	240 Vrms
HFAC link frequency	54 kHz
AC line frequency	60Hz
Leakage inductance	7 uH

Based on the system specifications and also considering the availability of the devices, 3 types of switches: IGBT, Si MOSFET and SiC MOSFET are picked for comparison. Their parameters are listed in Table II.

Table II. Switch parameters

	IGBT	Si MOSFET	SiC MOSFET
Part number	IKP15N65F5	IPP65R045C7	C2M0040120D
V_{DS}	650 V	700	1200 V
I_D (@25°C)	30 A	46	40 A
$R_{ds(on)}/V_F$	1.6 V	45 mΩ	40 mΩ
Q_{rr}	100 nC	13 µC	283 nC
t_{rr}	22 ns	725 ns	54 ns

Conduction losses in cycloconverter stage are calculated based on the parameters Table II. The conduction loss of Si MOSFET is calculated based on Fig. 3 (b). The calculation result is summarized in Fig. 5, which well demonstrates the superior performance of SiC MOSFET. Therefore, based on the above analysis, we proposed to use SiC MOSFET for this work in reason that the on-resistance is low and the reverse-recovery speed of the body diode is ultra-fast, which enable the SiC MOSFETs to work at a much higher frequency, i.e.,

50-100 kHz while achieving high power-density and high efficiency as well.

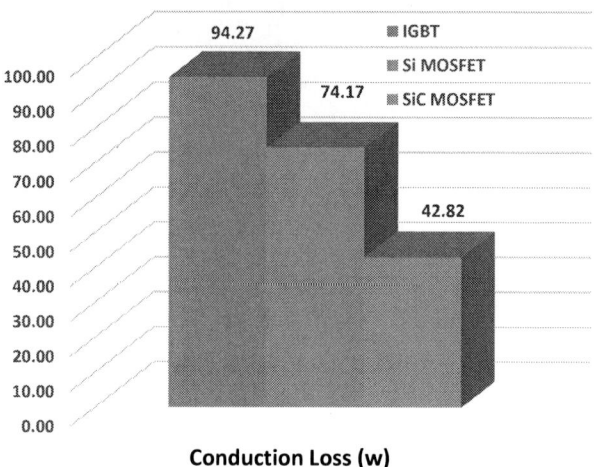

Conduction Loss (w)

Figure 5. Loss comparison between IGBT, Si and SiC MOSFETs

III. PROPOSED SOFT-SWITCHED HFAC LINK SiC DC-AC CONVERTER

In order to reduce the RMS current flowing through the HFAC link transformer, we proposed a unipolar-SPWM oriented modulation technique. The configuration of the proposed ZVZCS HFAC link DC-AC converter using SiC MOSFETs is shown in Fig. 1. For the FB stage, all the 4 switches Q1 to Q4, are equipped with a snubber capacitor to realize ZVS.

Fig. 6 shows the scheme of the SPWM generation for the cycloconverter stage. A count-down mode sawtooth waveform is used as the carrier. Under the proposed unipolar-SPWM oriented commutation strategy, half of the switches in cycloconverter: S_{1a}, S_{1b}, S_{3a} and S_{3b}, operate at line frequency which is 60 Hz instead of switching frequency which is 54 kHz in the design. This significantly reduces the switching loss to approximately 50% for cycloconverter. A 500 ns turn-off delay is added to all the HF AC switch gating for realizing ZCS. The essential reason for voltage spikes is because when the bus voltage changes its polarity, leakage inductance current is forced to be at the same value with the output inductor current, which means when output inductor is discharging, leakage inductance current is forced to be 0 and vice versa. This sudden change in leakage inductance current will cause high voltage spikes at the HFAC link. In the proposed commutation strategy, since there is a turn-off delay for the HF AC switches, i.e., S_{1a} and S_{4a}, the voltage polarity on the leakage inductance is changed before the current changes the direction, thus the leakage inductance current is naturally reduced to 0. The current natural commutation is realized. Furthermore, since there is a switching overlap period, the voltage spikes will be forced to 0.

978-1-4799-5494-0/14 $31.00 © 2014 IEEE

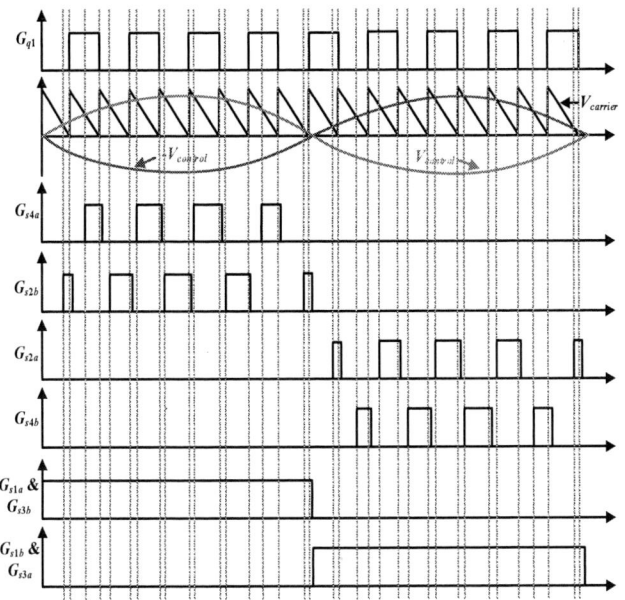

Figure 6. SPWM generation for cycloconverter

IV. SIMULATION AND EXPERIMENTAL VERIFICATION

A simulation model for the DC-AC inverter has been built in PSIM. Fig. 7 shows the gating for S_{2b}, AC link voltage, cycloconverter line-frequency AC (LFAC) voltage before the output filter, and filter inductor current. Fig. 8 is a zoomed-in of gating, FB output voltage and HFAC link current waveforms. As can be seen from Fig. 8, S_{1a}, S_{1b}, S_{3a} and S_{3b} operate at 60 Hz instead of switching frequency. S_{2a} and S_{4b} are not switching during this period of time because it is during the LFAC negative output interval. In other words, there are only 2 switches operating at one time for the cycloconverter.

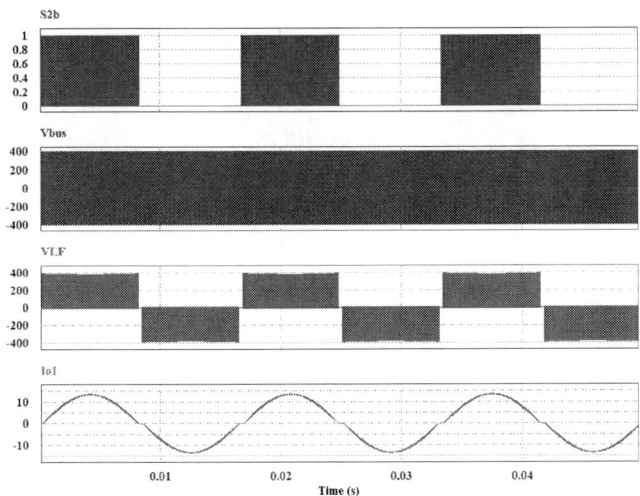

Figure 7. Simulation waveforms

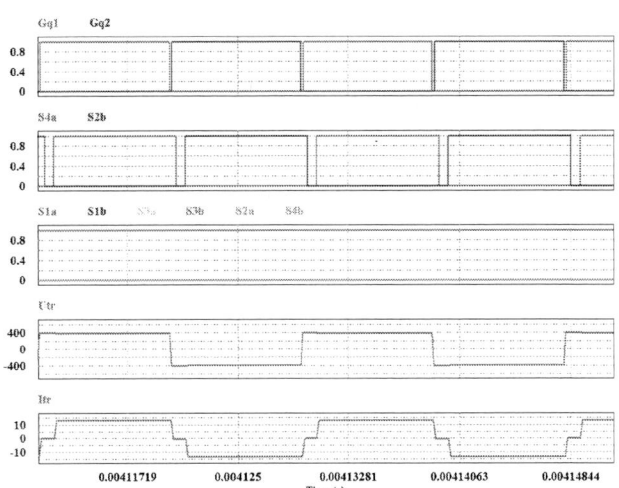

Figure 8. Zoomed-in simulation waveforms

A prototype for the FB stage and cycloconverter has been developed as shown in Fig. 9. DSP controller and the two stages are built in separate board for flexibility and scalability purposes. Each of the SiC MOSFET is equipped with an isolated driver circuit to mostly protect the gating signal from being intervened by the power stage. The major components used for the converter are listed in Table III. The current test conditions are 200 V DC input with 20 Ω load resistor. The HFAC link frequency is 54 kHz. Fig. 10 shows the experimental waveforms of gating, HFAC link voltage and current, and the output current, which demonstrate the functionality of the FB stage and cycloconverter. Fig. 11 (a) shows the zoomed-in experimental waveforms of FB and cycloconverter gating, HFAC link voltage and current, which match with the simulation results and theoretical analysis. Fig. 11 (b) and (c) further demonstrate the cycloconverter zero-current turning-off transition, and FB zero-voltage turning-on transition, respectively. The experimental results match with the simulation and well validate the proposed commutation strategy.

Table III. Component list

SiC MOSFET	CMF10120D
DSP controller	F28335
Isolated power supply	RP2424S
Digital isolator	ADuM1100
Gate driver IC	UCC27531

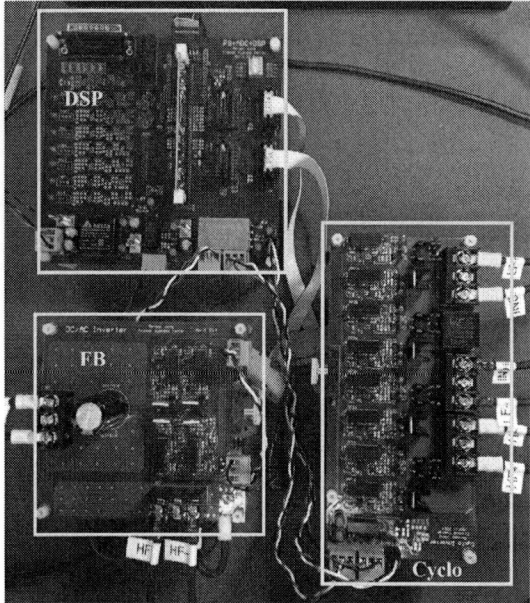

Figure 9. Prototype of the FB and cycloconverter

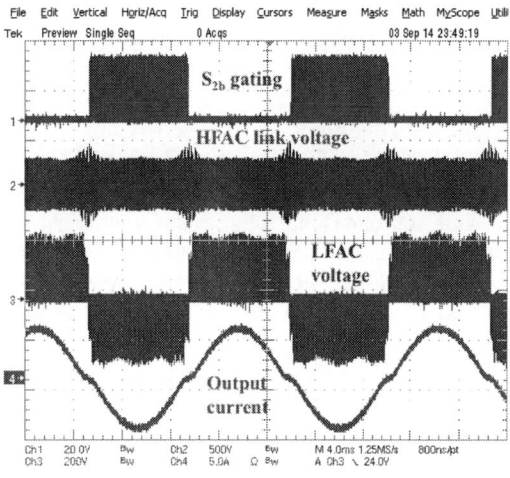

Figure 10. Experimental waveforms at 54 kHz

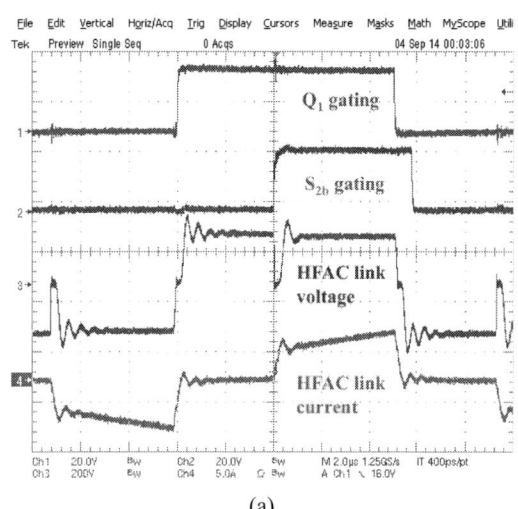

(a)

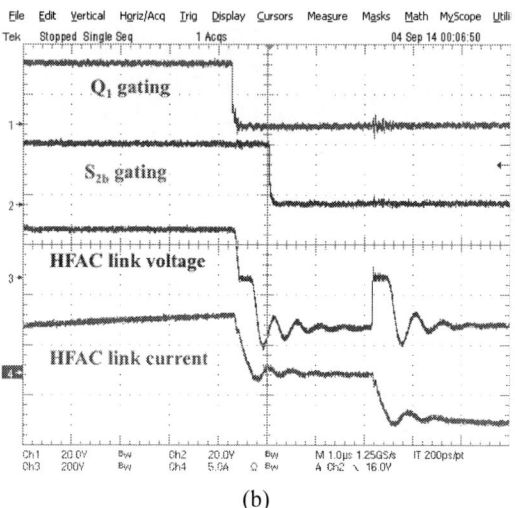

(b)

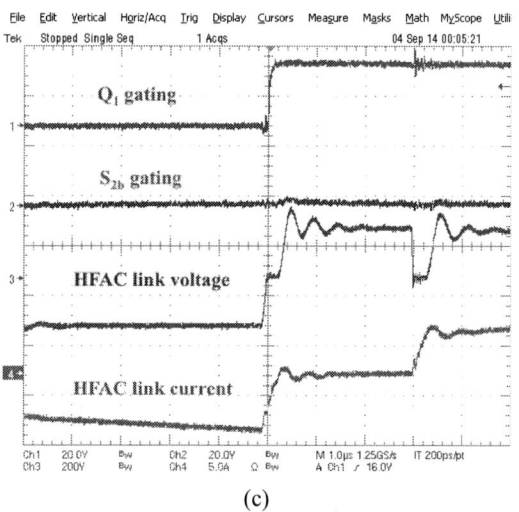

(c)

Figure 11. Zoomed-in experimental waveforms at 54 kHz. (a) zoomed-in waveforms in one switching cycle; (b) cycloconverter zero-current turning-off transition; (c) FB stage zero-voltage turning-on transition.

V. CONCLUSION

In order to increase the frequency of traditional DC-AC converter using FB cascaded with cycloconverter topology at 400 V or higher applications, SiC MOSFETs are utilized because of the low on-resistance and great switching performance under high voltage, high frequency conditions. A unipolar-SPWM-oriented modulation technique is proposed in this paper as well. The proposed modulation strategy enables the FB to realize ZVS and cycloconverter to realize ZCS, which largely reduces the voltage and current stresses on the device. Furthermore, the modulation strategy allows half of the switches in cycloconverter to work at 60 Hz instead of switching frequency, which significantly reduces the switching loss. Simulation and experimental results are provided to validate the

978-1-4799-5494-0/14 $31.00 © 2014 IEEE

performance of the SiC MOSFETs and modulation technique.

REFERENCES

[1] Ning, Puqi, Di Zhang, Rixin Lai, Dong Jiang, Fred Wang, Dushan Boroyevich, Rolando Burgos, Kamiar Karimi, Vikram D. Immanuel, and Eugene V. Solodovnik. "High-Temperature Hardware: Development of a 10-kW High-Temperature, High-Power-Density Three-Phase ac-dc-ac SiC Converter." *Industrial Electronics Magazine, IEEE*, vol. 7, no. 1, pp. 6-17, 2013.

[2] Zhang, Hui, Leon M. Tolbert, Burak Ozpineci, and Madhu S. Chinthavali. "A SiC-based converter as a utility interface for a battery system," In *Conference Record, 41st Industry Applications Conference IAS Annual Meeting*, vol. 1, pp. 346-350, 2006.

[3] L. M. Tolbert, B. Ozpineci, S. K. Islam, M. Chinthavali, "Wide bandgap semiconductors for utility applications," *IASTED International Conference on Power and Energy Systems (PES 2003)*, Palm Springs, California, pp. 317-321, February 24-26, 2003.

[4] T. Ericsen, "Future navy application of wide bandgap power semiconductor devices," *Proceedings of the IEEE*, vol. 90, Issue 6, pp. 1077-1082, June 2002.

[5] Matsui, Mikihiko and Yamagami, Masanori. "Asymmetric control of HF link soft switching converter for UPS and PV systems with bidirectional power flow," *Industry Applications Conference. Thirty-Third IAS Annual Meeting*, vol. 2, 1998.

[6] Norrga, Staffan. "Experimental study of a soft-switched isolated bidirectional ac–dc converter without auxiliary circuit," *IEEE Trans. Power Electron.*, vol. 21, no. 6, pp. 1580-1587, 2006.

[7] S. Manias, P. D. Ziogas, and G. Olivier, "Bilateral dc to ac convertor using a high frequency link," in *Proc. IEE B*, vol. 134, no. 1, pp. 15–23, 1987.

[8] Bhatnagar, Mohit and Baliga, B. Jayant. "Comparison of 6H-SiC, 3C-SiC, and Si for power devices," *IEEE Trans. Electron Devices.*, vol. 40, no. 3, pp. 645-655, 1993.

[9] Sree, Hari and Mohan, Ned. "Voltage sag mitigation using a high-frequency-link cycloconverter-based DVR." In *Industrial Electronics Society, IECON* 2000. 26th Annual Conference of the IEEE, vol. 1, pp. 344-349. IEEE, 2000.

[10] Chen, Daolian and Chen, Yanhui, "Step-up AC Voltage Regulators with High-Frequency Link," *IEEE Trans. Power Electron.*, vol. 28, no. 1, pp. 390-397, 2013.

[11] Norrga, Staffan, "Experimental study of a soft-switched isolated bidirectional AC-DC converter without auxiliary circuit," *IEEE Trans. Power Electron.*, vol.21, no. 6, pp. 1580-1587, 2006.

[12] Chen, Zhe, "Compensation schemes for a SCR converter in variable speed wind power systems", *IEEE Trans. Power Electron.*, vol. 19, no. 2, pp. 813-821, 2004.

978-1-4799-5494-0/14 $31.00 © 2014 IEEE

Dependence of Ti/C Ratio on Ohmic contact with TiC electrode for AlGaN/GaN structure

M. Okamoto, K. Kakushima, Y. Kataoka,
A. Nishiyama, N. Sugii, H. Wakabayashi,
K. Tsutsui, and H. Iwai, Frontier Research Center,
Tokyo Institute of Technology, Yokohama
226-8503, Japan, Email-Address:
okamoto.m.ae@m.titech.ac.jp

W.Saito
Semiconductor Company, Toshiba Corp. Kawasaki
212-8583, Japan, Email-Address:
wataru3.saito@toshiba.co.jp

Abstract - **The dependency of Ti atom composition in Ti-C mixed electrodes on Ohmic characteristics for AlGaN/GaN structure is examined by elucidating the role of both Ti and C atoms. Owing to AlGaN surface reduction by reaction with C atoms with thermal treatment, an enhanced reaction of Ti atoms and AlGaN layer has been confirmed. The border of reactive layer and the remaining AlGaN layer has shown rather uniform interface, which is contrast to conventional metallic spike formation. Higher Ti atom composition has revealed lower Ohmic contact resistance, especially 2.6 Ωmm with Ti:C of 5:1. The proposed reaction mechanism gives a new guideline to achieve uniform electron conduction for lower contact resistance.**

I. INTRODUCTION

III-nitride high-electron-mobility transistors (HEMTs) are focused for high-power and high-frequency device applications owing to its high electron velocity with high breakdown field [1]. Lateral AlGaN/GaN HEMTs are particularly promising, since the cost of the devices can be cut down by device scaling. In addition, the cost can also be further reduced as AlGaN/GaN layers can be epitaxially grown on Si (111) substrates through proper buffer layers [2]. One of the issues in AlGaN/GaN HEMTs is the presence of relatively high Ohmic contact resistance. A simple estimation based on ref. 3 indicates that power loss in the device can be reduced by 60% when specific Ohmic contact is reduced from 10^{-5} to 10^{-6} Ωcm^2 at a rate voltage of 600 V.

A common approach to achieve an Ohmic contact to 2 dimensional electron gas (2DEG) residing in GaN layer near AlGaN/GaN interface, is to use Ti-Al mixed metal with proper annealing [4]. By forming spikes of TiN at dislocations in AlGaN layer with thermal treatment, the metal layer and the 2DEG is brought into contact, achieving an Ohmic contact [5]. As the number of dislocation, typically above 10^8 cm^{-2}, strongly affects the contact resistance properties, this approach may pose an issue considering future dislocation-free wafers.

Recent study using Si-based layer in combination with Ti-Al mixed metal has shown another contact mechanism; extraction of nitrogen from AlGaN layer to create nitrogen vacancies (V_N) and form AlN at the surface of AlGaN layer [6]. As V_N act as a donor, band bending in AlGaN near metal interface reduces the tunneling distance to increase the electron transport [6]. Although this approach does not rely on dislocation sites in AlGaN layer, inhomogeneous AlN pyramid formation, originated from alloys in the metal layer, is still an issue to lead local electron conduction. Since Al forms alloys with other metal at relatively low temperature, one need to also consider other material selection that can effectively reduce the AlGaN layer.

It has been reported that reaction of Ti with AlGaN is completely different from Ti with GaN case. One of the main difference is that Ti can react with GaN layer to create electrically active V_N and can form TiN [8]. However, Ti is not able to create V_N in AlGaN layer from thermodynamic prediction. The fact that TiN spikes are formed at dislocations sites in AlGaN layer strongly indicates that the presence of grain boundaries or defects reduces the enthalpy of TiN formation. Based on the above backgrounds, an intuitive approach for low Ohmic contact with uniform electron conduction is to create V_N defects at the surface of AlGaN and let Ti to react with them.

In this study, we utilize C as a metal for reducing AlGaN layer. Reaction of C and AlGaN layer and creation of electrically active V_N have been characterized. Based on the role of C on AlGaN, the dependency of Ti atom composition in Ti-C mixed electrodes on Ohmic contact characteristics were examined.

II. DEVICE FABRICATION

Al$_{0.25}$Ga$_{0.75}$N and GaN layers were grown on Si (111) substrate with proper buffer layers to overcome the lattice mismatch. Figure 1 shows fabrication of the device. The sheet resistance (ρ_{sh}) of 2DEG layer was 472 Ω/sq. measured by Hall measurements. After forming mesa using plasma-enhanced chemical vapor deposited SiO$_2$ hard mask with Cl-based reactive ion etching, another SiO$_2$ was again deposited as a passivation layer followed by contact holes formations. C, Ti or Ti-C mixed metal layers with thickness of 20 nm, were deposited by RF magnetron sputtering. The composition ratio of Ti and C for mixed layer was set to 1:1, 3:1 and 5:1. TiN layer with thickness of 50 nm was deposited atop all the metal layers to avoid oxidation of

978-1-4799-5494-0/14 $31.00 © 2014 IEEE

the surface. Contact resistances were extracted from current-voltage (*I-V*) measurements through transmission line model (TLM).

III. RESULTS AND DISCUSSION
A Creation of V_N in AlGaN with C

The reaction of C with AlGaN layer was characterized by transmission electron microscope (TEM) and hard x-ray photoelectron spectroscopy (HXPES) with an x-ray energy of 7940 eV. For these characterizations, a carbon layer of 20 nm in thickness was used to efficiently collect photoelectrons under the layer. As the inelastic mean free path for photoelectrons Al $1s$ is 12 nm, all the photoelectrons arising from AlGaN layer can be collected [10]. Figure 1 shows Al $1s$ spectrum before and after annealing in 1075 °C in nitrogen ambient for 1 min. The binding energy was calibrated by each C $1s$ spectrum obtained from each sample. A positive shift in energy at the peak intensity from 1561.1 to 1561.6 eV is observed with the annealed sample. Moreover, a spectrum broadening from 1.16 to 1.26 eV in full width at half maximum (FWHM) can be confirmed. These two facts strongly suggests that the AlGaN layer is further bent down in additional to the effect of spontaneous and piezoelectric polarization. From deconvolution analysis, the spectrum change can be interpreted as V_N creation to form n$^+$-AlGaN at the surface of AlGaN, which is in good agreement with thermodynamic prediction of the reaction of C and AlGaN. We should note that downward bending of AlGaN layer is advantageous as it increases the density of 2DEG. From above results, the energy band diagram of C on AlGaN/GaN structure before and after annealing can be drawn as figure 2.

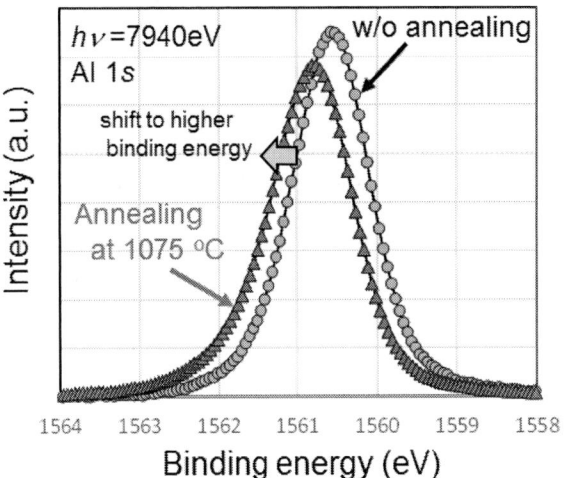

Figure 1 : Al $1s$ spectra arising from AlGaN layer before and after annealing with C metal. Energy shift at the peak intensity with spectrum broadening was observed

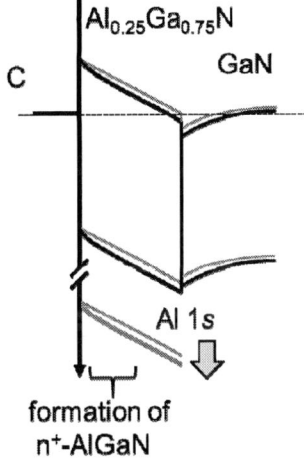

Figure 2 : Extracted energy band of C metal on AlGaN/GaN layer. Creation of V_N at the surface of AlGaN layer leads to the formation of n$^+$ layer.

Figure 3 shows TEM image of the sample after annealing. An atomically rough interface between C and AlGaN layer was confirmed, indicating reaction took place between the two layers. Out diffusion of Ga atoms into C layer as particles were confirmed. Here, it should be noted that the reaction of C and AlGaN is uniform and there is no local enhanced reaction at dislocations in AlGaN layer, presumably attributed to the amorphous structure of C layer. Therefore, one can conclude that C creates V_N at the surface of AlGaN layer by thermal treatment, and eventually reduces the surface.

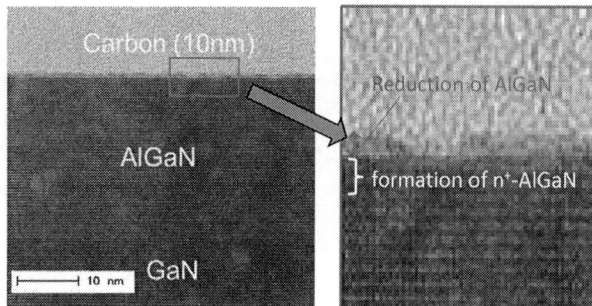

Figure 3 : Cross sectional TEM image of C metal on AlGaN/GaN after annealing. Reduction of AlGaN layer results in the creation of interface roughness.

B   Role of Ti and C in metal electrodes

The *I-V* characteristics of Ti, C and Ti-C (Ti:C=1:1) mixed electrodes are shown in figure 4(a). Although C creates V_N at the surface of AlGaN layer, no current has been observed when C electrode is used. On the other hand, the sample with Ti electrode showed non-linear IV curve, which is supposed to be local conduction at dislocation sites in AlGaN layer. When Ti-C mixed electrodes are used, an Ohmic conduction was obtained. TEM image of the sample, shown in figure 4(b) revealed reduction of AlGaN layer, and creation of an interface layer between Ti-C mixed layer and AlGaN layer. Schematic illustration of surface reaction is shown in figure 5.

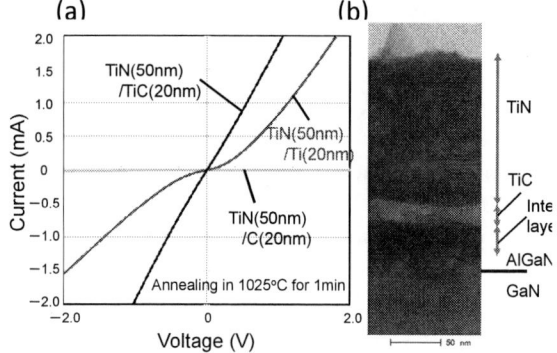

Figure 4(a): *I-V* characteristics of Ti-C mixed, Ti and C electrodes annealed at 1025°C. All the electrodes were capped with TiN layer. (b): Cross-sectional TEM image of TiN/TiC/AlGaN annealed at 1025 °C.

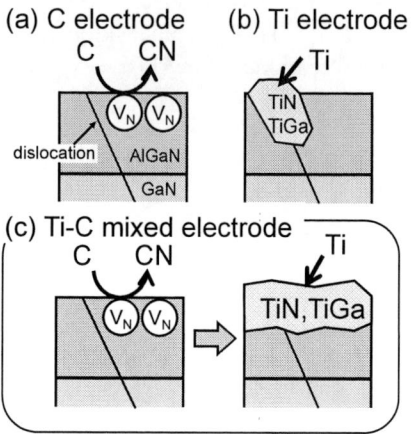

Figure 5 Schematic illustration of surface reaction of (a) C, (b) Ti and (c) Ti-C mixed electrodes with AlGaN layer.

C   Ti composition in Ti-C mixed electrodes

Contact resistances with different Ti composition in Ti-C mixed electrodes (Ti:C=1:1, 3:1 and 5:1) are shown in figure 6. With the annealing temperature step of 25 °C, Ti-C mixed electrodes with 1:1 showed Ohmic properties only when annealed at 1000 and 1025 °C. While increasing the Ti composition in the mixed electrode, Ohmic contact was achieved at lower temperature. Moreover, the annealing temperature process window required for Ohmic contact can be widen with large Ti composition. As for contact resistance, the sample with high concentration of Ti showed lower value of 2.6 Ωmm when annealed at 950 °C. Therefore, an optimum composition exist to utilize both reduction and reaction of AlGaN layer.

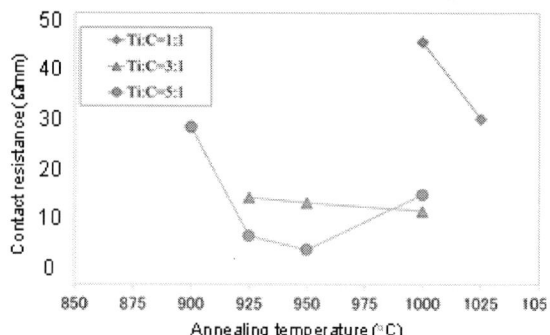

Figure 6: Contact resistance of Ti-C mixed electrodes on annealing temperature with different Ti composition.

IV. SUMMARY

The dependency of Ti atom composition in Ti-C mixed electrodes on Ohmic characteristics for AlGaN/GaN structure is examined by elucidating the role of both Ti and C atoms. Owing to AlGaN surface reduction by reaction with C atoms with thermal treatment, an enhanced reaction of Ti atoms and AlGaN layer has been confirmed. The border of reactive layer and the remaining AlGaN layer has shown rather uniform interface, which is contrast to conventional metallic spike formation. Higher Ti atom composition has revealed lower Ohmic contact resistance, especially 2.6 Ωmm with Ti:C of 5:1. The proposed reaction mechanism gives a new guideline to achieve uniform electron conduction for lower contact resistance.

V. REFERENCES

[1] T. P. Chow and R. Tyagi, "Wide bandgap compound semiconductors for superior high-voltage unipolar power devices", IEEE Trans. ED, Vol. 41, pp. 1481-1483 (1994).

[2] K. Cheng, M. Leys, S. Degroote, B. V. Daele, S. Boeykens, J. Derluyn, M. Germain, G. V. Tendeloo, J. Engelen, G. Borghs, "Flat GaN epitaxial layers grown on Si(111) by metalorganic vapor phase epiaxy using step-graded AlGaN intermediate layers", J. Elelctronic Materials, Vol. 35, pp. 592-598 (2006).

[3] W. Saito, I. Omura, T. Ogawa and H. Ohashi, "Theoretical limit estimation of lateral wide band- gap semiconductor power-switching device", Solid-State Electron., 48, pp. 1555-1562 (2004)

[4] L. Wang, F. M. Mohammed, "Differences in the reaction kinetics and contact formation mechanisms of annealed Ti/Al/Mo/Au Ohmic contacts on n-GaN", J. Appl. Phys., 101, 013702 (2007) .

[5] L. Wang, F. M. Mohammed, I. Adesidaa, "Dislocation-inducted nonuniform interfacial reactions of Ti/Al/Mo/Au ohmic contacts on AlGaN/GaN heterostructure", Appl. Phys. Lett, 87, 141915 (2005).

[6] B. Van Daele and G. Van Tenedeloo *et al*, "Mechanism for Ohmic contact formation on Si_3N_4", Appl. Phys. Lett., 89, 201908 (2006).

[7] D. C. Look, Z. Q. Fang, B. Claflin, "Identification of donors, acceptors and traps in bulk-like HVPE GaN", J. Cryst. Growth, Vo. 281, pp. 143-150 (2005).

[8] B. Van Daele, G. V. Tendeloo, W. Ruythooren, J. Derluyn, M. R. Leys, "The role of Al on Ohmic contact formation on n-type GaN and AlGaN/GaN", Appl. Phys. Lett., Vol. 87, 061905 (2005).

[9] K. Kakushima, K. Tachi, J. Song, S. Sato, H. Nohira, E. Ikenaga, P. Ahmet, K. Tsutsui, N. Sugii, T. Hattori, H. Iwai, "Observation of band bending of metal/high-k Si capacitor with high energy x-ray photoemission spectroscopy and its application to interface dipole measurement", J. Appl. Phys, Vol. 104, 104908 (2008).

[10] S. Tanuma, C. J. Powell, D. R. Penn, "Calculation of electron inelastic mean free paths (IMFPs) VII. Reliability of the TPP-2M IMFP predictive equation", Surf. Interface Anal. 35, 268 (2003).

Impact of Current Measurement on Switching Characterization of GaN Transistors

Jennifer Lautner and Bernhard Piepenbreier

Chair of Electrical Drives and Machines, University of Erlangen-Nuremberg, Germany

Email: jennifer.lautner@fau.de

Abstract—**Gallium nitride (GaN) power devices promise better performance and efficiency compared to silicon transistors. However, the switching characterization of GaN devices is challenging, because of the high switching speed. In this paper the switching behavior of a GaN transistor is investigated with special focus on the impact of current measurement. A circuit simulation model including the critical parasitic elements was built to analyze parasitic effects and their impact on the switching transients. To confirm the simulation results by experiments, a PCB board was designed and different current measurement methods are compared. It is shown that the current sensor adds additional inductance in the power loop and causes ringing and oscillations. Furthermore, it can be seen, that the current sensor resistance damps the oscillations and improves the switching transients. Therefor, both variables should be small in order to reproduce the real switching characteristics as accurate as possible.**

I. INTRODUCTION

Wide bandgap semiconductor material like gallium nitride (GaN) has higher bandgap, breakdown field and electron mobility compared to silicon (Si). This leads to higher temperature capability, higher blocking voltage, lower on-resistance and higher frequency operations of GaN transistors [1], [2]. The high switching speed reduces switching losses and offers great performance improvements of GaN devices. However, the switching characterization of fast switching devices is challenging [3]. Due to the faster rising and falling edges parasitic elements impact the switching transients of GaN more than of Si devices. Therefor, the parasitic inductance has to be minimized by proper PCB layout to avoid high ringing and oscillations [4]. However, to characterize the switching behavior of power devices the voltage and current have to be measured. The integration of the current sensor adds additional inductance (and resistance) in the power loop and impacts the switching waveforms.

In the last few years some papers about switching characterization of GaN devices have been published [5]–[9]. Nevertheless, the impact of the current sensor on the switching transients has not been investigated so far. The aim of this paper is to show the impact of the current sensor on the switching behavior of a GaN transistor by simulation and experimental results. The device under test is the commercially available GaN transistor EPC2001 from Efficient Power Conversion Cooperation (EPC). In Section II the double pulse measurement test setup is described and different current measurement methods are compared. Section III presents the simulation model of the circuit and evaluates the impact of the current sensor

analytically. The experimental results for two different current measurement methods are shown and discussed in section IV.

II. MEASUREMENT TEST SETUP

A. Double Pulse Tester

Commonly the switching characteristics are investigated by a double pulse tester (DPT) with clamped inductive load, as shown in Fig.1. The device under test (DUT) is switched with a double pulse. The first pulse duration determines the load current. The DUT's switching transients can be evaluate by measuring the drain-source voltage v_{ds} and the drain current i_d at the end of the first pulse and at beginning of the second pulse. By regulating the DC bus voltage and first pulse duration different voltage and current stresses of DUT can be observed. Since the DUT is only switched twice there is no heating of the device and the measurement of the switching transients is independent of junction temperature. The DUT

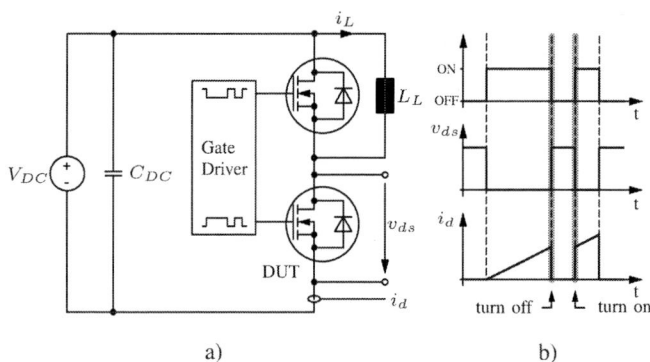

Fig. 1: a) Schematic of double pulse test circuit and b) typical waveforms.

is the commercially available 100V, 25A GaN transistor from EPC. This GaN device is a HEMT (High Electron Mobility Transistor) with a lateral structure and based on silicon [1]. In this paper two GaN transistors in a phase-leg configuration are used to imitate the behavior of a future converter. The transistors are driven complimentary by the special GaN half-bridge gate driver LM5113. In this paper the driver layout guidelines, as presented in [10], are used for PCB layout. Both GaN transistors are switched with 1Ω turn-on and turn-off gate resistance and without additional components to damp oscillations. The DC-link voltage is $V_{DC} = 50V$ and the switching current is $i_L = 10A$ for all simulations and experiments. For

978-1-4799-5494-0/14 $31.00 © 2014 IEEE

accurate measurement voltage and current probes with high bandwidth are necessary. The drain source voltage is measured with a 500MHz passive probe from LeCroy (PP008). The drain current measurement is discussed in the next section.

B. Current Measurement

The current measurement for characterization of fast switching devices is a particular challenge. Due to the high di/dt rates the current sensor needs high bandwidth and accuracy for precise measurement. Furthermore, the impact of the current sensor should be as small as possible. But the current sensor adds inductance in the power loop and worsens the switching waveforms. High oscillations and ringing in the measured voltage and current are the consequences. This additional inductance is mainly because of the larger trace on PCB board and not of the current sensor itself, as it is shown in Fig. 2. The inductance of the current sensor is very small and can be neglected in most cases. Fig. 2b shows a schematic of an optimal PCB-layout for the GaN half-bridge as presented in [4]. The current path of the power loop is illustrated in red color. In Fig. 2a the current sensor is inserted in the power loop. It can be seen, that the stray inductance of the power loop is significantly increased depending on the sensor size. This is a more serious problem for GaN devices with very

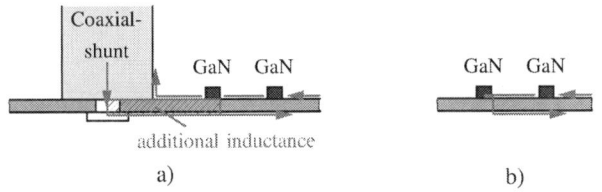

a)   b)

Fig. 2: Power loop current path on PCB board a) with and b) without current sensor.

small package inductance such as the GaN transistor from EPC because the inductance introduced by current sensor has a large portion of the overall power loop inductance. Hence, the size of the current sensor should be as small as possible to minimize the stray inductance in the power loop.
Commercially available current sensors with their respective advantages and disadvantages are listed in Tab. I. Due to the small size the rogowski coil would be suited. Nevertheless, the bandwidth of the coil is not sufficient for characterization of GaN devices as shown in [3]. The current transformer, e.g. used in [3], [8], has higher bandwidth as the current probe, but a limited maximum current. Both sensors are large and add consequently much inductance in the power loop. The coaxial shunt, which is used in [7], [9], [11], provides the highest bandwidth and the sensor construction is a bit smaller than with current transformer and current probe. The non-continuous current mode is no problem for double pulse test. However, a major disadvantage is the non-galvanic isolation. This has especially taken into account by using a passive probe for voltage measurement and can lead to grounding problems [11]. In some publications, e.g. [5], a custom made current sensor with 10x1Ω parallel resistors in 0201 package

TABLE I: Advantages and disadvantages of commercially available current sensors.

Sensor	Pros	Cons
Rogowski-coil	galvanic isolation size	low bandwidth and accuracy
Current probe	galvanic isolation continuous mode	size limited bandwidth
Current transformer	galvanic isolation high bandwidth	size current limitation
Coaxial shunt	high bandwidth and accuracy	non-galvanic isolation non continuous mode

TABLE II: Features of used current sensors for double pulse tester.

Sensor type	bandwidth	rise time	resistance
Fluke i50s	50MHz	7ns	–
T&M SDN-414-01	400MHz	1n	0.01Ω
T&M SDN-414-10	2GHz	0.18ns	0.1Ω

is used for switching characterization to reduce the parasitic inductance of the sensor construction in the power loop. However, this leads to current sharing problems.
Nevertheless, in this paper only commercially available current sensors are considered. A current probe from Fluke and two different coaxial shunts from T&M Research were used to compare different current measurement methods. The features of these sensors are listed in Tab. II. To fully exploit the potential of each current sensor two different PCB boards for current probe and coaxial shunt with optimal layout were designed.

III. SIMULATIONS

To investigate the impact of the current sensor on the switching behavior a simulation model of the double pulse test circuit including parasitic elements was built in LT-Spice. The simulation model is shown in Fig. 3. For the GaN devices manufacturer models are used, which includes parasitic capacitance and resistance. The parasitic inductance of the GaN device package is very small [1] and can be neglected. The drain inductance L_d includes all parasitic stray inductances of the power loop and is estimated with 1nH. The source inductances L_{s_1} and L_{s_2} are 100pH respectively. The gate resistor R_g is chosen to 1Ω. The inductive load and the DC-link capacitor contains parasitic elements. The current

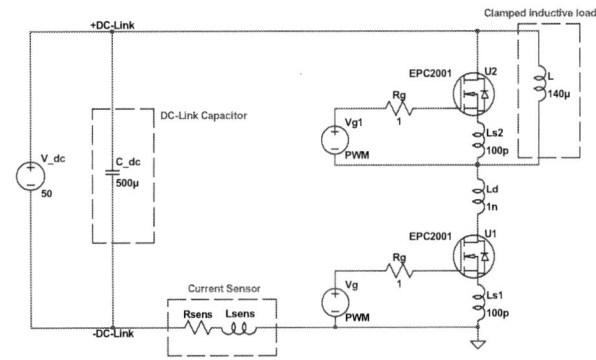

Fig. 3: Simulation circuit of double pulse tester including parasitic elements.

sensor is modeled by the resistance R_{Sens} and the inductance L_{Sens}. To evaluate the impact of the current sensor on the switching transients L_{Sens} and R_{Sens} are varied.

A. Influence of current sensor inductance L_{Sens}

The variable L_{Sens} describes the additional inductance introduced by the current sensor. Fig. 4 shows the turn-off and Fig. 5 the turn-on voltage v_{ds} and the drain current i_d for different values of L_{Sens}. The switching waveforms without current sensor are simulated with $L_{Sens} = 0$nH. $L_{Sens} = 5$nH represents approximately the behavior with coaxial shunt and $L_{Sens} = 25$nH the behavior with current probe.

The simulated turn-off transients show that L_{Sens} influences both the voltage and the current waveform by oscillations. The amplitude of the oscillation increases and the frequency of the oscillation decreases with increasing inductance L_{Sens}. Thus, the voltage stress increases with rising L_{Sens}. Moreover, it is obvious that the over-voltage is almost zero without current sensor.

The simulated turn-on transients show especially current stress due to charging current of the parasitic capacitance. The GaN transistors have no reverse recovery [1]. The current stress increases first for $L_{Sens} = 5$nH and decreases for higher sensor inductance. Furthermore, Fig. 5 shows that higher L_{Sens} slows down the current and the voltage drop is increased.

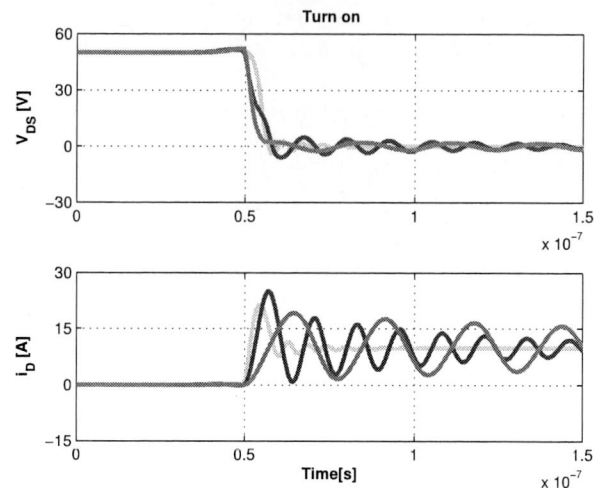

Fig. 5: Turn-on switching transients simulated with $L_{Sens} = 0$nH (—), $L_{Sens} = 5$nH (—) and $L_{Sens} = 25$nH (—).

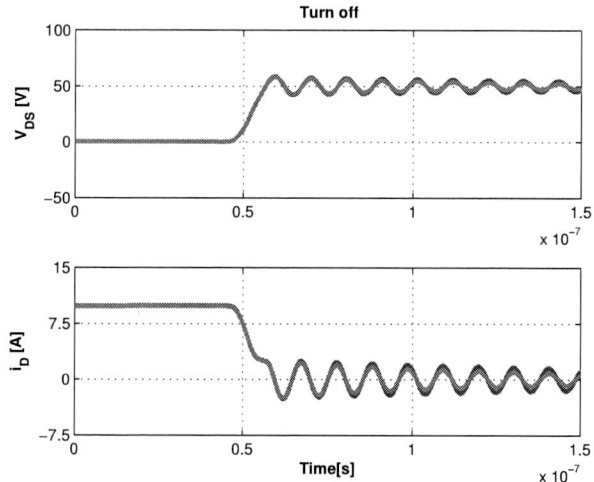

Fig. 6: Turn-off switching transients simulated with $R_{Sens} = 0\Omega$ (—), $R_{Sens} = 10$mΩ (—) and $R_{Sens} = 100$mΩ (—).

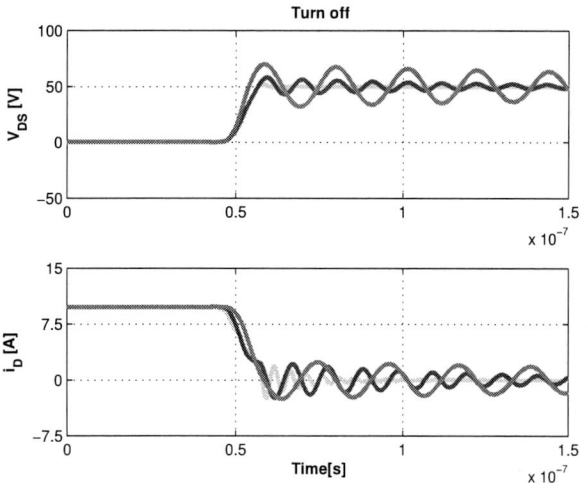

Fig. 4: Turn-off switching transients simulated with $L_{Sens} = 0$nH (—), $L_{Sens} = 5$nH (—) and $L_{Sens} = 25$nH (—).

B. Influence of current sensor resistance R_{Sens}

The influence of the current sensor resistance R_{Sens} on the switching transients is not as obvious as the influence of the sensor inductance, but should be nevertheless taken into account. Fig. 6 and Fig. 7 show the switching transients for different values of R_{Sens}. The sensor inductance L_{Sens} is chosen to nH. Simulated turn-off and turn-on transients show, that the resistor R_{Sens} damps the oscillation as expected. This effect is especially visible in the current turn-on waveform.

IV. EXPERIMENTAL RESULTS

To verify the simulations two double pulse test boards for measurement with current probe and with coaxial shunt, as described in Section II, were designed. Two coaxial shunts with different resistance were tested. The switching transients measured with the current probe are shown in Fig. 8 and with the coaxial shunt in Fig. 9 and Fig. 10. The current turn-off and turn-on waveforms measured with the current probe are delayed approximately by 10ns due to sensor properties. This delay is not compensated in the shown switching waveforms. The measurement with current probe shows higher overvoltage and lower frequency ringing in turn-off transients compared to measurement with coaxial shunt due to the larger power loop inductance introduced by current sensor construction. Furthermore, the turn-on current measured with current probe has smaller slope and reduced overcurrent. The comparison

978-1-4799-5494-0/14 $31.00 © 2014 IEEE

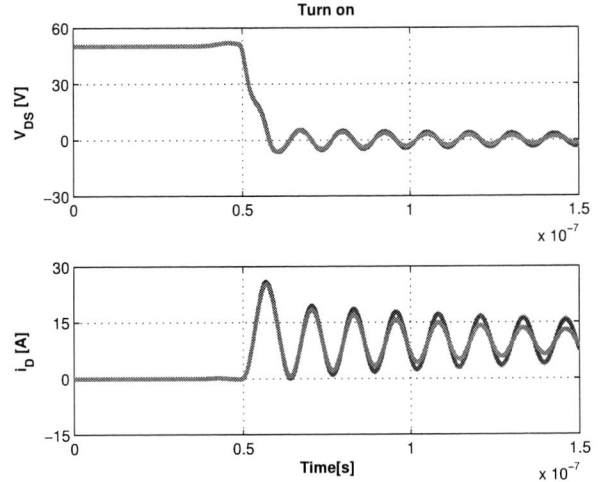

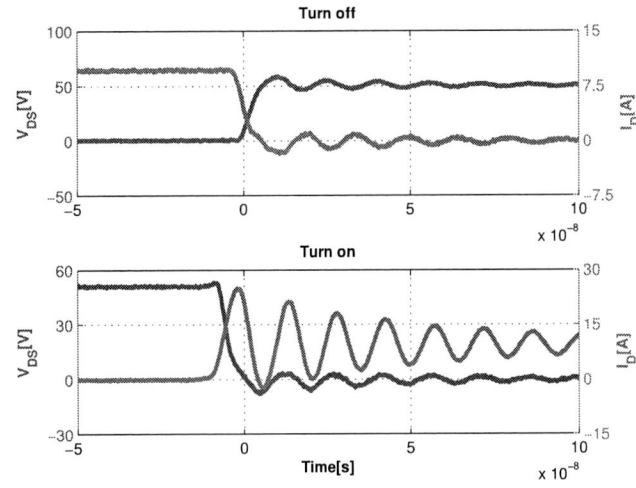

Fig. 7: Turn-on switching transients simulated with $R_{Sens} = 0\Omega$ (—), $R_{Sens} = 10\text{m}\Omega$ (—) and $R_{Sens} = 100\text{m}\Omega$ (—).

Fig. 9: Switching transients measured with coaxial shunt (R=100mΩ).

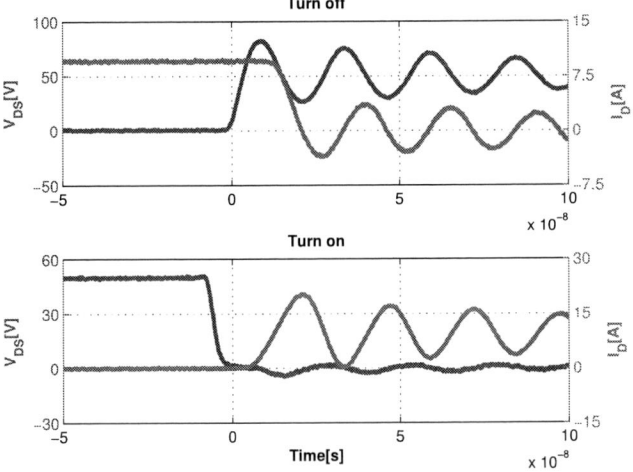

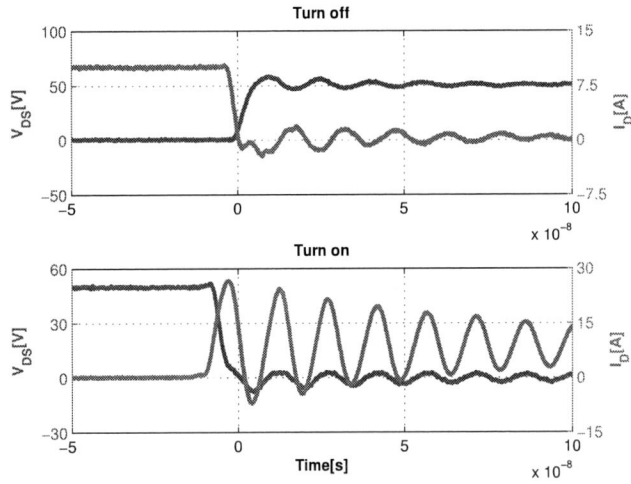

Fig. 8: Switching transients measured with current probe.

Fig. 10: Switching transients measured with coaxial shunt (R=10mΩ).

between the two coaxial shunts with different resistance shows damped waveforms especially the turn-on current for higher shunt resistance. Thus, the experimental results confirm the simulations of Section III.

V. CONCLUSION

In this paper the impact of the current sensor on the switching characterization of GaN transistors is analyzed in simulations and experiments. It has been shown, that the current sensor construction adds large additional inductance in the power loop and influences both current and voltage switching transients. This results in higher oscillations and ringing leading to large voltage and current stress. The best results were thereby achieved with the coaxial shunt. This sensor has a high bandwidth and can be inserted in the power loop with the least additional inductance. However, the impact of the current sensor is especially in the turn-on current visible

and worsens the switching transients. Nevertheless, the switching characterization is important and necessary, but additional inductance and resistance introduced by the current sensor should be as small as possible. Otherwise large deviations between expected characteristics and measured characteristics can occur.

REFERENCES

[1] A. Lidow, *GaN transistors for efficient power conversion: The eGaN FET journey continues.* El Segundo and CA: Power Conversion Publications, 2012.

[2] N. Kaminski, "State of the art and the future of wide band-gap devices," in *Power Electronics and Applications, 2009. EPE '09. 13th European Conference on,*, 2009, pp. 1–9.

[3] H. Li and S. Munk-Nielsen, "Challenges in switching sic mosfet without ringing," in *PCIM Europe 2014; International Exhibition and Conference for Power Electronics, Intelligent Motion, Renewable Energy and Energy Management; Proceedings of,* 2014, pp. 1–6.

[4] D. Reusch and J. Strydom, "Understanding the effect of pcb layout on circuit performance in a high-frequency gallium-nitride-based point of

load converter," *Power Electronics, IEEE Transactions on*, vol. 29, no. 4, pp. 2008–2015, 2014.

[5] M. Danilovic, Zheng Chen, Ruxi Wang, Fang Luo, D. Boroyevich, and P. Mattavelli, "Evaluation of the switching characteristics of a gallium-nitride transistor," in *Energy Conversion Congress and Exposition (ECCE), 2011 IEEE*, 2011, pp. 2681–2688.

[6] R. Mitova, R. Ghosh, U. Mhaskar, D. Klikic, Miao-Xin Wang, and A. Dentella, "Investigations of 600-v gan hemt and gan diode for power converter applications," *Power Electronics, IEEE Transactions on*, vol. 29, no. 5, pp. 2441–2452, 2014.

[7] Zheyu Zhang, Weimin Zhang, F. Wang, L. Tolbert, and B. Blalock, "Analysis of the switching speed limitation of wide band-gap devices in a phase-leg configuration," in *Energy Conversion Congress and Exposition (ECCE), 2012 IEEE*, 2012, pp. 3950–3955.

[8] N. Badawi, O. Hilt, E. Bahat-Treidel, S. Dieckerhoff, and H.-J. Würfel, "Switching characterisits of 200v normally-off gan hemts," in *PCIM Europe 2013; ; International Exhibition and Conference for Power Electronics, Intelligent Motion, Renewable Energy and Energy Management; Proceedings of*, 2013, pp. 319–324.

[9] Zhuxian Xu, Weimin Zhang, Fan Xu, F. Wang, L. Tolbert, and B. Blalock, "Investigation of 600 v gan hemts for high efficiency and high temperature applications," in *Applied Power Electronics Conference and Exposition (APEC), 2014 Twenty-Ninth Annual IEEE*, 2014, pp. 131–136.

[10] Youhao Xi, Min Chen, K. Nielson, and R. Bell, "Optimization of the drive circuit for enhancement mode power gan fets in dc-dc converters," in *Applied Power Electronics Conference and Exposition (APEC), 2012 Twenty-Seventh Annual IEEE*, 2012, pp. 2467–2471.

[11] Zheyu Zhang, Ben Guo, F. Wang, L. Tolbert, B. Blalock, Zhenxian Liang, and Puqi Ning, "Methodology for switching characterization evaluation of wide band-gap devices in a phase-leg configuration," in *Applied Power Electronics Conference and Exposition (APEC), 2014 Twenty-Ninth Annual IEEE*, 2014, pp. 2534–2541.

978-1-4799-5494-0/14 $31.00 © 2014 IEEE

The influence of SiC/SiO$_2$ interface morphology on the electrical characteristics of SiC MOS structures

L. Liu,[1,2] C. Jiao,[1] Y.Xu,[3] G.Liu,[3] L. C. Feldman[3] and S. Dhar[1]

1. Dept. of Physics, Auburn University, Auburn, AL, USA
2. School of Microelectronics, Xidian University, Xi'an, China
3. Institute for Advanced Materials, Devices and Nanotechnology, Rutgers University, Piscataway, NJ, USA

Abstract—The effect of roughness at the SiC/SiO$_2$ interface on electrical properties of 4H-SiC MOS devices has been investigated. Variations in surface roughness were generated by annealing 4H-SiC samples at high temperatures (1550°C-1650°C) with or without a graphitic cap layer. Subsequently, gate oxides were grown on these surfaces for n-type MOS capacitors and n-channel MOSFETs were fabricated. Although the interfaces demonstrated significantly different surface morphology, interface state density (D$_{it}$) measured on the capacitors were almost identical. This was reflected in the MOSFET characteristics where, to first order, no obvious difference in field-effect mobility was observed. This result verifies that long range roughness (in the micron scale) does not affect mobility of channel electrons where the mean free path is of the order of a ~ 1 nm due to the low inversion layer mobility.

Keywords— 4H-SiC; surface morphology; roughness; mobility; interface states

I. INTRODUCTION

Owing to its attractive physical properties such as a wide band gap, high critical electrical field, and high thermal conductivity, the 4H polytype of silicon carbide (4H-SiC) attracts great attention in the field of high performance power electronic devices. It is well known that high temperature annealing at temperatures greater than ~ 1400°C, required for post-implantation dopant activation, leads to out-diffusion of surface Si atoms and the formation of "step bunching" if the surface is not protected with a capping layer. It has been reported that the degradation in SiC surface morphology a impacts the interface state density and mobility of MOSFET[1][2]. Although many methods have been adopted to improve the surface roughness, such as using a protective layer of a graphitic cap, there still exists controversy about the influence of SiC/SiO$_2$ interface morphology after the post-implantation annealing on the interface state density and channel mobility. Fiorenza *et al.*[3], recently reported that the carbon cap used to protect SiC surface during 1650°C high temperature annealing results in an increased interface state density (D$_{it}$) and a decreased channel mobility compared to the devices without carbon cap. They attributed the increased D$_{it}$

to the different areal percentage exposing different basal planes before and after carbonization[4]. On the other hand, Haney *et al.*[5] found in their experiments that the presence of or lack of a carbon cap during high temperature annealing (1200°C ~1800°C) does not change the channel mobility and the D$_{it}$. Naik *et al.*[6] reported that by using a graphite layer during dopant activation annealing, both the D$_{it}$ and channel mobility decreased. They interpreted the decreased mobility as a consequence of increased roughness using graphitic cape, instead of Coulomb scattering. Masuda *et al.*[7] deposited a silicon film to restrain the desorption of Si from the surface during high-temperature annealing above the Si melting point, then removed Si by acid, and found that a macro-stepped structure was formed on the surface. Compared to the flat surface (20 cm^{2}V^{-1}s^{-1}), they obtained a high peak mobility (102 cm^{2}V^{-1}s^{-1}) on the surface in the presence of macro-step. Therefore, the effect of interface roughness on carrier transport in SiC MOS channels remains inconclusive. In this work, a systematic investigation is carried out by varying the interface roughness and studying its effect on electronic properties. Variations in surface roughness were generated by annealing 4H-SiC samples at high temperatures (1550°C-1650°C) with or without a graphitic cap layer. Roughness was characterized by AFM measurement. We find that the D$_{it}$ and field-effect mobility (μ$_{fe}$) show no obvious correlation with the surface roughness. This result verifies that long range roughness (on the micron scale) does not affect mobility of channel electrons where the mean free path is of the order of ~ 1 nm[8] due to the low inversion layer mobility.

II. EXPERIMENTAL

Long-channel lateral MOSFETs (W=200μm, L=210μm) are fabricated on the Si-face of 4H-SiC P$^-$/N$^+$ substrate. The p-type doping concentration of the SiC epilayer was 2×10^{16}cm^{-3}. Source and drain regions are formed by multiple ion implantation of N, with a uniform doping concentration 2~3×10^{19}cm^{-3}. The carbon capping layer was formed by spin coating a layer of photo-resist on the surface followed by baking at 600°C for about 30 minutes. Samples with or without the carbon cap process (as described above)

978-1-4799-5494-0/14 $31.00 © 2014 IEEE

experienced high temperature annealing in the temperature range 1550°C~1650°C for 30 minutes. Following this, O_2 plasma etching was used to remove the carbon cap. A sacrificial oxidation at 1150°C was performed to remove residual carbon. Subsequently gate oxides (~52nm) were grown by dry thermal oxidation at 1150°C followed by a 1175°C, 2-hour post-oxidation annealing in nitric oxide. Molybdenum, deposited by DC sputtering was used as the gate metal for the MOS capacitors and the MOSFETs. Sputtered Ni was used as source-drain contacts. The contacts were annealed at 800°C for 1 minute. The surface morphology was evaluated using atomic force microscopy (AFM) at the following processing steps (i) As received epitaxial surface (ii) After high temperature annealing (iii) After oxidation (iv)After removal of oxide. There was no significant difference in morphology between (ii), (iii) and (iv) but large difference changes was observed between (i) and (ii) depending on whether the surface was capped or not during the annealing. After the device fabrication, simultaneous high-low frequency capacitance-voltage (C-V) and current-voltage (I-V) characteristics were measured at room temperature. Low temperature C-V measurements were also carried out in order to evaluate fast interface states energetically located near the SiC conduction band-edge.

III. RESULTS

Fig.1 shows the surface morphology of the sample without a carbon cap during the processing. The measured RMS of the wavy surface is 6.1nm. As shown in Fig.2, it is evident that the carbon cap process results in a significantly smoother surface, reducing RMS to 2.6nm. Annealing carried out at a lower temperature of 1550°C, also shows the same trend (not shown here). This is consistent with previous observation that surface roughness increases with annealing temperature due to higher sublimation of Si atoms from the surface during the high temperature treatment [8].

To identify the relationship of mobility with surface roughness, the room temperature transfer characteristics (I_D-V_G) were measured at a small drain-to-source voltage (V_{ds} = 25mV). Typical results are shown in Fig. 3 and 4 respectively. The field effect mobility is calculated from I_D-V_G curve using the relation:

$$\mu_{FE} = \frac{L}{WC_{OX}V_{DS}}\frac{\partial I_{DS}}{\partial V_G} \qquad (1)$$

where L and W are the channel length and width respectively, C_{ox} is the gate capacitance per unit area, and $\partial I_D/\partial V_G$ is the derivative taken from I_D-V_G curve. As shown in these two figures, there is no obvious correlation between channel mobility and interface roughness. In Fig.3, the I_D-V_G curves of devices with and without carbon cap almost overlap, where the small difference can be attributed to the difference in contact resistance of these samples. This similar characteristic is reflected in comparison of the field-effect mobility curves in Fig.4. The peak mobility is 20 cm^{2}V^{-1}s^{-1} for device with carbon cap; 22 cm^{2}V^{-1}s^{-1}, without carbon cap, respectively. At high gate voltages, as the amount of inversion charge increases, the occupied traps are more effectively screened by the mobile inversion charge so Coulomb scattering diminishes, and the total mobility is dictated by surface roughness scattering mobility μ_{sr} and/or surface phonon scattering mobility μ_{ph}. The overlap of the tail part of these two curves obviously indicates that, although the correlation length L and root mean square deviation Δ are different between the devices processed with or without carbon cap, not much difference is caused by these two significantly different surface morphologies for electron transport along the channel.

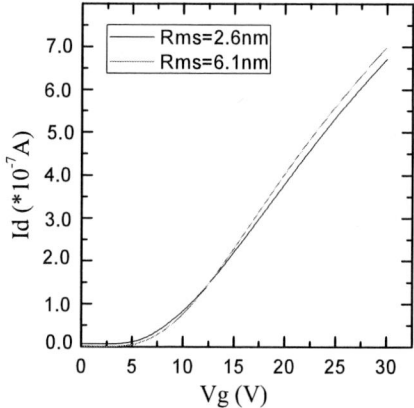

Fig.3. Comparison of I_d~V_g for different interface treatment (room temperature)

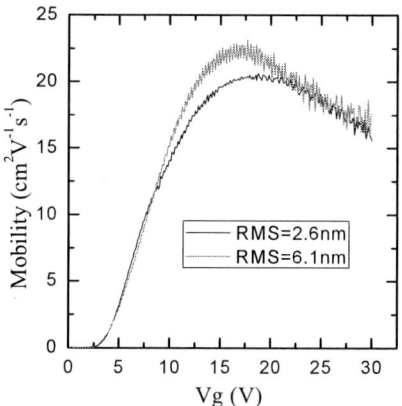

Fig.4. Comparison of mobility for different interface treatment (room temperature)

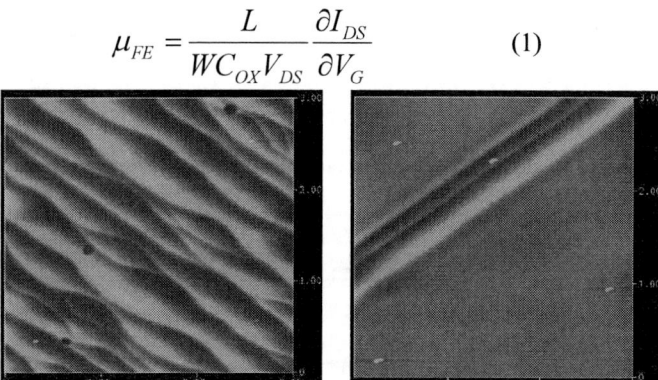

Fig.1. Surface morphology of device annealed without carbon under 1650°C (RMS=6.1nm)

Fig.2. Surface morphology of device annealed with carbon under 1650°C (RMS=2.6nm)

978-1-4799-5494-0/14 $31.00 © 2014 IEEE

The channel mobility of MOSFETs is proportional to electron mean free path λ, where λ is less than 1 nm, calculated from the low inversion channel mobility. On our sample surfaces, the dominant roughness spacing (peak to peak or valley to valley) is of the order of micron, much greater than λ, thus not defining the limiting factor for inversion mobility. However, the AFM scan is limited by spatial resolution in the order of nanometers. Thus the mobility may still be limited by a finer scale (nanoscale) roughness. Note that despite the large RMS roughness values, the high temperature anneal induced roughness is essentially step bunching, where the surface consists of relatively large flat terraces, with minimal crystal damage. This is distinctly different from the fine scale random roughness caused by reactive ion etching (RIE), which also involves significant crystal structure damage, in addition to large roughness. Other results from our laboratory show that SiC MOS devices fabricated on RIE roughened surfaces suffer from very early breakdown, due to roughness induced field crowding, as well as low mobility due to high defect density [10].

Fig.5 shows typical D_{it} profiles in the range of 0.2eV to 0.6eV from conduction band edge of the samples with different interface treatment, extracted from simultaneous high-low frequency C-V measurement on n-type MOS capacitors. It is evident from the results that the sample fabricated with carbon cap (solid circle) and without (solid triangle) exhibit identical D_{it} profile, which suggest that trapping at the interfaces with different morphologies is about the same.

H. Yoshioka et al.[11] pointed out that nitridation of the SiC/SiO₂ interface generates very fast states (higher than 1MHz). The response frequency of such very fast states increases when the energy level approaches to the conduction band edge. Hence, the conventional high-low CV method (usually with the maximum probe frequency of 0.1~1 MHz) cannot detect these fast states. However, at low temperature,

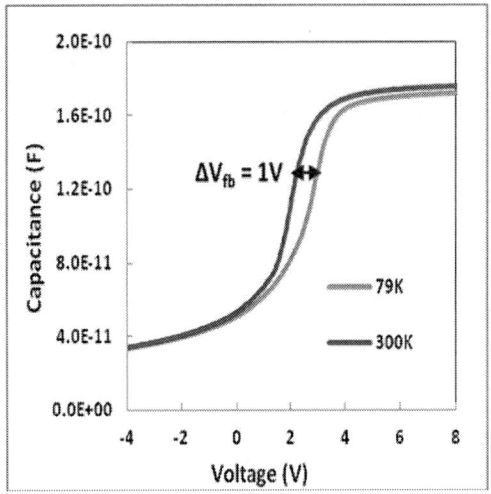

Fig.6. 100kHz C-V of NO annealed device (without interface roughness treatment) at room and low temperature

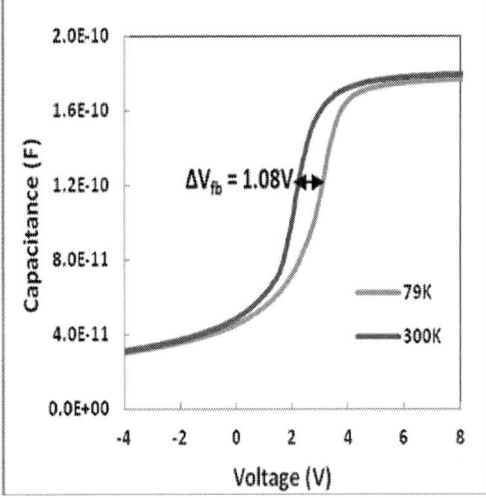

Fig.7. 100kHz C-V of NO annealed device (1650°C annealing without carbon cap, RMS=6.2nm) at room and low temperature

the emission times of interface states become longer, which slows down the overall trap response. This effect results in 'freezeout' of some of the fast traps near the conduction band-edge which are hard to detect at room temperature. These frozen in traps act like fixed charge and result in a shift of the CV curves to the right at the lower temperature. The flat-band voltage shift ΔV_{fb} between room temperature and low temperature CV provides a qualitative measure of fast interface states. In this experiment, the devices were cooled to 79K. Figures 6 and 7 show typical room temperature and low temperature CV characteristics of two different devices- (i) Device without any high temperature interface roughening treatment (smooth interface), (ii) Device annealed at 1650°C without carbon cap protection (rough interface, RMS=6.2nm). As evident from the figures the shift of the CV curves is about ~ 1V for both cases which indicates that the rough interface maintains the same D_{it} profile even taking the fast states into account.

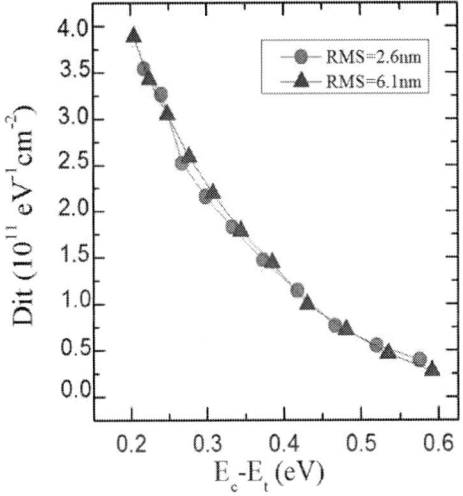

Fig.5. Comparison of D_{it} for different interface treatment

IV. CONCLUSION

The effect of SiC/SiO$_2$ interface roughness on electrical properties of 4H-SiC MOS capacitors and MOSFETs has been investigated in this work. Different surface roughness was generated by annealing 4H-SiC samples at high temperatures (1550°C-1650°C) with or without a graphitic cap layer. Following gate oxidation, the interfaces demonstrated significantly different surface morphology but D$_{it}$ is almost identical. This was reflected in the MOSFET characteristics and no obvious difference in field-effect mobility was observed between devices with different micro-scale roughness. This experiment verifies that micron-scale roughness (peak to peak or valley to valley) which is much greater than the electron mean free path (in nm scale) does not affect channel mobility due to the low collision probability of electrons near the SiC/SiO$_2$ interface.

ACKNOWLEDGMENT

This work was supported by the U.S. Army Research Laboratory (W911NF-07-2-0046, Program Manager: Dr. Aivars Lelis), and the II-VI Foundation Block-Gift Program.

REFERENCES

[1] W. Zhou, X. Zhong, K. Sheng, "High temperature stability and the performance degradation of SiC MOSFET", 2014, *IEEE Trans. On Power electronics*, 29(5):2329-2337.

[2] P. Fiorenza, L. K. Swanson, M. Vivona, *et al.*, "Comparative study of gate oxide in 4H-SiC lateral MOSFETs subjected to post-deposition-annealing in N$_2$O and POCl$_3$", 2013, *Appl. Phys. A: Mat. Sci. and Pro.*,:1-7.

[3] P. Fiorenza, F. Giannazzo, L. K. Swanson, *et al.*, "A look underneath the SiO$_2$/4H-SiC interface after N$_2$O thermal treatments", 2013, *Beilstein Journal of Nanotechnology*, 4:249-254.

[4] F. Roccaforte, P. Fiorenza, F. Giannazzo, "Impact of the Morphological and Electrical Properties of SiO$_2$/4H-SiC Interfaces on the Behavior of 4H-SiC MOSFET", 2013, *ECS Journal of Solid State Science and Technology*, 2(8): N3006-N3011.

[5] S. Haney, A. Agarwal, "The Effects of Implant Activation Anneal on the Effective Inversion Layer Mobility of 4H-SiC MOSFET", 2008, *J. Electron Mater.*, 37(5):666-670.

[6] H. Naik, K. Tang, T. P. Chow, "Effect of Graphite Cap for Implant Activation on Inversion Channel Mobility in 4H-SiC MOSFET", 2009, *Mater. Scie. Forum*, 615- 617:773-776.

[7] T. Masuda, S. Harada, T. Tsuno, *et.al.*, "High Channel Mobility of 4H-SiC MOSFET Fabricated on Macro-stepped Surface", *Mat. Sci.For*, 2009. 600- 603:695-698.

[8] M. A. Capano, S. Ryu, J. A. Cooper, *et al.*, "Surface roughening in ion implanted 4H-Silicon Carbide", *Journal of Electeronic Materials*, 1999, 28(3):214-219.

[9] Z. O. Wang, L. F. Mao, "Remote surface roughness effects on inversion electron density in nano-MOSFET", *Eur. Phys. J. Appl. Phys.* 2009, 48: 20301 - 5.

[10] G. Liu, Y. Xu, C. Xu, A. Basile, F. Wang, S.Dhar, E. Conrad, P. Mooney, T. Gustafsson, E. Garfunkel, L C. Feldman, "Effects and Mechanisms of RIE on SiC Inversion Layer Mobility and its Recovery", submitted.

[11] H. Yoshioka, T. Nakamura, and T. Kimoto, "Generation of very fast states by nitridation of the SiO2/SiC interface", *J. Appl. Phys.*, 112, 024520, 2012.

Optically-Switched High-Voltage Bipolar SiC Device

S.K. Mazumder and A. Mojab
Laboratory for Energy and Switching-Electronics Systems
University of Illinois at Chicago (UIC)
851 South Morgan Street, Chicago, IL: 60607
Emails: {mazumder, amojab2}@uic.edu

L. Cheng[1], A.K. Agarwal[2], and C.J. Scozzie[3]
[1]Cree Inc., 4600 Silicon Dr., Durham, NC 27703, USA
[2]U.S. Department of Energy, 1000 Independence Ave. SW, Washington DC 20585, USA
[3]U.S. Army Research Laboratory, 2800 Powder Mill Rd, Adelphi, MD

Abstract— **A new design of a very high-voltage, single-bias, and optically-controlled Emitter-Turn-Off (ETO) thyristor is introduced. Both the thyristor and the controlling switch are triggered optically using two optical sources. The silicon-carbide (4H-SiC) thyristor is triggered using a short wavelength laser source and the optical silicon switch in series with the anode contact of the thyristor is triggered with an 808-nm laser pulse. In the outlined structure, only a single power bias of 15 kV is required. The low-voltage control bias, typically used for triggering devices in an electrical ETO thyristor, is eliminated due to all-optical control. This eliminates the problem of susceptibility to electromagnetic interference (EMI) and yields complete electrical isolation between the power and the control stages.**

Keywords—**optical switching; bipolar; SiC; high voltage**

I. Approach

The patented single-biased all-optically-triggered 15-kV ETO, with one optical and two electrical terminals, is shown in Fig. 1 [1]. It has numerous potential applications ranging from pulsed power, fault isolation, electric traction, renewable energy, ac or dc power grid, motor drive to name a few [2]. In contrast, a conventional ETO [3] thyristor shown in Fig. 2 comprises three electrical terminals and requires low-voltage control as well as high-voltage power bias. The new optical ETO comprises an optically-gated 15-kV SiC thyristor [4]. Its device structure along with its doping densities and thickness of the different layers are shown in Fig. 3. The 15-kV voltage is blocked using a new (drift-layer) epitaxial structure methodology developed by Cree. Furthermore an optically-triggered power transistor (OTPT), shown in Fig. 4, is used instead of the PMOS transistor (used in conventional electrically-triggered ETO configuration) in order to make the triggering and control circuit of the optical ETO thyristor all-optically-triggered. This significantly enhances system reliability. Furthermore, the all-optical link between the device controller and the optical ETO thyristor provides significant immunity to electromagnetic interference (EMI), which is caused by high-power and high-voltage switching. In addition, the optical link of the optical ETO thyristor precludes any possibility of back propagation of power surge from the high-voltage-device based power stage to the controller.

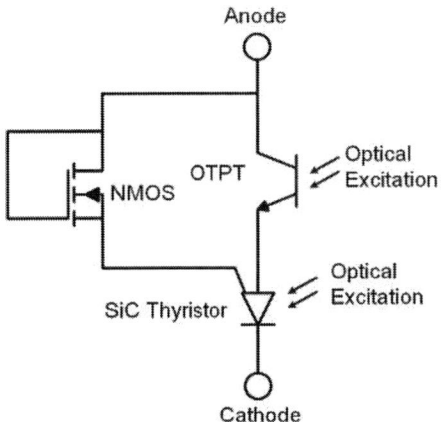

Fig. 1: Schematic diagram of the new proposed single bias all optically-triggered ETO thyristor. The NMOS transistor is connected in a diode configuration (gate and drain shorted) and will turn on only if a voltage drop of more than its threshold voltage is employed across it. This will occur during the ETO turn-off when the laser pulse on the OTPT is shut off. So the OTPT resistance rises rapidly and the voltage drop across the NMOS is increased. Almost immediately after that, all the current from the anode contact is commutated to the gate contact, forcing the optical SiC thyristor to turn off rapidly.

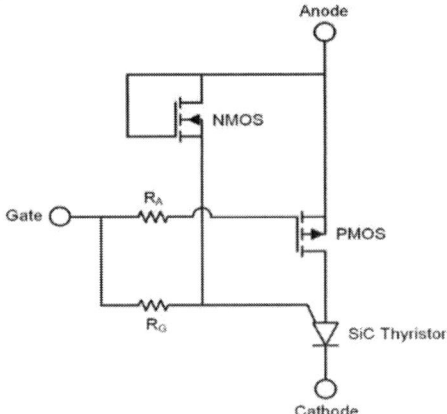

Fig. 2: Schematic diagram of a previously existing electrically-triggered ETO. In this configuration, the negative high voltage supply is connected to the cathode contact to reduce the stress on the switching circuit at the anode side. But, it is susceptible to EMI and isolation is not adequate. The turn-on process starts by setting the gate voltage to a negative level of about -15 V. This turns on both the PMOS transistor and the thyristor. The turn-off process starts by setting the gate voltage from -15 to zero. This turns off the PMOS and once the NMOS voltage exceeds its threshold voltage, it is turned on rapidly. Therefore a huge amount of anode current will commutate to the gate contact and force the thyristor to turn-off.

978-1-4799-5494-0/14 $31.00 © 2014 IEEE

II. RESULTS AND SIGNIFICANCE

A set of important and necessary physical models have been used in our Silvaco device-level simulations. In Fig. 5a, the voltage across the load and the optical control signal on the OTPT is shown. During the on-state, the voltage across the thyristor and across the OTPT is 4.6 V and 0.8 V, respectively, leading to a total on-state voltage drop of 5.4 V. During the ETO turn off of the optical ETO thyristor, the maximum transient voltage on the OTPT is less than 10 V (parasitic inductances are assumed to be small). Fig. 5b shows the results for the thyristor currents versus transient time for the optical ETO. A rise and fall times of 64 and 393 ns are achieved, respectively. Decreasing the defect in the epitaxial layer during its growth is an important factor in improving the performance of the optical SiC thyristor. Higher carrier lifetime leads to better thick drift layer conductivity (lower voltage drop) during the thyristor on-state (Fig. 6). The drawback of having longer lifetime is the high delay in the device turn-off, however this can be compensated by using higher thyristor driving currents.

ACKNOWLEDGEMENT

This work has been supported in part by U.S. NSF (award no. 1202384) and U.S. ARPA-E (award no. DE-AR0000336). However, any opinions, findings, conclusions, or recommendations expressed herein are those of the authors and do not necessarily reflect the views of NSF or ARPA-E.

REFERENCES

[1] S.K. Mazumder, "Photonically activated single bias fast switching integrated thyristor", USPTO Patent# US 8796728 B2, awarded on August 5, 2014.

[2] S.K. Mazumder, A. Mojab, and H. Riazmontazer, "Optically-switched wide-bandgap power semiconductor devices and device-transition control", *Physics of Semiconductor Devices, Springer*, pp. 57-65, 2014.

[3] J. Wang and A. Huang, "Design and characterization of high-voltage silicon carbide emitter turn-off thyristor," IEEE Transactions on Power Electronics, vol. 24, no. 5, pp. 1189-1197, May 2009.

[4] L. Cheng, A.K. Agarwal, C. Capell, M. O'Loughlin, K. Lam, J. Richmond, A. Burk, J. Palmour, A.A. Ogunniyi, H.K. O'Brien, and C. J. Scozzie, "15 kV, large Area (1 cm²) 4H-SiC p-type gate turn-off thyristors," Proceedings of the 9th European Conference on Silicon Carbide and Related Materials (ECSCRM'12), Mat. Sci. Forum, Vols. 740-742 (2013), pp. 978-981, Saint Petersburg, Rissia, Sept. 2-6, 2012.

[5] Q. Zhang, A. Agarwal, C. Capell, L. Cheng, M. O'Loughlin, A. Burk, J. Palmour, V. Temple, A. Ogunniyi, H. O'Brien, and C.J. Scozzie, "SiC super GTO thyristor technology development: Present status and future perspective," Pulsed Power Conference (PPC), pp. 1530-1535, 2011.

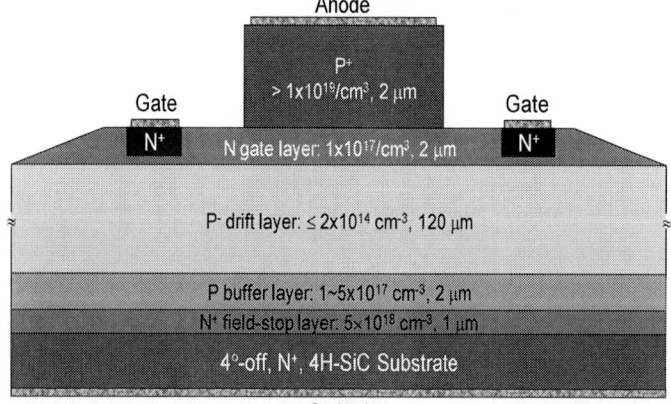

Cathode

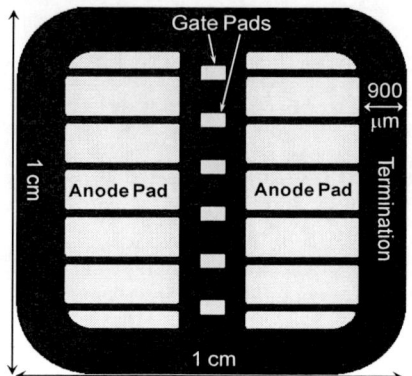

Fig. 3: Schematic cross-section and layout of 15 kV, 4H-SiC p GTO [4], [5].

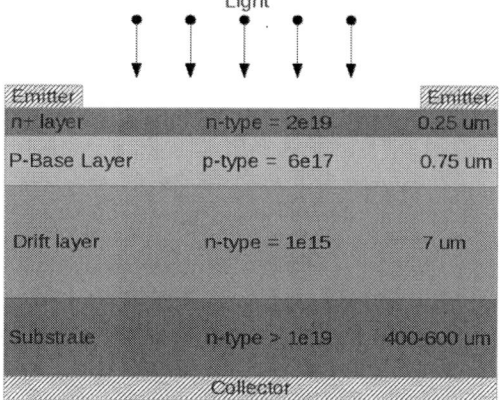

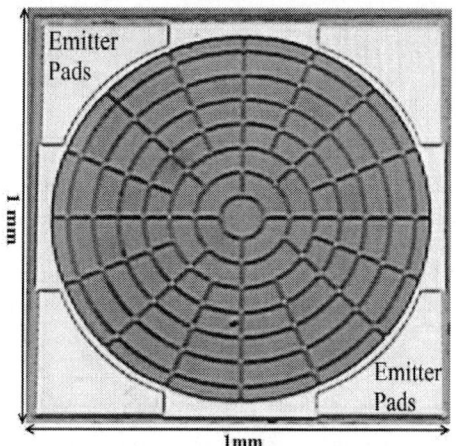

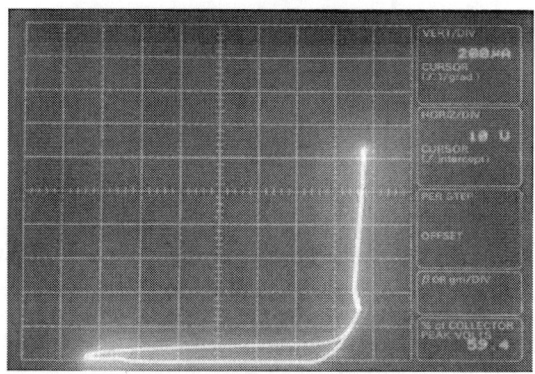

Fig. 4: Schematic cross-section and layout of the optical part of the Darlington OTPT. (Bottom) Breakdown test performed by Tektronix 371A on the OTPT.

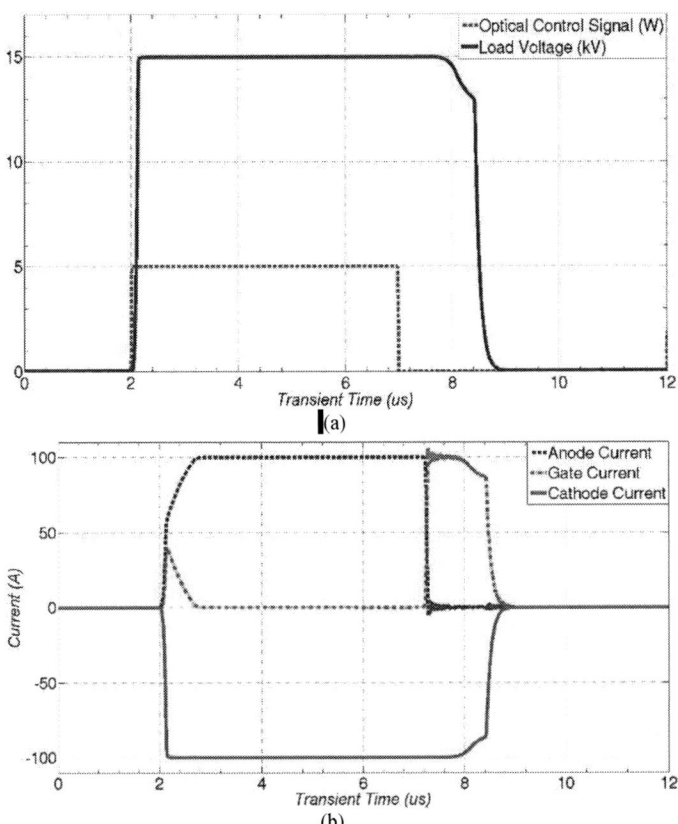

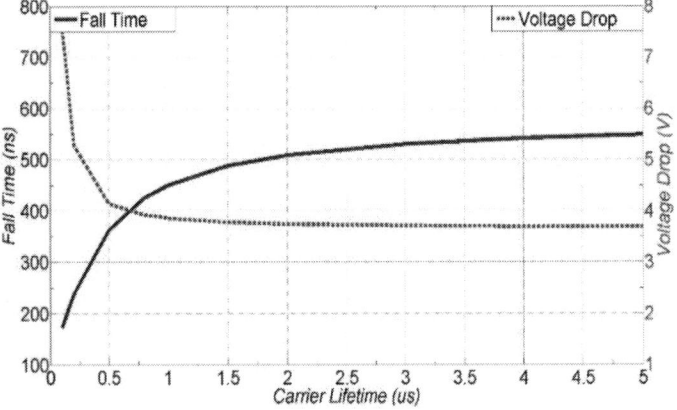

Fig. 5: Voltage across the load and the optical control signal on the OTPT (a), and Output currents for the thyristor in the optically-triggered ETO (b).

Fig. 6: Fall time and voltage drop vs. carrier lifetime for the optically-triggered ETO thyristor. Cree Inc. announced that they have achieved to increase the drift layer carrier lifetime up to about 2 μs and they are trying to increase that to 5 μs [6]. The advantage of having higher lifetime is the fast thyristor firing and low conduction voltage drop, which are big challenges in high-power electronic devices.

Enhanced Oxidation of SiC Substrates using La$_2$O$_3$ Capped Annealing and a Proposal for Uniform LaSiON Gate Dielectric Formation

Y. M. Lei, S. Munekiyo, T. Kawanago, K. Kakushima, K. Kataoka, H. Wakabayashi, K. Tsutsui, K. Natori, H. Iwai
Tokyo Institute of Technology
Yokohama 226-8503, Japan
Email address: yiming.l.aa@m.titech.ac.jp
M. Furuhashi, N. Miura

Mitsubishi Electric Corp.
Amagasaki 661-8661, Japan
Email
address:Furuhashi.Masayuki@ay.MitsubishiElectric.co.jp

Abstract—In this paper, Enhanced oxidation of SiC(0001) substrates using La$_2$O$_3$ capped annealing has been presented. Compared to thermal oxidation, lower oxidation temperature can be implemented to form SiO$_2$ layer, owing to higher oxidation rate by 10 times with La$_2$O$_3$ capped oxidation by catalytic effect of the film. Although oxidation kinetics are based on oxidation through step faces, the roughness of created SiO$_2$ has shown improved surface and interface over conventional dry oxidation, owing to oxygen radical oxidation induced by La$_2$O$_3$. Here, we propose a novel LaSiON dielectric formation based on the reaction of SiN and La$_2$O$_3$ layers. Owing to amorphous nature of the initial SiN$_x$ film, a uniform LaSiON film both thickness and composition parallel to the surface can be achieved. As oxidation of SiC substrates is inhibited, face dependent oxidation to cause thickness distribution can be eliminated.

Keywords—SiC, Oxidation, La-silicate

I. INTRODUCTION

As a wide band-gap semiconductor, Silicon Carbide (SiC) is a promising candidate for next generation power devices, the ability to grow silicon dioxide (SiO$_2$) by thermal oxidation is a unique advantage over other advantage over other wide band-gap semiconductor material. However, the properties of thermal oxidation grown SiO$_2$ on SiC still suffer from a lot problems. The one issue is that, high temperature dry oxidation causes large thickness variation in SiO$_2$ and also rougher SiO$_2$/SiC interface by enhanced oxidation at step bunches due to large facet dependent oxidation rate, which leads to local electric field crowding to degrade the reliability [1-2]. The issue can be solved by replacing dry oxidation by oxygen radical oxidation, which could reduce the oxidation reaction activation energy difference at step and terrace. Meanwhile, oxidation at lower temperature can also be achievable by the catalyst effect of radical oxygen. Here, La$_2$O$_3$ is known to create radical oxygen atoms from oxygen molecules, so that one can expect an enhanced oxidation even at low temperature [3]. 2nm La$_2$O$_3$ capped SiC substrate annealed in 5%-O$_2$ ambient has been examined and enhanced oxidation has been demonstrated. The presence of SiO$_2$ by 10 times in intensity and 10 nm in thickness can be obtained. Compared to dry oxidation in 100%-O$_2$ ambient, an enhance oxidation rate by 10

times can be achieved [4]. Moreover, supposing future trench devices, an enhanced oxidation of SiN$_x$ films using La$_2$O$_3$ to form LaSiON films is presented, which can avoid the issue of facet dependent oxidation rate.

II. SAMPLE PREPERATION

Three kind of sample has been prepared as follows: (a) SiC substrate deposited with La$_2$O$_3$; (b) SiC substrate deposited with SiN and La$_2$O$_3$; (c) SiC substrate without deposition. The sample structure are shown in Fig.1. All the sample substrate with a doping density of 1×10^{15} has been cleaned by SPM and HF treatment. A thin film of SiN with a thickness of 2nm has been deposited by Plasma-enhanced chemical vapor deposition (PECVD). La$_2$O$_3$ with thickness ranging from 2 to 4 nm was deposited by electron beam evaporation in an ultra-high vacuum chamber at a pressure of 10^{-8} Pa. The deposition rate of La$_2$O$_3$ was controlled to 0.2nm/min. Samples were then annealed for 30 min using a rapid thermal annealing (RTA) in 5%O$_2$/95%N$_2$ ambient from 500 °C to 1000 °C.

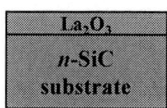 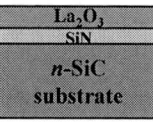 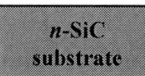

Fig.1 sample structure: (a) La$_2$O$_3$/SiC (b) La2O3/SiN/SiC   (c)SiC substrate only

The chemical bond state at substrate surface was analyzed by Attenuated total reflection-Fourier transform infrared spectroscopy (ATR-FTIR). Transmission electron microscopy (TEM) has also been utilized to analysis the cross-section image of near-SiC substrate surface. Surface Roughness comparison has been made by Atomic force microscopy (AFM). X-ray photoelectron spectroscopy (XPS) has also been used to analysis the element distribution on perpendicular orientation refer to the SiC substrate surface.

III. RESULTS AND DISCUSSION

A. Oxidation Effect Comparison between with and without La_2O_3

Fig. 2a and Fig. 2b shows the infrared spectra of SiC substrate without and with 2nm La_2O_3 after annealing at different temperature. Both are normalized by signal from SiC substrate Fig. 2a shows the ATR spectra of same SiC substrate annealed at different temperature. The absorption peak at around 1250 cm^{-1} with limited intensity was observed, which are attributed to the LO modes of the asymmetric stretching vibration of Si-O-Si bond absorption. The peak of Si-O-Si bond absorption appeared at 600 °C and increase with raising of temperature. Fig. 2b shows the ATR spectra of same SiC substrate with 2nm La_2O_3 deposited at different annealing temperature. The large absorption peak at around 1250 cm^{-1} and 1065 cm^{-1} are attributed to the LO modes of the asymmetric stretching vibration of Si-O-Si and La-O-Si-O bond. Both of the peaks appeared at 900 °C and increase significantly with the temperature raises. The huge increase of Si-O-Si bond absorption peak indicate that the oxidation of SiC substrate was enhanced by La_2O_3 capped annealing.

B. Sectional image of La_2O_3/SiC interface after annealing

Fig. 3 shows the cross section image of La_2O_3/SiC interface after annealed in 1000 °C for 30 min. The thickness of deposited La_2O_3 is 2nm. The white and black region on SiC are identified as SiO_2 and La-silicate, respectively. The thickness of SiO_2 layer is over 10nm, far more thicker than deposited La_2O_3. The formed La-silicate however, has not been a uniform layer, but separated into pieces distributing in the SiO_2 layer. Another thing should take into account is that the formed SiO_2 layer is not even. This thickness variation of SiO_2 shows that although with catalytic effect of La_2O_3, the facet dependent oxidation happened and induced oxidation fluctuation on SiC substrate.

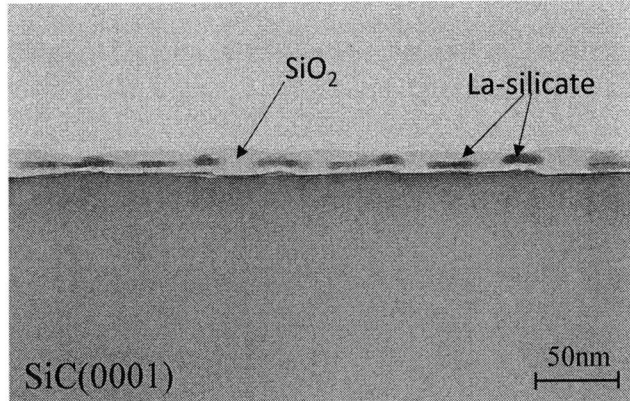

Fig. 3 Cross-Section TEM image of La_2O_3 capped SiC substrate after annealing

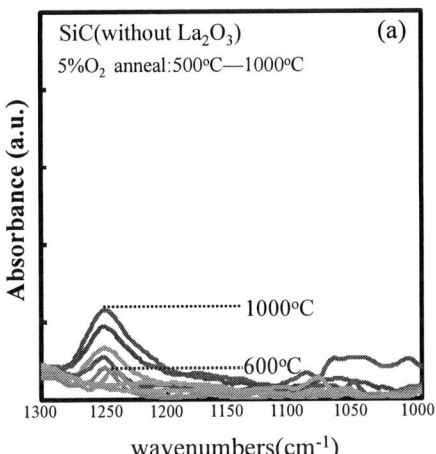

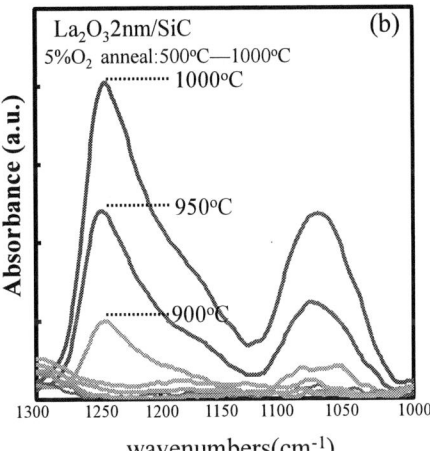

Fig. 2 infrared absorption spectra (a) without La_2O_3/SiC (b) La_2O_3 2nm/SiC

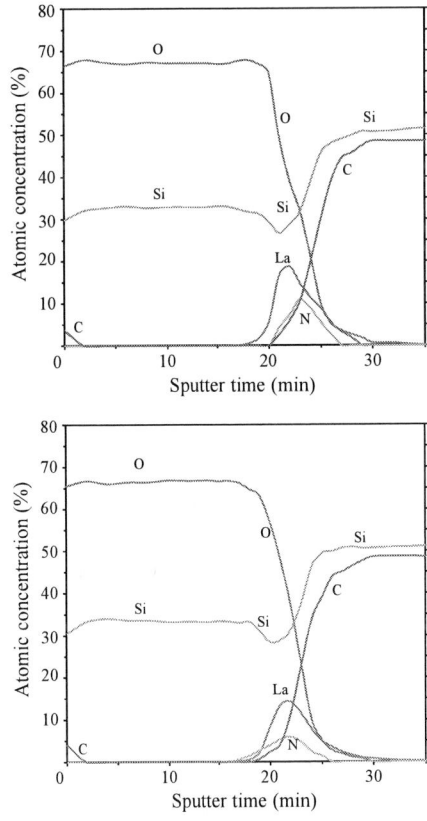

Fig. 4 Depth profiling of La_2O_3/SiN capped SiC substrate afer annealing by XPS

C. Depth profiling of La₂O₃/SiN capped SiC surface by XPS

C. Depth profiling of La$_2$O$_3$/SiN capped SiC surface by XPS

Since the facet dependent oxidation still happens at La$_2$O$_3$ coated SiC substrate, we propose a novel structure with SiN inserted between La$_2$O$_3$ and SiC. By the reaction of SiN with La$_2$O$_3$, facet dependent oxidation can be avoid and a fine La-silicate may form.

Two SiC substrate deposited with 2nm SiN and 4nm La$_2$O$_3$ was fabricated and then anneal at 500 °C and 800 °C for 30min. Fig. 4a and Fig. 4b shows the depth profiling of La$_2$O$_3$/SiN capped SiC annealed at 500 °C and 800 °C by XPS, respectively. The atomic concentration difference between Si and C at sputter time 20-25 min section shows a La-silicate region formed after 500 °C and 800 °C annealing. The expanded distribution of Nitrogen atomic concentration after annealed at 800 °C was observed, which indicate a LaSiON region has formed and grow with the temperature raises.

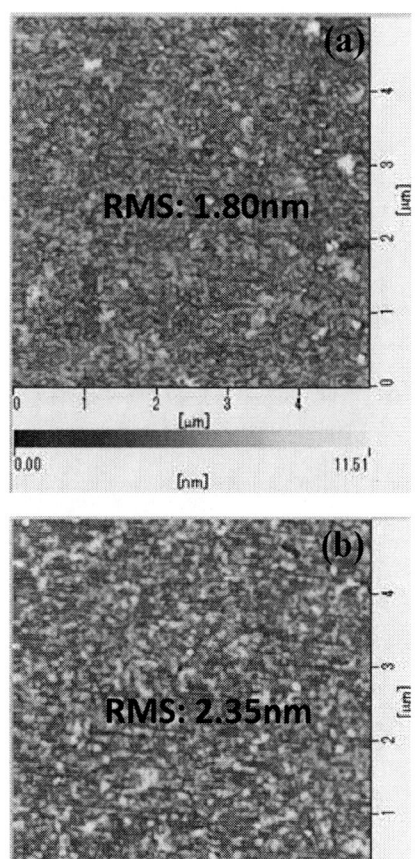

Fig. 5 AFM image of SiC substrate after annealing. Substrate structure: (a) La$_2$O$_3$ 4nm/SiN 2nm/SiC (b) La$_2$O$_3$ 4nm/SiC

D. Surface roughness of substrate with and without SiN deposition

D. Surface roughness of substrate with and without SiN deposition

Two SiC substrate deposited with 4nm La$_2$O$_3$ and one of the two deposited with 2nm SiN before La$_2$O$_3$ deposition was fabricated. Both the substrate were then annealed in 1000 °C for 30 min. Fig. 5a nad Fig. 5b shows the AFM image of SiC substrate with and without SiN deposition, respectively. For substrate with SiN deposition. A more flat surface has been observed. The Root Mean Square Roughness (RMS) of substrate with and without SiN insertion are 1.80nm and 2.35nm, respectively. This fact strongly indicated a more even surface has been achieved by SiN insertion. With the reaction by SiN and La$_2$O$_3$, facet dependent oxidation should be moderated and a more flat surface can be obtained.

E. Cross sectional image of La₂O₃/SiN/SiC interface after annealing

E. Cross sectional image of La$_2$O$_3$/SiN/SiC interface after annealing

Fig. 6 shows the cross section image of f La$_2$O$_3$/SiN/SiC interface after annealing in 1000 °C for 30 min. The thickness of La$_2$O$_3$ and SiN are 4nm and 2nm, respectively. The grey region are identified as SiO$_2$ and the black region are identified as La-silicate. A thick SiO$_2$ layer has been observed, indicating large oxidation has been taken place at SiC substrate. The formed La-silicate still distribute in SiO$_2$ layer as pieces, which should be related to the facet dependent oxidation. A more flat surface has been observed in comparison with substrate without SiN deposition, which indicated the moderation of oxidation variation by La$_2$O$_3$/SiN capped annealing.

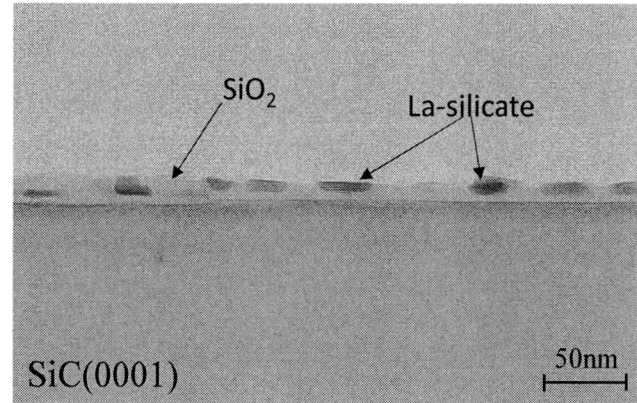

Fig.6 Cross-Section TEM image of La$_2$O$_3$/SiN capped SiC substrate after annealing

IV. CONCLUSION

Interface reaction of La$_2$O$_3$/SiN capped annealing has been examined by multiple experiment methods. An enhanced oxidation of SiC substrate has been achieved by La$_2$O$_3$ capped annealing. The reaction of La$_2$O$_3$ and SiN has been confirmed by X-ray photoelectron spectroscopy. Although facet dependent oxidation on SiC still occurred, a moderation of oxidation variation has been achieved by La$_2$O$_3$/SiN capped annealing.

REFERENCES

[1] T. Hosoi, K. Kozono, Y. Uenishi, et al., "Investigation of surface and interface morphology of thermally grown SiO$_2$ dielectircs on 4H-

SiC(0001(substarte," J. Materials Science Forum, vol. 679-680, pp. 342-345 (2011)

[2] W. Li, J. Zhao, Q. Zhu, and D. Wang, " Oxidation of setp edges on vicinal 4H-SiC(0001) surface," Appl. Phys. Lett. 103, 211603 (2013)

[3] K. Kita, T. Koyanagi, et al., "Characterization of flatband voltage roll-off and roll-up behavior in La_2O_3/silicate gate dielectirc," J. ECS Trans, vol. 61(2), p. 135-142 (2014)..

[4] K. Kita, R. H. Kikuchi, and H. Hirai, "Understranding of growth kinetics of thermal oxides on 4H-SiC (0001) for control of MOS characteristics," J. ECS Trans, vol. 61(2), p. 135-142 (2014)

Passivation of SiO$_2$/SiC Interface with La$_2$O$_3$ Capped Oxidation

S. Munekiyo, Y. M. Lei,
K. Natori, H. Iwai
Frontier Research Center, Tokyo
Institute of Technology,
Yokohama 226-8503, Japan
Email-Address :
munekiyo.s.aa@m.titech.ac.jp

T. Kawanago, K. Kakushima, K.
Kataoka, A. Nishiyama, N. Sugii,
H. Wakabayashi, K. Tsutsui
Department of Electronics and
Applied Physics,
Tokyo Institute of Technology,
Yokohama 226-8503, Japan
Email-Address :
kakushima@ep.titech.ac.jp

M. Furuhashi, N. Miura,
S. Yamakawa
Mitsubishi Electric Corp.
Amagasaki 661-8661, Japan
Email-Address :
Miura.Naruhisa@db.MitsubishiElectric.co.jp

Abstract - **Thermal oxidation of SiC(0001) substrates with La$_2$O$_3$ capped annealing has been performed. La$_2$O$_3$ capped oxidation has shown improvements in reduced hysteresis and interface state density (Dit) for MOS capacitors. La-silicate grains, agglomerated at the step bunches of SiC substrates, have been confirmed upon oxidation. We can anticipated that La-silicate grains are likely to passivate the charge trapping at step bunches and effectively suppresses the D$_{it}$.**

I. INTRODUCTION

Silicon carbide (SiC) has attracted attention as a power device material of next generation because it has excellent physical properties such as high breakdown field and high thermal conductivity [1]. In addition, SiC also has an ability that can be formed SiO$_2$ by thermal oxidation. However, the characteristics of 4H-SiC MOSFET using thermal oxidation SiO$_2$ as the gate dielectric exhibits a low inversion layer channel mobility due to large interface state density [2-3]. The possible origins of the poor MOS interface characteristics would be the carbon related defects in near-interface region of oxides and those on SiC surface caused by the biproducts of thermal oxidation [4]. Recent study on La-silicate interface layers for SiC substrates has shown improvements in interface states and reduction in electron trapping [5]. Besides these improvements, the basic physical reasons have not been clarified yet. Therefore, a systematic study to elucidate the effect of La atoms need to be clarified. For this purpose, we conducted La$_2$O$_3$ capped oxidation to form La-silicate interface layers and compared the electrical characteristics of MOS capacitors.

II. DEVICE FABRICATION

MOS capacitor were fabricated on n-type 4° off-axis 4H-SiC (0001) substrates. The substrates with ~1×10^{16}cm^{-3} doped n-type epitaxial layer were subjected to a solvent and

Fig. 1 Fabrication process and device structure for SiC-MOS capacitor with La$_2$O$_3$ capped oxidation.

diluted hydrofluoric acid surface cleaning. Next, 4nm and 10nm-thick La$_2$O$_3$ films, which act as capping layers for oxidation, were e-beam deposited (EBD) on substrates, followed by oxidation in 5%-O$_2$ ambient at 1000°C. Another SiO$_2$ layer is placed by plasma-enhanced chemical vapor deposition (PECVD) on top of the samples and annealed again at 1000°C. Additionally, post oxidation anneal (POA) was also carried out in 5%-O$_2$ ambient at 800°C. Tungsten (W) gate electrodes were deposited by RF sputtering, thereby producing a MOS capacitor. Finally, samples were subjected to annealing in forming gas (3%-H$_2$) at 420°C. Note that no annealing in NO or N$_2$O ambient was conducted in all the samples. A sample without La$_2$O$_3$ layer and without POA were also fabricated as a reference. The measurement was performed Capacitance-Voltage characteristics of devices, it was studied the electrical characteristics due to the insertion of La$_2$O$_3$ layer.

978-1-4799-5494-0/14 $31.00 © 2014 IEEE

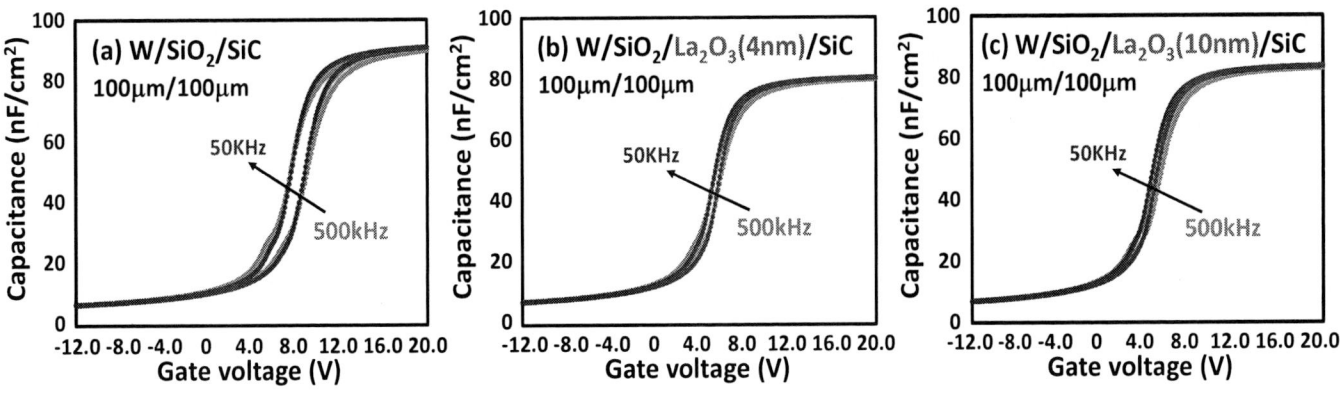

Fig. 2 CV curves (a) w/o, (b) 4-nm and (c) 10-nm-thick La$_2$O$_3$ layers.

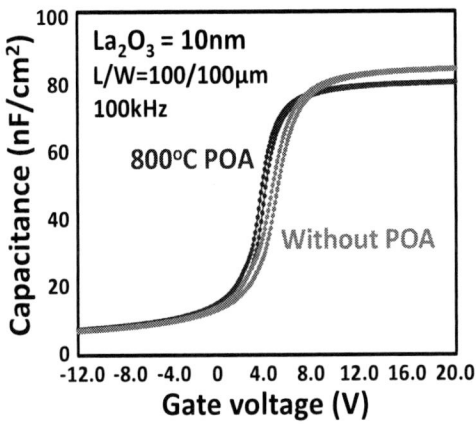

Fig. 3 Effect of post oxidation anneal in CV characteristics.

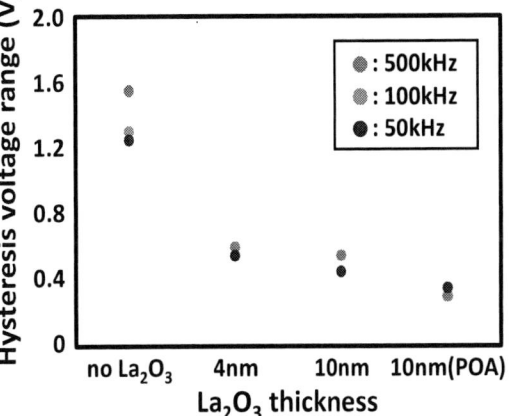

Fig. 4 Hysteresis voltage range.

III. RESULT AND DISCUSSION

A. Thickness dependence of La$_2$O$_3$ layer in CV characteristics

Fig. 2 shows capacitance-voltage (CV) characteristics of samples (a) without La$_2$O$_3$, (b) 4-nm-thick La$_2$O$_3$ and (c) 10-nm-thick La$_2$O$_3$ insertion. First, flat-band voltage is shifted to the negative side with increasing insertion La$_2$O$_3$ thickness. This result indicate that positive fixed charge density in gate dielectrics is decreased by La$_2$O$_3$ capped oxidation. Secondly, hysteresis of the CV curve was reduced in the case of thicker La$_2$O$_3$ layer. These results suggest the quality improvement of the gate dielectrics on 4H-SiC substrate can be expected by the La$_2$O$_3$ capped oxidation.

B. Effect of post oxidation anneal

Fig. 3 shows CV characteristics (100kHz) of samples that with and without POA which carried out in 5%-O$_2$ ambient at 800°C. From this result, stretch-out behavior in the CV curve was improve by POA. Stretch-out affect the interface state density (D$_{it}$) of MOS interface. Therefore, this result suggest that D$_{it}$ is reduced by the POA process. For this reason, it was anticipated that the strain of

oxide/SiC interface was mitigated by La atom in La-silicate move as network modifier during POA.

C. Suppression of hysteresis voltage range

Fig. 4 shows hysteresis voltage range of all samples. Clockwise hysteresis was effectively suppressed with La$_2$O$_3$ capped oxidation, and further improvements was observed with the sample having thicker La$_2$O$_3$ layer. Hysteresis affect the electron trap in gate dielectrics. Hence, it was considered that those electron trap has been reduced by insertion thicker La$_2$O$_3$ layer. In addition, hysteresis was also reduced by carrying out the POA, smallest hysteresis was obtained in that case (<0.4V).

D. TEM image of La$_2$O$_3$ capped oxidation

Cross sectional transmission electron microscope (TEM) image showed formation of La-silicate grains, with composition of La$_2$SiO$_5$, agglomerated at step bunches at the surface of SiC substrates (Fig. 5). From the observed change in the interface and obtained electrical characteristics, one can anticipated that La-silicate grains are likely to passivate the charge trapping at step bunches and effectively suppresses the D$_{it}$.

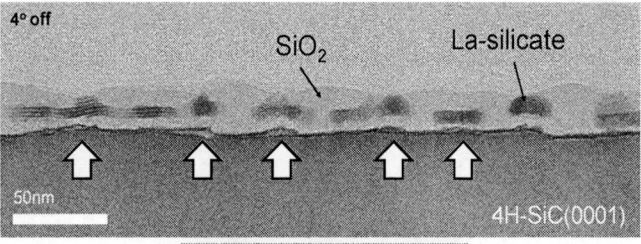

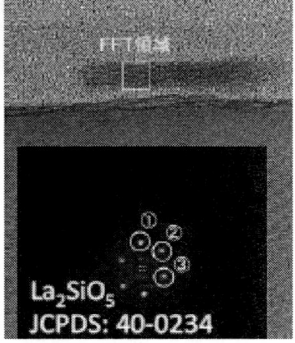

Fig. 5 TEM image of La$_2$O$_3$ capped oxidation indicate formation of La-silicate grains at step bunches.

IV. SUMMARY

We evaluated electrical characteristics of 4H-SiC MOS capacitor with La$_2$O$_3$ capped oxidation. Flat-band voltage is shifted to the negative side with increasing insertion La$_2$O$_3$ thickness. In addition, hysteresis voltage range of the CV curve was reduced in the case of thicker La$_2$O$_3$ layer. Furthermore, stretch-out and hysteresis were more improved by performing the POA, the best electrical characteristics obtained in this case. La-silicate grains agglomerated at step bunches at the surface of SiC substrates, one can anticipated that La-silicate grains are likely to passivate the charge trapping at step bunches. From the above results, it can be expected to obtain even better interfacial and electrical properties by La$_2$O$_3$ capped oxidation.

ACKNOWLEDGMENT

This work is supported by Mitsubishi Electric Corporation.

REFERENCES

[1] J. A.Cooper, Jr. and A. K.Agarwal, "SiC Power-Switching Devices-The Second Electronics Revolution?," *Proc. IEEE,* 90, 956 (2002).

[2] S. Sridevans, B. J. Baliga, "Lateral N-Channel Inversion Mode 4H-SiC MOSFET's," *IEEE Electron Device Lett.* 19, 228 (1998).

[3] R. Schörner, P. Friedrichs, and D. Peters, "Detailed Investigation of N-Channel Enhancement 6H-SiC MOSFET's," *IEEE Trans. Electron Devices* 46, 533.

[4] K. Kita, R. H. Kikuchi, and H. Hirai, "Understranding of growth kinetics of thermal oxides on 4H-SiC (0001) for control of MOS characteristics," J. ECS Trans, vol. 61(2), p. 135-142 (2014)

[5] X. Yang et al., "High Mobility 4H-SiC MOSFETs Using Lanthanum Oxide Interfacial Engineering and ALD Deposited SiO2," ICSCRM Th-2B-5 (2013).

Effect of Post Deposition Annealing for High Mobility 4H-SiC MOSFET Utilizing Lanthanum Silicate and Atomic Layer Deposited SiO$_2$

Xiangyu Yang, Bongmook Lee, and Veena Misra
Department of Electrical and Computer Engineering
North Carolina State University
Raleigh, NC 27606, USA
xyang2@ncsu.edu

Abstract—**We have demonstrated high mobility Si-face 4H-SiC MOSFET results using a novel lanthanum silicate (LaSiO$_x$) interface engineering and Atomic Layer Deposited (ALD) SiO$_2$. In this work, the impact of post deposition annealing (PDA) conditions on the mobility of MOSFET with LaSiO$_x$ is investigated. The sample received 900 °C PDA in nitrous oxide (N$_2$O) ambient shows the highest mobility and higher PDA temperature reduces the peak mobility. Mobility results measured at elevated temperatures show that the electron mobility of the La-containing devices is limited by the phonon scattering as opposed to the coulombic scattering, indicating improved interface properties.**

Keywords— Atomic layer deposition (ALD); lanthanum silicate; mobility; MOSFET; SiC.

I. INTRODUCTION

With superior material properties, silicon carbide (SiC) is one of most promising substrate materials for future power switches [1]. Although Si-face 4H-SiC metal oxide semiconductor field effect transistors (MOSFETs) have been studied for decades, the typical inversion electron mobility values are still low [3-5]. It has reported that the poor mobility of MOSFET using thermally grown SiO$_2$ as the gate dielectric is caused by high density of interface states (D$_{it}$) at SiC/SiO$_2$ interface, which are believed to be defects associated with carbon atoms released from the substrate during thermal oxidation, near interface traps, and dangling bonds [1]. To improve the device mobility, various techniques have been developed to passivate the interface states, including incorporations of nitrogen using nitrous oxide (N$_2$O) and nitric oxide (NO) and phosphorous into the thermally grown oxide through high temperature anneals [3-5]. The presence of nitrogen or phosphorous near the SiC/SiO$_2$ interface effectively reduces the interface state density and thus improves the MOSFET mobility. However, the achieved mobility values remain low compared to the bulk 4H-SiC mobility and the introduction of nitrogen and phosphorous species results in low threshold voltage and even normally-on devices, which leads to the safety and reliability concerns of power MOSFETs [1]. This phenomenon is known as the trade-off between mobility and threshold voltage. To solve this problem, deposited dielectrics can be used to replace the thermal oxide [6-8]. With

deposited gate dielectrics, the substrate consumption is minimized and thus the carbon related defects associated with thermal oxidation can be avoided. Using deposited dielectrics also enables interface engineering to independently control the SiC/dielectrics interface properties. SiO$_2$ remains the most suitable gate dielectric due to its large bandgap, as well as a high conduction band offset to SiC which suppresses electron tunneling current [9]. SiO$_2$ deposited by Atomic Layer Deposition (ALD) has shown promising electrical properties on 4H-SiC substrate, due to its good overall film quality and minimal substrate consumption, as well as other benefits from using ALD, including superior film uniformity, precise thickness control, conformal deposition, and low process temperature [7-10]. However, even without thermal oxidation and the carbon related defects, the SiC/SiO$_2$ interface requires interface treatment to improve the interface properties [6-8]. We have demonstrated a novel interface engineering technique to improve the MOSFET mobility while maintaining a sufficiently positive threshold voltage by introducing an ultra thin lanthanum silicate (LaSiO$_x$) layer at the SiC/dielectric interface [11]. LaSiO$_x$, which exhibits high dielectric constant and good thermal stability, has been studied as potential high-k gate dielectric for advanced Si technology [12, 13]. The LaSiO$_x$/SiO$_2$ stack can be created by annealing a lanthanum oxide (La$_2$O$_3$)/SiO$_2$ stack to high temperatures (>800 °C), due to La$_2$O$_3$'s scavenging effect [12, 13]. Incorporation of a La-rich layer has been proven to be effective for the suppression of interfacial layer growth on Si substrate [12, 13]. We have shown that the presence of a La-rich layer at SiC/dielectric interface can also prevent low quality interfacial layer formation during the processing [11]. In this work, we have further explored the impact of the PDA conditions on SiC/LaSiO$_x$ interface properties.

II. EXPERIMENTAL PROCEDURES

MOS capacitors were fabricated on n-type 8° off-axis Si-face (0001) 4H-SiC substrates with nitrogen doping of 6×10^{15} cm^{-3}. The wafers were first subjected to a solvent and diluted hydrofluoric (HF) acid cleaning. A 1nm thick La$_2$O$_3$ layer was deposited using a molecular beam epitaxy (MBE) system through reactive evaporation of lanthanum metal in oxygen ambient. Then, the samples were immediately transferred to an

This work is supported by Toyota Motor Corporation.

978-1-4799-5494-0/14 $31.00 © 2014 IEEE

ALD system for 30 nm SiO_2 deposition at 150 °C substrate temperature [7]. The samples received a post deposition annealing (PDA) in a rapid thermal annealing (RTA) tool in N_2O ambient at various temperatures. RF sputtered tantalum nitride (TaN) capped with tungsten (W) was deposited as the gate electrode. The capacitors received a post metallization annealing (PMA) in N_2 to remove the sputter damage.

The n-channel lateral MOSFETs were fabricated on aluminum doped ($N_A=5\times10^{15}$ cm^{-3}) 8° off-axis Si-face 4H-SiC wafers. To form the ohmic contacts, source and drain phosphorus implantation and the activation anneal were performed. The samples went through the same surface cleaning and La_2O_3, ALD SiO_2, and gate electrode depositions as the MOS capacitors. Finally, nickel silicide was created as the source and drain ohmic contact by annealing sputtered nickel after the gate stack fabrication. The nickel silicide formation annealing also serves as the PMA for the gate electrode. The fabrication processes and the schematic of the cross-section of La-containing MOSFET are shown in Fig. 1 (a) and (b) respectively. The MOS capacitor capacitance-voltage (C-V) characteristics were measured using an HP 4284A LCR meter. MOSFET DC characteristics were measured with a Keithley 4200 semiconductor parameter analyzer. The carrier transport mechanisms of MOSFETs with and without $LaSiO_x$ were studied by evaluating the peak field effect mobility's dependence on device operating temperatures.

III. RESULTS AND DISCUSSIONS

Fig. 2 shows the C-V characteristics of MOS capacitor using 1nm La_2O_3/ALD SiO_2 stack after 900 °C N_2O PDA. The C-V curves show sharp C-V transitions and minimal hysteresis. The presence of the La-rich layer reduces the interface state density and gate leakage current [11, 14].

(a)

- Source and drain implantation and activation
- Field oxide deposition and patterning
- Surface cleaning
- 1nm La_2O_3 deposition by MBE
- 30 nm SiO_2 deposition by ALD
- PDA in N_2O at various temperatures
- TaN/W gate sputtering and lift-off
- Contact hole etching
- Source and drain contact Ni deposition and lift-off
- Ni-silicide formation anneal at 950 °C

(b)

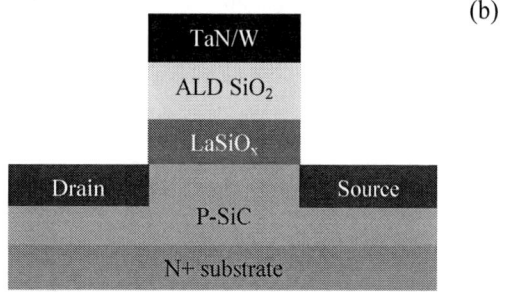

Fig. 1. (a) MOSFET fabrication process flow; (b) the schematic of fabricated La-containing lateral MOSFETs. The LaSiOx layer was created by annealing the La_2O_3/ALD SiO_2 stack. The physical thickness of as-deposited La_2O_3 and ALD SiO_2 are 1nm and 30nm respectively.

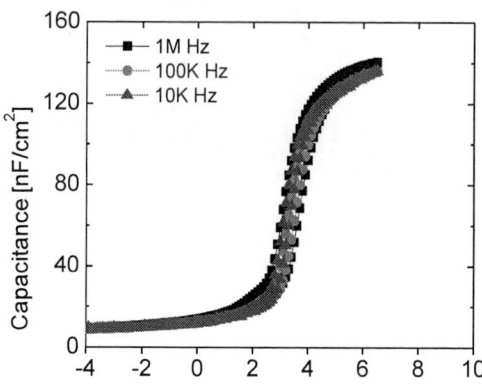

Fig. 2. C-V characteristics of La-containing MOS capacitors measured at various frequencies. The sample received 900 °C PDA in N_2O. The C-V curves show sharp C-V transitions and minimal hysteresis.

As shown in Fig. 3, an abrupt interface between SiC and the La-rich layer is observed from the Z-contrast scanning transmission electron microscopy (STEM) image of the SiC/$LaSiO_x$/ALD SiO_2 stack. The $LaSiO_x$ layer, which is about 2.5 nm thick, remains amorphous after 900 °C N_2O PDA. If the PDA temperature is increased, it is expected that the $LaSiO_x$ thickness and composition may change due to substrate oxidation or additional reaction between the silicate and ALD SiO_2 layer, which should result in lower La concentration at the SiC/$LaSiO_x$ interface.

Fig. 4 shows the field effect mobility results of MOSFETs with and without $LaSiO_x$. It clearly shows at least 4 times improvement in peak mobility with the ultrathin $LaSiO_x$ layer compared to the device using ALD SiO_2 as the gate dielectric.

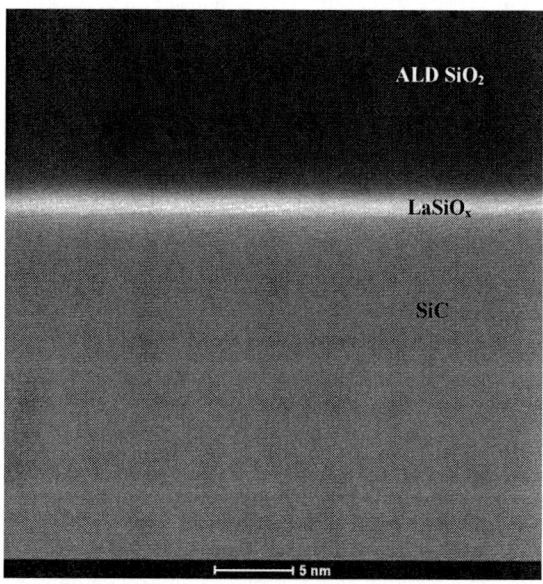

Fig. 3. Z-contrast STEM image of the $LaSiO_x$/ALD SiO_2 stack after 900 °C N_2O PDA. An abrupt interface between SiC and LaSiOx is observed. The specimen was lift-out from the gate area of the La-containing MOSFET with 900 °C PDA using focused ion beam.

978-1-4799-5494-0/14 $31.00 © 2014 IEEE   118

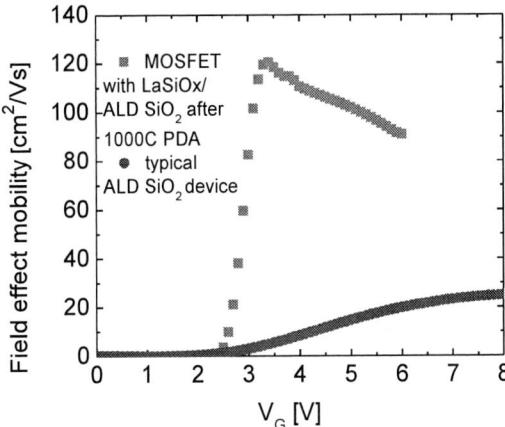

Fig. 4. Field effect mobility of the La-containing MOSFET with 1000 °C PDA and a typical ALD SiO_2 device with N_2O PDA. The MOSFET with $LaSiO_x$ after 1000 °C PDA shows at least 4 times higher peak mobility.

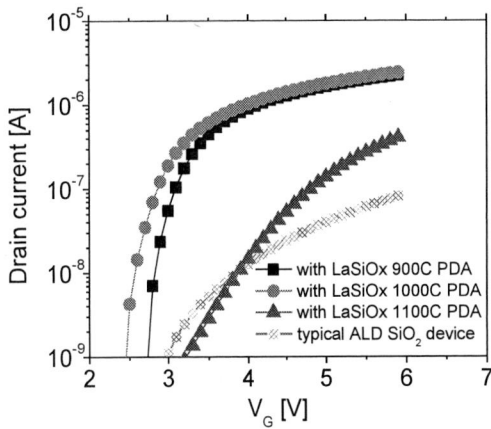

Fig. 6. Sub-threshold characteristics of La-containing MOSFETs and a typical ALD SiO_2 device with N_2O PDA. La-containing MOSFETs with 900 °C and 1000 °C N_2O PDA show dramatically improved the sub-threshold swing, compared to the device without $LaSiO_x$.

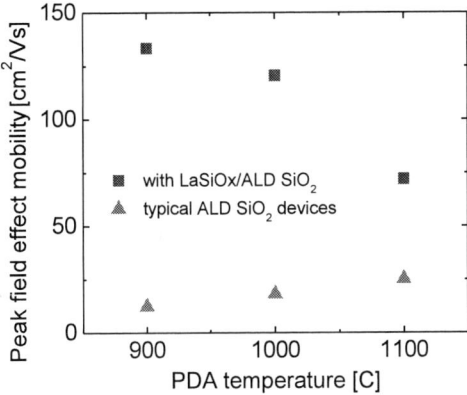

Fig. 5. Peak field effect mobility of La-containing MOSFETs and ALD SiO_2 devices with various N_2O PDA temperatures. The peak mobility decreases with the N_2O PDA temperature for the La-containing MOSFETs, whereas the peak mobility increases with the N_2O PDA temperature for the devices without $LaSiO_x$.

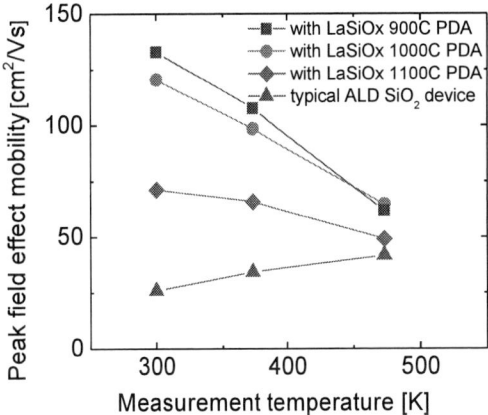

Fig. 7. Peak mobility of La-containing MOSFETs and a typical ALD SiO_2 device at various measurement temperatures. Higher peak mobility is observed at elevated temperatures for the MOSFET using ALD SiO_2 as the gate dielectric. The peak mobility decreases with the measurement temperature for the La-containing MOSFETs

Fig. 5 shows the extracted peak mobility results as a function of N_2O PDA temperatures for MOSFETs with and without $LaSiO_x$. The peak mobility of MOSFETs using only ALD SiO_2 as the gate dielectric increases with the N_2O PDA temperatures due to the nitridation effect of N_2O at higher temperatures [7]. The MOSFETs with $LaSiO_x$ display clearly different trend. The sample received 900 °C PDA shows the highest mobility and higher PDA temperature reduces the peak mobility. Compared to the device with 900 °C N_2O PDA, the 1000 °C PDA device shows slightly reduced peak mobility value whereas the device with 1100 °C PDA significantly decreases the peak mobility. The mobility reduction after 1100 °C PDA in N_2O can be attributed to thermal oxidation at the SiC/$LaSiO_x$ interface or further reaction between the silicate and ALD SiO_2, which may lead to lower La concentration. Therefore, with 1nm La_2O_3, the N_2O PDA temperature should be limited to around 1000 °C to avoid reduction of La concentration near the SiC/$LaSiO_x$ interface.

The sub-threshold characteristics of the La-containing MOSFETs are shown in Fig. 7. As indicated in the mobility results in Fig. 6, MOSFETs with $LaSiO_x$ after 900 °C and 1000 °C PDA show similar sub-threshold swings, which are significantly improved compared to the device without $LaSiO_x$. The sub-threshold swing increases significantly after 1100 °C N_2O PDA indicating the interface properties has been degraded. The sub-threshold swing results are consistent with the mobility results shown in Fig. 6. The result confirms that, with 1nm La_2O_3, the 1100 °C N_2O PDA weakens the effect of La incorporation due to potentially lower La concentration at the SiC/$LaSiO_x$ interface. It is noted that the PDA time also have strong impact on the amount of thermal oxidation. Thus, RTA offers clear advantages over conventional furnace anneal due to its precise annealing time control. In this work, the PDA time at the target temperatures is limited to 40 s for all the samples. Additionally, with thicker La_2O_3, the peak

978-1-4799-5494-0/14 $31.00 © 2014 IEEE   119

mobility and sub-threshold swings' dependence on the N_2O PDA temperature may be different from the results using 1nm La_2O_3 shown in Fig. 6 and Fig. 7. Therefore, to achieve optimal La concentration at the SiC/dielectric interface, the La_2O_3 thickness, annealing time, and temperature must be carefully optimized.

To study the carrier transport mechanism, MOSFET mobility at elevated temperatures were measured by heating the samples on a heating chuck to 100 °C and 200 °C and the peak mobility's evolution from room temperature (20 °C) to 200 °C was evaluated. Based on Matthiessen's rule, the lowest component of the bulk, coulombic, phonon, and surface roughness mobility, limits the total low field mobility and thus the corresponding scattering process dominates [15]. It is shown that the bulk and surface phonon mobility decrease with measurement temperature, whereas coulombic mobility increases with temperature and the surface roughness mobility does not have explicit temperature dependence [15]. Therefore, if mobility value is increased with temperature, the mobility is limited by the coulombic scattering, which is associated with interface states and fixed charges. On the other hands, if the peak mobility decreases with temperature, the total mobility is limited by the bulk mobility and/or the phonon scattering processes. Since the bulk 4H-SiC mobility is significantly higher than other mobility components at temperatures of interest, the bulk mobility can be ignored in this study [15]. Thus, if the peak mobility decreases with measurement temperature, the total low field mobility is limited by scattering processes dominate.

In Fig. 7, MOSFETs with $LaSiO_x$ after 900 °C and 1000 °C PDA show similar change in measured peak mobility when the measurement temperature is raised from room temperature to 200 °C. The device received 1100 °C N_2O PDA shows mild mobility reduction from room temperature to 100 °C and the mobility decreases further at 200 °C. Compare to the devices with $LaSiO_x$, higher peak mobility is observed at elevated temperatures from the MOSFET using ALD SiO_2 as the gate dielectric. Based on the methodology discussed above, it is clear that the mobility of MOSFET with ALD SiO_2 only is limited by coulombic scattering. However, the mobility of MOSFETs with $LaSiO_x$ after 900 °C and 1000 °C PDA is limited by phonon scattering processes, indicating significantly improved interface properties compared to the devices without $LaSiO_x$. Although lower peak mobility is observed at elevated temperatures from the La-containing MOSFET with 1100 °C PDA, the slight change from room temperature to 100° C may suggest that, due to possibly lower La concentration, the mobility may be determined by phonon scattering or more than one mobility components.

IV. CONCLUSION

In summary, we have further explored the high mobility and high threshold voltage interface engineering technique using $LaSiO_x$ and SiO_2 deposited by ALD. The impact of PDA conditions on MOSFET mobility and sub-threshold characteristics has been investigated. The device received 900 °C PDA and 1000 °C PDA in N_2O ambient show at least 4 times higher peak mobility than device with ALD SiO_2 only.

1100 °C N_2O PDA significantly lowers the peak mobility, possibly due to lower La concentration at SiC/dielectric interface, but the device still show higher mobility than MOSFETs without $LaSiO_x$. The $LaSiO_x$ layer after 900 °C and 1000 °C PDA substantially improves sub-threshold swing compared to the device without La. The PDA conditions should be carefully selected to maintain the optimal La concentration for mobility enhancement. To achieve the optimal La concentration at the SiC/dielectric interface, the La_2O_3 thickness, PDA time, and PDA temperature must be optimized. Peak mobility results measured at elevated temperatures show that the electron mobility of the La-containing devices after 900° C and 1000° C PDA is limited by the phonon scattering as opposed to the coulombic scattering, indicating improved interface properties.

REFERENCES

[1] G. Pensl et al, "Alternative techniques to reduce inteface traps in n-type 4H-SiC MOS capacitors," Phys. Stat. Sol. (b), vol. 245, no. 7, pp. 1378–1389, 2008.

[2] V.V. Afanasev, M. Bassler, G. Pensl, and M. Schulz, "Intrinsic SiC/SiO interface states," Phys. Stat. Sol. (A), vol. 162, pp. 321–337, 1997.

[3] G. Y. Chung et al., "Improved inversion channel mobility for 4H-SiC MOSFETs following high temperature anneals in nitric oxide," IEEE Electron Device Lett. , vol. 22, no. 4, pp. 176–178, Apr. 2001.

[4] P. Jamet and S. Dimitrijev, "Physical properties of N2O and NO-nitrided gate oxides grown on 4H SiC," Appl. Phys. Lett., vol. 79, pp. 323–325, 2001.

[5] D. Okamoto, H. Yano, K. Hirata, T. Hatayama, and T. Fuyuki, "Improved inversion channel mobility in 4H-SiC MOSFETs on Si face utilizing phosphorus-doped gate oxide," IEEE Electron Device Lett. , vol. 31, no. 7, pp. 710–712, Jul. 2010.

[6] S. K. Haney, "Investigation of low temperature, atomic-layer-deposited oxides on 4H-SiC and their effect on the SiC/SiO₂ interface," Ph.D. dissertation, MSE Dept., NC State Univ., Raleigh, NC, 2012.

[7] X. Yang, B. Lee, and V. Misra, "Electrical properties of 4H-SiC devices using SiO₂ deposited by Atomic Layer Deposition," unpublished.

[8] C. Kim et al., "Comparison of thermal and atomic-layer-deposited oxides on 4H-SiC after post-oxidation-annealing in nitric oxide," Appl. Phys. Lett., vol. 100, 082112, 2012.

[9] A. K. Agarwal, S. Seshadri, and L. B. Rowland, "Temperature dependence of Fowler–Nordheim current in 6H and 4H SiC MOS capacitors," IEEE Electron Device Lett., vol. 18, no. 2, pp. 592–594, 1997.

[10] S. M. George, "Atomic layer deposition: An overview," *Chem. Rev.*, vol. 110, no. 1, pp. 111–131, Jan. 2010.

[11] X. Yang, B. Lee, and V. Misra, "High mobility 4H-SiC MOSFETs using lanthanum silicate interface engineering and ALD deposited SiO₂", Material Sci. Forum, vol. 778-780, pp. 557-561, 2014.

[12] B. Lee, S. R. Novak, D. J. Lichtenwalner, X. Yang and V. Misra, "Investigation of the origin of V_T/V_{FB} modulation for NMOS application: role of La diffusion, effect of host high-k layer, and interface properties, " IEEE Trans. on Electronic Devices, 58, pp. 3106-3115, 2011.

[13] D. J. Lichtenwalner et al., "Lanthanum silicate gate dielectric stacks with subnanometer equivalent oxide thickness utilizing an interfacial silica consumption reaction," J. Appl. Phys. 98, 024314, 2005.

[14] X. Yang, B. Lee, and V. Misra, "High mobility 4H-SiC lateral MOSFET using lanthanum silicate and atomic layer deposited SiO₂", unpublished.

[15] S. K. Powell et al., "Physics-based numerical modeling and characterization of 6H-silicon-carbide metal–oxide–semiconductor field-effect transistors," J. Appl. Phys. , vol. 92, no. 7, pp. 4053–4061, Oct. 2002.

Wide Bandgap Power Devices Based High Efficiency Power Converters for Data Center Application

Weimin Zhang, Ben Guo, Fan Xu, Yutian Cui, Yu Long,

Fred Wang, Leon M. Tolbert, Benjamin J. Blalock, Daniel J. Costinett

Department of Electrical Engineering and Computer Science
The University of Tennessee
Knoxville, USA
wzhang29@vols.utk.edu

Abstract— Wide band gap (WBG) power devices, such as Silicon Carbide (SiC) and Gallium Nitride (GaN) devices, have been innovatively applied in the data center power converters, which are based on the high voltage DC (HVDC) power distribution architecture, to evaluate the potential efficiency improvement. For the front-end AC-DC rectifier, a buck rectifier using SiC devices was implemented. The SiC devices were tested at first to obtain the static and switching characteristics. The number of devices in parallel, the switching frequency and the input/output filters were investigated. A prototype of 7.5 kW, 3 phase 480 VAC input, 400 VDC output front-end rectifier was built and tested. The peak efficiency reaches up to 98.55%, and the full load efficiency is 98.54%. For the intermediate DC-DC bus converter, the impact of the GaN devices on the LLC resonant converter efficiency was evaluated and compared with the Si counterparts. Based on the device loss analysis and the FEA simulation on the transformer winding loss, the GaN devices exhibited the reduced device loss, and also the capabilities to reduce the transformer winding loss. A 300 W, 400 VDC input, 12 VDC output GaN device based DC-DC bus converter was built and tested by 96.3% peak efficiency and 96.1% full load efficiency.

Keywords—SiC; GaN; high efficiency; power converter

I. INTRODUCTION

In the United States, about 67 TWh, or 2.2% of all generated electricity, was consumed by data centers in 2010; this equates to $3.3 billion in electricity bills [1]. The power required to support the computing infrastructure in large data centers can reach up to several tens of MW [2], but the power efficiency is very low. Fig. 1 shows an example of power flow in a server. Nearly half of the power is lost in the power delivery path, resulting in an end-to-end efficiency below 50% [3]. The main reasons are that there are many cascaded power conversion stages and the efficiency of each power stage is low. Because all power dissipation results in the generation of heat, additional power is needed to operate an active cooling system to remove the excess heat, further reducing the systems power usage effectiveness.

There are many power distribution architectures proposed or employed in data centers. The widely applied architecture in datacenter nowadays is the AC power distribution architecture, shown in Fig. 2 [4]. It contains Uninterruptible Power Supply (UPS), power distribution unit (PDU), power supply unit (PSU) and voltage regulator (VR). The overall efficiency of the AC

power distribution architecture applied in the data center nowadays is about 67% in average [5].

Recently, a 400 V high voltage DC (HVDC) power distribution architecture is attracting increasing attention due to the greatly improved efficiency. As shown in Fig. 3, by removing several power conversion stages from the AC power distribution architecture, the overall efficiency improves from 67% to 77% [5]. The power converters are simplified to three power stages: the three phase 480 VAC to 400 VDC front-end rectifier, the 400 VDC to 12 VDC intermediate bus converter (IBC), and the 12 VDC to 1 VDC point of load (POL) converter.

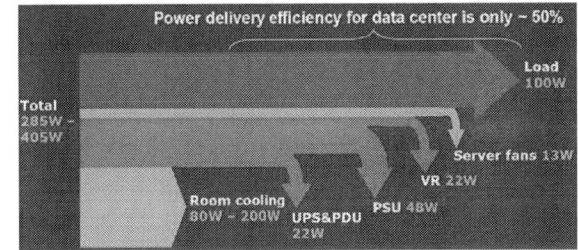

Fig. 1. Power flow in a server.

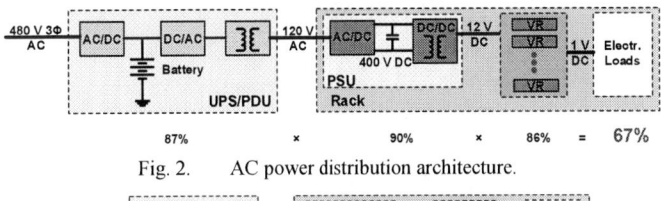

Fig. 2. AC power distribution architecture.

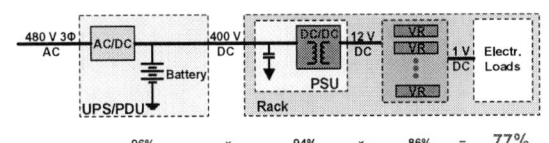

Fig. 3. 400 V high voltage DC power distribution architecture.

Wide band gap (WBG) power semiconductor devices, such as silicon carbide (SiC) and gallium nitride (GaN), have become commercially available in recent years. Due to the superior material properties, WBG devices feature higher blocking voltage, lower on-state resistance and faster switching speed compared with the Si device counterparts, therefore facilitate higher switching frequency and lower device loss of the power converter. It is expected that the application of WBG power devices in the data center power converters will bring improved efficiency to the entire power conversion system. On the other hand, the reduced device loss also leads to the reduced cooling

978-1-4799-5494-0/14 $31.00 © 2014 IEEE

system requirement, which saves the converter size and cost. Therefore, WBG power semiconductor devices have great potential benefits to improve the power conversion efficiency in the data center power infrastructure.

This paper discusses the design and implementation of high efficiency data center power converters based on HVDC power distribution architecture using WBG power devices, and mostly focuses on the 480 VAC to 400 VDC front-end rectifier, the 400 VDC to 12 VDC intermediate bus converter. The efficiency improvement by utilizing the WBG devices is investigated. The paper is organized as following sections: Section II discusses the design, optimization and implementation of the front-end buck rectifier. Section III analyzes the benefits of GaN device in intermediate LLC resonant converter on device loss and transformer winding loss. Section IV is the conclusion of the paper.

II. FRON-END RECTIFIER

A. Topology and Specifications

A 7.5-kW all-SiC based three-phase buck rectifier is designed, fabricated, and tested as the front-end rectifier [6]. The converter specification and topology are shown in Table I and Fig. 4. The switching element Sx (x= 1, 2, …, 6) consists of a MOSFET and a diode in series. The input filter consists of the AC inductor L_{ac}, AC capacitor C_{ac} and the damping resistor R which is paralleled with L_{ac}. The output filter includes the DC capacitor C_{dc} and the DC-link inductor L_{dc}. The efficiency target for front-end rectifier is 99%. To reach such a high efficiency, the loss of each part in the converter should be minimized.

TABLE I. THREE-PHASE BUCK RECTIFIER SPECIFICATIONS

Power rating	7.5 kW
Input voltage	3 phase 480 Vac
Input range	± 10%
Input current	9 A
Output voltage	400 Vdc
Output current	18.75 A
Input power factor	> 0.99
Current total harmonic distortion	< 5%
Operating temperature	50 °C

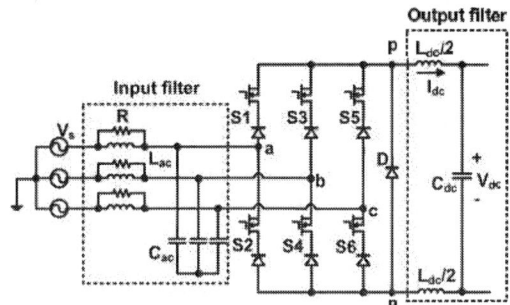

Fig. 4. Topology of three-phase buck rectifier.

B. Semiconductor Devices

As the dominant portion of total loss in a converter, the power device loss are considered first. With 680 V amplitude value of input line-to-line voltage, the buck rectifier is built using 1200 V SiC MOSFETs, CMF20120D, as active switches and 1200 V SiC Schottky diodes, SDP60S120D, as series diodes and freewheeling diode.

The static characteristics of SiC MOSFET were obtained using Tektronix 371B curve tracer at various temperatures. The on-state resistance of the SiC MOSFET used in the buck rectifier design was obtained over the temperature from 25 °C to 125 °C, at a gate-source voltage V_{gs} of 20 V, as shown in Fig. 5. The on-state resistance increases as the junction temperature gets higher. The switching performance of the switches was measured in the double pulse test, as shown in Fig. 6. The commutation between two switches, as well as the one between one switch and the freewheeling diode were both be measured in the test.

Fig. 7 and Fig. 8 show the SiC MOSFET switching waveforms at 680 V and 20 A. The turn-off time t_{off} is 40 ns and the turn-on time t_{on} is 55 ns. Here, t_{off} is defined from the time I_{ds} falls to 90% until V_{ds} rises to 90%. Similarly, t_{on} is defined as the time from I_{ds} rising to 10% until V_{ds} falls to 10%. The turn-off energy E_{off} and turn-on energy E_{on} are measured at different DC voltage values and constant DC current of 20 A, as shown in Fig. 9.

In order to reduce the conduction loss, several devices are paralleled. But the switching loss may increase as more devices are paralleled. As shown in Fig. 10, 4 SiC MOSFETs and 2 SiC Schottky diodes are paralleled considering the trade-off between conduction loss and switching loss.

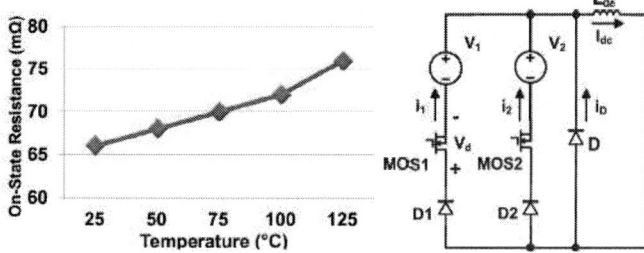

Fig. 5. SiC MOSFET on-state resistance versus temperature.

Fig. 6. Double pulse test circuit.

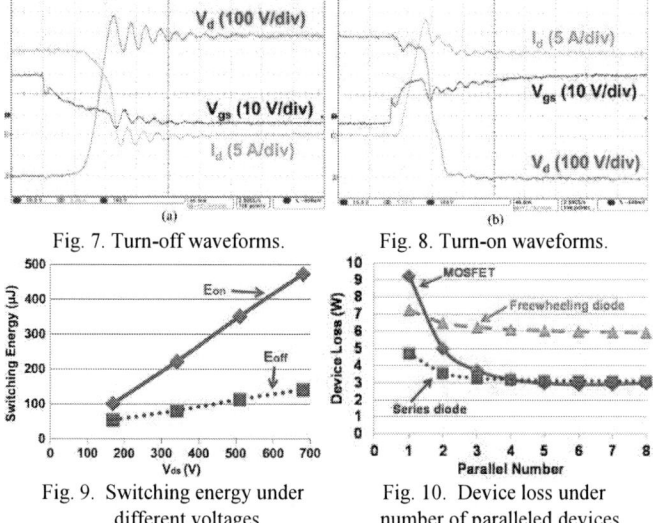

Fig. 7. Turn-off waveforms.

Fig. 8. Turn-on waveforms.

Fig. 9. Switching energy under different voltages.

Fig. 10. Device loss under number of paralleled devices.

C. Switching Frequency and Filter Design

Switching frequency is an important factor for converter efficiency. High switching frequency is helpful to reduce the size of the passive components, but causes high switching loss of power devices. Fig. 11 compares the losses of the SiC based buck rectifier at different switching frequencies, including power devices loss, passive components loss, and auxiliary circuit loss. As shown in Fig. 11, the converter achieves the highest efficiency at 28 kHz.

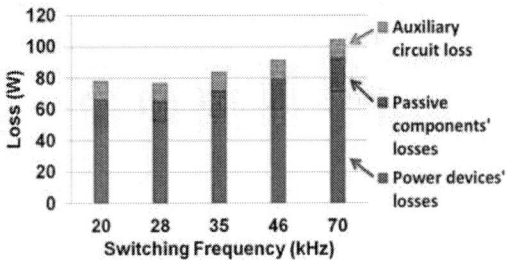

Fig. 11. Buck rectifier loss versus switching frequency.

The input filter is an L-C filter to meet the input current harmonics requirement. In each phase, the input capacitor consists of four paralleled 1.5 μF/330 V_{ac} film capacitors, whose loss can be neglected. The inductor cores with high permeability and low core loss are suitable to build the ac side inductors. Ferrite R, EE cores are selected to build the ac inductors. The total power loss of the input filter with 180 Ω damping resistors R, paralleled with ac inductors, is 10.57 W.

The dc capacitor C_{dc} is designed to limit the output voltage ripple less than 0.5% of V_{dc} at full load, and the dc inductor L_{dc} is designed to limit the inductor current ripple less than 20% of I_{dc} at full load. Three 50 μF/450 V_{dc} film capacitors are paralleled to build the dc capacitor. The nanocrystalline cut cores are used to reduce the core loss of dc inductor. The copper foil with 0.254 mm thickness and 50.8 mm width is used as the conductor instead of solid wire due to its smaller copper loss at high frequency. The loss of the dc inductor is calculated to be 8.4 W.

D. Design Results and Prototype

According to the design and loss calculation above, the full load efficiency of SiC based 7.5 kW three-phase buck rectifier with 28 kHz switching frequency is 98.79%. The auxiliary circuit loss of 12.4 W, which is used to power the DSP system, signal processing circuits, and sensors, is included in the total loss of 91.52 W. The loss distribution is shown in Fig. 12, where 2/3 of the total loss is the device loss and around 50% of the device loss is series diode conduction loss.

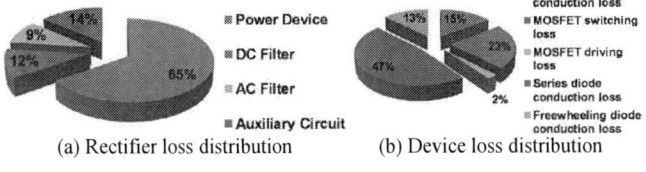

(a) Rectifier loss distribution   (b) Device loss distribution

Fig. 12. Calculated loss distribution.

To verify the design of the buck rectifier, a 7.5 kW prototype is built in Fig. 13. The experimental waveform under full output power is shown in Fig. 14. The input current THD is 2.9% and in phase with the input voltage. With Yokogawa power analyzer PZ4000, the efficiency curve is measured under different output power levels, as shown in Fig. 15. With more devices in parallel,

the efficiency of the rectifier increases under high output power levels. But under low output power, the switching loss will increase with more paralleled devices. The full load efficiency is 98.54%.

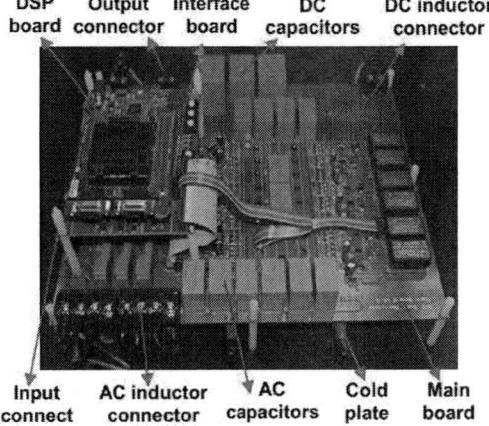

Fig. 13. Prototype of 7.5 kW buck rectifier.

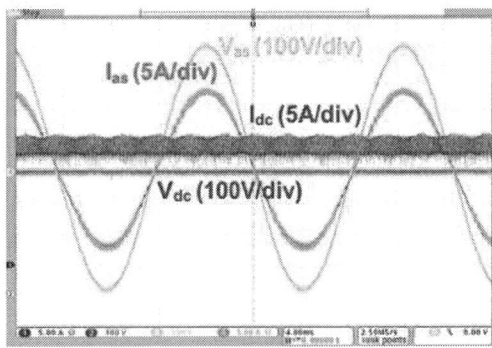

Fig. 14. Buck rectifier test waveforms under 7.5 kW output power. (Time: 4 ms/div).

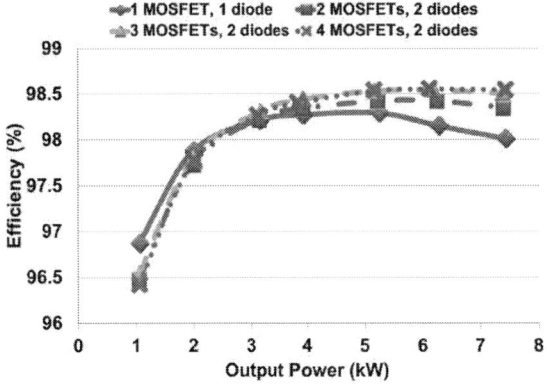

Fig. 15. Measured efficiency of the buck rectifier.

III. INTERMEDIATE BUS CONVERTER

A. Topology and Specifications

An Intermediate Bus Converter which converts 400 VDC voltage to 12 VDC voltage is applied in the HVDC power distribution architecture. The unregulated LLC resonant converter is selected as the topology, and operates as a DC-DC transformer to step down the voltage efficiently [7]. The converter specification and topology are shown in Table II and Fig. 16.

978-1-4799-5494-0/14 $31.00 © 2014 IEEE

TABLE II.   LLC RESONANT CONVERTER SPECIFICATIONS

Power Rating	300 W
Input Voltage	400 V_{dc} (±1%)
Minimum Input Voltage (for momentary loss of ac voltage (20 ms))	200 V_{dc}
Output Voltage (from full load to no load)	10.8 V_{dc} ~ 12.6 V_{dc}
Hold-up Time Requirement	20 ms

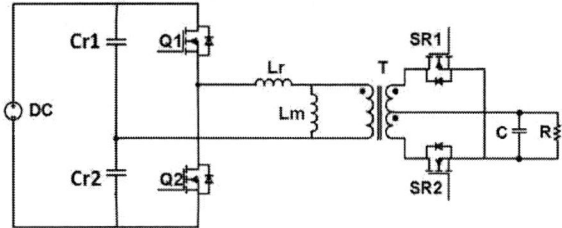

Fig. 16.   Topology of LLC resonant converter.

B. GaN Benefits on Device Loss

The LLC resonant converter can achieve zero voltage switching (ZVS) for the primary side switches and zero current switching (ZCS) for the secondary side switches when it operates at resonant frequency. Both the Si devices and GaN devices are considered in the design comparison. 1 MHz switching frequency is selected to evaluate the GaN device advantages in high frequency operation.

One tradeoff in the design of the LLC resonant converter is between dead time and peak magnetizing current. During the dead time, the output capacitance of both primary side and secondary side devices need to be discharged to achieve ZVS turn-on. Part of the peak magnetizing current is necessary to charge/discharge the winding capacitance as well. High peak magnetizing current implies high RMS value for the magnetizing current, which leads to high device conduction loss and high transformer winding loss, especially at light load or no load condition. Larger dead time results in smaller effective energy transfer time from input to load, so as to the high RMS current at full load. Therefore, in order to achieve high converter efficiency at high frequency, it is better to have small magnetizing current and small dead time. The devices with low output capacitance (low output charge) are preferable.

Fig. 17 shows the primary side RMS current versus the dead time for Si devices and GaN device. As is shown that, low output capacitance leads to smaller dead time and lower primary side RMS current. However, when the output capacitance is decreasing, the on-state resistance is increasing. Therefore, it is interesting to investigate the Si and GaN device loss with the same on-state resistances and different junction capacitances. Table III and Table IV show the device parameters of the primary side and secondary side devices, including the on-state resistance, the output capacitance, the output charge and the gate charge. The GaN devices exhibit lower output capacitance/charge and gate charge, which, therefore, inevitably show lower conduction loss compared with the Si device.

In the unregulated LLC resonant converter, the switching frequency is equal to the resonant frequency. The SR device driving signal is synchronized with the primary device driving

signal. The total loss of the primary side device mainly includes conduction loss, turn-off loss and driving loss. The total loss of SR device mainly includes conduction loss and driving loss. Figs. 18 and 19 show the loss breakdown of both Si-based design and GaN-based design. Conduction loss is the dominant on the primary side devices. On the secondary side devices, the driving loss is comparable with the conduction loss due to four devices in parallel.

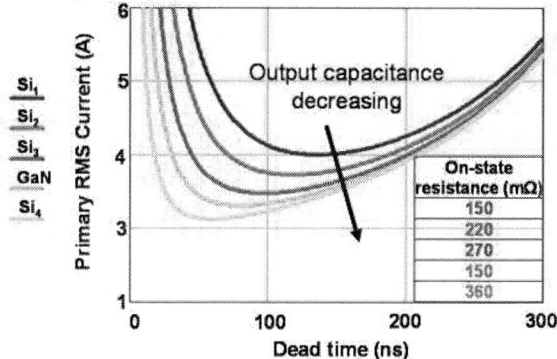

Fig. 17.   RMS current versus dead time.

TABLE III.   PRIMARY SIDE DEVICE PARAMETERS

Primary side devices	Voltage rating (V)	R_{dson} (mΩ) @25°C	Effective output capacitance (pF)	Gate charge (nC)
IPP60R165CP (Si)	600	150	220	26
TPH3006PS (GaN)	600	150	110	9.3

TABLE IV.   SECONDARY SIDE DEVICE PARAMETERS

Secondary side devices	Voltage rating (V)	R_{dson} (mΩ) @25°C	Output charge (nC)	Gate charge (nC)
BSC027N04LS (Si)	40	3.3	40	25
EPC2015 (GaN)	40	3.4	18.5	9.5

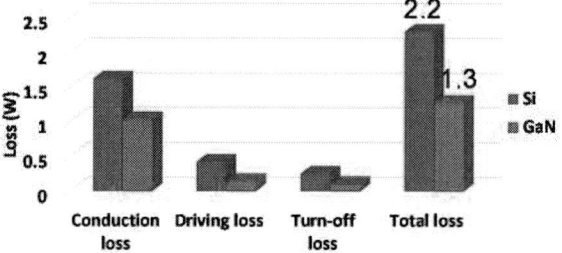

Fig. 18.   Primary side device loss breakdown and comparison.

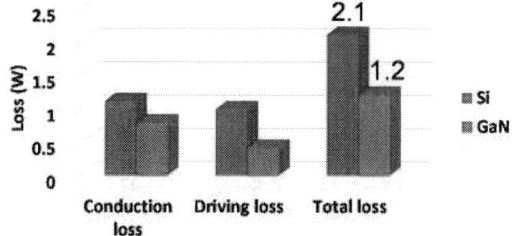

Fig. 19.   Secondary side device loss breakdown and comparison.

C. GaN Benefits on Transformer Winding Loss

Transformer loss accounts for a large part of the total converter loss in the LLC resonant converter. Since the

978-1-4799-5494-0/14 $31.00 © 2014 IEEE

transformer cores for both Si and GaN based converters are designed under the same power rating and same operation frequency, the core size and loss are the same. This section will investigate the impact of GaN device on the transformer winding loss.

The design of interleaving winding structure usually helps to reduce the winding loss by cancelling the magnetic field between the windings where the transformer primary side current and secondary side current exist in the same amplitude and in the opposite direction. However, in the LLC resonant converter, since the magnetizing current is part of the primary side current, there is a phase angle between the primary side and secondary side current, resulting in the noncancelable magnetic field even in the interleaving winding structure. The phase angle ϕ_d is determined by

$$\phi_d = arcsin(\frac{-I_{Lmpeak}}{\sqrt{2}I_{priRMS}}) \qquad (1)$$

where I_{Lmpeak} is the peak magnetizing current; I_{priRMS} is the primary side RMS current.

In order to investigate the impact of ϕ_d on the winding loss, a 16:1 center-tapped transformer winding is modeled in the Maxwell 2D simulation, shown in Fig. 20. There are sixteen turns in series as the primary side winding, and four turns in parallel as one secondary side winding. The geometry model is axisymmetric about Z axis (or RZ plane). Each PCB winding is a 70 μm or 2oz copper. The width of the primary side winding is 1 mm, and the secondary winding is 5 mm. The space between the PCB windings is 0.127 mm. The magnetic transient solver is applied in the simulation. Fig. 21 and Fig. 22 show the simulation results when the ϕ_d is equal to 0 and 0.7. Fig. 23 shows simulation results of the winding loss versus ϕ_d. The loss becomes higher when the ϕ_d is large. As is depicted in (1), ϕ_d is determined by the peak magnetizing current and primary side RMS current. According to the discussion in the sector B, the selection of ϕ_d is fundamentally determined by the device parameters.

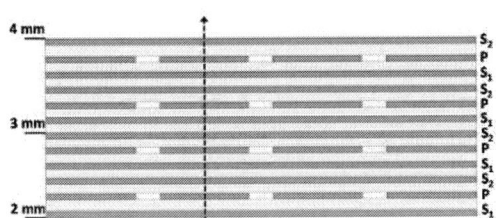

Fig. 20.   Transformer winding structure in simulation.

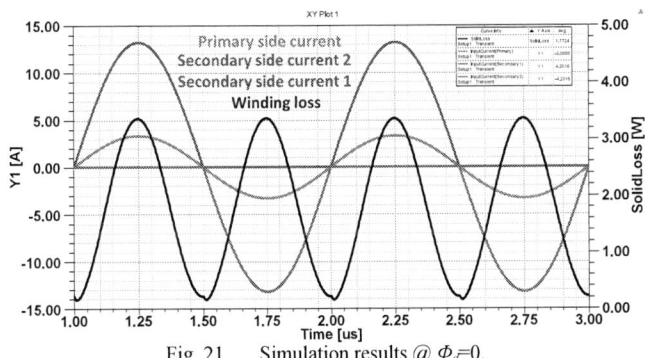

Fig. 21.   Simulation results @ ϕ_d=0.

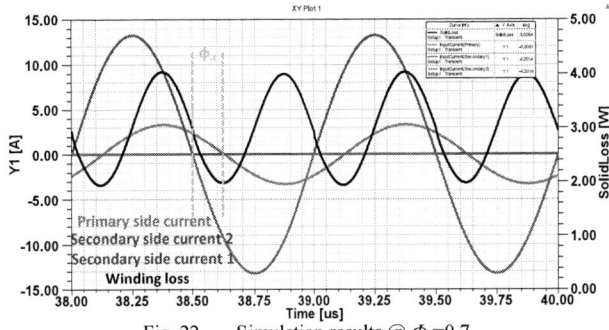

Fig. 22.   Simulation results @ ϕ_d=0.7.

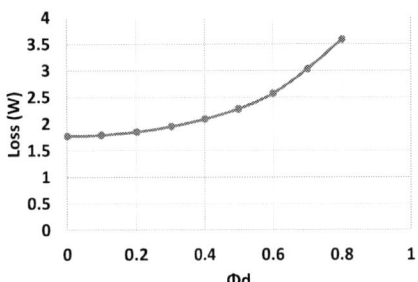

Fig. 23.   Winding loss versus ϕ_d.

Fig. 24 plots the winding loss versus the dead time for both Si and GaN cases. When the dead time is large, the required peak magnetizing current is low and therefore the ϕ_d is approaching to zero. The winding loss is also approaching its lowest value, which satisfies the curve shown in Fig. 23. Consequently, the winding loss is T_d depended. Fig. 25 and Fig. 26 plot the sum of the device and winding loss with constant winding loss and dead time depended pure winding loss respectively. In the Scenario I in the Fig. 25, the winding losses in two cases are derived when ϕ_d is equal to zero, since usually ϕ_d is not considered in the winding loss calculation, while in the Scenario II, the winding losses are obtained from Fig. 24 at the points where dead time equals to 80 ns for GaN and 125 ns for Si. The loss in Scenario I is apparently underestimated.

In fact, no matter how much the winding loss is considered in Fig. 25, optimal dead time point is always around 80 ns for the GaN case and 125 ns for the Si case. However, if the dead time depended winding loss is considered during the design, the dead time can be further extended and the sum of the loss is lower than that in Fig. 25 Scenario II, as depicted in Fig. 26. As is mentioned before, larger dead time results in smaller effective energy transfer time from input to load, so as to the higher RMS current and higher loss. But the winding loss reduction caused by the increase of T_d (decrease of I_{Lmpeak} and ϕ_d) overcomes that. As a result, the real optimal dead time point is about 120 ns for GaN case and 150 ns for Si case.

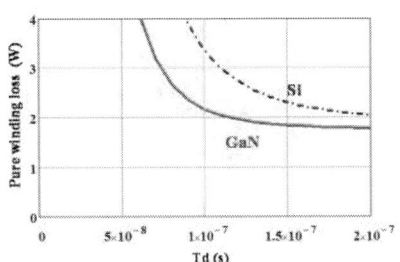

Fig. 24.   Winding loss versus dead time.

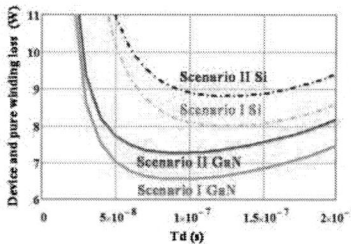

Fig. 25.   Device and winding loss with constant winding loss.

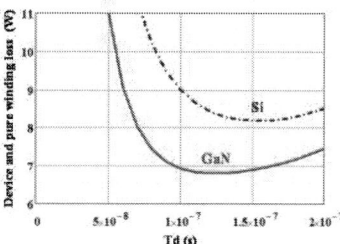

Fig. 26.   Device and winding loss with dead time depended winding loss.

D. Converter Test Results

Both Si-based and GaN-based 400 V-12 V/300 W/1 MHz LLC resonant converter prototype are built, shown in Fig. 27. The waveforms of the two converters at full load are shown in Fig. 28. The GaN-based converter demonstrate smaller drain-source voltage falling time, therefore the dead time is reduced. On the other hand, the peak magnetizing current of the GaN-based converter is also lower than that of the Si-based converter. As a result, the RMS current of the GaN-based converter is lower at the same output power.

The efficiencies were measured for Si-based and GaN-based converter under different load condition, shown in Fig. 29. The GaN-based converter demonstrates 96.3% peak efficiency and 96.1% full load efficiency, which is about 0.6% higher than the Si-based converter. Compare the test results with the loss analysis results shown in Fig. 26, there are about 5 W extra loss in the test results. This is mainly because of the transformer winding terminal loss and fringe effect loss. These losses are not influenced by the different type of the devices, therefore they are not discussed in the above sections.

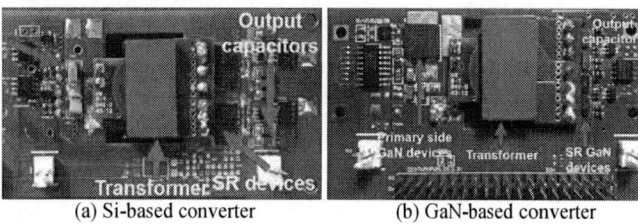

(a) Si-based converter   (b) GaN-based converter

Fig. 27.   Intermediate bus converter prototypes.

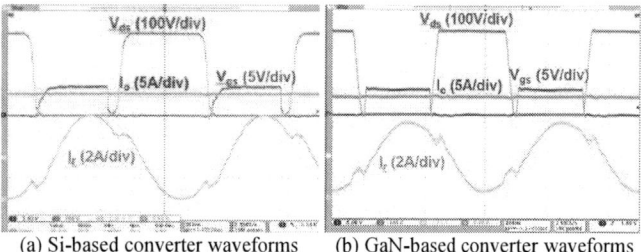

(a) Si-based converter waveforms   (b) GaN-based converter waveforms

Fig. 28.   Test waveforms at full load.

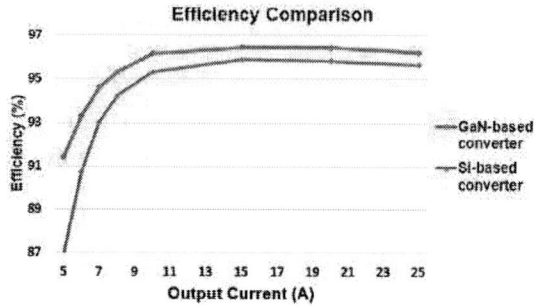

Fig. 29.   Measured efficiency of Si and GaN based converter.

IV.   CONCLUSIONS

This paper discusses the application of SiC and GaN devices in the high efficiency data center power supplies. A 7.5 kW, 3 phase 480 V AC input, 400 V DC output front-end rectifier was designed and optimized with the SiC devices. The device characteristics, the device paralleling, the switching frequency, and the input/output filter are investigated to minimize the loss. The converter test results show the peak efficiency reaches up to 98.55%, and the full load efficiency is 98.54%. For the intermediate bus converter, both the Si and GaN based 300 W, 400 V DC input, 12 V DC output LLC resonant converters are designed and compared. The GaN device benefits of the low output capacitance on the device loss and transformer winding loss are analyzed. The test results show 96.3% peak efficiency and 96.1% full load efficiency for the GaN-based converter.

ACKNOWLEDGMENTS

This work made use of Engineering Research Center Shared Facilities supported by the Engineering Research Center Program of the National Science Foundation and DOE under NSF Award Number EEC-1041877 and the CURENT Industry Partnership Program.

REFERENCES

[1]   J. Koomey, 2011, "Growth in Data center electricity use 2005 to 2010," Oakland, CA: Analytics Press. August 1.

[2]   http://en.wikipedia.org/wiki/Data_center

[3]   A. Pratt, P. Kumar, K. Bross, T. Aldridge, "Powering Compute Platforms in High Efficiency Data Centers," IBM Technology Symposium, 2006.

[4]   A. Pratt, P. Kumar, T. V. Aldridge, "Evaluation of 400V DC Distribution in Telco and Data Centers to Improve Energy Efficiency," Proc. 29th International Telecommunications Energy Conf. (INTELEC 2007), Rome, Italy, Oct. 2007, pp. 32-39.

[5]   D. Geary, T. C. Lai, T. Martinson, "New Ideas & New Equipment for Energy Efficient Data Centers", www.delta-americas.com, Jan. 2010.

[6]   F. Xu, B. Guo, L. M. Tolbert, F. Wang, B. J. Blalock, "Design and performance of an all-SiC three-phase buck rectifier for high efficiency data center power supplies," in IEEE Energy Conversion Congress and Exposition (ECCE), 15-20 Sept. 2012, pp. 2927-2933.

[7]   W. Zhang, Z. Xu, Z. Zhang, F. Wang, L. M. Tolbert, and B. J. Blalock, "Evaluation of 600 V cascode GaN HEMT in device characterization and all-GaN-based LLC resonant converter," in Proc. IEEE Energy Conversion Congress and Exposition (ECCE), 2013, pp. 3571 – 3578.

125 W Multiphase GaN/Si Hybrid Point of Load Converter for Improved High Load Efficiency

Luke L. Jenkins, Benjamin K. Rhea, William Abell, Frank T. Werner, Christopher G. Wilson, Robert N. Dean, and
Daniel K. Harris
Electrical and Computer Engineering
Auburn University
Auburn, AL 36849 USA
llj0005@tigermail.auburn.edu

Abstract—**This work presents experimental results of an optimized five-phase GaN/Si hybrid synchronous buck converter designed to improve low voltage point of load (POL) converter efficiency while meeting increasing load demands typical in high performance computing. The best enhancement-mode GaN HEMTs have very low gate charge, switch much faster than Si, and enable higher frequency operation which can reduce size. However, GaN HEMTs are not yet available with extremely low $R_{DS(ON)}$, like comparable Si MOSFETs. For this reason, GaN-based systems are susceptible to poor efficiency at high loads. To mitigate this consequence, the appropriate combination of GaN and Si is utilized in a 12 to 1 V POL converter with peak efficiency above 95% and 125 W rated load. The results of this work submit that optimal performance can be achieved by combining two semiconductor technologies, and the design is a novel approach to improve upon current state-of-the-art POL converters.**

I. INTRODUCTION

WIDE bandgap (WBG) semiconductors are becoming more popular in power electronic systems due to their better performance over Silicon (Si) devices of similar voltage ratings [1]. The low voltage WBG market is predominately GaN at this time, while the high-voltage WBG market is predominately SiC. The overlapping area between these two materials is around 600 V. There are three basic types of GaN power semiconductors: depletion-mode devices, cascode devices, and enhancement-mode devices. Depletion-mode devices are normally on and require more complex start-up procedures. In most switch-mode power supply (SMPS) applications, depletion-mode devices are not ideal. Cascode devices use a low voltage Si MOSFET to force a depletion-mode GaN FET to operate like an enhancement-mode device, but cascode devices must make some sacrifice performance to accomplish this. Lastly, the first commercially available enhancement-mode GaN devices were introduced in 2009 [3][4]. Current enhancement-mode GaN devices are heterostructure field effect transistors (HFETs) or high electron mobility transistors (HEMTs), and they are lateral devices

where conduction occurs in the two dimensional electron gas (2DEG). Vertical conduction enhancement-mode GaN FETs are not currently available. A typical lateral GaN on Si device structure is shown in Fig. 1. These lateral GaN HEMTs switch faster, have a wider bandgap, superior theoretical on-resistance to breakdown voltage, higher electron mobility, and higher electron velocity than comparable Si MOSFETs [3][5]-[7]. GaN technology is still young and shows promise for the future of efficient and high frequency power conversion [3]-[5], but the selection of GaN devices is still limited—forcing compromises in drain-to-source on-resistance ($R_{DS(ON)}$) or requiring parallel devices to reduce effective $R_{DS(ON)}$. Since Si is a mature technology, there are many comparable Si MOSFETs with significantly lower $R_{DS(ON)}$.

Commercial state-of-the-art low voltage POL converters typically operate between 200 and 600 kHz to accomplish between 80 and 90% efficiency [5][8]. Poor POL converter efficiency leads to shorter battery life in portable and telecommunication systems [9], generates heat near the load that may require cooling [10], and can be detrimental to operating cost in large scale systems like data center and large computing facilities [11]. Today, the cost of maintaining digital cloud storage and cloud computing is growing quickly, and these systems are heavily dependent upon POL converters which are the final conversion stage that reduces voltage levels to approximately 1 V, suitable for modern processors. High performance computing (HPC) and increased cloud applications are increasing energy consumption [12], and projections indicate that cloud computing will grow rapidly for many years [13]. Modern processors continue to operate at lower voltages and higher currents. Consequently, POL converters suffer from poor efficiency due to high conduction loss [14], and the POL is typically the least efficient conversion stage in the data center power supply chain [15]. The development of efficient, high power density and high current (> 100 A) POL converters is fundamental to meeting future computational and data storage demands.

I. TOPOLOGY & TRANSISTORS

Non-isolated POL converters, like the synchronous

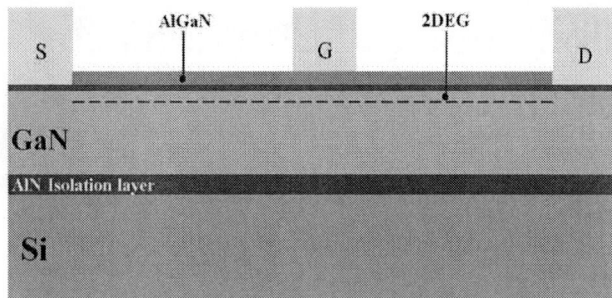

Fig. 1. Typical GaN on Si lateral HEMT structure.

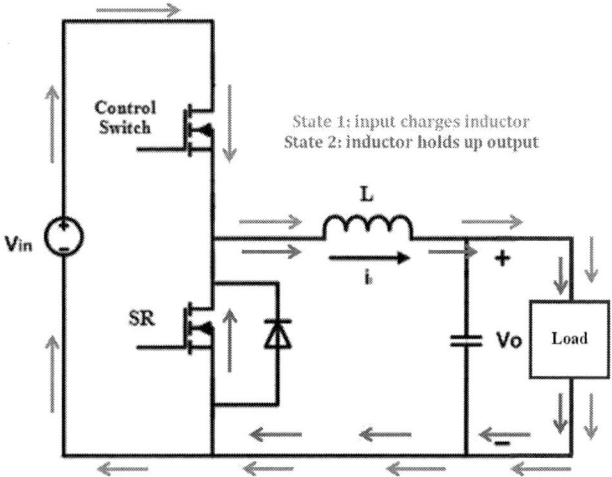

Fig. 2. Synchronous buck converter topology in its two primary states.

buck converter as seen in Fig. 2, charge an inductor during state 1 and allow the inductor to support the load during state 2 (see Fig. 2). The smaller the ratio V_{OUT}/V_{IN}, the smaller the duty cycle (D) and the more free-wheeling current that circulates through the synchronous rectifier (SR). The POL converters in this work are developed for efficient operation at loads exceeding 100 A at 12 to 1 V power conversion – a common voltage transformation ratio in servers and HPC. At a conversion ratio of 12 to 1 V, the duty cycle will be slightly below but approximately 10%. This means the control switch will only conduct for 10% of the period, while the SR will conduct for 90% of the period. Thus, switching losses will likely be the dominating loss mechanism in the control switch, and conduction losses will be the dominating loss mechanism in the SR. The switching loss of a semiconductor can be approximated by

$$P_{SW} = \frac{1}{2} V_{in} * I_{out} * (t_r + t_f) * f_S \qquad (1)$$

where the rise time (t_r) and the fall time (t_f) are approximated by

$$t_r \approx t_f \approx \frac{Q_G}{I_G} + \frac{L_{LUMP}*I_G}{V_{GS}-V_{th}}. \qquad (2)$$

A semiconductors gate charge, Q_G, holds the largest influence on how fast a transistor can turn on or off. This is the case because Q_G must be charged and discharged by the gate driver every period. The Q_G/I_G term dominates the rise and fall time. Therefore, switching loss is approximately proportional to Q_G, and Q_G is a parameter used to compare the switching time of two transistors. The conduction loss of the control switch can be calculated using

$$P_{C,SW} = I_{out}^2 * R_{DS(ON)} * D, \qquad (3)$$

and the conduction loss of the SR can be calculated using

$$P_{C,SR} = I_{out}^2 * R_{DS(ON)} * (1 - D). \qquad (4)$$

Conduction loss is exactly proportional to $R_{DS(ON)}$ in the control switch and SR; thus, $R_{DS(ON)}$ is also a common parameter to compare transistors. The third and last loss mechanism is the gate losses given by

$$P_G = Q_G * V_{GS} * f_S * \left(\frac{R_{G\_sink}}{R_{G\_sink}+R_g+R_{gi}} + \frac{R_{G\_source}}{R_{G\_source}+R_g+R_{gi}} \right). \qquad (5)$$

The gate loss simply refers to the power expended charging and discharging the transistor, and it is proportional to Q_G, gate-to-source voltage (V_{GS}), and switching frequency (f_S). Except for very light load operation, P_G is usually small relative to switching and conduction losses. More details on the parameters in these equations and complete synchronous buck converter loss models can be found in [3] and [16]-[17].

The power semiconductors and inductor create the majority of power loss in the non-isolated POL converter, so choosing the proper semiconductors is essential to efficient power conversion. Many Si and GaN power transistors are compared using a figure of merit (FOM) defined as

$$FOM = Q_G \times R_{DS(ON)}. \qquad (6)$$

This FOM provides an understanding of the transistor's switching and conduction loss. The lower the FOM, the better the device should perform. In many cases, Q_G can be sacrificed to lower $R_{DS(ON)}$ or vice versa, so the application can govern whether low conduction loss or low switching loss is more valuable. Many GaN and Si transistors have been compared leading up to this work; one GaN HEMT and one Si MOSFET performed significantly better than the others: the EPC2015 GaN HEMT by Efficient Power Conversion (EPC) and the BSC010NE2LS Si MOSFET by Infineon. Table 1 compares the key characteristics of these two power semiconductors. The

978-1-4799-5494-0/14 $31.00 © 2014 IEEE

transistors were utilized to design three similar converters: a GaN POL converter, a Si POL converter, and a GaN/Si hybrid POL converter. The GaN HEMT switches faster and thus reduces switching loss. This is ideal for the control switch, since it only conducts for about 10% of the period. The Si MOSFET switches slower than the GaN HEMT, but has nearly one fourth the $R_{DS(ON)}$ and reduces conduction loss. It would require four parallel GaN HEMTs to accomplish this low $R_{DS(ON)}$, and that would become a physically large design with an effective gate charge $Q_{G,eff} = 42$ nC, which will likely switch slower than the Si devices. Therefore, the application, load requirements, and frequency will determine the appropriate transistor—a GaN HEMT for efficient switching or a Si MOSFET for efficient conduction.

II. POINT OF LOAD CONVERTER DESIGN

Commercial POL converters have been employing Si for over twenty-five years, and recent development of WBG semiconductors has led to development of high frequency and high power density GaN-based converters. In order to improve the high load performance (> 100 A), the design in this work combines a fast switching GaN HEMT to reduce switching loss in the control switch and a low $R_{DS(ON)}$ Si MOSFET to reduce conduction loss in the SR. A single phase of this design is presented in Fig. 3.

The POL converter uses the Texas Instruments (TI) LM5113 gate driver, which has been designed specifically to drive enhancement-mode GaN HEMTs in a half bridge configuration. The driver has an internal 2.1 mΩ pull-up gate resistance to reduce gate overvoltage, and it uses a bootstrap technique to internally clamp the gate at 5.2 V, which prevents the GaN HEMT from exceeding its gate-to-source voltage limit ($V_{GS,MAX}$) of 6 V [18]. The layout of the GaN HEMT is essential for performing fast switching because the device has very low package inductance; therefore, poor layout creates parasitic inductance that can be larger than the package inductance. Layout techniques described in [11] were used to reduce layout parasitics near the GaN control switch (Q1 in Fig. 3). The Si SR (Q2) is about 150% the size of the GaN HEMT but is still quite small compared to similar Si devices. The Si device is also driven with the LM5113 at $V_{GS} = 4.5$ V. Since Si MOSFETs are capable of being driven with a wide range of gate voltages, there is a direct trade-off between V_{GS}, $R_{DS(ON)}$, and Q_G. This design drives the Si MOSFET with a low gate voltage that results in slightly higher $R_{DS(ON)}$ and lower Q_G. External gate resistors (R1-R4) are 0 Ω resistors because it was not necessary to reduce I_G to prevent gate overvoltage (see Fig. 3). C2 – C5 are input capacitors. C9 – C12 are output capacitors. The inductor footprint was designed to accommodate several commercial off the shelf (COTS) inductors such as this 2 µH, 0.9 mΩ Coilcraft SER1410-202 inductor, which was the largest inductance value and most efficient inductor used in this design.

Table 1. GaN and Si transistor comparison with $V_{GS} \approx 4.5$ V.

Manufacturer	Part Number	Material	V_{DS} (V)	I_D (V)	$R_{DS(ON)}$ (mΩ)	Q_G (nC)	FOM (mΩ x nC)
EPC	EPC2015	GaN	40	33	4	10.5	42
Infineon	BSC010NE2LS	Si	25	100	1.1	27	29.7

Fig. 3. A single phase GaN/Si hybrid POL converter.

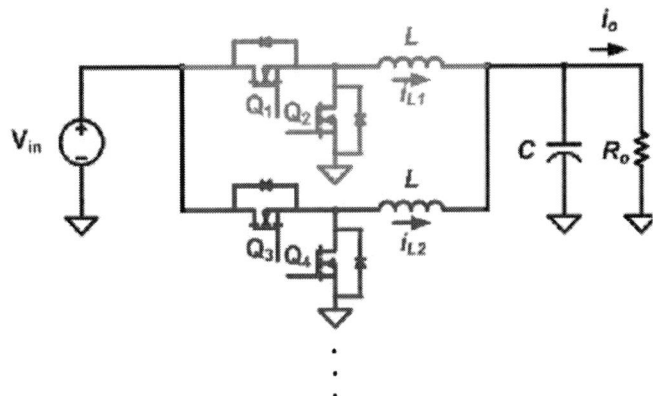

Fig. 4. Multiphase synchronous buck converter example

III. MULTIPHASE POINT OF LOAD CONVERTER DESIGN

The synchronous buck converter is an easy topology to implement in parallel. Thus, the system can be designed and optimized with a single phase, and expanding to multiple phases is easy—as long as the controller is designed efficiently, is fast enough to manage all active phases, and manages proper spacing of the PWM signals for interleaving applications. Multiphase synchronous buck converters are implemented by connecting V_{IN} and V_{OUT}, and the output capacitance are shared among all phases (see Fig. 4). Each phase has a gate driver but is driven by the same controller. The phases can be synchronous or interleaving. Synchronous phases turn on and off at the same time, while interleaving phases are staggered evenly throughout the period. Interleaving phases have many advantages such as raising the effective output frequency by

$$f_{eff} = f_s \times N, \qquad (7)$$

where f_{eff} is the effective frequency or frequency of the POL converter output; f_s is the switching frequency of each phase, and N is the number of phases in an interleaved arrangement. Thus, five phases switching at 200 kHz results in a 1 MHz converter. The higher output frequency significantly reduces output voltage ripple, which can reduce the passives required to meet design specifications. Additionally, multiphase POL converters respond better to transient loads because the higher effective frequency means that at least one phase is never far from turning on; this does however require a controller with enough bandwidth to sense the change in load and increase the duty cycle to accommodate the change. Fig. 5 compares the steady-state 5 A output voltage and current waveforms of a single phase and an interleaved five phase POL converter. Notice the five phase POL converter has significantly less output voltage ripple. There are many other great advantages of multiphase POL converters such as distributing current to minimize I_2R loss, cleaner output signal, higher load capability, and phase shedding. Phase shedding promotes high efficiency by dropping phases to operate the remaining phases at the load that yields peak efficiency.

I. EXPERIMENTAL RESULTS

The GaN/Si hybrid POL converter significantly improves efficiency beyond similar Si-based designs, and it improves high load efficiency beyond GaN-based designs. Fig. 6 compares experimental efficiency of a Si-based, a GaN-based, and a GaN/Si hybrid POL converter. These three converters are all interleaved five phase synchronous buck converters capable of more than 125 A load current. All three converters utilize the same controller, inductors, capacitors, and gate drivers. The only changes between them are the SR. The SR of the GaN-based POL converter is two EPC2015s in parallel to yield an effective on-resistance $R_{DS(ON),EFF} = 2\ m\Omega$, which is still twice that of the Si SR. Additionally the two parallel GaN HEMTs in the SR have an effective gate charge $Q_{G,EFF} = 21\ nC$.

Each GaN/Si hybrid POL phase is approximately 0.74 in^2 and can be nearly half this size with the inductor mounted on the backside of the printed circuit board (PCB). Fig. 7 includes thermal images of the GaN/Si hybrid POL converter at 25 and 100 A and without any forced cooling. The thermal images are upside-down relative to Fig. 3. In both designs, "A" represents the GaN control switch and "B" denotes the Si SR. Both the control switch and SR are similar in temperature, even though the SR conducts for nine times

as long as the control switch; with that said, the SR also has a larger surface area to dissipate heat.

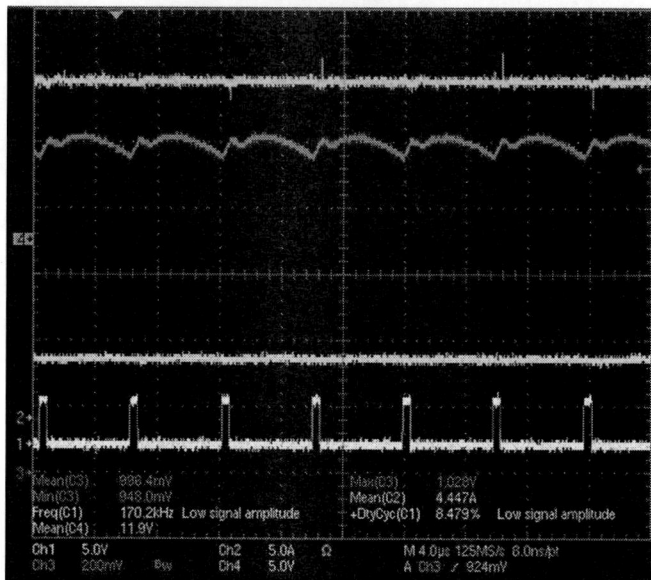

(a) Single phase POL converter, 5 A output.

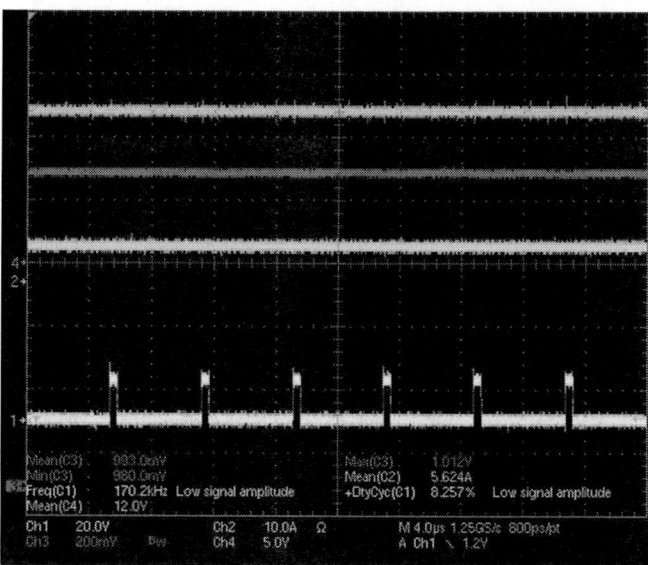

(b) Interleaved five phase POL converter, 5 A output.

Fig. 5. Comparison of output voltage waveforms (purple) between a single phase POL converter and interleaved five phase POL converter.

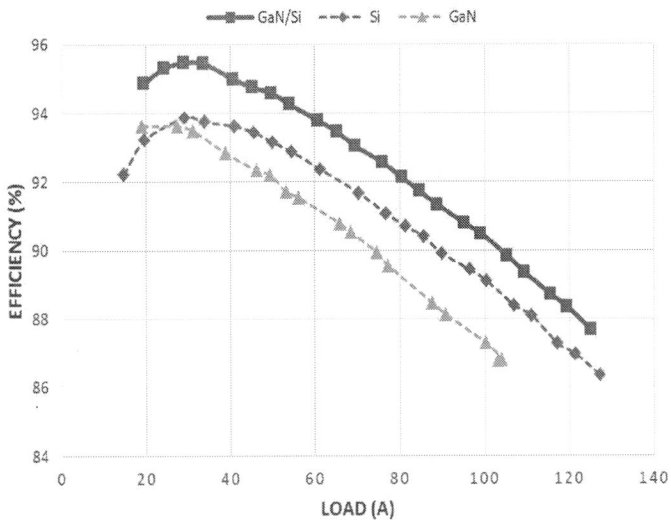

Fig. 6. Efficiency comparison of a Si-based, a GaN-based, and a GaN/Si hybrid POL converter (each an interleaved five phase synchronous buck converter topology)

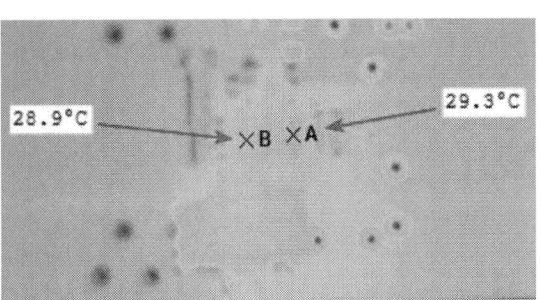

(a) 25 A load current

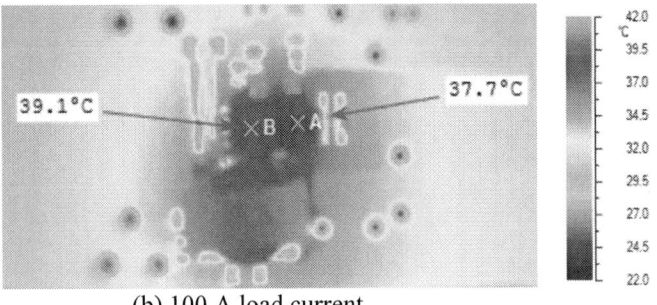

(b) 100 A load current

Fig. 7. Thermal images of one GaN/Si POL converter phase. No forced cooling and $T_{ambient}$ = 23.3 °C.

II. CONCLUSION

GaN certainly has some remarkable characteristics—particularly switching around three times faster than Si. There is certainly a bright future for GaN semiconductors, and they will continue to improve. Nevertheless, there are still some applications best suited for Si, and perhaps some applications that are best when utilizing GaN and Si. In typical POL applications where $V_{IN} = 12 - 48$ V_{DC} and $V_{OUT} = 0.9 - 3.3$ V_{DC}, the control switch of the synchronous buck converter conducts for a very short time while the SR conducts for the majority of the period. For this reason it is logical to combine fast switching GaN with low $R_{DS(ON)}$ Si. This design is compared to similar Si-based and GaN-based synchronous buck converters to demonstrate how a GaN HEMT can reduce switching loss in the control switch while a very low on-resistance Si MOSFET can lessen conduction loss in the SR. Experimental results show that the GaN POL converter is dominated by conduction loss at high loads, whereas the Si POL converter is dominated by switching loss at light loads. However, the GaN/Si hybrid approach yields improved efficiency over a large load range and maintains comparable package size. Implemented as an interleaved five phase POL converter, the GaN/Si hybrid design enables over 125 W output power at 12 to 1 V DC-DC conversion. Peak efficiencies above 95% occur between 20 and 40 A load current, and this surpasses today's state-of-the-art POL converters that rarely reach 90% peak efficiency. Efficiency is still above or near 90% up to full load, and the output waveforms of this 1 MHz effective frequency POL converter has very low steady-state voltage and current ripple. The 1.1 mΩ Si transistor is certainly the best device for the SR. Since the EPC GaN FET still proved to be the best applicant for the control switch, the conclusion is that, with the current semiconductor market, a combination of GaN and Si can produce better results than either technology can independently, and efficiency of low voltage POL converters are being raised by several percent with this GaN/Si hybrid POL converter. The expectation is that eventually, as GaN transistors mature, they will surpass Si with very low $R_{DS(ON)}$; however, at this time, the GaN/Si hybrid POL converter approach improves overall performance.

III. REFERENCES

[1] Scott, M.J.; Ke Zou; Inoa, E.; Duarte, R.; Yi Huang; Jin Wang, "A Gallium Nitride switched-capacitor power inverter for photovoltaic applications," *Applied Power Electronics Conference and Exposition (APEC), 2012 Twenty-Seventh Annual IEEE* , vol., no., pp.46,52, 5-9 Feb. 2012

[2] Zhang, W.; Long, Y.; Zhang, Z.; Wang, F.; Tolbert, L.; Blalock, B.; Henning, S.; Wilson, C.; and Dean, R., "Evaluation and Comparison of Silicon and Gallium Nitride power Transistors in LLC Resonant Converters," in *Energy Conversion Congress and Exposition (ECCE), 2012 IEEE,* 2012, pp. 1362–1366.

[3] A. Lidow, J. Strydom, M. Rooij, and Y. Ma, "GaN Transistors for Efficient Power Conversion," Power Conversion Publications, El Segundo, 2012

[4] Reusch, D.; Strydom, J., "Understanding the effect of PCB layout on circuit performance in a high frequency gallium nitride based point of load converter," *Applied Power Electronics Conference and Exposition (APEC), 2013 Twenty-Eighth Annual IEEE* , vol., no., pp.649,655, 17-21 March 2013

[5] Shu Ji; Reusch, D.; Lee, F.C., "High frequency high power density 3D integrated Gallium Nitride based point of load module," *Energy Conversion Congress and Exposition (ECCE), 2012 IEEE* , vol., no., pp.4267,4273, 15-20 Sept. 2012 doi: 10.1109/ECCE.2012.6342242

[6] Reusch, D.; Lee, F.C.; Gilham, D.; Yipeng Su, "Optimization of a high density gallium nitride based non-isolated point of load module," *Energy Conversion Congress and Exposition (ECCE), 2012 IEEE* , vol., no., pp.2914,2920, 15-20 Sept. 2012

[7] Efficient Power Conversion. (2013). *Is it the End of the Road for Silicon in Power Conversion?*[Online]. Available: http://epc-co.com/epc/documents/product-training/Appnote_Si_endofroad.pdf

[8] Jenkins, Luke L.; Wilson, Christopher G.; Moses, Justin D.; Aggas, Jeffrey M.; Abell, William; Dean, Robert N., "Performance Comparison of Multiphase GaN vs. GaN/Si Hybrid 12-1 V Point of Load Converters," *Government Microcircuit Applications and Critical Technology Conference (GOMACTech) 2014*, March 31 – April 3 2014

[9] Mandyam, G.D., "Improving battery life for wireless web services through the use of a mobile proxy," *Personal Indoor and Mobile Radio Communications (PIMRC), 2010 IEEE 21st International Symposium on* , vol., no., pp.2763,2768, 26-30 Sept. 2010

[10] Iyengar, M.; David, M.; Parida, P.; Kamath, V.; Kochuparambil, B.; Graybill, D.; Schultz, M.; Gaynes, M.; Simons, R.; Schmidt, R.; Chainer, T., "Server liquid cooling with chiller-less data center design to enable significant energy savings," *Semiconductor Thermal Measurement and Management Symposium (SEMI-THERM), 2012 28th Annual IEEE* , vol., no., pp.212,223, 18-22 March 2012

[11] Jenkins, Luke L.; Wilson, Christopher G.; Moses, Justin D.; Aggas, Jeffrey M.; Rhea, Benjamin K.; Dean, Robert N.; "Optimization of a 96% Efficient 12-1 V Gallium Nitride Based Point of Load Converter," *Applied Power Electronics Conference (APEC) 2014*, March 16-20 2014

[12] Warkozek, G.; Drayer, E.; Debusschere, V.; Bacha, S., "A new approach to model energy consumption of servers in data centers," *Industrial Technology (ICIT), 2012 IEEE International Conference on* , vol., no., pp.211,216, 19-21 March 2012

[13] "World Energy Projections Plus," Energy Information Administration (EIA), 2009

[14] Jenkins, Luke L.; Wilson, Christopher G.; Moses, Justin D.; Aggas, Jeffrey M.; Rhea, Benjamin K.; Dean, Robert N., "The impact of parallel GaN HEMTs on efficiency of a 12-to-1 V buck converter," *Wide Bandgap Power Devices and Applications (WiPDA), 2013 IEEE Workshop on* , vol., no., pp.197,200, 27-29 Oct. 2013

[15] Cui,Yutian; Xu, Fan; Zhang, Weimin; Guo, Ben Tolbert, Leon M.; Wang, Fred; Blalock, Benjamin J.; Jenkins, Luke L.; Wilson, Christopher G.; Aggas, Jeffrey M.; Rhea, Benjamin K.; Moses, Justin D.; Dean, Robert N.; "High Efficiency Data Center Power Supply Using Wide Band gap Power Devices," *Applied Power Electronics Conference (APEC) 2014*, March 16-20 2014

[16] Mappus, Steve, "Optimizing MOSFET Gate Driver Voltage", Texas Instruments Inc.

[17] Peter Markowski, "Estimating MOSFET Switching Losses Means Higher Performance Buck Converters," Planet Analog.com, December 18, 2002

[18] Texas Instruments (TI), LM5113 5A, 100V Half-Bridge Gate Driver for Enhancement-Mode GaN FETs. Data sheet available at: http://www.ti.com/product/lm511.

A Full-Bridge Current-Source Isolated DC/DC Converter with Reduced Number of Switches and Voltage Stresses for Photovoltaic Applications

Feng Guo, Lixing Fu, He Li, Mohammed Alsolami,
Xuan Zhang, and Jin Wang
Department of Electrical and Computer Engineering
The Ohio State University
Columbus, Ohio, U.S.A.

Jie Zhang
School of Electrical and Electronic Engineering
Hubei University of Technology
Wuhan, China

Abstract— A full-bridge current-source isolated dc/dc converter for Photovoltaic applications is proposed in this paper. The converter utilizes a Quasi-Switched-Capacitor circuit as the secondary side, which features a reduced number of switches and voltage stresses, and an additional boost function. Through the proposed control algorithm, soft-switching can be realized for all switches. A 1.2 kW, 1 MHz, 40 V/ 400 V prototype is built and tested in the lab. To increase the efficiency at high switching frequency, GaN HEMT from Transphorm is implemented, and a peak efficiency of 92.7 % at 500 kHz and 89.0% at 1 MHz is achieved.

Keywords— current-source converter; Quasi-Switched-Capacitor circuit; Photovoltaics; GaN HEMT

I. Introduction

The increasing concern about the energy crisis and the rapid improvement of technology have stimulated the growth of the Photovoltaic (PV) energy conversion systems. In the PV system, a line frequency transformer is commonly installed. It can boost the output voltage of the dc/ac stage, so the system can be connected to the medium or high voltage utility grid. This transformer also can eliminate the ground leakage current. However, this low frequency transformer is bulky and costly. It significantly increases the system cost and maintenance fee. To eliminate this low frequency transformer, a front stage dc/dc converter with high frequency transformer can be utilized. Voltage-source isolated dc/dc converters have been widely used in this application, such as flyback, dual active bridge, and resonant topologies [1] [2].

At dc input side of these converters, several PV panels are usually put in series to increase the voltage level, which brings the partial shading issue to the system. This issue can be completely solved by utilizing micro inverter or micro converter, which only use one PV panel at input side. Alternatively, the partial shading effect can be greatly reduced by putting PV panels in parallel [3]. However, in either method, a high boost ratio is required for the front side dc/dc converter. For voltage-source isolated dc/dc converters, the efficiency is decreased with the high boost ratio or large input current. In the meantime, the input current ripple is usually

large, which will influence the tracking of maximum power point and shorten the lifetime of PV panels. In contrast, the current-source converters feature inherent boost capability, lower input current ripple, and easy current controllability, which become good candidates in this application.

Current-source isolated dc/dc converters have been widely used in high power or high boost ratio applications [5]-[8]. At secondary side of these converters, a bridge structure for rectification is implemented. The semiconductor devices in the bridge need to sustain the full output dc voltage, which is a fairly high value. Furthermore, this bridge structure cannot offer any boost function. Another practical issue with the current-source isolated dc/dc converter is that the current in the leakage inductor can introduce large voltage spike when the switch turns off. Hence, certain soft-switching strategies need to be implemented.

A full-bridge current-source isolated dc/dc converter based on the Quasi-Switched-Capacitor (QSC) circuit is proposed in this paper, which features: 1) The high voltage side QSC circuit has less number of switches and voltage stresses [4]; 2) The QSC circuit has additional boost function, with which a high boost ratio can be easily achieved. In addition, because of this boost function, the voltage stress on the transformer is also reduced; 3) For primary side switches, ZCS turn-on and ZVS turn-off are realized. For secondary side switches, both ZVS turn-on and turn-off are realized; and 4) GaN HEMT is implemented in the prototype to achieve a high switching frequency, high power, high boost ratio, and high efficiency operation.

II. Proposed Circuit Topology

A. Secondary side Quasi-Switched-Capacitor circuit

The topology of the secondary side QSC circuit is shown in Fig. 1. The circuit is derived by applying the traditional voltage tripler ac/dc circuit and replacing the three diodes with three controllable switches. In the circuit, S_a and S_b are a pair of switches that turn on and off simultaneously. The control signal for S_c is complementary to that of $S_a \& S_b$ with a certain amount of dead time inserted in between. In contrast to [4], which fixed the duty ratio of S_a&S_b to be 0.5, a variable duty ratio D can be implemented for S_a&S_b. C_1 and C_2 have identical capacitance. In one switching cycle, they will be charged in parallel and discharged in series. L_m represents the transformer magnetizing inductance.

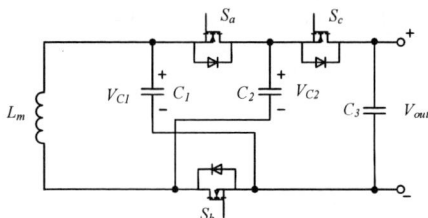

Fig. 1. The topology of the QSC circuit.

Based on the voltage-second balance of L_m, it can be derived that:

$$V_{C1} = V_{C2} = \frac{1-D}{2-D} V_{out} . \qquad (1)$$

From (1), it is noted that, by utilizing the magnetizing inductor of the transformer, the output voltage can be boosted higher than the capacitor voltage, so an additional boost function can be realized by the QSC circuit. Furthermore, compared to other circuit topologies, where the voltage stress on the switches equals V_{out}, the voltage stress on the switches in the QSC circuit is reduced to:

$$V_{Sa\_max} = V_{Sb\_max} = V_{Sc\_max} = V_{out} - V_{C1} = \frac{1}{2-D} V_{out} . \qquad (2)$$

The voltage across the transformer winding, which is V_{Lm}, is also reduced from V_{out} to the following value:

$$V_{Lm\_max} = \max\{\frac{1-D}{2-D} V_{out}, \frac{D}{2-D} V_{out}\} . \qquad (3)$$

B. Full dc/dc converter circuit

The proposed full dc/dc converter circuit topology is presented in Fig. 2. The QSC circuit is used on the secondary side. The primary side of the converter is a full-bridge current-source circuit, and the reverse blocking diode is not needed. The primary side input inductor limits the current ripple flowing into the PV module, so the lifetime can be extended and a more accurate MPPT can be achieved. A high-frequency transformer is utilized to connect the primary and secondary sides, and in the meantime provides galvanic isolation in the circuit. The ground leakage current of the PV modules can be greatly reduced by this configuration.

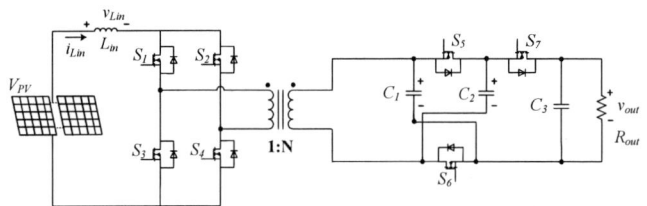

Fig. 2. Circuit topology of the proposed isolated dc/dc converter.

III. The Operation Principle

The key waveforms in the circuit are shown in Fig. 3. There is a total of six states in one switching cycle. On the primary side, S_1&S_4 are one pair of switches that turn on and off at the same time. S_2&S_3 are the other pair of switches. On the secondary side, S_5&S_6 are the third pair of switches. S_5&S_6 and S_7 are controlled as complementary pairs. There are totally six states in one switching cycle. In contrast to traditional control methods, the control signals of primary side switches are synchronized with the secondary side QSC circuit and in most cases are asymmetrical.

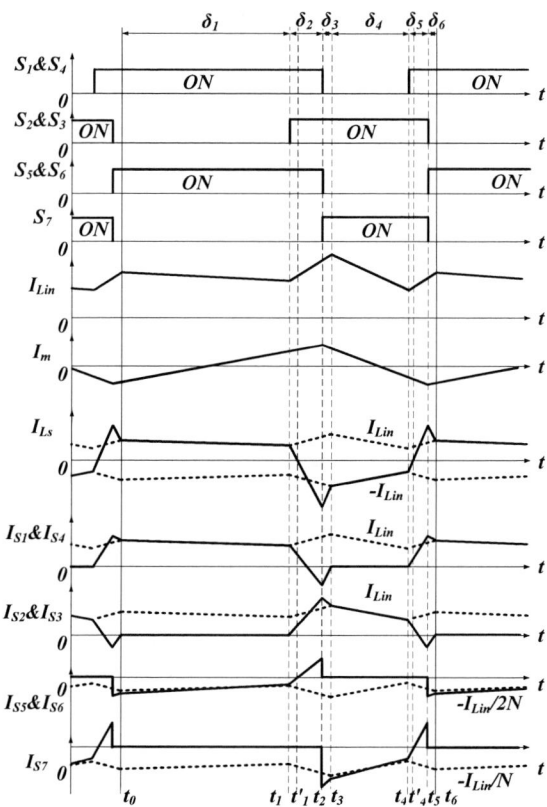

Fig. 3. Key waveforms of the circuit.

A. State I (t_0-t_1), C_1&C_2 charging.

In this state, C_1&C_2 are connected in parallel and charged by the input inductor. The mutual inductor current is increasing. C_3 and load are disconnected from the main circuit and the load voltage is held by C_3. The current flow in State I is shown in Fig. 4(a).

B. State II (t_1-t_2), leakage inductor current changing direction.

This state starts when S_2&S_3 are turned on. The input inductor is charged by the input voltage source. C_1&C_2 are connected in parallel to charge the leakage inductor and the mutual inductor. The current in the leakage inductor drops to zero at t'_1, and then continues to increase at the reverse direction. The mutual inductor current keeps increasing. The current flow in State II is shown in Fig. 4(b) and (c).

978-1-4799-5494-0/14 $31.00 © 2014 IEEE

Ideally, this state should end when the current in the leakage inductor has the same magnitude but opposite direction as the input current. In practice, the leakage inductor current is charged a little higher than the input current to leave some margin and avoid a voltage overshoot.

At t_1, when S_2&S_3 are turned on, the currents in the input inductor and the leakage inductor are the same. With the decrease of the leakage inductor current, the current in S_2&S_3 will begin to increase from zero. Since L_s limits the switch current increase rate while the voltages across the switches drop instantly, S_2&S_3 realize zero-current switching (ZCS) during the turn-on transition.

C. State III (t_2-t_3), leakage inductor current freewheeling.

This state starts when S_1&S_4 and S_5&S_6 are turned off, and S_7 is turned on. Since the magnitude of the current in the leakage inductor is larger than the input current, the energy at the input side cannot be transferred to the load. Instead, the input inductor remains charged by the input voltage source in this state. In the meantime, C_1&C_2 are connected in series with the load. Since the body diodes of S_1&S_4 are conducting, the current in the leakage inductor is decreasing rapidly. This state ends when the magnitude of the current in the leakage inductor

is equal to the input current. The mutual inductor current starts to decrease in this state. The current flow in State III is shown in Fig. 4(d).

At t_2, when S_1&S_4 are turned off, the magnitude of the current in the leakage inductor is larger than the input current, and the body diodes of S_1&S_4 are conducting. Therefore, S_1&S_4 realize zero-voltage switching (ZVS) during turn-off transition.

In addition, since the current in leakage inductor is a negative value and larger than the mutual inductor current, the body diode of S_7 will conduct first considering the dead time between S_5&S_6 turning-off and S_7 turning-on. When S_7 is turned on, it will operate under synchronous rectification mode. Therefore, S_7 realizes ZVS during turn-on transition.

The turning-off of S_5&S_6 is hard switching. However, by adding an additional small capacitor between the drain and source of S_5-S_7, the voltage across S_5&S_6 will increase slowly during the turn-off transition, and the switch current will decrease to zero before the drain-to-source voltage is built up to the steady state value. Therefore, a ZVS also can be realized during the turning off of S_5&S_6.

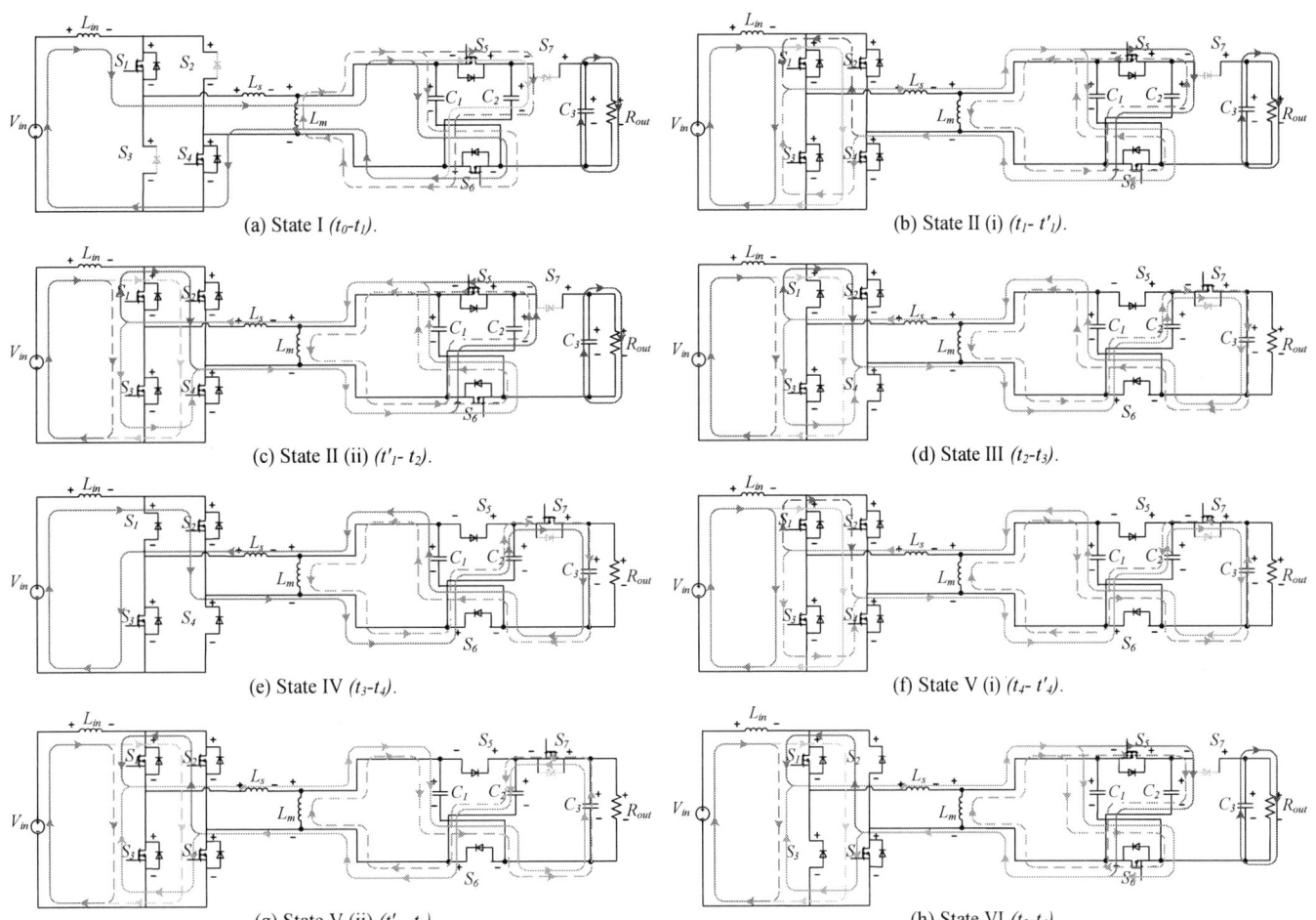

(a) State I (t_0-t_1).

(b) State II (i) (t_1- t'_1).

(c) State II (ii) (t'_1- t_2).

(d) State III (t_2-t_3).

(e) State IV (t_3-t_4).

(f) State V (i) (t_4- t'_4).

(g) State V (ii) (t'_4- t_5).

(h) State VI (t_5-t_6).

Fig. 4. Operation principle.

D. State IV (t_3-t_4), energy transferring to the load.

This state starts when the body diodes of S_1&S_4 are turned off at t_3. C_1&C_2 and input source are connected in series with the load. C_3 is charged and the energy is transferred from the input side to the load. C_1&C_2 are also discharging energy to the load. Mutual inductor current keeps decreasing. The current flow in State IV is shown in Fig. 4(e).

E. State V (t_4-t_5), leakage inductor current changing direction.

This state starts when S_1&S_4 are turned on. The input inductor is charged by the input voltage source. C_1&C_2 and the load are connected in series to charge the leakage inductor and the mutual inductor. The current in the leakage inductor increases to zero at t'_4, and then continues to increase in the positive direction. The mutual inductor current keeps decreasing. The current flow in State V is shown in Fig. 4(f) and (g).

Ideally, this state should end when the current in the leakage inductor has the same magnitude and direction as the input current. In practice, the leakage inductor current is charged a little higher than input current to leave some margin and avoid a voltage overshoot.

Similar to State II, at t_4, S_1&S_4 realize ZCS during the turn-on transition.

F. State VI (t_5-t_6), leakage inductor current freewheeling.

This state starts when S_2&S_3 and S_7 are turned off, and S_5&S_6 are turned on. Since the magnitude of the current in the leakage inductor is larger than the input current, the energy on the input side cannot be transferred to the load. Instead, the input inductor is still charged by the input voltage source in this state. In the meantime, C_1&C_2 are connected in parallel. Since the body diodes of S_2&S_3 are conducting, the current in the leakage inductor is decreasing rapidly. This state ends when the magnitude of the current in the leakage inductor is equal to the input current. The mutual inductor current starts to increase in this state. The current flow in State VI is shown in Fig. 4(h).

Similar to State III, at t_5, S_2&S_3 and S_7 realize ZVS during the turn-off transition. S_5&S_6 realize ZVS during the turn-on transition.

After State VI, the circuit will operate from State I again.

IV. CIRCUIT ANALYSIS

Based on Fig. 3, define the duty ratio of S_5&S_6 as D, then the duty ratio of S_7 is $(1-D)$. Therefore:

$$\begin{cases} D = \delta_1 + \delta_2 + \delta_6 \\ 1 - D = \delta_3 + \delta_4 + \delta_5 \end{cases}, \tag{4}$$

where δ_1 is the duty ratio of State I, δ_2 is the duty ratio of State II, etc.

Based on the voltage-second balance of the mutual inductor, it can be derived that:

$$V_{out} = \frac{1 + \delta_3 + \delta_4 + \delta_5}{\delta_3 + \delta_4 + \delta_5} V_{C1} = \frac{2 - D}{1 - D} V_{C1}. \tag{5}$$

Comparing (5) with (1), it is noted that the relationship between V_{c1} and V_{out} does not change in the full circuit.

In the meantime, based on the voltage-second balance of the input inductor, it can be derived that:

$$V_{in} = \frac{V_{C1}}{N} (\delta_1 + \delta_4 \frac{D}{1 - D}). \tag{6}$$

Assume that the current ripple in the input inductor is negligible, and the average current in the input inductor is I_{in}. Then at the end of State III, the current in the leakage inductor is $-I_{in}$. Therefore:

$$\delta_3 = \frac{V_{C1}}{(V_{out} - V_{C1} - V_{C2})} \delta_2 - \frac{2 I_{in} N L_s}{(V_{out} - V_{C1} - V_{C2}) T_s}. \tag{7}$$

Similarly, at the end of State VI, the current in the leakage inductor is I_{in}. Therefore:

$$\delta_6 = \frac{(V_{out} - V_{C1} - V_{C2})}{V_{C1}} \delta_5 - \frac{2 I_{in} N L_s}{V_{C1} T_s}. \tag{8}$$

Notice that based on energy conservation:

$$V_{in} I_{in} = \frac{V_{out}^2}{R_{out}}. \tag{9}$$

To simplify the control and ensure that the peak leakage inductor current at the end of State II and State V are the same, the following relationship should be satisfied:

$$\frac{V_{C1}}{N L_s} \cdot \delta_2 = \frac{(V_{out} - V_{C1} - V_{C2})}{N L_s} \cdot \delta_5 \Rightarrow \delta_5 = \frac{1 - D}{D} \cdot \delta_2. \tag{10}$$

Combine (4) to (10), and it can be derived that:

$$\frac{4 L_s}{R_{out} T_s} (\frac{V_{out}}{V_{in}})^2 + \frac{2(1 - D)}{N(2 - D)} (D - 2\delta_2) \frac{V_{out}}{V_{in}} - 1 = 0. \tag{11}$$

It can be observed from (11) that by controlling D, δ_2, and T_s, the boost ratio can be controlled.

Because of the asymmetrical control signal, there can be dc offset current in the mutual inductor of the transformer. To calculate it, assume that the current ripple in the mutual inductor is negligible and the average value is I_m. Notice that the average current of S_7 is the same as the average load current, so it can be derived that:

$$\begin{aligned} I_m = \frac{1}{1 - D} (N \frac{V_{out}}{R_{out}} &- (1 - D - 2\delta_3) \cdot I_{in} \\ &- \frac{\delta_2 \delta_3 (1 - D) T_s}{2 - D} \cdot \frac{V_{out}}{N L_s} + \frac{(\delta_3^2 + \delta_5^2) D T_s}{2 - D} \cdot \frac{V_{out}}{2 N L_s}) \end{aligned}. \tag{12}$$

V. SIMULATION AND EXPERIMENTAL RESULTS

A. Simulation verification

A simulation model of the circuit is first built in PSIM to verify the circuit analysis. The circuit parameters are summarized in TABLE I. In the simulation, the parasitic components are not included.

TABLE I.   CIRCUIT PARAMETERS OF THE SIMULATION MODEL.

Transformer turns ratio, N	1:3
Transformer mutual inductance, L_m	6 uH
Transformer leakage inductance, L_s	70 nH
Input inductance, L_{in}	2.3 uH
Capacitors, C_1 and C_2	10 uF
Capacitor, C_3	4.4 uF
Switching frequency, f_s	500 kHz
Input voltage, V_{in}	40 V
Output resistor, R_{out}	133 Ohm

The simulation results are shown in Fig. 5. Comparing the simulation results with the theoretical waveforms in Fig. 3, it can be concluded that they are similar to each other. The simulation results also show that: 1) voltage stress on the secondary side switches is smaller than the output voltage; 2) voltage stress on the secondary side winding of the transformer is smaller than the output voltage; 3) The primary side switches realize soft switching at both turn-on and turn-off transition; and 4) The dc offset current of the transformer can be reduced to a very small value by a proper selection of control parameters.

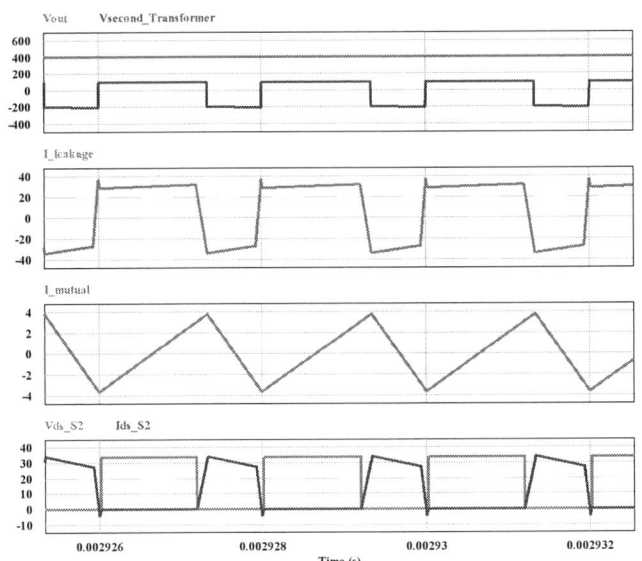

Fig. 5.   Simulation results of the proposed converter.

B. Prototype design and experimental results

A 1.2 kW prototype is built with the parameters in TABLE I, as shown in Fig. 6. At primary side, Si OptiMOSFETs (IPP200N25N3) from Infineon are applied for each switch. At secondary side, Gallium Nitride (GaN) HEMTs (TPH3006P) from Transphorm are applied for each switch. Ferret cores E42/21/20 and E32/16/9 are selected for the transformer and

inductor, respectively. Ceramic capacitors are selected for C_1, C_2, and C_3 in the QSC circuit. Since the same prototype is also served as a dual-input dc/dc converter with a higher power level, the design of the passive components in the circuit are not optimized [9] for this application. In the experiments, a NHR 9200 dc power supply is utilized at the PV input, and resistor banks are used as the load. The control is implemented in a Texas Instruments DSP (TMS320F2808). The efficiency is measured with a Yokogawa WT3000 power meter with two 701933 current probes.

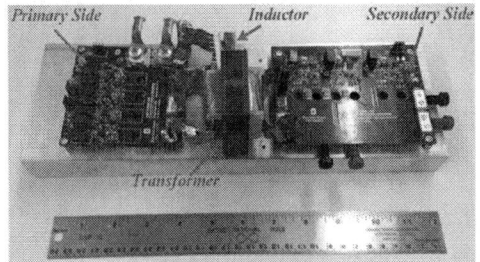

Fig. 6.   Converter prototype.

Fig. 7 shows the experimental waveform at the operation point where f_s=500 kHz, V_{in}=45 V, V_{out}=400 V, and P_{out}=1.2 kW. It can be noted that the voltage stress on the secondary side switches is reduced to 300 V, which matches the theoretical analysis. By a proper control of the shoot through time, the overshoot current on the leakage inductor is minimized. Compared to the simulation result, the slightly higher input voltage is caused by the power loss in the circuit. The ringing on the drain-to-source voltage of S_1 is caused by the resonance between the transformer leakage inductor and the parasitic capacitor of the switch.

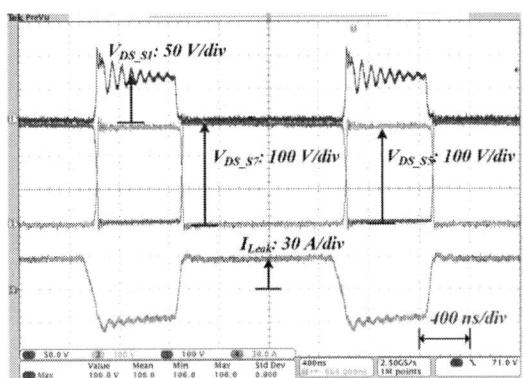

Fig. 7.   Experimental result at switching frequency of 500 kHz.

Fig. 8 shows the experimental waveform at the operation point where f_s=1 MHz, V_{in}=43 V, V_{out}=400 V, and P_{out}=1.0 kW. A similar operation condition is achieved compared to the 500 kHz case.

The switching waveforms to show the soft switching characteristics of the proposed converter are displayed in Fig. 9. It can be seen that the primary side switch S_1 realizes ZVS off, while the secondary side switch S_7 realizes both ZVS on and off. Since the switch current cannot be measured directly in the prototype, the ZCS on of the primary side switches cannot be observed directly.

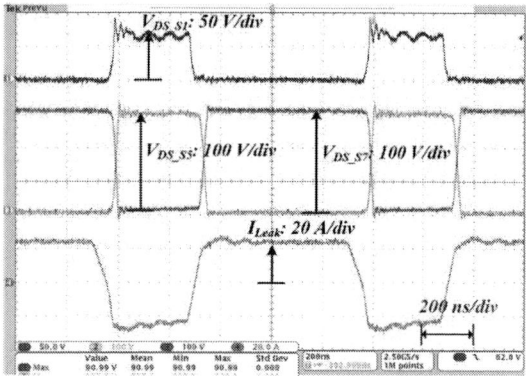

Fig. 8. Experimental result at switching frequency of 1 MHz.

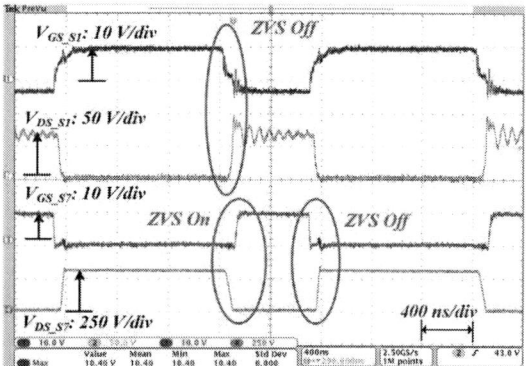

Fig. 9. Soft switching waveforms of the proposed converter.

The efficiency curves of the converter under different switching frequencies are shown in Fig. 10. The peak efficiency is 92.7% at 600 W with a switching frequency of 500 kHz. With a switching frequency of 1 MHz, the peak efficiency is 89.0% at 750 W. Compared to other current-source isolated dc/dc converters in the literature, a comparable efficiency is achieved with a much higher switching frequency.

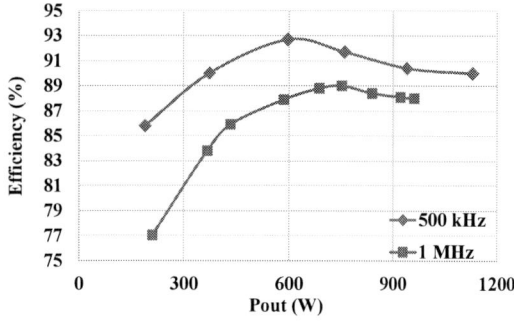

Fig. 10. Efficiency curves at different switching frequencies.

The calculated power loss distribution of the converter at 1 kW, 500 kHz is shown in Fig. 11. It can be seen that most of the loss (76.5%) is from the switch conduction loss. The switching loss (1.11 W) is from the energy loss in the output capacitor of the primary side switches during turn-on transition. There is around 40 W differences between the calculated power loss and the measured power loss, which is possibly caused by the partial fulfillment of ZCS turn-on of primary side switches. Since the leakage inductance is

extremely small and the turn-on speed of the Si switches is not fast enough, the leakage inductor current has increased to a large value before the switch voltage drops to zero. If the turn-on switching loss is considered (around 35 W in total), the calculated power loss will closely match the measured power loss. In future, GaN devices can also be utilized in the primary side switches to eliminate this loss.

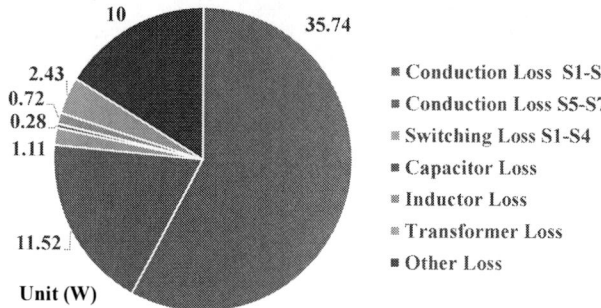

Fig. 11. Calculated power loss distribution at 1 kW, 500 kHz.

VI. CONCLUSIONS

In this paper, a full-bridge current-source isolated QSC dc/dc converter is proposed for PV applications. The operation principle is analyzed in detail. A 1.2 kW, 1 MHz, 40 V/ 400 V prototype utilizing GaN switching devices is built in the lab. Theoretical analysis is verified by both simulation and experimental results. A peak efficiency of 92.7% is achieved at 500 kHz, and 89.0% at 1 MHz.

REFERENCES

[1] M. Rehman, R. Hassan, and N. Zaffar, "High efficiency modified dual-active bridge converter for photovoltaic integration," *IEEE PowerTech*, Jun. 2013, pp. 1-5.

[2] Z. Liang, R. Guo, J. Li, and A. Huang, "A high-efficiency PV module-integrated dc/dc converter for PV energy harvest in FREEDM systems," *IEEE Trans. Power Electronics*, vol. 26, no. 3, pp. 897-909, Mar. 2011.

[3] L. Gao, R. A. Dougal, S. Liu, and A. P. Iotova, "Parallel-connected solar PV system to address partially and rapidly fluctuating shadow conditions," *IEEE Trans. Ind. Electron.*, vol. 56, no. 5, pp. 1548–1556, May 2009.

[4] X. Zhang, C. Yao, C. Li, L. Fu, F. Guo, and J. Wang, "A Wide Bandgap Device-Based Isolated Quasi-Switched-Capacitor DC/DC Converter," *IEEE Trans. Power Electron.*, vol. 29, no. 5, pp. 2500-2510, May 2014.

[5] K. Wang, F. C. Lee, and J. Lai, "Operation principles of bi-directional full-bridge DC/DC converter with unified soft switching scheme and soft starting capability," in *Proc. IEEE Applied Power Electronics Conference and Exposition (APEC)*, 2000, pp. 111–118.

[6] S. Jalbrzykowski and T. Citko, "Current-fed resonant full-bridge boost DC/AC/DC converter," *IEEE Trans. Ind. Electron.*, vol. 55, no. 3, pp. 1198-1205, Mar. 2008.

[7] X. Pan, and A.K. Rathore, "Novel bidirectional snubberless naturally commutated soft-switching current-fed full-bridge isolated DC/DC converter for Fuel Cell vehicles," *IEEE Trans. Industrial Electronics*, vol. 61, no. 5, pp. 2307-2315, May 2014.

[8] F. Peng, H. Li, G. Su, and J. Lawler, "A new ZVS bidirectional DC-DC converter for fuel cell and battery application," *IEEE Trans. Power Electronics*, vol. 19, no. 1, pp. 54-65, Jan. 2004.

[9] F. Guo, L. Fu, X. Zhang, and J. Wang, "A family of dual-input dc/dc converters based on Quasi-Switched-Capacitor circuit," in *Proc. IEEE Energy Conversion Congress and Exposition (ECCE)*, to be published.

Reliability and stability of SiC power MOSFETs and Next-Generation SiC MOSFETs

B. Hull, S. Allen, Q. Zhang, D. Gajewski,
V. Pala, J. Richmond, S. Ryu, M. O'Loughlin,
E. VanBrunt, L. Cheng, A. Burk,
J. Casady, D. Grider, and J. Palmour

Cree, Inc.
4600 Silicon Drive, Durham, NC 27703, USA
Brett_hull@cree.com

Abstract— In this paper, we present reliability and stability data based on a large body of data accumulated from high volume production of SiC power MOSFETs. The SiC MOSFETs (Gen2, C2M) showed excellent body diode and threshold voltage stability after 1000 hours of accelerated stressing tests. Results from next generation SiC power MOSFET development efforts are also presented. A significant reduction in specific on-resistance was demonstrated, and a wide range of blocking voltages, from 900 V to 15 kV, has also been demonstrated.

Keywords—power MOSFETs; threshold voltage; stability; body diode

I. INTRODUCTION

While the merits of 4H-SiC power MOSFETs have been known for decades [1], the technological and engineering barriers have only recently been overcome to allow manufacturing on a scale that allows the SiC MOSFET to become competent commercial power switching devices. In early 2013 Cree commercially introduced its 2nd Generation 25 mΩ and 80 mΩ, 1200 V planar SiC MOSFETs (C2M series). Innovation on SiC MOSFET device design and manufacturing processes continues to drive reliability and performance improvements and cost reductions. These ongoing refinements are being implemented into current- and next-generation SiC MOSFETs and have lead to a significant improvement in long-term stability and reliability, as well as a reduction in the specific on-resistance ($R_{on,sp}$) of 40% in a 1200 V device compared to Cree's C2M product family. The reductions in $R_{on,sp}$ allows for smaller chips for a given device rating, leading to more chips per wafer and to higher device yields, all leading to reduced manufacturing cost per chip. Furthermore, this next generation technology is easily scalable to a large range of voltage ratings, and we have scaled the designs and demonstrated SiC MOSFETs employing this next generation technology for voltage ratings ranging from 900 V to 15 kV.

II. PLANAR SiC POWER MOSFET STRUCTURE

Figure 1 shows a schematic cross-section of the SiC MOSFET unit cell employed for the fabrication of all planar MOSFETs from 900 V-rated to 15 kV-rated, along with the component resistances of the device. The total on-resistance ($R_{DS,on}$) consists of the sum of the channel resistance (R_{CH}), JFET gap resistance (R_{JFET}), spreading resistance (R_{SP}), drift layer resistance (R_{Drift}) and substrate resistance (R_{SUB}). In low voltage devices R_{CH} is the dominant resistive component, but as the voltage rating of a device increases, the drift layer epi doping decreases and thickness increases, so R_{Drift} increases, becoming the dominant resistive component in devices rated to higher biases. Theoretically the trench MOSFET structure [2] provides the potential for increasing the cell packing density, decreasing R_{CH}, and eliminating R_{JFET}, making the trench MOSFET a very desirable device for the lower voltage ratings. However, planar MOSFET technology is maturing rapidly, and while the trench MOSFET structure provides these potential benefits over the planar MOSFET, we have determined that an optimized planar MOSFET structure that pushes the limit for blocking layer design results in $R_{on,sp}$ that is the lowest in the industry without any compromise in reliability that arises from the trench MOSFET structure. Furthermore, as described above, as the voltage rating of a SiC MOSFET increases above about 2 to 3 kV, the R_{CH} becomes a relatively small portion of the overall device resistance, R_{Drift} begins to dominate, and cell packing density becomes less significant, negating the benefits of the trench MOSFET design for these higher voltage MOSFETs.

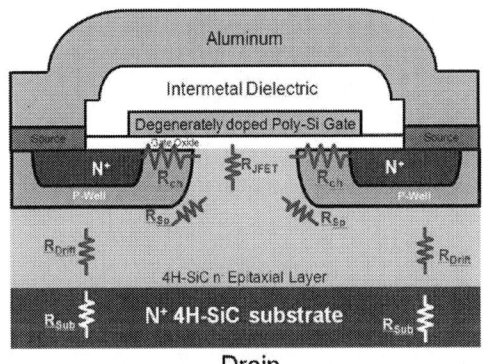

Fig. 1. Schematic cross-section of planar DMOSFET.

III. THRESHOLD VOLTAGE AND BODY DIODE STABILITY OF 4H-SiC POWER MOSFETs

One of the most critical challenges in SiC MOSFET device reliability has been the threshold voltage (V_T) stability, which has been examined in numerous reports [3,4]. As part of the development of the next generation SiC MOSFET at Cree, Inc., we have placed a great deal of emphasis on refining the device design and processing steps to achieve V_T stability. Figure 2 shows the threshold voltage of 1200V rated MOSFETs during the course of a 1000 hour high temperature gate bias (HTGB) stress test at V_G = -15 V and T = 150°C. Threshold voltage was measured periodically (at 1 minute intervals for the first hour of testing and at 1 hour intervals thereafter) *in situ* at 150°C by halting the V_G stress, applying a drain-to-source bias and modulating V_G until 2 mA of Drain current was measured. The measurement algorithm takes less than 500 msec per device. At 1000 hours, the average V_T shift for the 15 devices under stress was -50 mV, with a maximum shift of -90 mV. The maximum negative V_T shift occurs at about 4 hours into the stress (with a maximum shift of any device of -140 mV), and after that time, V_T begins shifting back in the positive direction. Figure 3 shows V_T shift under a +20V gate bias at 175°C, with a maximum V_T shift of 280 mV at 1000 hours of stress. The majority of the shift occurs within the first 5 to 10 hours, and the ΔV_T reverses rapidly upon removal of the positive bias. In a power switching application, since the MOSFETs are constantly switching from positive gate bias to zero or negative gate bias, V_T shift in actual usage is minimal.

The stability of the body diode of SiC MOSFETs is also of concern, based on demonstrations of PiN diode instability under minority carrier injection conditions [5]. Figure 4 shows the shift in body diode V_F and MOSFET $V_{DS,on}$ following a 1000 hour DC stress of the body diodes of 20 1200V-rated MOSFETs. These metrics were measured at room temperature

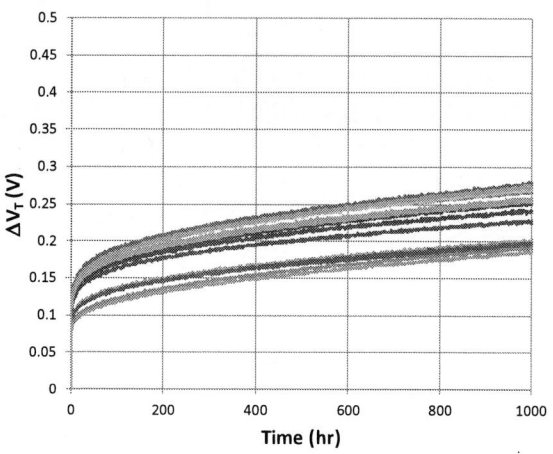

Fig. 3. Threshold voltage shift as a function of time for a 1000 hour HTGB stress at +20 V and 175°C.

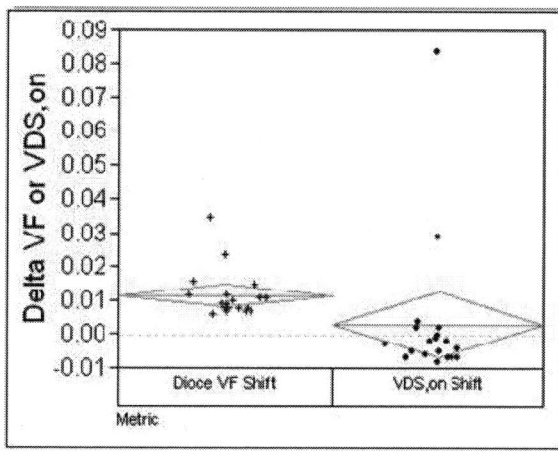

Fig. 4. Threshold voltage shift as a function of time for a 1000 hour HTGB stress at +20 V and 175°C.

before and after the 1000 hour stress. During stressing, the devices were mounted to a water-cooled plate and 22 A was forced through the body diode with the MOSFET channel pinched fully off with V_{GS} = -5V. Through the course of stressing, the maximum increase in diode V_F was 0.035V (or 0.8%) and the maximum increase in MOSFET $V_{DS,on}$ was 0.084V (or 5.4%). Furthermore, all parts initially blocked 1200 V with a typical leakage current of 10 nA, and no parts increased by more than a few nA of current at V_{DS} = 1200 V following the 1000 hour stress.

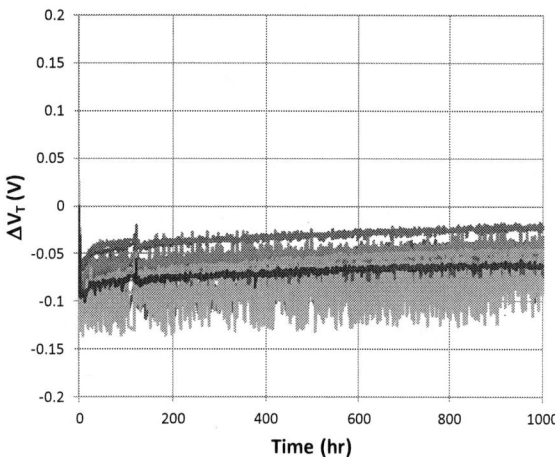

Fig. 2. Threshold voltage shift as a function of time for a 1000 hour HTGB stress at -15 V and 150°C.

IV. NEXT GENERATION SiC POWER MOSFETs

The MOSFETs were fabricated on 100 mm 4H-SiC substrates, cut 4° off of the [0001] axis. 4H-SiC epitaxial blocking layers were first grown on these substrates, with the thickness and doping dependent on the desired blocking voltage rating. The source and p-wells were formed by ion implantation and were activated at 1600°C. The gate oxides are grown thermally in O_2 with a post-oxidation anneal in NO, and the gate electrode was degenerately doped Poly-Si. Following deposition of a passivation dielectric, vias were formed down to the SiC and gate, into which ohmic contacts were formed. Table 1 details the chip design parameters, including $R_{on,sp}$, measured avalanche voltage, and measured total switching energy normalized to the active area of the device (details on the switching measurements are provided below). Since the 900 V- and 1200 V-rated MOSFETs have a $R_{on,sp}$ of less than 3 mΩ·cm², R_{SUB} is a fairly large contribution to $R_{DS,on}$ of these MOSFETs. To reduce this parasitic resistance, the substrates for MOSFETs of these voltage ratings were thinned from the backside during fabrication, to a thickness of about 180 μm. The thinning process reduced $R_{on,sp}$ by about 0.5 mΩ·cm², which allows for a chip of a given resistance and current rating to be reduced in size by about 15 to 20%. As the voltage rating increases to 1700 V and above, there is little to no benefit to thinning since R_{SUB} becomes a negligible fraction of the total chip resistance.

As shown in Fig. 5, at low breakdown voltages (V_{BR}), $R_{on,sp}$ diverges from the theoretical limit of the $R_{on,sp}/V_{BR}$ trade-off since R_{CH} is dominant when R_{Drift} is low. Despite the high R_{CH}, by optimizing device layout and drift layer epitaxy (and minimizing R_{Drift}, R_{SP}, R_{JFET} and R_{SUB}), we have achieved $R_{on,sp}$ in these planar MOSFETs that are competitive with reported values for trench MOSFETs of similar voltage ratings [6,7]. With increasing V_{BR} to 3.3 kV and higher, these next generation planar MOSFETs achieve close to the theoretical limit on the $R_{on,sp}/V_{BR}$ trade-off curve.

TABLE I. DESIGN AND PERFORMANCE METRICS OF NEXT GENERATION POWER MOSFETs

Voltage Rating [kV]	Drift Layer Doping [cm⁻³]	Specific on-Resistance [mΩ·cm²]	Avalanche Voltage [kV]	Specific Switching Energy (mJ/cm²)
0.9	1.1×10^{16}	2.3	1.1	Not Tested
1.2	8×10^{15}	2.7	1.6	12
1.7	6×10^{15}	3.4	1.9	Not Tested
3.3	2.8×10^{15}	10.6	4.1	16
6.5	1.3×10^{15}	40	7.2	Not Tested
10	6×10^{14}	125	11	22
15	3.6×10^{14}	250	16	66

V. MOSFET SWITCHING CHARACTERISTICS

Switching performance of a series of these next generation MOSFETs was examined with clamped inductive load, double-pulsed setups. For MOSFETs rated at 3.3 kV and lower, the inductor in the switching circuit was 856 μH, external gate resistance was 6.8Ω, and the rectifier was a SiC JBS diode with a rating comparable to the MOSFET under test. The switching setup of the 10 kV and 15 kV MOSFETs consisted of a 14 mH air-core inductor, two 10 kV-rated SiC JBS diodes connected in series, and an external gate resistance of 6.8Ω. In all tests, the MOSFETs were switched on to V_{GS} = +20 V and off to V_{GS} = -5 V. The switching performance, normalized to the active area of each device, is summarized in Table 1. To be consistent for comparison purposes, the switching losses shown in Table 1 were collected for switching at 70% of the rated voltage and the full-rated DC current (for T_J = 150°C) of a given device. With increasing voltage-rating, the specific switching energy losses ($E_{Tot,sp}$) increase from 12 mJ/cm² for the 1200 V MOSFET to 66 mJ/cm² for the 15 kV MOSFET under these test conditions.

SUMMARY

Next generation SiC MOSFETs with voltage ratings from 900 V to 15 kV were demonstrated. Using planar MOSFET technology, we have optimized device layout and fabrication to approach the theoretical limit of performance, and we have demonstrated planar MOSFETs at 900 V and 1200 V that rival the performance of trench MOSFETs. Switching performance of the MOSFETs over the range of voltage ratings was demonstrated, with the switching energies increasing as expected with increasing voltage rating and voltage being commutated. Long-term reliability was also demonstrated, with quite stable threshold voltages for a V_{GS} = -15 V stress of 1000 hours at 150°C, and for a V_{GS} = +20 V stress of 1000 hours at 175°C demonstrated. Furthermore, we have shown stable body diode performance under a 1000 hour DC body diode stress of 22 A.

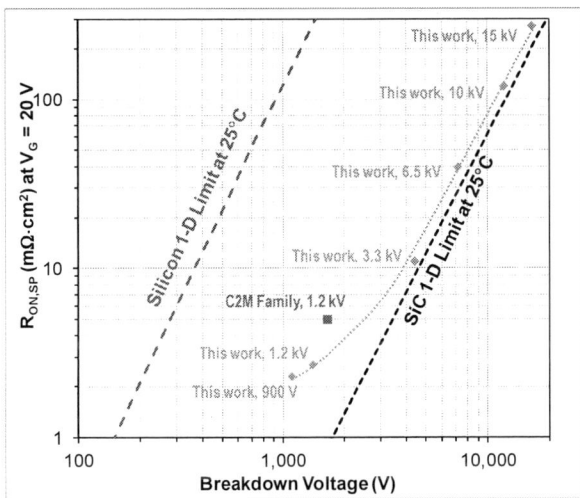

Fig. 5. Specific on-resistance ($R_{on,sp}$) of next generation SiC MOSFETs, measured at a gate bias of 20 V, as a function of breakdown voltage at 25°C

ACKNOWLEDGMENT

This research was sponsored, in part, by the Army Research Laboratory under Cooperative Agreement Numbers W911NF-12-2-0064 and W911NF-10-2-0038, and by the

Office of Naval Research and Penn State EOC under Contract No. N00014-10-D-0145. The views and conclusions contained in this document are those of the authors and should not be interpreted as representing the official policies, either expressed or implied, of ARL, NRL, EOC or the U.S. Government.

REFERENCES

[1] K. Shenai, R.S. Scott, and B.J. Baliga, *IEEE Trans. Elect. Dev.*, **36** (9), pp. 1811-1823 (1989).

[2] B. J. Baliga, *Fundamentals of Power Semiconductor Devices*, Springer Science, New York, NY (2008), p. 279.

[3] A. J. Lelis, D. Habersat, G. Lopez, J. M. McGarrity, F. B. McLean, and N. Goldsman, *Mater. Sci. Forum*, vol. 527–529 (2006), pp. 1317–1320.

[4] M. J. Tadjer, K. D. Hobart, E. A. Imhoff, and F. J. Kub, *Mater. Sci. Forum*, vol. 600-603 (2009), pp. 1147-1150.R. Nicole, "Title of paper with only first word capitalized," J. Name Stand. Abbrev., in press.

[5] H. Lendenmann, F. Dahlquist, N. Johansson, R. Söderholm, P. A. Nilsson, J. P. Bergman, and P. Skytt, *Mater. Sci. Forum*, vol. 353-356 (2001), p. 727.

[6] Y. Kagawa, N. Fujiwara, K. Sugawara, R. Tanaka, Y. Fukui, Y. Yamamoto, N. Miura, M. Imaizumi, S. Nakata, and S. Yamakawa, *Mater. Sci. Forum*, vol. 778-780 (2014), pp. 919-922.

[7] Y. Saitoh, M. Furumai, T. Hiyoshi, K. Wada, T. Masuda, K. Hiratsuka, Y. Mikamura, and T. Hatayama, *Mater. Sci. Forum*, vol. 778-780 (2014), pp. 931-934.

6.5 kV SiC Normally-Off JFETs – Technology Status

J. L. Hostetler, P. Alexandrov, X. Li, L. Fursin, A. Bhalla

United Silicon Carbide, Inc.
7 Deer Park Drive, Monmouth Junction, NJ, 08852
USA

Abstract— Large area 6.5 kV normally-off JFETs and JBS diodes have been developed for high DC-link voltage applications. The basic performance characteristics are examined along with the switching behavior using double pulse testing at 3 kV, 11A with an inductive load. In addition, a performance comparison between the single 6.5 kV JFET and a 6.0 kV super-cascode approach, built from stacking five normally-on 1.2 kV JFETs, is presented. The technology readiness level of each approach is discussed.

Keywords—Silicon Carbide; JFETs; JBS Diodes; high DC-link voltage; super-cascode;

I. INTRODUCTION

While 1.2 kV SiC JFETs are now commercially available from at least two suppliers [1, 2] and MOSFETs up to 1.7 kV from several vendors, there is significant development effort underway worldwide in bringing higher voltage unipolar devices to market. At voltages between 3.3 to 6.5 kV, Si-IGBT solutions exist, but the huge advantage of faster switching speeds offered by SiC unipolar devices, makes them very attractive to system developers seeking higher power densities, cost reduction of magnetics and improved efficiency.

Medium voltage SiC switches rated from 3.3 to 6.5 kV have become particularly attractive to motor drives and traction systems. Electricity accounts for ~ 43% of the world energy consumption, and out of that, ~45% of the electricity used is accounted for by electrical motors. In addition it has been shown that SiC devices can increase the efficiency of motor drive systems by 20 to 30% over the incumbent Si-IGBT switches, making the application high on the list for widespread adoption to address global energy consumption and climate control initiatives [3].

Higher DC-link voltage applications, such as modern electric railway systems, utilize 6.5 kV devices as they are able to operate directly from the 3 kV overhead lines, can accommodate transients up to 5 kV, and enable very simple, robust multi-system concepts. In addition, transformerless topologies are sought for huge weight reduction in trains by cascading 6.5 kV devices to directly access 15 kV lines [4]. The advantages of 6.5 kV rated unipolar SiC devices over Si devices lies in the improved switching losses where SiC devices exhibit shorter turn-on and turn-off times that are invariant with temperature, thus allowing faster switching and higher power handling capabilities. Furthermore, SiC devices can operate at much higher temperatures than Si devices, allowing for reduction of cooling apparatus. Both of these aspects enabled by SiC, can greatly increase power density, while increasing system efficiency as well [5, 6].

II. DEVICE FABRICATION

The 6.5 kV normally-off JFETs and JBS's were fabricated on 100 mm diameter 4H-SiC 4° offcut wafers utilizing a 75 μm drift layer N doped at 8E14 cm^{-3} where the JFET utilizes a vertical structure as seen in Fig 1(a) & (b).

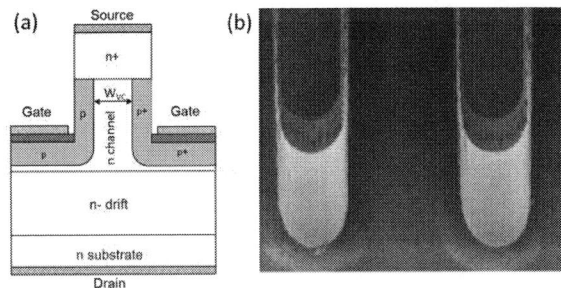

Fig. 1. Vertical JFET device crosssection (a) and SEM of source mesas for 6.5 kV devices (b).

Both devices utilize a guard ring edge termination structure designed to block ~10 kV at 250°C and the die size was 6x6 mm to accommodate a 15 A rating each. The JFET and JBS die geometries are shown in Fig. 2 where a split gate bus was utilized for low JFET gate resistance.

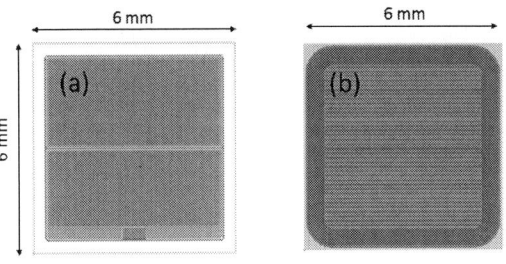

Fig. 2. 6.5 kV JFET geometry (a) and 6.5 kV JBS geometry (b).

For switching evaluation, the 6.5 kV enhanced mode JFET was soldered to an Al_2O_3 DBC using Au eutectic solder. The super-cascode evaluation used USCi's discrete TO-247 package, where the 1.2 kV normally on JFET devices used in are described in USCi's datasheet and an earlier report [1, 7].

The authors would like to thank the DOE Energy Storage Program for partial funding of this project

978-1-4799-5494-0/14 $31.00 © 2014 IEEE

III. DEVICE PERFORMANCE

A. 6.5 kV Enhanced Mode SiC JFET

Figures 3(a) and 3(b) show the testing results for forward conduction and blocking characteristics at room temperature for the normally-off JFET. RDSon values ~350 mΩ with a Vg= +3.0 V are observed. The leakage current, with Vg=0 V remains under 500 μA at 6.5 kV. The typical gate threshold voltage of 1.7 V is seen in Fig. 4.

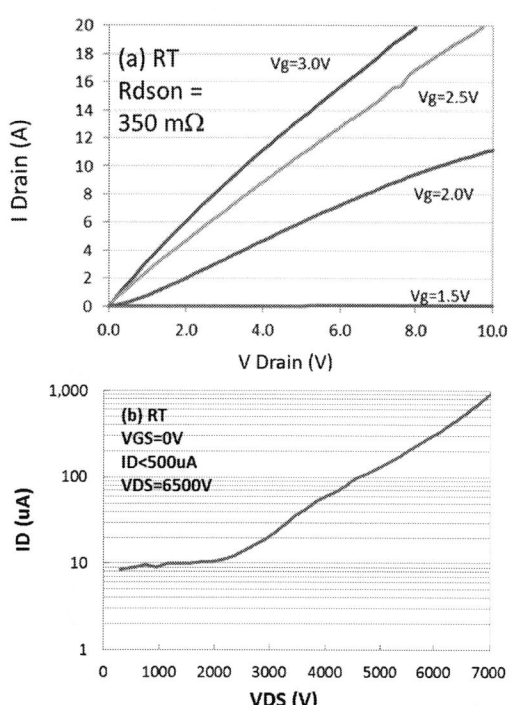

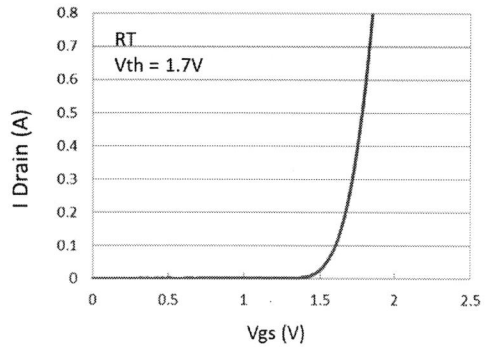

Fig. 3. 6.5 kV normally-off JFET forward conduction showing RDSon=350 mΩ (a) and reverse leakage at Vg=0 V showing I_D< 500 μA at 6.5 kV.

Fig. 4. Transfer curve showing gate threshold of 1.7 V for the 6.5 kV JFETs.

B. 6.5 kV JBS Diodes

Figures 5(a) and 5(b) show the testing results for the 6.5 kV JBS diodes. The forward voltage V_f=3.8 V for the rated current of 15 A, corresponds to an RDSon=250 mΩ. The

reverse blocking shows a current leakage, Ir <150 μA for the rated voltage of 6.5 kV.

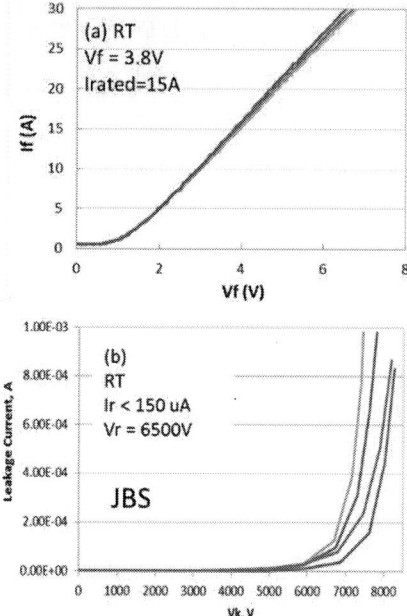

Fig. 5. 6.5 kV JBS diode forward conduction showing V_f=3.8 V at 15 A rated current (a) and reverse leakage for 4 typical diodes showing Ir < 150 μA at 6.5 kV.

C. 1.2 kV Depletion Mode Discrete JFETs

The normally-on 1.2 kV JFET packaged in a TO-247 used in the super-cascode demonstration is described elsewhere in more detail [1, 7], where the RDSon=45 mΩ and the devices exhibit a pinch-off voltage of -5 V. The characteristic performance of the stacked devices in the super-cascode is presented in Section IV.

IV. SINGLE DEVICE VS. SUPER-CASCODE DISCUSSION

For the advancement of 6.5 kV switches, it is appropriate to address the technology readiness levels of different approaches. As this falls well outside the voltage blocking level of GaN/Si devices, the competing technologies are straight forward single SiC 6.5 kV unipolar devices, 6.5 kV Si-IGBTs, and a stacking approach where lower voltage SiC devices are stacked serially to form a high voltage switch [8.9.10]. While the technology readiness level is mature for 6.5 kV Si-IGBTs, the traditional Si bipolar devices are plagued with high switching losses and excessive cooling requirements, which limit switching frequency, and subsequently have a negative impact on the magnetic components of power conversion systems. As next generation inverters are requiring higher power densities and extended reliability over that capable with Si devices, SiC unipolar devices have gained considerable attraction in the last decade. The well-known material, processing and cost issues associated with SiC devices has begun to subside, and with the development of 6" SiC substrates, this trend will continue. The SiC MOSFET is gaining acceptance, however the SiC MOS interface still suffers from reliability issues as compared to the SiC JFET

978-1-4799-5494-0/14 $31.00 © 2014 IEEE   144

[11]. For or this study, a normally-off 6.5 kV SiC JFET is compared to a super-cascode, where five normally-on 1.2 kV rated SiC JFETs are stacked serially with a low voltage Si-MOSFET to form a 6.0 kV rated switch exhibiting the desired normally-off behavior. The technology readiness level of fabricating single devices is compared to the multiplexing of the more mature 1.2 kV devices. Figure 6 (a) shows the traditional switch with matching JBS in antiparallel for freewheeling. Figure 6(b) shows a simplified super-cascode circuit, where a low voltage Si-MOSFET is in series with five 1.2 kV normally-on SiC JFETs. Avalanche diodes are placed in between each JFET to clamp the voltage at or below the rated 1.2 kV for each device.

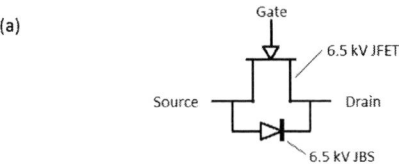

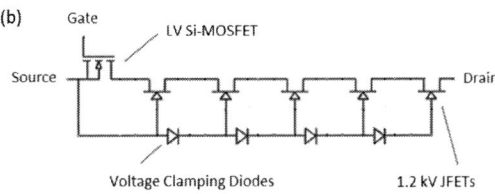

Fig. 6. Traditional arrangement of a switch with matching SiC JBS in antiparallel (a) and the super-cascode approach with five 1.2 kV JFETs in series with a low voltage Si-MOSFET (b).

In the super-cascode, when the low voltage Si-MOSFET is on, only a small voltage drop exists across the gate and source of the normally-on JFETs and hence they are all in forward conduction mode, where the total RDSon is the sum of the 5 JFETs. When the Si-MOSFET is turned off, the source-gate bias of the first JFET increases until pinch off occurs driving the first JFET into off-state. The drain voltage of the first JFET continues to increase until the first clamping diode avalanches, tying the second JFET gate to a predetermined voltage, driving it into off-state, and likewise for the remaining 3 JFETs in a domino fashion [8]. Figure 7 shows the super-cascode in forward conduction mode where a total RDSon of 226 mΩ is exhibited for a gate-source voltage of V_{GS}=10 V.

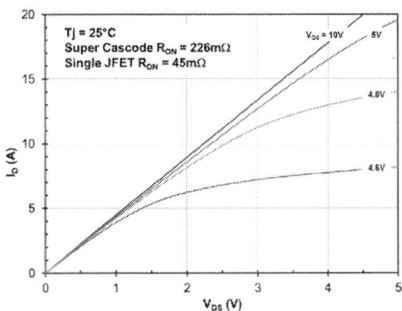

Fig. 7. Super-cascode forward conduction exhibiting a total RDSon of 226 mΩ for a V_{GS}=10 V.

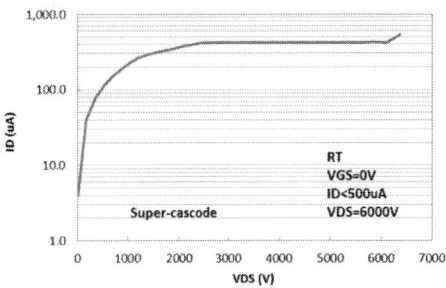

Fig. 8. Super-cascode blocking behavior exhibiting <500 µA at rated voltage of 6 kV.

Figure 8 shows the blocking behavior of the super-cascode using clamping diodes with an avalanche voltage of 1.2 kV, where the leakage above 6 kV demonstrates the last JFET entering avalanche mode. Unlike, the single 6.5 kV JFET, the leakage of the super-cascode is higher at lower voltages (2-3 kV) due to the bias current of the avalanche diodes.

V. SWITCHING PERFORMANCES

The switching performance of the single 6.5 kV JFET and the 6.0 kV super-cascode were evaluated using double pulse testing at 3 kV, 11A (33 kW pulse power) using an inductive load. Figure 9 shows the turn-on and turn off waveforms for both approaches. Both exhibit excellent characteristics.

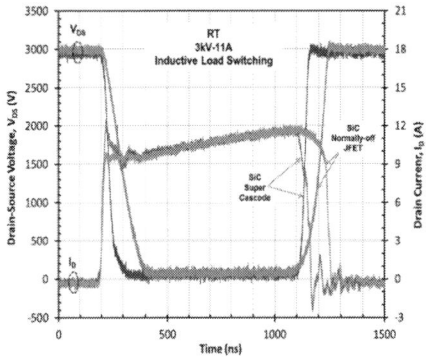

Fig. 9. Turn-on and turn-off waveforms for the single enhanced mode 6.5 kV JFET (pink=voltage&green=current) and the 6.0 kV super-cascode (blue=voltage & red=current) for 3 kV switching at 11 A with an inductive load.

The single JFET device exhibits a rise and fall time of 150 ns and 100 ns respectively and turn-on and turn-off energies of 2.7 and 1.5 mJ, respectively. However, the super-cascode exhibits a much faster rise time of only 67 ns, and a fall time of 35 ns, which results in lower turn-on energy of 1.25 mJ and a lower turn-off energy of 0.5 mJ as well. The slower switching of the single device is due to the fact that during turn-on and turn-off the single switch Gate-Source and Gate-Drain capacitances are charged and discharged by a limited current capability gate driver. In the case of the super-cascode, the JFET switch's Gate-Source and Gate-Drain capacitances are charged and discharged through ground, allowing higher displacement current and faster switching.

978-1-4799-5494-0/14 $31.00 © 2014 IEEE   145

VI. Discussion

Both approaches for achieving a 6.5 kV rated SiC switch have advantages and disadvantages. There is no significant inherent advantage for each approach with respect to RDSon. In this experiment, the total RDSon of the super-cascode is lower, however, this is due to device design differences and is not an inherent advantage. There is an advantage with respect to blocking, however, as the single device has lower leakage at switching voltages around 3 to 4 kV as seen in Fig 3(b). In the super cascode, voltage clamping components requires a biasing current in order to work stably, which will likely increase the leakage current of the super cascode (see Fig 8).

While both approaches exhibit excellent switching behavior, the slower switching nature of the single 6.5 kV device can be advantageous when dv/dt's are a critical issue, however it is clear that the super-cascode can switch faster with nearly half of the switching losses.

With the respect to the device fabrication of the two approaches, the super-cascode has an advantage primarily arising from yield. The epitaxial growth of the thin layer (~10 um) for 1.2 kV device, is far more mature resulting in less embedded defects, and furthermore, the device size is 1/5 the size of the single 6.5 kV device. The total area of SiC used is roughly the same between the two approaches, however, the smaller device combined with the thinner drift layer will have a significant yield improvement over the very large 6x6 mm single device utilizing thick epitaxy (~75 um).

The advantages and disadvantages of module assembly for both approaches are not so clear. Thermally, having the devices separated, as in the case with the super-cascode, is advantageous, which can allow for higher current densities, but also requires more heat sink area. The super-cascode requires more devices, i.e. avalanche diodes and other components for parasitic control, however this approach does not require the use of anti-parallel SiC diodes rated the same as the SiC JFET, as the body diode of the super-cascode also exhibits low reverse recovery losses. The Si-devices are very cheap and assembly straightforward for the super cascode, where the bulk of the work lies in the design, and not the assembly. The only inherent issue is with so many Si components, the temperature rating of a super-cascode module would be limited to 175°C, while the single 6.5 kV JFET and anti-parallel JBS diode can operate with much higher temperatures (~300°C) limited only by the packaging materials. For current-scaling using multiple devices within one module, the single 6.5 kV device has the advantage of simple paralleling, while the super cascode, must parallel arrays of super-cascodes, which need more balancing and parasitic control. The super-cascode can, however, use discrete devices far more efficiently than the single device approach making it attractive for addressing multiple platforms and applications based on the same modular component. To assess the advantages and disadvantages with respect to module packaging, clearly the end application will drive the result.

Finally, to address the technology readiness level of both approaches, the issue of reliability should be discussed. Since all of the components of the super-cascode have already been tested under JEDEC and even more harsh qualifications, a module built from these components should inherently be reliable. However, multi-component integration of the super-cascode requires further qualification. The 6.5 kV single devices, while expected to be as reliable as the 1.2 kV devices, require a high kV infrastructure and are currently in qualification. With continuous improvements in thick epitaxial growth, SiC device processing, and with the introduction of 6" substrates, the single device approach will certainly mature in the near future and enable a whole class of next generation power converters. The super-cascode can also address many applications requiring >5 kV switches, gain the advantages of SiC material properties while relying on a more mature 1.2 kV device.

VII. Conclusion

The technology readiness level of SiC 6.5 kV rated JFET switches was considered. Single device enhanced mode 6.5 kV SiC JFETs were compared to a super-cascode approach where five 1.2 kV rated normally-on devices were stacked in series with a Si-MOSFET to form a 6.0 kV rated switch with normally-off behavior. Results show both approaches are valid with advantages and disadvantages associated with each, however, with respect to reliability, the technology readiness level of the super-cascode is further along and ready for medium voltage applications such as motor drives and traction systems.

References

[1] USCI datasheet UJN1205K (www.unitedsic.com)

[2] Infineon datasheet IJW120R100T1 (www.infineon.com)

[3] International Energy Agency, "World Energy Outlook", 2011.

[4] A. Steimel, "Power-Electronics Issues of Modern Electric Railway Systems", 10th Int. Conf. on Development & Application Systems, Suceava, Romania, May 27-29, 2010.

[5] R. Mallwitz, C. Althof, S. Buchhold and E. Kiel, "First 99% PV Inverter with SiC JFETs on the Market-Future Role of SiC", *Proc. of the PCIM Europe*, May 2012.

[6] Burak Ozpineci, Leon Tolbert , Silicon Carbide: Smaller, Faster, Tougher, IEEE spectrum, 27 Sep 2011.

[7] J. Hostetler, X. Li, P. Alexandrov, L. Fursin, A. Bhalla, F. Hoffmann, R. Myers-Ward, P. Klein, B. Stahlbush, and D. Kurt Gaskill, "SiC JFETs Reduce the Balance-of-system for Stationary Storage Power Conversion Systems", EESAT 20132, San Diego, CA, Oct. 2013.

[8] P. Friedrichs, H. Mitlehner, R. Schorner, K. Dohnke, R. Elpelt, and D. Stephani, "Stacked high voltage switch based on Sic VJFETs", ISPSD, April 14-17, Cambridge, UK, 2003.

[9] Biela, J.; Aggeler, D.; Bortis, D.; Kolar, J.W., "5kV/200ns Pulsed Power Switch based on a SiC-JFET Super Cascode," *IEEE Int. Power Mod. and High Voltage Conf*, vol., no., pp.358,361, 27-31 May 2008.

[10] Aggeler, D.; Biela, J.; Kolar, J.W., "A compact, high voltage 25 kW, 50 kHz DC-DC converter based on SiC JFETs," *APEC*, pp.801-807, 24-28 Feb. 2008.

[11] J. Flicker, D. Hughart, R. Kaplar, S. Atcitty, and M. Marinella, "Performance and Reliability Characterization of 1200 V Silicon Carbide Power MOSFETs and JFETs at High Temperatures" HiTEC, Albuquerque, 2014.

Silicon Carbide Transient Voltage Suppressor for Next Generation Lightning Protection

Avinash S. Kashyap, Peter Sandvik, James McMahon, Alexander Bolotnikov, Jeffrey Erlbaum, Emad Andarawis

General Electric Global Research Center
Niskayuna, NY, USA
Kashyap@ge.com

Abstract— Transient voltage suppressors (TVS) fabricated in silicon carbide (SiC) and subjected to extensive surge testing is presented. The SiC TVS devices can work at high temperatures, have high current density capability, are smaller and have lower capacitance than comparable Si devices. They are also rugged and reliable, capable of withstanding multiple back-to-back lightning hits. These devices are expected to provide enhanced surge protection for composite aircrafts and enable high temperature power electronics circuits.

Keywords—silicon carbide, transient voltage suppressor, DO-160, composite airframe, surge suppressor

I. INTRODUCTION

Wide bandgap discrete power semiconductors [1], [2] and integrated circuits (ICs) [3], [4] capable of extreme environment operation and/or miniaturization have seen tremendous progress over the last few years. Nearly all electronic circuits require protection from over-voltages due to transient events, but currently no solution exists in the market for surge protection of high temperature electronics. This is because Si devices are extremely leaky beyond 150 °C. Silicon-on-insulator (SOI) technology is not conducive for designing lightning protection components (as vertical structures cannot be made in SOI), while gallium nitride (GaN) is not widely available in native substrates, and not as mature as SiC. Therefore, TVS devices were designed and fabricated in SiC [5] especially for the aviation, power electronics and down-hole markets that value extreme environment survivability & reduction of weight, volume and board area.

II. DEVICE STRUCTURE AND PHYSICS

The SiC TVS consists of an n-p-n structure with an angled mesa as shown in Fig. 1. The TVS is designed to work as a punchthrough device. When a positive voltage is imposed upon the top electrode, the upper p-n junction becomes reverse biased and the depletion region present in the junction starts increasing in size. As the positive voltage is increased, the depletion region grows and at a certain voltage (punchthrough voltage), it reaches the depletion region of the lower junction and punchthrough ensues, along with current flow. Due to the symmetrical nature of the structure, these devices are inherently bidirectional.

The ability of SiC TVS' to work at high temperatures (compared to Si) is linked to two material properties – (a) low

intrinsic carrier concentration (about 18 orders of magnitude lower then Si at room temperature), which keeps the leakage current low even at highly elevated temperatures and, (b) higher thermal conductivity. This can be seen in Fig. 2, where the SiC TVS is compared with a commercial off-the-shelf (COTS) Si device at 225 °C (breakdown voltage ~50V). The SiC devices were tested successfully up to 300 °C, with low leakage currents.

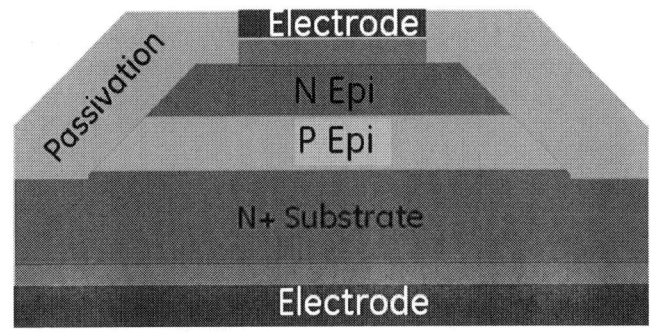

Fig. 1. Cross-section of the SiC TVS device.

On the other hand, the Si device in Fig. 2 becomes practically unusable due to the excessive leakage current, while the same for the SiC TVS remains about 6 orders of magnitude lower.

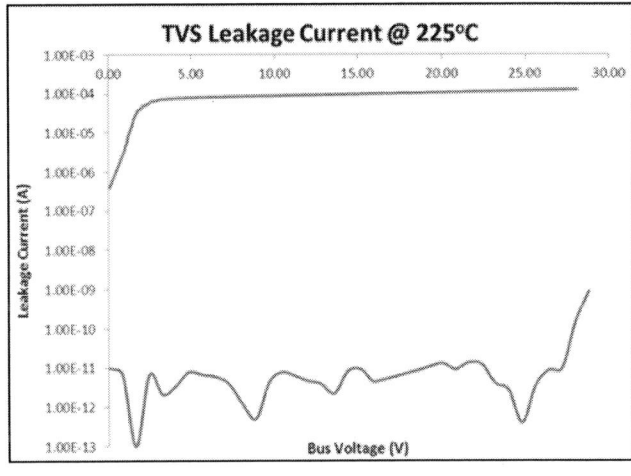

Fig. 2. Comparison of leakage currents of Si (blue) and SiC (red) TVS @ 225 °C. Note the several orders of magnitude lower leakage for the GE SiC device.

The higher thermal conductivity helps in dissipating the heat generated in the die during surge events, which usually occur for short durations (10s to 100s of microseconds). This enables the wide bandgap device to withstand multiple hits even at elevated temperatures.

III. DEVICE PERFORMANCE

SiC TVS devices of various sizes were fabricated on 4" wafers (Fig. 3). The wafers were then diced and the TVS devices were placed in hermetic packages capable of withstanding high temperatures and surge currents. These packaged parts were initially subjected to DC tests on a curve tracer. The current density (J) - voltage characteristics of the device as seen in Fig. 4 clearly demonstrates the capability of the SiC material and the device design as it conducts over 8 kA/cm^2, for a voltage drop that is less than 2X the breakdown voltage. It is to be noted that unlike a surge test, the J-V sweep is a cumulative test, consisting of dozens of long duration (250 us pulses), causing significant self-heating and degrading the device performance. The low voltage drop across the device as it conducts extremely high current densities makes the SiC TVS an ultra-rigid clamper, especially when compared to metal-oxide varistors. The sharp J-V curves (or lower voltage drops across the device) ensure that greater protection is afforded to downstream electronics. In other words, active components that are being protected by the SiC TVS can be de-rated to a lesser amount, allowing for a more optimized or even aggressive design.

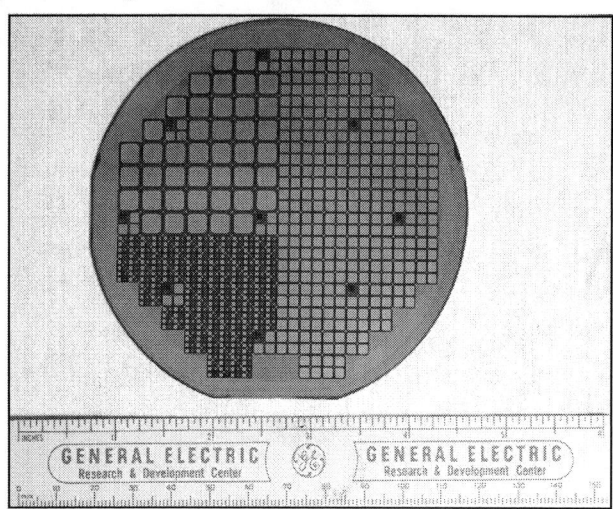

Fig. 3. A 4" SiC wafer with TVS devices of various sizes (current rating).

Also, in COTS Si TVS, it may not be possible to find exact combinations of breakdown voltages and power ratings. Circuit designers typically connect two or more devices in series or parallel to mitigate that problem.

The higher current density in SiC allows dies to be smaller than Si devices for the same current rating, and they can be custom designed for any breakdown voltage. This obviates the need to use multiple parts, thereby saving space and improving reliability. Fig. 5 shows the SiC replacement for three Si devices that need to be connected in series for a certain application. The smaller size and weight provides a tremendous benefit in terms of board area, weight and volume for space

critical applications such as aerospace, oil & gas, military hardware or high performance power electronics.

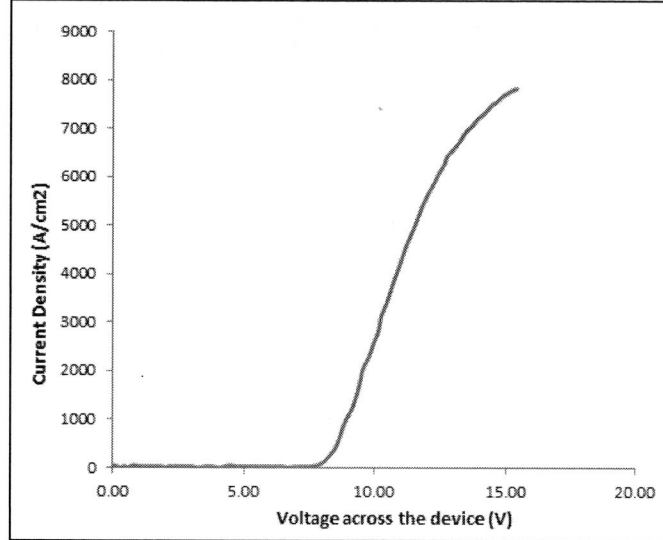

Fig. 4. Voltage vs. current density of the SiC device. The device is shown to conduct over 8 kA/cm^2 for a drop of ~2X the break-down voltage, demonstrating low resistance.

Another advantage of having smaller dies is the consequent reduction of capacitance of the device due to a smaller area. CV measurements of the SiC devices have yielded capacitance values at least 2-3X lower (~17 nF/cm^2) than the comparable Si devices. This helps because - (a) there is less loading of the downstream electronics and, (b) in certain instances, p-n junction diodes connected in series to TVS devices to reduce capacitance can be eliminated.

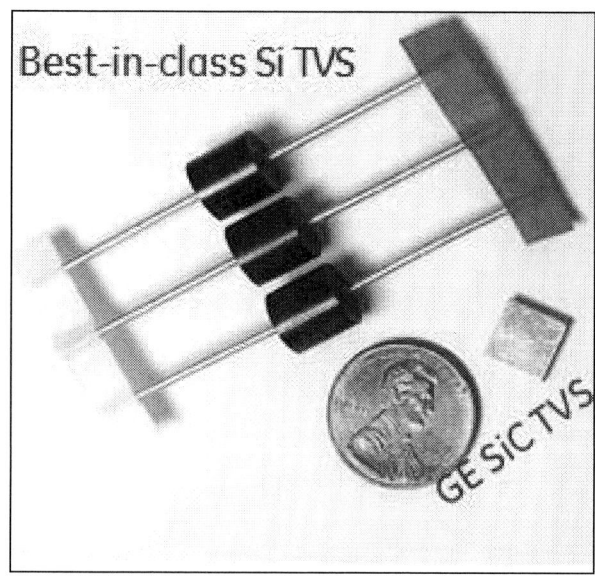

Fig. 5. Three Si TVS devices in series can be replaced with a single, smaller SiC device (in a hermetic package) for a certain application.

IV. SURGE TESTING

SiC based Zener diodes [6] have been demonstrated previously with high current densities. But those results were only from small devices conducting less than 0.5 A of current. To be used in real-life applications, TVS devices have to be scaled to much larger areas and be able to conduct hundreds of Amperes of current or more.

For surge testing, the DO-160 standard [7] (environmental conditions and test procedures for airborne equipment) was followed to simulate lightning conditions in aircraft electronics. Voltage & current waveforms such as 6 us/69 us (rise/fall times), 40 us/120 us, 10 us/1000 us were imposed upon the device to understand the clamping properties and current conduction entitlement. They are grouped into "levels" and "waveform types" according to the duration and amplitude of the signals. Typically these waveforms have shapes as shown in Fig. 6.

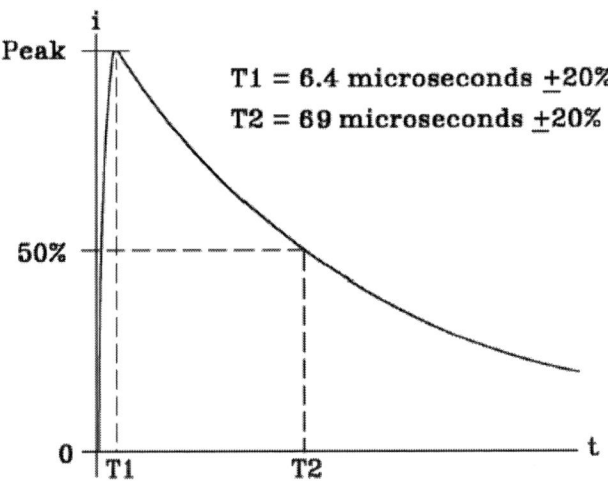

Fig. 6. A DO-160 6 us/69 us waveform [7].

Extensive surge testing demonstrated that the SiC TVS devices are able to conduct at current densities of up to 12 kA/cm^2 (for an 8 us/20 us waveform). This is consistent with the DC measurement performed previously. The surge tests were also performed at 170 °C and 225 °C in order to understand their operation at extreme temperatures. The SiC TVS devices were able to pass the Waveform 5A, Level 5 (1600 V/1600 A) test successfully even at 225 °C.

These higher level waveforms have become critical requirements for electronics used in composite aircrafts. In terms of COTS parts, there are few options that can pass these current levels and certainly none at elevated temperatures without drastic derating.

Fig. 7 shows a SiC TVS device successfully clamping a Waveform 5A/Level 5 surge at 170 °C (dissipating 40.5 kW of power). The best-in-class Si TVS device is derated to 0 W at this temperature.

The ruggedness of the device was tested by performing multiple back-to-back hits in the forward and reverse directions in one minute intervals. A sampling of some of the surge tests

are presented below. As explained previously, the higher thermal conductivity (~3.5X that of Si) is the enabler for this enhanced performance. When a surge current hits a TVS device, the heat initially remains in the die because the pulse duration is much shorter than the time required for heat spreading. From the die, the heat is slowly transferred to the package. The higher thermal conductivity of SiC allows the heat to spread faster across the die and towards the package compared to Si devices, and therefore it is able to withstand multiple hits even at elevated temperatures without encountering problems associated with thermal runaway.

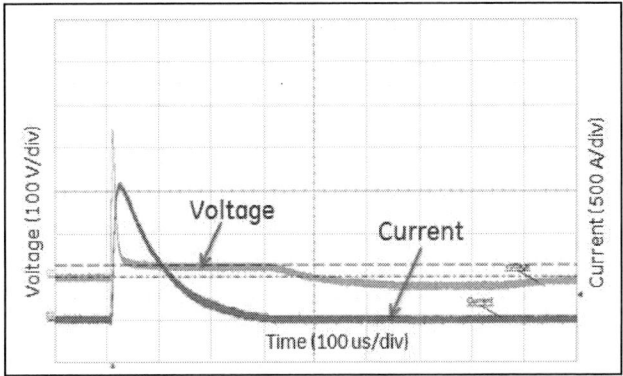

Fig. 7. SiC TVS successfully clamping a waveform 5/level 5 surge at 170 °C.

TABLE 1. DO-160 surge test results for SiC TVS devices.

Test	V/I Levels	Result
Level3, Waveform 5A	300V/300A	Pass
Level 4 Waveform 5A	750V/750A	Pass
Level 5, Waveform 4	1600V/320A	Pass
Level 5, Waveform 5A	1600V/1600A	Pass
Level 5, Waveform3 @ 1MHz	3200V/128V	Pass
Level 4, Waveform 4	750V/150V	Pass

V. SUMMARY

The first known punchthrough physics based SiC TVS is designed and fabricated. They are capable of operating at high temperatures (>200 °C) with low leakage currents. The devices are also shown to be several times smaller than their Si counterparts with lower capacitance, and high surge current capability with little or no derating. Due to the material properties of SiC, it was also shown that these TVS devices are capable of withstanding multiple back-to-back lightning hits. Long term reliability testing (HTRB - high temperature reverse bias) to understand lifetime is currently underway. These devices are expected to play a vital role in enabling future aviation and downhole electronics.

978-1-4799-5494-0/14 $31.00 © 2014 IEEE

ACKNOWLEDGMENT

The authors would like to thank Lightning Technologies Inc. for performing the DO-160 tests on the devices and Sensitron Semiconductor for packaging the SiC TVS devices.

REFERENCES

[1] C. DiMarino, Z. Chen; M. Danilovic, D. Boroyevich, R. Burgos, P. Mattavelli, "High-temperature characterization and comparison of 1.2 kV SiC power MOSFETs," *IEEE Energy Conversion Congress and Exposition (ECCE)*, pp.3235-3242, 15-19 Sept. 2013.

[2] J. S. Glaser, J. J. Nasadoski, A. S. Kashyap, P. A. Losee, K. S. Matocha, L. D. Stevanovic, "Direct Comparison of Silicon and Silicon Carbide Power Transistors in High-Frequency Hard-Switched Applications,"

IEEE Applied Power Electronics Conference (APEC 2011), Ft. Worth, TX, March 6-10, 2011.

[3] A. S. Kashyap, C. Chen, and R. Ghandi, A. Patil, E. Andarawis, L. Yin, D. Shaddock, P. Sandvik, K. Fang, Z. Shen, W. Johnson, "Silicon carbide integrated circuits for extreme environments," *IEEE Workshop on Wide Bandgap Power Devices and Applications*, Columbus, Ohio, USA, Oct. 27-29, 2013.

[4] S. L. Garverick, C. Soong, M. Mehregany, "SiC JFET Integrated Circuits for Sensing and Control at Temperatures up to 600°C," *IEEE Energytech Conference*, Cleveland, OH, May 29-31, 2012.

[5] A. S. Kashyap, D. Shaddock, E. Andarawis, P. Sandvik, S. Arthur, V. Tilak, "System for Transient Voltage Suppressors," U. S. Patent 8530902, Sept. 10, 2013.

[6] K. Vassilevski, K. Zekentes, A. B. Horsfall, C. M. Johnson, N. G. Wright, "Low Voltage Silicon Carbide Zener Diode," *Material Science Forum*, vol. 457-460, pp. 1029-1032.

[7] Environmental Conditions and Test Procedures for Airborne Equipment, RTCA/DO-160G, RTCA Inc. December 8, 2010.

Temperature Dependent Design of Silicon Carbide Schottky Diodes

Rahul Radhakrishnan[1], Tony Witt and Richard Woodin

Global Power Technologies Group
20692 Prism Place, Lake Forest, CA- 92630 USA
Email: rahulrad@gmail.com[1]

Abstract—A well designed Silicon Carbide (SiC) Schottky-Barrier Diode (SBD) has to optimize reverse leakage current in addition to the familiar silicon (Si) device trade-off of avalanche breakdown voltage (BV) and forward voltage drop (V_F). Reverse leakage current is a strong function of temperature and we show experimentally that this sensitivity of reverse leakage to temperature is a function of the electric field at the Schottky interface at rated voltage. We model this phenomenon using Hatakeyama's approximations of electron emission across the metal-SiC barrier. This model and our results point to a previously un-reported design trade-off between BV margin over rated voltage and temperature dependence of SiC SBD rated reverse leakage current. Using this trade-off, we show that optimum epitaxial design of a SiC SBD is different for a given rated voltage than what mere BV vs R_{on} considerations indicate. Optimized epitaxy is also different for different JBS geometries.

Keywords—SiC, Diode, Schottky, JBS, MPS, Modeling, Design

I. INTRODUCTION

Over the last decade, wide band gap semiconductors have been replacing Silicon (Si) as the material of choice for switching power devices. In particular, Silicon Carbide (SiC) based rectifiers have already found wide commercial success, particularly in power applications where they are paired with traditional Silicon switches. Since SiC has a high built-in voltage, P-N diodes have on-state voltage > 3 V. To reduce voltage drop and to prevent minority carrier storage (and consequent switching speed and energy loss) in on-state, Schottky diodes (SBDs) are the preferred SiC rectifiers. The Schottky barrier is chosen to be ~ 1 V, within the flexibility offered by the fabrication process, to keep room temperature forward voltage drop ~ 1.5 V.

Si P-N junction rectifiers traditionally used in power switching circuits have almost negligible reverse leakage current before the reverse voltage is high enough to set-off avalanche breakdown. So, the design trade-off for Si rectifiers is simply between high reverse breakdown voltage and low forward on-resistance (R_{on}). Since SiC rectifiers are SBDs, their reverse leakage current is non-negligible well before breakdown. In addition, this leakage current is very sensitive to temperature. Besides reverse bias static power loss, the leakage current at rated voltage and high temperatures is also limited by power dissipation considerations of the package. Consequently, it is well understood that design trade-off for SiC includes

reverse leakage current at rated voltage, in addition to BV and V_F. So, device designs like Junction-Barrier Schottky (JBS) [1] are used to reduce electric field at the Schottky junction, which reduces leakage at the expense of V_F.

In this work, we investigate the dependence of reverse leakage current on temperature as a function of the margin of avalanche breakdown voltage over rated voltage. Depending on the avalanche breakdown voltage, the electric field at rated voltage at the metal-SiC interface changes. We analyze the impact of this change on reverse leakage current and its temperature dependence; both experimentally using GPTG's 1200 V JBS diodes and theoretically using an analytical model [2]. We then use the analytical I-V models of SiC SBDs to find the optimum epilayer design for a 1200 V SiC rectifier and contrast it with the optimum epilayer predicted by a Silicon device-like BV v/s R_{on} optimization.

II. TEMPERARURE DEPENDENT LEAKAGE

A 1200 V rated JBS diode (fig. 1) was first designed to have ~2100 V avalanche breakdown voltage (BV) and with a Schottky barrier height of 1.08 eV. This epitaxy was doped $6.1E15cm^{-3}$ and 12.5 µm thick, and the p+ JBS implants were 0.35 µm deep (R in fig. 1) and spaced 2µm (2k in fig. 1) apart. The reverse leakage performance at different temperatures for a device of active area 0.063 cm^2, rated for 15 A forward current, is shown in fig. 2. Leakage current at room temperature is < 10 µA but, at 175 °C, it is > 1 mA; which not only increases static power loss but also might lead to thermal runaway. The high temperature leakage current can be reduced by increasing barrier height or designing JBS implants to be deeper or narrower together. However, both of these techniques have significant costs in increased V_F. So, we explored the impact of changing the BV on temperature sensitivity of leakage current at rated voltage.

To continuously model the variation of device performance with various parameters, analytical models were used. The leakage current of JBS diodes was calculated by a 2-step analytical model which at first used a closed form expression for maximum electric field at the Schottky contact in reverse bias as a function of epitaxy, mask and process parameters [3]; and then used another closed form solution relating the leakage current to the electric field at the Schottky contact. The latter model combines thermionic emission across the Schottky barrier and electron tunneling between metal and SiC. Closed

978-1-4799-5494-0/14 $31.00 © 2014 IEEE

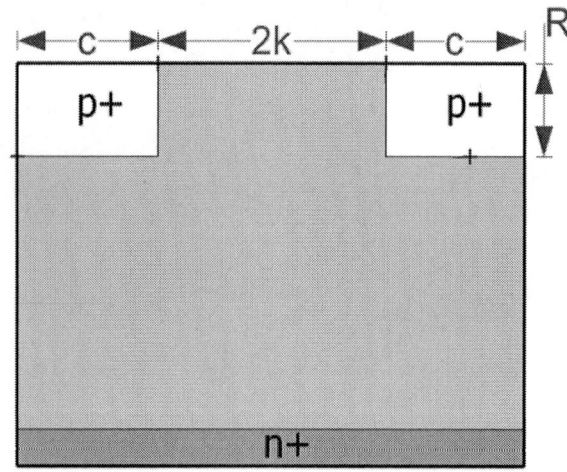

Fig. 1. Cross-section of a JBS diode figure.

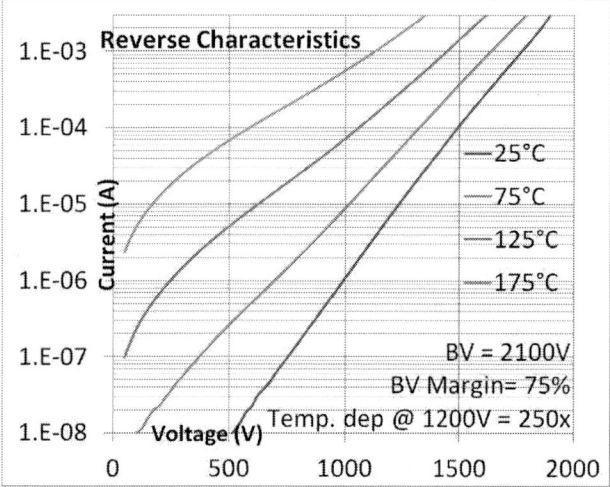

Fig. 2. Reverse I-V of a 1200V rated device showing high leakage at elevated temperatures.

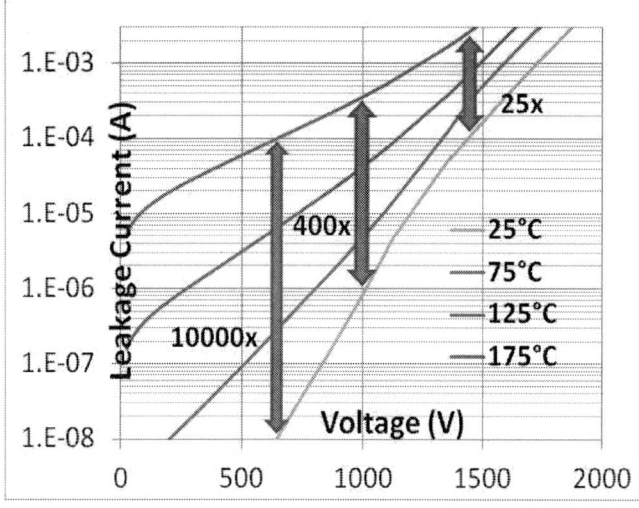

Fig. 3. Model predictions showing reduction in temperature variation of reverse leakage with increasing voltage/electric field at contact.

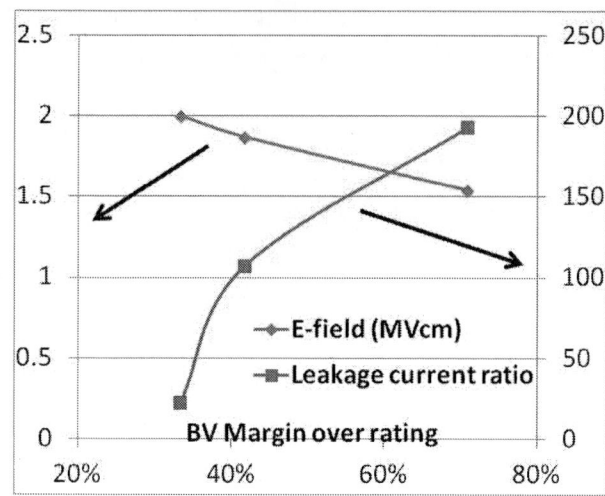

Fig. 4. Variation of the ratio of leakage current at 175 °C vs leakage at 25 °C as a function of BV margin over rated voltage.

form expressions for electron tunneling were derived by Hatakeyama [2] based on the more involved differential equations published by Padovani and Stratton [4]. Fig. 3 shows the predictions of this combined model by adding thermionic emission and tunneling together. For SiC operating at high electric fields near the rated voltage, tunneling swamps thermionic field emission current. It can be seen from the figure that temperature sensitivity of leakage current reduces with increasing voltage. At ~600 V, the leakage at 175 °C is 10000 x the leakage at 25 °C but this sensitivity of leakage to temperature reduces to ~25 x at ~1400 V. From the model [2], this trend in leakage could be due to two reasons-

- At low voltages and hence low electric field at the Schottky barrier, leakage is due to thermionic field emission where electrons are thermally excited to the energy levels from which they tunnel across the junction. As electric field increases, more corresponding energy levels in the metal and SiC are occupied and hence capable of tunneling. So the number of carriers tunneling is less sensitive to temperature and the consequent thermal excitation.

- At higher electric fields, leakage mechanism shifts to field emission, which is less sensitive to temperature. This leads to the leakage curves at high voltages bunching together.

Fig. 3 from the analytical model showed reduction in temperature sensitivity of leakage current to reverse voltage, because of the change of electric field at the contact with voltage. Changing the epitaxy can also change electric field at the contact at the same voltage. Since avalanche breakdown for a particular device happens at nearly the same peak field (called the critical field), the electric field at the anode contact at rated voltage can be increased by reducing BV of the epitaxy. By reducing the BV margin over rated voltage, the sensitivity of leakage current to temperature can be reduced. This is illustrated in fig. 4 which shows the model prediction for increase of electric field as BV is reduced. It also shows experimental reduction in the leakage current ratio for different device designs as BV margin over rated reverse voltage is reduced. For example, a BV margin of 40 % for a 1200 V rated device means BV is 1200 X 1.40 = 1680 V.

III. OPTIMUM SiC SBD DESIGN

The margin of designed BV over rated voltage has a lower bound in that there should be sufficient yield of devices that have a safety margin of BV above the rated voltage. From the traditional BV vs R_{on} trade-off, increasing BV increases R_{on}. From the previous section, it shows that increasing BV will also increase the high temperature leakage of the diode. For a 1200 V diode, we chose BV to be ~1650 V, which allows for a 10% BV margin over 1200 V (1320 V) even with 25 % variation in BV due to epitaxial doping and thickness variations.

The additional design constraint we introduce is high temperature leakage current, which is set not to exceed 100 µA at 175 °C, the highest rated junction temperature of operation. It is interesting to note that this is a current limit rather than a current density limit, which means that this method will yield different optimum designs for different chip sizes. Instead of absolute leakage, a similar constraint could be the maximum power dissipation for the given package that doesn't lead to thermal runaway. Although it was found that a higher Schottky barrier allows the diode to be designed for lower V_F at the same leakage current and BV, we fix the Schottky barrier as a constant (1.23 eV) since the process to achieve any given barrier cannot be easily designed. With these constraints, we find the optimum JBS spacing and depth as well as epitaxial doping and thickness to minimize V_F of the device. The forward and reverse I-V curves of the optimized device at different temperatures are shown in figs. 5 and 6 respectively.

The optimization curves of different epitaxial thicknesses and doping are shown in fig. 7 for a JBS diode, from which a design can be chosen based on the leakage constraint. The star marker in fig. 7 shows the epitaxial design that has the lowest V_F while still having leakage not more than 100 µA. Table 1 shows the comparison of such optimized epitaxial designs under different cases along with the epitaxial designs chosen just by BV v/s R_{on} trade-off for SiC [5]. Since the latter is independent of Schottky barrier, the leakage current obtained ins not constrained; indeed in the case with barrier of 1.23 eV, high temperature leakage is > 1 mA. However, when we constrain the high temperature leakage to < 100 µA through the

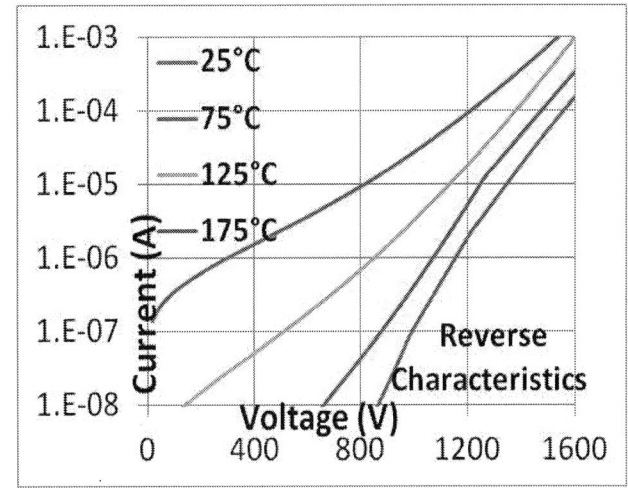

Fig. 6. **Reverse I-V of the optimized design with reverse leakage < 100 µA.**

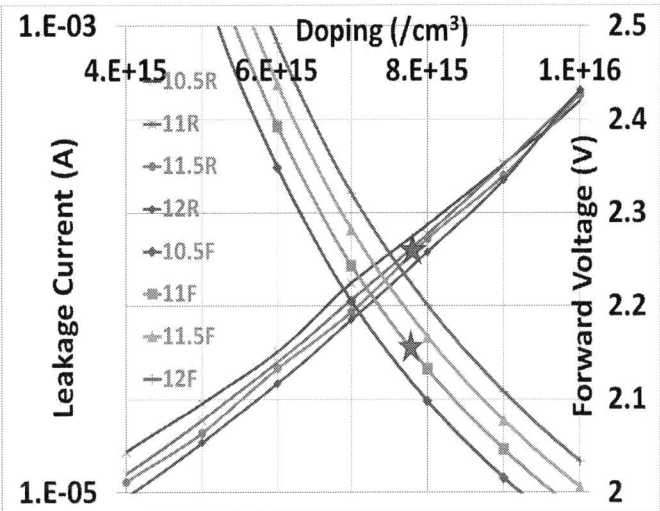

Fig. 7. **Optimization curves at 175°C showing the JBS diode design with lowest V_F at high temperature.**

TABLE I.   COMPARISON OF MODELS AT 25C AND 175C

	Epitaxy		BV	Leakage	V_F (V)
	/cu.cm	µm	(V)	(uA)	
Schottky	5.6E15	10.5	1885	100	2.33
				2	1.60
JBS	7.6E15	11	1680	100	2.17
				2	1.54
BV/ R_{on}	9.1E15	10.5	1650	3800	2.08
				111	1.62

added constraint in the first two rows of table 1, we are able to optimize to a different epitaxy that still yields a lower V_F both for Schottky and JBS diodes. It can also be seen that the optimum epitaxy for JBS diodes is different from corresponding Schottky diodes and, as expected, the JBS

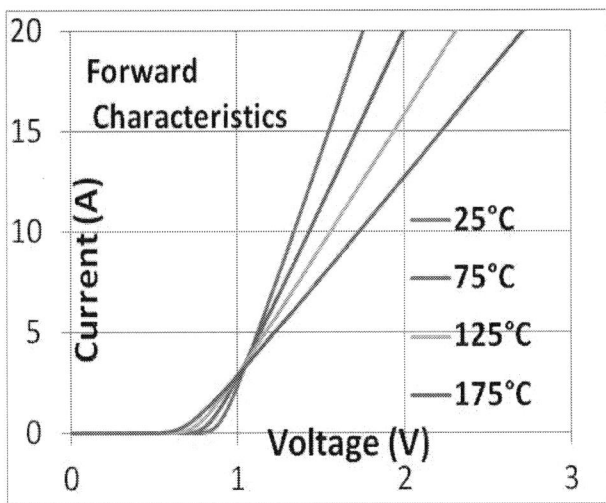

Fig. 5. **Forward I-V of the optimized design showing 175 °C V_F < 2.3V**

diodes can be designed to offer a better V_F vs leakage current trade-off.

IV. CONCLUSION

This paper proposes the margin of avalanche breakdown voltage over rated reverse voltage as an important design parameter in wide band-gap SBD design with impact on temperature dependence of leakage current. Limiting BV has conventionally been recognized to help limit V_F but we show that it also has impact on limiting the variation of leakage current at rated voltage with temperature. While the absolute value of leakage current can be limited by changing barrier height at the cost of V_F, BV margin is shown to be an additional knob in tuning its temperature dependence.

We then used the analytical models for forward and reverse I-V curves of the SBD and JBS diodes in an optimization scheme to find the epitaxial design that minimizes V_F. This optimization of epitaxy is found to be different from that obtained from BV v/s R_{on} trade-off and also sensitive to JBS geometry. So, we conclude that the epitaxial design at a given rated reverse voltage has to be designed specifically for each Schottky barrier and JBS geometry to optimize performance and we propose an optimization scheme for the same.

ACKNOWLEDGMENT

The authors would like to acknowledge stimulating discussions with Prof. James Cooper of Purdue University as fruitful to this work.

REFERENCES

[1] J. B. Baliga, "Analysis of junction-barrier-controlled schottky (JBS) rectifier characteristics," Solid State Electronics, vol. 28, pp. 1089-93, 1955.

[2] T. Hatakeyama and T. Shinohe, "Reverse Characteristics of a 4H-SiC Schottky Barrier Diode," Material Science Forum, Vol. 389-93, pp. 1169-72, 2002.

[3] R. Radhakrishnan and J. H. Zhao, "A 2-dimensional fully analytical model for design of high voltage junction barrier Schottky (JBS) diodes," Solid State Electronics, vol. 63, pp. 167-76, 2011.

[4] F. A. Padovani and R. Stratton, "Field and thermionic field emission in Schottky barriers," Solid State Electronics, vol. 9, pp. 695-707 (1966).

[5] J. B. Baliga, Power Semiconductor Devices,.PWS Publishing Company, Boston MA, fig. 3.7 p. 76.

A 12 to 1 V Five Phase Interleaving GaN POL Converter for High Current Low Voltage Applications

Benjamin K. Rhea, Luke L. Jenkins, William E. Abell, Frank T. Werner,
Christopher G. Wilson, Robert N. Dean, Daniel K. Harris
Electrical and Computer Engineering Department
Auburn University
Auburn, Alabama 36849
Email: bkr0001@auburn.edu

Abstract—**As applications like high performance cloud computing grow exponentially, there is an immediate need for high efficiency high current point of load (POL) converters. Interleaving phases is a common technique to reduce conduction loss, lower output ripple, improve high load performance, and reduce stress on power switching devices. However, multiphase POLs are primarily limited by size. The development of Gallium Nitride (GaN) power semiconductors is enabling higher power density in power supply applications. This paper demonstrates this with a 120 W five-phase interleaving GaN synchronous buck converter that achieves a peak efficiency above 96% and maintains over 90% efficiency up to 105 A.**

I. Introduction

Modern microprocessors typically operate between 0.9 - 1.2 V and current demand is on the rise [1], [2]. The Point of Load (POL) is often the lowest efficiency power conversion stage due to high conduction loss and low output voltage. Additionally, conduction loss between the POL converter and the load dissipates additional heat near the processor. The solution lies in the development of efficient converter design and small size that can be physically located near the load. Thus, the application of GaN is pivotal to increase the power density and efficiency of high performance switch-mode power supplies [3], [4]. The effectiveness of GaN is further enhanced in buck converter topology with the application of parallel synchronous rectifiers (SR) [5]. One of the primary drawbacks to GaN technologies is the reluctance of the industry to adopt a new unproven device technology over silicon (Si); however, there are promising results after $125\,^{\circ}\text{C}$ reliability tests and over a year of continuous operation [6], [7].

One approach to increasing efficiency is the expansion to multiple phases. This reduces the overall inductor current ripple which improves the reliability of the devices as well as the load that is being driven by the converter [8]. The multiphase converter also increases the effective output switching frequency which results in smaller size and higher power density [9]. Multiphase converters in previous efforts have demonstrated operating temperatures on the power devices of approximately $55\,^{\circ}\text{C}$ up to $60\,^{\circ}\text{C}$ at 40 A (10 A per-phase) output current with thermal management control techniques and forced air flow [10]. The presented design looks to lessen the need for advanced control techniques and minimize the need for forced air.

II. Interleaving

One problem with operating multiple conversion phases in parallel is that if they are driven by the same driving signal then the magnitude of the output voltage ripple increases. This is fixed by even spacing of the PWM signals, called interleaving. The PWM control was implemented on a TI F28335 control-Card which was chosen for its five easily programmable pulse width modulation (PWM) signals.

These five PWM signals are interleaved and switched at 160 kHz as seen in Fig 1. The resulting current output waveform is seen in Fig. 2. The interleaving results in a reduced output voltage ripple [11]. This can be seen in the steady state results of a single and five phase board. Typical peak to peak values of a single phase range from approximately 0.95 to 1.03 V when operating at 5 A, as seen in Fig. 3. Five phases range from approximately 0.98 to 1.01 V, as seen in Fig. 4. The output voltage ripple decreases from $\pm$ 80 mV to $\pm$ 32 mV peak to peak due to the interleaving.

The expansion to multiple phases also introduces a new term, called the effective switching frequency, which is the frequency of the output signal which is seen by the load. With multiple phases interleaved, the load sees a higher frequency then each individual phase. The effective switching frequency (F_{eff}) is defined as the switching frequency (F_s) multiplied by the number of active phases (N) Eq. 1. Increasing the frequency reduces the need for addition passives required by the converter.

$$F_{eff} = F_s \times N \qquad (1)$$

III. Results

This multiphase POL converter significantly improves high load performance upon commercial state-of-the-art POLs, minimizes output voltage ripple by interleaving phases, and increases output frequency from 800 kHz up to 2 MHz. Each synchronous buck phase is configured with a single GaN EPC2015 switch and two parallel GaN synchronous rectifiers (SRs) located on the same PCB, as seen in Fig. 5. Due to the design operating as a 12 to 1 V converter, the switch is on for roughly 8% of the time and the SR is on for 92% of the time.

978-1-4799-5494-0/14 $31.00 © 2014 IEEE

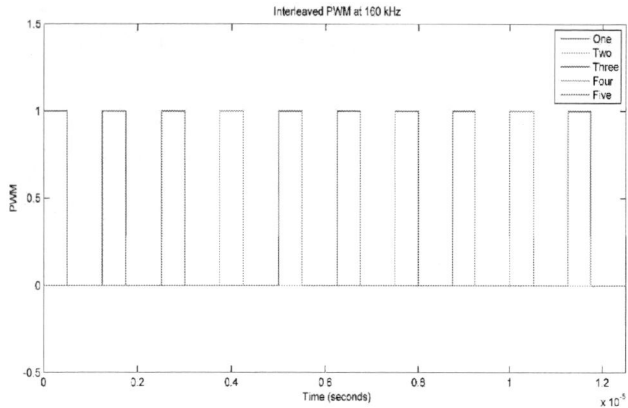

Fig. 1. Five interleaved PWM signals used in the five phase design. Fs = 160 kHz Feff = 800 kHz

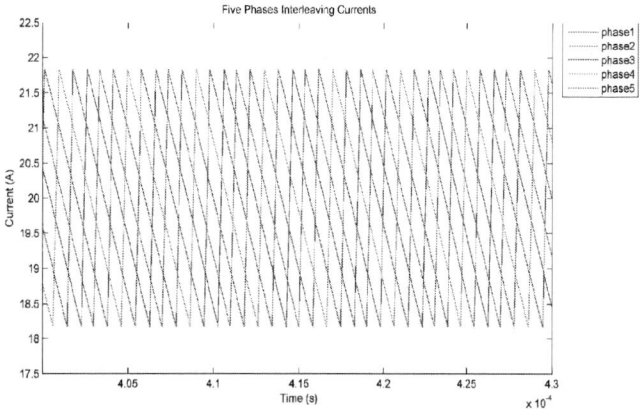

Fig. 2. Five current phases resulting from the interleaving PWM signals.

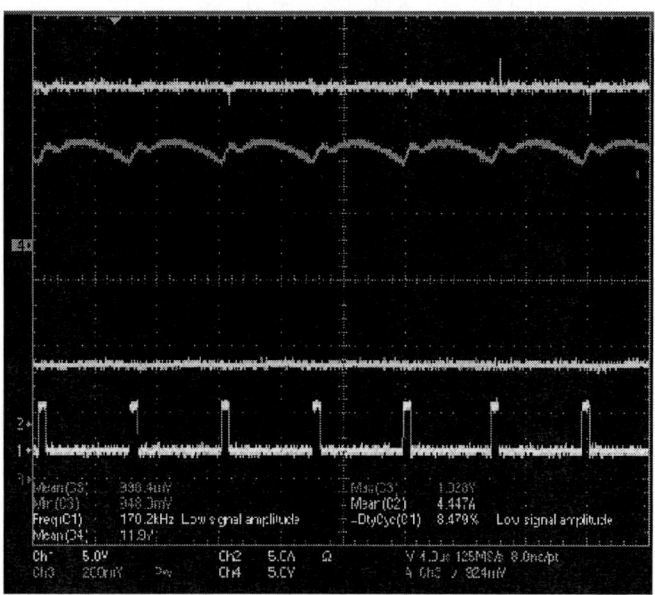

Fig. 3. Steady state operation of single phase at 5 A. VOUT in magenta

To increase the efficiency of the system, two SRs are used in parallel since this reduces the on-resistance of the combined SR which conduct a majority of the time.

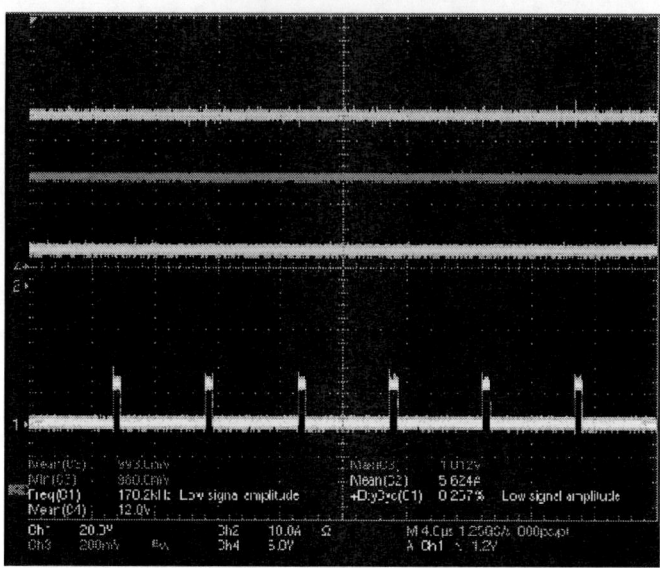

Fig. 4. Steady state operation of five phases at 5 A. VOUT in magenta

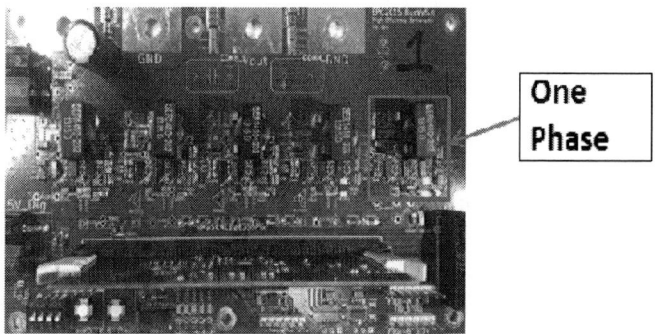

Fig. 5. The five phase synchronous buck converter using GaN HEMTs with peak efficiency over 96%.

The POL is designed with a wide input and output operating range: VIN = 2 - 24 V and VOUT = 0.5 - 5 V. With removal of developmental test points, the converter has a power density of approximately 270 W/in^2, a feat enabled by GaN power semiconductors.

With five phases sharing the load current, the heat produced next to the load can be decreased, which decreases the reliance on external cooling. Thermal images of a single and multiphase converter are taken with the same output current of 20 A. The current sharing between the five phase converter means that each individual phase is effectively operating at 4 A each. Both of the these designs have the same individual buck layout with only the phase count increasing and no forced air cooling. The single phase results can be seen in Fig. 6 and the five phase results in Fig. 7. Here a 18.4 °C temperature difference can be seen where the single phase peaks at 46.6 °C and the five phase at 28.2 °C on the GaN power switching device.

The five phase design also showed superior thermal performance when each phase carried the same amount of current. The single phase board operating at 10 A output current reached a peak temperature of 37.6 °C Fig. 8. The five phase board operating with 10 A per-phase (50 A total) peaked at only 31.9 °C Fig. 9. This shows a 5.7 °C improvement

978-1-4799-5494-0/14 $31.00 © 2014 IEEE

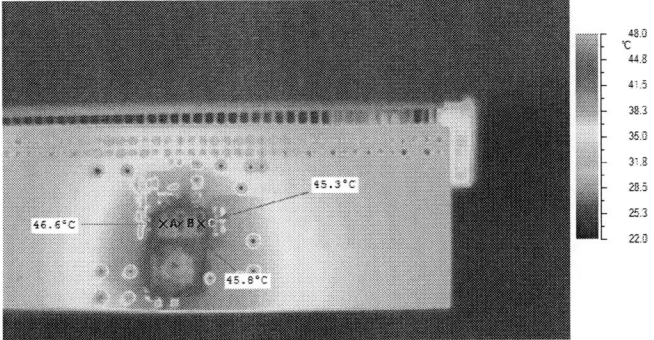

Fig. 6. Thermal image of a single phase converter operating at 20 A output current showing high heat dissipation near the load

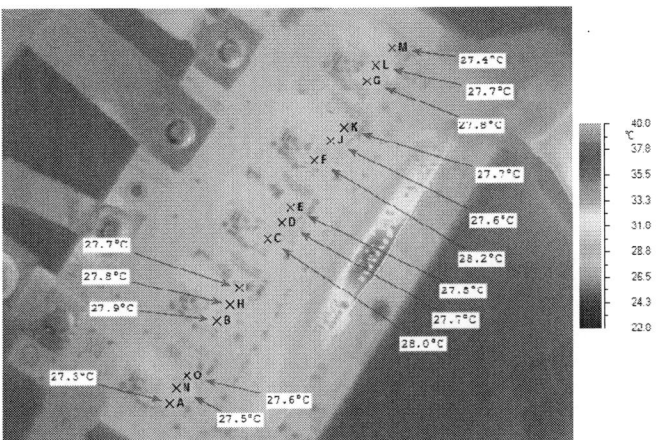

Fig. 7. Thermal image of a five phase converter operating at 20 A output current showing the reduced heat dissipation near the load

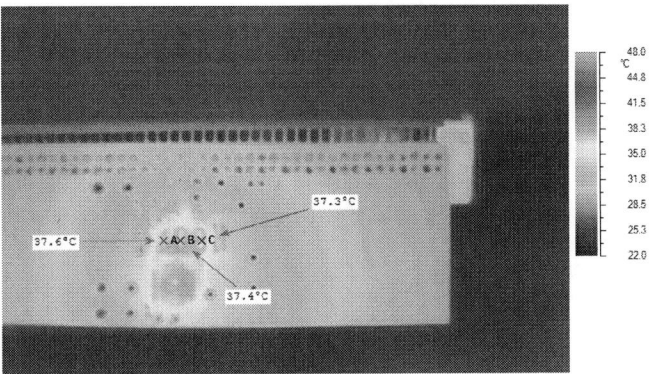

Fig. 8. Thermal image of the single phase converter operating at 10 A load and no forced air cooling and a peak temperature of 37.6 °C

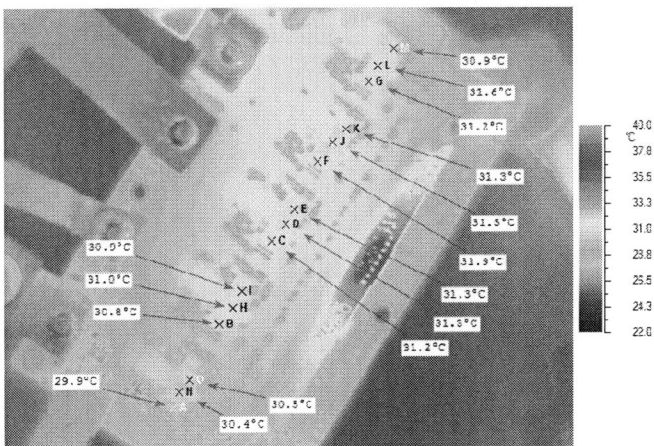

Fig. 9. Thermal image of the five phase converter operating at 50 A load and no forced air cooling and a peak temperature of 31.9 °C

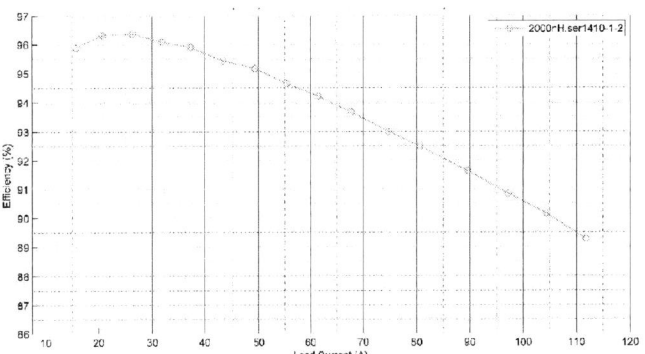

Fig. 10. Load vs efficiency for the five phase converter which maintains over 90% efficiency up to 105 A. Fs = 160 kHz Feff = 800 kHz

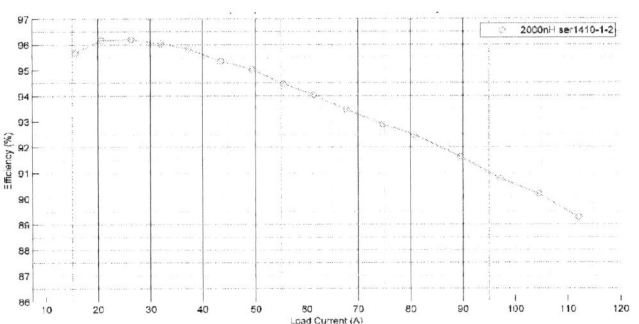

Fig. 11. Load vs efficiency for the five phase converter which maintains over 90% efficiency up to 100 A. Fs = 200 kHz Feff = 1 MHz

in the five phase design over the single phase design for comparable operating currents per-phase. Although this design lacks complex thermal management controller techniques, the difference in temperatures between the center and outside GaN switches is only 2 °C at 50 A due in part to an optimized layout.

Various switching frequencies were tested where 160 kHz was determined to be most efficient for the five phase converter. The POL achieves a peak efficiency over 96% at 25 A output current. While operating at 25 A, the converter never gets over 28.4 °C, minimizing the need for external cooling. The converter maintains this high performance with over 90% efficiency up to 105 A, as seen in Fig. 10. Although the system peaks at an effective switching frequency of 800 kHz, the system still performs well at an effective output switching frequency of 1 MHz maintaining over 96% efficiency over 30 A and maintains over 90% efficiency over 100 A Fig. 11. Increasing the effective switching frequency to 2 MHz, the system peaks efficiency peaks over 94% and maintains over 90% efficiency up to 95 A, as seen in Fig. 12.

978-1-4799-5494-0/14 $31.00 © 2014 IEEE

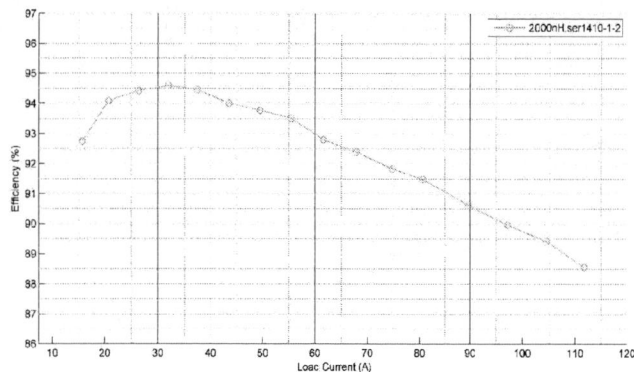

Fig. 12. Load vs efficiency for the five phase converter which maintains over 90% efficiency up to 95 A. Fs = 400 kHz Feff = 2 MHz

IV. CONCLUSION

A five phase synchronous buck converter has been developed to maximize efficiency for modern microprocessors that demand low voltage and high current operation. A new term for multiphase converters, called the effective switching frequency was introduced describing the frequency seen by the load. This results in a significantly reduced output voltage ripple as well as optimizing efficiency at higher loads.

The presented POL reaches a peak efficiency of 96% at 25 A and maintains over 90% efficiency up to 105 A output current with an effective switching frequency of 800 kHz. This design also performs well at an effective switching frequency, of 1 MHz and 2 MHz with over 90% efficiency maintained over 100 A and 95 A, respectively. The 120 W design features two parallel SR GaN power devices for each individual phase to minimize the on-resistance.

A detailed thermal analysis was performed on both a single and five phase design. The converter showed superior thermal performance over the single phase design with a 5.7 °C lower operating temperature at a comparable 10 A per-phase output current. The five phase converter design minimizes the heat dissipation near the load. All of the thermal results presented are without any external forced air cooling.

Overall, the 12-1 V five phase converter has demonstrated the effectiveness of GaN used in switched mode power converters. This design boasts over 96% efficiency, and an effective switching frequency up to 2 MHz which results in minimal heat dissipation. With GaN still considered a young technology, this leaves room for improvements of future designs.

REFERENCES

[1] Yi-Chung Wang; Ying-Yu Tzou, "Design and Realization of a Digital Multiphase-Interleaved VRM Controller Using FPGA," Industrial Electronics Society, 2007. IECON 2007. pp.1978-1982, 5-8 Nov. 2007.

[2] Pattnaik, S.; Panda, AK.; Mahapatra, K.K., "An Improved Multiphase converter for New Generation Microprocessr," Industrial and Information Systems, 2008. ICIIS 2008. IEEE Region 10 and the Third international Conference on , vol., no., pp.1,5, 8-10 Dec. 2008

[3] Scott, M.J.; Jinzhu Li; Jin Wang, "Applications of Gallium Nitride in power electronics," Power and Energy Conference at Illinois (PECI), 2013 IEEE , pp.1,7, 22-23 Feb. 2013.

[4] Reusch, D.; Lee, F.C.; Gilham, D.; Yipeng Su, "Optimization of a high density gallium nitride based non-isolated point of load module," Energy Conversion Congress and Exposition (ECCE), 2012 IEEE , vol., no., pp.2914,2920, 15-20 Sept. 2012

[5] Jenkins, Luke L.; Wilson, Christopher G.; Moses, Justin D.; Aggas, Jeffrey M.; Rhea, Benjamin K.; Dean, Robert N., "The impact of parallel GaN HEMTs on efficiency of a 12-to-1 V buck converter," Wide Bandgap Power Devices and Applications (WiPDA), 2013 IEEE Workshop on , vol., no., pp.197,200, 27-29 Oct. 2013

[6] Jenkins, Luke L.; Aggas, Jeffrey M.; Rhea, Benjamin K.; Abell, William; Moses, Justin D.; Wilson, Christopher G.; Dean, Robert N., 1,000 Hour GaN and SiC Power Semiconductor Reliability Study, Reliability Science of Advanced Materials and Devices (RSAMD) 2014, 7-9 Sep. 2014

[7] Jenkins, Luke L.; Rhea, Benjamin K.; Aggas, Jeffrey M.; Abell, William; Moses, Justin D.; Wilson, Christopher G.; Dean, Robert N., One Year Reliability Validation of GaN Power Semiconductors in Low-Voltage Power Electronic Applications, Reliability Science of Advanced Materials and Devices (RSAMD) 2014, 7-9 Sep. 2014

[8] Panov, Y.; Jovanovic, M.M., "Stability and dynamic performance of current-sharing control for paralleled voltage regulator modules," Applied Power Electronics Conference and Exposition, 2001. APEC 2001. Sixteenth Annual IEEE , vol.2, no., pp.765,771 vol.2, 2001

[9] Shu Ji; Reusch, D.; Lee, F.C., "High-Frequency High Power Density 3-D Integrated Gallium-Nitride-Based Point of Load Module Design," Power Electronics, IEEE Transactions on , vol.28, no.9, pp.4216,4226, Sept. 2013

[10] Cao, P.; Ng, J.C.W.; Trescases, O., "Thermal management for multiphase current mode buck converters," Applied Power Electronics Conference and Exposition (APEC), 2011 Twenty-Sixth Annual IEEE , vol., no., pp.1124,1129, 6-11 March 2011

[11] Rhea, Benjamin K.; Wilson, Christopher G.; Jenkins, Luke L.; Dean, Robert N., "The impact of inductor selection on a 12-1 V GaN POL converter with over 94% peak efficiency and higher load optimization," Wide Bandgap Power Devices and Applications (WiPDA), 2013 IEEE Workshop on , vol., no., pp.28,31, 27-29 Oct. 2013

High-Frequency Wireless Charging System Study Based on Normally-off GaN HEMTs

Yongsheng Fu, Lei Shi, Kevin (Hua) Bai, *IEEE Member*

Department of Electrical and Engineering, Kettering University

1700 University Ave, Flint, MI 48504, USA

E-mail: hbai@kettering.edu

Abstract—The Gallium nitride high electron mobility transistor (GaN HEMT) has become popular in the power electronics industry as it offers the possibility to reach lower switching loss and higher switching frequency compared to Si devices. This paper adopts a normally-off GaN HEMT in the wireless power transfer (WPT) system to charge a 48V battery pack on the E scooter. 178W power was delivered to the battery pack with 813 kHz switching. A simulation model of GaN HEMT is also built in LTspice to itemize the system loss. At the end some thoughts of improving the system efficiency were proposed.

Keywords— GaN HEMT, SiC, CoolMOS, wireless power transfer, zero voltage switching

I. INTRODUCTION

High efficiency and high compactness are the two major specifications for the power supply designs. Reduction of the product body weight and size requires the increment of the switching frequency, which brings higher switching loss again. Wide-bandgap (WBG) devices, e.g., GaN HEMTs are attracting more and more attention due to high electron velocity and high electron mobility, which allows GaN HEMT to switch faster transition and lower switching loss. It has been reported that by using GaN HEMT the switching frequency can be pushed to ~MHz [1-3]. Consequently passive components could be dramatically reduced.

Although SAE standard J2954 claims 80kHz as the optimal switching frequency for the wireless charger, this paper still designed a ~MHz WPT system using WBG devices to familiarize the characteristics of such devices and obtain the first-hand data of its effectiveness on reducing the system size and weight. The WPT system designed in this paper is based on the resonance theory, i.e., utilizing the receiver/transceiver leakage inductance to resonate with the external capacitors [4-5]. Such system could be used in charging the electric scooters, where 48V battery packs are used to drive the in-wheel hub-motor with the charging power of ~200W. The clearance between the scooter basis and the ground is ~10cm.

The Series-Series connection (SSC) resonant topology is used for this charger design with the equivalent circuit shown in Fig. 1(a)~(b). Q_1-Q_4 are GaN HEMTs. Diodes D_1-D_4 are silicon carbide SDP30S120. C_p and L_p are the primary-side resonant capacitor and inductor, respectively. C_s and L_s are the secondary-side resonant capacitor and inductor, respectively. L_m is the mutual inductance. R_1 and R_2 are primary and secondary coil internal resistance, respectively. R_L is the equivalent load resistance.

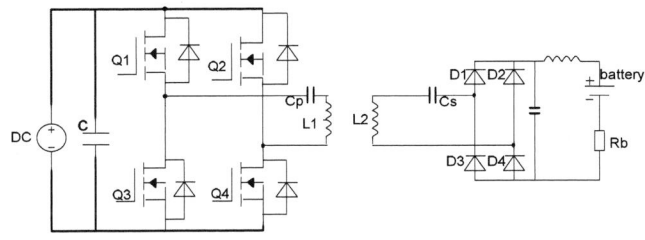

(a)

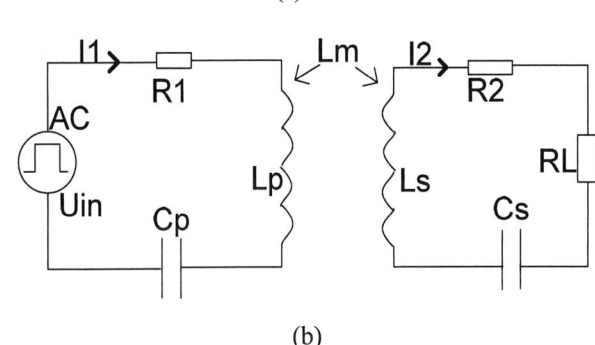

(b)

Fig. 1. (a) SSC topology based WPT system and (b) Equivalent circuit

II. GAN HEMT CHARACTERIZATION

Fig. 2(a) shows the used cascode GaN-HEMT structure. L_1~L_5 are parasitic inductances inside the package. The cascode GaN HEMT takes advantage of both Si MOSFET and GaN FET. It is more compatible with the existing gate-drive design for MOSFETs.

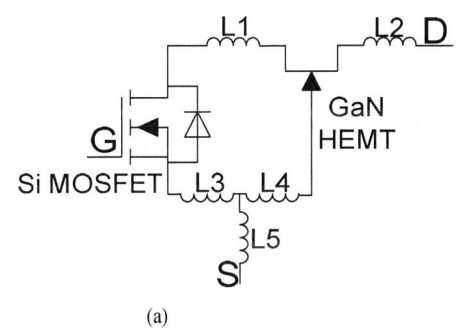

(a)

Sponsor of this project is Transphorm, Inc

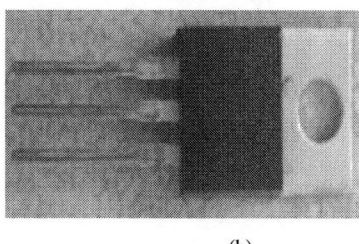

(b)

Fig. 2. (a) Cascode GaN HEMT structure and (b) TO-220 GaN FEMT

In this system four 600V GaN HEMTs TPH3002PS from Transphorm, Inc have been used. Its functional model is built based on LTSpice with the performance-evaluation circuit shown in Fig. 3. M_1 and M_2 are GaN HEMTs with M_1 driven by a 800kHz PWM signal at 50% duty cycle while M_2 off for free-wheeling. L_1 is the common-source inductance (CSI) shared by the power loop and the gate-drive loop. L_2 is the parasitic inductance of the driving loop. The CSI will slow down turn-on/off transition thereby increase the switching loss.

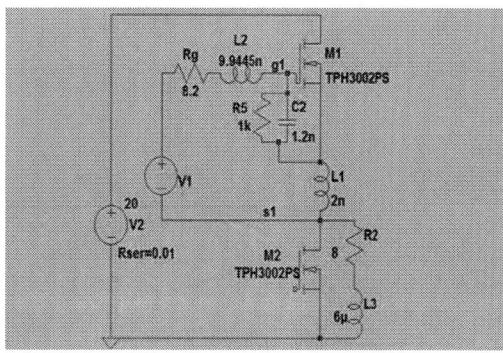

Fig. 3. Performance evaluation circuit

In the above evaluation circuit, V_2=20V, R_g=4Ω, R_2=25Ω, R_5=1k, L_1=2nH, L_2=9.9445nH, L_3=6uH, C_2=1.2nF, and the gate voltage of turn-on and turn-off are 8.1v and -3.9v, respectively. Fig. 4(a)~(b) shows simulated rising and falling times of Vgs are 24.25ns and 24.4ns, respectively. Fig. (c)~(d) show experimental rising and falling times of Vgs are both 24ns.

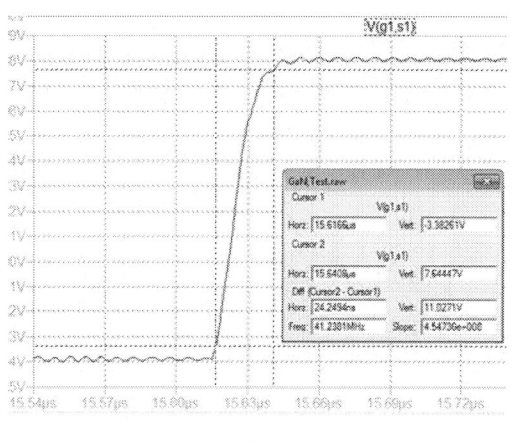

(a)

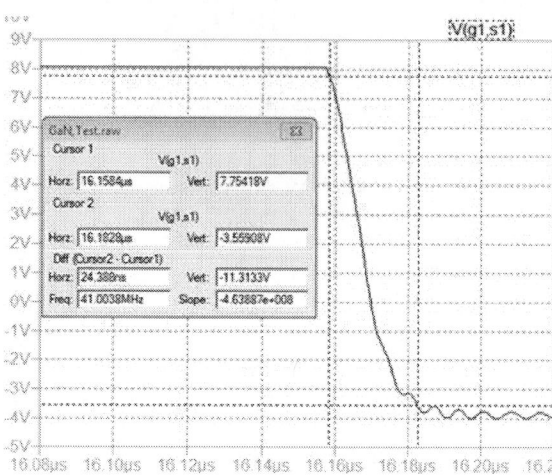

(b)

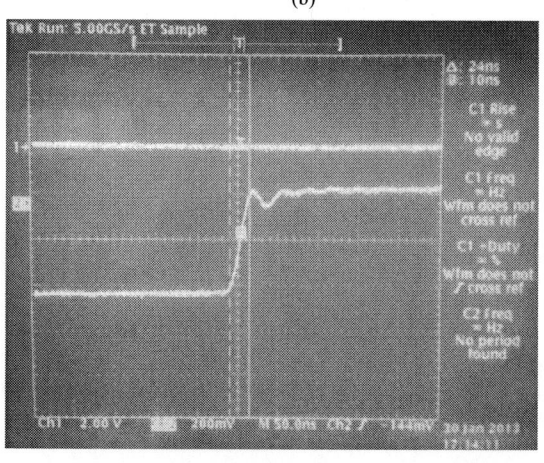

(c)

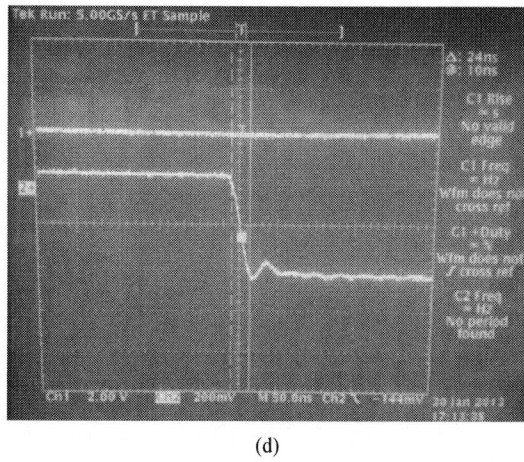

(d)

Fig. 4. Vgs waveform at Rg=4Ω(a) Rising edge (simulation) (b) Falling edge (simulation) (c) Rising edge (experiments, X-axis: 50ns/div, Y-axis: 4v/div) and (d) Falling edge (experiments, X-axis: 50ns/div, Y-axis: 4v/div)

With Rg = 8.2Ω, Fig. 5(a)~(b) show simulation results where the rising and falling time of Vgs are 70ns and 56ns, respectively. Fig. 5(c)~(d) show experimental rising and falling times of Vgs are 70ns and 56ns, respectively.

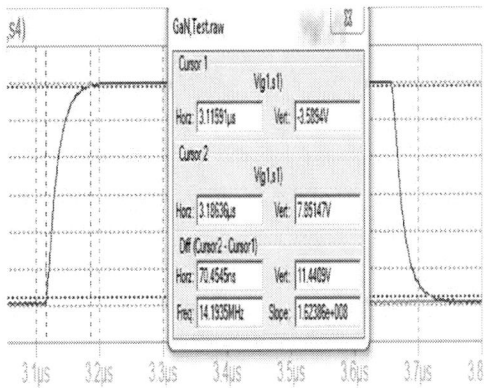

(a)

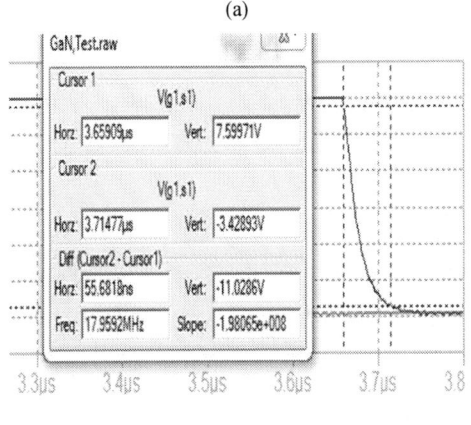

(b)

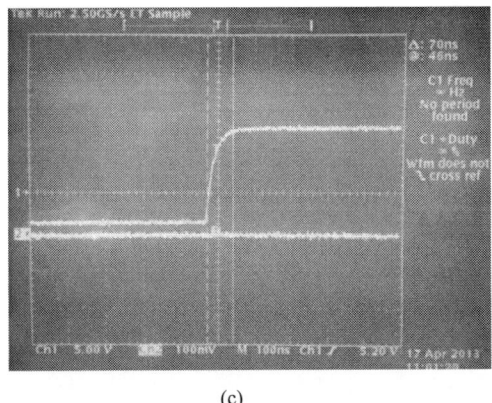

(c)

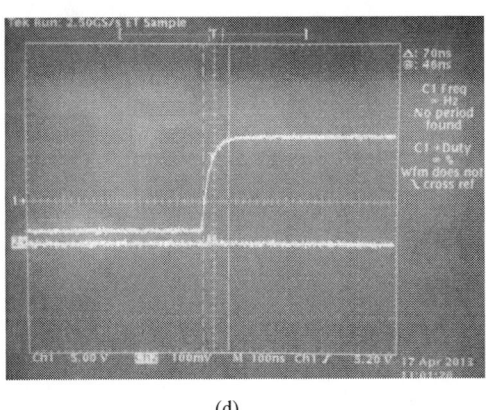

(d)

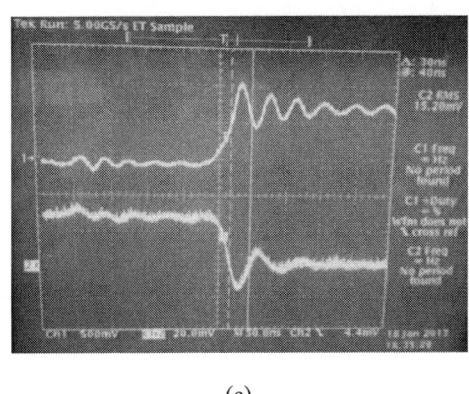

(e)

Fig. 5. Vgs waveform at Rg=8.2Ω (a) Rising edge (simulation) (b) Falling edge (simulation) (c) Rising edge (experiments, X-axis: 100ns/div, Y-axis: 5v/div) (d) Falling edge (experiments, X-axis: 100ns/div, Y-axis: 5v/div) and (e) GaN HEMT turn-off waveform (1-Vds, 2-Ids, X-axis: 50ns/div, Y-axis: 2 A/div)

Fig. 5(e) shows GaN HEMT turn-off waveform at 2A, which indicates that the reverse recovery time is 30ns and turn-off time is ~20ns. R_g=8.2Ω. Ambient temperature is 30°C.

In order to reveal the advantage of using GaN HEMT instead of MOSFET at the power loss aspect, a CoolMOS with the same voltage rating and current rating of GaN HEMT TPH3002PS is selected. Four CoolMOSs replace four GaN HEMTs in Fig. 1(a). Fig. 6(a)~(c) show the comparisons of switching loss, conduction loss and the sum of conduction and switching loss for each GaN and CoolMOS under the same output powers from LTspice simulation results, respectively. The system is under ZCS condition with the output power of 600W, otherwise, it is under ZVS condition.

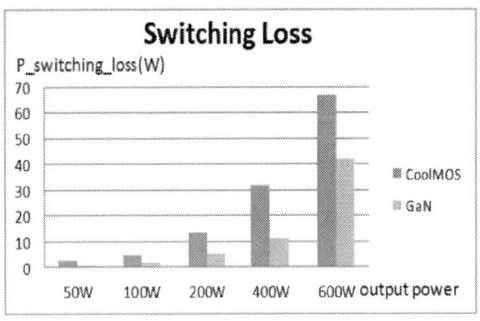

(a)

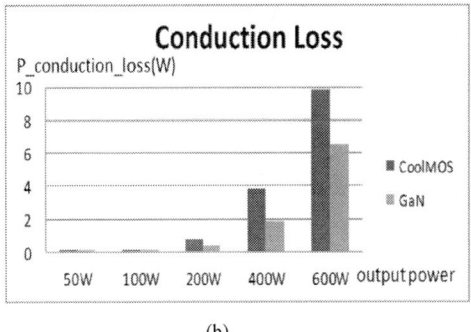

(b)

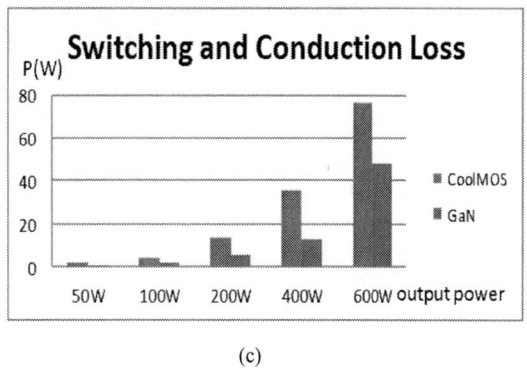

(c)

Fig. 6. (a) Switching loss comparison (b) Conduction loss comparison and (c) total loss comparison

III. COIL COUPLING MODEL AND SIMULATION

In this study the air-gap between primary (power transmitter) and secondary (power receiver) is 10cm, secondary-side coil has 15cm radius, particularly used for electric scooters. The dimension of primary side is decided by the maximum coupling coefficient between the primary and secondary coils. Fig. 7 shows the coupling coefficient when primary side radius changes from 0cm to 20cm, which indicates that the coupling coefficient reaches the maximum at the 18cm radius for the primary coil.

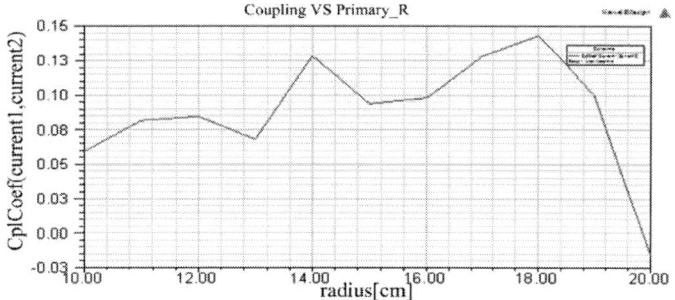

Fig. 7. Coupling coefficient VS primary coil radius

The designed coils are shown in Fig. 8 (a) with the wire diameter as 1.68mm and the loop distance as 1mm. There are 7 turns for both coils. The primary and secondary coils' radius are 15cm and 18cm, respectively.

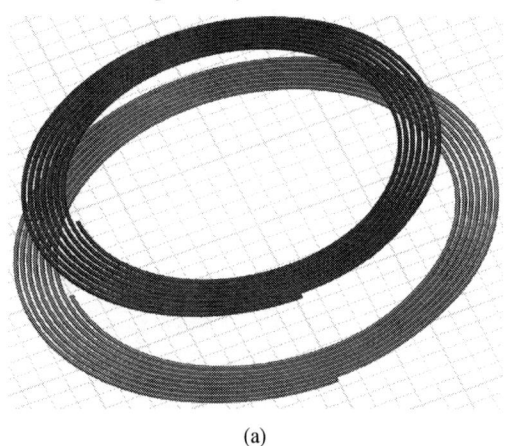

(a)

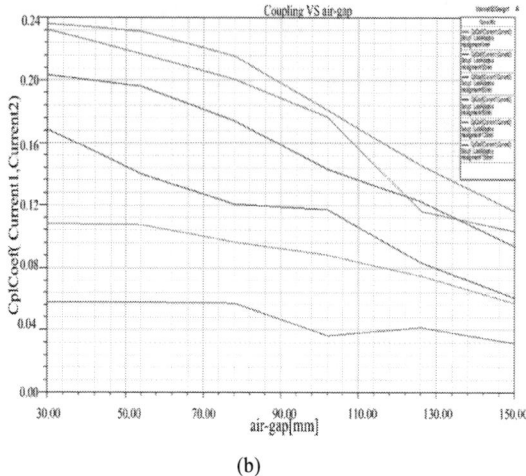

(b)

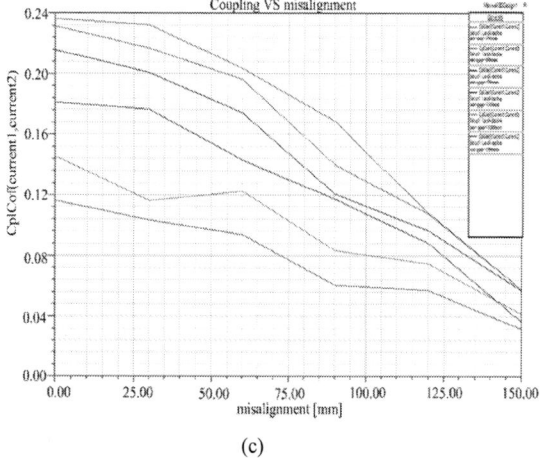

(c)

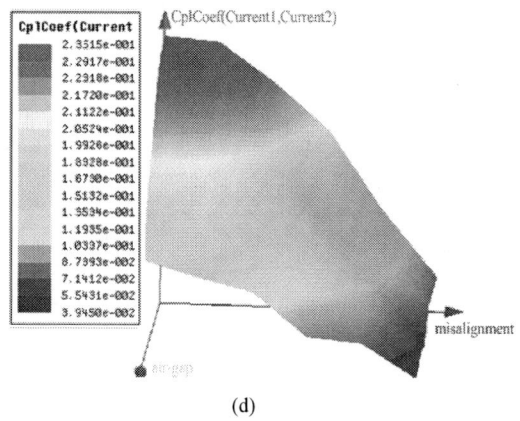

(d)

Fig. 8. (a) Wireless charging coil model (b) Coupling VS air-gap (c) Coupling VS misalignment and (d) 3D Coupling VS misalignment and air-gap

Coil inductance was extracted from Ansys/Maxwell. The self-inductances of primary and secondary side are 30.868uH and 25.619uH, respectively. The mutual inductance is 5.5156uH. Fig. 8(b) and (c) show the coupling coefficient at different air-gap and misalignments, respectively. Fig.8(d) shows the coupling coefficient 3D view. In order to maintain the zero-voltage switching (ZVS) at 813 kHz switching frequency, the resonant capacitance of primary and secondary side are chosen 1.77nF and 2nF, respectively. The simulation

978-1-4799-5494-0/14 $31.00 © 2014 IEEE

model and results are show in Fig. 9(a)~(b) when charging a 48V battery pack. R_5 is the primary coil's internal resistance as 210mΩ, and R_6 is the secondary coil's internal resistance as 160mΩ.

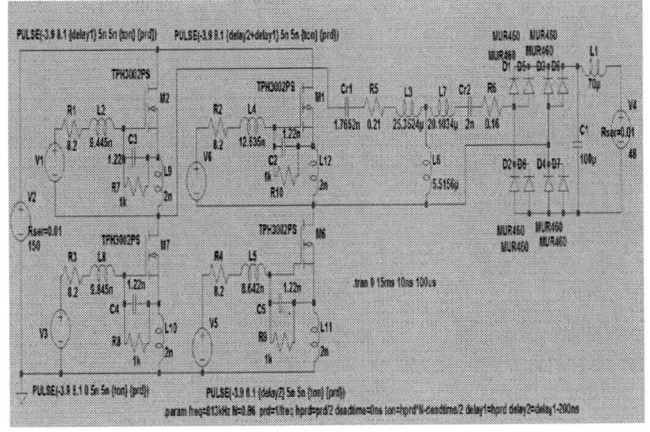

(a)

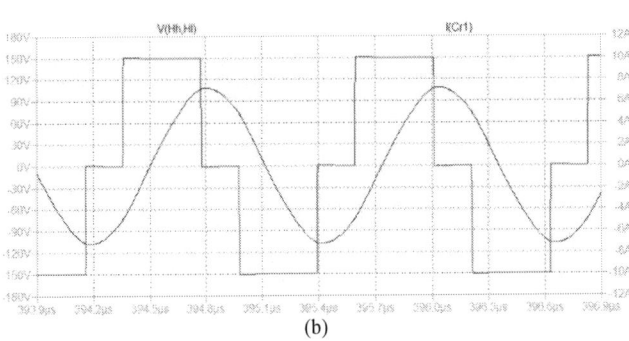

(b)

Fig. 9. (a) LTspice simulation model and (b) H –bridge waveform (red-H - bridge voltage, blue-H -bridge current)

IV. EXPERIMENTAL VALIDATIONS AND LOSS ALYNASIS

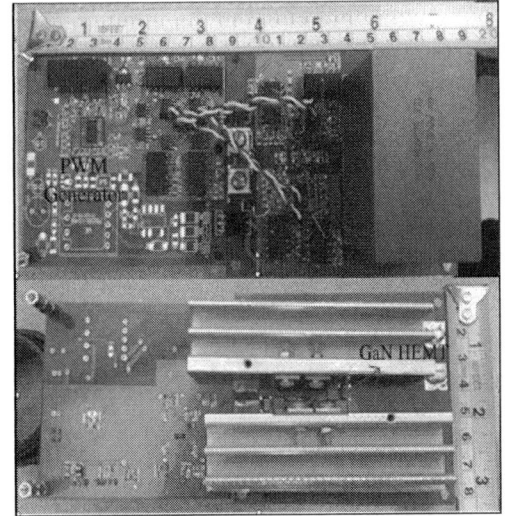

Fig. 10. Hardware prototype

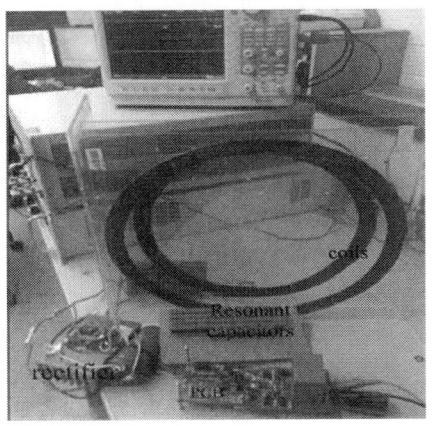

Fig. 11. Test bench of WPT system

Fig.10-11 shows hardware prototype. The PWM signals were generated from UCC28950. An open loop control is used with the fixed duty cycle. Four LM5114 gate drivers were used to drive four GaN HEMTs. Table 1 shows the system parameters.

TABLE. 1 TESTING PARAMETERS OF WPT

Parameter	Value	Unit
Switching frequency	813	kHz
Dead time	90	ns
PWM duty cycle	43%	
Turn on/off resistor for GaN HEMT	8.2	kHz
Self-inductance of primary side	30.87	uH
Self-inductance of secondary side	25.62	uH
Mutual inductance	5.52	uH
Input voltage	150	V
Input average current	1.34	A
Output voltage	48	V
Output current	3.7	A

The experimental data indicates 89% overall system efficiency with the 178w output power. Fig. 12(a) shows the H –bridge output voltage waveform. The primary and secondary coil current waveforms are shown in Fig. 12(b) and (c), respectively.

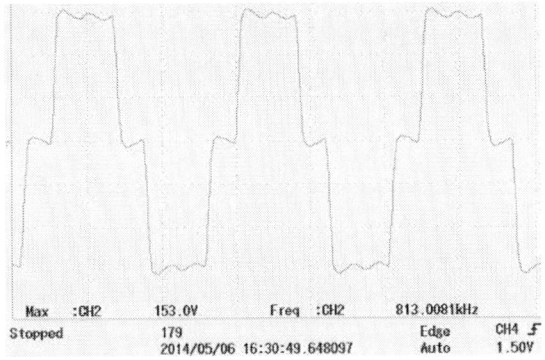

(a)

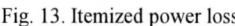

Fig. 13. Itemized power loss

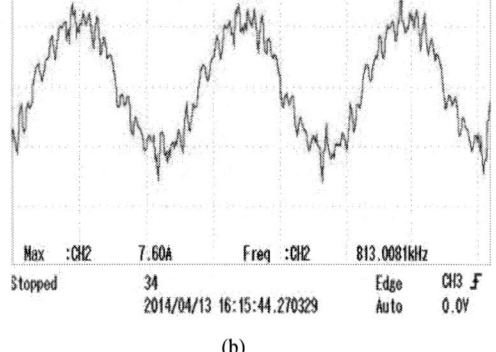

(b)

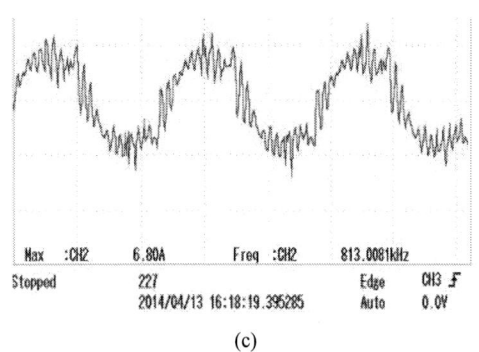

(c)

Fig. 12. (a) H –bridge voltage waveform, (b) Current waveform of primary coil and (c) Current waveform of secondary coil

From Fig. 12(b) and (c) the RMS current of primary and secondary coil are 5.4A and 4.8A, respectively. Thus the conduction loss of primary and secondary coil is 6.1W and 3.7W, respectively. Note this loss calculation considers the skin effect. Fig. 13 shows the itemized power loss.

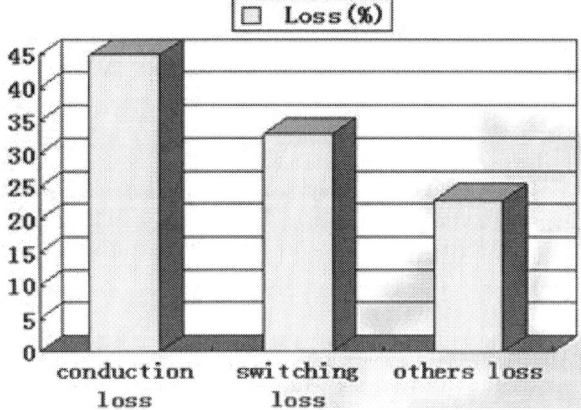

V. CONCLUSION

This paper validates the SSC topology based high-frequency wireless charging system in electric scooter application using Normally-off GaN HEMT. 178W charging power and 89% system efficiency are achieved with 813 kHz switching frequency. The switching performance of GaN HEMT has been evaluated by the simulation and experiments with power loss analysis given as well. Future optimization will be focused on improving system efficiency by reducing the conduction loss and PCB parasitic inductance.

978-1-4799-5494-0/14 $31.00 © 2014 IEEE

REFERENCES

[1] Z.Y. Liu, X.C. Huang, Fred C. Lee, Q. Li, "Simulation Model Development and Verification for High Voltage GaN HEMT in Cascode Structure", 2013, pp. 1-4.

[2] X.C. Huang, Q. Li, Z.Y. Liu, Fred C. Lee, "Analytical Loss Model of High Voltage GaN HEMT in Cascode Configuration", 2013, IEEE, pp.1-8.

[3] W.M. Zhang, Z.X. Xu, Z.Y. Zhang, Fred Wang,. Leon M, "Evaluation of 600V Cascode GaN FEMT in Device Characterization and All-GaN-Based LLC Resonant Converter", 2013 European Union, pp. 1-8.

[4] H.Bai, C.Duan, C.G.Jiang and A.Taylor, "Design of a zero-voltage-switching large-air-gap wireless charger with low electric stress for electric vehicles", IET Power Electronic, 2012, pp. 1-9

[5] S.Q.Li, N.Tommy, W.H.Li and C.Mi, "Wireless Charging for Safe and Economic Future Transportation", 2013, pp.12-14

[6] Z.R. Jia, M.Chen, X.F. Lv, Z.M. Qian, "Gallium Nitride Based High Power-Density Automotive HID Electronic Ballast", 2013 IEEE, pp. 1-8.

[7] Y.L. Xiong, S.Sun, H.W. Jia, John. Shen, "New Physical Insights on Power MOSFET Switching Losses", 2009 IEEE Transactions on Power Electronics, VOL, 24, No.2. pp. 1-6.

[8] X.C. Huang, Q. Li, Z.Y. Liu, Fred C. Lee, "Analytical Loss Model of High Voltage GaN FEMT in Caecode Configuration", 2014 IEEE Transactions on Power Electronics, VOL.29, No. 5. pp. 1-6

[9] J. Fu, Z. Zhang, Y. Liu, and P.C. Sen, "MOSFET switching loss model and optimal design of a current source driver considering the current diversion problem", IEEE Trans. Power Electron., VOL. 27, No.2, pp. 98-1012, Mar. 2008.

Reflected Wave Phenomenon in Motor Drive Systems Using Wide Bandgap Devices

Mark J . Scott, Jared Brockman, Boxue Hu, Lixing Fu,
Longya Xu, and Jin Wang*
The Ohio State University
Department of Electrical and Computer Engineering
Columbus, Ohio, United States of America
*: wang.1248@osu.edu

Rachid Darbali Zamora
The University of Puerto Rico
Department of Electrical and Computer Engineering
Mayaguez, PR 00681

Abstract—**Fast switching transients from power devices can produce large voltage spikes in motor drive systems with long feeder cable lengths. Referred to as the reflected wave phenomenon, it has been previously demonstrated that the maximum cable length for a given per unit overshoot decreases with increasing switching speed. With the emergence of higher switching speed components based on wide bandgap (WBG) semiconductors, this issue will present itself at shorter feeder cable lengths than comparably rated silicon (Si) devices. This work explores the reflected wave phenomenon for a gallium nitride (GaN) device and compares it to a Si MOSFET. A PSPICE model is created to predict the behavior of this issue. It is validated with experimental results. Tests were performed on a 600 V GaN device, a 600 V Si CoolMOS MOSFET, and a 650 V SiC MOSFET.**

Keywords—Reflected wave phenomenon, motor drive, wide bandgap devices, silicon carbide, gallium nitride.

I. INTRODUCTION

Research in the field of power electronics continuously aims to increase power density, improve operating efficiencies, and lower manufacturing costs. A growing consensus among researchers is that silicon (Si) based power devices have matured to a point where these types of improvements are becoming increasingly more difficult [1], [2]. For this reason, wide bandgap (WBG) based power devices are actively being explored as potential replacements for Si power devices in existing and emerging applications. Due to the merits of their atomic structure, WBG devices have the potential to switch at faster speeds, with lower conduction losses, and operate at higher temperatures than equivalently rated Si devices [1]–[4]. Already, manufacturers are offering gallium nitride (GaN) and silicon carbide (SiC) based devices.

Motor drive systems are an area where significant gains in performance are expected with WBG based implementation. Already, a three phase inverter has been presented with an efficiency exceeding 99% [4]. The benefits of higher switching frequency were shown to shink the size of the output filtering and improve the operating efficiency in a GaN based motor drive system [3]. WBG based power devices also have the ability to operate at higher junction temperatures. This is an area of active research as a higher temperature inverter could

enable the automotive industry to eliminate one of the cooling loops in their hybrid electric vehicles [2].

WBG devices, GaN in particular, are able to switch at much higher speeds than comparably rate Si components. In [5], switching speeds as high as 140 V/ns were achieved. The benefits of faster switching transitions favor lower switching losses and higher commutation frequencies. However, they do present other design issues such as electromagnetic interference (EMI) [6]. Additionally, they can cause mistrigger issues as transients propagate through the miller capacitance [7]. The correlation between switching speed, feeder cable length, and voltage spikes at machine terminals has been explored by several research groups [8]–[11]. During the late 1980's, as motor drives started to incorporate insulated gate bipolar transistors (IGBTs), failures began to appear in drive systems with long feeder cable lengths (see Fig. 1) [11]. Motor insulation and cable dielectric failures were occurring as a result of high voltage transients. These voltage spikes are due to the cable behaving like a transmission line as a result of the higher speed IGBT devices, a process that is referred to as the *reflected wave phenomenon* [8]–[11].

The impact of high switching speed WBG devices on the reflected wave phenomenon has not been explored in great detail. Yet, it can produce transients at the load terminal that exceed the DC bus voltage by two times. The goal of this work is to understand the severity of this issue through testing and simulation. The structure of the rest of the paper is as follows: section II reviews the theoretical background of the reflected wave phenomenon. The development of a PSPICE model, and the simulation results are provided in section III. Section IV contains experimental results for a 600 V rated GaN device, a 600 V rated Si device, and a 650 V SiC device. Two types of cables are evaluated, one is shielded and the other is

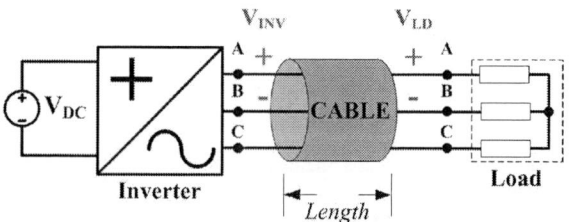

Fig. 1.   Motor drive system with long feeder cables.

978-1-4799-5494-0/14 $31.00 © 2014 IEEE

unshielded. A 4.5 ft. cable and a 13.5 ft. cable of each type were tested. For every configuration, tests were performed on a ground plane and floating on top of a non-conductive benchtop. The paper concludes with a summary of this research and the proposed future work.

II. BACKGROUND

The reflected wave phenomenon occurs in motor drive systems with long feeder cables. As the switching speed of power devices increases, the feeder cables will behave like transmission lines at smaller lengths [8], [9]. Without proper consideration, transients exceeding twice the DC bus voltage can be produced. This exposes the electric machine to undue stress on the winding insulation and can lead to its premature failure. It can also lead to the breakdown of the cable dielectric. Researchers using wide bandgap (WBG) devices in motor drive systems should be aware of this issue because it will present itself at shorter feeder cable lengths than comparably rated silicon (Si) drives.

The velocity (v) of a voltage waveform travelling down a lossless cable is related to the characteristic impedance (Z_O) of the cable by the following relationship:

$$v = \frac{1}{\sqrt{L_x C_x}}, \tag{1}$$

$$Z_O = \sqrt{\frac{L_X}{C_X}}, \tag{2}$$

where L_X is the per unit length inductance and C_X is the per unit length capacitance. Using (1), the time (τ) it takes the voltage to travel from the inverter to the motor terminals is calculated as:

$$\tau = \frac{l}{v}, \tag{3}$$

where l is the cable length. If τ is longer than 1/3 of the device's switching time (t_r or t_f), a full reflection will occur [9]. The peak magnitude ($V_{LD_{PEAK}}$) of the overshoot at the load will be equal to:

$$V_{LD_{PEAK}} = (1 + \Gamma_L) V_{DC}, \tag{4}$$

$$\Gamma_L = \frac{Z_L - Z_O}{Z_L + Z_O} \tag{5}$$

where Γ_L is the reflection coefficient at the motor and Z_L is the impedance of the load (the motor in this case). In the case of an electric machine, the inductive nature makes it appear as an open circuit to the fast switching transient and Γ_L is close to one (approximately 0.9 for small machines [9]). The inverter terminals, on the other hand, will appear as a short circuit due to the DC bus capacitance (i.e. Γ_S is close to -1). Given this

relationship, the critical cable length (l_c), or the length at which a full reflection occurs, is given by:

$$l_c = v \cdot \frac{t_r}{3}, \text{ or } l_c = v \cdot \frac{t_f}{3}. \tag{6}$$

This process is illustrated in Fig. 2 where it is assumed that $\tau \gg$. During the first interval, $t < \tau$, the voltage waveform travels from the inverter terminals to the load as shown in Fig 2b, and the load voltage (V_{LD}) equals zero. At $t = \tau$, V_{LD} increases to $V_{LD_{PEAK}}$. The theoretical rise time for the transient at the load is

$$\frac{dV_{LD}}{dt} = 2 \cdot t_r \tag{7}$$

because V_{LD} is equal to the incident waveform plus the reflected waveform. During the second interval, the reflected wave travels back to the inverter. When $t = 2\tau$, the wave is incident at the inverter terminals. Since Γ_S is negative, the wave is inverted and reflected back to the load as shown in Fig 2b. At $t = 3\tau$, V_{LD} decreases to $(1 + \Gamma_S \Gamma_L^2)V_{IN}$. Another reflected wave is produced at the load and sent back to the load terminals (not shown in Fig 2b). It returns from the inverter terminals with a positive amplitude. At $t = 5\tau$, the voltage begins to rises to $(1 + \Gamma_S^2 \Gamma_L^3)V_{IN}$ due to the third incident waveform. This process continues until the load voltage stabilizes at V_{DC}. From Fig 2, it can be observed that the oscillation frequency (f_{rw}) of the reflecting wave is equal to

$$f_{rw} = \frac{1}{4\tau}. \tag{8}$$

III. MODELLING

A model of the test setup was created in PSPICE. The feeder cables were simulated as distributed transmission lines. Their parameters were extracted using an impedance analyzer and incorporated into the model. The schematic for the test setup is shown in Fig. 3. A double pulse test setup that was

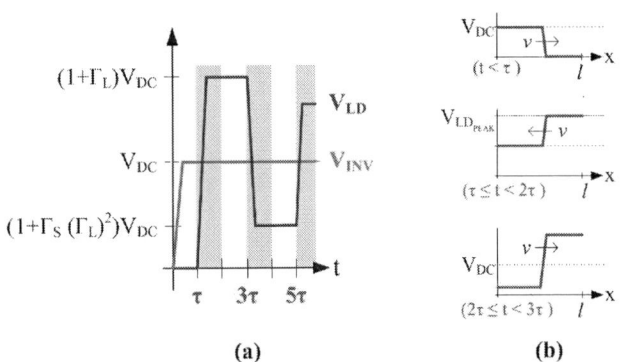

Fig. 2. The (a) large voltage transients at the load terminals in Fig. 1 results from the switching transient being (b) reflected between the inverter and load terminals.

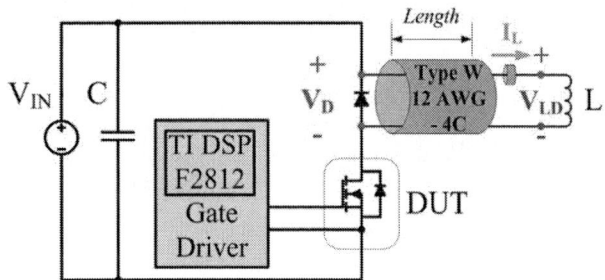

Fig. 3. Schematic of the test setup used to evaluate the reflected wave phenomenon.

used for characterizing switching devices was augmented to conduct this evaluation. The device under test (DUT) was one of the three switching devices. It was feed with two pulses; the first to control the current (I_L) in the load inductor (L), and the second pulse was used to examine the switching transients. A long cable was inserted between the freewheeling diode and inductor. The voltage across the diode (V_D) and the inductor (V_{LD}) were measured to determine how the transient propagated through the cable.

Two types of cables were evaluated during this research. Each was a four conductor bundle of 12 AWG wire; one was shielded and another was unshielded. This gauge of wire was selected because it corresponds with the current capabilities of the switching devices being tested. A 4294A impedance analyzer was used to perform the measurements on a one foot unit length of each cable. The distributed resistance (R_X) and inductance (L_X) were measured by shorting the wires at one end and performing the measurements at the other end [9]. The capacitance (C_X) and conductance (G_X) were measured with the terminals at the other end open. Measurements were made from 10 kHz to 10 MHz.

Using the data from the impedance analyzer and (2), the cable velocity was calculated for each type of cable. The data shown in Fig. 4 indicates that the shielded cable has a higher velocity then the unshielded one. This is due to L_X of the unshielded cable being higher than that of the shielded cable. The cable velocity results at 1 MHz were used to calculate

expected delays (τ) for cable lengths of 4.5 ft. and 13.5 ft. The results are listed in Table I. Therefore, it is expected that the reflected wave will be more pronounced with the shielded cable then the unshielded one. As a cross check for the measurements, the characteristic impedance (Z_O) was also calculated at 1 MHz. These values are in accordance with other values listed in literature [8].

TABLE I.   ESTIMATED CABLE PARAMETERS

Cable Type	Z_O^a (Ω)	Cable Velocity^a (m/s)	Estimated Delay (ns)	
			1.37 m *(4.5 ft)*	*4.12 m* *(13.5 ft)*
Shielded	102.3	1.69×10^8	8.1	24.4
Unshielded	121.8	1.55×10^8	8.9	26.6

a. Parameters estimated at 1 MHz

The per unit length measurements of R_X, G_X, L_X, and C_X extracted at 1 MHz were used in the transmission line model in PSPICE. Two types of stimulus were applied to the model; initially a pulse source was used with rise and fall times equal to $\dfrac{dV}{dt}$ values in the datasheet. It was found that the sharp edges of the pulse function were not a good approximation to the circuit's behavior. Therefore, this source was replaced with a piecewise linear voltage source where the experimental data was exported to a file and used to create the data points for the simulation.

The simulation results are shown in Fig. 5. The pink waveform, V_D, is the voltage transient measured across the freewheeling diode during the 13.5 ft, unshielded cable test of the GaN device. The simulated result (V_{LD} (SIM)), shown in blue, is in close agreement with the experimental waveform (V_{LD} (EXP)), shown in green.

IV. EXPERIMENTAL RESULTS

The experimental test setup is show in Fig 6. The voltages at the terminal of diode (V_D) and at the load inductor (V_{LD}) were measured using a Tektronix MSO5204 and two differential probes (THDP0200). The data points were exported to a file so that the results could be evaluated using

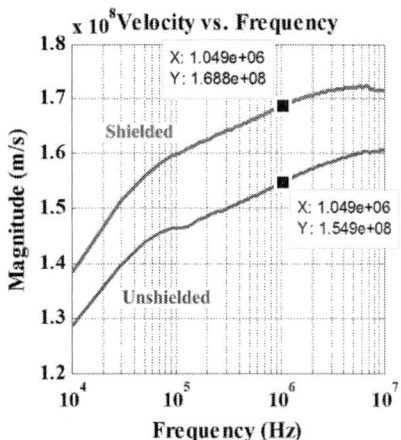

Fig. 4. Estimated voltage waveform velocitys for a shielded and unshielded cable.

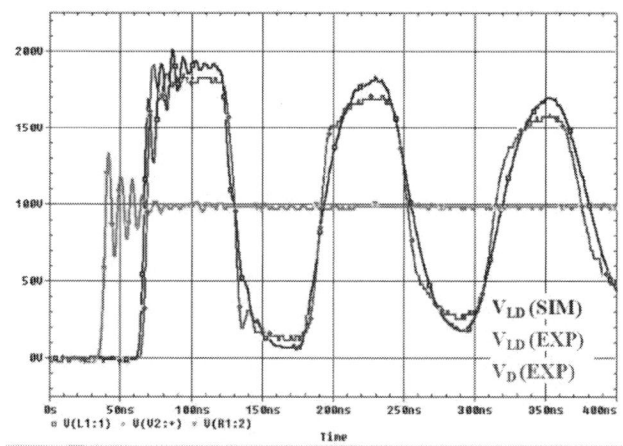

Fig. 5. PSPICE simulation results compared with experimental results obtained for the 13.5 ft. unshielded cable test (second series of tests).

978-1-4799-5494-0/14 $31.00 © 2014 IEEE

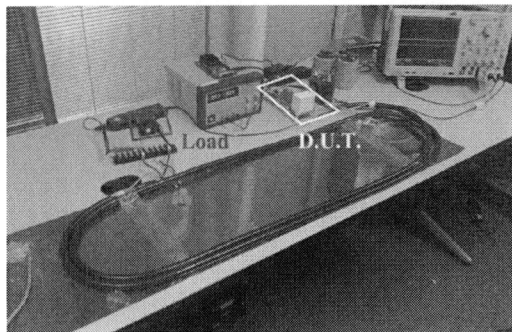

Fig. 7. Experimental test setup showing the 13.5 ft. shielded cable resting on top of the ground plane.

Matlab and PSPICE. Three different devices were tested, a 600 V GaN transistor, a 600 V Si MOSFET - CoolMOS, and a 650 V SiC MOSFET. Four different cables were used during the experiments. Two were 4.5 ft long, one was shielded and the other was not. The other two were 13.5 ft long; again one was shielded and the other was not. Measurements were conducted with the cable resting on top of the bare table top and with the cable placed on a ground plane (as shown in Fig. 6). The copper sheet was connected to a copper rod cemented into the build floor. For the experiments involving the shielded cable, the shield was connected to the ground plane. The DC bus voltage was set to 100 V for each of the following tests.

Each device had its own PCB for these tests with nearly identical layout. A single channel gate driver from Texas Instruments (UCC27511) was used to drive the devices. The gate voltages were 10 V, 15 V, and 20 V, for the GaN, Si, and SiC devices, respectively. A turn on resistor of 10 Ω and a turn off resistor of 2.4 Ω were used for all of the devices.

The results are shown in appendix. It can be observed that the rise time for the GaN device is about 20 % faster than the Si device. Both of these devices are nearly four times faster than the SiC device. In all cases, the GaN device had the largest overshoot and the SiC device had the lowest overshoot. This can be attributed to the switching speeds.

The impact of cable length can also be seen. At 13.5 ft, the delay times are well over 20 ns with the shielded cable being faster than the unshielded cable in every test. Validating the conclusion made with the test results from the impedance analyzer. The delay times are also close to the predictions given in Table I.

For the SiC test, it can be seen that the rise time at the load is twice as fast as the rise time at the diode; confirming (7). This device best illustrates the impact of switching speed on the magnitude of the reflected wave. In both tests, the peak voltage on the load was less than twice the voltage on the free-wheeling diode. Additionally, the peak voltage was smaller for the 4.5 ft. test as compared to the 13.5 ft. test.

To explore this fact further, a second series of tests were performed with the GaN and Si devices. In this case, the gate voltages were fixed at 10 V. The impact of switching speed can clearly be seen by looking at Fig. 7. The results for the 4.5 ft tests are shown in Fig. 7a. The peak overshoot for the Si device is about 50 %, while the GaN device is roughly 100 %. Contrast this with Fig. 7b, the results for the 13.5 ft tests. In this experiment both devices yield nearly 100 % overshoot. The difference can be attributed to the slower switching speed of the Si device, which was 22.4 ns for the 4.5 ft. shielded cable, and 22.8 ns for the 13.5 ft. shielded cable. Such switching speeds are similar to those of Si IGBTs. Therefore, it can be expected that the reflected wave phenomenon will present a greater challenge when migrating from IGBTs to WBG switching devices.

Finally, it can be seen that the presence of the ground plane had a minimal impact on the results during the testing. For the unshielded cable, in more than one instance, either the peak overshoot, or the delay time, or both were the same for corresponding tests, with and with the ground plane.

V. CONCLUSION AND FUTURE WORK

The following work explored the reflected wave phenomenon for motor drive systems based on WBG power devices. Three different devices were evaluated in this work; a GaN transistor, a Si MOSFET, and a SiC MOSFET. The

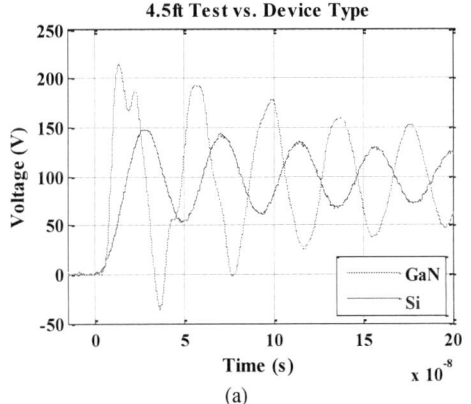
(a)

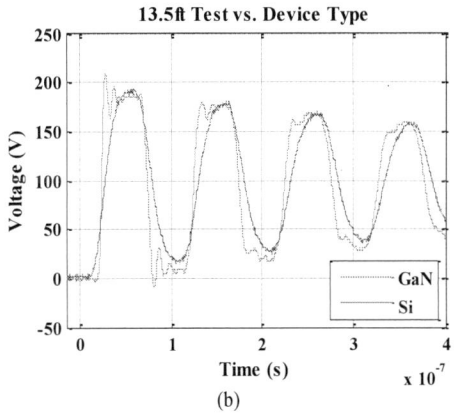
(b)

Fig. 6. Load Waveforms (V_{LD}) for GaN and Si during (a) 4.5 ft. and (b) 13.5 ft unshielded cable test

overshoot at the load was most severe with the GaN device due to its higher switching speed. The SiC device experience the slowest turn on time and best illustrated the impact of switching speed on the magnitude of the reflected wave. This test was repeated with the turn on speed of Si deliberately slowed down. For the 13.5 ft test, nearly 100% overshoot was seen. However, the 4.5 ft. test was closer to 50 %.

Additionally, a PSPICE model was developed using the distributed parameters measured with an impedance analyzer. Good agreement was shown between the experimental results and simulated results. The developed model is intended for future studies of EMI produced from WBG based motor drive system. The results of this effort will appear in future publications.

VI. APPENDIX A – EXPERIMENTAL DATA

TABLE II.   GAN – 4.5 FT. TEST RESULTS

Parameters	Unshielded Floating	Shielded Floating	Unshielded Ground Plane	Shielded Ground Plane
V_D Peak	122.8 V	122 V	122 V	122.8 V
V_L Peak	255.6 V	233.2 V	255.6 V	228.4 V
Delay	9.4 ns	7.4 ns	9.4 ns	8 ns
V_D rise time	3.4 ns	3.4 ns	3.6 ns	3.6 ns
V_{LD} rise time	3.6 ns	3.6 ns	3.4 ns	3.6 ns

TABLE III.   SI – 4.5 FT. TEST RESULTS

Parameters	Unshielded Floating	Shielded Floating	Unshielded Ground Plane	Shielded Ground Plane
V_D Peak	102 V	100.4 V	100.4 V	99.6 V
V_L Peak	191.6 V	193.2 V	190 V	193.2 V
Delay	8.8 ns	6.6 ns	8.8 ns	7 ns
V_D rise time	4.8 ns	3.8 ns	4.4 ns	3.8 ns
V_{LD} rise time	4.2 ns	4.4 ns	4 ns	4 ns

TABLE IV.   SIC – 4.5 FT. TEST RESULTS

Parameters	Unshielded Floating	Shielded Floating	Unshielded Ground Plane	Shielded Ground Plane
V_D Peak	102.2 V	103.2 V	96.4 V	94.8 ns
V_L Peak	179.2 V	174.8 V	178 V	176.4 ns
Delay	10.6 ns	8.8 ns	12 ns	11.4 ns
V_D rise time	14.2 ns	15.4 ns	15.8 ns	13 ns
V_{LD} rise time	7.2 ns	7.4 ns	7.2 ns	7.2 ns

TABLE V.   GAN – 13.5 FT. TEST RESULTS

Parameters	Unshielded Floating	Shielded Floating	Unshielded Ground Plane	Shielded Ground Plane
V_D Peak	121.6 V	120 ns	120.8 ns	120.8 ns
V_L Peak	231.6 V	239.6 ns	231.6 ns	241.2 ns
Delay	29.6 ns	23 ns	27.2 ns	22.6 ns
V_D rise time	3.6 ns	3.6 ns	3.6 ns	3.4 ns
V_{LD} rise time	3.4 ns	2.8 ns	3.6 ns	3.0 ns

TABLE VI.   SI – 13.5 FT. TEST RESULTS

Parameters	Unshielded Floating	Shielded Floating	Unshielded Ground Plane	Shielded Ground Plane
V_D Peak	103.6 V	106 V	103.6 V	106.8 V
V_L Peak	191.6 V	194.8 V	191.6 V	194.8 V
Delay	27.8 ns	22 ns	28.2 ns	22.6 ns
V_D rise time	4.4 ns	4.6 ns	4.4 ns	5 ns
V_{LD} rise time	4.4 ns	4.4 ns	4.4 ns	4 ns

TABLE VII.   SIC – 4.5 FT. TEST RESULTS

Parameters	Unshielded Floating	Shielded Floating	Unshielded Ground Plane	Shielded Ground Plane
V_D Peak	101.2 V	106.8 V	110.4 V	114 V
V_L Peak	189.6 V	189.2 V	187.6 V	189.2 V
Delay	27.4 ns	23.8 ns	29.6 ns	24.4 ns
V_D rise time	18.4 ns	16.2 ns	15 ns	14.2 ns
V_{LD} rise time	7.4 ns	7.4 ns	7.2 ns	7.4 ns

REFERENCES

[1] L. M. Tolbert, et al., "Power electronics for distributed energy systems and transmission and distribution applications," Oak Ridge National Laboratory, Oak Ridge, TN, ORNL/TM-2005/230, Dec. 2005.

[2] K. Boutros, R. Chu, and B. Hughes, "Gan power electronics for automotive application," in *Energytech, 2012 IEEE*, May 2012, pp. 1–4.

[3] K. Shirabe, M. Swamy, J. K. Kang, M. Hisatsune, Y. Wu, D. Kebort and J. Honea, "Advantages of high frequency PWM in AC motor drive applications," *IEEE Energy Conversion Congr. and Expo. (ECCE), 2012 IEEE*, Raleigh, NC, Sept. 15-20, 2012, pp. 2977-2984.

[4] T. Morita, S. Tamura, Y. Anda, M. Ishida, Y. Uemoto, T. Ueda, T. Tanaka and D. Ueda, "99.3% Efficiency of three-phase inverter for motor drive using GaN-based Gate Injection Transistors," *26th Annu. IEEE Applied Power Electronics Conf. and Expo. (APEC), 2011*, Fort Worth, TX, March 6-11, 2011, pp. 481-484.

[5] B. Hughes, Y. Y. Yeong, D. M. Zehnder, and K. S. Boutros, "A 95% efficient normally-off GaN-on-Si HEMT Hybrid-IC boost-converter with 425W output power at I MHz," in *2011 IEEE Compound Semiconductor Integrated Circuit Symp.*, Waikolowa, HI, Oct. 16-19, 2011, pp. 154-156.

[6] X. Gong and J.A. Ferreira, "Comparison and Reduction of Conducted EMI in SiC JFET and Si IGBT-Based Motor Drives," *IEEE Trans. Power Electron.*, vol. 29, no. 4, pp. 1757-1767, April 2014.

[7] M. J. Scott, et al., "Merits of gallium nitride power conversion," in *Semiconductor Science and Technology*, vol. 28, no. 7, pp. 1-10, June 2013.

[8] L. A. Saunders, et al., "Riding the reflected wave – IGBT drive technology demands new motor and cable considerations", *IEEE Ind. Appl. Soc. 43rd Annu. Petroleum and Chemical Ind. Conf.*, Philadelphia, PA, Sept. 23-25, pp. 75-88.

[9] A. von Jouanne, et al., "Application Issues for PWM adjustable speed AC motor drives", *IEEE Ind. Appl. Mag.*, pp. 10-18, Sept/Oct. 1996.

[10] J. M. Bentley, and P. J. Link, "Evaluation of Motor Power Cables for PWM AC Drives," in *IEEE Trans. Ind. Appl.*, vol. 33, no. 2, pp. 342-358, Mar./Apr. 1997.

[11] E. Persson, "Transient Effects in Application of PWM Inverters to Induction Motors," in *IEEE Trans. Ind. Appl.*, vol. 28, no. 5, pp. 1095-1101, Sept./Oct. 1992

Driving and Characterization of Wide Bandgap Semiconductors for Voltage Source Converter Applications

Zheyu Zhang and Fred Wang
University of Tennessee, Knoxville

October 13, 2014
Knoxville, TN

Outline

I. **Gate Driver for Wide Bandgap Power Semiconductors**
   - ❑ Gate Driver Fundamentals
   - ❑ Gate Driver Related Characterization of Power Semiconductors
   - ❑ Gate Driver Design Basics
   - ❑ Protections for Voltage Source Converter Applications
   - ❑ Summary and Key Message

II. **Switching Characterization of Wide Bandgap Power Semiconductors**
   - ❑ Double Pulse Test Fundamentals
   - ❑ Unique Challenges and Considerations for Wide Bandgap Power Semiconductors
   - ❑ Summary and Key Message

THE UNIVERSITY *of* TENNESSEE
KNOXVILLE

Outline

I. **Gate Driver for Wide Bandgap Power Semiconductors**

   ❑ Gate Driver Fundamentals
   - Main Functions of a Gate Driver
   - Basic Functional Blocks of a Gate Driver

   ❑ Gate Driver Related Characterization of Power Semiconductors

   ❑ Gate Driver Design Basics

   ❑ Protection for Voltage Source Converter Applications

   ❑ Summary and Key Message

THE UNIVERSITY *of* TENNESSEE
KNOXVILLE

Fundamentals of a Good Gate Driver

Static requirement

❑ **Keep the switch in ON state**
- Minimize ON state voltage and corresponding conduction losses

❑ **Safely keep the switch in OFF state**
- Minimize leakage current
- Prevent spurious change of the switch state due to external or internal disturbances

Dynamic requirement

❑ **Drive the switch from ON to OFF and OFF to ON state, with**
- Low switching losses
- Acceptable EMI
- Low commutation over-voltage

Protection (Advanced function)

❑ **Protect the switch in case of any hazardous situation**
- Over-current
- Over-voltage
- Over-temperature

THE UNIVERSITY *of* TENNESSEE
KNOXVILLE

Basic Functional Blocks of a Gate Driver

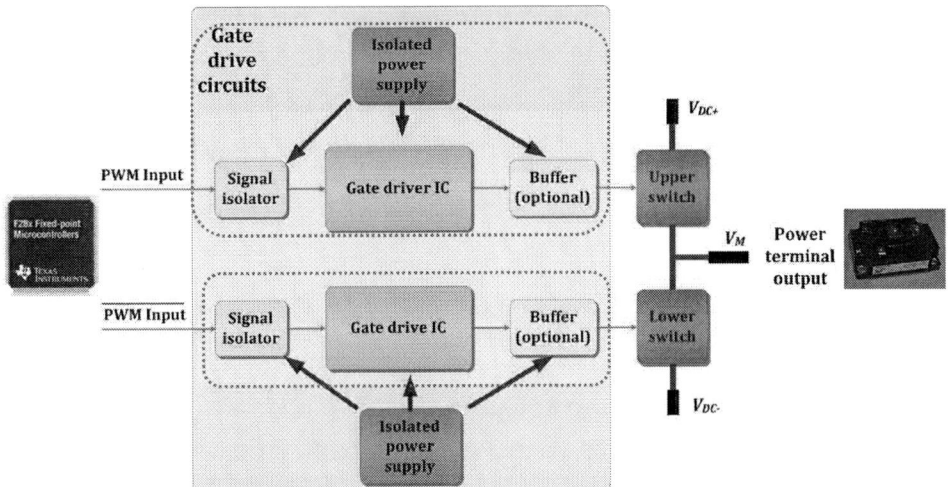

❑ **Function of each gate driver block**
- Gate driver IC & Buffer: switch the power device with sufficient driving capability
- Signal Isolator: provides galvanic isolation between the control loop and power loop
- Isolated Power Supply: power secondary side of isolator, gate driver IC & buffer

THE UNIVERSITY of TENNESSEE
KNOXVILLE

Outline

I. Gate Driver for Wide Band-gap Power Semiconductors

❑ Gate Driver Fundamentals

❑ Gate Driver Related Characterization of Power Semiconductors

- Static Characteristics

- Dynamic Characteristics

❑ Gate Driver Design Basics

❑ Protection for Voltage Source Converter Applications

❑ Summary and Key Message

THE UNIVERSITY of TENNESSEE
KNOXVILLE

Static Characteristics

❑ **Output characteristics**
  ▪ Minimize the on-state voltage drop and conduction loss

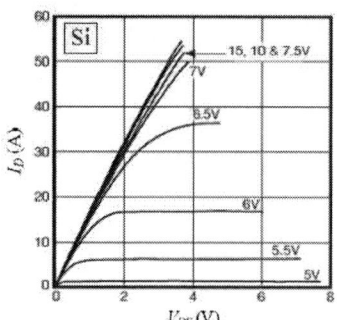

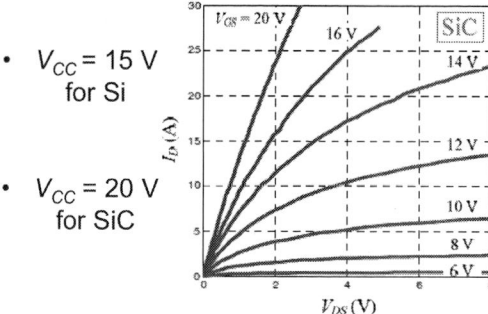

- $V_{CC} = 15$ V for Si

- $V_{CC} = 20$ V for SiC

❑ **Gate voltage maximum ratings**
  ▪ Control gate-source voltage within the required range

Type	Manufacturer	Model	Gate voltage maximum ratings
Si MOSFET	Microsemi	APT34M120J	+30 V / -30 V
SiC MOSFET	CREE	C2M0080120D	+25 V / -10 V

CURENT

THE UNIVERSITY of TENNESSEE
KNOXVILLE

Switching Commutation — Load Current Flows In

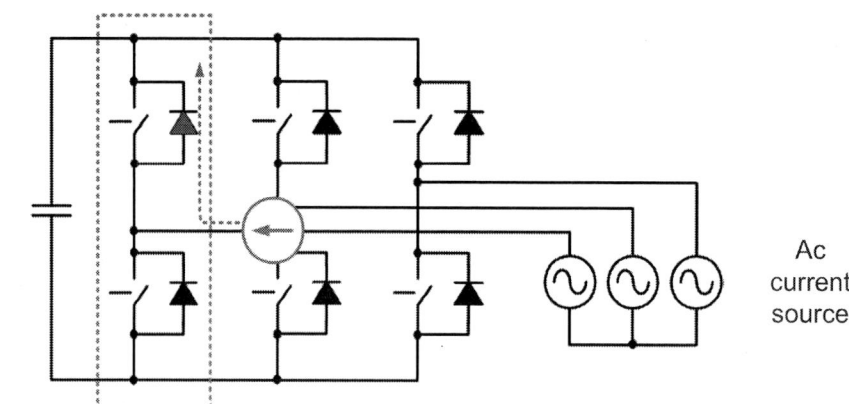

Ac current source

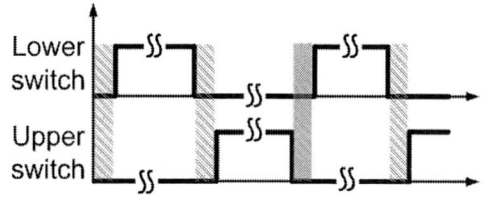

❑ If current flows into middle point
  ▪ Commutation between lower switch & upper diode
  ▪ Upper switch operates as a synchronous switch

CURENT

THE UNIVERSITY of TENNESSEE
KNOXVILLE

Switching Commutation — Load Current Flows Out

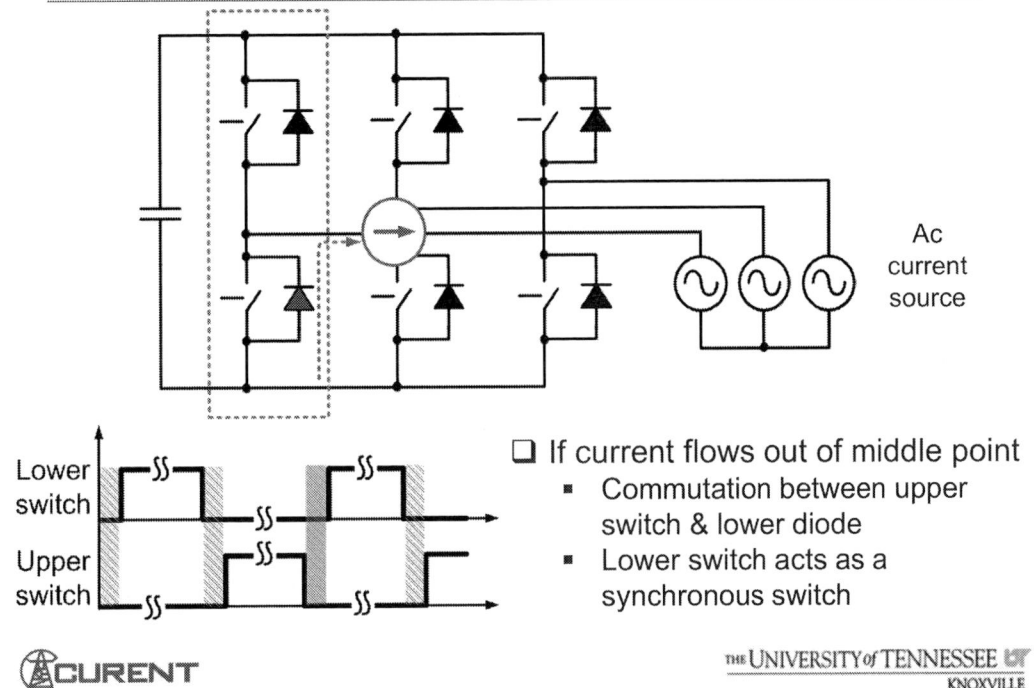

❑ If current flows out of middle point
- Commutation between upper switch & lower diode
- Lower switch acts as a synchronous switch

Equivalent Circuit of Switching Commutation

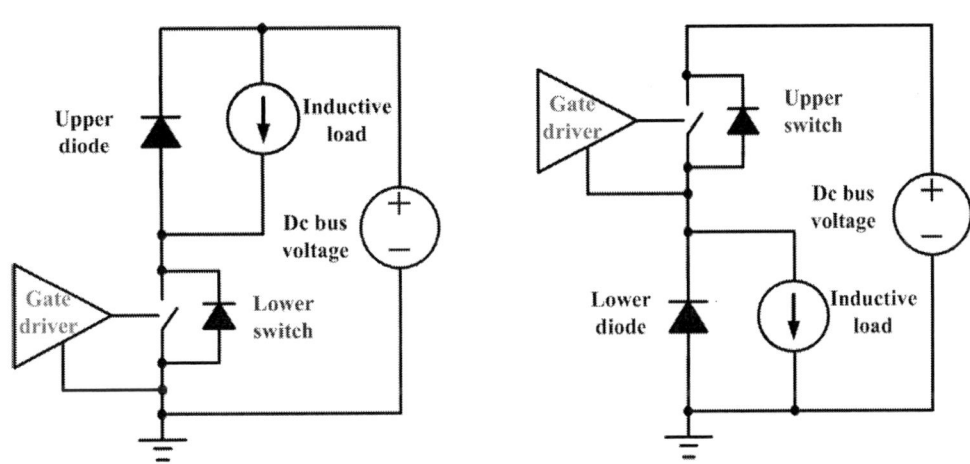

Load current flows into the middle point   Load current flows out of the middle point

❑ **Switching performance is impacted by**
- Power semiconductors
- Gate driver
- Operating Conditions

978-1-4799-5494-0/14 $31.00 © 2014 IEEE   173

Dynamic Characteristics

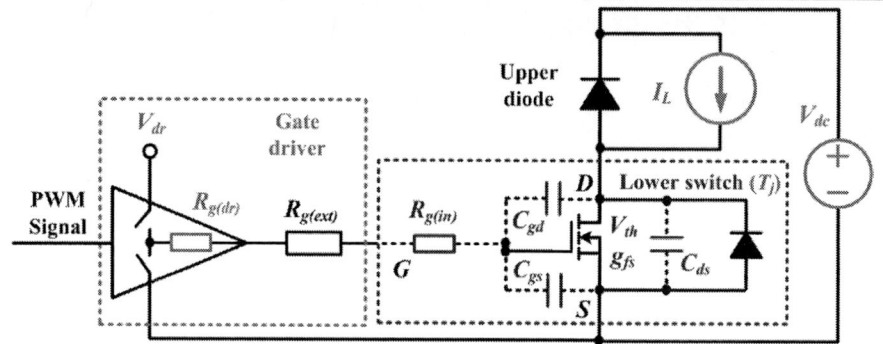

❏ **Power semiconductors**
- C_{rss}: Miller capacitance (i.e., C_{gd})
- C_{iss}: input capacitance (i.e., sum of C_{gd} & C_{gs})
- C_{oss}: output capacitance (i.e., sum of C_{gd} & C_{ds})
- $R_{g(in)}$: internal gate resistance of device
- V_{th}: threshold voltage
- g_{fs}: transconductance

❏ **Gate driver**
- $R_{g(ext)}$: external gate resistance
- $R_{g(dr)}$: internal resistance of gate driver
- V_{dr}: output voltage of gate driver

❏ **Operating conditions**
- V_{dc}: dc bus voltage
- I_L: inductive load current
- T_j: junction temperature

Outline

I. Gate Driver for Wide Band-gap Power Semiconductors

❏ Gate Driver Fundamentals

❏ Gate Driver Related Characterization of Power Semiconductors

❏ Gate Driver Design Basics

- Gate Driver Configuration

- Gate Driver Isolations

- Gate Driver IC & Gate Resistor

- Case Study

❏ Protection for Voltage Source Converter Applications

❏ Summary and Key Message

Gate Driver Configuration

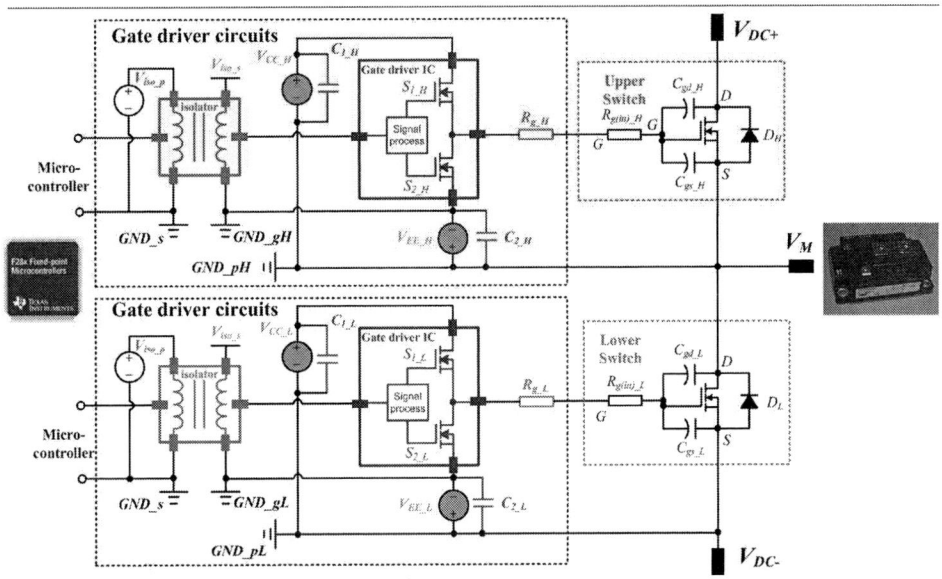

❏ **Gate driver mainly consists of**
- Signal isolator
- Isolated PS
- Gate driver IC
- Gate resistor
- Decoupling cap

THE UNIVERSITY of TENNESSEE
KNOXVILLE

Signal Isolator

❏ Isolate the ground of micro-controller and gate driver IC and safely transfer the control and error signal from the input to the output of the gate driver

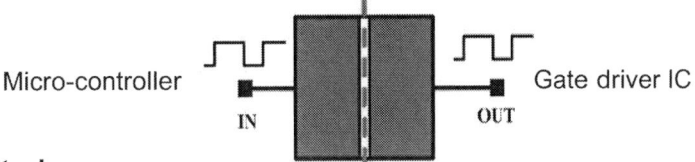

Selection criteria:

❏ Galvanic isolation capability: greater than voltage rating of power switches
❏ CM transient immunity: greater than *dv/dt* during switching transients
❏ Maximum/minimum frequencies: cover required switching frequency range
❏ Propagation delay: determined by switching frequency & control accuracy

Galvanic isolation	> 1500 V	Propagation delay	< 100 ns
CM transient immunity	> 50 kV/µs	Input to output capacitance	1..10 pF
Minimum frequency	0 Hz (DC)	Maximum frequency	10 kHz…1MHz

THE UNIVERSITY of TENNESSEE
KNOXVILLE

978-1-4799-5494-0/14 $31.00 © 2014 IEEE

CM Transient Immunity for Signal Isolator

❏ **CM transient immunity: source & effects**

- Switching transitions cause high *dv/dt* across the signal isolator
- Coupling capacitances offer the parasitic paths
- *dv/dt* coupled through the parasitic paths leads isolator to lose control by inadvertently triggering a function or causing false feedback.

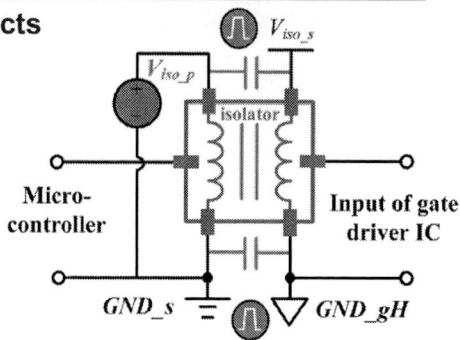

❏ **Typical *dv/dt* for wide bandgap switches**

- SiC discrete switch (CREE CMF20120D): ~ 30 kV/µs
- SiC power module (CREE CPM212000025B): ~ 80 kV/µs
- GaN transistor (Transphorm TPH3006PD): ~ 140 kV/µs

❏ **Commercial available signal isolator**

Types	Opto-coupler	Magnetically-coupler	Capacitive-coupler
CMTI	30 kV/µs	35 kV/µs	50 kV/µs

THE UNIVERSITY *of* TENNESSEE
KNOXVILLE

Isolated Power Supply

❏ Power supply is the "gas tank" of a gate driver.

- Provides necessary voltage for operation of the gate driver circuits
  - Signal isolator
- Provides necessary voltage for driving the switch into ON and OFF state
  - The output stage supply

Selection Criteria:

❏ Galvanic isolation capability: greater than voltage rating of power switches

❏ CM transient immunity: greater than dv/dt during switching transients

❏ Output power: greater than power dissipation by secondary side of signal isolator, gate driver IC, and output stage

$$P_{out} > P_{iso} + P_{gd} + P_{sw} \quad (1) \qquad P_{sw} = (V_{CC} - V_{EE}) \times Q_g f_s \quad (2)$$

P_{iso}: power dissipated by secondary side of isolator, obtained from the datasheet of the isolator
P_{gd}: power dissipated by gate driver IC, obtained from the datasheet of the gate driver IC
P_{sw}: power dissipated during switching transition, calculated by (2)
V_{CC} / V_{EE}: positive/negative output voltage of gate driver IC
Q_g: gate charge, obtained from the datasheet of the power device

❏ Output voltage: static & dynamic requirements, maximum ratings

THE UNIVERSITY *of* TENNESSEE
KNOXVILLE

Output Voltage for Isolated Power Supply

❑ Static requirement
  ▪ Minimize ON state resistance and conduction losses

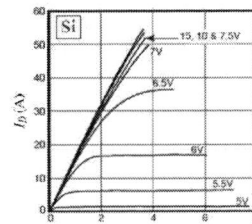

  • V_{CC} = 15 V for Si

  • V_{CC} = 20 V for SiC

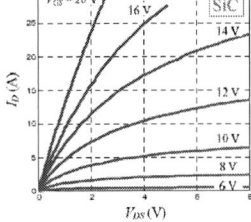

  ▪ Safely keep power switches in OFF state
    • Minimize leakage current
    • Sufficient margin to prevent spurious change of switch state due to the external or internal disturbance

❑ Dynamic requirement
  ▪ Fast turn-on and turn-off power switches
  ▪ Maintain sufficient margin to not exceed gate voltage maximum ratings of power switches

THE UNIVERSITY of TENNESSEE
KNOXVILLE

Non-Isolated Solution: Bootstrap Power Supply

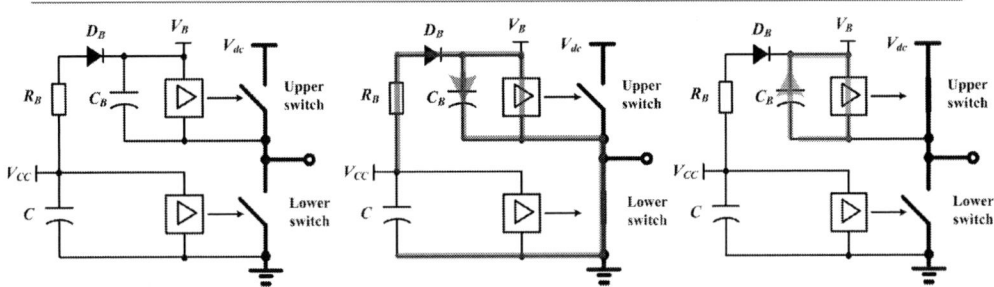

❑ **The simplest solution, used in low cost low power application**
  ▪ The low side gate driver supplied directly from *Vcc*
  ▪ The high side gate driver supplied from *Vcc* via a bootstrap diode D_B damping resistor R_B and capacitor C_B
    • When the lower switch is conducting, D_B is conducting, charging C_B and feeding the upper gate driver
    • When the upper switch is conducting, D_B is reverse polarized, high side gate driver is disconnected from V_{CC}, C_B is being discharged, feeding the upper gate driver

❑ **Not possible to provide negative gate voltage or maintain duty cycle of the upper switch close to 1**

THE UNIVERSITY of TENNESSEE
KNOXVILLE

Gate Driver IC

❑ Gate driver IC is the "engine" of a gate driver.

- Provides necessary voltage for driving the switch into ON and OFF state
- Provides sufficient current to charge and discharge the switch input capacitance
- Provides low impedance loop with quick response for fast switching

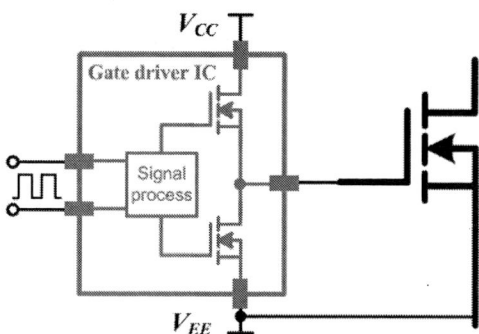

Selection criteria:

❑ Operating voltage range: greater than $V_{CC} - V_{EE}$

❑ Peak source / sink drive current: greater than $(V_{CC} - V_{EE})/R_g$

❑ Propagation delay: determined by switching frequency & control accuracy

THE UNIVERSITY of TENNESSEE
KNOXVILLE

Gate Driver IC (cont'd)

Selection criteria (cont'd):

❑ Rise / fall time of gate driver IC output voltage: shorter than switching delay time

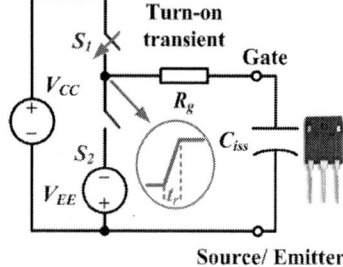

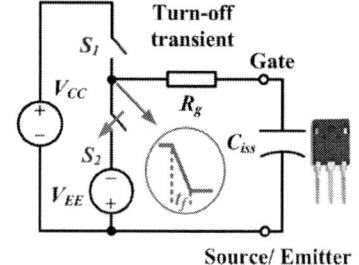

❑ On state resistance of S_1 and S_2: smaller than desired gate resistance

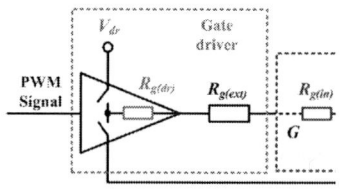

THE UNIVERSITY of TENNESSEE
KNOXVILLE

Gate Resistor

❑ Gate resistor is the "gas pedal" of a gate driver, controlling the speed of switching transients.

❑ Different turn-on & turn-off gate resistance

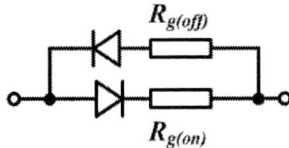

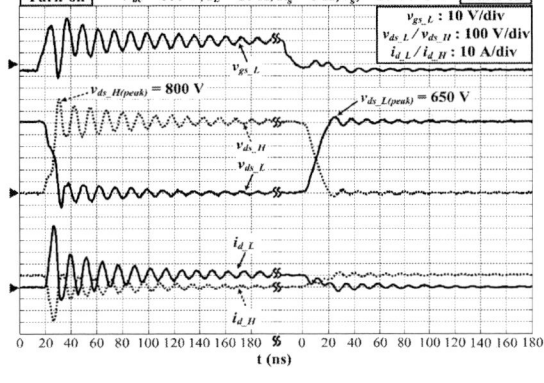

Selection criteria:
❑ Resistance
  • Datasheet
  • Analytical calculation
  • Finally, tuned by set of experiments

❑ Power rating
$$P_{Rg} = (V_{CC} - V_{EE}) \times Q_g f_s$$

THE UNIVERSITY of TENNESSEE
KNOXVILLE

Decoupling Capacitor

❑ Decoupling capacitor is the "fuel injector" of the gate driver, providing the current pulse during switching transients

Selection criteria:

❑ Voltage ratings: $V_{C1} > V_{CC} \quad V_{C2} > |V_{EE}|$

❑ Capacitance: $C_1 > Q_g/\Delta V_{CC} \quad C_2 > Q_g/\Delta V_{EE}$

❑ Types: ceramic (low equivalent series inductance)

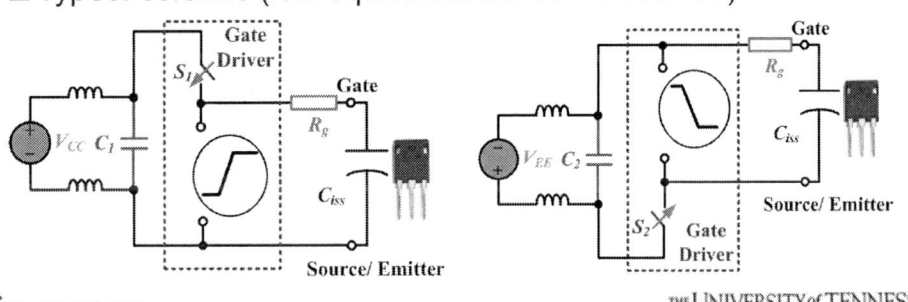

THE UNIVERSITY of TENNESSEE
KNOXVILLE

Outline

I. **Gate Driver for Wide Bandgap Power Semiconductors**

  ❏ Gate Driver Fundamentals

  ❏ Gate Driver Related Characterization of Power Semiconductors

  ❏ Gate Driver Design Basics

  ▪ Gate Driver Configuration

  ▪ Gate Driver Isolations

  ▪ Gate Driver IC & Gate Resistor

  ▪ **Case Study**

  ❏ Protection for Voltage Source Converter Applications

  ❏ Summary and Key Message

CURENT

THE UNIVERSITY of TENNESSEE
KNOXVILLE

Gate Driver Design — Case Study

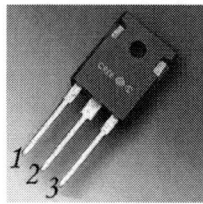

Parameters of device under evaluation

Model	Voltage rating	Current rating	$R_{ds(on)}$
C2M0080120D	1200 V	20 A @ 100°C	80 mΩ

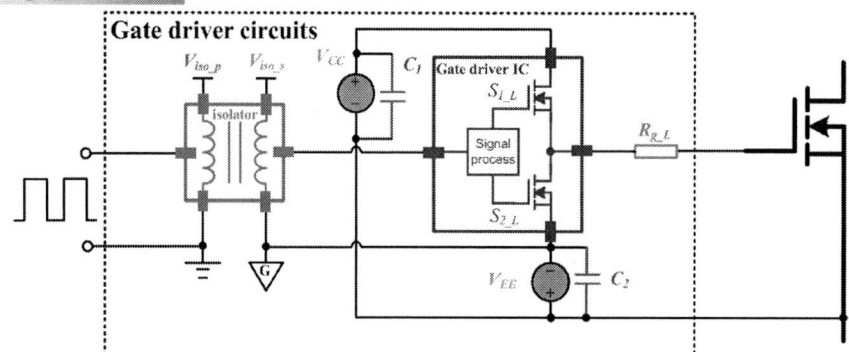

1. Signal Isolator 2. Isolated PS 3. Gate driver IC

4. Gate resistor 5. Decoupling capacitor

CURENT

THE UNIVERSITY of TENNESSEE
KNOXVILLE

Gate Driver Design — Signal Isolator

❑ **Signal isolator**
- Galvanic isolation capability: > 1200 V [based on datasheet]
- CM transient immunity: > 50 V/ns [based on datasheet]

Fall time	Rise time	Test conditions
18.4 ns	13.6 ns	V_{DD} = 800 V
dv/dt (on)	dv/dt (off)	I_D = 20 A
35 V/ns	47 V/ns	

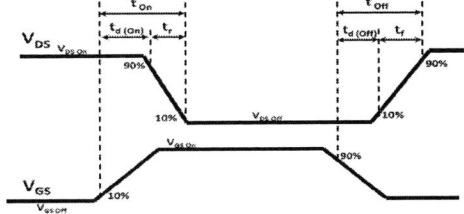

- Maximum/minimum frequencies: > 100 kHz [based on applications]
- Propagation delay: < 100 ns (1 % of minimum switching period)

❑ **Transformer based isolator ADuM 5240 is selected**

Insolation voltage	2500 Vrms	Maximum frequency	1 MHz
CM transient immunity	35 V/ns	Propagation delay	< 75 ns
Secondary side power dissipation		13 mW	

THE UNIVERSITY of TENNESSEE
KNOXVILLE

Gate Driver Design — Isolated PS

Isolated power supply
❑ Galvanic isolation capability: > 1200 V [based on datasheet]
❑ CM transient immunity: > 50 V/ns [based on datasheet]
❑ Output voltage: 20 V to -5 V [based on datasheet]
❑ Output power: > 0.159 W [based on datasheet and applications]

$$P_{out} > P_{sw} + P_{iso} + P_{gd} \ (1) \quad P_{sw} = (V_{CC} - V_{EE}) \times Q_g f_s \ (2)$$

V_{CC}	20 V	V_{EE}	-5 V	Q_g	49.2 nC	f_s	100 kHz

P_{iso}: power dissipated by secondary side of isolator	13 mW
P_{gd}: power dissipated by gate driver IC	23 mW
P_{sw}: power dissipated during switching transition, calculated by (2)	123 mW

❑ **Traco power THB 3 series DC/DC converter is selected**

Isolation voltage	3000 Vrms	Input-Output capacitance	13 pF
Power	3 W	Output voltage	24 V / 5 V

THE UNIVERSITY of TENNESSEE
KNOXVILLE

Gate Driver Design — Gate Drive IC

Gate drive IC
- ❑ Operating voltage range: > $(V_{CC} - V_{EE})$ = 30 V [based on datasheet]
- ❑ Peak source / sink drive current:
  = $(V_{CC} - V_{EE})/R_g < (V_{CC} - V_{EE})/R_{g(in)}$ = 6.5 A [based on datasheet]
- ❑ Propagation delay: < 100 ns (1 % of minimum switching period)
- ❑ Rise / fall time of the output voltage of gate driver IC:
  < min $(t_{d(on)}, t_{d(off)})$ = 12 ns [based on datasheet]
- ❑ Pull-up / pull-down resistance: < $R_{g(desire)}$ or << $R_{g(in)}$ = 4.6 Ω

$R_{g(in)}$	4.6 Ω	$t_{d(on)}$	12 ns	$t_{d(off)}$	23.2 ns	C_{iss}	950 pF

❑ IXYS IXDN_609 gate driver IC is selected

Operating voltage	4.5 V to 35 V	Peak source/sink current	9 A
Rise / fall time	7 / 5 ns (@ 1.5 nF)	Propagation delay	40 ns (@ 25 V)
Pull-up / pull-down resistance	0.5 / 0.33 Ω (@ 25 V)	Quiescent power dissipation	23 mW (@ 20 ºC)

Gate Driver Design — Gate Resistor

Gate resistor
- ❑ Resistance: based on datasheet and finally tuned by set of experiments
- ❑ Different turn-on & turn-off gate resistance
  - ▪ Using diode
  - ▪ Usually, turn-off gate resistance is smaller than turn-on
- ❑ Power rating: > 0.123 mW [based on datasheet and applications]

$$P_{Rg} = (V_{CC} - V_{EE}) \times Q_g f_s$$

V_{CC}	20 V	V_{EE}	-5 V	Q_g	49.2 nC	f_s	100 kHz

- ❑ It is preferred to use several gate resistors in parallel to tune the resistance

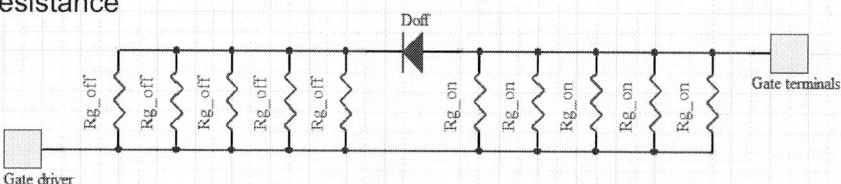

Gate Driver Design — Decoupling Capacitor

Decoupling capacitor

❑ Voltage ratings [based on datasheet]

$$V_{C1} > V_{CC} = 20\,V \qquad V_{C2} > |V_{EE}| = 5\,V$$

❑ Capacitance [based on datasheet]

$$C_1 > Q_g/\Delta V_{CC} = Q_g/(k_{GS} \times V_{CC}) = 0.246\,\mu F$$

$$C_2 > Q_g/\Delta V_{EE} = Q_g/(k_{GS} \times V_{EE}) = 0.984\,\mu F$$

V_{CC}	20 V	V_{EE}	-5 V	Q_g	49.2 nC
k_{GS}: Gate voltage ripple coefficient during switching transient					1%

❑ Types: surface mount ceramic capacitor

THE UNIVERSITY *of* TENNESSEE
KNOXVILLE

Outline

I. Gate Driver for Wide Bandgap Power Semiconductors

❑ Gate Driver Fundamentals

❑ Gate Driver Related Characterization of Power Semiconductors

❑ Gate Driver Design Basics

❑ Protection for Voltage Source Converter Applications

- Cross-Talk

- Over-Current

❑ Summary and Key Message

THE UNIVERSITY *of* TENNESSEE
KNOXVILLE

Mechanism Causing Cross-Talk (Turn-On)

Lower switch as the device under test

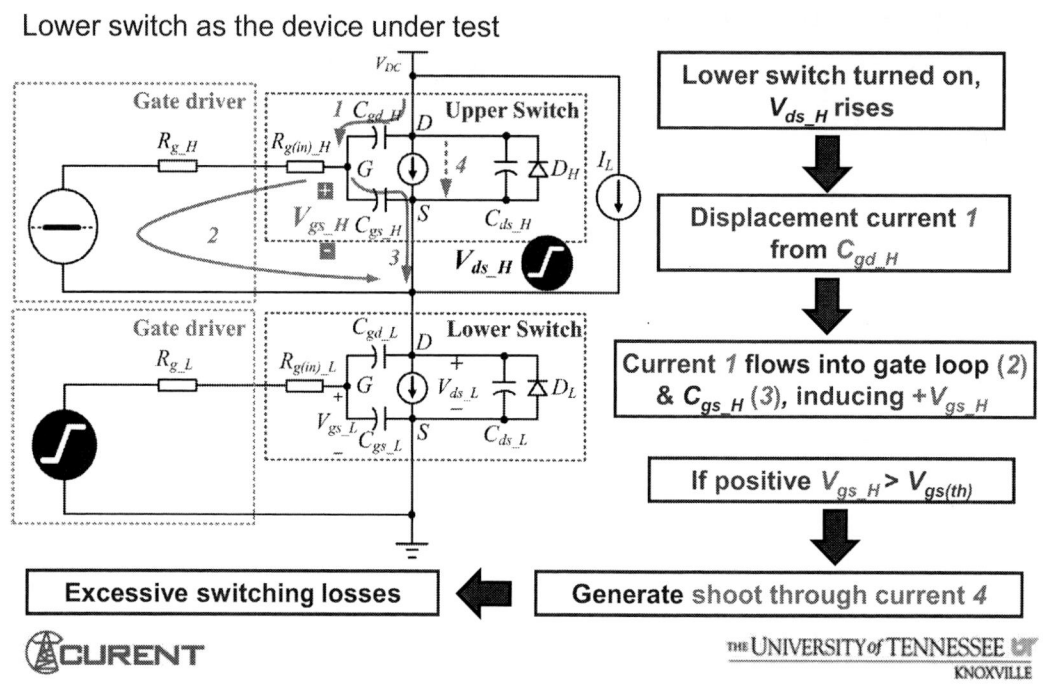

Mechanism Causing Cross-Talk (Turn-Off)

Lower switch as the device under test

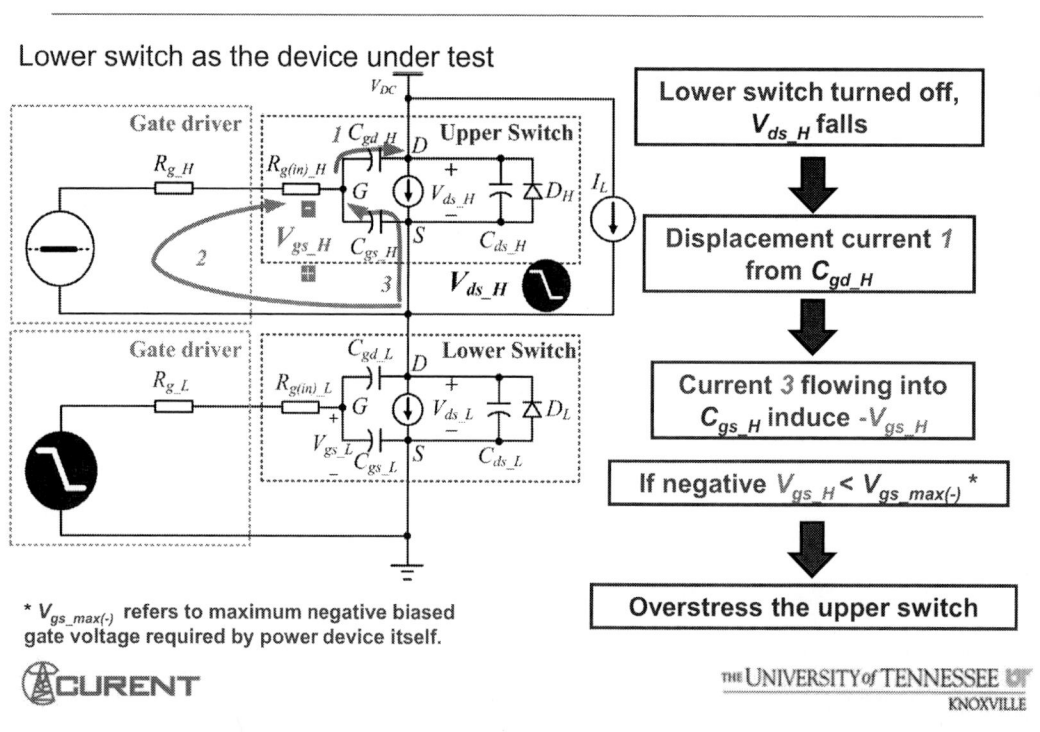

* $V_{gs\_max(-)}$ refers to maximum negative biased gate voltage required by power device itself.

Cross-Talk for WBG Switches (SiC as an Example)

Characteristics of Several Comparable Si / SiC Power Devices

Type	Manufacturer	Model	V_{DS}/I_D (100 °C)	Q_{gs}	$V_{gs(th)}$ (25 °C)	$V_{gs\_max(-)}$
Si IGBT	IR	IRGP20B120U	1200 V / 20 A	169 nC	4.5 V	-20 V
Si MOSFET	Microsemi	APT34M120J	1200 V / 22 A	560 nC	4.0 V	-30 V
SiC MOSFET	CREE	C2M0080120D	1200V / 20 A	49.2 nC	2.2 V	-10 V

❑ **Properties of high voltage SiC devices**

- Faster switching speed
- Lower threshold voltage
- Lower maximum allowable negative gate voltage

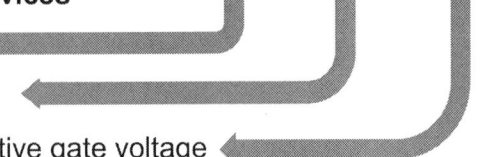

❑ **SiC devices in a phase-leg configuration are easily affected by cross-talk, leading to**

- Extra switching losses & reliability issues

THE UNIVERSITY of TENNESSEE
KNOXVILLE

Basic Ideas for Cross-Talk Suppression

❑ **To suppress cross-talk, we need to minimize spurious V_{gs_H}**

- Reduce gate loop impedance during the switching transient
  - Gate impedance regulation (GIR) assist circuit
- Pre-charge the gate-source capacitance before the switching transient
  - Gate voltage control (GVC) assist circuit

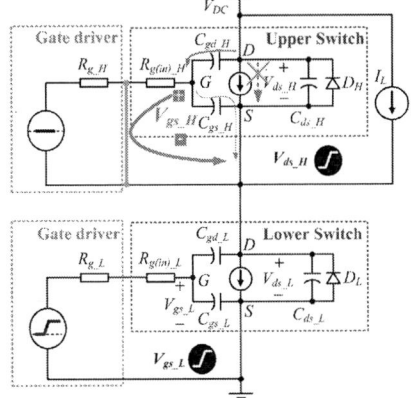 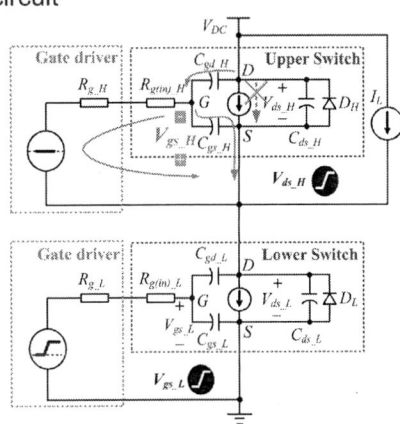

Turn-on transient of lower switch as an example

UNIVERSITY of TENNESSEE
KNOXVILLE

Gate Impedance Regulation (GIR) Assist Circuit

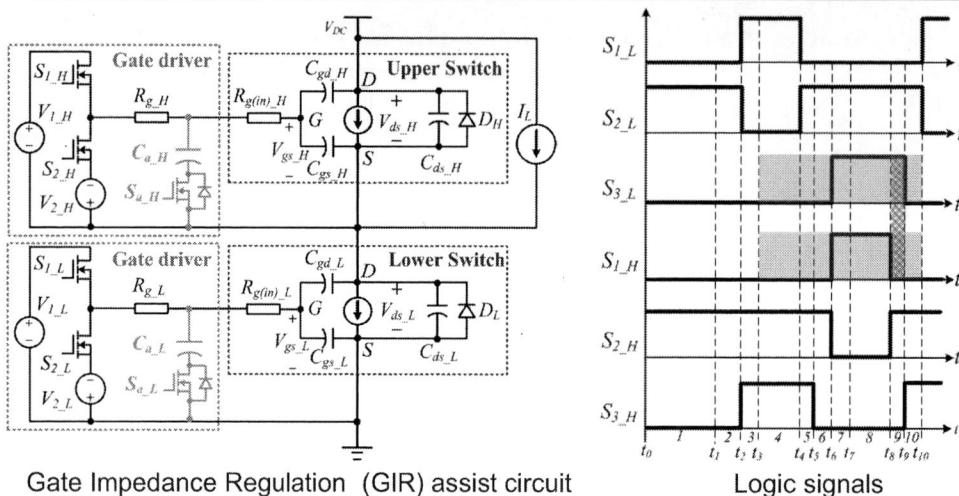

Gate Impedance Regulation (GIR) assist circuit

Logic signals

☐ **Compared with conventional gate driver, GIR assist circuit adds**

- One auxiliary transistor (S_{a_H} or S_{a_L}) in series with one capacitor (C_{a_H} or C_{a_L}) for each device in a phase-leg.

THE UNIVERSITY of TENNESSEE
KNOXVILLE

Operating Principle of GIR Circuit (Turn-On)

Lower switch as the device under test

Lower switch turned on; S_{a_L} remained off; S_{a_H} turned on

Lower switch

| C_{a_L} disconnected |

⬇

Turn-on performance of lower switch not affected

Upper switch

| C_{a_H} connected; gate impedance of the upper switch minimized |

⬅ **Cross-talk mitigated; turn-on energy loss reduced**

THE UNIVERSITY of TENNESSEE
KNOXVILLE

978-1-4799-5494-0/14 $31.00 © 2014 IEEE

Operating Principle of GIR Circuit (Turn-Off)

Lower switch as the device under test

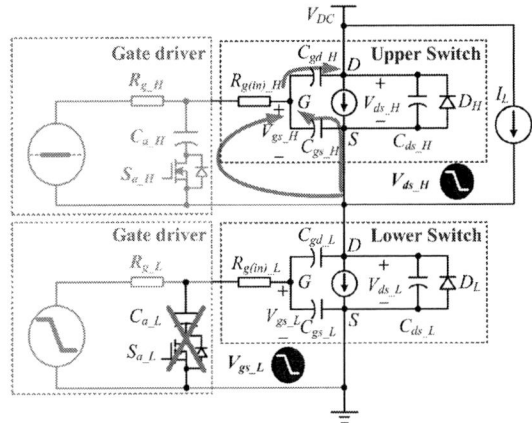

Lower switch turned off; S_{a_L} remained off; S_{a_H} remained on

Lower switch

C_{a_L} disconnected

Turn-off performance of lower switch not affected.

Upper switch

C_{a_H} connected; gate impedance of the upper switch minimized

Cross-talk mitigated; negative spurious gate voltage minimized

THE UNIVERSITY *of* TENNESSEE
KNOXVILLE

Parameter Design Criterion

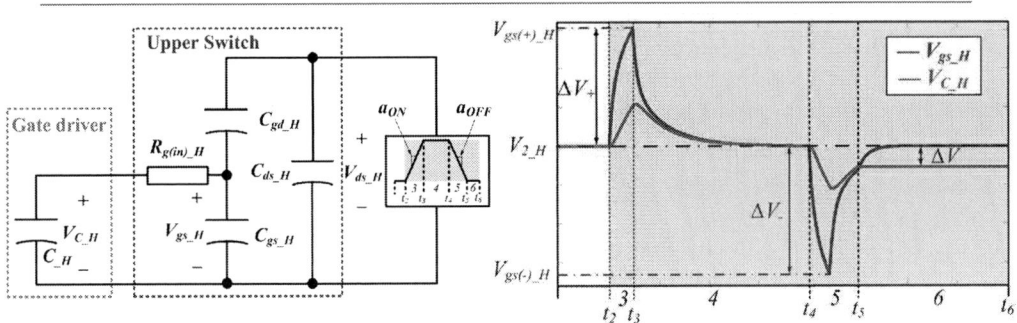

Simplified equivalent circuit of upper switch during switching transient of lower one

Spurious gate voltage & auxiliary capacitor voltage

❏ To avoid cross-talk

- Spurious gate voltage $\in (V_{gs\_max(-)}, V_{gs(th)})$

- $\Delta V_+ + |\Delta V_-| \leq V_{gs(th)} - V_{gs\_max(-)}$, where ΔV_+ & ΔV_- are related to C

- $V_{gs\_max(-)} + \Delta V_- \leq V_2 (i.e., V_{EE}) \leq V_{th} - \Delta V_+$

where ΔV_+, ΔV_-, and V_2 refer to positive spurious gate voltage, negative spurious gate voltage, and negative turn-off gate voltage, respectively. C is the auxiliary capacitor

THE UNIVERSITY *of* TENNESSEE
KNOXVILLE

GIR Assist Circuit Design — Case Study

Parameters of device under evaluation

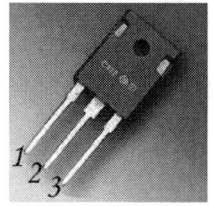

C_{gd}	13 pF	C_{gs}	1900 pF
$V_{gs(th)}$	2.5 V	$V_{gs\_max(-)}$	-5 V
V_{dc}	800 V	a_{ON}^{*}	27 V/ns
$R_{g(in)}$	5 Ω	a_{OFF}^{*}	23 V/ns

* determined according to test results.

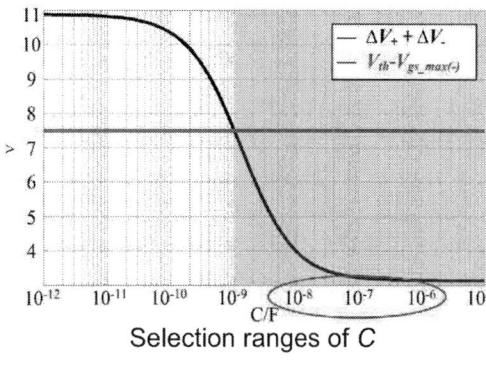

Selection ranges of C

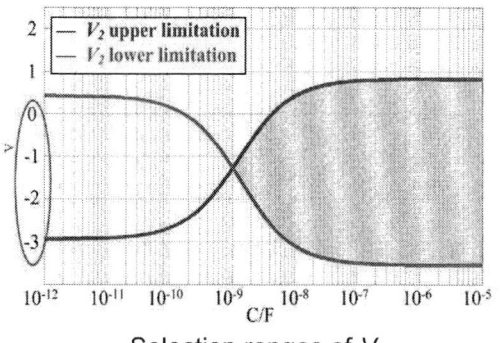

Selection ranges of V_2

CURENT

THE UNIVERSITY of TENNESSEE
KNOXVILLE

Turn-on Transient of the Lower Switch

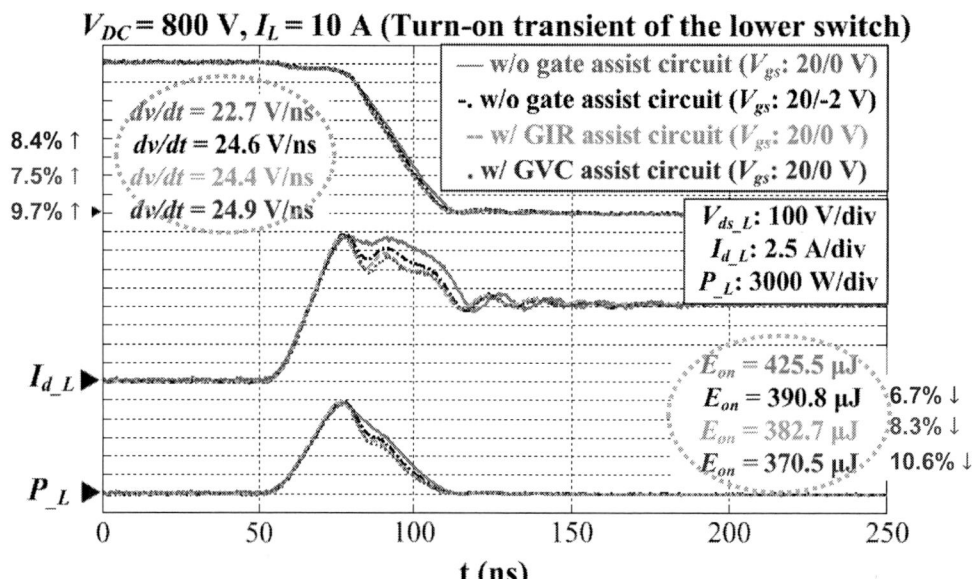

[1] Zheyu Zhang, "Active gate driver for cross talk suppression of SiC power devices in a phase-leg configuration", in *Proc. IEEE Trans. Power Electronics*, vol. 29, no. 4, 2014.

CURENT

THE UNIVERSITY of TENNESSEE
KNOXVILLE

Turn-on Transient of the Lower Switch (cont'd)

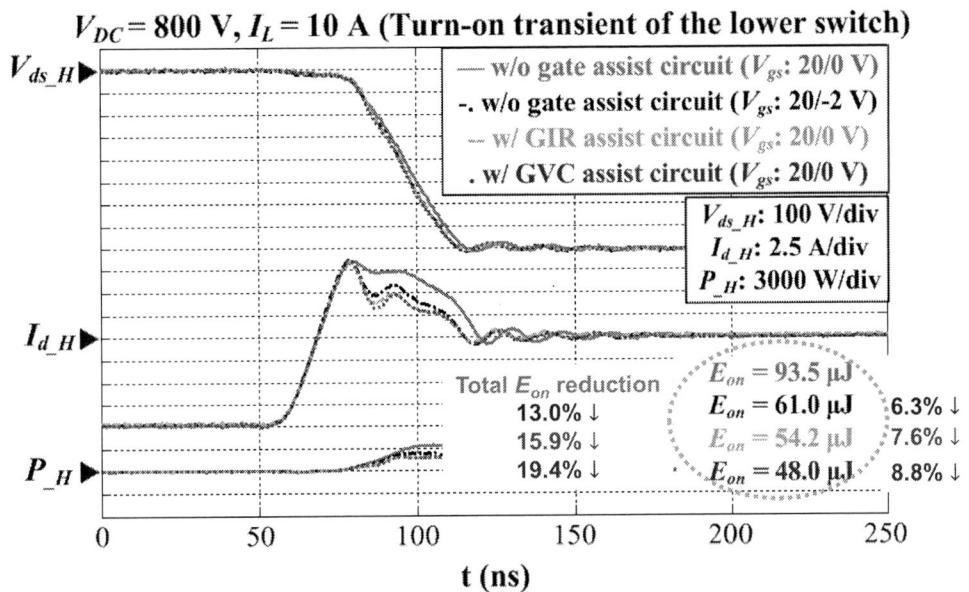

Turn-off Transient of the Upper Switch

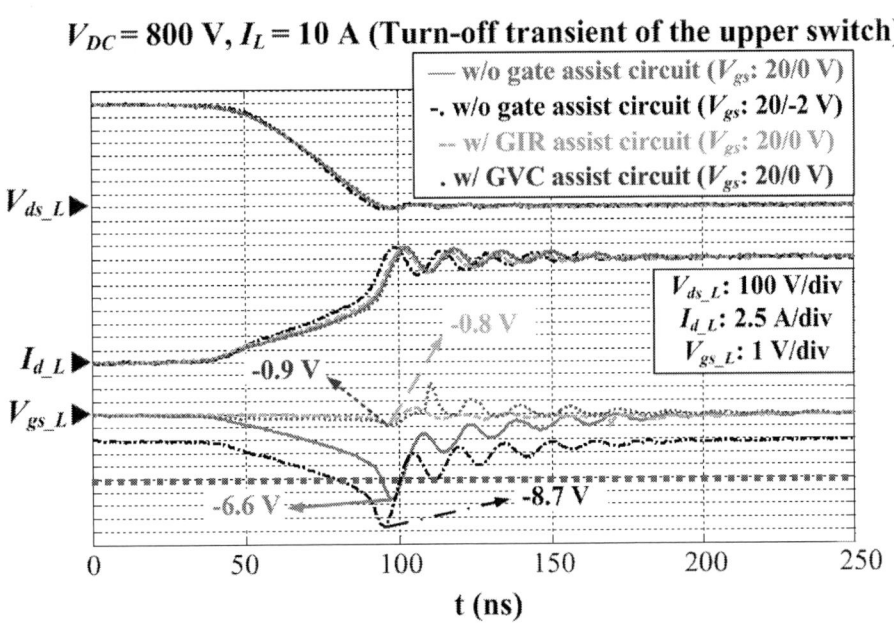

Outline

I. **Gate Driver for Wide Bandgap Power Semiconductors**

   ❏ Gate Driver Fundamentals

   ❏ Gate Driver Related Characterization of Power Semiconductors

   ❏ Gate Driver Design Basics

   ❏ Protections for Voltage Source Converter Applications

   ▪ Cross-Talk

   ▪ Over-Current

   ❏ Summary and Key Message

THE UNIVERSITY of TENNESSEE
KNOXVILLE

Short Circuit Modes & Causes

Arm Short Circuit ← *Transistor or diode destruction*

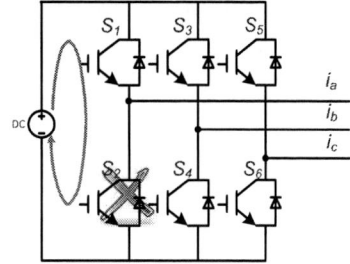

Series Arm Short Circuit ← *Faulty gate drive signal*

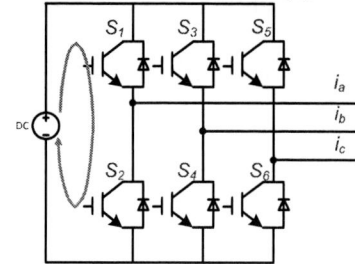

Short in Load ← *Miswiring or load short circuit*

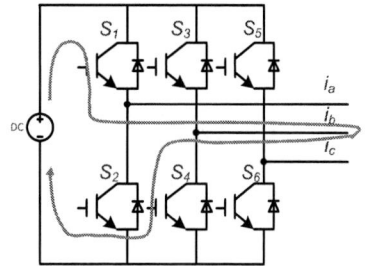

Ground Fault ← *Miswiring or dielectric breakdown*

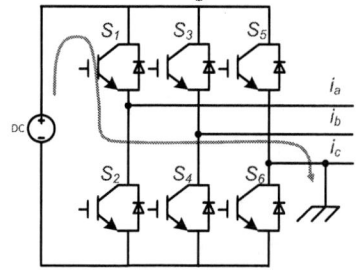

[1]:Fuji IGBT modules application manual, 2004.

THE UNIVERSITY of TENNESSEE
KNOXVILLE

Fault Type-Hard Switching Fault (HSF)

❑ HSF--Short circuit fault at turn-on switching transient

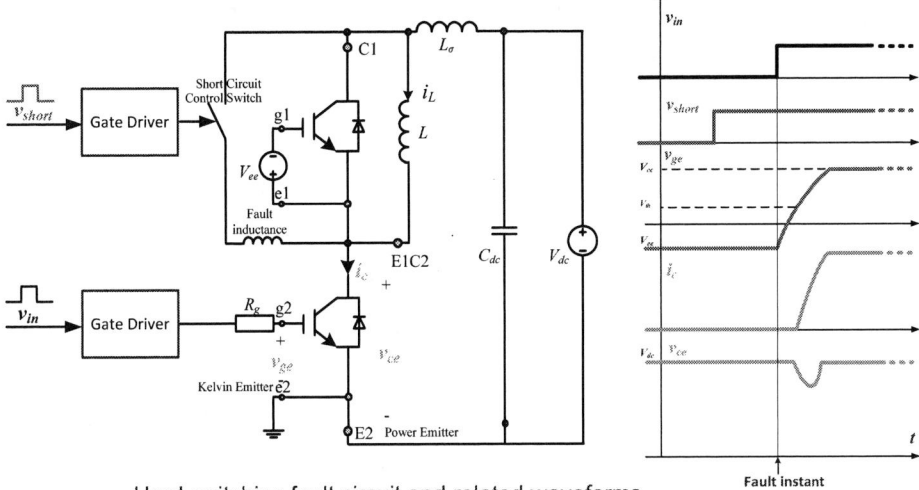

Hard switching fault circuit and related waveforms

Fault Type-Fault Under Load (FUL)

❑ FUL--Short circuit fault during the on-state condition

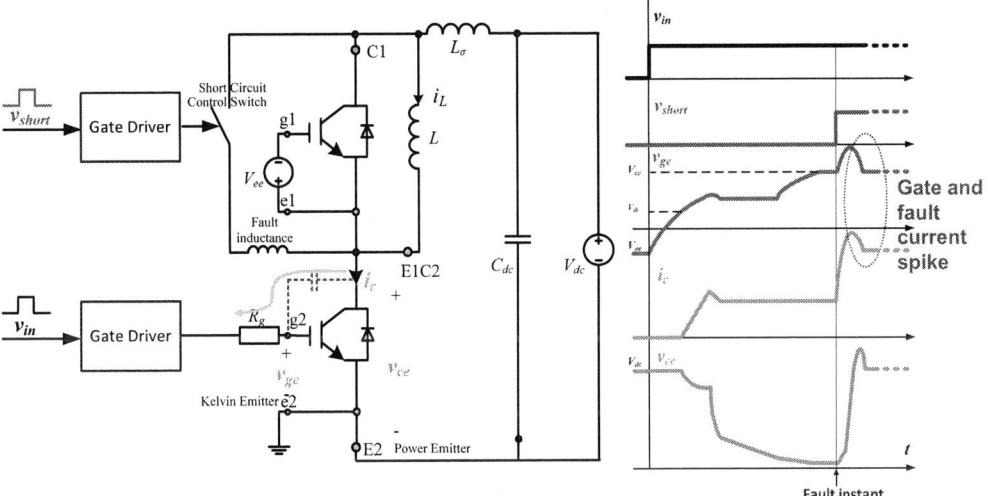

Fault under load circuit and related waveforms

Fault Type - Example

❑ A protection circuit (e.g., desaturation protection) should be activated within the short circuit withstand time t_{sc}

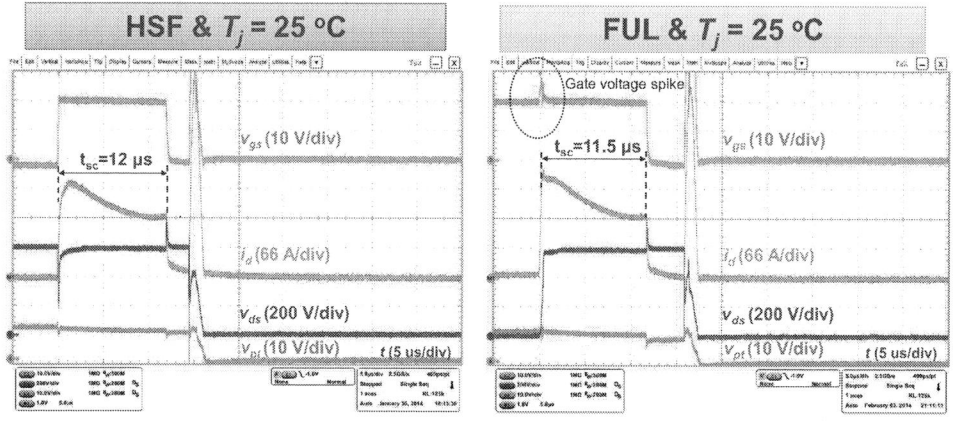

Test Condition: V_{gs} = +20/-2V, V_{dc} = 600 V, CREE 1G SiC MOSFETs (1200 V / 24 A)

THE UNIVERSITY of TENNESSEE
KNOXVILLE

IGBT Desaturation Protection

❑ Under steady-state conduction, IGBT operates in saturation region

❑ Under short circuit condition, IGBT operation point moves from saturation region to active region, so-called "Desaturation" or "Desat"

$$I_c \uparrow \quad \rightarrow \quad V_{ce} \uparrow$$

❑ A short circuit protection can be triggered by the increased collector-emitter voltage Vce

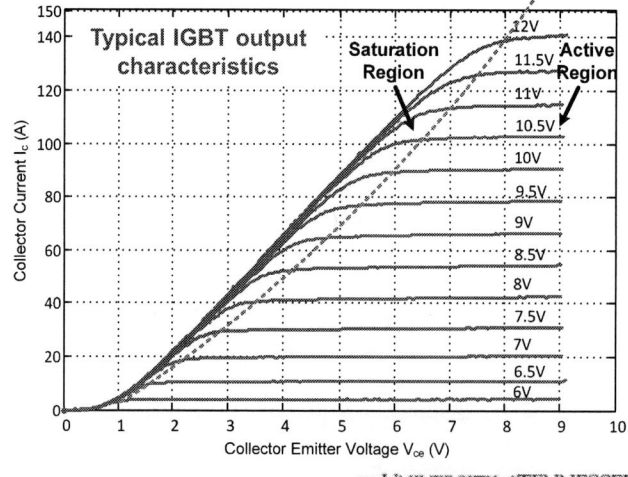

THE UNIVERSITY of TENNESSEE
KNOXVILLE

Desaturation Protection for SiC MOSFET

Challenges :

❑ Commercial IGBT/MOSFET gate drivers usually have slow fault response time (>3 μs) versus short short-circuit withstand time (SCWT) of SiC devices

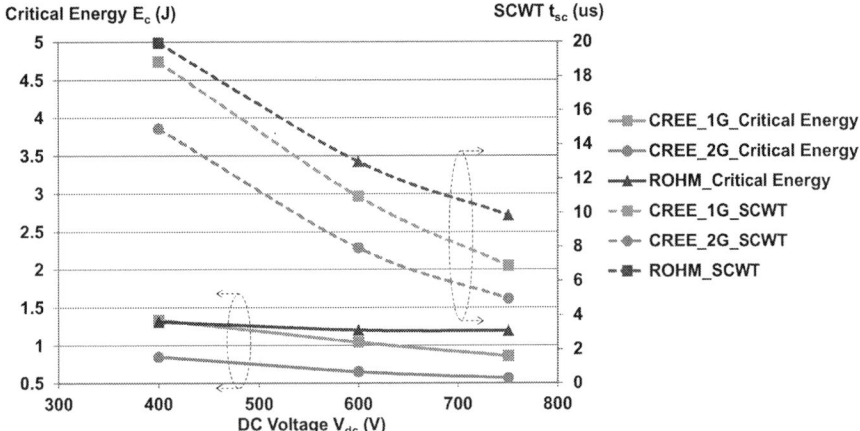

Desaturation Protection for SiC MOSFET

Challenges (cont'd):

❑ Protection threshold for SiC MOSFETs is not straightforward due to unclearly defined active region

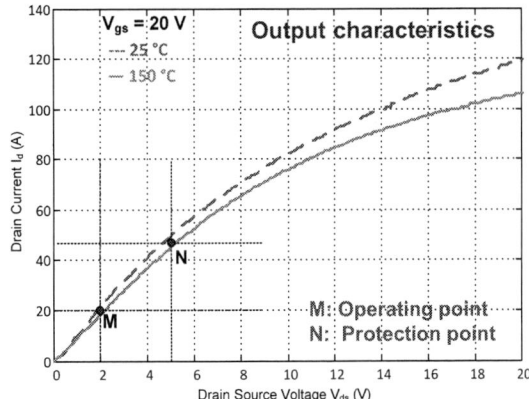

Blanking Time vs. Noise Immunity

Challenges (cont'd):

❑ The noise immunity and fault response time become sharp contradictions

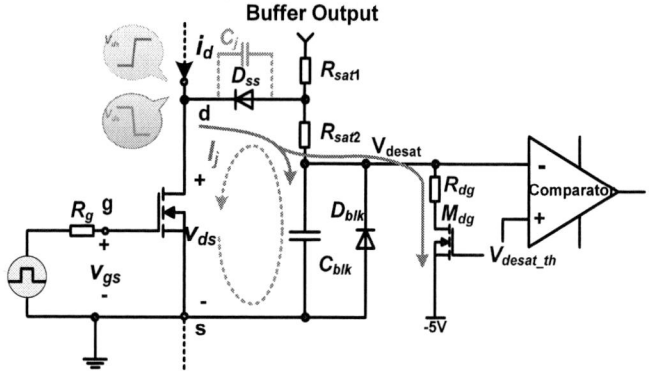

❑ Protection circuit can be falsely triggered due to high *dv/dt* during turn-on/turn-off transients

❑ Large C_{blk}/C_j, large R_{sat2} could more effectively suppress the impact of *dv/dt* on V_{desat}, while blanking time will increase

 THE UNIVERSITY of TENNESSEE
KNOXVILLE

Blanking Time Setting

Solution

❑ Reduction of fault response time with acceptable noise immunity capability so that fault response time << short-circuit withstand time

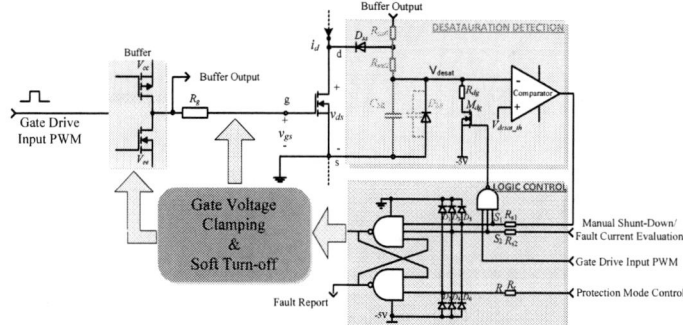

❑ Blanking time is determined by the RC network (R_{sat1}, R_{sat2}, and C_{blk}) :

- The blanking time is: $t_{blk} = \tau \ln \dfrac{V_{cc}}{V_{cc} - V_{desat\_th}}, \quad \tau = (R_{sat1} + R_{sat2}) \cdot C_{blk}$

- Blanking time $t_{blanking}$ > Turn-on switching time t_{on}

➡ $t_{blanking}$ =100~200 ns is preferable

 THE UNIVERSITY of TENNESSEE
KNOXVILLE

Desaturation Technique – Testing Results

V_{dc} = 750 V, 200 °C, V_{gs} = +20/-2V, Blanking time:100 ns, CREE 2G SiC MOSFET

Hard Switching Fault	**Fault Under Load**

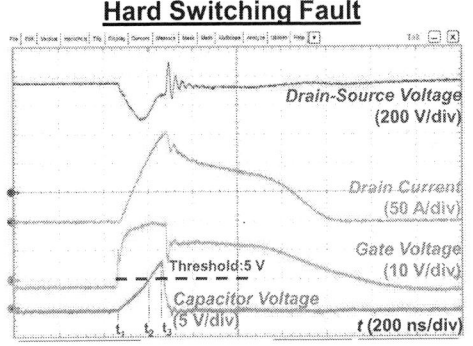 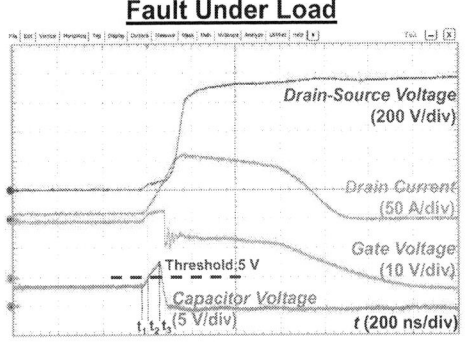

☐ Total response time: 195 ns
- Blanking time delay ($t_1 \sim t_2$): 130 ns
- Comparator response delay ($t_2 \sim t_3$): 65 ns

☐ Total response time: 85 ns
- Blanking time delay ($t_1 \sim t_2$): 20 ns
- Comparator response delay ($t_2 \sim t_3$): 65 ns

Outline

I. Gate Driver for Wide Bandgap Power Semiconductors

☐ Gate Driver Fundamentals

☐ Gate Driver Related Characterization of Power Semiconductors

☐ Gate Driver Design Basics

☐ Protections for Voltage Source Converter Applications

- Cross-Talk

- Over-current

☐ Summary and Key Message

Summary & Key Message

❑ **No power conversion without power semiconductors**

❑ **Power semiconductors is NOTHING without a gate driver!**

- **The gate driver will properly drive a power semiconductor and bring the maximum performance. For WBG devices,**

  - Driving capability of gate driver IC (rise/fall time, pull-up/pull-down resistance) & CM transient immunity of gate driver isolation are special requirements.

- **The gate driver will protect a power semiconductor and entire converter if something goes wrong. For WBG devices,**

  - Cross-talk is easily induced, leading to potential hazard of shoot-through failure and gate terminal reliability issues. A gate assist circuit was introduced for cross-talk suppression.

  - Short circuit capability is limited. The desaturation protection circuit with < 200 ns response time was described for device reliability enhancement.

THE UNIVERSITY of TENNESSEE
KNOXVILLE

Outline

II. Switching Characterization of Wide Bandgap Power Semiconductors

❑ Double Pulse Test Fundamentals

- Main Function of Double Pulse Test

- Double Pulse Test Configuration

- Double Pulse Test Circuit Design Basics

❑ Unique Challenges and Considerations for Wide Bandgap Power Semiconductors

❑ Summary and Key Message

THE UNIVERSITY of TENNESSEE
KNOXVILLE

Switching Characterization

❑ **Why** do we perform switching characterization?
- Important for high power high frequency applications
  - Switching losses for efficiency estimation & thermal management
  - Switching time for dead-time setting & switching frequency selection

❑ **How** do we perform switching characterization?
- The Double Pulse Test (DPT) is the typical method.

THE UNIVERSITY *of* TENNESSEE
KNOXVILLE

Double Pulse Test Configuration

❑ Configuration

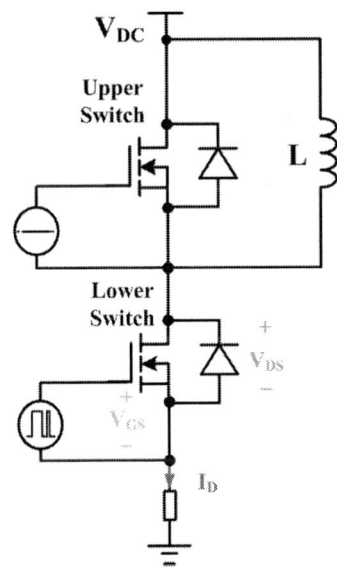

❑ Typical switching waveforms
- 1st pulse to establish desired current
- V_{DC} to establish desired voltage
- The end of the 1st pulse for turn-off switching characterization
- The beginning of the 2nd pulse for turn-on switching characterization

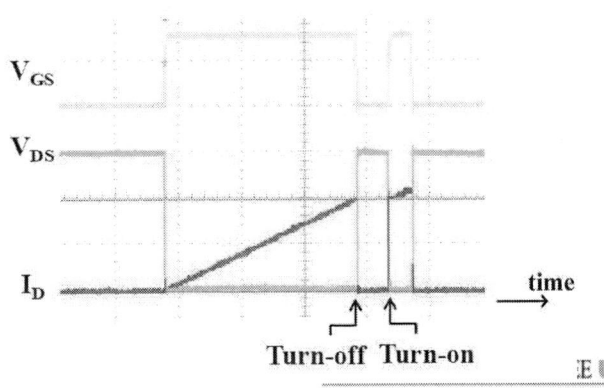

KNOXVILLE

Double Pulse Test Setup

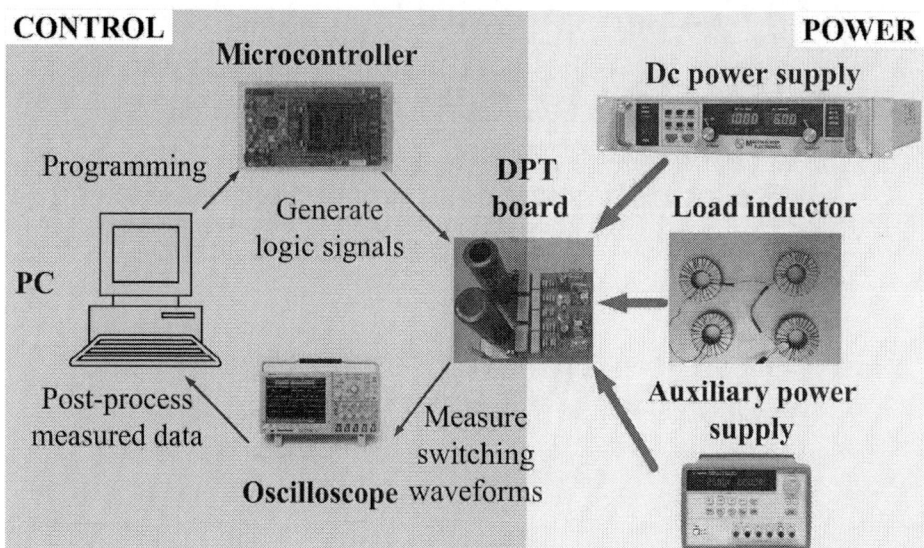

 THE UNIVERSITY of TENNESSEE
KNOXVILLE

Load Inductor Design

❑ Inductance

- Ideally, the inductive current during switching transient stays nearly constant

- Practically, during switching transient, inductive current variation ΔI_L should be slight

$$\Delta I_L = \frac{V_{DC}}{L}\, t_{sw} \leq k_{\Delta I_L} \times I_L$$

Where V_{DC}, I_L are the operating voltage and current; t_{sw} is the switching time; L is the load inductance; $k_{\Delta IL}$ is the current ripple coefficient

Model	$V_{DC(max)}$	$I_{L(min)}$	$t_{sw(max)}$	$k_{\Delta IL}$	L_{min}
1200-V / 20-A C2M0080120D	800 V	5 A	50 ns	2%	400 µH

❑ Fabrication

- Several inductors in series to minimize equivalent parallel capacitance (EPC)

 THE UNIVERSITY of TENNESSEE
KNOXVILLE

978-1-4799-5494-0/14 $31.00 © 2014 IEEE   198

Double Pulse Test Board

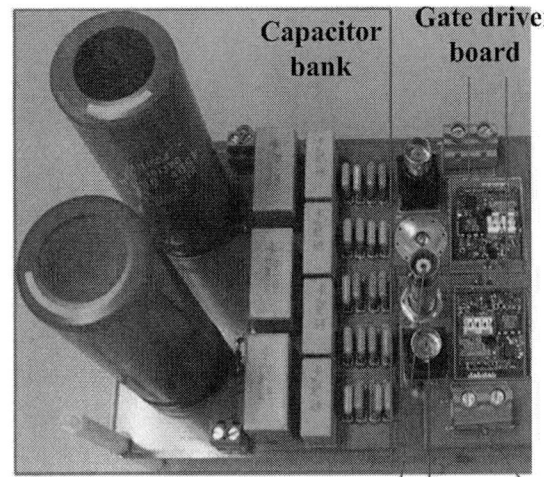

Capacitor bank · **Gate driver board** · **Switching current & voltage sensor** · **Load inductor connector**

❑ Capacitor bank
  ▪ Electrolytic or film capacitor: supply current during the pulse interval
  ▪ Decoupling capacitor: supply current during switching transient
❑ Gate driver board
  ▪ Device under test
  ▪ Gate driver circuits
❑ Measurement connector
  ▪ Drain current
  ▪ Drain-source voltage
❑ Load inductor connector

THE UNIVERSITY *of* TENNESSEE
KNOXVILLE

Dc Capacitor Bank Design

❑ Capacitance

  ▪ During the 1st pulse, the energy to establish the inductive current come from the dc capacitor

  $$\frac{1}{2}CV_{DC}^2 - \frac{1}{2}C(V_{DC} - \Delta V_{DC})^2 = \frac{1}{2}LI_L^2$$

  ▪ Practically, during the 1st pulse, dc capacitor voltage variation ΔV_{DC} should be slight

  $$\Delta V_{DC} \leq k_{\Delta VDC} \times V_{DC}$$

  where $k_{\Delta VDC}$ is the voltage ripple coefficient

Model	$V_{DC(min)}$	$I_{L(max)}$	L	$k_{\Delta VDC}$	C_{min}
1200-V / 20-A C2M0080120D	200 V	20 A	400 µH	1%	200 µF

❑ Type

  ▪ Electrolytic or film capacitor

THE UNIVERSITY *of* TENNESSEE
KNOXVILLE

Decoupling Capacitor Design

❏ Capacitance

- During the switching transient, decoupling capacitor is used to supply transient current & mitigate over-voltage

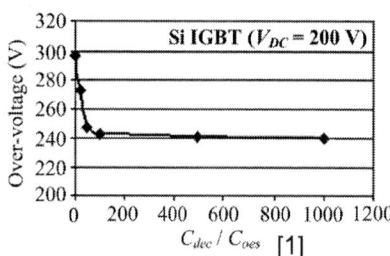
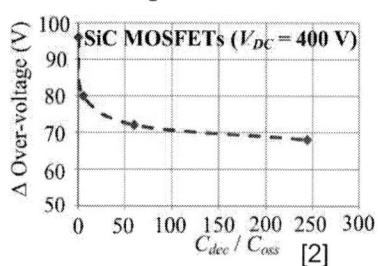

- Practically, C_{dec} should be designed > 250X of the output capacitance

❏ Type

- Ceramic capacitor

Model	C_{oss}	$C_{dec(min)}$
C2M0080120D	950 pF	237.5 nF

[1] Qian Liu, 2007 [2] Zheng Chen, 2013

Outline

II. Switching Characterization of Wide Bandgap Power Semiconductors

❏ Double Pulse Test Fundamentals

❏ Unique Challenges and Considerations for Wide Bandgap Power Semiconductors

- Measurement

- Grounding

- Data Processing

❏ Summary and Key Message

Measurement — Overview

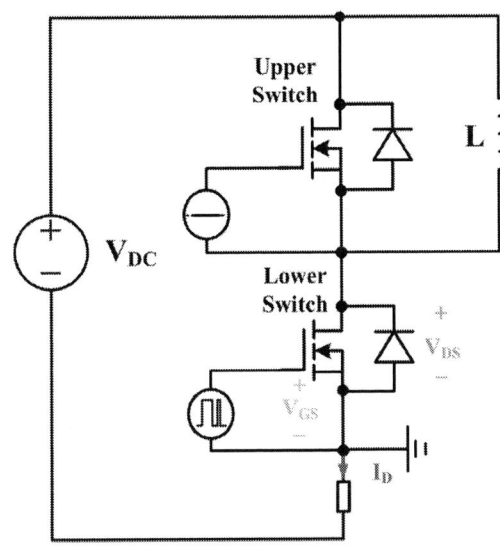

- Waveforms typically measured in Double Pulse Test:
  - Gate-source voltage, V_{GS}
  - Drain-source voltage, V_{DS}
  - Drain current, I_D
- Requirements for probes & scope:
  - Dynamic range
  - Bandwidth
  - Accuracy

CURENT

THE UNIVERSITY of TENNESSEE
KNOXVILLE

Bandwidth Requirement

- Bandwidth requirement:
  - A square-wave consists of many harmonics
  - Sine-wave to determine a rising edge's equivalent frequency
  - $f_{SW} = 0.35/(\min(t_r, t_f))$

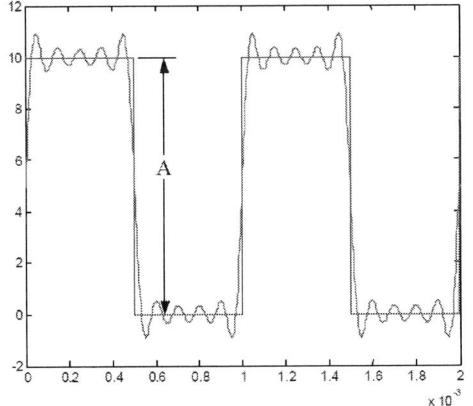 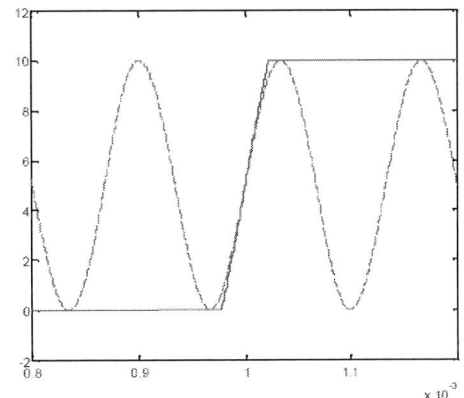

[1] Joseph B. Witcher, M.S thesis, 2002

CURENT

THE UNIVERSITY of TENNESSEE
KNOXVILLE

Switching Voltage Detection

❑ Voltage measurement requirement:
  ▪ High bandwidth ▪ High dynamic range

Types	Models	Maximum voltage	Band width	Rise time	Pros	Cons
Differential probes	THDP0200	± 1600 V	200 MHz	8.75 ns	Galvanic isolation	Limited bandwidth, longer connection leads to device
	TMDP0200	± 750 V	200 MHz	8.75 ns		
Passive probes	P5150	2,500 V	500 MHz	3.50 ns	High bandwidth	Non-galvanic isolation
	TPP0850	2,500 V	800 MHz	2.19 ns		

THE UNIVERSITY of TENNESSEE
KNOXVILLE

Switching Current Detection

❑ Current measurement requirement:
  ▪ High bandwidth ▪ High accuracy

Types/Models	Current type	Band-width	Rise time	Pros	Cons
Coaxial shunt SSDN-10	dc/ac	2,000 MHz	0.88 ns	High bandwidth, high accuracy	Non-galvanic isolation, non-continuous operation
Current transformer Pearson 2877	ac	200 MHz	8.75 ns	High bandwidth, galvanic isolation	Saturation for large current, non-continuous operation
Current probe TCP0030A	dc/ac	120 MHz	14.58 ns	Galvanic isolation, continuous operation	Low accuracy, limited bandwidth
Rogowski Coil CWT Mini	ac	17 MHz	102.9 ns	Galvanic isolation	Low bandwidth and accuracy

THE UNIVERSITY of TENNESSEE
KNOXVILLE

Voltage & Current Alignment

❏ Why is alignment necessary?
- Each of the voltage and current probes has its own characteristic propagation delay.
- The difference in these two delays will cause serious timing misalignment, eventually leading to the inaccurate switching losses.
- For WBG, switching loss is significantly sensitive to V-I alignment.

$V_{DC} = 600$ V, $I_L = 10$ A, $R_g = 2$ Ω

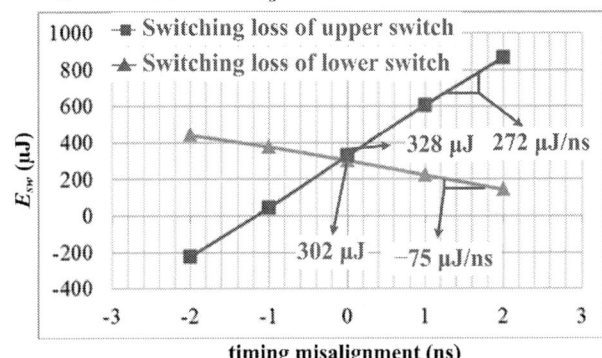

SENSITIVITY OF V-I ALIGNMENT ON SWITCHING LOSSES

R_g	10 Ω	5 Ω	2 Ω
$\Delta E_{sw\_L}/\Delta t$	26 μJ/ns	132 μJ/ns	272 μJ/ns
$\Delta E_{sw\_H}/\Delta t$	-16 μJ/ns	-47 μJ/ns	-75 μJ/ns

THE UNIVERSITY of TENNESSEE
KNOXVILLE

Voltage & Current Alignment (cont'd)

❏ Method
1. Use the power measurement deskew and calibration fixture [1]
   - Only current probes from Tektronix can be deskewed due to the connector constraints.
   - Maximum allowable voltage of the fixture is 8 V$_{rms}$.
2. Use the Probe Compensation output of the scope as a standard square waveform signal source for V−I alignment [2]
   - Only coaxial shunt can be used due to the connector constraints.
   - Output voltage from scope is 2.5 V.
3. Modify the DPT by removing the inductor and replacing one device with a low inductance 100 Ω resistor [3]
   - The voltage across a pure resistor is 100X larger than the current through it without any phase shift.
   - Maximum voltage is determined by resistor (typically > 150 V).

[1] Tektronix, power measurement deskew & calibration fixture instructions.
[2] C. Zhen, M.S thesis, 2009.
[3] Cree, SiC MOSFET double pulse fixture

THE UNIVERSITY of TENNESSEE
KNOXVILLE

Outline

II. **Switching Characterization of Wide Bandgap Power Semiconductors**

☐ Double Pulse Test Fundamentals

☐ Unique Challenges and Considerations for Wide Bandgap Power Semiconductors

- ▪ Measurement
- ▪ **Grounding**
- ▪ Data Processing

☐ Summary and Key Message

THE UNIVERSITY of TENNESSEE
KNOXVILLE

Grounding

☐ **The test point connects with grounds via passive probes**

- ▪ CM noise path includes both auxiliary and power circuits.
- ▪ The parasitic inductance and capacitance of this loop causes common-mode ringing in the current measurement

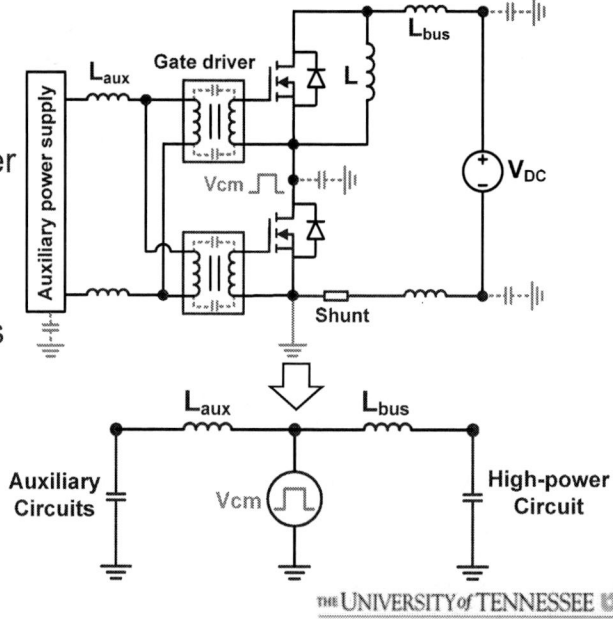

THE UNIVERSITY of TENNESSEE
KNOXVILLE

Grounding (cont'd)

❑ **To alleviate this issue:**

- A common-mode choke with high impedance at the common-mode ringing frequency can be employed.
- Ground of the oscilloscope can be floated.
- The grounding point in the double pulse circuit cannot be the middle point in a phase leg .

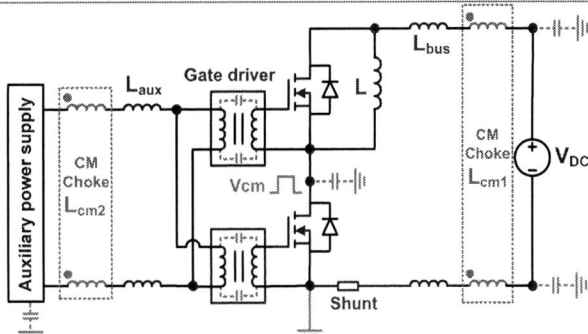

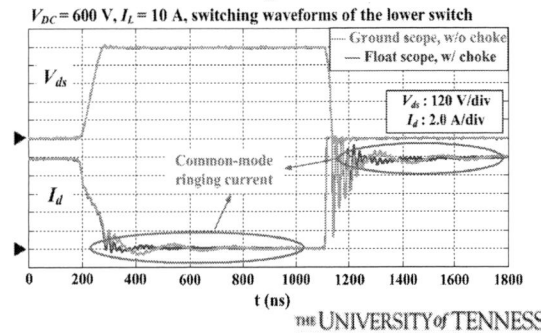

V_{DC} = 600 V, I_L = 10 A, switching waveforms of the lower switch

Tip-to-Ground

❑ **Probe ground leads are inductors in series with the measurement**

❑ **In the presence of high *dv/dt* and large electromagnetic fields, this wire will create errors in the voltage measurement**

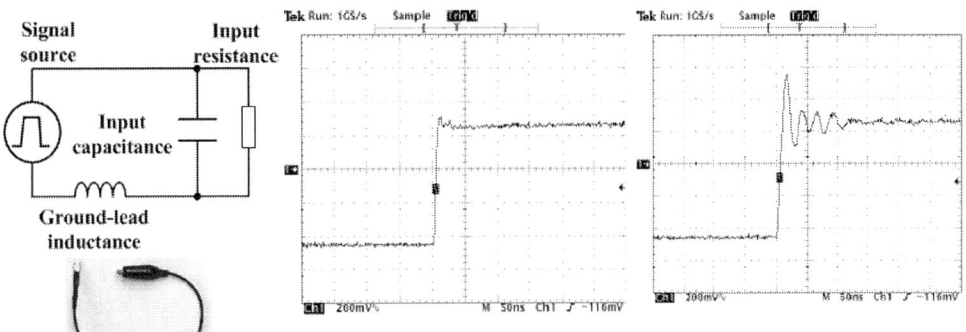

6.5-inch probe ground clip   28-inch probe ground clip [1]

[1] Tektronix, ABCs of Probes

Tip-to-Ground (cont'd)

❑ **Probe-tip adaptor is recommended for tip-to-ground reduction**

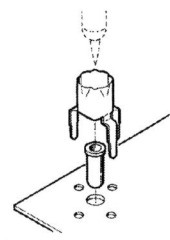

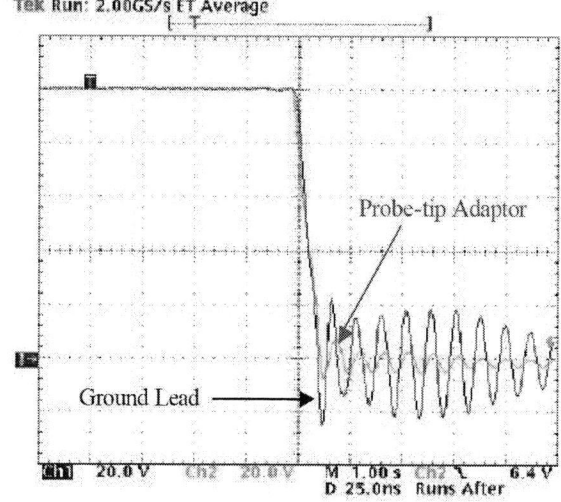

[1] Joseph B. Witcher, M.S thesis, 2002

Outline

II. **Switching Characterization of Wide Bandgap Power Semiconductors**

❑ Double Pulse Test Fundamentals

❑ Unique Challenges and Considerations for Wide Bandgap Power Semiconductors

- Measurement
- Grounding
- Data Processing

❑ Summary and Key Message

Switching Data Processing

❑ Traditional method of processing switching data is [1]

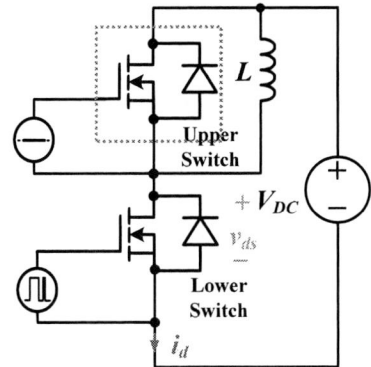 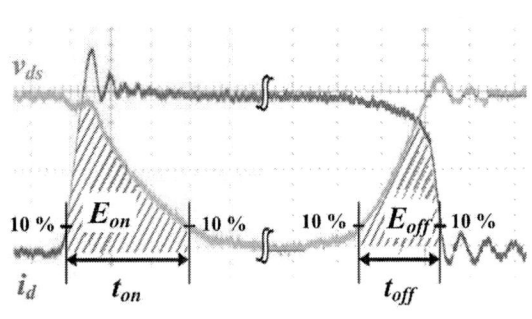

❑ Two factors need to be taken into account for fast switching WBG devices

- V-I alignment
- Switching loss of the upper switch

[1] C. Zhen, M.S thesis, 2009.

THE UNIVERSITY *of* TENNESSEE
KNOXVILLE

Impact of Cross-Talk on E_{sw}

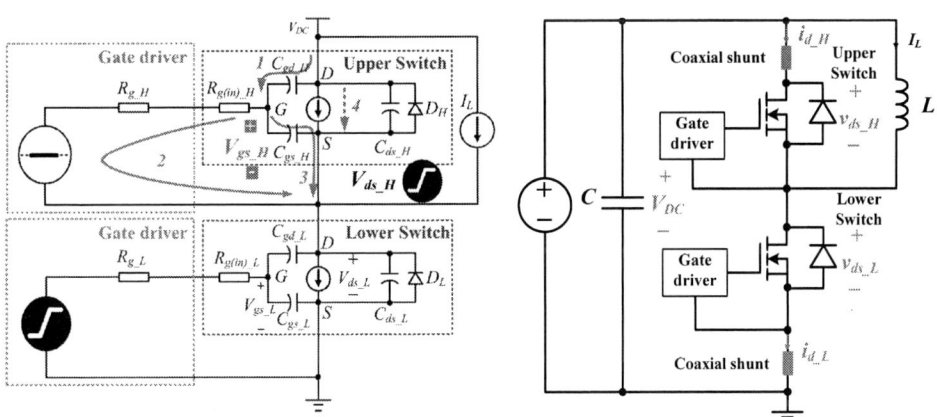

❑ Cross-Talk may induce additional switching loss at the opposite switch during the switching transient of the operating one in a phase-leg

$$E_{sw} = \int (v_{ds\_L} i_{d\_L} + v_{ds\_H} i_{d\_H}) dt = \int (V_{DC} i_{d\_L} - I_L v_{ds\_H}) dt$$

THE UNIVERSITY *of* TENNESSEE
KNOXVILLE

Double Pulse Test — Case Study

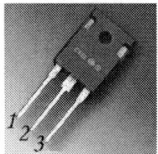

Model	Voltage rating	Current rating	$R_{ds(on)}$
C2M0080120D	1200 V	20 A @ 100°C	80 mΩ

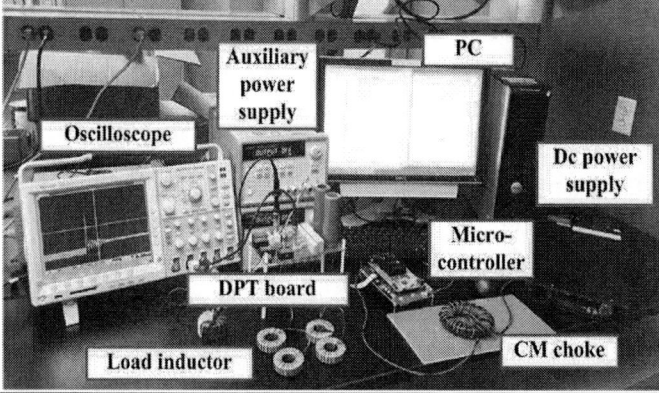

Load Inductor	460 μH / 20 A (4 in series, 115 μH per each)
Dc cap bank	270 μF / 900 V electrolytic cap (2 in series, 540 μF / 450 V per each)
Decoupling cap	3 μF / 1 kV ceramic cap (30 in parallel, 100 nF / 1 kV per each)

THE UNIVERSITY of TENNESSEE
KNOXVILLE

Case Study — Measurement & Data Processing

Scope	1GHz, Tek DPO 401
V_{gs}	1 GHz, Tek TPP 1000 passive probe
V_{ds}	800 MHz, Tek TPP0850 HV single-ended probe
I_d	2 GHz, T & M Research SSDN-10 coaxial shunt

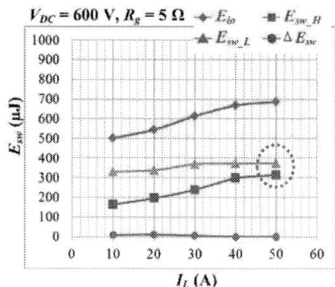

$$E_{io} \approx E_{sw} = E_{sw\_L} + E_{sw\_H}$$

V_{DC} = 600 V, I_L = 30 A, R_g = 10 Ω (Lower switch as DUT)

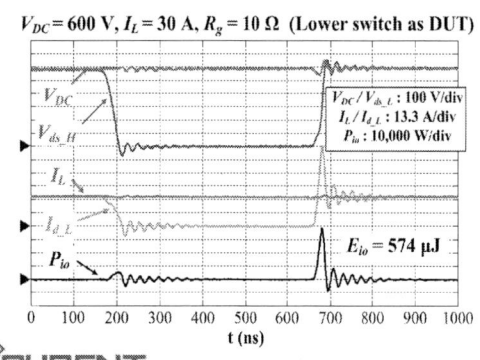

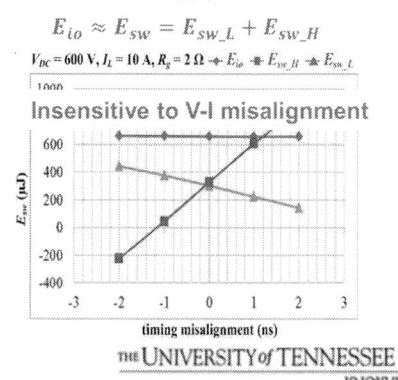

Insensitive to V-I misalignment

THE UNIVERSITY of TENNESSEE
KNOXVILLE

Outline

II. Switching Characterization of Wide Bandgap Power Semiconductors

❏ Double Pulse Test Fundamentals

❏ Unique Challenges and Considerations for Wide Bandgap Power Semiconductors

❏ Summary and Key Message

THE UNIVERSITY *of* TENNESSEE
KNOXVILLE

Summary & Key Message

❏ **Four factors need to be taken into account for DPT of WBG devices**

Factors	Design considerations	Recommendations
1. Scope & probe	• Satisfy bandwidth, dynamic range, accuracy	• HV passive probe • Coaxial shunt
2. V-I alignment	• High sensitive on switching loss • Must be carefully deskewed	• Resistive based DPT
3. Probe induced grounding	• Affect measurement accuracy • Must be mitigated	• CM choke • Probe-tip adaptor
4. Cross-talk	• Extra switching loss @ non-operating device	• Following method

❏ **A practical method is introduced for switching loss assessment**

- $E_{sw} = \int (V_{DC} i_{d\_L} - I_L v_{ds\_H}) dt$

- **Considers cross-talk, insensitive to V-I alignment**

THE UNIVERSITY *of* TENNESSEE
KNOXVILLE

Acknowledgements

The Lecturer would like to thank the II-VI Foundation and the Engineering Research Center Shared Facilities supported by the Engineering Research Center Program of the National Science Foundation and DOE under NSF Award Number EEC-1041877 and the CURENT Industry Partnership Program.

The lecturer would also like to thank Zhiqiang Wang, Ben Guo, Weimin Zhang, Edward A. Jones for their contributions to this tutorial.

Thank you for your attention !

Questions?

Reliability and High Field related issues in GaN-HEMT devices – PART I

Gaudenzio Meneghesso
Enrico Zanoni, Matteo Meneghini,
University of Padova,
Department of Information Engineering,
via Gradenigo 6/B, 35131 Padova, Italy
gaudenzio.meneghesso@unipd.it

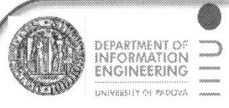

University of Padova – Department of Information Engineering

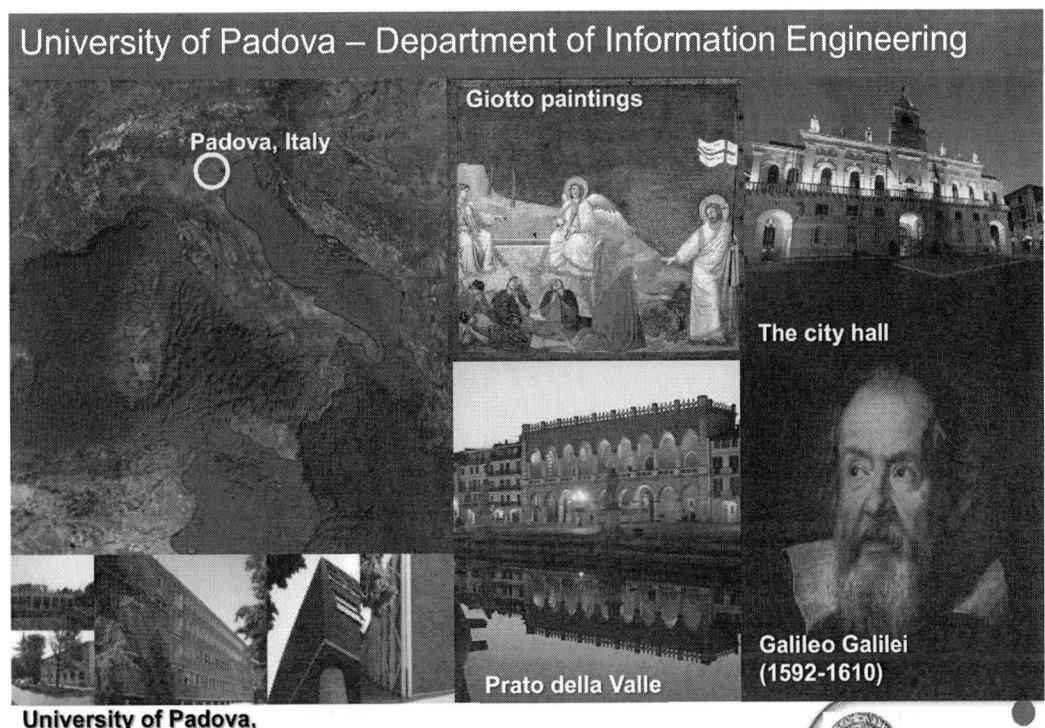

Microelectronics Laboratory Staff

- **3 Full professors**
  - E. Zanoni, A. Paccagnella, G. Meneghesso

- **2 Associate professors**
  - A. Neviani, A. Gerosa

- **5 Assistant professors**
  - A. Cester, A. Bevilacqua, D. Vogrig, S. Gerardin, M. Meneghini

- **6 post-docs**
  - N. Wrachien, N. Trivellin, F.A. Marino, A. Stocco, C. Desanti, I. Rossetto

- ≈15 Ph.D. students

- 1 technician

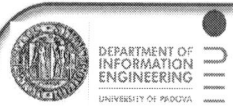

RESEARCH TOPICS

- Characterization, modeling and reliability of compound semiconductor devices
  - Study of short-channel and trapping effects in HEMTs
  - Reliability and parasitic effects on GaN HEMTs
  - Characterization, modeling and reliability of high-efficiency GaN LEDs and Blu-Ray Laser diodes
- Development of RF-MEMS switches
  - Characterization and charge trapping phenomena
  - Cycling and reliability issues
  - ESD events on RF-MEMS
- Characterization and reliability of Organic Devices
  - Organic TFT – IMEC & Slovak University of Technology in Bratislava.
  - Organic LED: Universal Display Corporation
  - Organic Solar cells, IMEC & University of Rome Tor Vergata
- Reliability of advanced CMOS devices
  - Gate oxides/Soft error/Advanced memories, Radiation hardness
  - DNA Recognition, Development of ESD protection circuits
- Design of integrated circuits for analog signal processing
  - Analog implementation of turbo decoders, Ultra wide band receivers
  - Integrated circuits for implantable biomedical devices

Acknowledgments

  http://www.alinwon-fp7.eu/fp7/

  http://www.hiposwitch.eu/

  http://www.e2cogan.eu/

 ONR project N000141010608, monitor: Paul Maki

Outline

❑ Introduction & Motivations
❑ Parasitic effects in GaN-HEMTs (Part I)
  ❑ Current collapse
    ❑ Definitions & Characterization
  ❑ Study of parasitic and dispersion effects: gate-lag, drain-lag, current collapse
  ❑ V_{th} instabilities in Power GaN HEMTs
  ❑ Controlling trapping phenomena
  ❑ Hot electrons
  ❑ Reliability issues in GaN-HEMTs (Part II)
  ❑ Conclusions

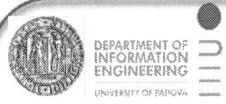

IEEE TED October 2013 issue

IEEE TRANSACTIONS ON
ELECTRON DEVICES

A PUBLICATION OF THE IEEE ELECTRON DEVICES SOCIETY

Call for Papers
for a special issue in
IEEE Transactions on Electron Devices
"Wide Bandgap Semiconductor Power Devices for Energy Efficiency and Renewable Energy Utilization"

Please submit papers by using the website: http://mc.manuscriptcentral.com/ted
BE SURE TO MENTION THE SPECIAL ISSUE WITHIN THE COVER LETTER
Submission Deadline: **June 30, 2014**   Scheduled Publication Date: **February 2015**

Guest Editors:
Krishna Shenai Argonne National Laboratory, kshenai@anl.gov
Mietek Bakowski Acreo, Mietek.Bakowski@acreo.se
Noboru Ohtani Kwansei Gakuin University, ohtani.noboru@kwansei.ac.jp

OCTOBER 2013   VOLUME 60   NUMBER 10   IETDAI   (ISSN 0018-9383)

SPECIAL ISSUE ON GaN ELECTRONIC DEVICES

GUEST EDITORIAL
Foreword .... *K. J. Chen, G. Meneghesso, R. Quay, and T. Egawa*

REGULAR PAPERS
Fabrication and Characterization of GAN Based Devices
Scaling of GaN HEMTs and Schottky Diodes for Submillimeter-Wave MMIC Applications ....
.... *K. Shinohara, D. C. Regan, Y. Tang, A. L. Corrion, D. F. Brown, J. C. Wong, J. F. Robinson,*
H. H. Fung, A. Schmitz, T. C. Oh, S. J. Kim, P. S. Chen, R. G. Nagele, A. D. Margomenos, and M. Micovic
Current Stability in Multi-Mesa-Channel AlGaN/GaN HEMTs ....
.... *K. Ohi, J. T. Asubar, K. Nishiguchi, and T. Hashizume*
p-Channel Enhancement and Depletion Mode GaN-Based HFETs With Quaternary Backbarriers ....
.... *H. Hahn, B. Reuters, A. Pooth, B. Holländer, M. Heuken, H. Kalisch, and A. Vescan*
High-Performance GaN-Based Nanochannel FinFETs With/Without AlGaN/GaN Heterostructure ....
.... *K.-S. Im, C.-H. Won, Y.-W. Jo, J.-H. Lee, M. Bawedin, S. Cristoloveanu, and J.-H. Lee*
DC and RF Performance of Gate-Last AlN/GaN MOSHEMTs on Si With Regrown Source/Drain

---

IOP PUBLISHING

SEMICONDUCTOR SCIENCE AND TECHNOLOGY

Semicond. Sci. Technol. 28 (2013) 070301 (1pp)

doi:10.1088/0268-1242/28/7/070301

PREFACE

Gallium nitride electronics

Siddharth Rajan
The Ohio State University, USA

Debdeep Jena
University of Notre Dame, USA

In the past two decades, there has been increasing research and industrial activity in the area of gallium nitride (GaN) electronics, stimulated first by the successful demonstration of GaN LEDs. While the promise of wide band gap semiconductors for power electronics was recognized many years before this by one of the contributors to this issue (J Baliga), the success in the area of LEDs acted as a catalyst. It set the field of GaN electronics in motion, and today the technology is improving the performance of several applications including RF cell phone base stations and military radar. GaN could also play a very important role in reducing worldwide energy consumption by enabling high efficiency compact power converters operating at high voltages and lower frequencies.

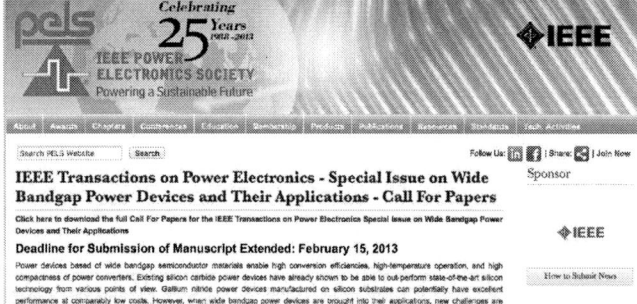

IEEE Transactions on Power Electronics - Special Issue on Wide Bandgap Power Devices and Their Applications - Call For Papers

Click here to download the full Call For Papers for the IEEE Transactions on Power Electronics Special Issue on Wide Bandgap Power Devices and Their Applications

Deadline for Submission of Manuscript Extended: February 15, 2013

Power devices based of wide bandgap semiconductor materials enable high conversion efficiencies, high-temperature operation, and high compactness of power converters. Existing silicon carbide power devices have already shown to be able to out-perform state-of-the-art silicon technology from various points of view. Gallium nitride power devices manufactured on silicon substrates can potentially have excellent performance at comparably low costs. However, when wide bandgap power devices are brought into their applications, new challenges are faced. These are for instance cost, reliability, or EMI issues.

Prospective authors are invited to submit original contributions, or survey papers for review for publication in this special issue on Wide Bandgap Power Devices and Their Applications. Topics of interest include, but are not limited to:

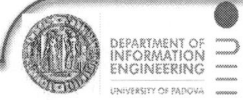

Introduction to the GaN-HEMTs

	Si	GaAs	4H-SiC	6H-SiC	GaN/AlGaN
Band gap energy E_g (eV)	1.1 ind.	1.43 dir.	3.26 ind.	3.0 ind.	3.42 dir.
Electron mobility μ_e (cm²/Vs)	1500	8500	1000	500	1300 >2000 (2DEG)
Electric breakdown field E_{crit} (10⁶ V/cm)	0.3	0.4	2.0	2.4	3.3
Saturation velocity v_{sat} (10⁷ cm/s)	1.0	2.0	2.0	2.0	2.7
Thermal conductivity κ (W/Kcm)	1.5	0.46	4.9	4.9	2.4
Johnsons Figure of Merit ($\sim V_{Br}^2 \times v_{sat}^2$)	1	7	180	260	760
Maximum operation temperature T_{max} (°C)	200	300	500	500	500

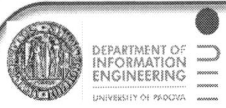

.... but fully usable only if heat dissipation can be managed

→ trade-off

Joachim Würfl, "GaN Power Devices (HEMT): Basics, Advantages and Perspectives", ECPE Workshop 2013

DEPARTMENT OF
INFORMATION
ENGINEERING
UNIVERSITY OF PADOVA

Introduction to the GaN-HEMTs

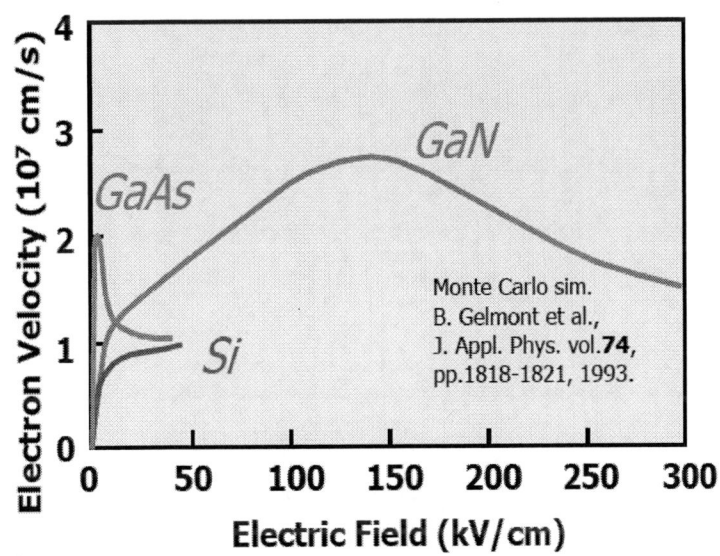

Monte Carlo sim.
B. Gelmont et al.,
J. Appl. Phys. vol.**74**,
pp.1818-1821, 1993.

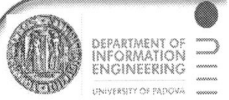

DEPARTMENT OF
INFORMATION
ENGINEERING
UNIVERSITY OF PADOVA

Introduction & Motivations

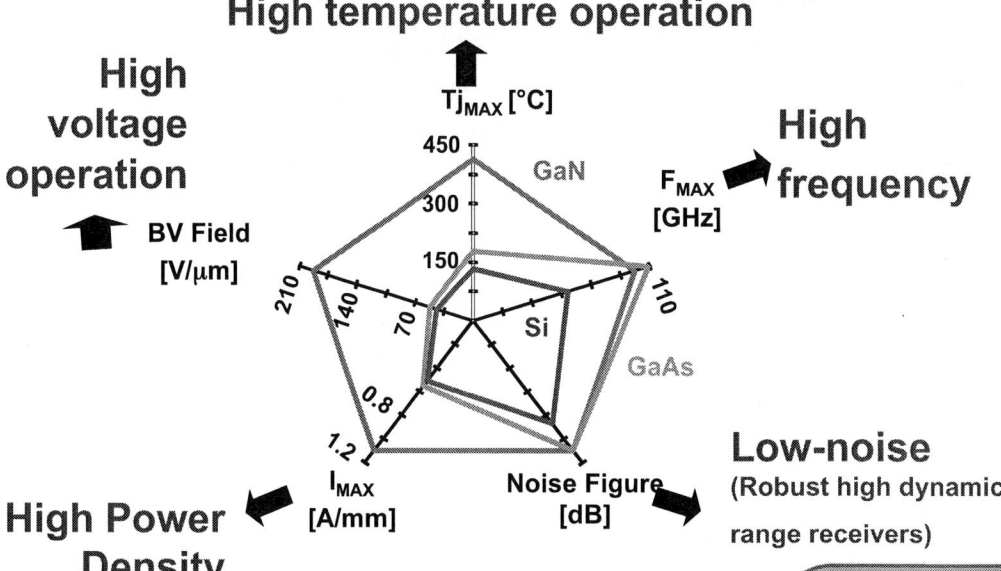

Introduction to the GaN-HEMTs

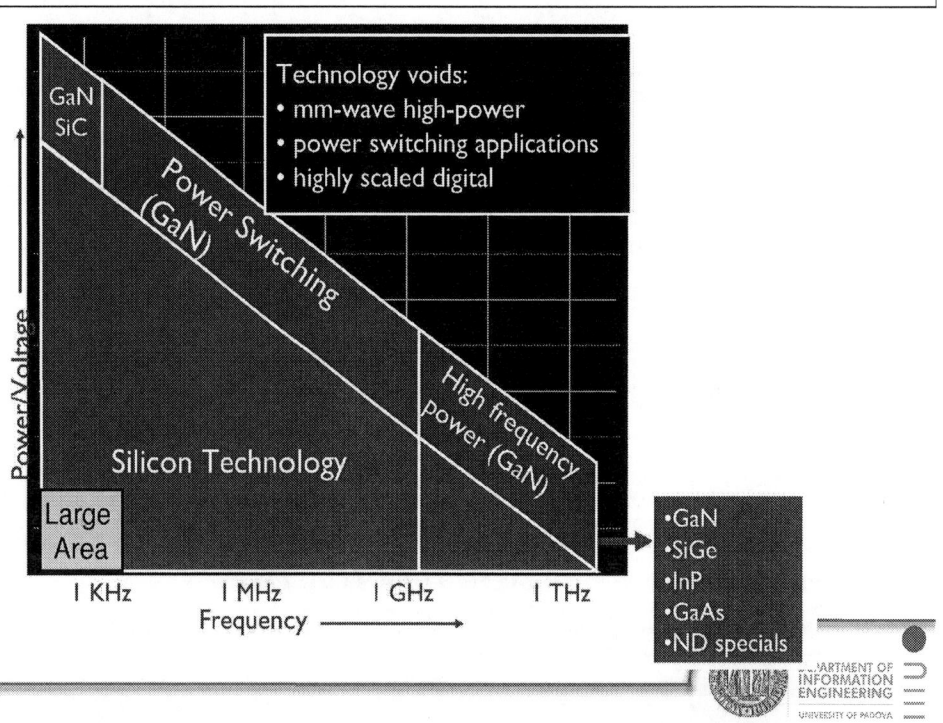

GaN: Polar material

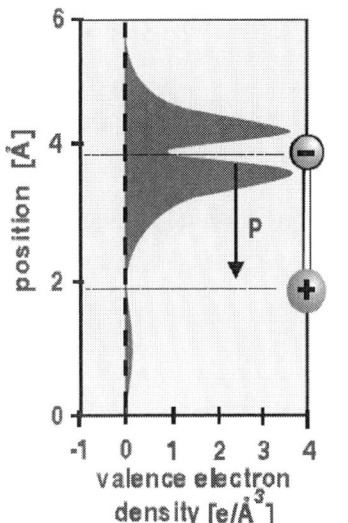 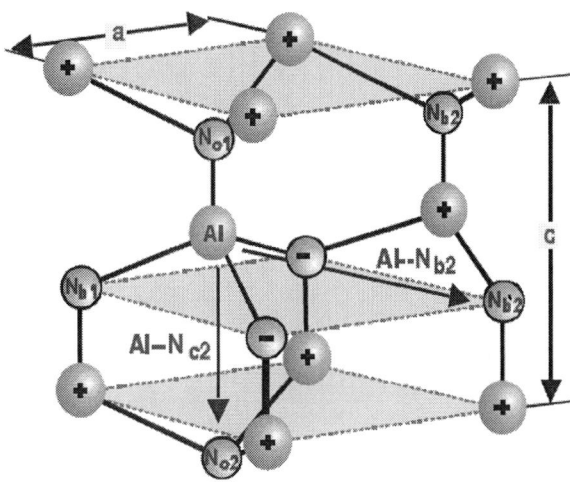

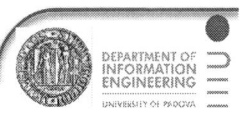

O. Ambacher et al JAP 87, 334 (2000)
O. Ambacher et al. J.Appl. Phys. 85, 3222 (1999)

GaN: Polar material

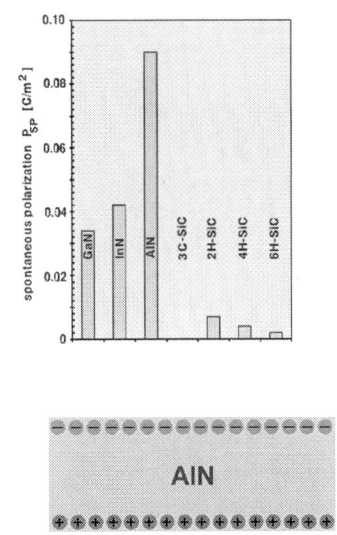 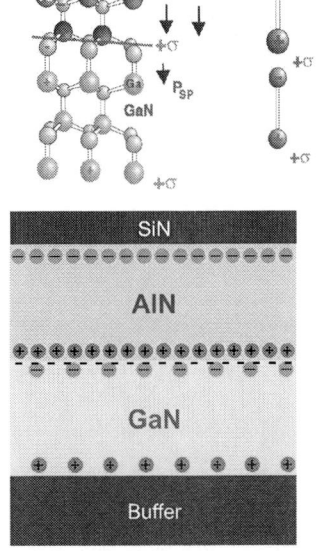 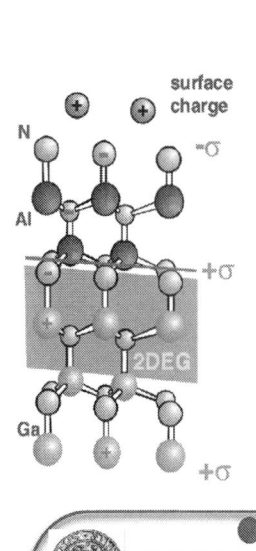

978-1-4799-5494-0/14 $31.00 © 2014 IEEE   217

GaN: Polar material

GaN/AlGaN

OK the lattice constant

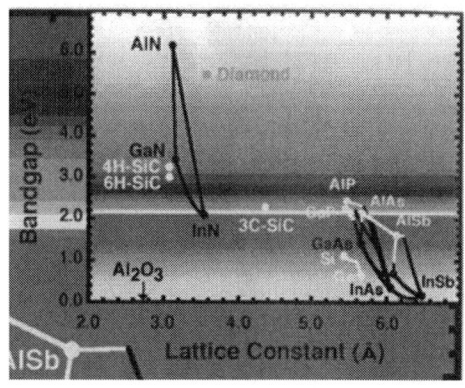

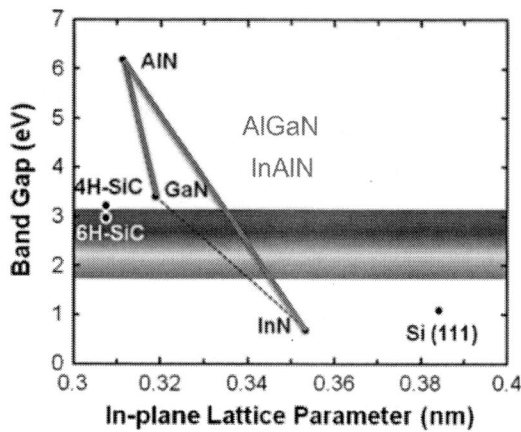

Typically 20-25% Al composition

GaN HEMTs: 2DEG w/o doping!

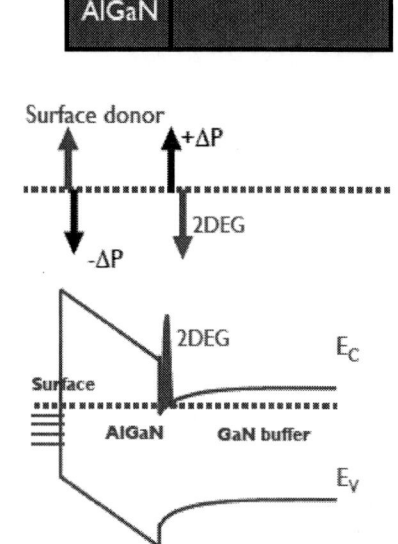

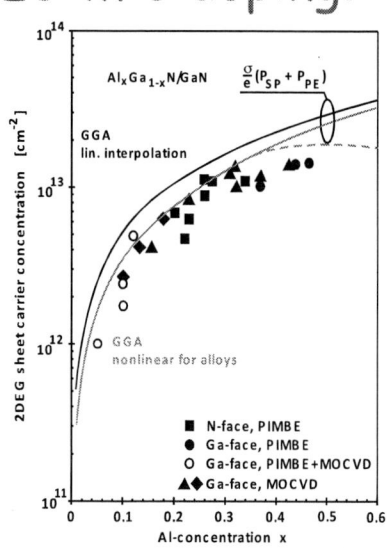

O. Ambacher et al JAP 87, 334 (2000)
O. Ambacher et al. J.Appl. Phys. 85, 3222 (1999)

Basic HEMTs design $I_{Dmax}= 2.1$ A/mm

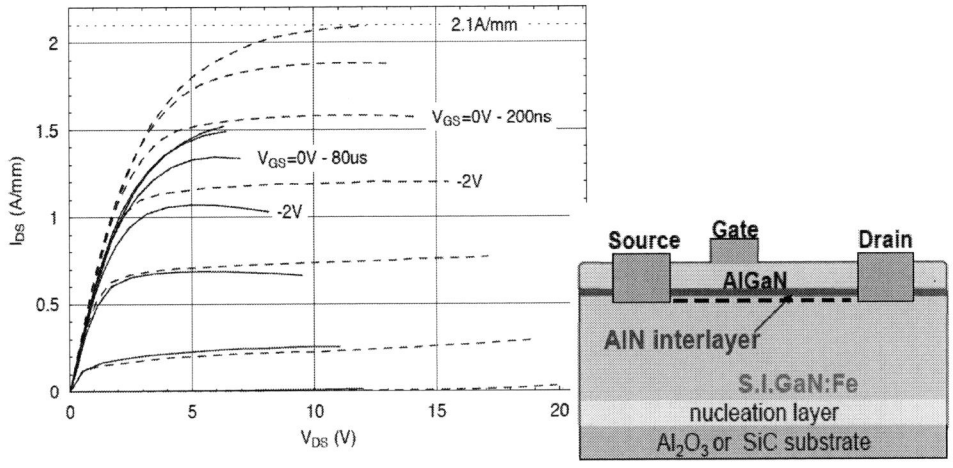

A. Chini, R. Coffie, G. Meneghesso, E. Zanoni, D. Buttari, S. Heikman, S. Keller, and U. K. Mishra
"A 2.1A/mm Current Density AlGaN/GaN HEMT"
IEE Electronics Letters, vol. 39, N. 7, April 2003, pp. 625-626

GaN-HEMTs capabilities: proven!

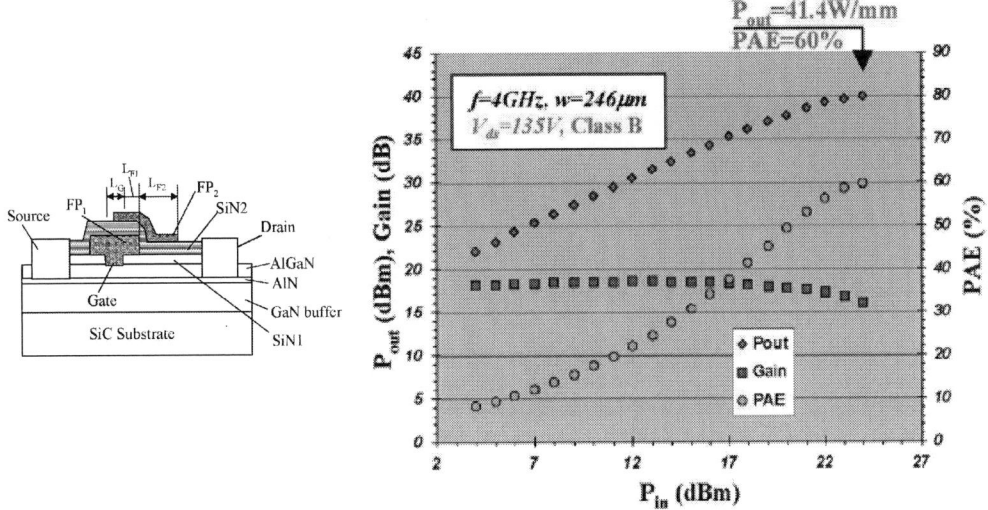

[3] Y.-F. Wu, M. Moore, A. Saxler, T. Wisleder and P. Parikh "40-W/mm Double Field-plated GaN HEMTs", Device Research Conference, . 151, 2006

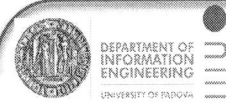

978-1-4799-5494-0/14 $31.00 © 2014 IEEE   219

Field-plate Optimization

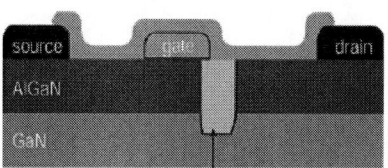

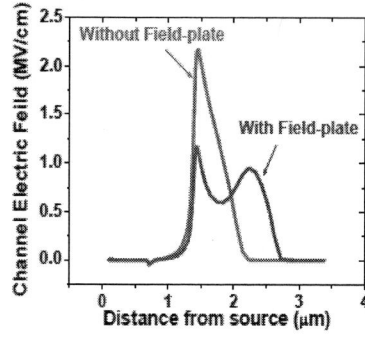

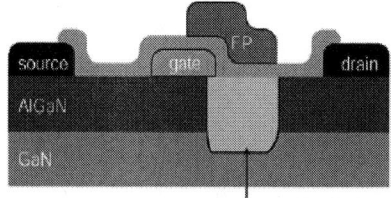

Field-plate reduces electric field

- Increase breakdown voltage
- lower electron injection into traps ➡ less dispersion

courtesy: Umesh Mishra, UCSB

Very high blocking voltage with compact size

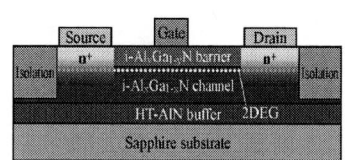

Fig.1   Schematic structure of HEMTs with the hetero-structure consisted of the $Al_xGa_{1-x}N$ channel layer and the $Al_yGa_{1-y}N$ barrier layer ($Al_yGa_{1-y}N/Al_xGa_{1-x}N$, y > x).

Takuma Nanjo et al, "Remarkable Breakdown Voltage Enhancement in AlGaN Channel HEMTs", **IEDM 2007**

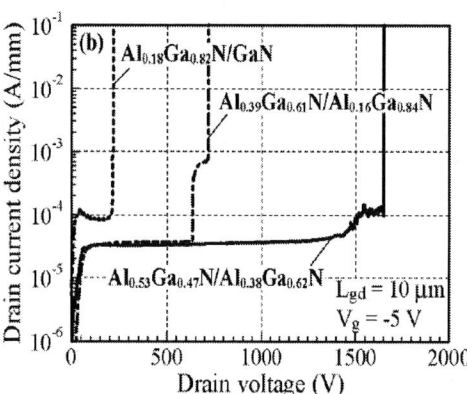

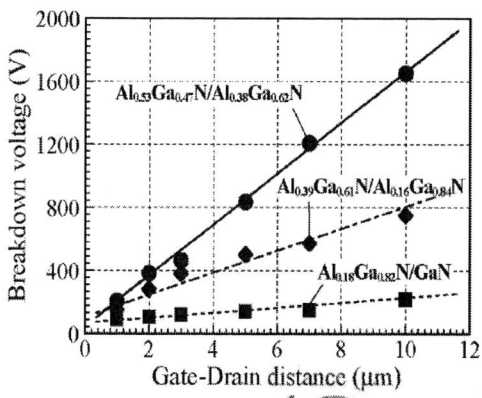

978-1-4799-5494-0/14 $31.00 © 2014 IEEE

Figure of Merit in semiconductors

Table 1 Figures of merit of various semiconductors

	Si	GaAs	4H-SiC	GaN
JFM	1	11	410	790
KFM	1	0.45	5.1	1.8
BFM	1	28	290	910
BHFM	1	16	34	100

JFM : Johnson's figure of merit for high frequncy devices = $(EbVs/2\pi)^2$

KFM : Keyes's figure of merit considering thermal limitation= $\kappa(EbVs/4\pi\varepsilon)^{1/2}$

BFM : Baliga's figure of merit for power switching = $emEg^3$

BHFM : Baliga's figure of merit for high frequency power switching = μEb^2

POWER APPLICATIONS

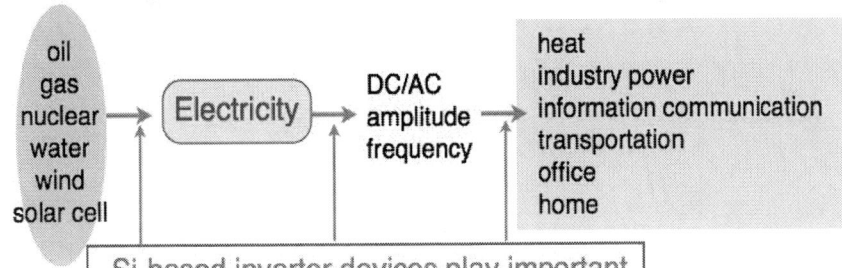

Efficiency of present inverter : 80~ 90%
10~20% loss still remains !!
      mainly due to the limitation of material properties of Si

Ultra-low loss inverter is a key device for next-generation
energy saving society

GaN HEMTs for high efficiency power electronics

99.3% Efficiency of three-phase inverter for motor drive using GaN-based Gate Injection

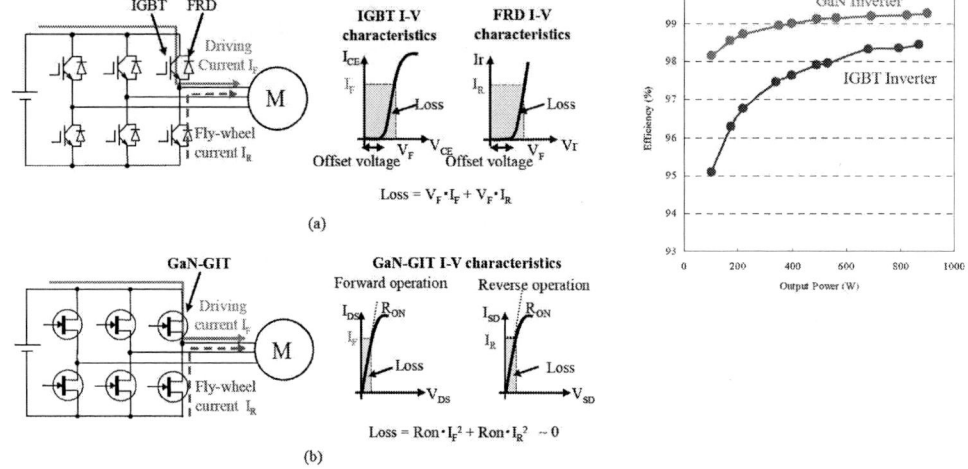

(a)

(b)

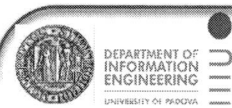

Morita, T.; Tamura, S.; Anda, Y.; Ishida, M.; Uemoto, Y.; Ueda, T.; Tanaka, T.; Ueda, D.
99.3% Efficiency of three-phase inverter for motor drive using GaN-based Gate Injection
Transistors Applied Power Electronics Conference and Exposition (APEC),
2011 Twenty-Sixth Annual IEEE 2011, Page(s): 481 – 484

IEEE ELECTRON DEVICE LETTERS, VOL. 29, NO. 8, AUGUST 2008

A 97.8% Efficient GaN HEMT Boost Converter With 300-W Output Power at 1 MHz

Yifeng Wu, Matt Jacob-Mitos, Marcia L. Moore, and Sten Heikman

@ 300-W output power

@ 1 MHz

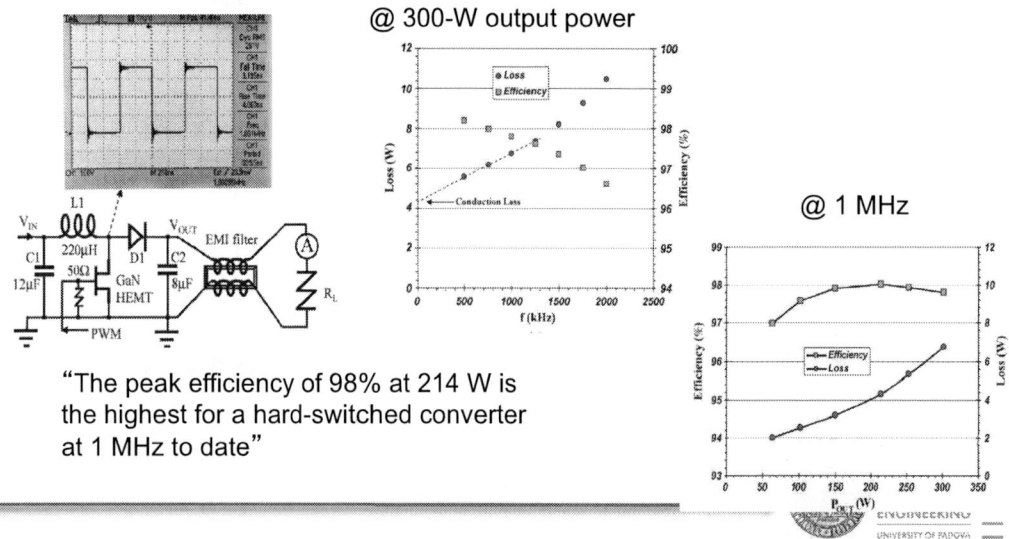

"The peak efficiency of 98% at 214 W is the highest for a hard-switched converter at 1 MHz to date"

978-1-4799-5494-0/14 $31.00 © 2014 IEEE   222

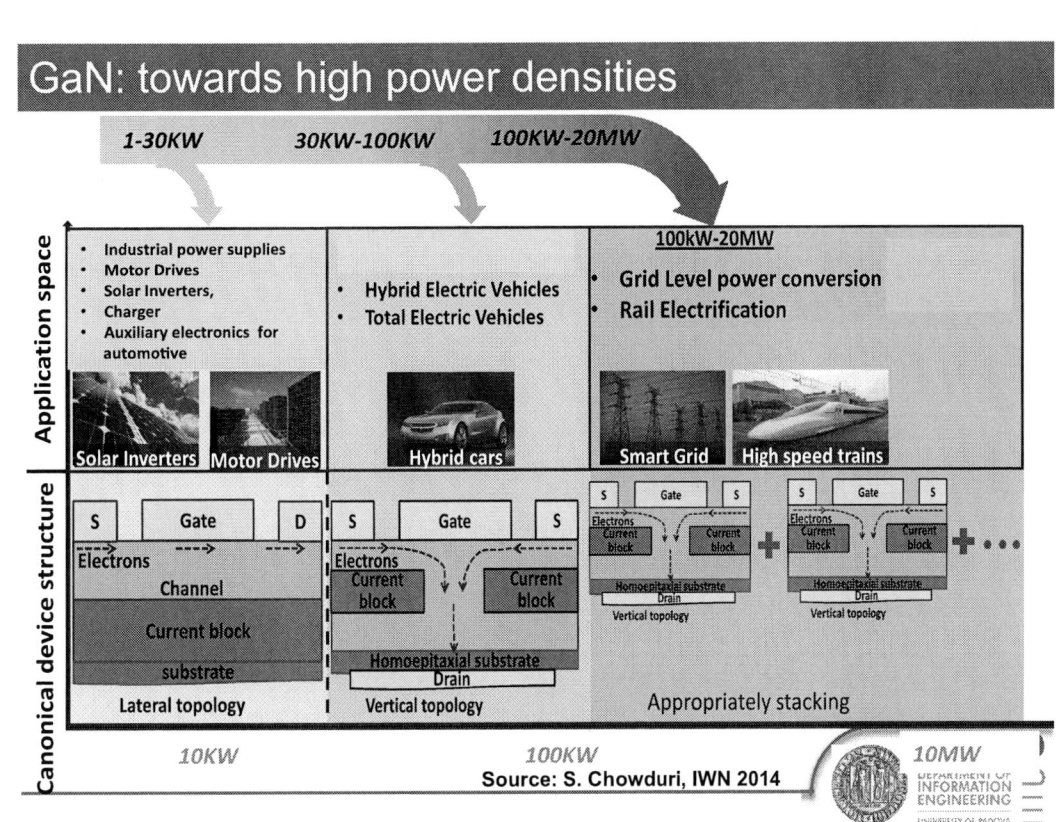

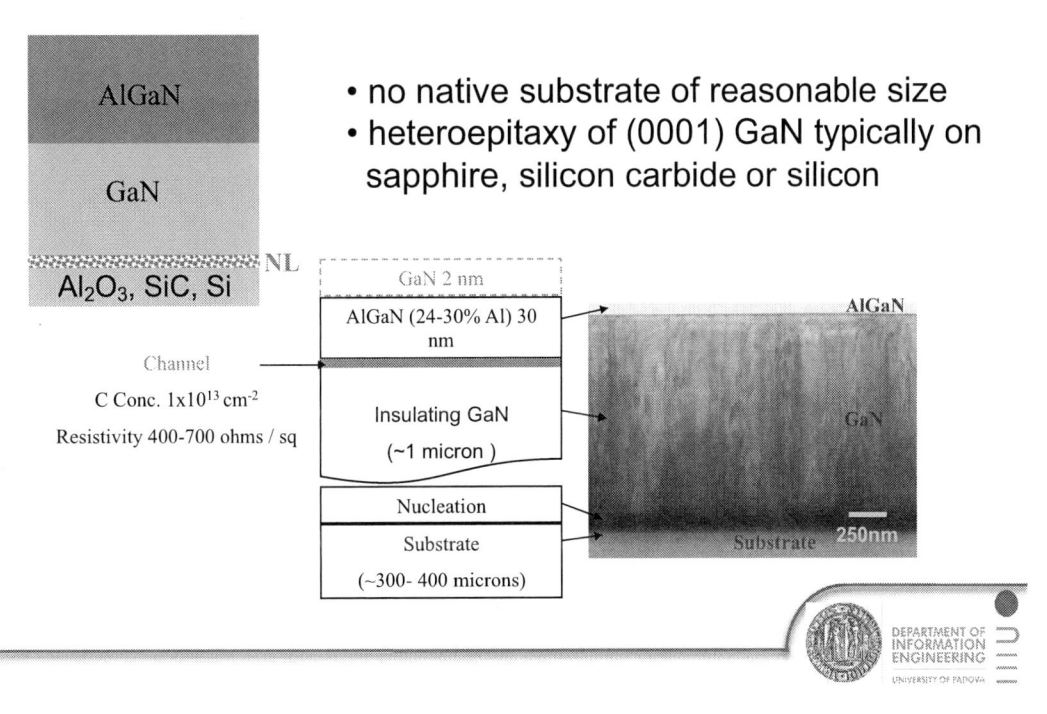

GaN: cost vs performance

GaN economically possible on foreign substrates only
* GaN on SI-SiC, would be the best in terms of performance, but cost is high

GaN on Si
* very cost efficient
* Scalable to large wafer diameters 8" → mass production in CMOS line feasible
* Performance trade-off? Thermal management must be optimized

	n-type SiC	s.i. SiC	GaN bulk	Si
Lattice mismatch (%)	3.1	3.1	0	17
Availability / Price (4", €)	700	2500	6000	80
Thermal conductivity (W/cmK)	4	4	1.3	1.48

Best performance   **versus**   Cost efficient realization

Joachim Würfl, "GaN Power Devices (HEMT): Basics, Advantages and Perspectives", ECPE Workshop 2013

Insulated Gate GaN HEMT

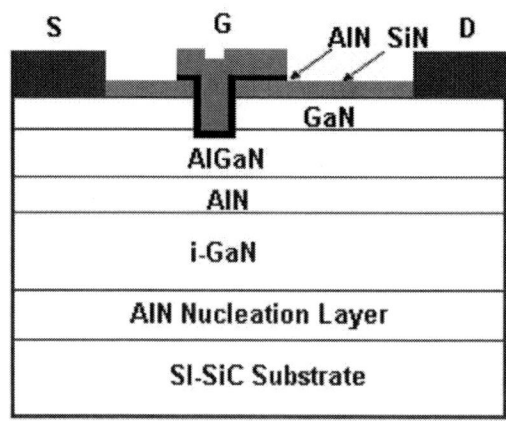

Fig.1 Schematic cross-section of the developed AlGaN/GaN MIS HEMT with AIN dielectric

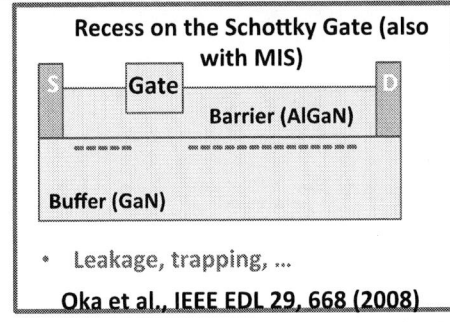

Recess on the Schottky Gate (also with MIS)

- Leakage, trapping, …

Oka et al., IEEE EDL 29, 668 (2008)

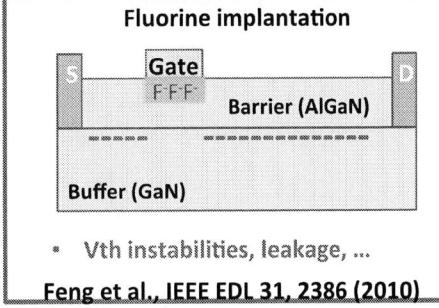

Fluorine implantation

- Vth instabilities, leakage, …

Feng et al., IEEE EDL 31, 2386 (2010)

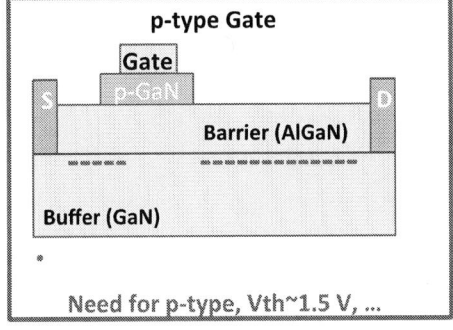

p-type Gate

Need for p-type, Vth~1.5 V, …

Uemoto et al., IEEE TED 54, 3393 (2007)

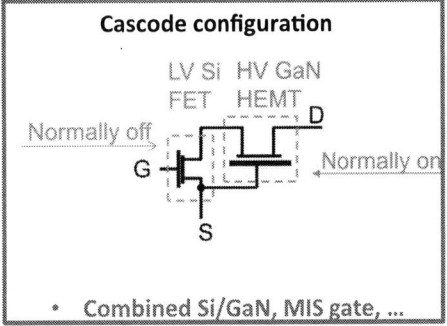

Cascode configuration

- Combined Si/GaN, MIS gate, …

Eg. Transphorm,

IOP PUBLISHING

SEMICONDUCTOR SCIENCE AND TECHNOLOGY

Semicond. Sci. Technol. 28 (2013) 074014 (8pp)

doi:10.1088/0268-1242/28/7/074014

INVITED REVIEW

Current status and scope of gallium nitride-based vertical transistors for high-power electronics application[*]

Vertical vs Lateral GaN HEMT

Srabanti Chowdhury[1], Brian L Swenson[2], Man Hoi Wong[3] and Umesh K Mishra[4]

Invited Review

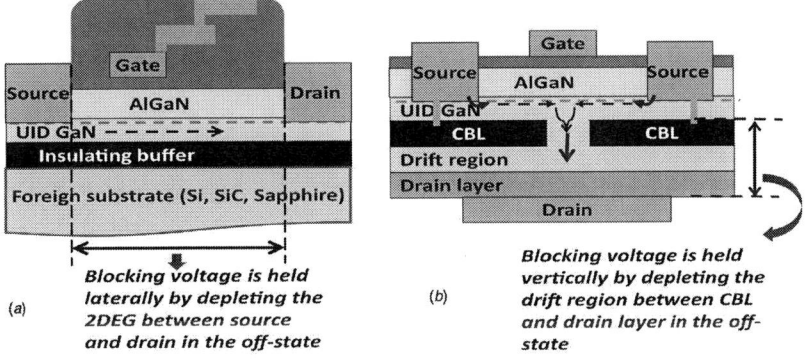

Figure 4. (*a*) A lateral AlGaN/GaN power HEMT (*b*) A vertical transistor using AlGaN/GaN layer structure on bulk GaN drift layer and substrate.

Outline

❑ Introduction & Motivations
❑ Parasitic effects in GaN-HEMTs (Part I)
  ❑ Current collapse
    ❑ Definitions & Characterization
  ❑ Study of parasitic and dispersion effects: gate-lag, drain-lag, current collapse
  ❑ V_{th} instabilities in Power GaN HEMTs
  ❑ Controlling trapping phenomena
  ❑ Hot electrons
❑ Reliability issues in GaN-HEMTs (Part II)
❑ Conclusions

DEPARTMENT OF
INFORMATION
ENGINEERING
UNIVERSITY OF PADOVA

Current collapse

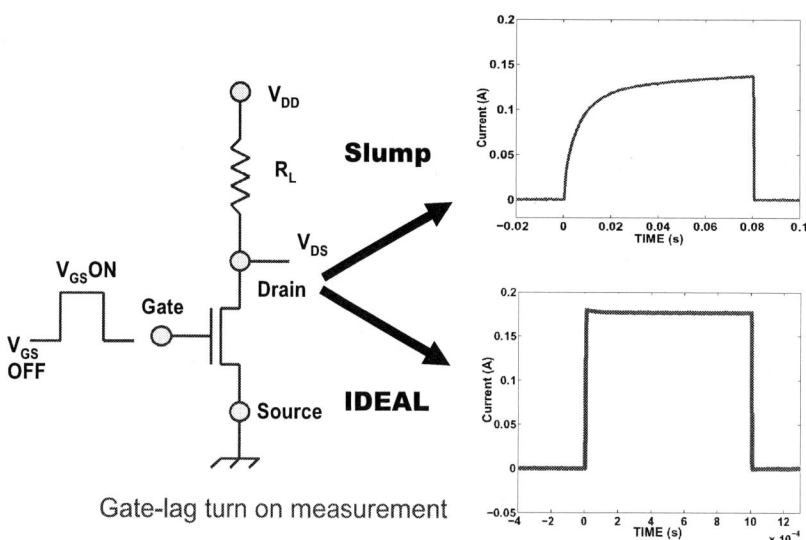

Gate-lag turn on measurement

[7] STEVEN C. BINARI, et al., "Trapping Effects in GaN and SiC Microwave FETs", PROCEEDINGS OF THE IEEE, VOL. 90, NO. 6, . 1048- 1058, JUNE 2002

DEPARTMENT OF
INFORMATION
ENGINEERING
UNIVERSITY OF PADOVA

Current collapse

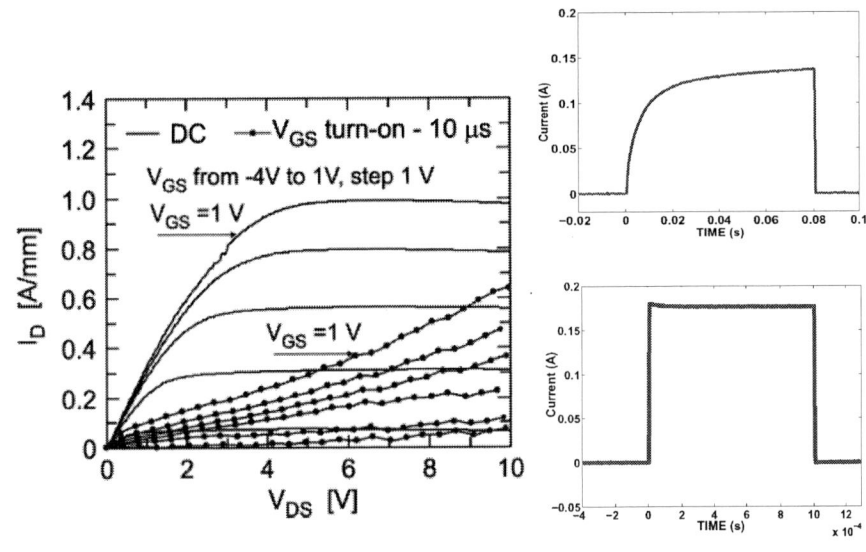

Current collapse

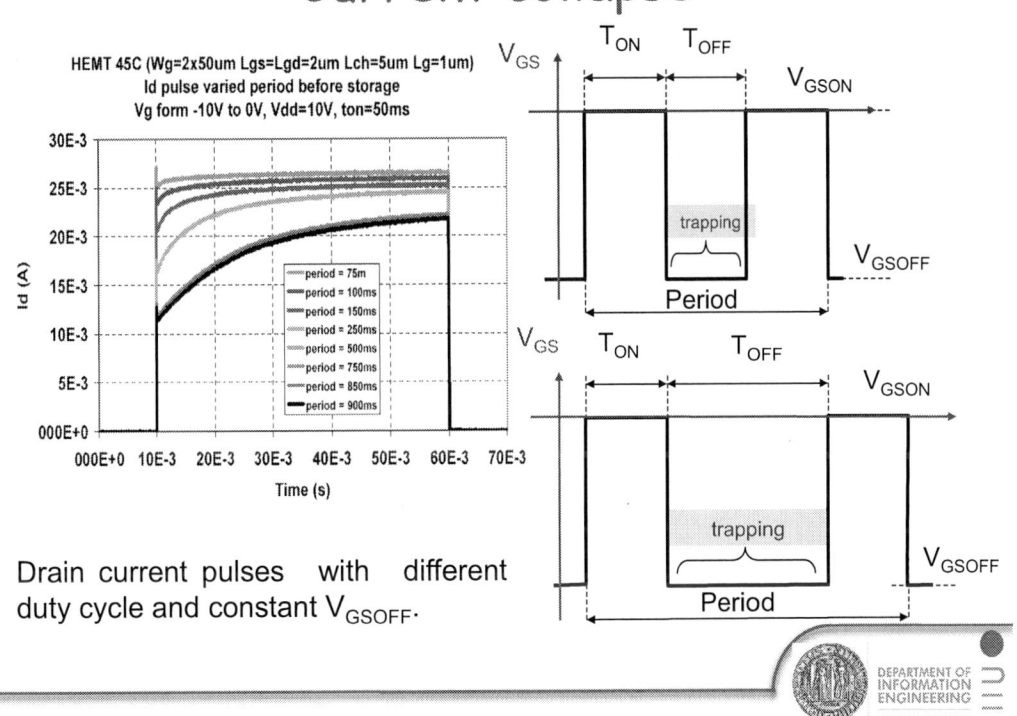

Drain current pulses with different duty cycle and constant V_{GSOFF}.

Current collapse

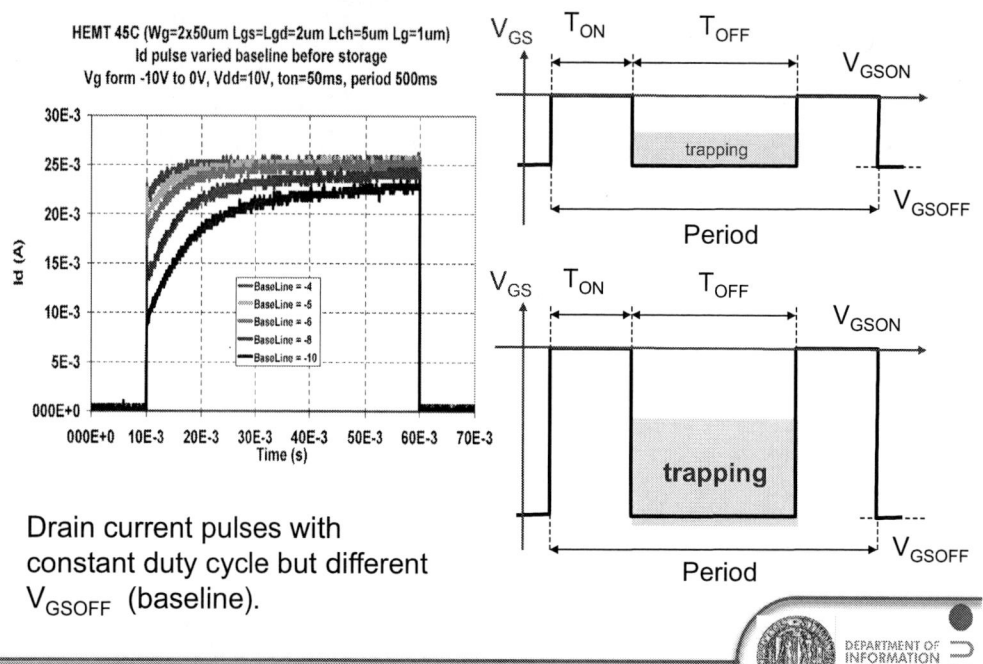

HEMT 45C (Wg=2x50um Lgs=Lgd=2um Lch=5um Lg=1um)
Id pulse varied baseline before storage
Vg form -10V to 0V, Vdd=10V, ton=50ms, period 500ms

Drain current pulses with constant duty cycle but different V_{GSOFF} (baseline).

DEPARTMENT OF
INFORMATION
ENGINEERING
UNIVERSITY OF PADOVA

Outline

❏ Introduction & Motivations
❏ Parasitic effects in GaN-HEMTs (Part I)
   ❏ Current collapse
      ❏ Definitions & Characterization
   ❏ Study of parasitic and dispersion effects: gate-lag, drain-lag, current collapse
   ❏ V_{th} instabilities in Power GaN HEMTs
   ❏ Controlling trapping phenomena
   ❏ Hot electrons
❏ Reliability issues in GaN-HEMTs (Part II)
❏ Conclusions

DEPARTMENT OF
INFORMATION
ENGINEERING
UNIVERSITY OF PADOVA

Double-Pulse Characterization

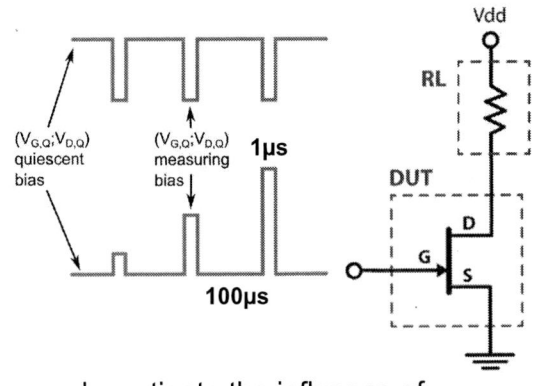

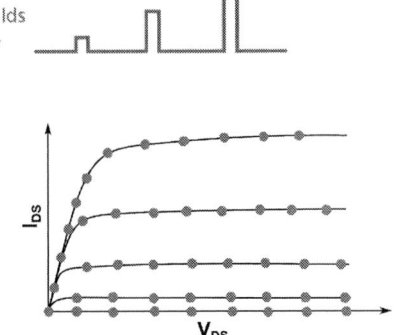

- Investigate the influence of **trapping phenomena** on I_D-V_D and I_D-V_G **dynamic characteristics**.
- Identify the Ids current dispersion in terms of dynamic Vth shift and dynamic Ron collapse.

Current collapse

Current collapse is (in general) due to both V_{th} shift and g_{mpeak} decrease.

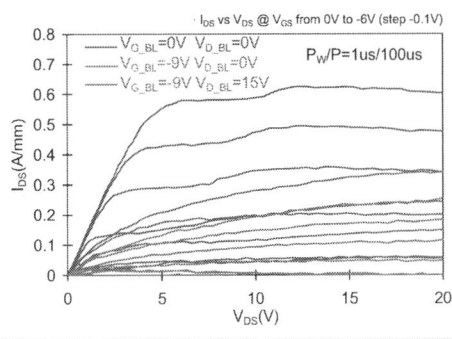

Current collapse is due mainly to surface traps and can be corrected by SiN passivation.

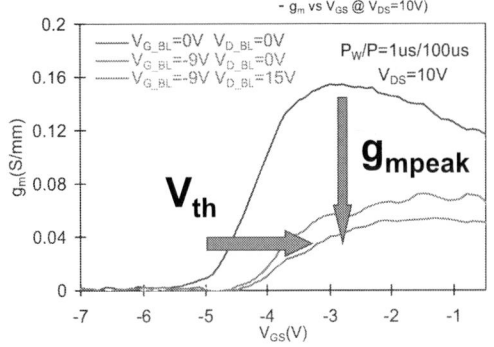

Epi traps can also induce collapse through V_{th} shift

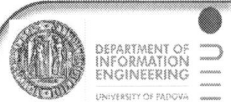

Current collapse

No - Traps

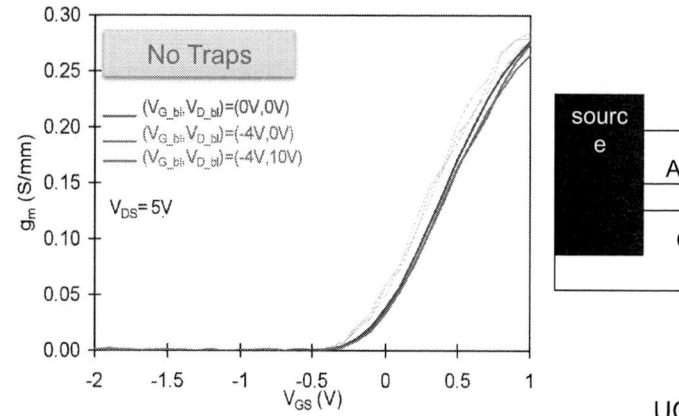

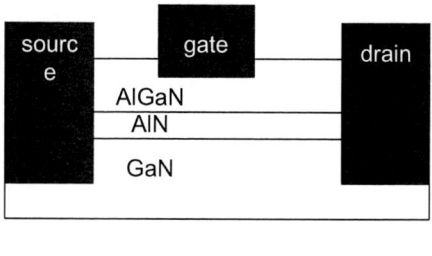

UCSB - Collaboration

Current collapse

Traps "under" the gate

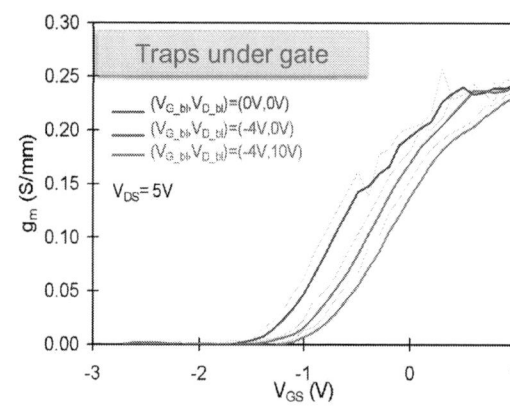

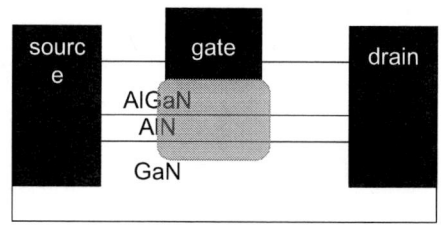

UCSB - Collaboration

Current collapse

Traps in the access regions

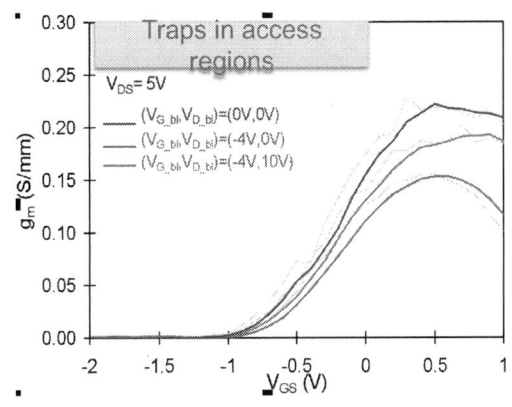

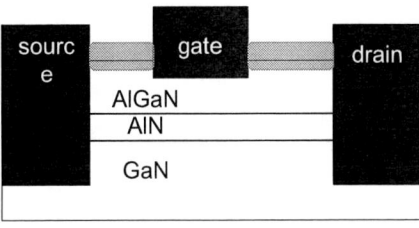

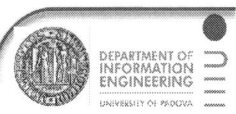

UCSB - Collaboration

Current collapse

Traps "under" the gate and in the access regions

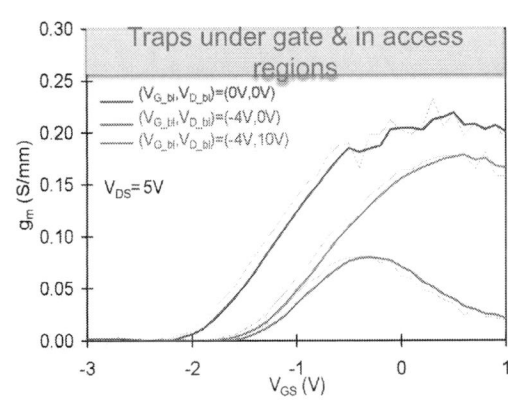

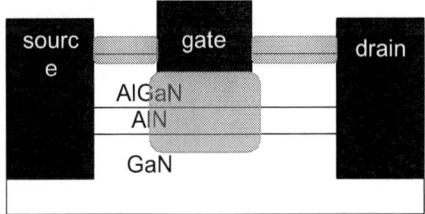

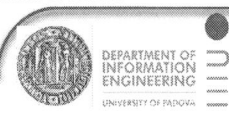

UCSB - Collaboration

978-1-4799-5494-0/14 $31.00 © 2014 IEEE

Traps localization

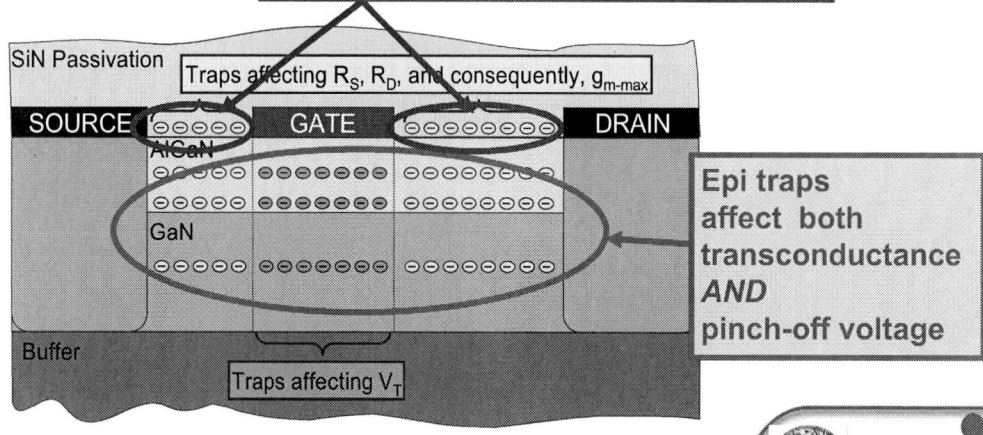

surface traps affect transconductance through series resistance increase they can not affect pinch-off voltage, unless Lg is fairly short

SiN Passivation

Traps affecting R_S, R_D, and consequently, g_{m-max}

SOURCE GATE DRAIN

AlGaN

GaN

Epi traps affect both transconductance *AND* pinch-off voltage

Buffer

Traps affecting V_T

Introduction

- **Characterization Techniques for 3-terminal HEMT structures**
  - **Double-Pulse**
    - Investigate the influence of trapping phenomena on I-V dynamic characteristics.
    - Identify the Ids current disperion, dynamic Vth shift and dynamic Ron collapse.
  - **Drain Current Transients**
    - Pulsed technique; test trapping phenomena in the real switching conditions with loadline.
    - Allows to separate the different defect-states involved for the overall current disperion, and to gather information on the trapping mechanisms.
  - **Gm(f)**
    - Quasi-equilibrium technique; investigate the deep-levels affecting the small-signal dynamic behavior.
    - Allows to map the depth (in vertical direction) of the detected defect-states.
- **Characterization Techniques for 2-terminal FATFET structure**
  - **DLTS**
    - Span a very large area of the Arrhenius plot.
- **Deep-Levels Database**

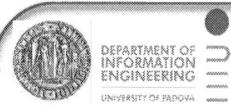

Drain Current Transient Analysis

Drain Current Transient Technique

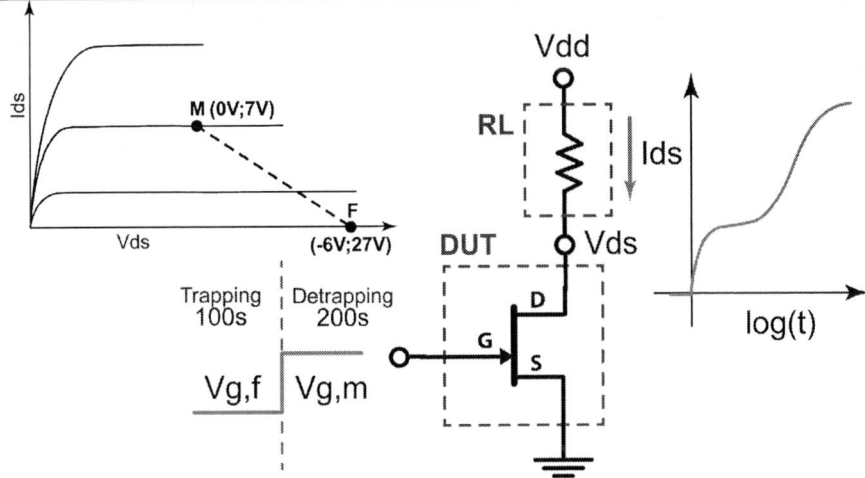

Conversely to Double-Pulse characterization, the Drain Current Transient analysis comprehensively investigate the **time evolution of carrier (de)trapping processes**. The **deep-levels** signatures – **activation energies** and **capture cross-sections** – and **their localization** can be achieved by performing the measurements under different bias conditions and different base-plate temperatures.

Drain Current Transient: Data Analysis

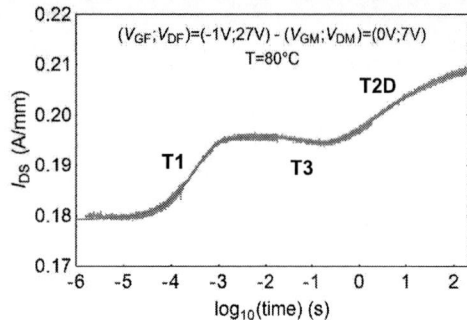

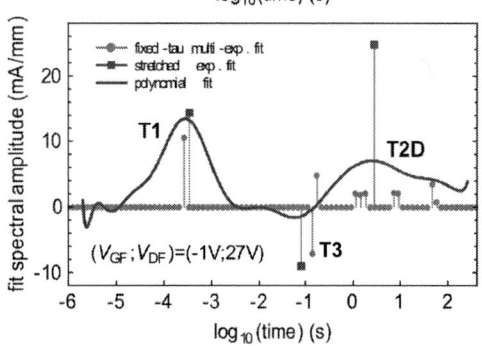

Since the thermal emission of trapped-carriers is governed by exponential law, the following algorithms have been adopted to fit the experimental data, and to extrapolate the emission rate of the observed (de)trapping processes:

- **Stretched multi-exponential**

$$I_{DS}(t) = I_{DS,\infty} - \sum_{i}^{N} A_i e^{-\left(\frac{t}{\tau_i}\right)^{\beta_i}}$$

- **Multi-exponential with fixed-τ-set**

$$I_{DS}(t) = I_{DS,\infty} - \sum_{i}^{100} A_i e^{-\frac{t}{fixed,\tau_i}}$$

- **Peak of the transient derivative**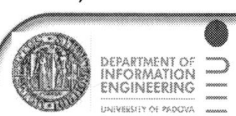
(fitted by polynomial function)

Drain Current Transient: Filling Bias Comparison

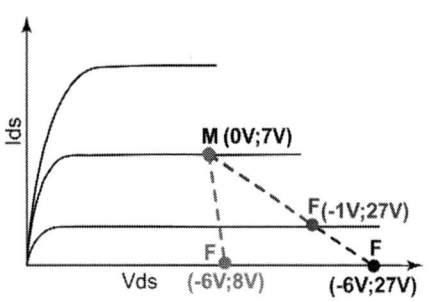

Properly selecting gate- and drain- voltage during the trapping phase promotes trapping phenomena in different device region:

- **Semi-on-state** (high V_{DG} and relatively high I_{DS}) promote charge **trapping into buffer region**, likely caused by **hot-electrons** mechanisms.

- **Gate-reverse-bias** promotes **trapping below the gate region**, where the free-carriers available for capturing are likely supplied by the **gate-leakage current**.

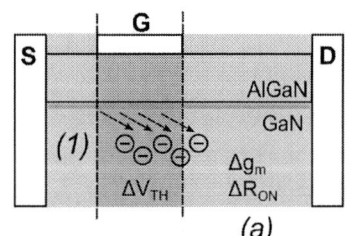

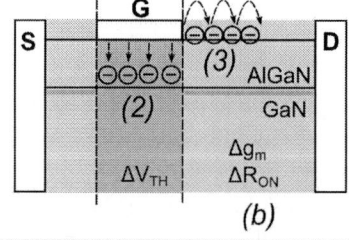

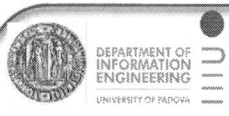

Drain Current Transient: Filling Bias Comparison

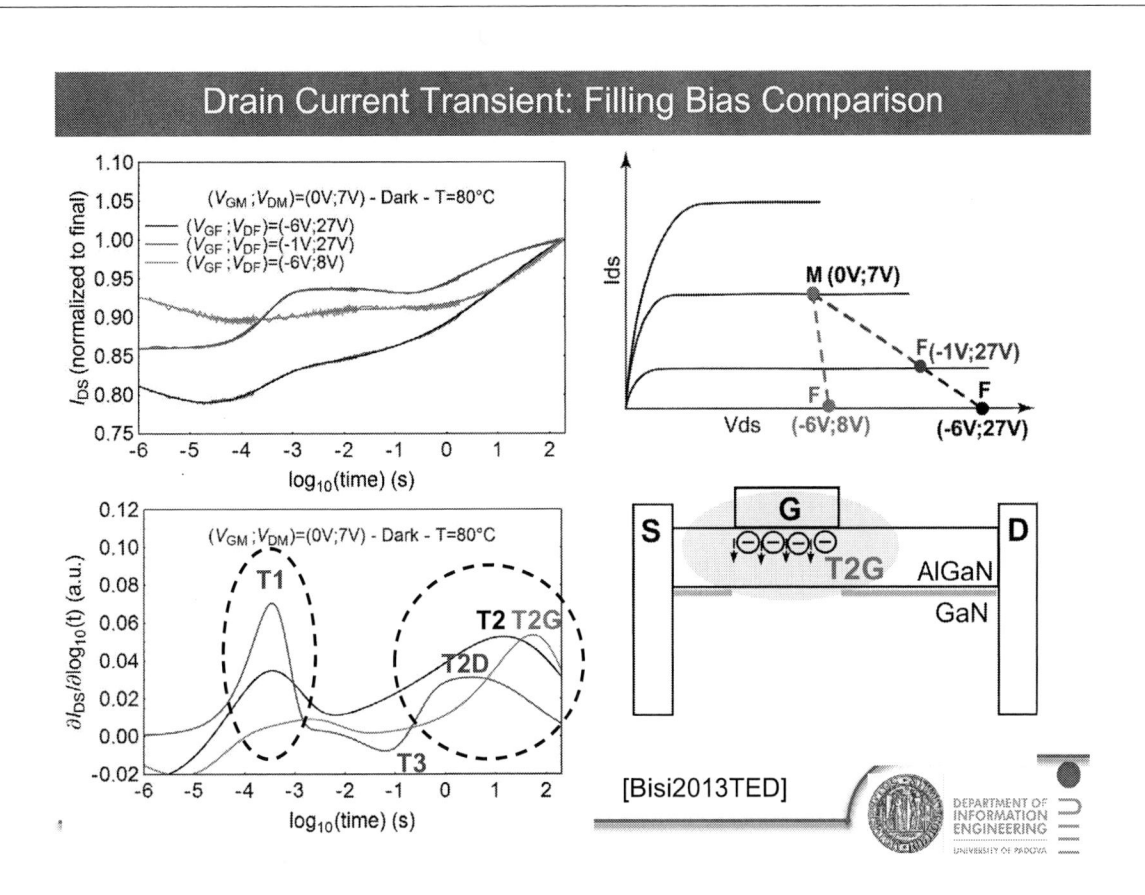

[Bisi2013TED]

Drain Current Transient: Thermal Activation

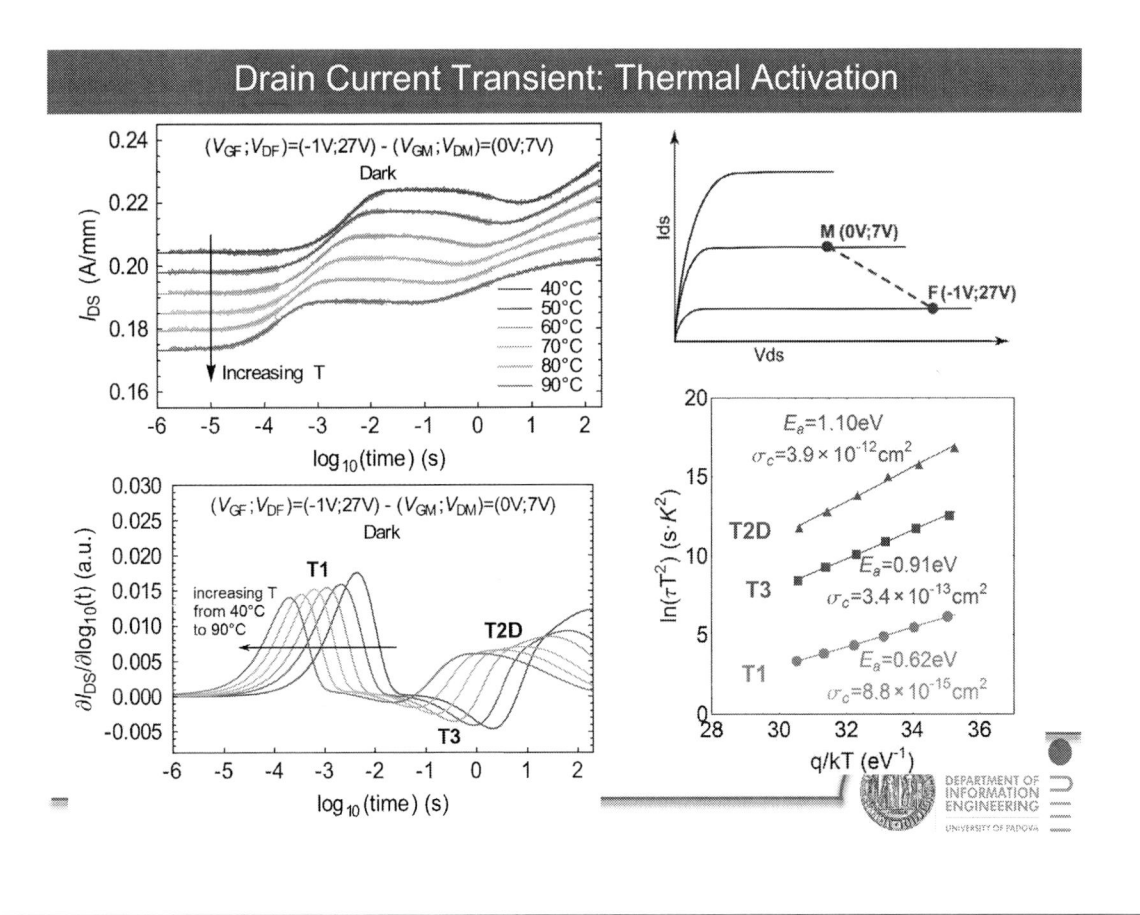

Drain Current Transient: Thermal Activation

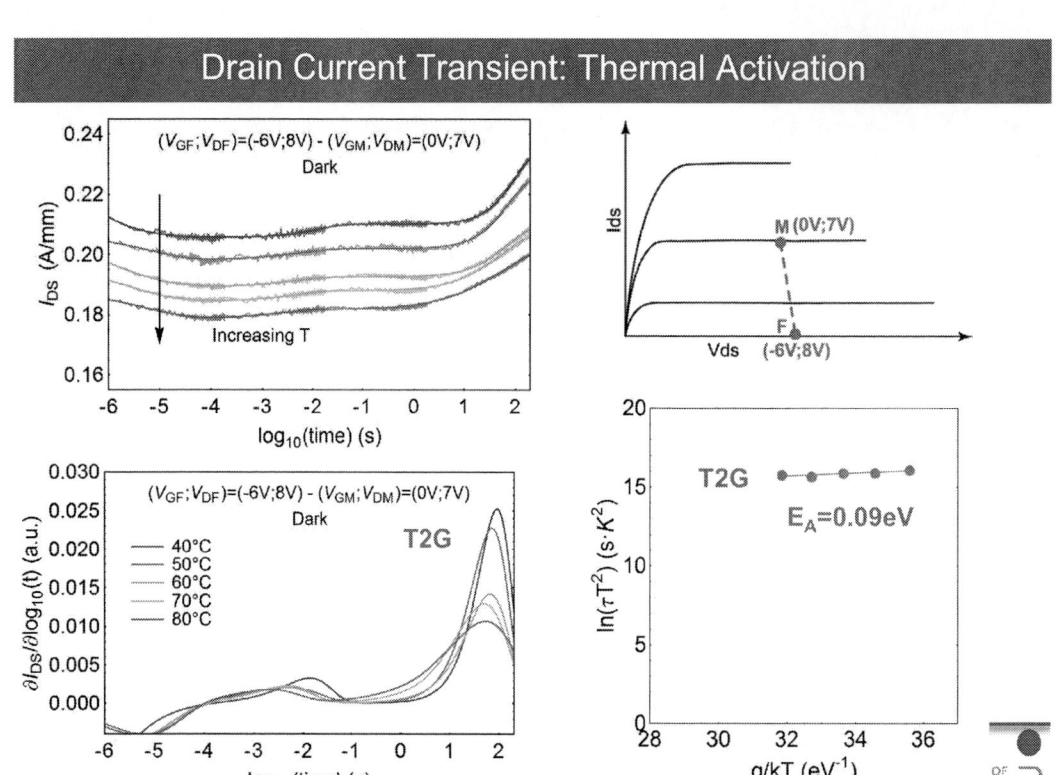

Non-thermally-activated Detrapping Processes

The weakly- or non-thermally-activated de-trapping transients could be dominated by **direct-** or **trap-assisted-tunneling** or **hopping mechanisms**, either at superficial defect-states or within AlGaN barrier mid-gap states.

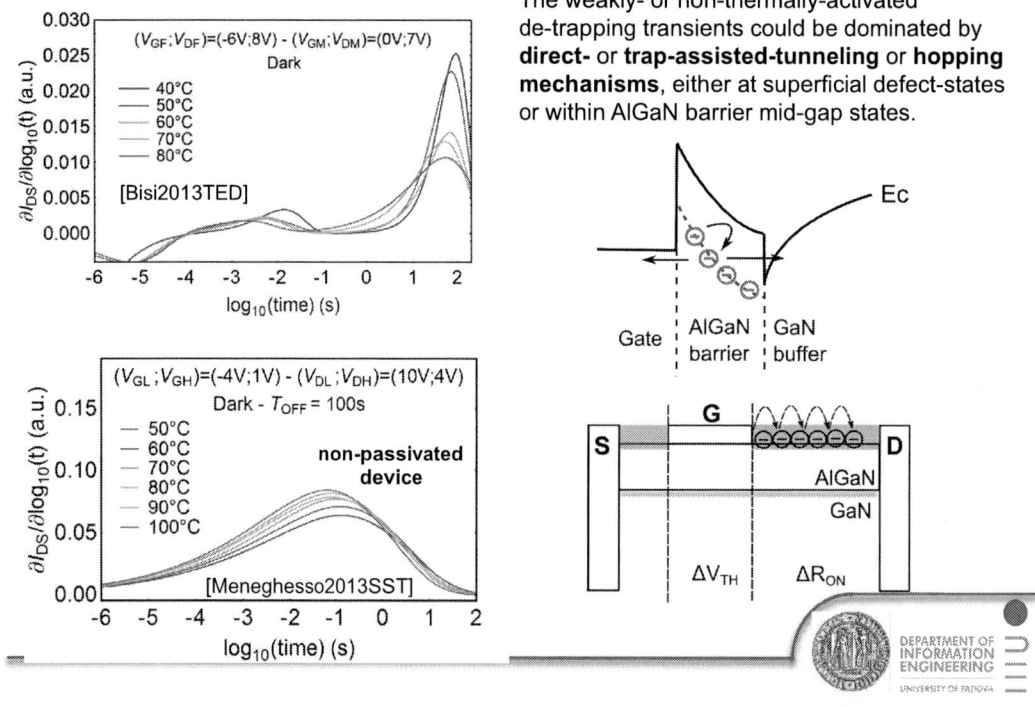

Self-heating effects and channel temperature estimation

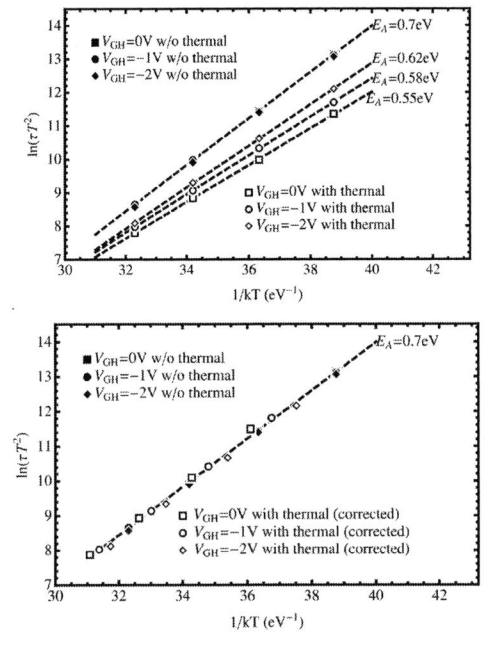

Since Drain Current Transients are acquired in **non-tracurable self-heating bias conditions**, the actual channel temperatures, employed in the Arrhenius Plot, have been estimated by means of thermal resistance and dissipated power.

[Chini2013TED]

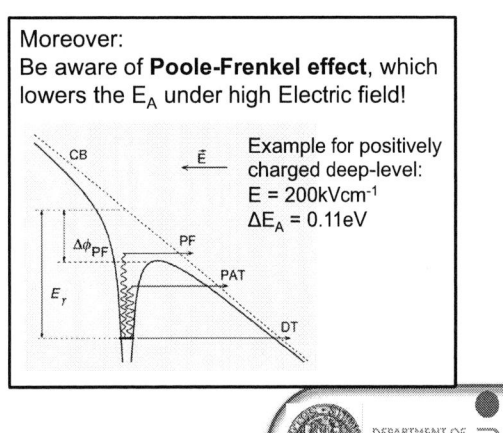

Moreover:
Be aware of **Poole-Frenkel effect**, which lowers the E_A under high Electric field!

Example for positively charged deep-level:
E = 200kVcm^{-1}
ΔE_A = 0.11eV

Non-exponential Stretched Behaviour

Non-exponential stretched emission processes are often detected.

Likely causes could be related to:

- **superposition** of different deep-levels emission processes;

- **high density** of defect states, as in the case of **extended line defects**, that lead to trapped **electron delocalization** and formation of **mini-bands**;

- presence of **defect-states in alloy materials**, such as $Al_xGa_{(1-X)}N$, which lead to **deep-level splitting** due to band discontinuity;

- **non-uniform distribution** of **electric-field** and **temperature** within the complex HEMT structure;

- (de)trapping kinetics are not governed by thermal emission, but by other mechanisms such as **hopping** or **trap-assisted-tunneling**.

Stretching factor:
- β=1: Ideal exponential,
- 0.5<β<1: commonly observed stretched behaviour,
- B<0.5: rarely observed highly-stretched-exponential.

$$I_{DS}(t) = I_{DS,final} - Ae^{-\left(\frac{t}{\tau}\right)^{\beta}}$$

978-1-4799-5494-0/14 $31.00 © 2014 IEEE   237

Trapping and how to identify buffer traps?

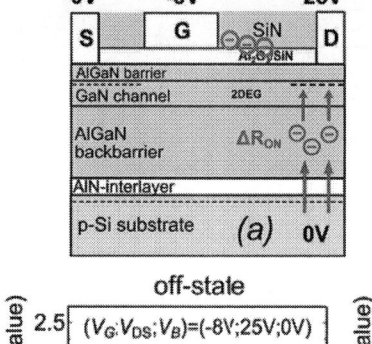
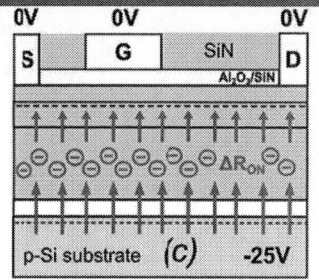

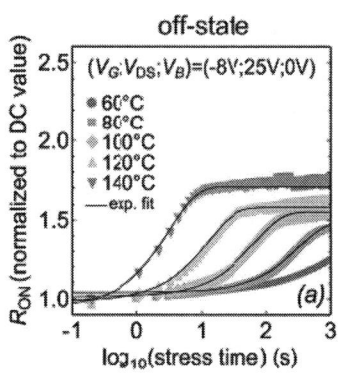
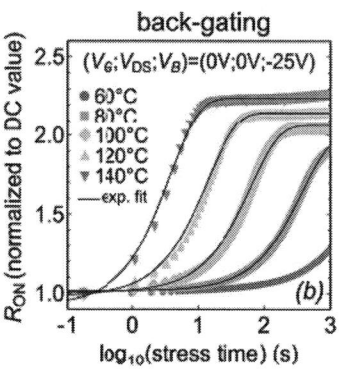

- **Trapping time in power devices require long time (100 s)**
- **Temperature accelerate the trapping mechanisms (this is why one can see larger trapping at higher T)**
- **Buffer or surface traps?**

EDL 35, 1004 (2014)

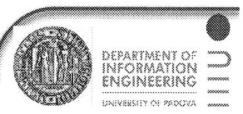

Pre- and Post-Stress Characterization

DCT characterization during stress-tests gets an insight on defects evolution throughout the device degradation. **Electrical-stress may promote**:

- The **increase of pre-existing defects density**,
  e.g., the increase of dislocations density under thermo-mechanical stress.

- The **generation of new electrically active defect-states**,
  e.g., Oxigen in diffusion under gate-edge-degradation.

- The **strengthening of trapping mechanisms**,
  e.g., the generation of parasitic paths which leads to higher leakage currents, thus higher density of free-electrons available for trapping.

[Tapajina2010APL]   [Caesar2012IRPS]   [Rossetto2013ESREF]

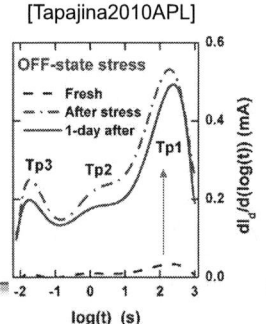
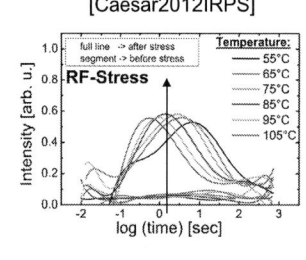
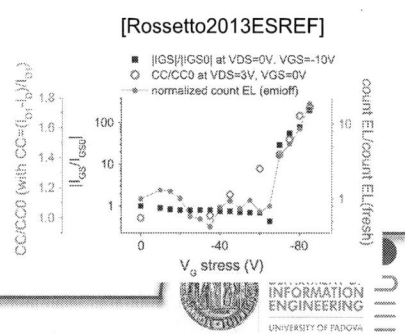

Deep-Levels Database

Deep-Levels Database

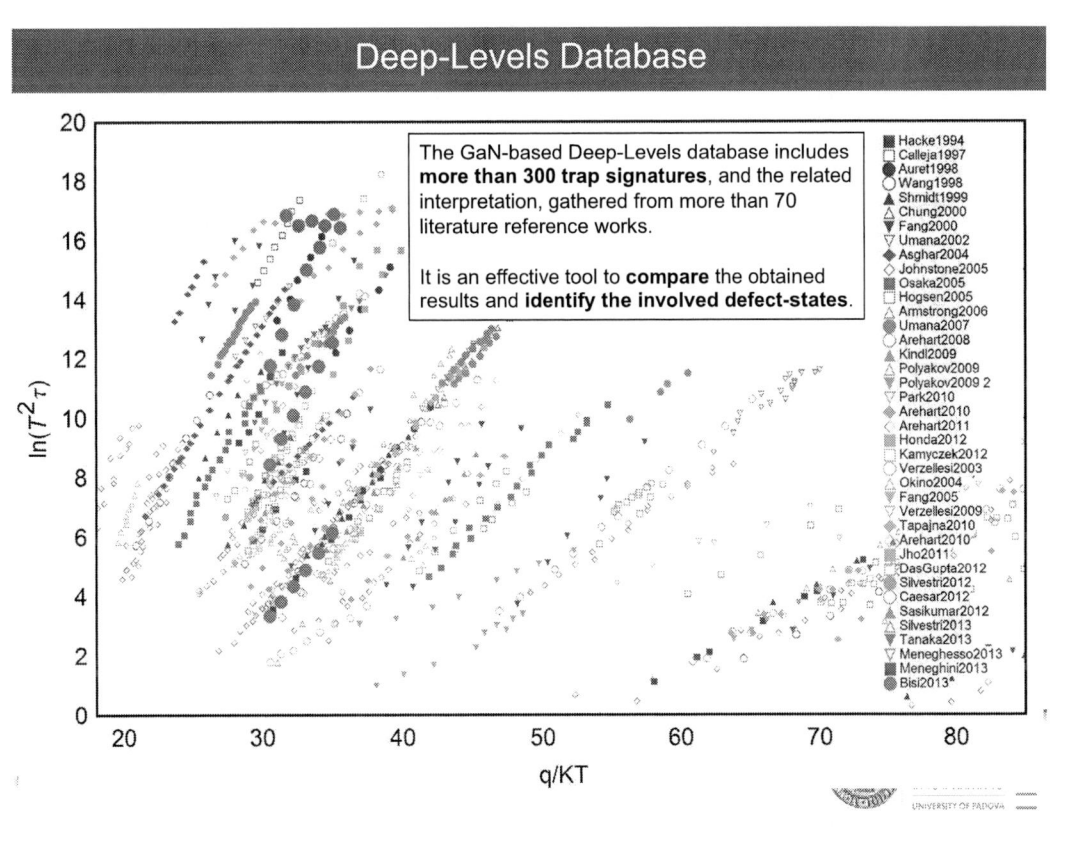

The GaN-based Deep-Levels database includes **more than 300 trap signatures**, and the related interpretation, gathered from more than 70 literature reference works.

It is an effective tool to **compare** the obtained results and **identify the involved defect-states**.

Deep-Levels Database

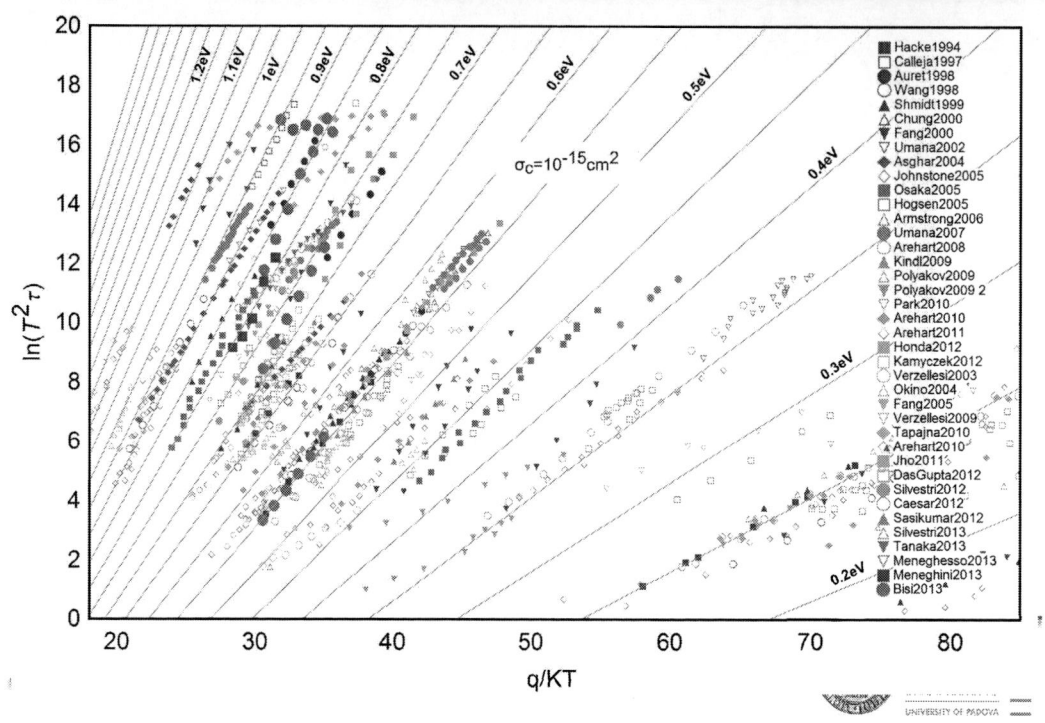

Deep-Levels causing Dynamic Current Dispersion

The list below reports the **electrically-active deep-levels** detected in our labs on several GaN-based HEMTs families. Their tentative, **still under debate origins** are also reported.

	E_A	σ_c (cm^2)	Comments
Ec -	0.62eV	9×10^{-15}	• Commonly-observed E2/0.6eV level. • Iron-doping influenced
Ec -	0.50eV	2×10^{-19}	• Reported as Oxigen-related defect, or open-core edge dislocations.
Ec -	0.64eV	2×10^{-16}	
Ec -	0.83eV	1×10^{-15}	• Reported as AlGaN-related defect.
Ec -	0.85eV	1×10^{-14}	• Observed mainly in Carbon-doped devices.
Ec -	1.1eV	4×10^{-12}	• Tentatively ascribed to dislocations. • Also reported as Gallium-interstitial, or Gallium-antisite defect.
Ev +	0.91eV	3×10^{-13}	• Gallium-Vacancy. • Carbon in Nitrogen substitutional-position.
Non-thermally-activated and highly-stretched behavior			• Detected in non-passivated devices. • Ascribed to superficial trap-states.
Non-thermally-activated			• AlGaN barrier /GaN cap (via tunneling).

Bisi, IEEE TED Oct. 2013

Outline

- ❏ Introduction & Motivations
- ❏ Parasitic effects in GaN-HEMTs (Part I)
    - ❏ Current collapse
        - ❏ Definitions & Characterization
    - ❏ Study of parasitic and dispersion effects: gate-lag, drain-lag, current collapse
    - ❏ V_{th} instabilities in Power GaN HEMTs
    - ❏ Controlling trapping phenomena
    - ❏ Hot electrons
- ❏ Reliability issues in GaN-HEMTs (Part II)
- ❏ Conclusions

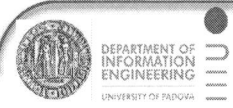

Samples

- Depletion-mode AlGaN/GaN HEMTs on Si

- GaN buffer intentionally doped with C (GaN:C)

- Undoped GaN channel (GaN:UID),

- $Al_{0.22}Ga_{0.78}N$ barrier

- Insulated gate and double field-plate structure.

- L_G=3 μm, L_{GS}=1.5 μm, L_{GD}=12 μm, source-connected field-plate overhang of 6 μm.

- Large periphery, packaged devices.

G. Meneghesso et al., IEEE IRPS 2014

Double Pulse

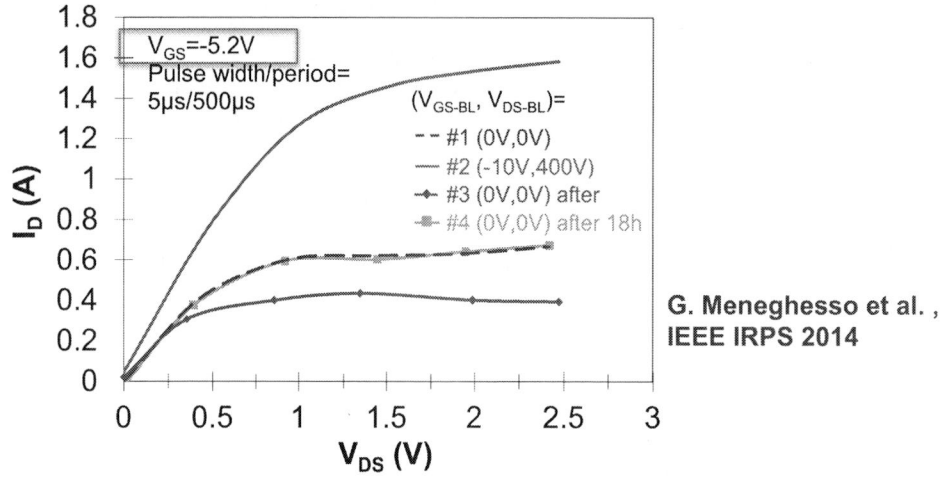

V_{GS}=-5.2V
Pulse width/period= 5μs/500μs

(V_{GS-BL}, V_{DS-BL})=
- #1 (0V,0V)
- #2 (-10V,400V)
- #3 (0V,0V) after
- #4 (0V,0V) after 18h

G. Meneghesso et al. ,
IEEE IRPS 2014

Drain current instabilities are triggered by pulsed measurements with an off-state quiescent point.

DEPARTMENT OF INFORMATION ENGINEERING
UNIVERSITY OF PADOVA

Double Pulse

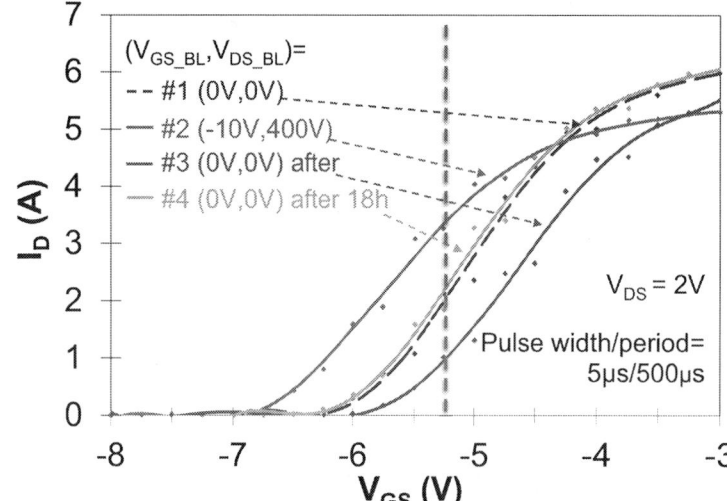

(V_{GS_BL}, V_{DS_BL})=
- #1 (0V,0V)
- #2 (-10V,400V)
- #3 (0V,0V) after
- #4 (0V,0V) after 18h

V_{DS} = 2V

Pulse width/period= 5μs/500μs

G. Meneghesso et al. ,
IEEE IRPS 2014

I_D instabilities are due to dynamic V_{TH} shifts of both polarities: initial state → negative ΔV_{TH} → positive ΔV_{TH} → negative ΔV_{TH} (recover to initial state)

DEPARTMENT OF INFORMATION ENGINEERING
UNIVERSITY OF PADOVA

DC Gate Bias Stresses

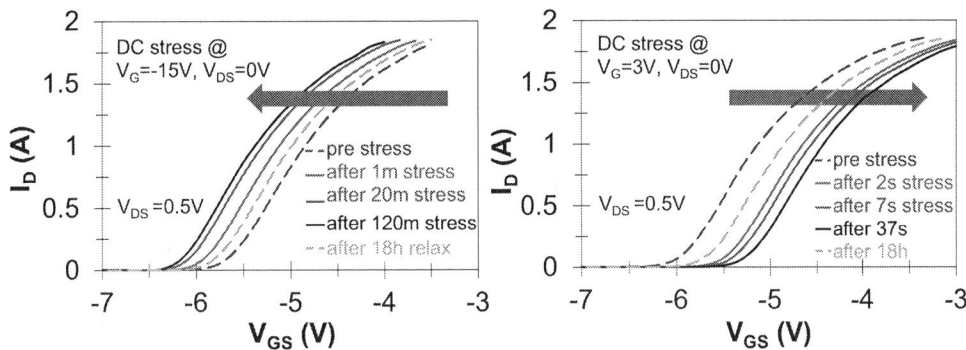

Negative gate bias stresses cause a negative V_{TH} shift, while positive gate bias stresses induce a positive V_{TH} shift.
These shifts are only partially recoverable (depending on stress time and magnitude).

G. Meneghesso et al. ,
IEEE IRPS 2014

Transient Behavior

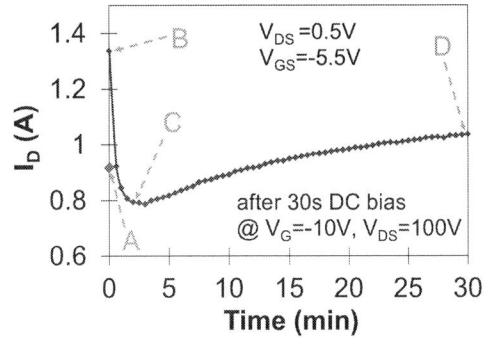
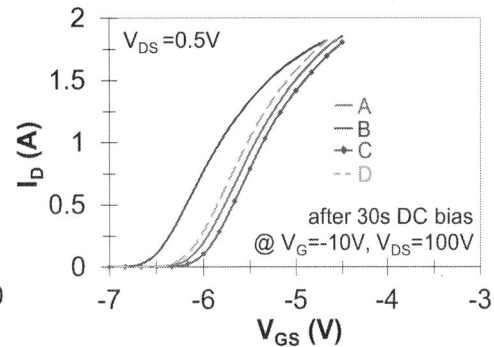

A: pre-stress ($I_D = I_{PRE}$)

B: $I_D \gg I_{PRE} \Leftrightarrow \Delta V_{TH} < 0$

C: $I_D < I_{PRE} \Leftrightarrow \Delta V_{TH} > 0$

D: $I_D \approx I_{PRE} \Leftrightarrow \Delta V_{TH} < 0$

B to C: $\tau \approx 10^2$ s

C to D: $\tau \approx 10^3$ s

G. Meneghesso et al. ,
IEEE IRPS 2014

Negative V_{TH} – mechanisms

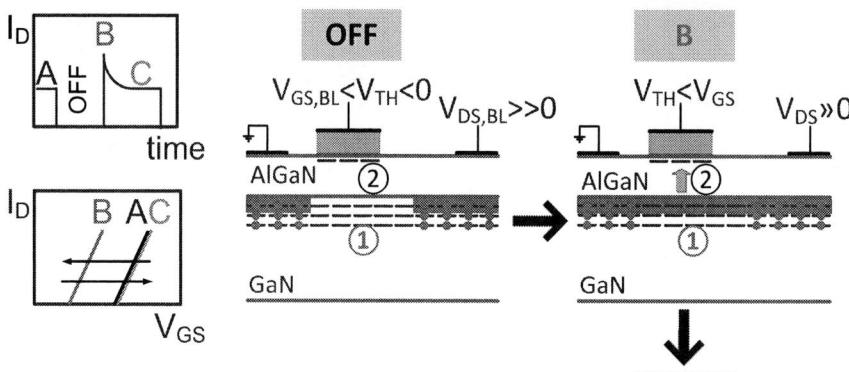

① e- emission from channel traps during OFF, followed by 2DEG reformation and e- trapping into channel traps:

- basic mechanism always present unless negligible trap density in GaN channel;
- supported by observation that ΔV_{TH} vanishes with C doping optimization and same processing.

G. Meneghesso et al. , IEEE IRPS 2014

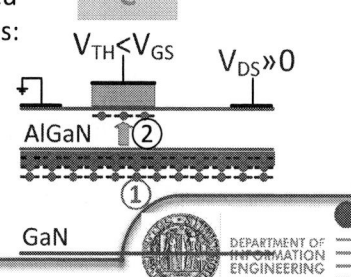

Positive V_{TH} – mechanisms

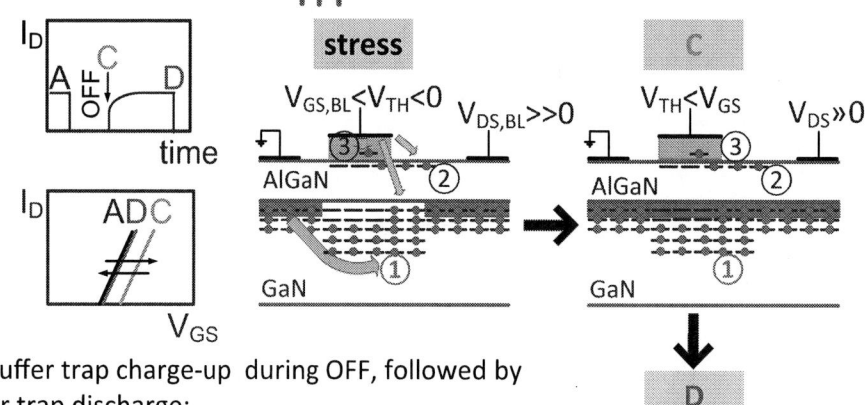

① buffer trap charge-up during OFF, followed by buffer trap discharge:

- trap charge/discharge can occur by e- capture/ emission (e- traps in the GaN:UID channel?) and/or to h+ emission/capture (h+ traps in the GaN:C buffer?);
- supported by observation that ΔV_{TH} vanishes with C doping optimization and same processing.

Numerical Simulations

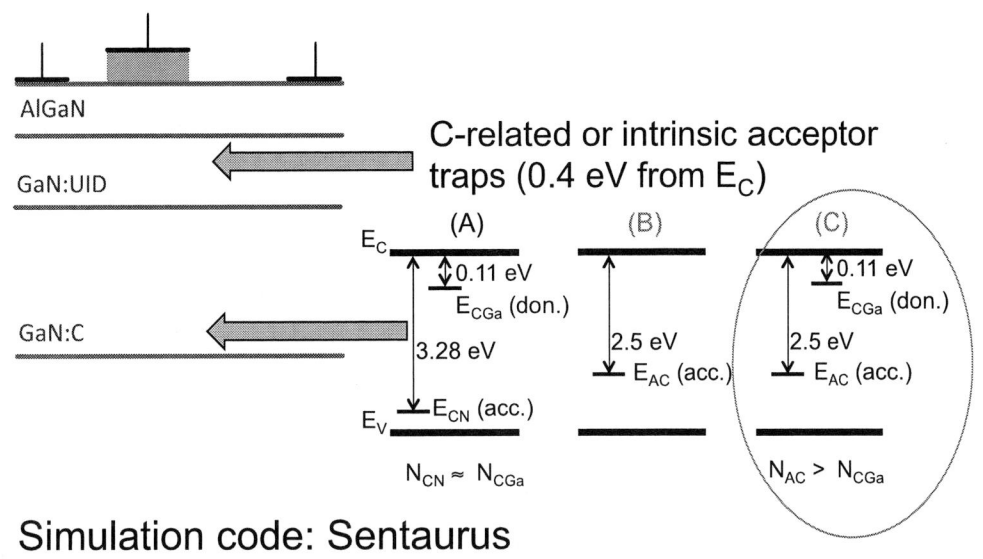

C-related or intrinsic acceptor traps (0.4 eV from E_C)

Simulation code: Sentaurus Device™ (version G-2012).

Positive V_{TH} – mechanisms

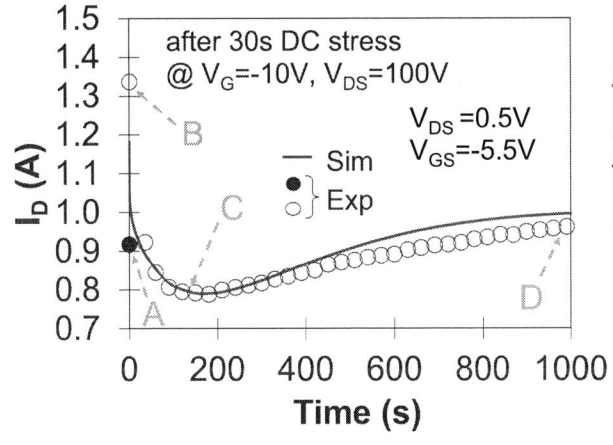

after 30s DC stress @ V_G=-10V, V_{DS}=100V

V_{DS} =0.5V
V_{GS} =-5.5V

— Sim
● Exp

A: pre-stress
B: neutral GaN:UID traps (no e⁻) & neg. GaN:C traps (no h⁺)

G. Meneghesso et al., IEEE IRPS 2014

B to C: e⁻ capture into GaN:UID traps
C: neg. GaN:UID traps & neg. GaN:C traps
C to D: h⁺ capture into GaN:C traps

Outline

❑ Introduction & Motivations
❑ Parasitic effects in GaN-HEMTs (Part I)
 ❑ Current collapse
  ❑ Definitions & Characterization
 ❑ Study of parasitic and dispersion effects: gate-lag, drain-lag, current collapse
 ❑ V_{th} instabilities in Power GaN HEMTs
 ❑ Controlling trapping phenomena
 ❑ Hot electrons
❑ Reliability issues in GaN-HEMTs (Part II)
❑ Conclusions

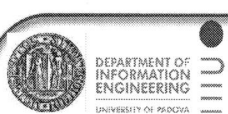

Material Quality

**Wafer Quality Target for Current-Collapse-Free GaN-HEMTs
in High Voltage Applications**

Hidetoshi Fujimoto, Wataru Saito, Akira Yoshioka, Tomohiro Nitta, Yorito Kakiuchi and Yasunobu Saito

Semiconductor Company, Toshiba Corp.

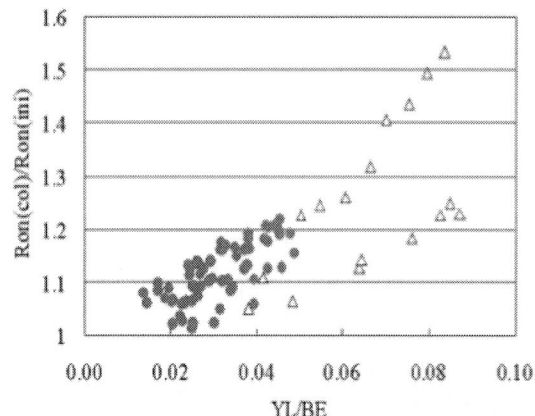

Correlation between YL-to-BE intensity ratio (YL/BE) and on-resistance ratio Ron(col)/Ron(ini) for conventional (triangles) and improved (circles) wafers

[9] CS MANTECH Conference, April 14-17, 2008, Chicago, Illinois, USA

Passivation

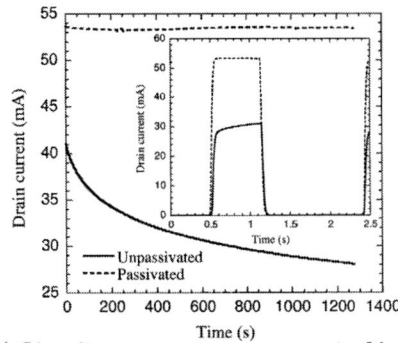 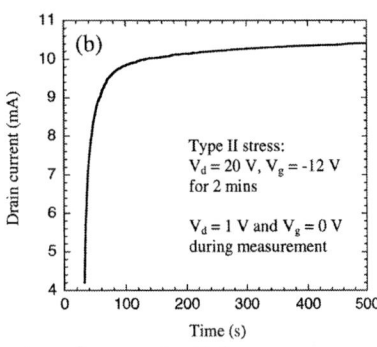

✓ Simultaneous measurement of I_{DS} and surface potential

✓ Degradation attributed to electron accumulation into deep surface traps.

✓ **SiN$_x$ passivation results in surface-trap-density reduction or in charge stabilization**

[10] Koley, G.; Tilak, V.; Eastman, L.F.; Spencer, M.G., "Slow transients observed in AlGaN/GaN HFETs: effects of SiNx passivation and UV illumination", IEEE Transactions on Electron Devices, vol. 50, pp. 886-893, 2003

Cap Layer

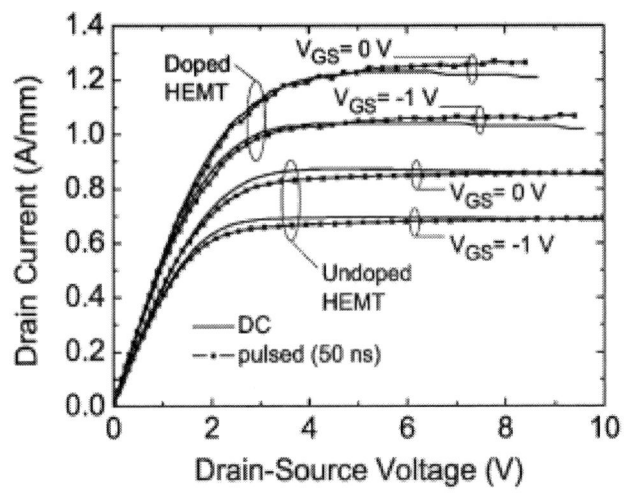

✓ **GaN cap layer can successfully remove the current collapse**

✓ **However some gate-leakage issues can rise.**

[11] G. Meneghesso, F. Rampazzo, P. Kordos, G. Verzellesi, E. Zanoni, "Current Collapse and High-Electric-Field Reliability of Unpassivated GaN/AlGaN/GaN HEMTs, IEEE Transaction on Electron Devices, Vol. 53, No. 12, pp. 2932-2941, 2006

deep gate recess

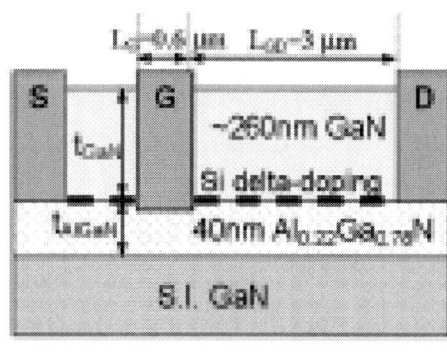
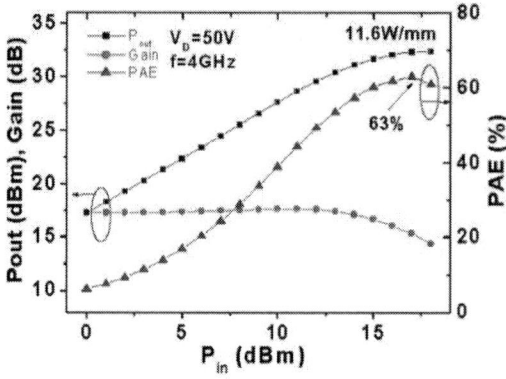

Principle:

Keep surface (traps) far away from the channel as much as possible

[12] L. Shen, Y. Pei, L. McCarthy, C. Poblenz, A. Corrion, N. Fichtenbaum S. Keller, S. P. Denbaars, J. S. Speck and U. K. Mishra, IEEE Microwave Theory and Techniques Symposium, 2007

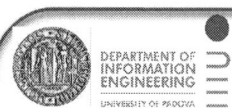

Outline

❑ Introduction & Motivations
❑ Parasitic effects in GaN-HEMTs (Part I)
    ❑ Current collapse
        ❑ Definitions & Characterization
    ❑ Study of parasitic and dispersion effects: gate-lag, drain-lag, current collapse
    ❑ V_{th} instabilities in Power GaN HEMTs
    ❑ Controlling trapping phenomena
    ❑ Hot electrons
❑ Reliability issues in GaN-HEMTs (Part II)
❑ Conclusions

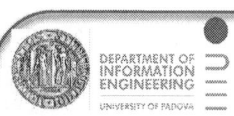

978-1-4799-5494-0/14 $31.00 © 2014 IEEE   248

I.I. in GaAs Based HEMTs

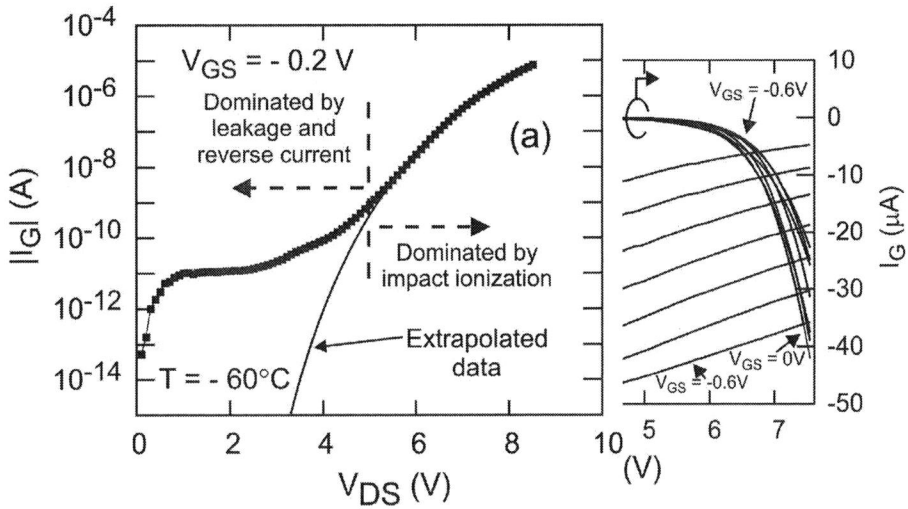

[21] G. Meneghesso, et al, "Hot carriers effects in AlGaAs/InGaAs High Electron Mobility Transistors: failure mechanisms induced by hot-carrier testing", Journal of Applied Physics, Vol. 82, No. 11, pp. 5547-5554, 1997

I.I. induced Gate Current and light emission in GaAs Based HEMTs

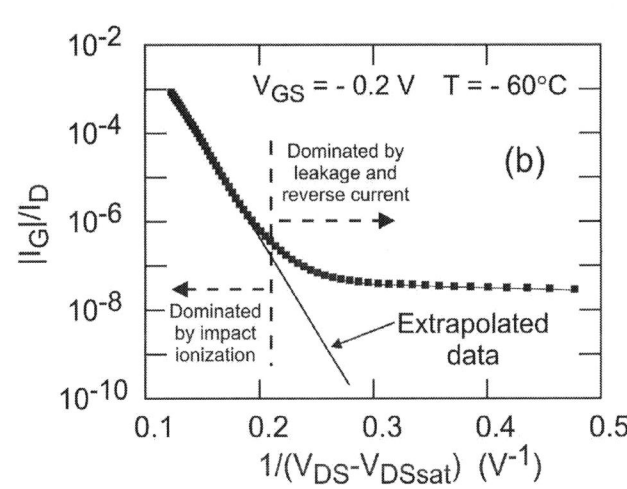

$$|I_G|/I_D \simeq \alpha_n \cdot L_{eff} \simeq L_{eff} \cdot \exp\left(\frac{1}{E}\right)$$

$$\simeq L_{eff} \cdot \exp\left(\frac{-L_{eff}}{V_{DS} - V_{DSsat}}\right)$$

Chynoweth's law:
A.G. Chynoweth, Phys. Rev., 109, pp. 1537-1540, 1958
and:
K. Hui et al, IEEE Electron Device Lett. 11, p. 113 1990.

978-1-4799-5494-0/14 $31.00 © 2014 IEEE   249

I.I. induced Gate Current and light emission in GaAs Based HEMTs

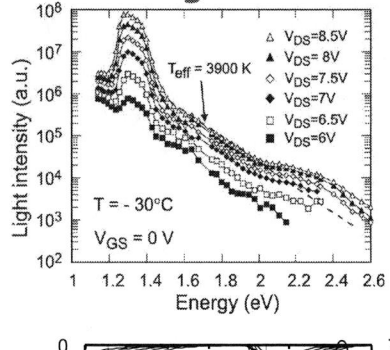

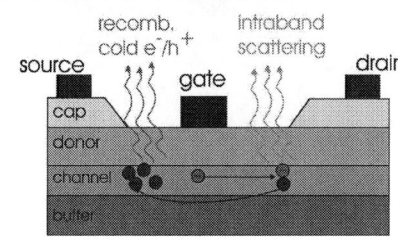

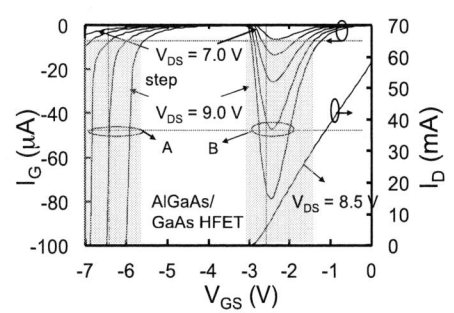

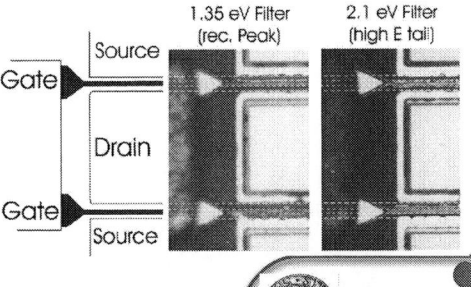

What happens in GaAs HEMTs??

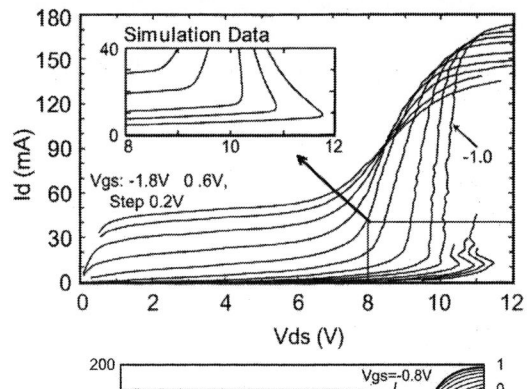

In GaAs HEMTs:

- **Impact Ionization**
- **Hole accumulation**
- **Parassitic BJT**
- **Snap-back**
- **+ thermal phenomena**

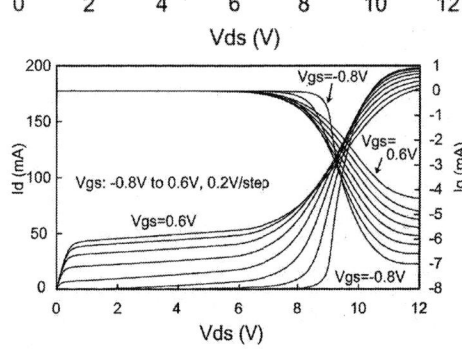

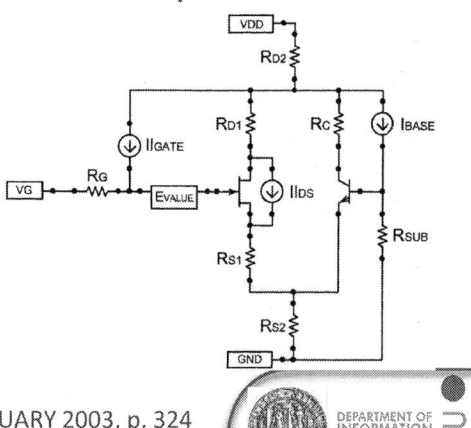

G. Meneghesso et al, IEEE TED vol. 50, NO. 2, FEBRUARY 2003, p. 324

If e-h pare are present, recombination peak is present

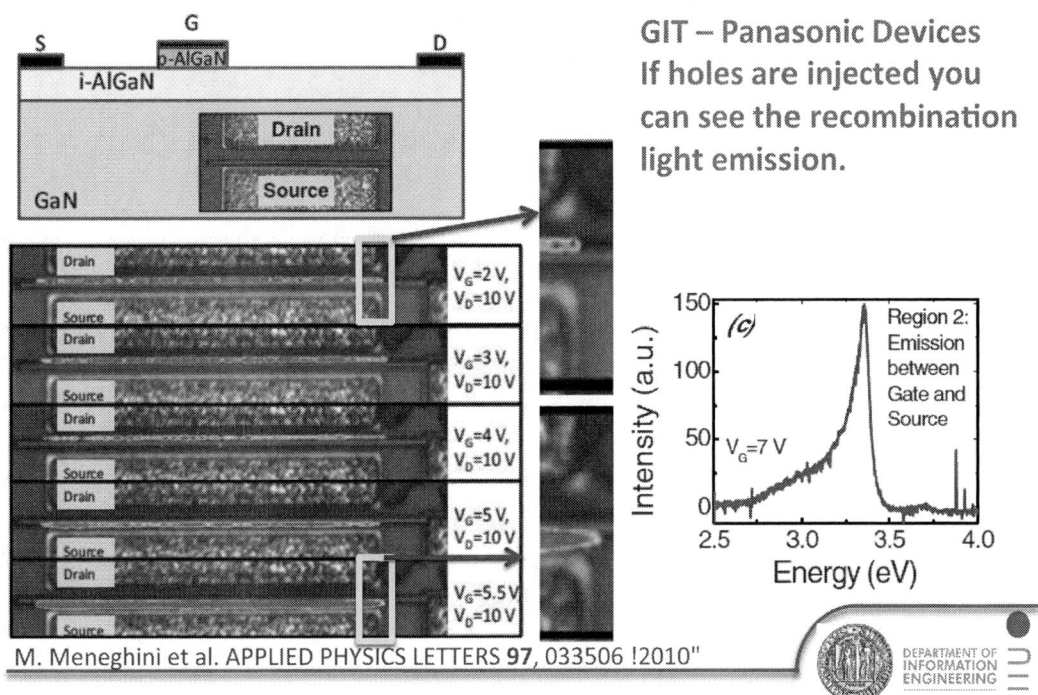

GIT – Panasonic Devices
If holes are injected you can see the recombination light emission.

M. Meneghini et al. APPLIED PHYSICS LETTERS **97**, 033506 !2010"

I.I. induced Gate Current and in GaAs Based HEMTs

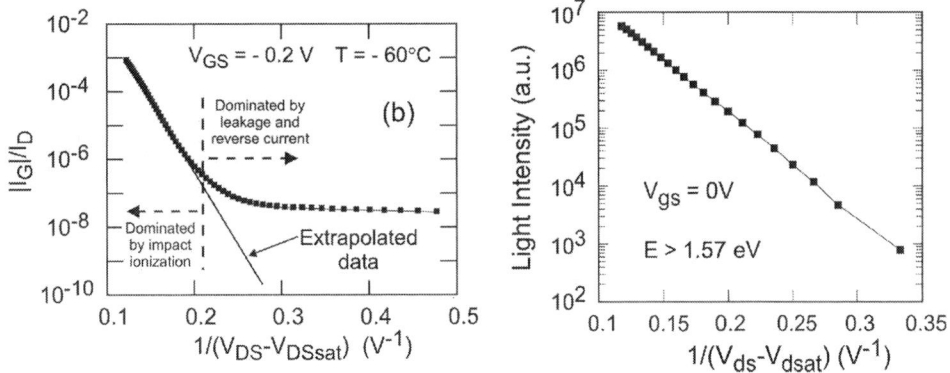

The total (integrated) light emission follow the same behavior as the impact ionization gate current i.e. the Chynoweth's law!

I.I. induced Gate Current
in GaN-Based HEMTs

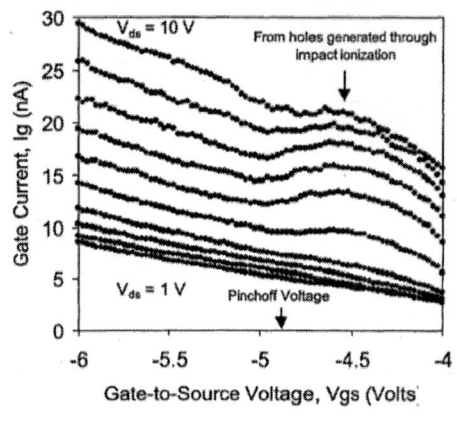

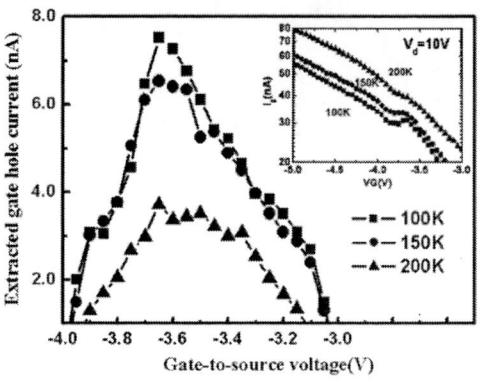

[22] B. Brar et al. Proc. IEEE Lester Eastman Conf., pp. 487-491, 2002.

[23] Ching-Hui Lin et al. IEEE Electron Device Letters, 2005 pp. 710 - 712

The impact ionization gate current is always much **lower** than the leakage current!

DEPARTMENT OF
INFORMATION
ENGINEERING
UNIVERSITY OF PADOVA

False I.I. induced Gate Current
in GaN-Based HEMTs

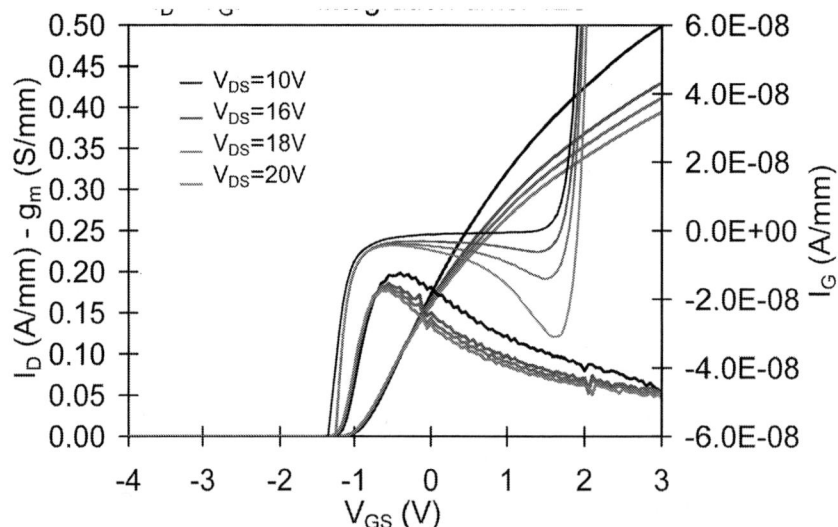

INFORMATION
ENGINEERING
UNIVERSITY OF PADOVA

978-1-4799-5494-0/14 $31.00 © 2014 IEEE   252

False I.I. induced Gate Current in GaN-Based HEMTs

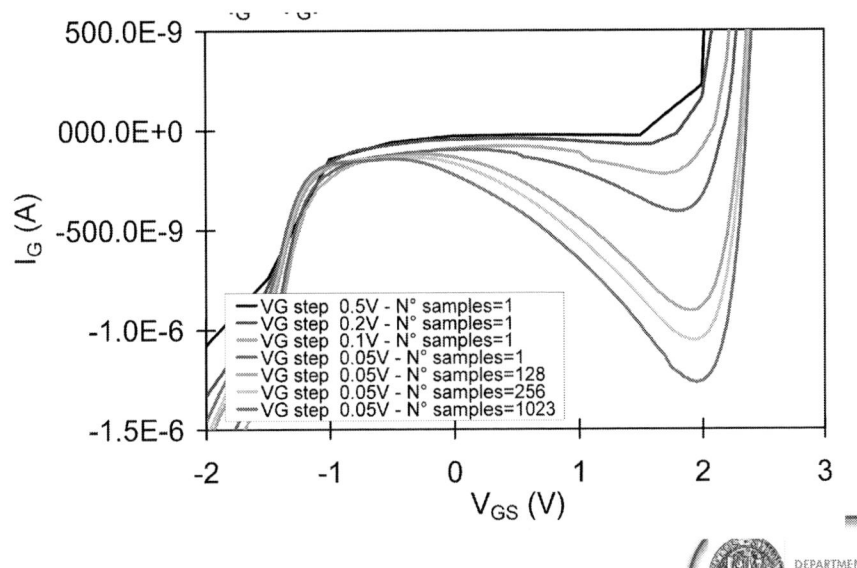

I.I. induced Gate Current GaAs vs GaN Based HEMTs

The gate leakage current at high V_{DS} can not be used as a good probe for the presence of impact-ionization in GaN-based HEMTs since:

a) the contribution of holes generated by impact ionization gate current is always much **lower** than the leakage current

b) A different scenario is present in GaN: impact-ionization is negligible in GaN, holes have a much lower life-time, ...

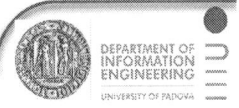

Light emission in **GaN** Based HEMTs

SOURCE

GATE

DRAIN

DRAIN

SOURCE

What about using the light emission as a good probe for the presence of hot electrons in the channel?

OK, lets analyze the light emission!

DEPARTMENT OF
INFORMATION
ENGINEERING
UNIVERSITY OF PADOVA

Light emission in **GaN** Based HEMTs

Gate leakage current, NO "bell shape" -> bad probe for hot electrons

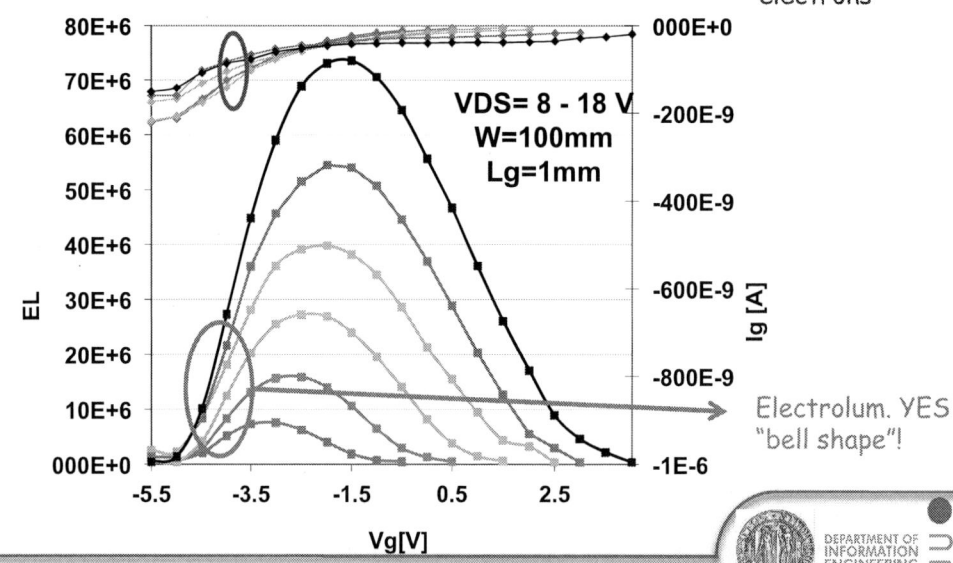

VDS= 8 - 18 V
W=100mm
Lg=1mm

Electrolum. YES "bell shape"!

DEPARTMENT OF
INFORMATION
ENGINEERING
UNIVERSITY OF PADOVA

Light em.: a different scenario

AlGaN/GaN HEMT, Lg = 0.5 μm   **AlGaAs/GaAs HEMT, Lg = 0.25 μm**

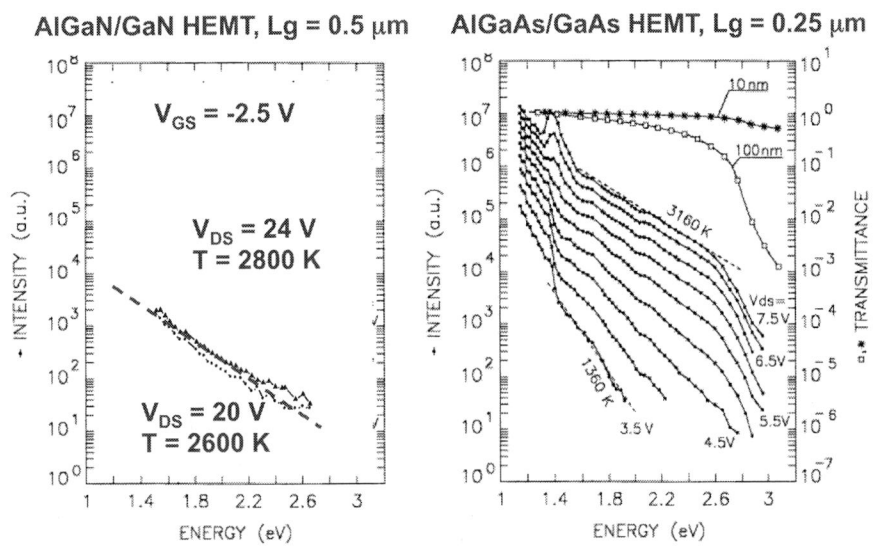

No evidence of e/h pair generation and recombination
Intensity relatively low compared to GaAs-based devices

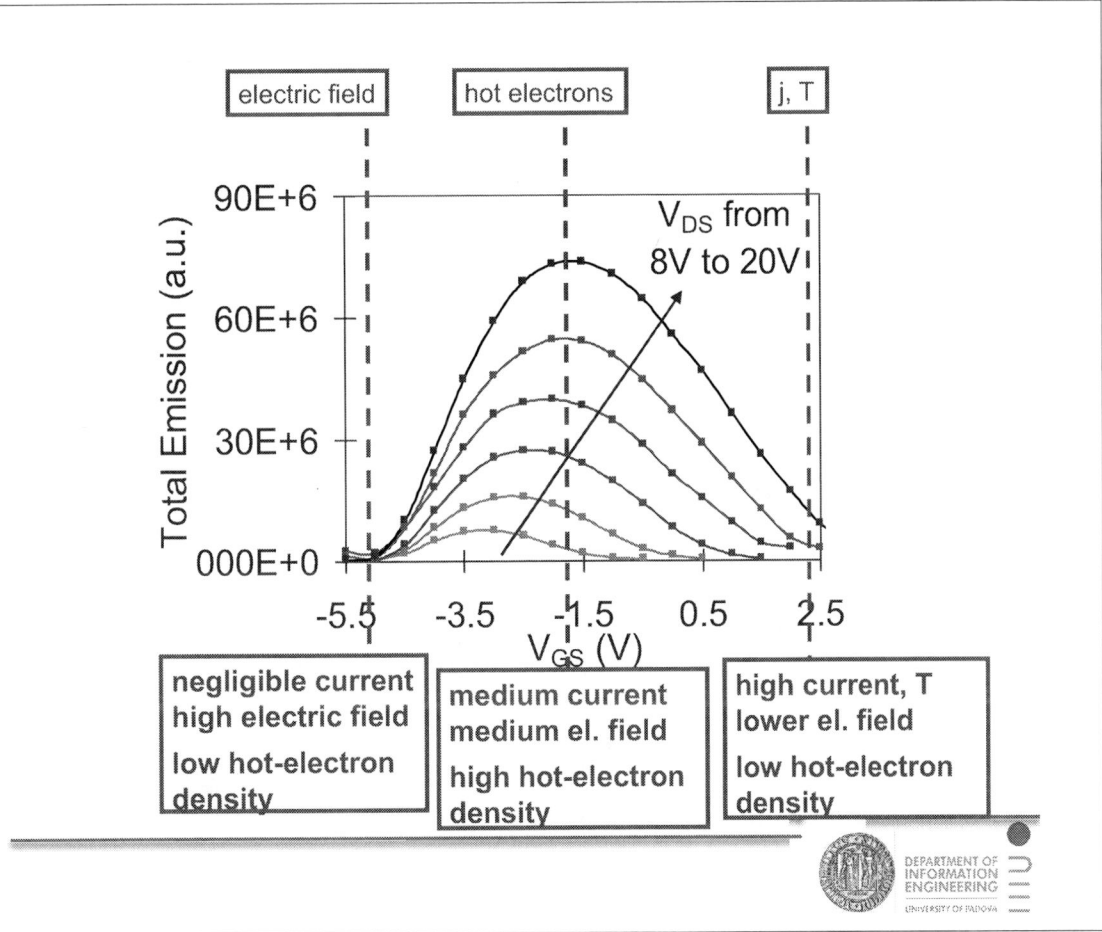

Conclusions – Part I

- An overview of the more common parasitic effects in GaN has been presented: current slump, gate leakage; kink in the output I-V, hot electrons;
- Both material (substrate, epitaxial layers, passivation ....) and processing play crucial role in the game against parasitic phenomena;
- A rather complex scenario has been highlighted due to the complex nature of the GaN

By means of detailed analysis of material quality and of adequate device processing it will be possible, with the help of careful device characterization and simulations, to identify and localize parasitic phenomena and possibly to develop technological countermeasure (passivation, FP, gate recess, ...) for the improvement of the GaN-based devices characteristics.

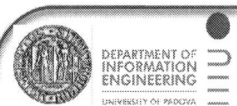

DEPARTMENT OF
INFORMATION
ENGINEERING
UNIVERSITY OF PADOVA

Outline

- ❏ Introduction & Motivations
- ❏ Parasitic effects in GaN-HEMTs (Part I)
    - ❏ Current collapse
        - ❏ Definitions & Characterization
    - ❏ Study of parasitic and dispersion effects: gate-lag, drain-lag, current collapse
    - ❏ V_{th} instabilities in Power GaN HEMTs
    - ❏ Controlling trapping phenomena
    - ❏ Hot electrons
- ❏ Reliability issues in GaN-HEMTs (Part II)
- ❏ Conclusions

Reliability and High Field related issues in GaN-HEMT devices – PART II

Gaudenzio Meneghesso

Enrico Zanoni, Matteo Meneghini,
University of Padova,
Department of Information Engineering,
via Gradenigo 6/B, 35131 Padova, Italy
gaudenzio.meneghesso@unipd.it

"HEMT reliability" papers 1987-2012 (>25 years !)

Scopus

Search | Alerts | My list | Settings

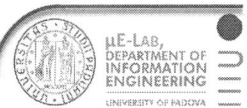

Your query: TITLE(((gallium nitride) OR (gan)) AND ((reliability) OR (degradation) OR (trapping)))

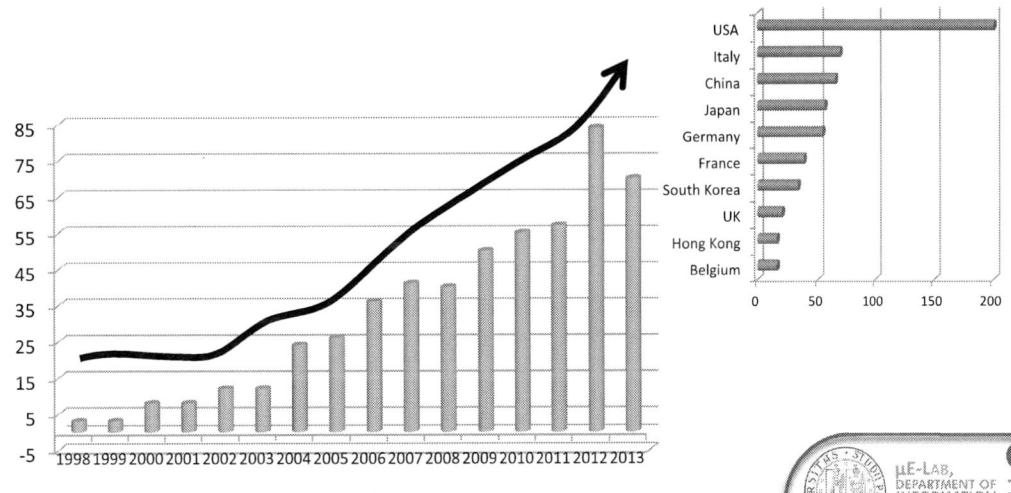

Failure mechanisms of AlGaN/GaN HEMTs

Schottky contact degradation
Thermally activated Au-Ni interdiffusion
Gate leakage current increase

Reverse bias or off-state degradation
Converse piezoelectric effect
Deep levels generation
Time dependent breakdown (point defects percolation)
Electrochemical surface degradation (GaN oxidation)
GaO dissolution, oxygen indiffusion
Groove formation, surface pitting
Increase of gate leakage current, decrease of drain current, increased current collapse

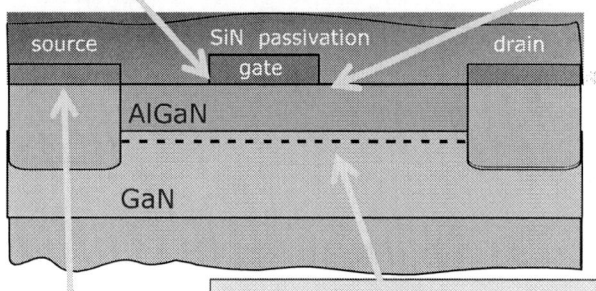

Ohmic contact degradation
Id and gm decrease

On state degradation
Hot electron effects
Charge trapping in the SiN
Dehydrogenation of point defects
Deep levels generation
Transconductance degradation
Drain current decrease
Threshold voltage shifts

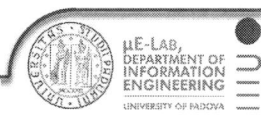

Outline 1

- Introduction
- Degradation mechanisms induced by reverse-bias or off-state testing of GaN HEMTs
  - converse piezoelectric effect
  - generation of deep levels
  - time-dependent AlGaN breakdown
  - donor trap generation at the AlGaN/GaN interface
  - electrochemical surface degradation
  - indiffusion of impurities
  - controversial effects
- Effects of charge trapping on the gate-drain access region
- Competing degradation effects (off-state and on-state)
  - gate leakage degradation (walk-in), gate-drain charge trapping (walk-out)
  - DC vs rf degradation

Outline 2

- Hot electron induced degradation (on-state degradation)
  - diagnostic of hot electrons using electroluminescence
  - dependence on gate and drain bias
  - acceleration law
  - model based on defects dehydrogenation

- Thermally-induced gate Schottky contact metallurgical degradation
  - NiSi formation at gate edges
  - Au indiffusion at sidewall and through the Ni layer

- Conclusions

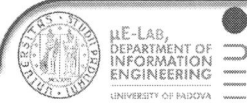

Converse piezoelectric effect

- Off-state degradation
- Gate leakage increase
- Converse piezoelectric effect
- Critical voltage concept
- Multiphysics modeling of piezoelectric effects

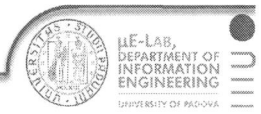

Step-stress gate reverse bias experiments

HEMTs may degrade when submitted to reverse-bias stress

Joh et al. suggested a step-stress method to identify the degradation

There is a "critical (gate-drain) voltage" beyond which the device starts degrading, showing a permanent increase in gate leakage current.

Piezoelectric effect model

Joh et al., EDL 29, 287, 2008

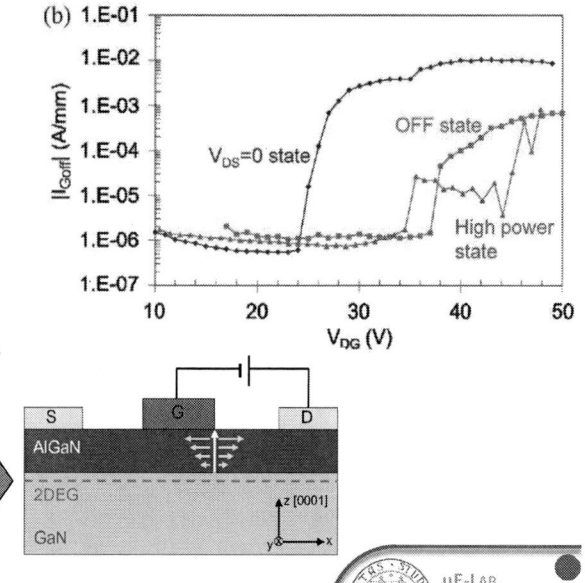

Off-state gate degradation: current collapse increase

Deep levels are generated, causing current collapse
Pulsed current slump and gate leakage current increase are correlated. Trap generation with activation energy 0.56 eV

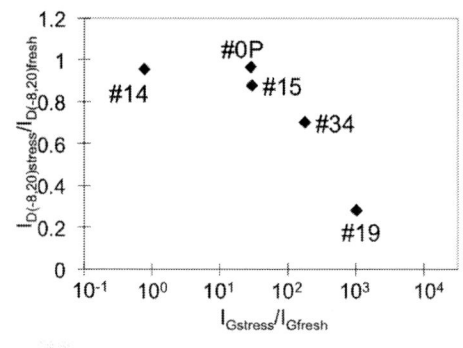

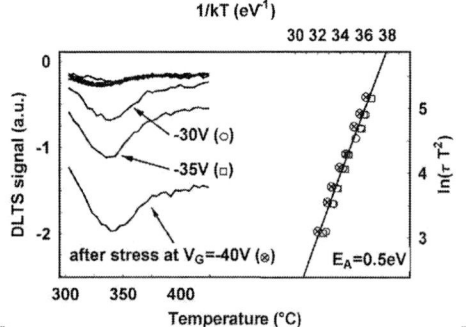

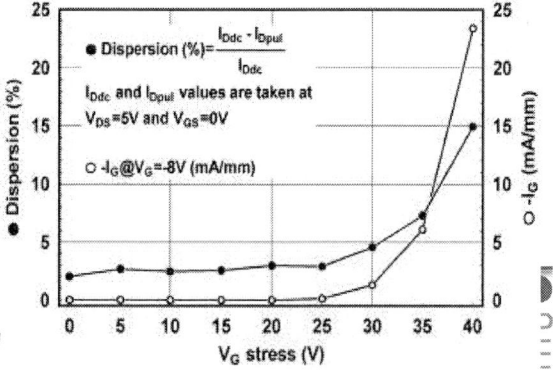

Multiphysics modeling of converse piezoelectric effect

Multi-dimensional device simulator: fully-coupled

- * equations of electroelasticity
- * drift-diffusion carrier transport
- * linear heat conduction

hp:
- * steady-state conditions
- * biaxial in-plane strain subtracted → from 3D to 2D
- * thick substrate modeled as thermal resistance

Analysis of converse piezoelectric effects in AlGaN/GaN HEMTs

Mechanical failure criteria

Simulations : 0.3 µm gate HEMT 30% 25 nm AlGaN, 1.6 µm gate-drain spacing
junction temperature 400°C, peak E ~11.5 MV/cm stress 4.6 GPa

2011 Ancona et al. SISPAD

Multiphysics modeling of converse piezoelectric effect

Simulations : 0.3 µm gate HEMT 30% 25 nm AlGaN, 1.6 µm gate-drain spacing
junction temperature 400°C, peak E ~11.5 MV/cm stress 4.6 GPa

INTRINSIC DEVICE (NO PIT) : failure due to converse piezoelectric effect
is unlikely (piezoelectric contribution to stress 0.5 GPa, thermal 9.6 GPa)

DAMAGED DEVICE (AFTER PIT FORMATION) : once a pit is formed
the strain under pit (2nm x 3nm) is much higher (13 GPa) and increases
if a crack is formed (35 GPa if crack transversed the AlGaN)

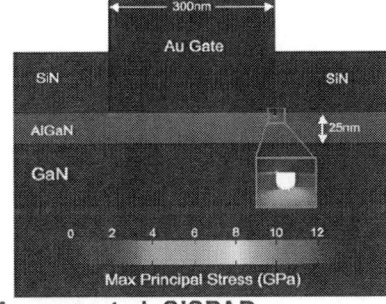

2011 Ancona et al. SISPAD

EL as a tool for analyzing the degradation process

Stress induces the generation of leaky paths under the gate → Generation of lattice defects

Leakage paths can be identified by electroluminescence microscopy

A direct relationship between EL hot spots, surface pits, and gate current leakage has been demonstrated

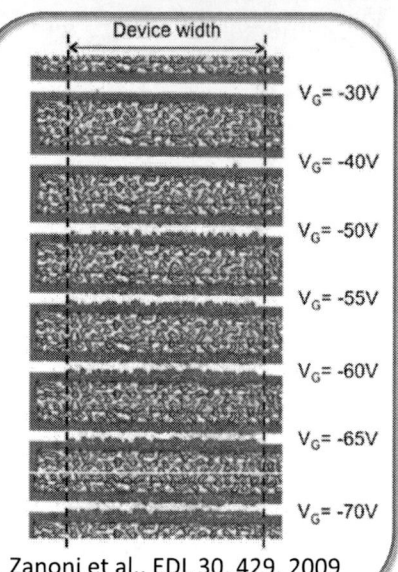

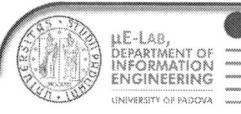

Zanoni et al., EDL 30, 429, 2009

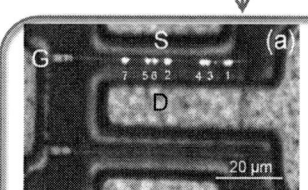 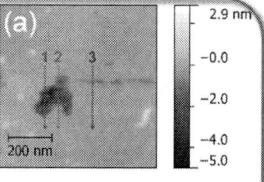

Montes Bajo et al., Appl. Phys. Lett. 101, 033508 (2012)

EL as a tool for analyzing the degradation process

Electroluminescence microscopy images of a representative HEMT sample. Images were taken at V_G = -10 V (with V_S=V_D=0V) after each step-stress

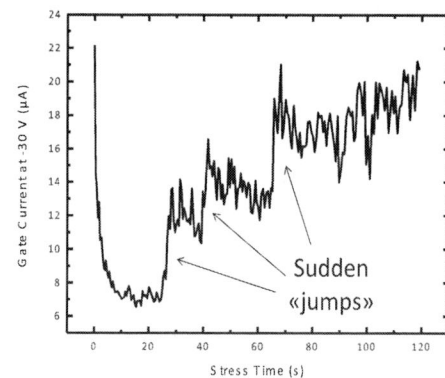

Evolution of gate current I_G measured during stress on a device biased at V_G = -30 V and V_S = V_D = 0 V.

Reverse bias testing – Gate edge degradation

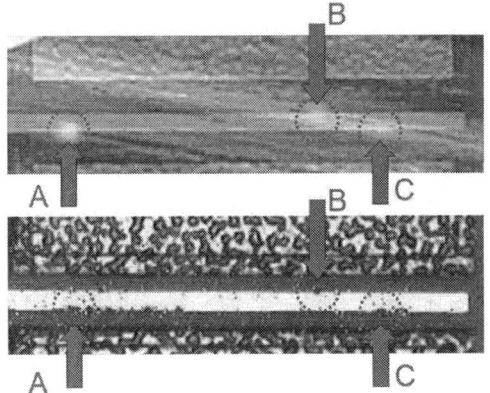

Gate-edge degradation is possibly triggered by pre-existing defects within the epitaxial layers, or extending from the substrate up to the surface. These defects enhance the gate tunneling current,

The electric field at the gate edge is the key variable governing the degradation; temperature also possibly contributes to accelerate failure

Meneghesso et al. 2010JMWT

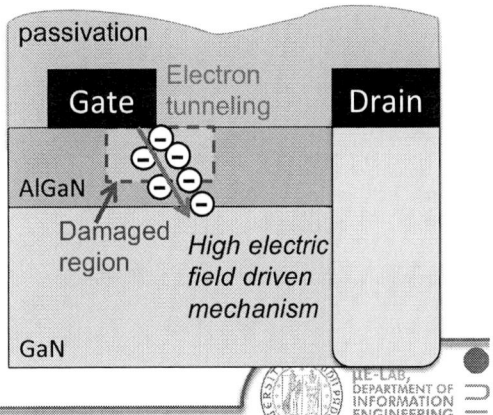

Correlation between degradation, trapping and EKL

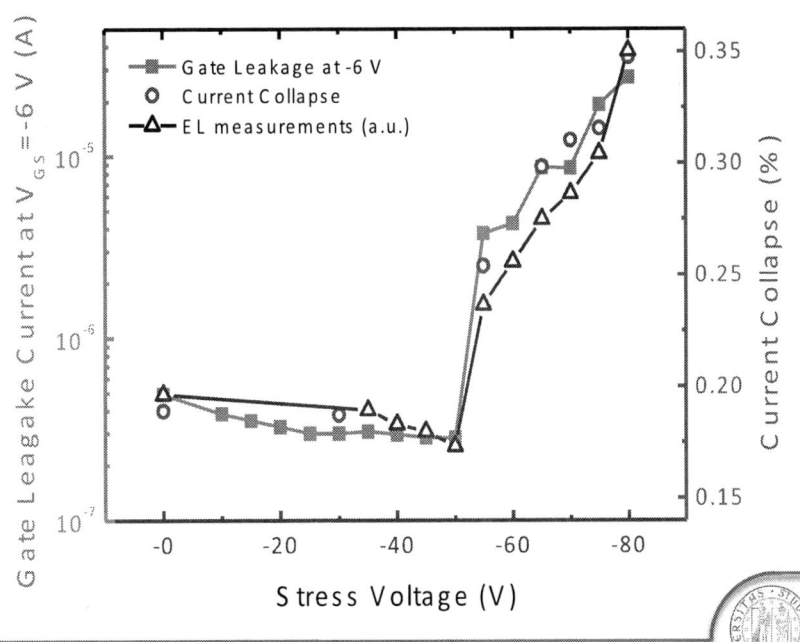

978-1-4799-5494-0/14 $31.00 © 2014 IEEE

Common failure mode for different device design

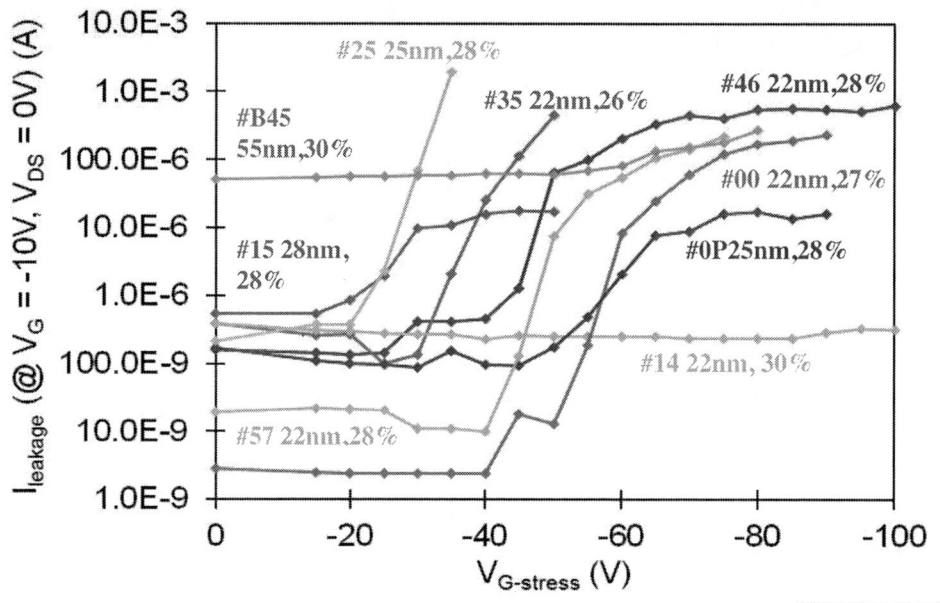

D. Cullen et al. IEEE Trans. Dev. Mat. Reliability March 2013

TEM investigation → Generation of defects under the gate

Chang et al, 2011: TEM analysis of degraded devices

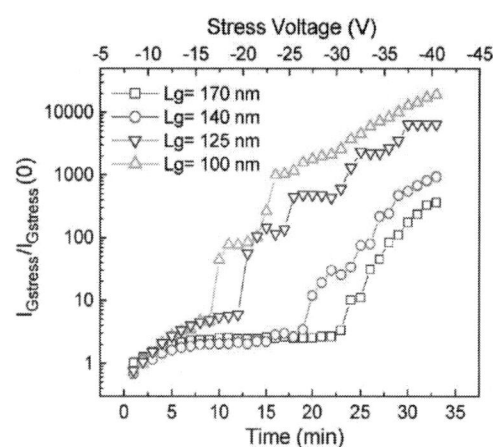

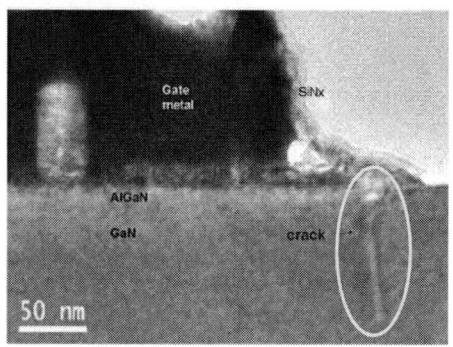

[Chang et al., TDMR 11, 187, 2011]
Univ. of Florida

Chang et al. indicated that degradation is dependent on the electric field; TEM analysis shows generation of defects under the gate

978-1-4799-5494-0/14 $31.00 © 2014 IEEE   264

Degradation of gate leakage current

- time-dependent AlGaN breakdown
- donor trap generation at the AlGaN/GaN interface
- electrochemical surface degradation
- indiffusion of impurities

Time-dependent AlGaN breakdown

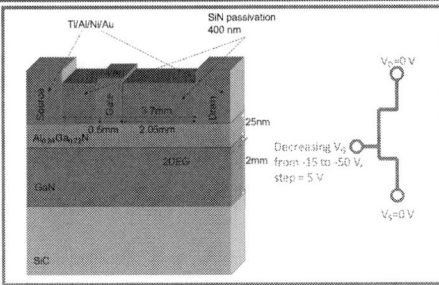

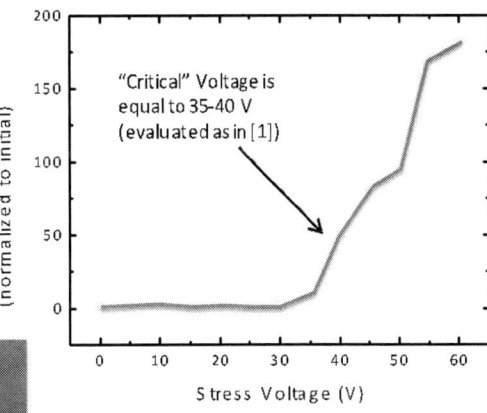

Leakage current measured at different stages of a step-stress experiment

Degradation is a time dependent process, that may occur even below the "critical voltage"

M. Meneghini et al., Appl. Phys. Lett. 100, 033505 (2012)

Time-dependent AlGaN breakdown

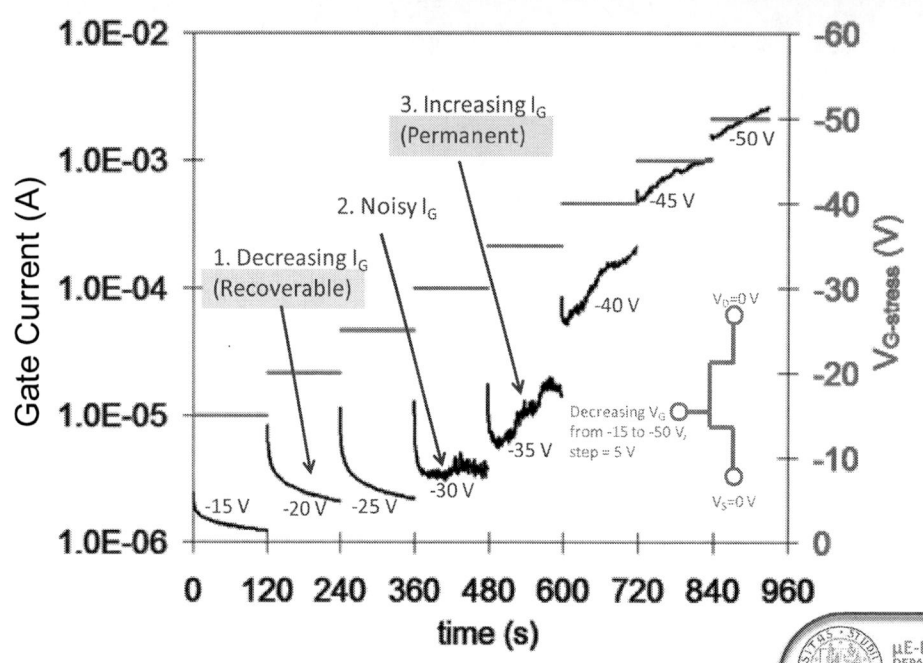

Time-dependent AlGaN breakdown

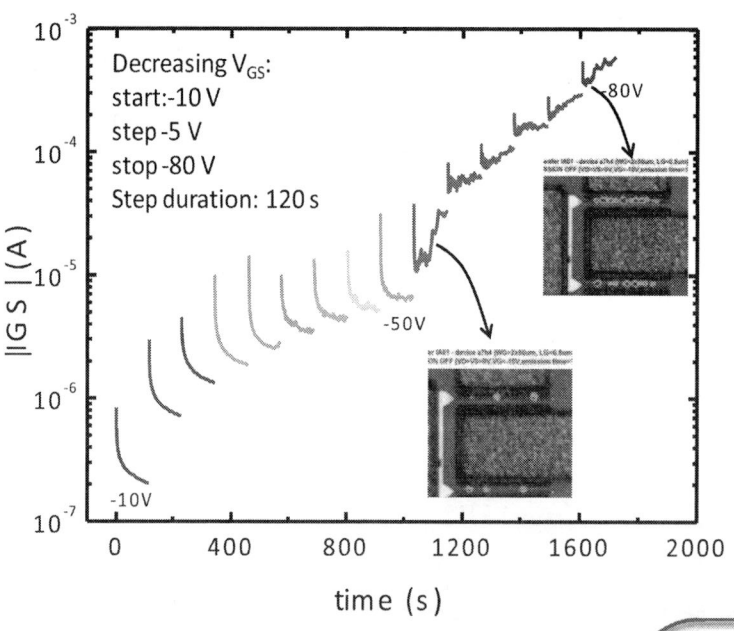

Constant-Voltage Stress Tests

What happens below the "critical voltage"? → Constant Voltage Stress

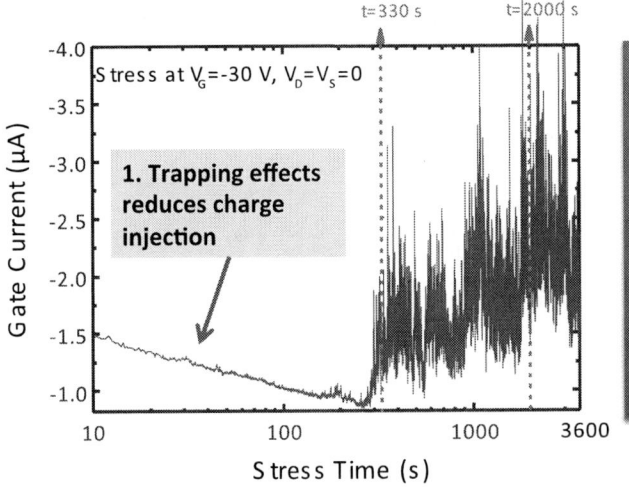

Stress voltage is below the "critical voltage"

Stress induces

→Recoverable degradation → gradual decrease in gate current

Constant-Voltage Stress Tests

What happens below the "critical voltage"? → Constant Voltage Stress

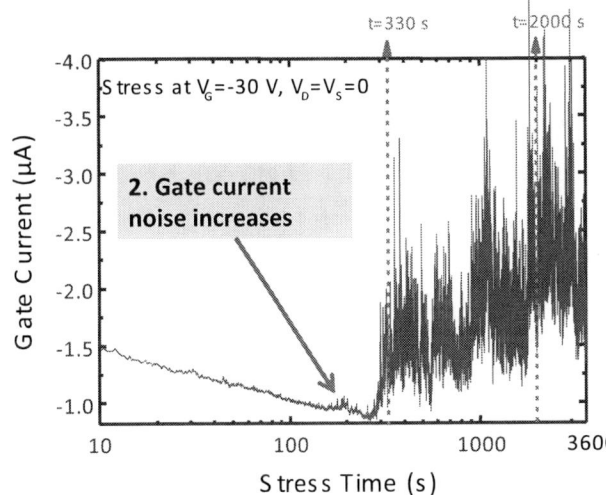

Stress voltage is below the "critical voltage"!

Stress induces

→Recoverable degradation → gradual decrease in gate current

Constant-Voltage Stress Tests

What happens below the "critical voltage"? → Constant Voltage Stress

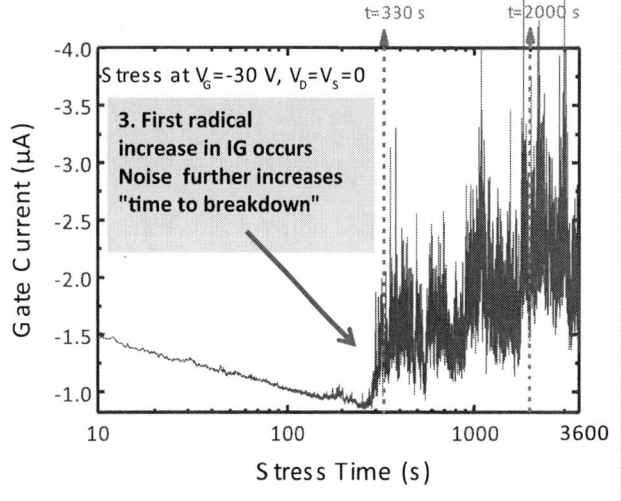

3. First radical increase in IG occurs Noise further increases "time to breakdown"

Stress at V_G=-30 V, V_D=V_S=0

Stress voltage is below the "critical voltage"!

Stress induces

→Recoverable degradation → gradual decrease in gate current

→Permanent degradation → sudden increase in IG

EL Signal during stress

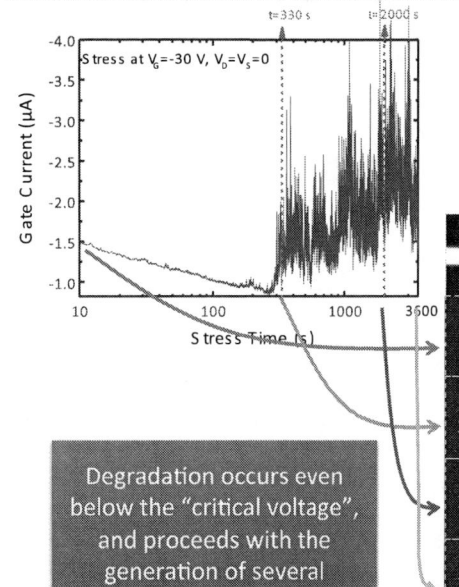

Stress at V_G=-30 V, V_D=V_S=0

Degradation occurs even below the "critical voltage", and proceeds with the generation of several leakage paths

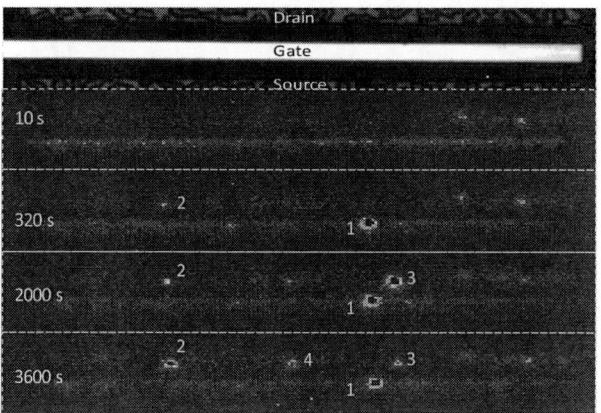

Stress conditions: VG=-30 V, VD=VS=0 V

Drain
Gate
Source

10 s

320 s

2000 s

3600 s

Permanent degradation occurs even below V_{crit}

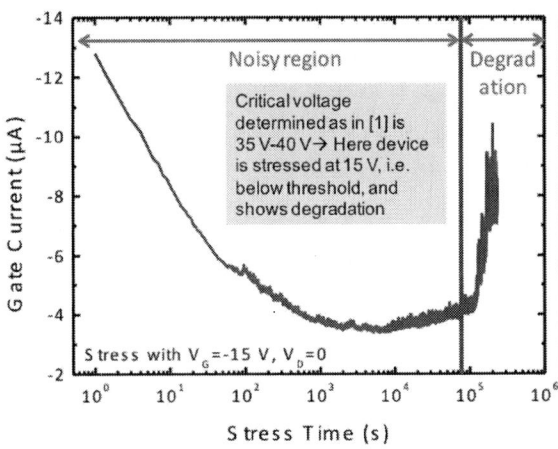

Stress with $V_G = -15$ V, $V_D = 0$

• Degradation can occur even for stress voltage levels smaller than the "critical" voltage → In this case (stress at VG=-15 V!)

• Gate current becomes noisy well before the failure, which occurs after 10^5 s (27 hours), due to the increase in trap density under the gate (conduction becomes randomly possible)

Degradation is ascribed to a defect generation and percolation process, that eventually results in the generation of a permanent leakage path between gate and channel

M. Meneghini et al., Appl. Phys. Lett. 100, 033505 (2012)

µE-Lab,
DEPARTMENT OF
INFORMATION
ENGINEERING
UNIVERSITY OF PADOVA

Model: trap-induced percolative conduction path

Permanent degradation: facts

• Gate current becomes noisy before degradation

• Generation of leaky paths (EL measurements!)

• There is a "time to breakdown" (even below V_{crit})

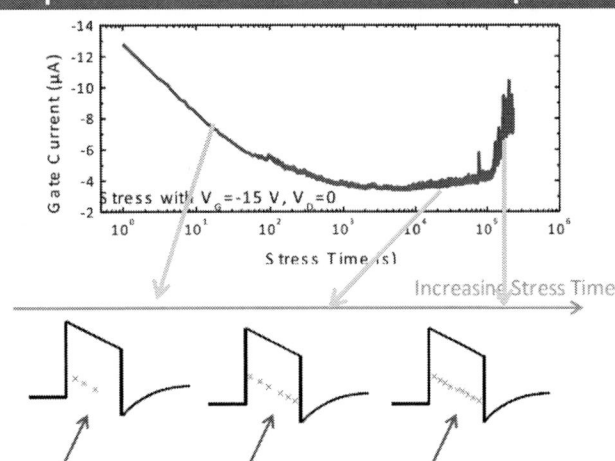

Stress with $V_G = -15$ V, $V_D = 0$

Increasing Stress Time

Generation of trap states in the AlGaN layer → Randomly distributed

Density of trap states increases → If neighbouring traps temporarily overlap, conduction becomes possible. Gate current becomes

Breakdown occurs when traps form a conductive path between gate and the channel region → Sudden increase in gate current

Defect generation and percolation process

µE-Lab,
DEPARTMENT OF
INFORMATION
ENGINEERING
UNIVERSITY OF PADOVA

978-1-4799-5494-0/14 $31.00 © 2014 IEEE

Generation of non-radiative recombination centers

Reverse bias testing of AlGaN/GaN HEMT
comparison of electroluminescence and photoluminescence

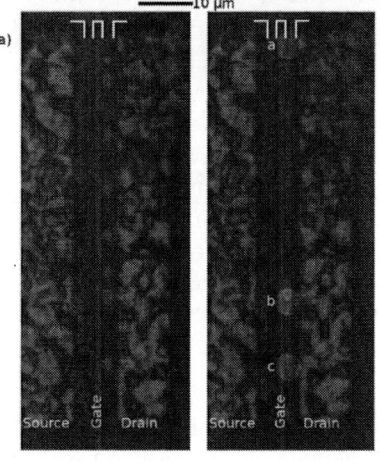

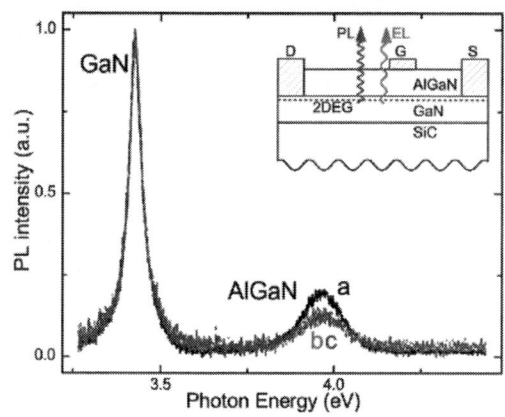

Generation of non-radiative defects within the AlGaN layer →
indirect confirmation of percolative paths formation

Hodges et al. 2012 "Optical inv..." App.Phys. Lett.

Dependence of time to breakdown on stress voltage

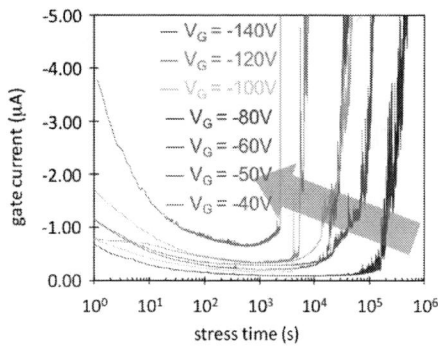

- Time to Breakdown strongly depends on the applied electric field (degradation even below the "critical voltage")

•Stress tests carried out at different (negative) voltage levels show that degradation kinetics are strongly accelerated at high electric fields

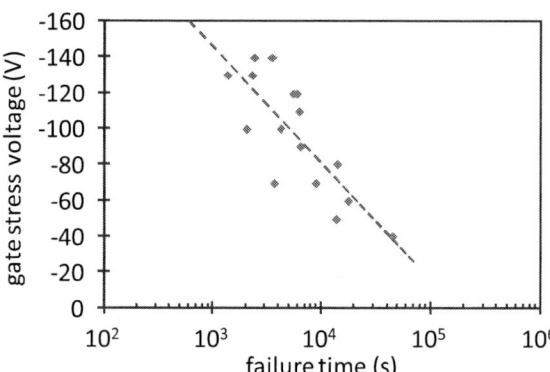

M. Meneghini et al., Appl. Phys. Lett. 100, 033505 (2012)

Dependence of time to breakdown on initial leakage

- Time to Breakdown measured at $V_G = -30$ V
$V_D = V_S = 0$ V

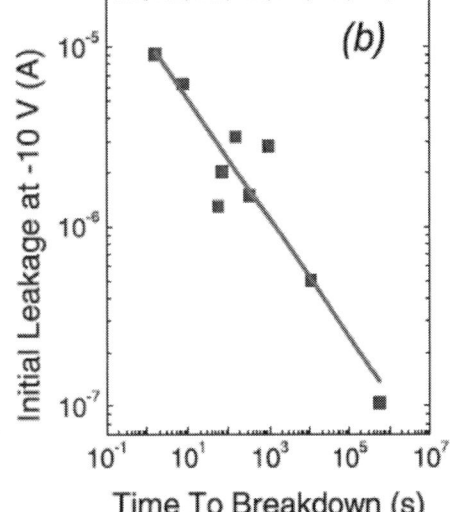

M. Meneghini et al., Appl. Phys. Lett. 100, 033505 (2012)

µE-LAB,
DEPARTMENT OF
INFORMATION
ENGINEERING
UNIVERSITY OF PADOVA

Time to breakdon vs. Critical Voltage

$V_{CRITICAL}$ **strongly depends on the t_{STEP} used during test**

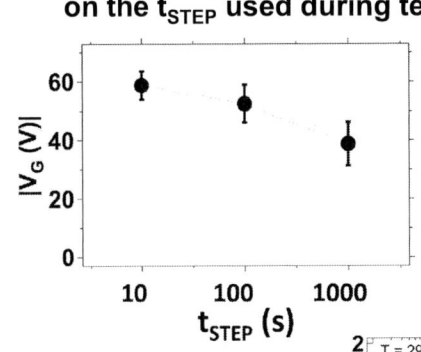

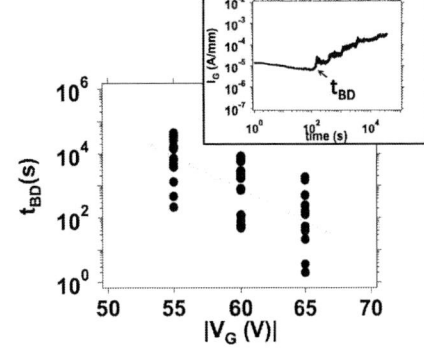

Similar to the dielectric degradation on CMOS devices the t_{BD} for GaN devices can also be fitted to the Weibull distribution.

D. Marcon et al.
IEEE TED October 2013

µE-LAB,
DEPARTMENT OF
INFORMATION
ENGINEERING
UNIVERSITY OF PADOVA

Life time extrapolation

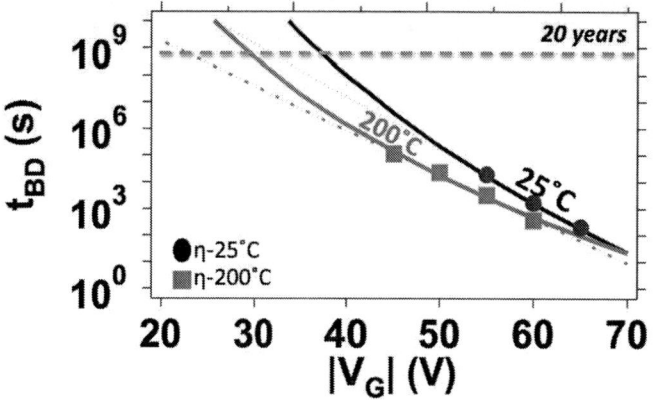

D. Marcon et al.
IEEE TED October 2013

Extrapolation of the t_{BD} towards low bias conditions at 25°C (298K) (black) and at 200°C (437K) (red) for 63.2% failure level. Data are fitted by means of exponential law (dotted line) and of power law (continuous line).

Degradation versus device width

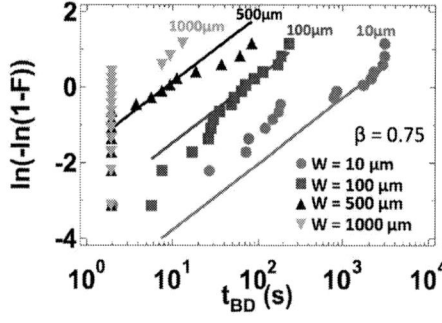 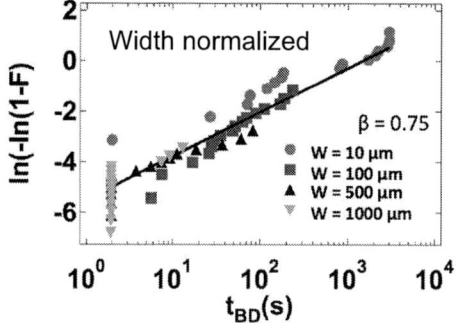

Assuming that the breakdown site responsible for leakage current increase occurs at a random position along the device width, then a long device should statistically show faster t_{BD} than shorter device. Therefore, the probability to form a percolation path along the longer device is n times larger. Nevertheless, the failure distribution and β should remain the same because the nature of the failure does not change with the device width.

$$\eta_1 = \eta \left(\frac{W_2}{W_1} \right)^{\frac{1}{\beta}}$$

D. Marcon et al.
IEEE TED October 2013

978-1-4799-5494-0/14 $31.00 © 2014 IEEE

Random telegraph noise as signature of trap generation

H. P. Rao et al. IEEE TDMR 12(1) 31 2012; Point defects located at around 4.5 nm from the gate metal-semiconductor interface; 0.9 eV below the AlGaN conduction band

P. Marko et al. APL 100, 143507 2012; Micr. Reliab. 52, 2194, 2012 Modulation of current along an intrinsic or stress-induced percolation path across the AlGaN barrier by electron capture and emission on a trap within the barrier

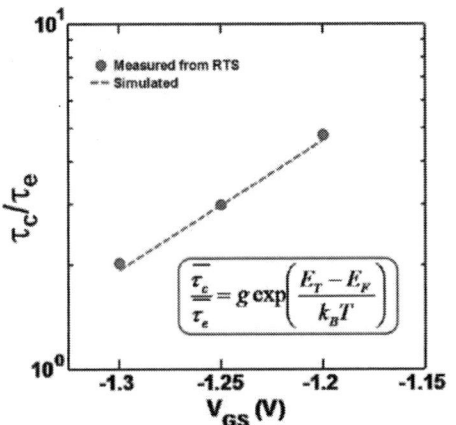

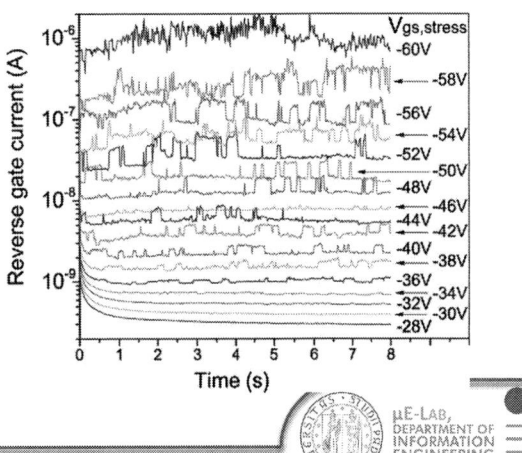

Generation of defects at the AlGaN/GaN interface

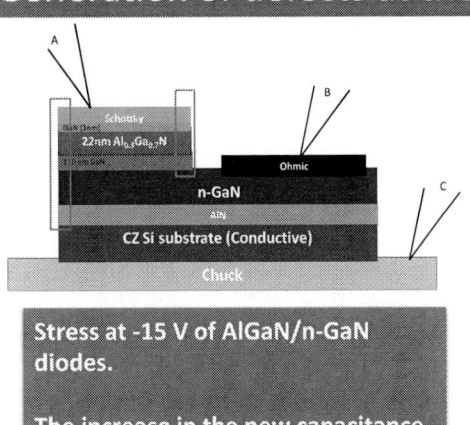

Stress at -15 V of AlGaN/n-GaN diodes.

The increase in the new capacitance peak is strongly correlated to the increase in gate leakage current

Trap activation energy extracted from C(V) vs T measurements: Ea~0.1-0.3 eV far from the conduction band

M. Meneghini et al. IEEE IEDM 2011 , 19.5.1

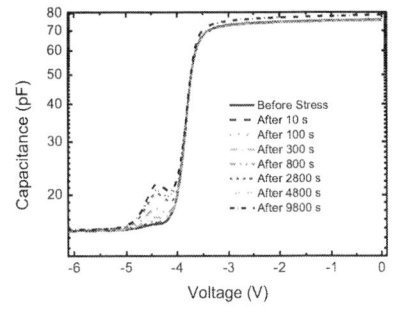

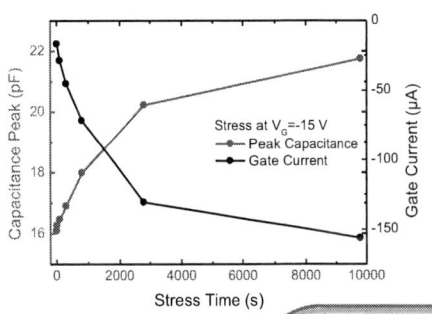

Interface state generation : dependence on the electric field

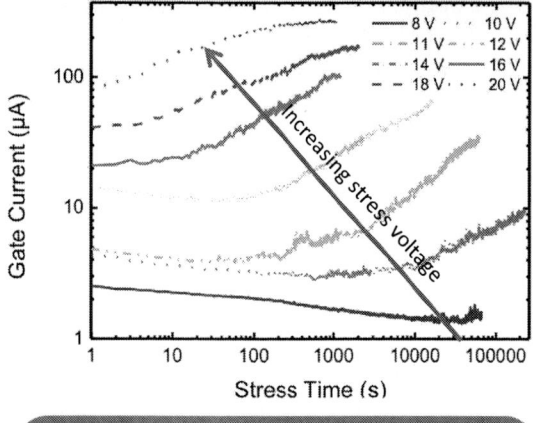

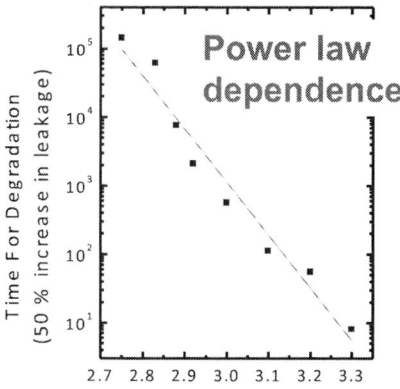

High electric fields can significantly accelerate the degradation process

Power law dependence

Vertical structures → Good control of the electric field, lifetime extrapolation becomes possible

M. Meneghini et al. IEEE IEDM 2011 , 19.5.1

DEPARTMENT OF INFORMATION ENGINEERING
UNIVERSITY OF PADOVA

Trap generation at AlGaN/GaN interface: degradation model

1. During a constant-voltage stress, electrons are injected through the AlGaN layer due to the high electric field → Leakage current

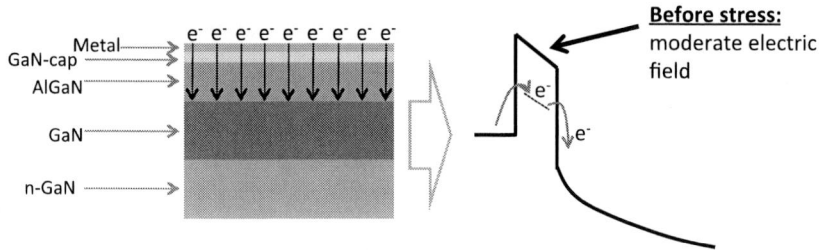

Before stress: moderate electric field

2. Generation of donor traps in the GaN-buffer (new peak in the C-V curves), due to the high electric field and/or to the injection of highly accelerated electrons

3. Further increase in current injection due to the generated positive charge

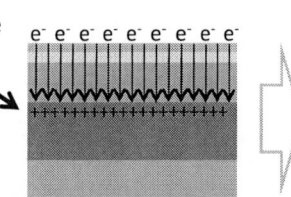

After stress: increased electric field

Donor traps generated after stress

M. Meneghini et al. IEEE IEDM 2011 , 19.5.1

µE-LAB, DEPARTMENT OF INFORMATION ENGINEERING
UNIVERSITY OF PADOVA

978-1-4799-5494-0/14 $31.00 © 2014 IEEE

Structural Degradation in GaN HEMTs

Cross-section

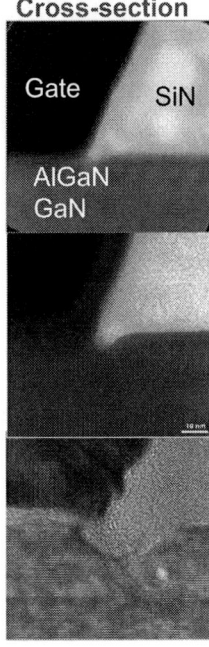

Gate   SiN

AlGaN
GaN

1. Below and around V_{crit}:
Fast groove formation in GaN cap

2. Beyond V_{crit}:
Pit formation in AlGaN barrier

3. Pit growth (to AlGaN/GaN interface) and merge

Joh, ROCS 2010

Makaram, APL 2010

Plan-view

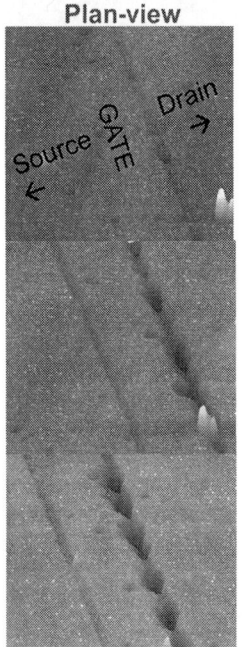

Source   GATE   Drain

DEPARTMENT OF
INFORMATION
ENGINEERING
UNIVERSITY OF PADOVA

AFM analysis → Generation of grooves/pits under the gate

Stress at 150 °C, up to V_{DG}=50 V

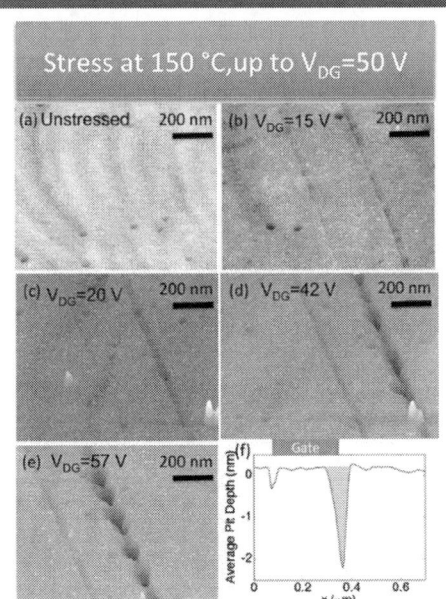

(a) Unstressed   200 nm   (b) V_{DG}=15 V   200 nm

(c) V_{DG}=20 V   200 nm   (d) V_{DG}=42 V   200 nm

(e) V_{DG}=57 V   200 nm   (f) Gate

Makaram et al. (MIT), Appl. Phys. Lett. 96, 233509 (2010)

Three-steps process:
1. a groove forms in the GaN cap layer due to field-induced oxidation or electrochemical etching
2. pit formation and growth
3. subsequent crack formation

Mechanism thermally enhanced, but requiring an electric field to occur

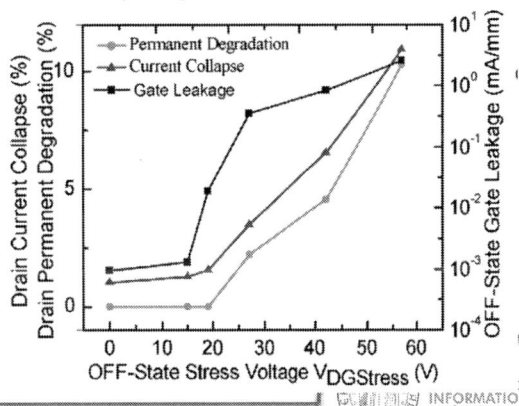

GaN HEMTs Tutorial – G. Meneghesso- 2°WiPDA October 13, 2014   38

INFORMATION
ENGINEERING
UNIVERSITY OF PADOVA

978-1-4799-5494-0/14 $31.00 © 2014 IEEE   275

Oxidation of the surface

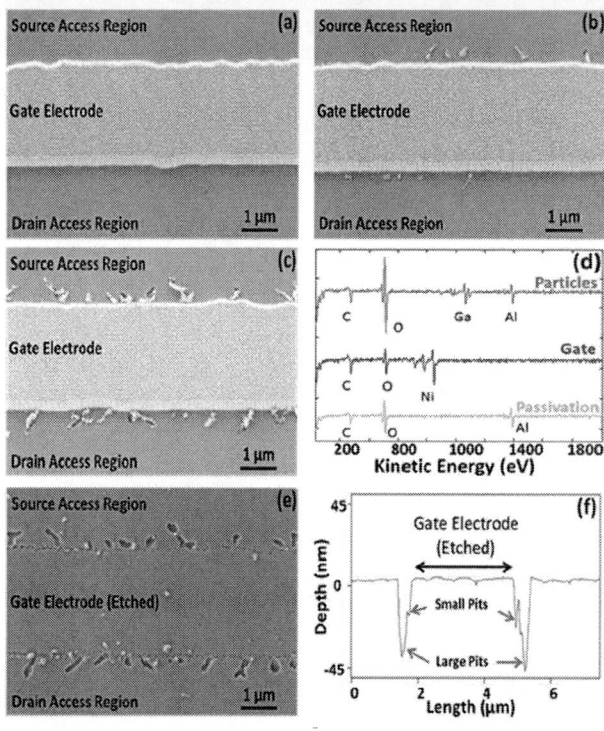

- Oxide particles were found to form along the gate edge of stressed devices (off-state tests 40 V 60 s, 600s, 6000 s)
- When the gate electrode is removed, pits are seen to have formed underneath each particle" in the area where electric field is maximum
- Reverse-bias degradation can be due to the chemical oxidation of the nitride semiconductor surface
- GaN is decomposed to Ga^{3+} and nitrogen gas. The Ga^{3+} then reacts with oxygen ions or oxyanions to form Ga_2O_3. The supply of oxygen ions and oxyanions is therefore key for this reaction to happen

Gao, Palacios et al. , Appl. Phys. Lett. 99, 223506 (2011)

Redox in the off-state degradation of AlGaN/GaN HEMTs

Redox in the off-state degradation of AlGaN/GaN HEMTs

The surface electrochemical reactions

- *Oxidation* of $Al_xGa_{1-x}N$ (or GaN)
  $$2Al_xGa_{1-x}N + 6h^+ = 2xAl^{3+} + 2(1-x)Ga^{3+} + N_2\uparrow$$
  $$2xAl^{3+} + 2(1-x)Ga^{3+} + 6OH^- = xAl_2O_3 + (1-x)Ga_2O_3 + 3H_2O$$

- *Balance* of H^+ and OH^-
  $$H^+ + OH^- = H_2O$$

- *Reduction* of hydrogen
  $$2H^+ + 2e^- = H_2\uparrow$$

F. Gao, C. V. Thompson, and T. Palacios, MIT (unpublished)

Complete balanced electrochemical reaction:
$$2Al_xGa_{1-x}N + 6h^+ + 6e^- + 3H_2O = xAl_2O_3 + (1-x)Ga_2O_3 + N_2\uparrow + 3H_2\uparrow$$

Effect of atmosphere on reverse bias degradation

❖ **Pits and Grooves in different environment but in the same stress conditions**

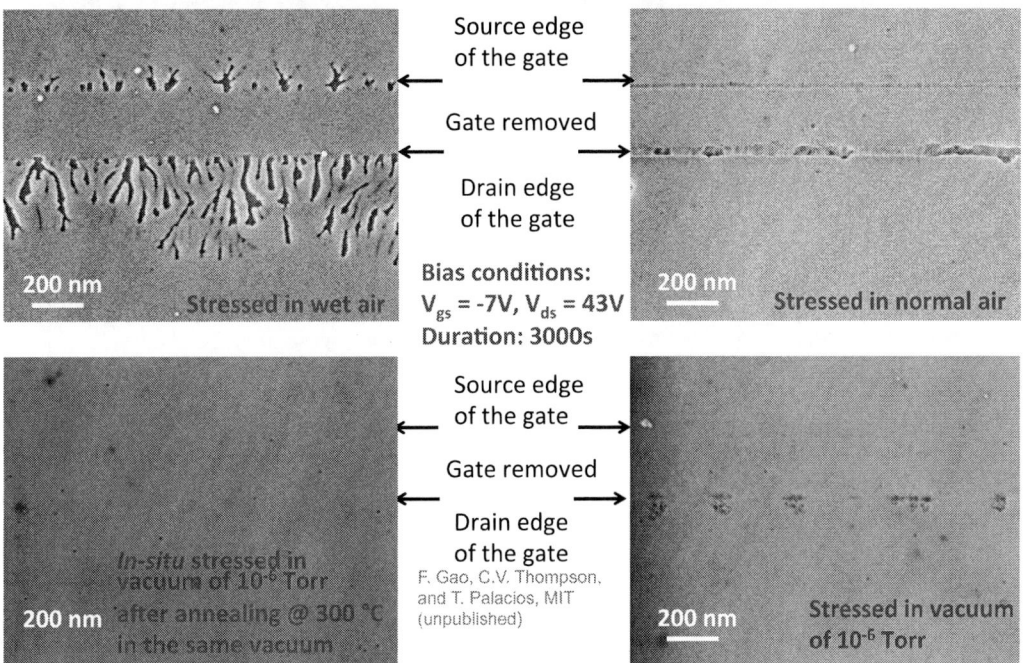

Source edge of the gate

Gate removed

Drain edge of the gate

Bias conditions:
V_{gs} = -7V, V_{ds} = 43V
Duration: 3000s

200 nm — Stressed in wet air

200 nm — Stressed in normal air

In-situ stressed in vacuum of 10^{-6} Torr after annealing @ 300 °C in the same vacuum

200 nm

F. Gao, C.V. Thompson, and T. Palacios, MIT (unpublished)

200 nm — Stressed in vacuum of 10^{-6} Torr

Effect of test time

OFF-state stress $V_{DGstress}$=50 V (>V_{crit}), T_{base}=150 C

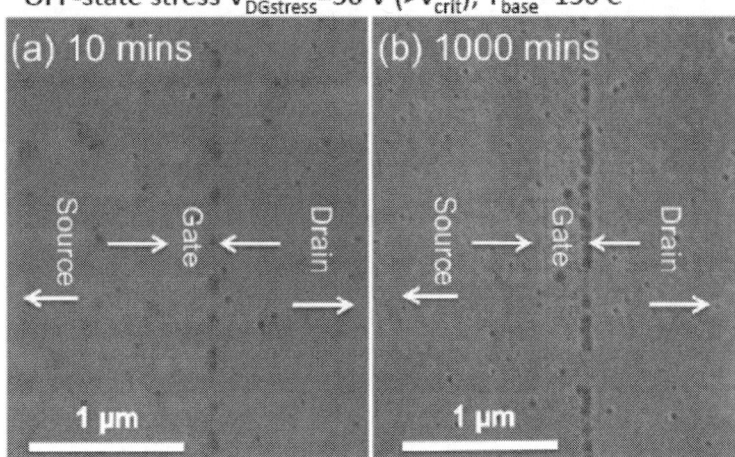

(a) 10 mins

(b) 1000 mins

Source → Gate ← Drain — 1 µm

Source → Gate ← Drain — 1 µm

M. Makaram , J. Joh et al., Formation of structural defects in AlGaN/GaN High Electron Mobility Transistors under Electrical Stress Proc. ROCS 2010

µE-LAB, DEPARTMENT OF INFORMATION ENGINEERING UNIVERSITY OF PADOVA

Effect of temperature distribution

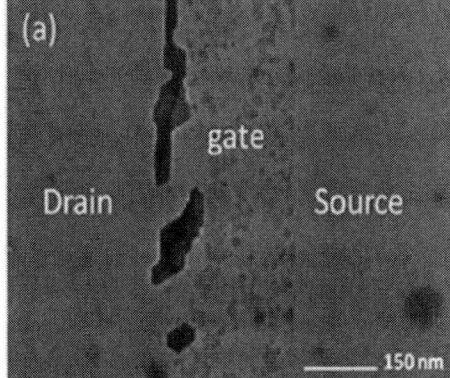 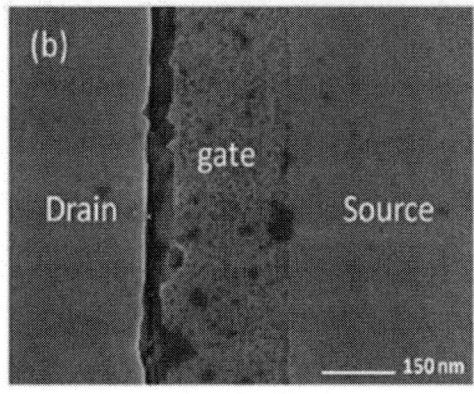

outer part of the finger
(cooler, less degradation)

inner part of the finger
(hotter, more degradation)

High-power state at Vds = 40 V, IDQ = 250 mA/mm, Tbaseplate = 150°C.

L. Li, J. Joh et al., APL 100, 172109 (2012)

GaN oxidation in reverse-biased LEDs

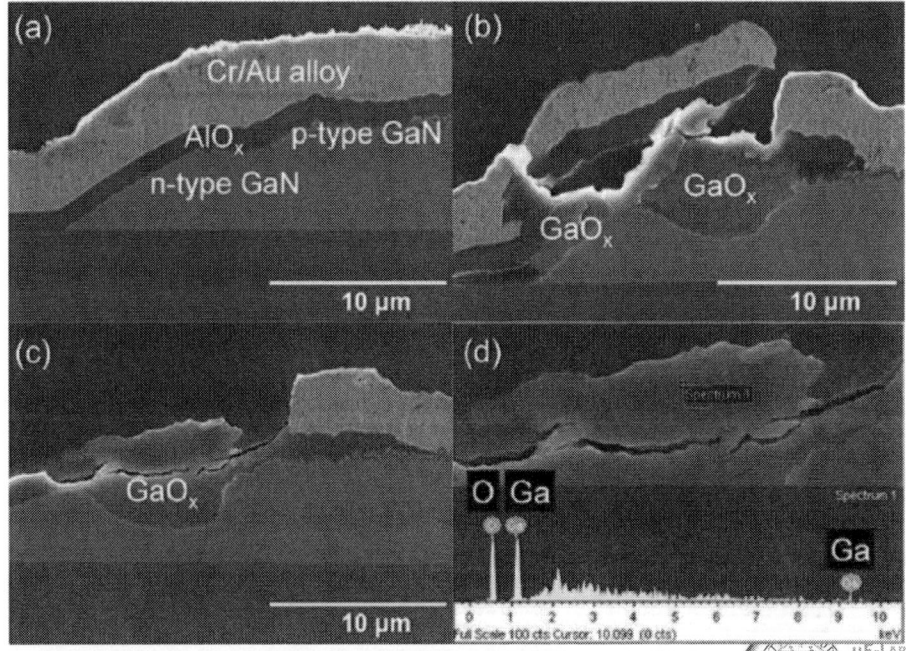

H-H. Yen et al., IEEE Photonics Tech. Lett. 22 (15), 1168, 2010.

Controversial results

- Role of temperature

- Correlation between gate leakage increase and drain current degradation

- Crack and pit formation

Off-state test 0.25 μm gate AlGaN/GaN HEMT on SiC

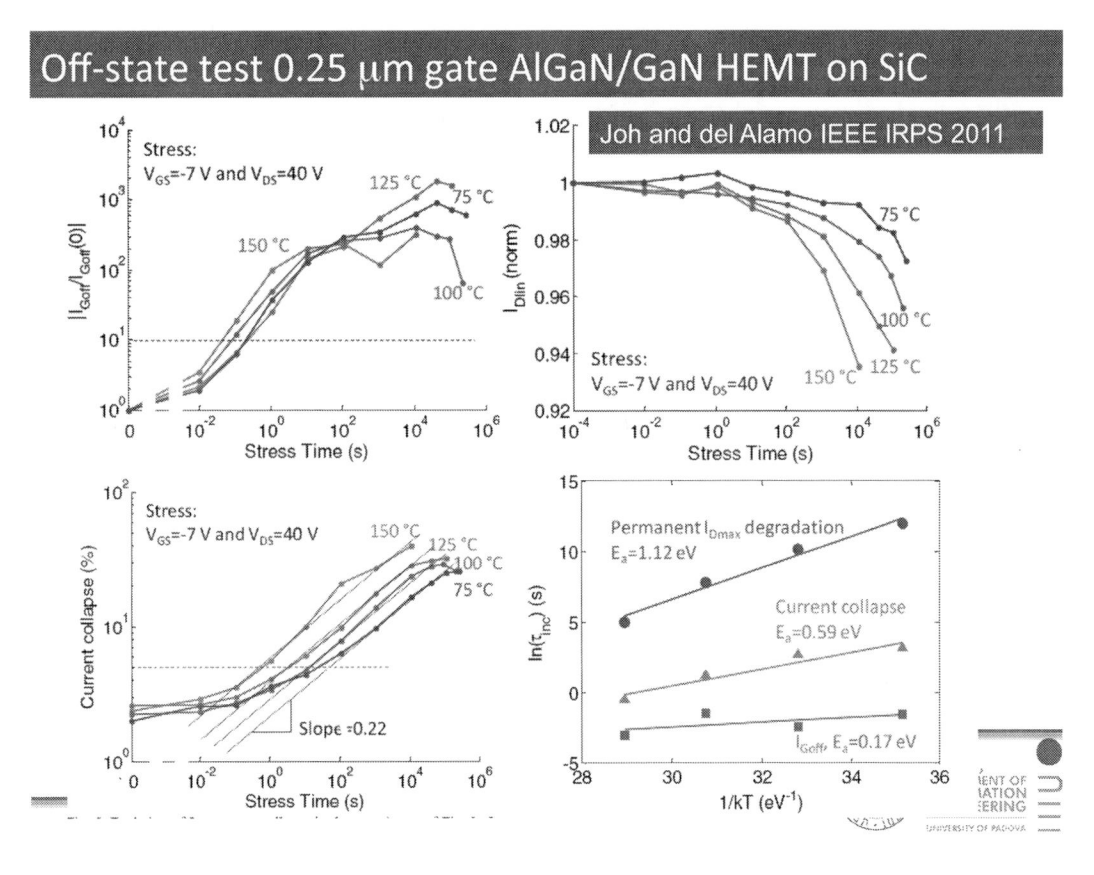

Joh and del Alamo IEEE IRPS 2011

Off-state test 0.25 μm gate AlGaN/GaN HEMT on SiC

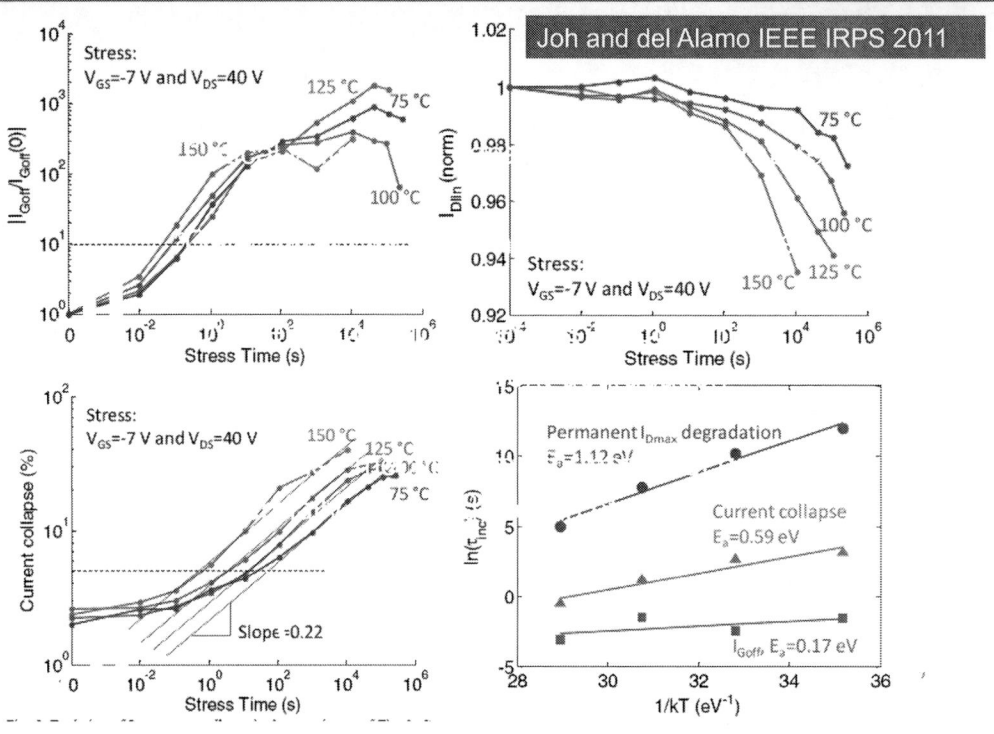

Joh and del Alamo IEEE IRPS 2011

Critical voltage decreasing with temperature

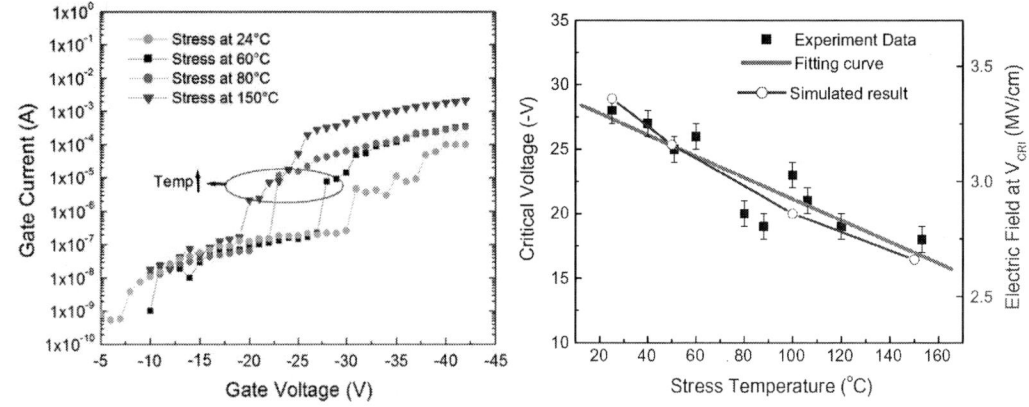

Both the critical voltage and the electric field value required to induce the abrupt increase of leakage decrease with temperature. Failure is attributed to GaO dissociation and NiO formation.

Douglas et al. Microel. Reliab. 52 (2012) 23-28

Comparison of off-state and on-state stress results

Off-state Step stress experiment -20 V to 100 V 5 V steps, tstep = 10 s, RT
0.5 μm AlGaN/GaN HEMT, Lgd = 4 μm, Lgs = 1 μm 22 nm 26% AlGaN
Degradation of Ig and Id occurs at different values of critical voltage

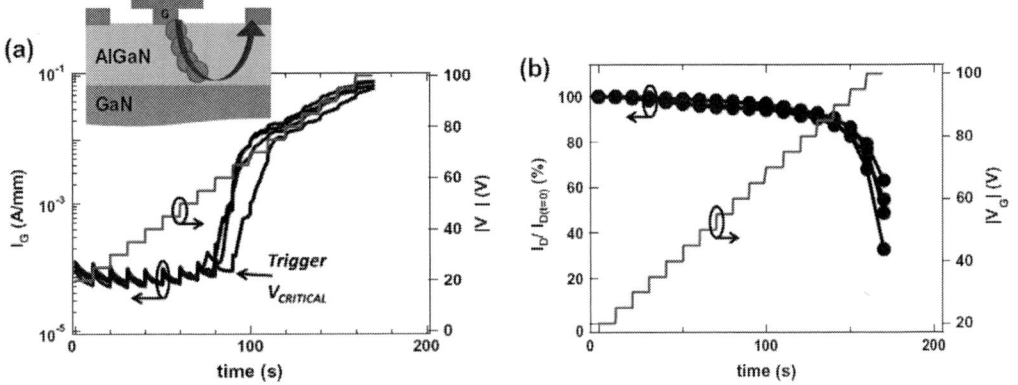

No pitting detected in TEM microsections
for devices tested in off-state

D. Marcon et al., Microelectronics Reliability 52 (2012) 2188

No increase in gate leakage in on-state experiments

On-state DC stress experiment VDS = 50 V, Tj = 280°C, 300°C, 330°C
NO INCREASE IN GATE LEAKAGE CURRENT →
DIFFERENT FAILURE MECHANISM

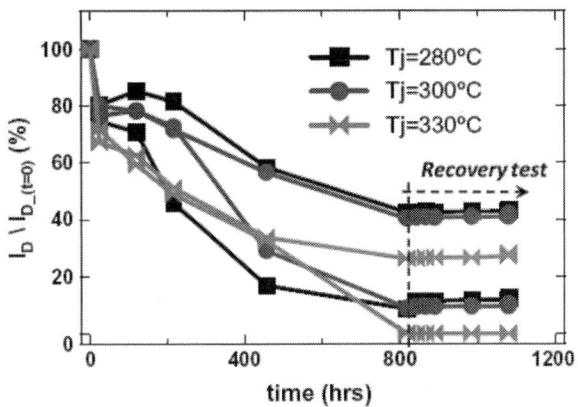

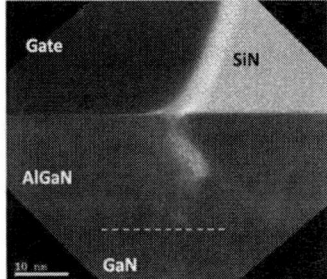

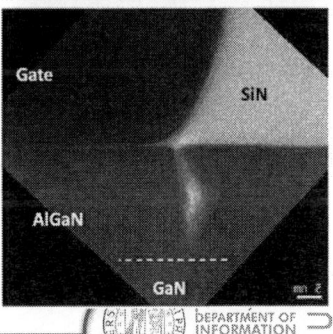

Significant pitting, oxide detected in the crack (possible
contamination during sample preparation ?)

D. Marcon et al., Microelectronics Reliability 52 (2012) 2188

Generation of non-radiative recombination centers

Reverse bias testing of AlGaN/GaN HEMT

Depth resolved cathodoluminescence, Kelvin probe force microscopy surface photovoltage spectroscopy

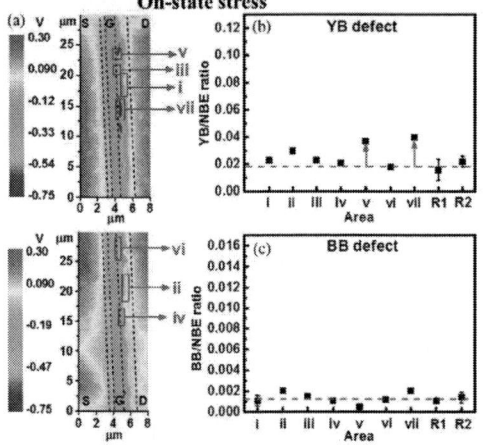

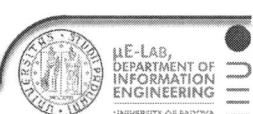

Defects 1.2 eV above GaN valence band accumulate with degradation

Compatible with converse piezoelectric degradation (but also other mechanisms)

OFF-state: YB and BB defects generation

ON-state: only YB defects.

Correlated with surface potential decrease

C.-L. Lin et al., IEEE Trans. El. Dev. 2667, 2012

Oxygen interdiffusion ?

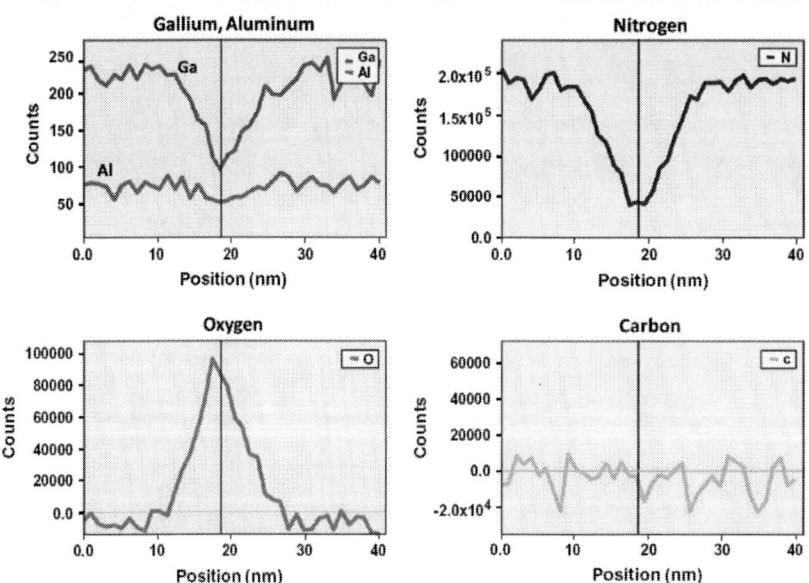

STEM scans in the region of the crack

D. Marcon et al., Microelectronics Reliability 52 (2012) 2188

978-1-4799-5494-0/14 $31.00 © 2014 IEEE

Experiments at RFMD

Tests on 0.5 μm AlGaN/GaN HEMTs on SiC under different stressors in nitrogen:
- high voltage (V_{DS} = 60 V and 100 V) low current
- high dc power > 11 W/mm

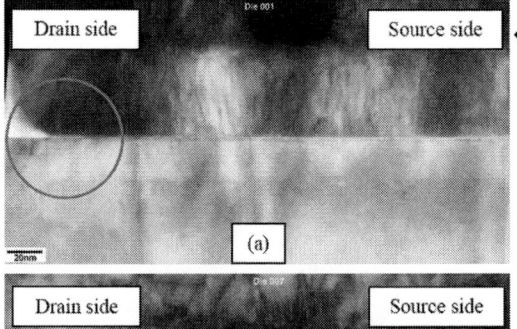

← high voltage tested device

High voltage tested devices did not show pits or cracks, which were present in devices tested at high power

← high power tested device

2011 Christiansen et al. IEEE IRPS 2011

Indiffusion of impurities

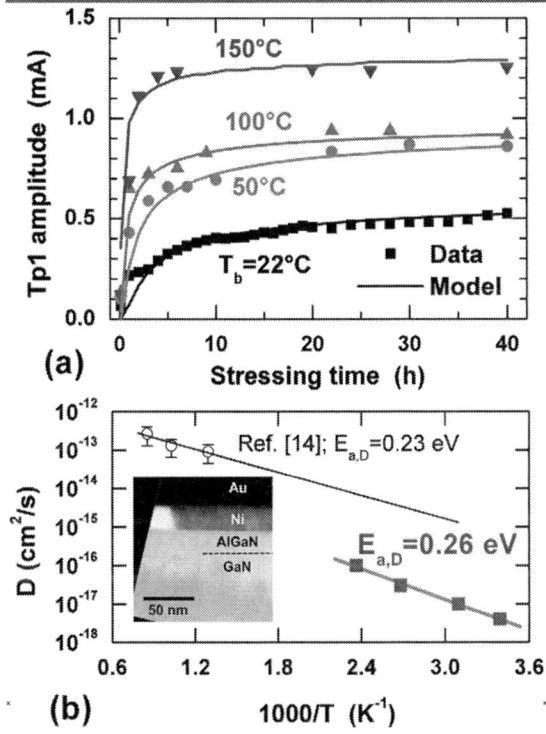

2 μm gate devices
On-state tests Vds = 30 V, Vgs = 0 V
at Tj = 300°C
Off-state tests Vds = 30 V

Drain current decrease and threshold voltage shift
Thermally activated trap generation or activation
Oxygen indiffusion along dislocations ?
No cracks at the drain edge

Tapajna, Kuball et al
Applied Physics Letters
97, 023503 2010

978-1-4799-5494-0/14 $31.00 © 2014 IEEE   283

Indiffusion of impurities

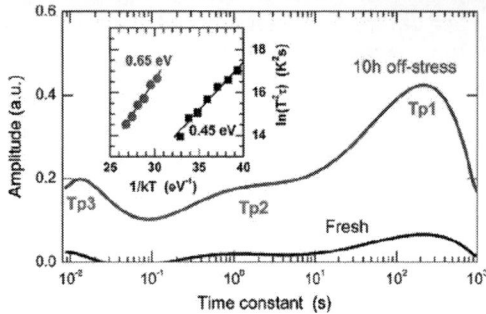

20 nm 26% AlGaN/GaN HEMT on SiC
On-state tests Vds = 30 V, Vgs = 0 V
at Tj = 300°C
Off-state tests Vgs = 0 V Vds = 30 V

Generation of traps in the AlGaN (Tp1)
possible diffusion of impurities along
dislocation, like oxygen or carbon

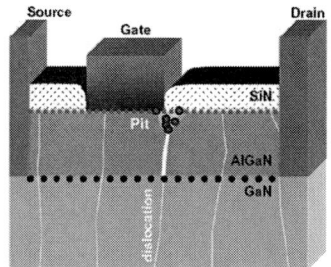

M. Kuball et al Microelectronics
Reliability 51 2011 195

RF testing of 0.25 µm AlGaN/GaN HEMTs on SiC

0.25 µm 22nm 22% AlGaN/GaN HEMT on SiC
RF stress at VDS = 42 V f = 10 GHz compression 2dB to 4 dB
Peak channel temperature 260°C to 300°C t > 1000 hours
Failure criterion 1 dB loss in gain

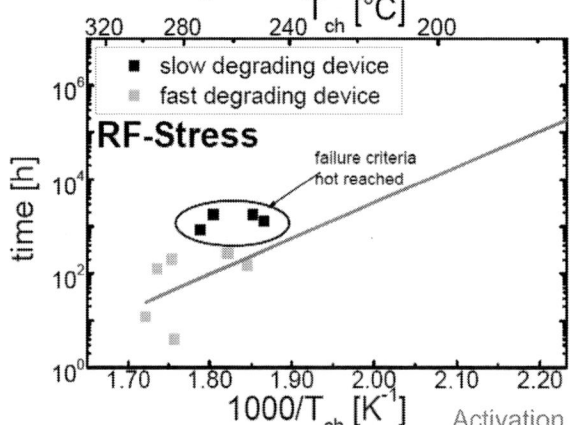

Activation energy 1.7 eV
Extrapolated lifetime 2 x 10^5 hours

M. Dammann et al. IEEE IIRW 2011
M. Caesar et al. IEEE IRPS 2012

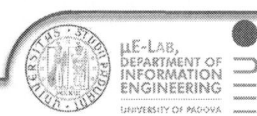

RF testing of 0.25 μm AlGaN/GaN HEMTs on SiC

0.25 μm 22nm 22% AlGaN/GaN HEMT on SiC
DC stress at VDS = 42 V Peak channel temperature 260°C to 310°C t > 1000 hours

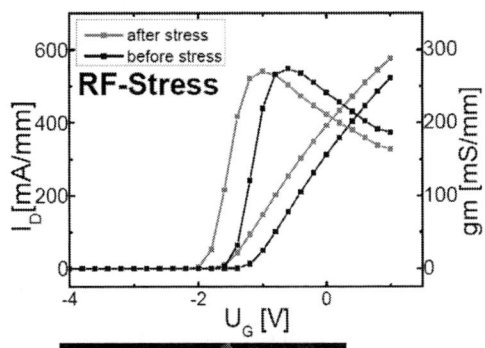 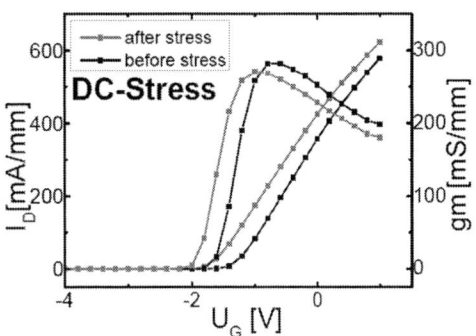

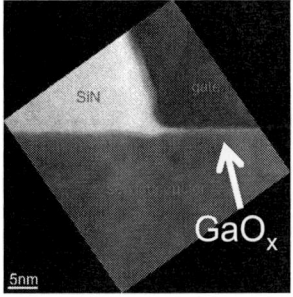

Generation of donor-like traps under the gate:
Decomposition of GaOx followed by
indiffusion of oxygen into the GaN cap layer
or even deeper

M. Dammann et al. IEEE IIRW 2011
M. Caesar et al. IEEE IRPS 2012

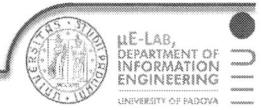

Interaction of different failure mechanisms

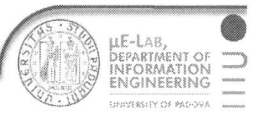

978-1-4799-5494-0/14 $31.00 © 2014 IEEE   285

Charge trapping in the access region

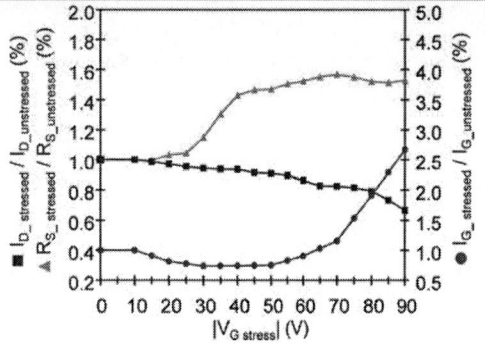

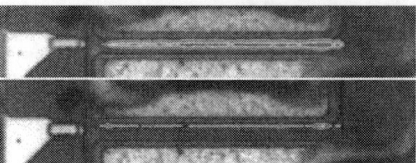

EL –on state Before stress

EL –on state After stress

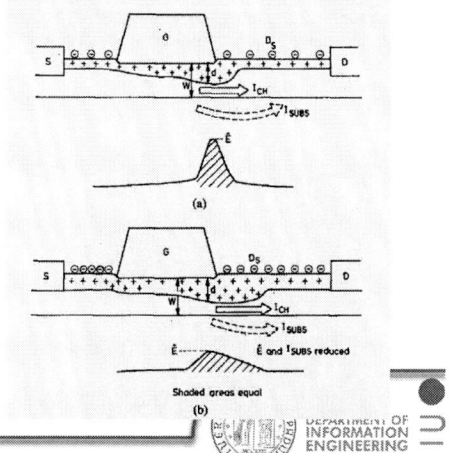

Another effect induced by the reverse gate biasing is related to electron trapping in the gate-drain access region

As far as electrons remain trapped, they create a "virtual gate" effect resulting in
- decrease of transconductance
- increase of parasitic resistance
- reduction of electric field
- breakdown walkout
- decrease of electroluminescence and EL/ID

Photoionization of traps by UV radiation

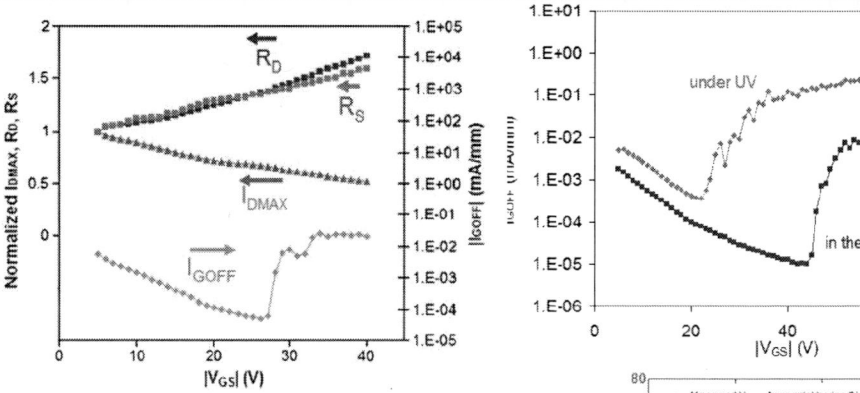

Photoionization of traps by shining UV radiation, restoring a high value of electric field. As a consequence, critical voltage is reduced

Demirtas2010MR, Demirtas2010IRPS

Competition between failure mechanisms

Charge trapping in G-D region
Off-state and on-state

Reduces drain current
Reduces electric field
Reduces gate leakage Ig
Reduces electroluminescence EL
Improves breakdown voltage Vbr

Requires hot electrons, HE, either
coming from gate (leakage)
or from device channel (semi-on
state)

Fast process, can follow rf signal
enhanced by rf compression
(higher electric field with HE)
WALK-OUT
Ig ↓ EL/Id ↓ Vbr ↑

Gate degradation
Mostly off-state: very high
electric field

Increases gate leakage Ig
even without reducing Id
Induces injection paths for HE
→ EL from percolation paths

Triggered by surface pitting,
gate leakage current or electric
field alone, even in off-state

Slow time-dependent process,
can hardly follow rf signal; more
evident if compression is avoided
WALK-IN
Ig ↑ EL/Id ↑ Vbr ↓

On-state device degradation

Hot electron degradation of AlGaN/GaN HEMTs submitted to on-
state stress tests

- exposure to on-state stress induces a remarkable decrease in
  drain current

- degradation rate strongly depends on the intensity of the EL
  signal emitted by the devices during stress, while it has a
  negligible dependence on temperature

- degradation is ascribed to **electron trapping in the gate-drain
  access region, induced by** hot electrons **(acceleration law is
  derived)**

Experimental Details

- Samples grown by MBE 0.25 µm gate

- Preliminary off-state stress tests

- Stress plan: on-state stress tests (14 h, V_{DS}= 30 V, several V_G levels)

- Degradation was monitored by electrical and EL measurements

- Critical voltage > 200 V

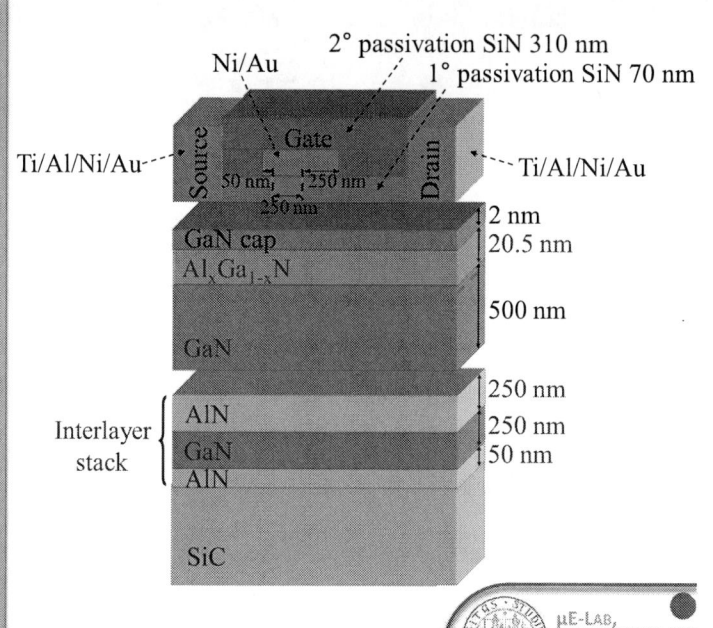

Hot electron degradation

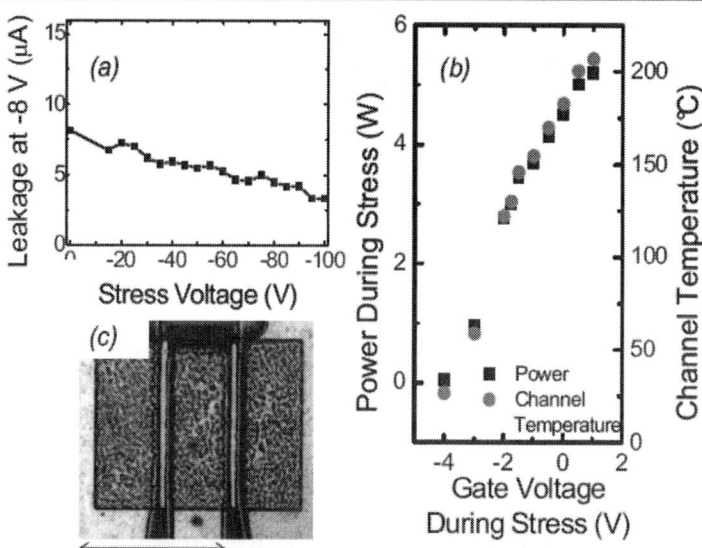

Analyzed devices do not degrade in off-state up to V_{GD}=-100 V →
on-state stress tests at VDS=30 V, various gate voltages and power dissipation
M. Meneghini et al., Appl. Phys. Lett. 100, 233508 2012

Electroluminescence as a probe for hot electrons.

Gate leakage current, not related to channel hot electrons

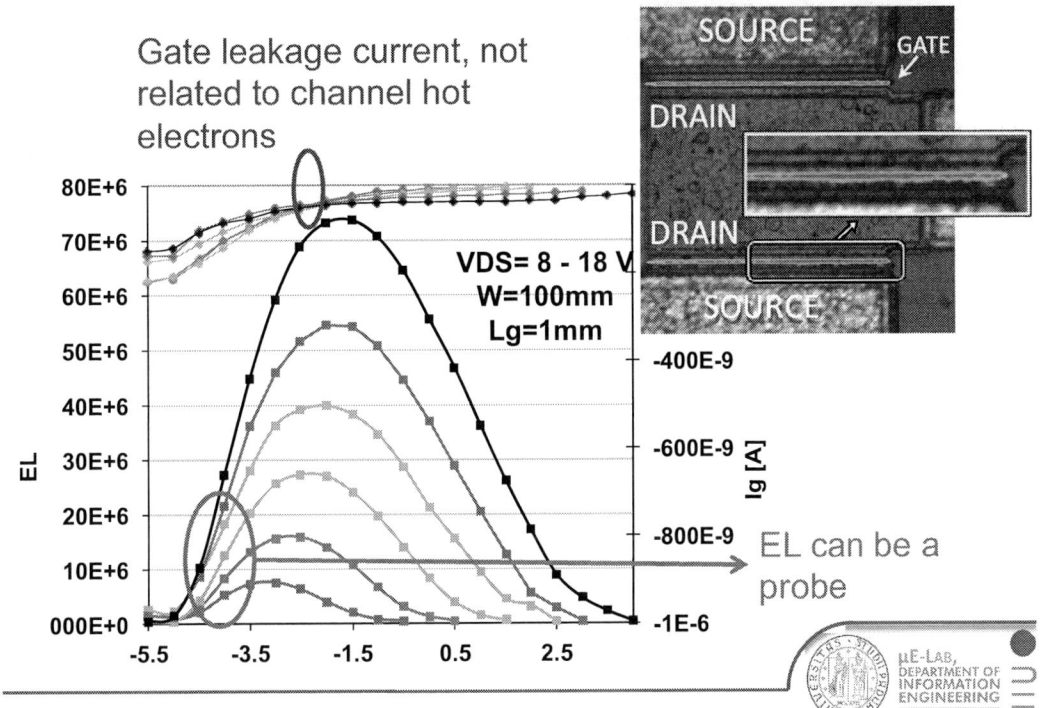

VDS= 8 - 18 V
W=100mm
Lg=1mm

EL can be a probe

μE-LAB, DEPARTMENT OF INFORMATION ENGINEERING
UNIVERSITY OF PADOVA

Before Stress → EL analysis

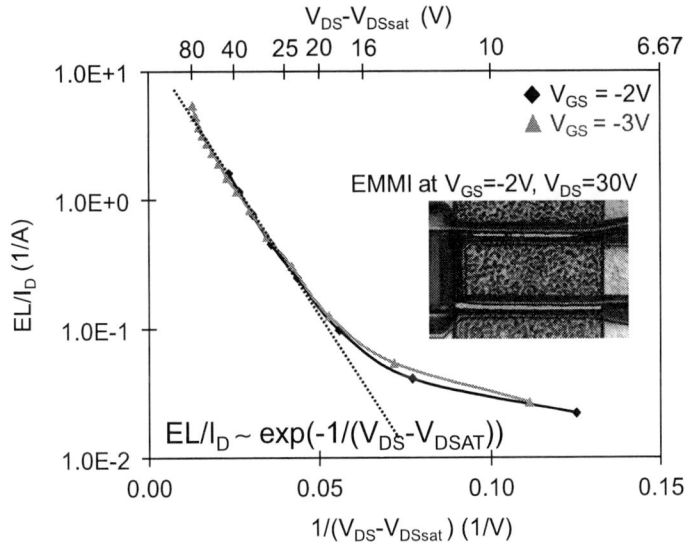

EMMI at V_{GS}=-2V, V_{DS}=30V

$EL/I_D \sim \exp(-1/(V_{DS}-V_{DSAT}))$

Under on-state conditions → Weak EL signal, localized at the edge of the drain

EL intensity follows the Chynoweth's law (hot electrons!)

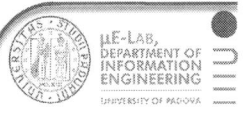

μE-LAB, DEPARTMENT OF INFORMATION ENGINEERING
UNIVERSITY OF PADOVA

978-1-4799-5494-0/14 $31.00 © 2014 IEEE

Electrical Characteristics before/after stress

Stress conditions →14 h on-state stress at V_{DS} = 30 V, V_{GS} = -1 V, W=200 µm, Lg=0.25 µm

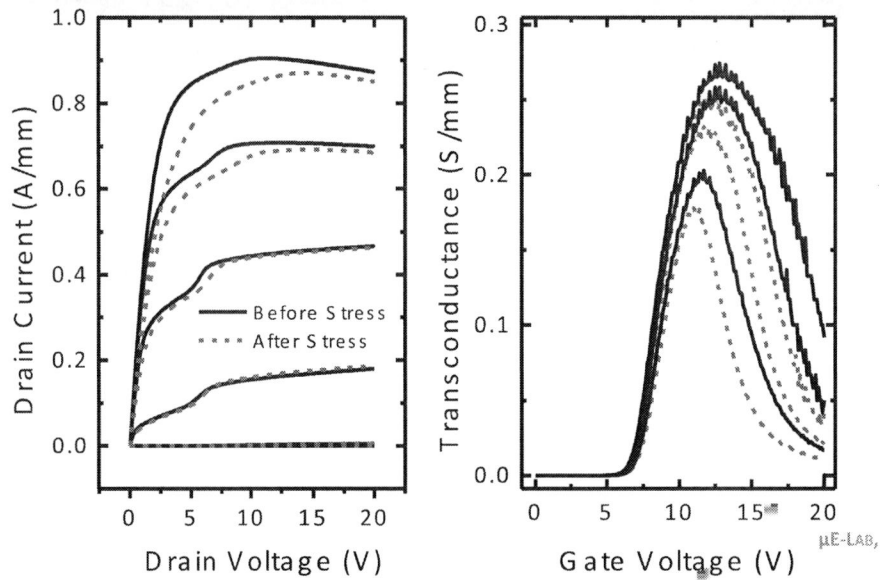

Degradation of drain current and electroluminescence

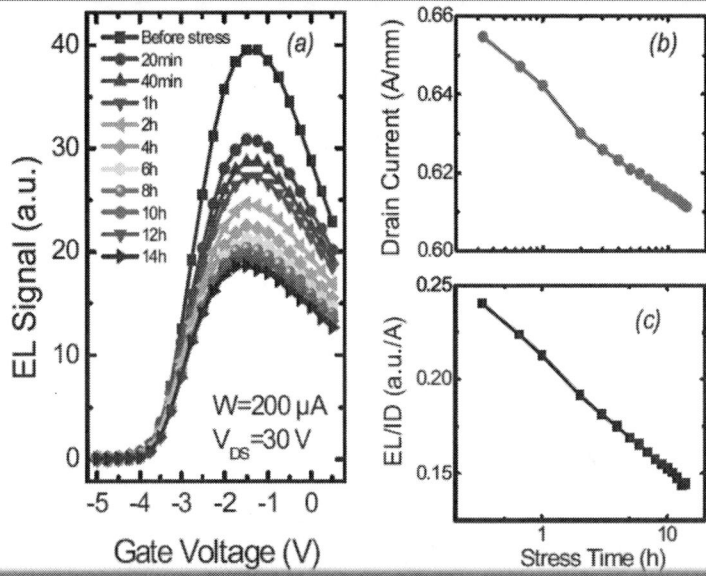

After stress →
Decrease in the EL/ID signal, correlated to the decrease in ID
This indicates a decrease in the electric field
(= trapped charge in the GD access region)

HOT Electron Degradation

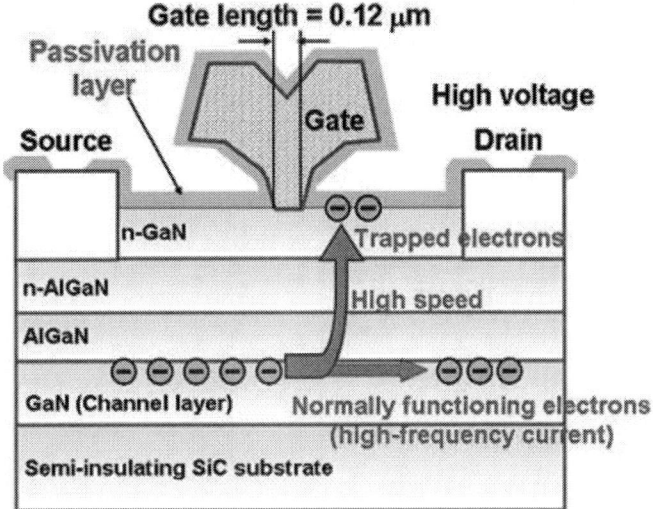

Fujitsu hot-electron trapping scheme:

http://www.fujitsu.com/global/news/pr/archives/month/2010/20101004-01.html

---

Degradation vs VGS level - constant voltage

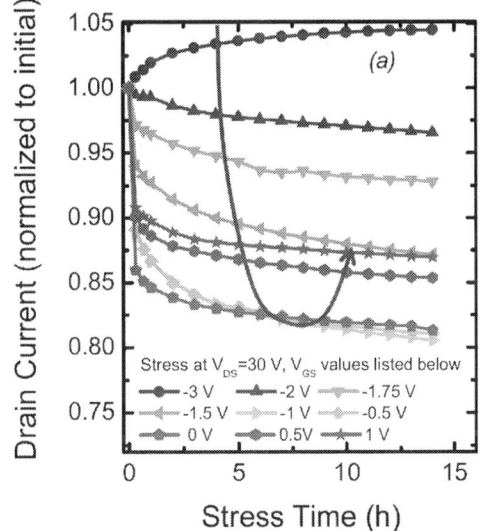

To achieve a more detailed description of the degradation process, we studied the degradation as a function of V_{GS} at constant V_{DS} level of 30 V and found a NON MONOTONIC BEHAVIOUR, like that of electroluminescence

Degradation Rate vs VGS level: effect of temperature

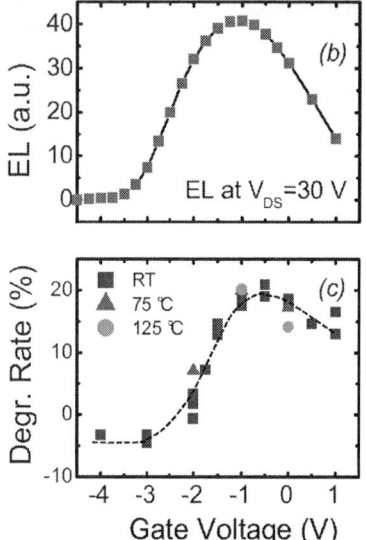

For low V_{GS} values → increase in I_D, due to a small negative shift in the threshold voltage

For higher V_{GS} levels → ΔI_D reaches a maximum and then decreases (despite the highest dissipated power), similarly to the EL signal

Tests at T_{AMB} = 75°C and 125°C showed similar degradation rates, highlighting the marginal role of the temperature as accelerating factor

Degradation Rate vs VGS level

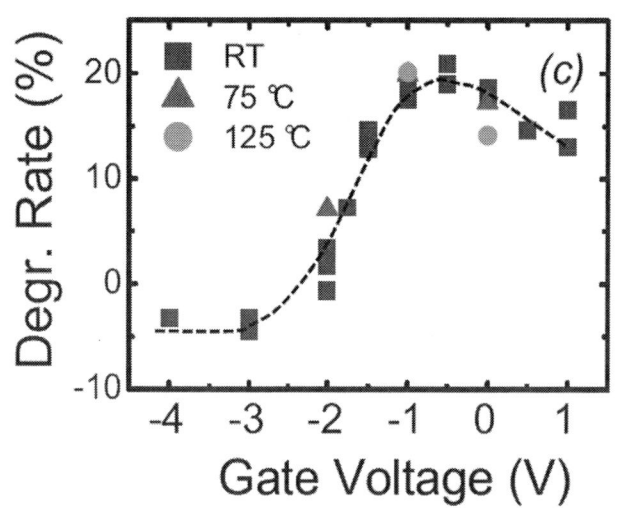

Interpretation

At high V_{DS} → Hot electrons can achieve sufficient energy to be injected in the AlGaN, at the SiN/AlGaN interface, in the SiN or in the GaN buffer, eventually remaining trapped there

Trapped charge induces an increase in Ron and a decrease in the electric field

Tests at T_{AMB} = 75°C and 125°C showed similar degradation rates, highlighting the marginal role of the temperature as accelerating factor

978-1-4799-5494-0/14 $31.00 © 2014 IEEE

Hot electron tests at constant power dissipation

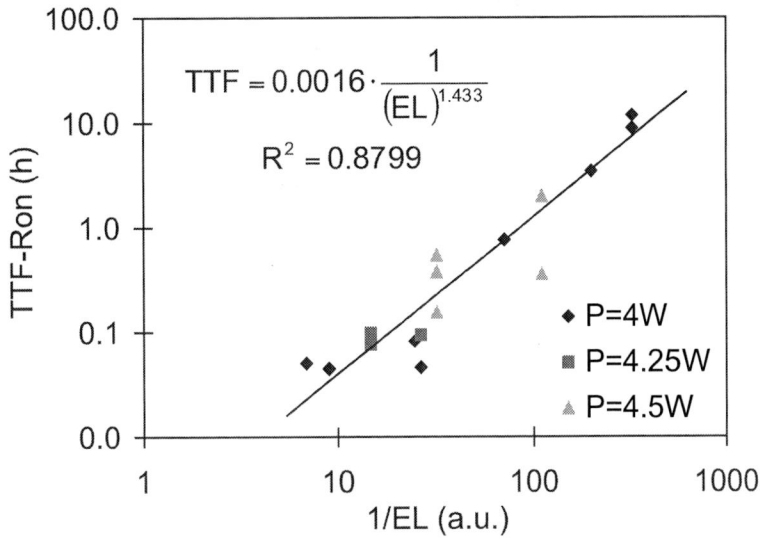

$$TTF = 0.0016 \cdot \frac{1}{(EL)^{1.433}}$$

$$R^2 = 0.8799$$

- ◆ P=4W
- ▪ P=4.25W
- ▲ P=4.5W

- •TTF was defined as the time for a 10% increase of R_{on}
- •A linear relationship was verified between log(TTF) and 1/EL

Hot-electron degradation: defects dehydrogenation

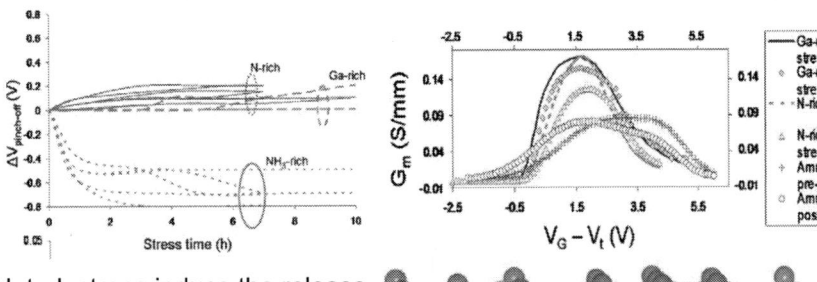

Hot-electrons induce the release of hydrogen which passivates point defects leading to:

* transconductance degradation
* increase in yellow luminescence

FIG. 1. (Color online) Atomic configurations of triply hydrogenated (a) gallium vacancy, (b) nitrogen antisite, and (c) divacancy.

Density functional theory model: HE dehydrogenation of the V_{Ga}-H_3 defect (gm decrease, YL)

Dehydrogenation of the nitrogen antisite (MOCVD, MBE ammonia-rich growth)→ decrease in the acceptor concentration → negative pinch-off shift)

Dehydrogenation of Ga vacancies → positive pinch-off shift

Y.S. Puzyrev et al. Jour. Appl. Phys. 109, 034501 2011

Thermal storage @325°C: Ni/Au metal interdiffusion

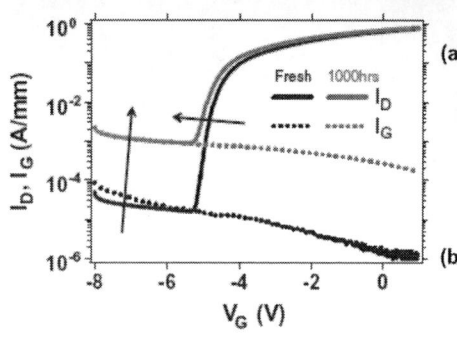

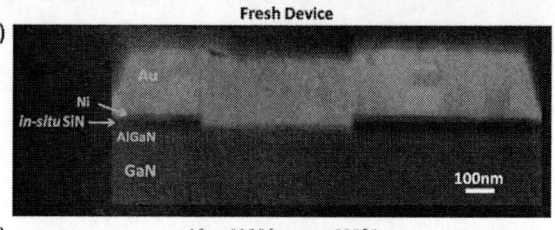

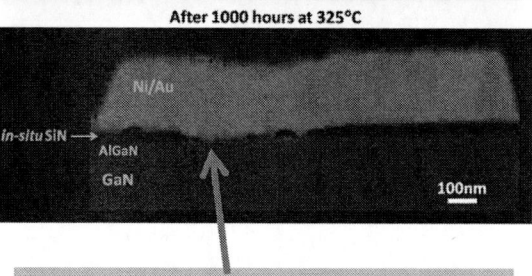

Ni/Au T-gates on AlGaN/GaN HEMT on Si in-situ Si_3N_4 passivation etched before deposition

After 1000 hours of storage in ambient:
- negative threshold voltage shift (fluorine removal), slight increase of I_D
- significant increase in drain and gate leakage current

Severe interdiffusion between Au and Ni, consuming the Ni layer, creating a Au contact on the AlGaN, and possibly diffusing into AlGaN itself

D. Marcon et al. Microel. Reliability 51 (2011) 1717.
Similar effects in F. Vitobello et al. IEEE IRPS 2012 2C.4.1

µE-LAB,
DEPARTMENT OF
INFORMATION
ENGINEERING
UNIVERSITY OF PADOVA

Ni/Au metal interdiffusion after 700 h life test @300°C

Commercially available AlGaN/GaN HEMT on Si submitted to 700 h rf test ad V_{DS} = 45 V, T_j =300°C
10% decrease of output power and drain current, 150% increase of gate current

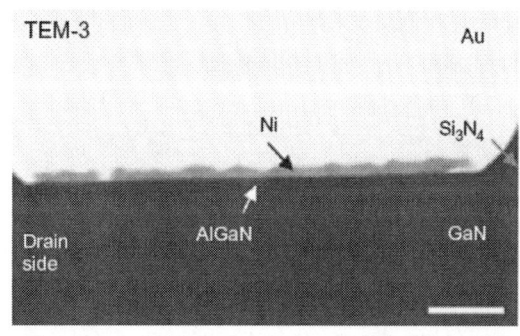

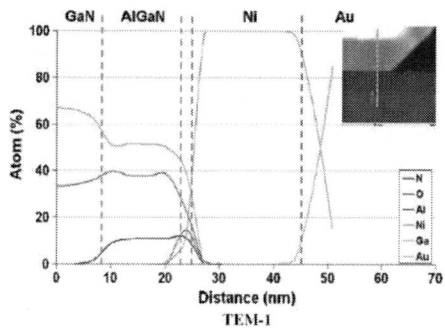

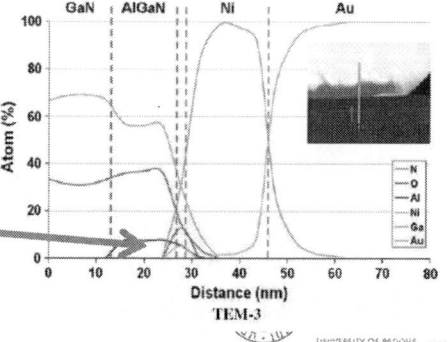

Au interdiffusion changes the contact from Ni/NiOx/AlGaN to Ni/Au/NiOx/AlGaN

J.-B. Fonder et al., Microel. Reliability 52 (2012) 2205

UNIVERSITY OF PADOVA

Conclusions

AlGaN/GaN technology for microwave and switching power applications has reached significant levels of maturity, with off-state critical voltages in excess of 200 V for 0.25 μm gate devices, and extrapolated lifetimes exceeding 2×10^5 hours at 175°C, 42 V, 10 GHz

Several open issues on failure mechanisms remain, in particular:

*Identification of thermally-activated failure mechanisms responsible for long-term degradation

* Further studies on electrochemical surface degradation mechanisms

*Accelerated methods for the extrapolation of time-dependent AlGaN breakdown

* More data on hot electron degradation

* More data on Ni/Au gate metallization stability and possible alternatives

Acknowledgments

  **http://www.alinwon-fp7.eu/fp7/**

  **http://www.hiposwitch.eu/**

  **http://www.e2cogan.eu/**

 **ONR project N000141010608, monitor: Paul Maki**

Acknowledgements

Post PhD, PhD and MS Students at University of Padova:
A. Stocco, C. De Santi, I. Rossetto, S. Dalcanale, R. Silvestri, …

Umesh Mishra and his group (UCSB),

Alessandro Chini, Giovanni Verzellesi (Univ. of Modena and Reggio Emilia),

G. Pozzovivo, G. Curatola, H. Haebeler, Infineon

P. Moens On Semiconductors

J. Sonsky - NXP

Oliver Hilt, Joachim Würfl (Ferdinand Braun Institut)

Dyoniz Pogany (Vienna University of Technology)

Antonio Cetronio, Claudio Lanzieri, Marco Peroni (Selex SI),

2nd IEEE Workshop on Wide Bandgap Power Devices and Applications (WiPDA), Oct 13, 2014

Wide-Band Gap (WBG) WBG Devices Enabled MV Power Converters for Utility Applications – Opportunities and Challenges

Subhashish Bhattacharya

Dept. of ECE

FREEDM Systems Center

North Carolina State University

Performance Evaluation and Converter Design with 1200V / 100A SiC MOSFET Module

Performance Evaluation and Converter Design with 1200V / 100A SiC MOSFET Module

2-Level 3ph Converter with 800V DC Bus

- 1200V/100A half bridge module for each leg

- Converter with forced air cooling can be designed for 50kW

- The designed converter is tested up to 35kW

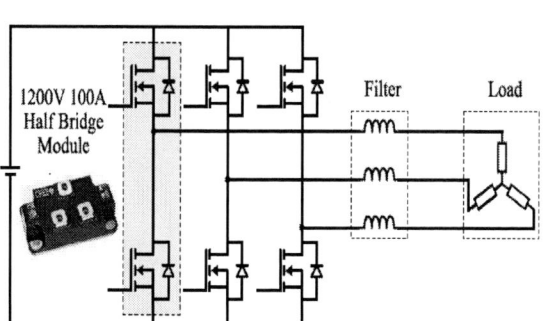

1200 V / 100 A SiC MOSFET based voltage source converter

1200V / 100A SiC MOSFET Module Structure and Forward Characteristics

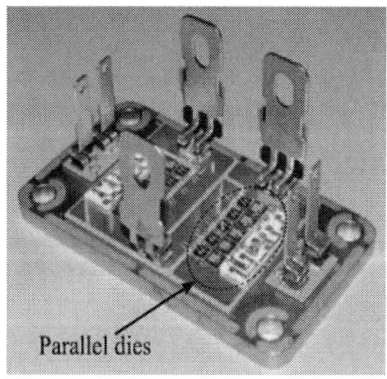

1200 V, 100 A SiC MOSFET module made from parallel dies

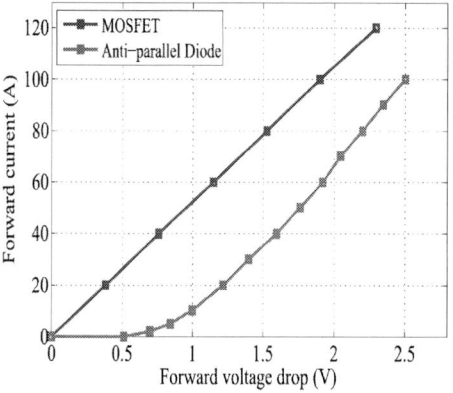

V-I characteristics of SiC MOSFET and anti-parallel SiC Schottky diode at 125 C and gate voltage of +20/-5 V

1.2kV and 1.7kV, 50A SiC MOSFET Test Circuit

Test Circuit Schematic

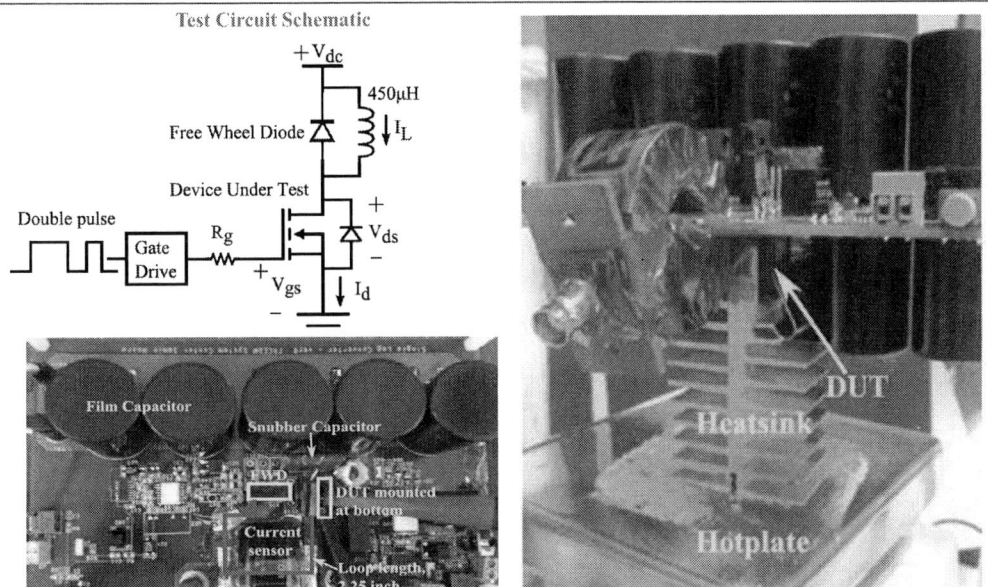

Hard-switching characterization circuit board

Device Heating Arrangement

1200V / 100A SiC MOSFET Switching transient
T_j=125°C, R_g=15Ω, V_d=800V, I_d=100A V_{gs}=+20/-5V

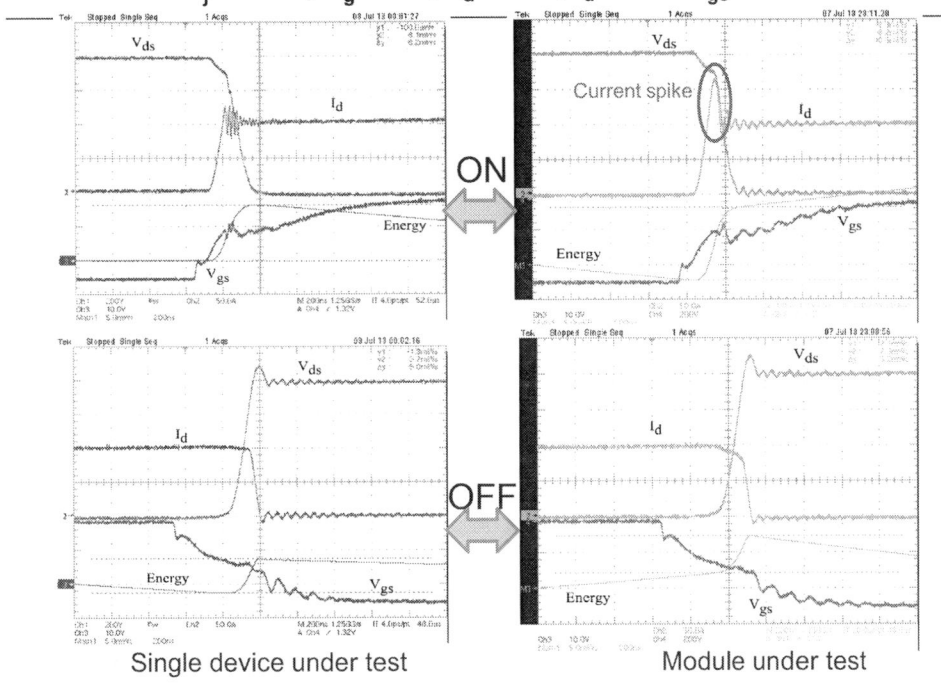

Single device under test

Module under test

978-1-4799-5494-0/14 $31.00 © 2014 IEEE

1200V / 100A SiC MOSFET Loss Dependents

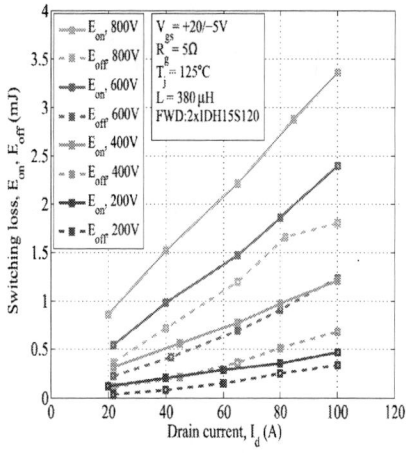

Loss variation with current
(single device)

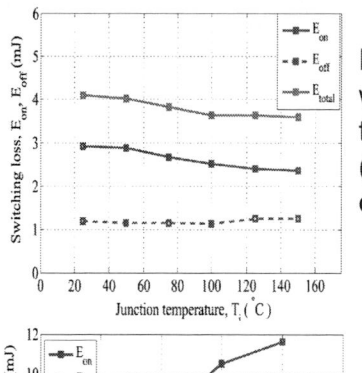

Loss variation with temperature (single device)

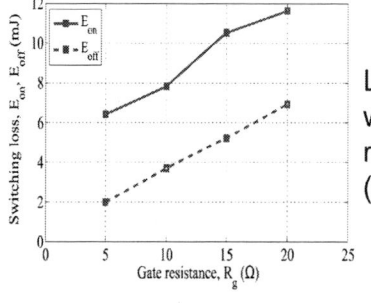

Loss variation with gate resistance (module)

1200V / 100A SiC MOSFET switching
Effect of gate resistance

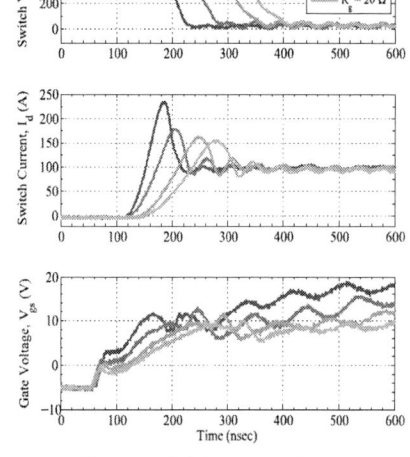

Turn ON transient

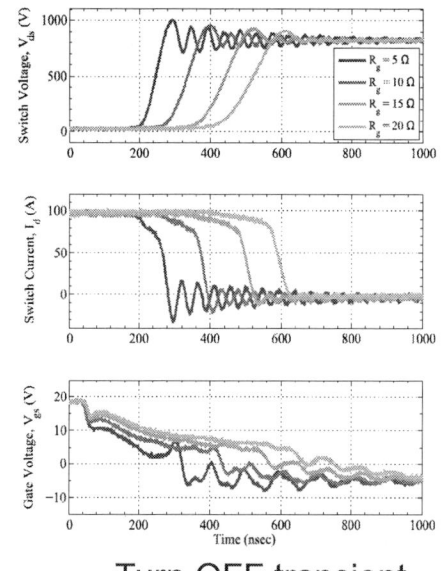

Turn OFF transient

978-1-4799-5494-0/14 $31.00 © 2014 IEEE   300

1200V / 100A SiC MOSFET
Loss and Efficiency compared to Si-IGBT

❖ IGBT - Semikron SK100GB12T4T, 1200 V, 100 A

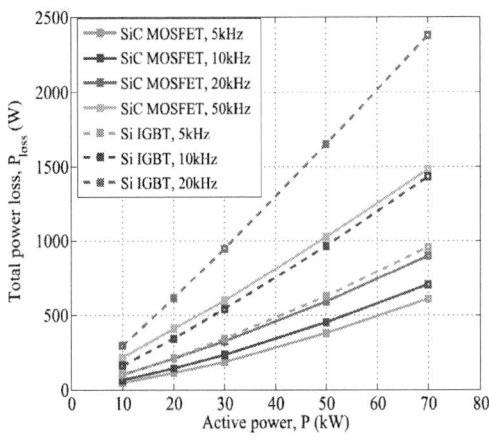

Power loss compared at different switching frequency

Converter efficiency compared at different switching frequency

1200V / 100A SiC MOSFET
3-phase Voltage Source Converter results

 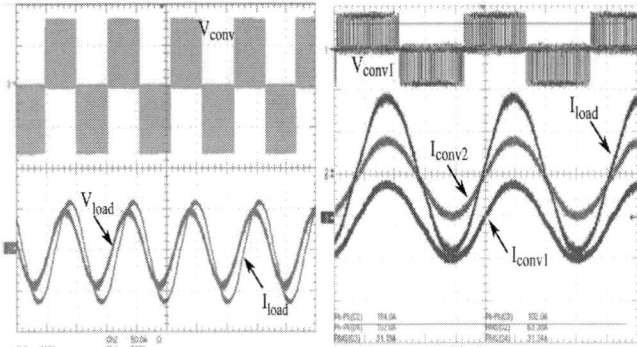

Packaged 3-phase Voltage Source converter
High Power Density

Converter supplying load of 35 kVA, Scale: Line current, I_{load} 50A/div, converter line voltage, V_{conv} 500V/div, Load line voltage, V_{load} 500V/div Switching Frequency = 20kHz

50kW loading, two parallel converter

50 kW 1200V/100A SiC-MOSFET Inverter

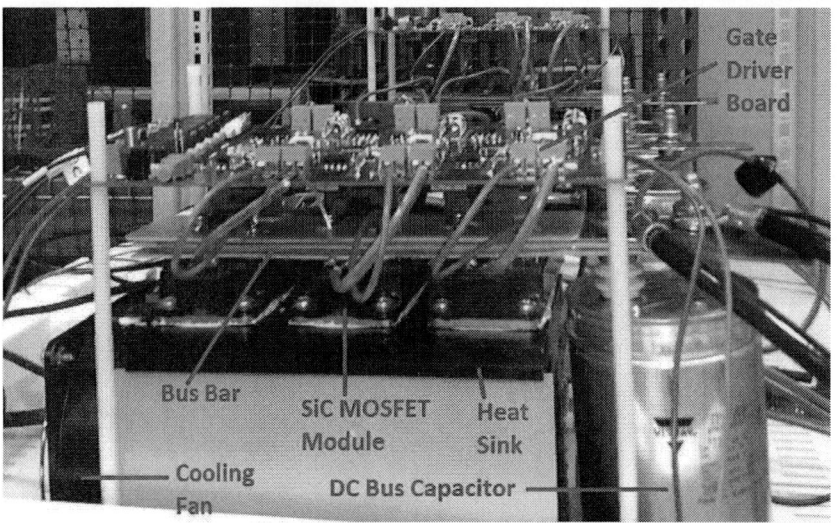

50 kW drive system, using 1200 V/100A SiC-Mosfet.
The SiC devices can be switched at 20-50 kHz, Vdc = 800V.

Dual Active Bridge (DAB) Isolated DC-DC Converter with SiC MOSFET

Why DAB & SiC?

• DAB – most suitable topology for high power density

• Si devices are limiting further increase in density (at higher power levels)

• DAB with SiC MOSFETs – Considerable increase in power density with high efficiency!

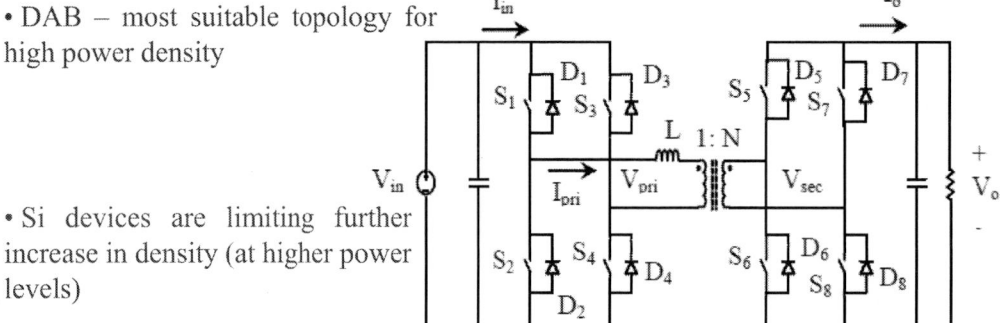

DAB Circuit Representation

978-1-4799-5494-0/14 $31.00 © 2014 IEEE   302

Dual Active Bridge (DAB) Isolated DC-DC Converter
with SiC 1200V / 20A MOSFET

 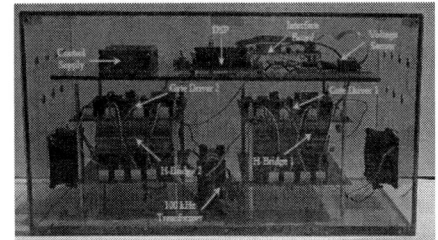

10 kVA DAB Prototype with SiC 1200V / 20A MOSFET devices

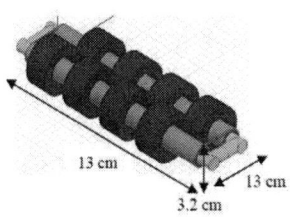 

Core material	Vitroperm 500F
Bac [T]	0.35
Cross section area [cm^2]	8.96
Core mass [g]	520
Number of turns (Pri, Sec.)	8, 4
Primary winding	Litz 216/36
Secondary winding	split copper tube
Mag. inductance	1.5 mH
Leakage inductance	9.3 uH
Winding loss	2 W
Core loss	45 W
Total loss	47 W

10 kVA 100 kHz Transformer

Duality of Solenoid vs Coaxial Winding Transformer (CWT)

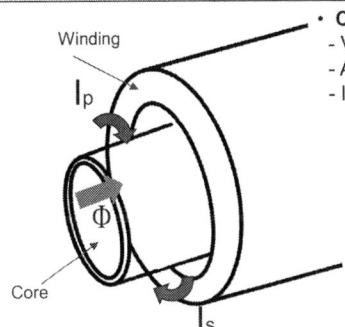

- **Coaxial type transformer**
  - Very low parasitic components
  - Accurately predictable high frequency response
  - Ideal geometry from electrical and thermal insulation point of view

 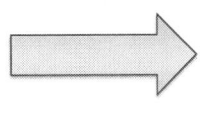 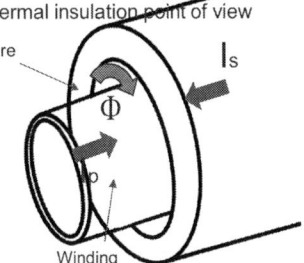

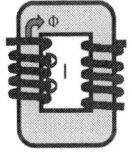     

Dual Active Bridge (DAB) Isolated DC-DC Converter with SiC 1200V / 20A MOSFET – experimental results

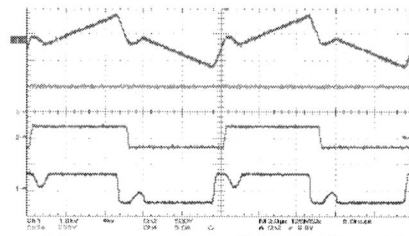

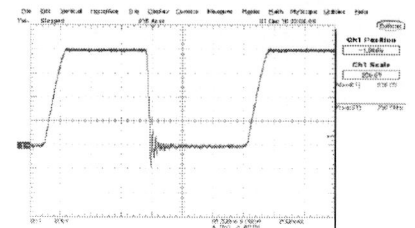

Closed loop operation of the DAB at 100 kHz
V_{in} - 600 V; V_o - 200 V; 4.8 A I_{pri} (peak)

750 kHz Switching at 800 V, 4.8 A

Switching times ~ 30ns

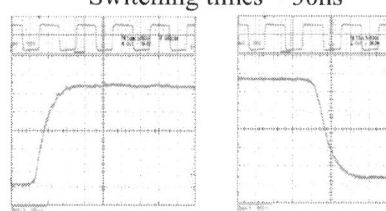

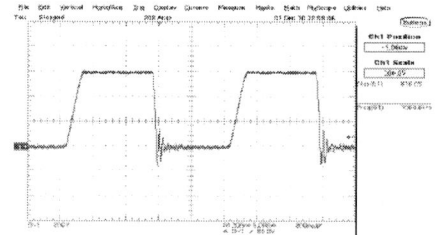

Turn-off and Turn-on transients at 500 V, 4 A

1 MHz Switching at 600 V, 6 A

Performance Characterization of 1700V / 50A SiC MOSFET

1.7kV, 50A SiC MOSFET Switching Characterization

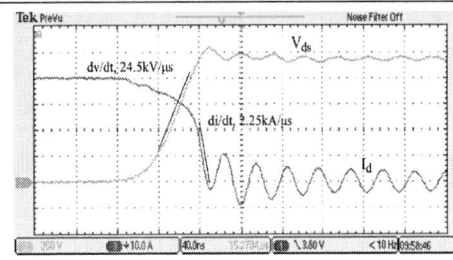

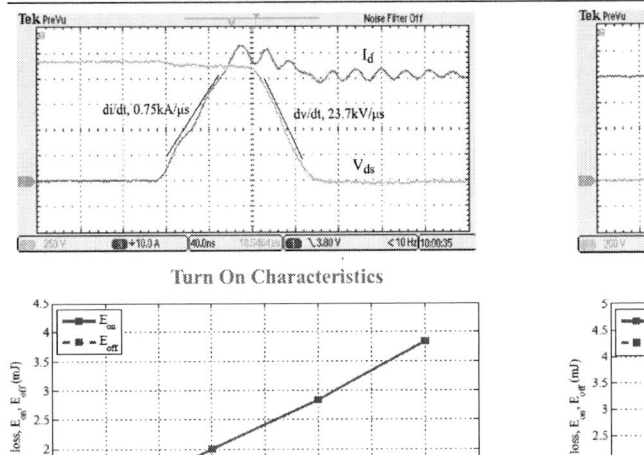

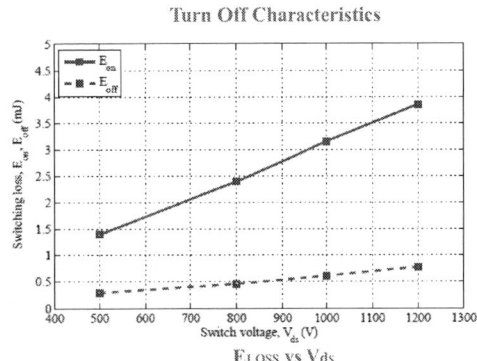

Turn On Characteristics

Turn Off Characteristics

E_{LOSS} vs I_d

E_{LOSS} vs V_{ds}

1.7kV, 50A SiC MOSFET Characterization

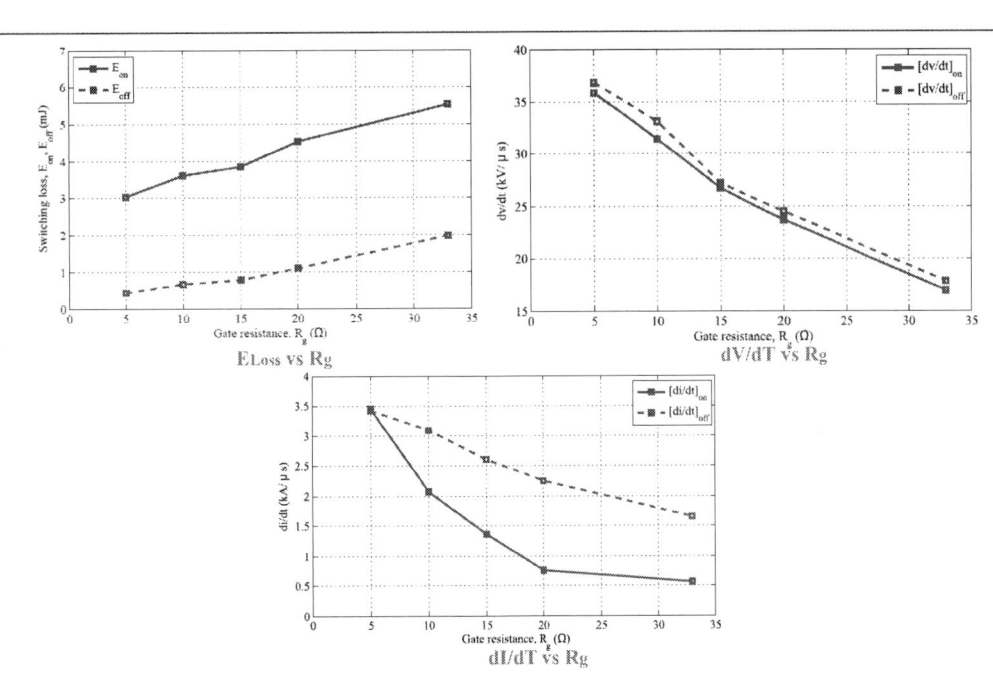

E_{LOSS} vs R_g

dV/dT vs R_g

dI/dT vs R_g

978-1-4799-5494-0/14 $31.00 © 2014 IEEE

1.7kV, 50A SiC MOSFET Characterization

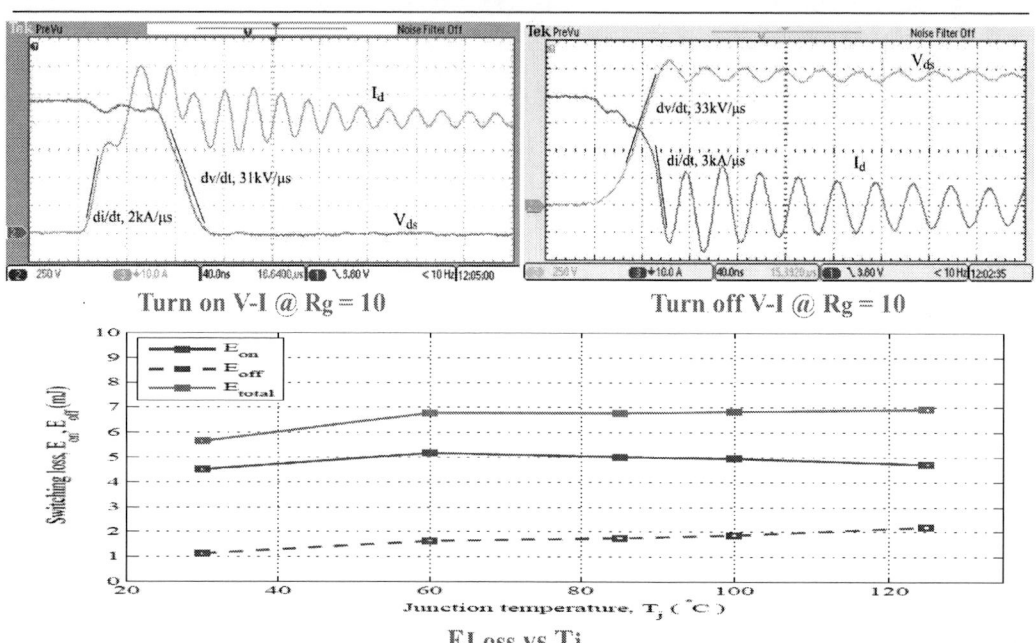

Turn on V-I @ Rg = 10   Turn off V-I @ Rg = 10

ELoss vs Tj

1.7kV, 50A SiC MOSFET Performance Comparison

Performance Comparison of 1.7kV/50A SiC MOSFET, 1.7kV/50A Si IGBT, 1.7kV/42A BiMOSFET

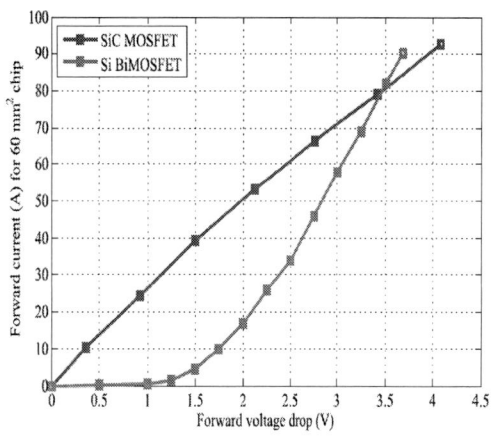

SiC MOSFET & Si BiMOSFET
Forward characteristics
Comparison

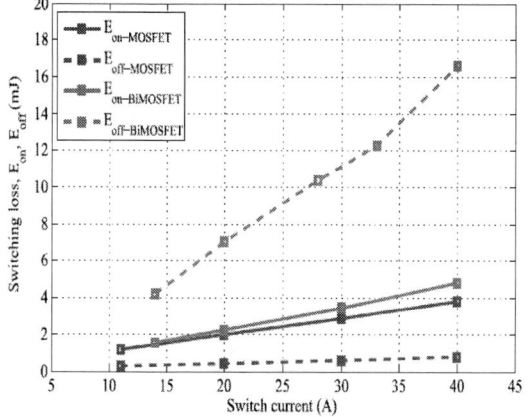

SiC MOSFET & Si BiMOSFET
Switching characteristics
Comparison

978-1-4799-5494-0/14 $31.00 © 2014 IEEE   306

Performance Characterization of 1200V/45A SiC JFET

1200V / 45A SiC JFET Module - FF45R12W1J1_B11

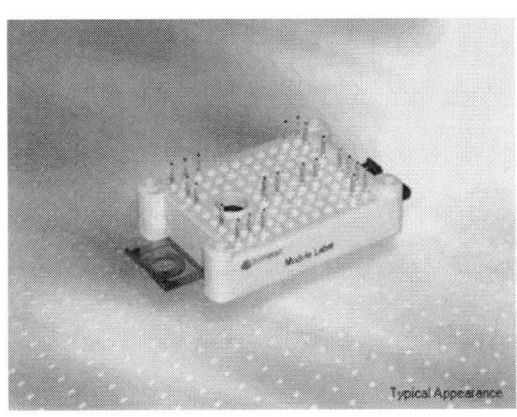

Infineon's Easy1B Package

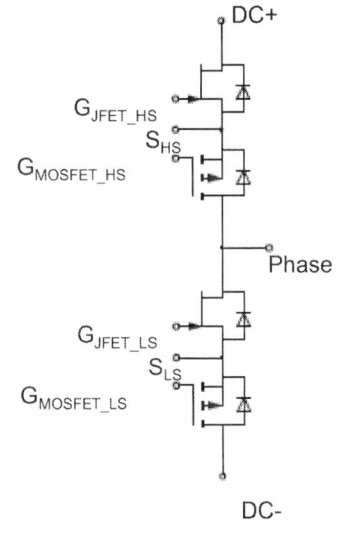

1200V / 45A SiC JFET Module

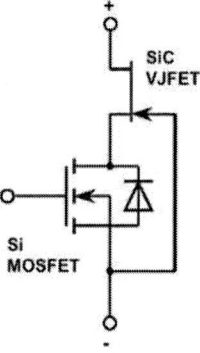

Classic Cascode Topology

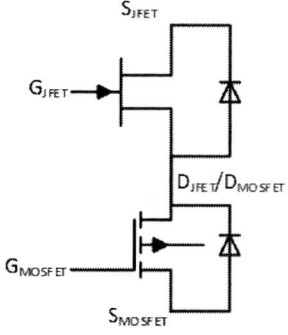

**Novel Cascode Topology with P-MOSFET
and Anti-Parallel Diodes**

1200V/45A SiC JFET Module Switching Results: Turn-On, Turn-Off

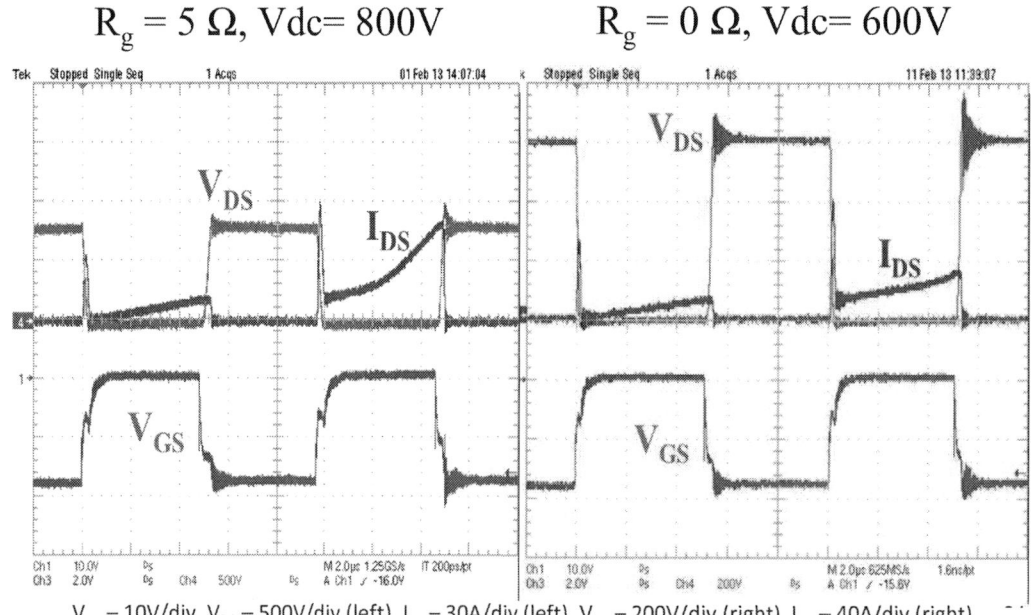

$V_{GS} - 10V/div$, $V_{DS} - 500V/div$ (left), $I_{DS} - 30A/div$ (left), $V_{DS} - 200V/div$ (right), $I_{DS} - 40A/div$ (right)

1200V / 45A SiC JFET Module Switching Results: Turn-On

$$R_g = 5 \; \Omega$$

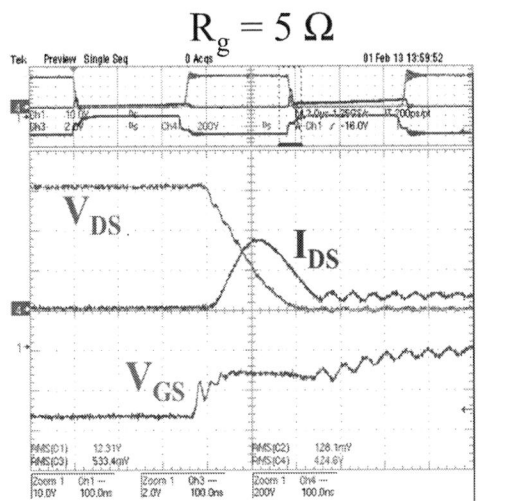

$$R_g = 0 \; \Omega$$

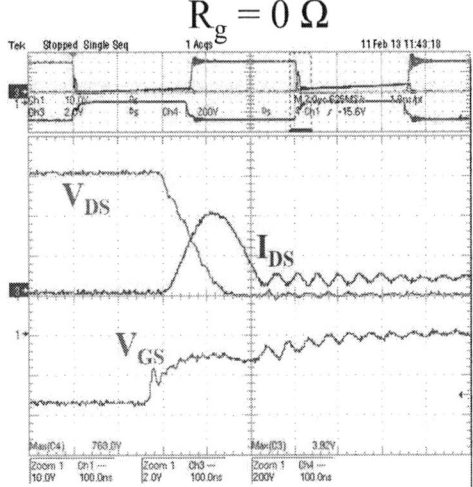

$V_{GS} - 10V/div$, $V_{DS} - 200V/div$, $I_{DS} - 40A/div$; $Vdc = 600V$

1200V / 45A SiC JFET Module Results: Eon, Eoff, C-V curve

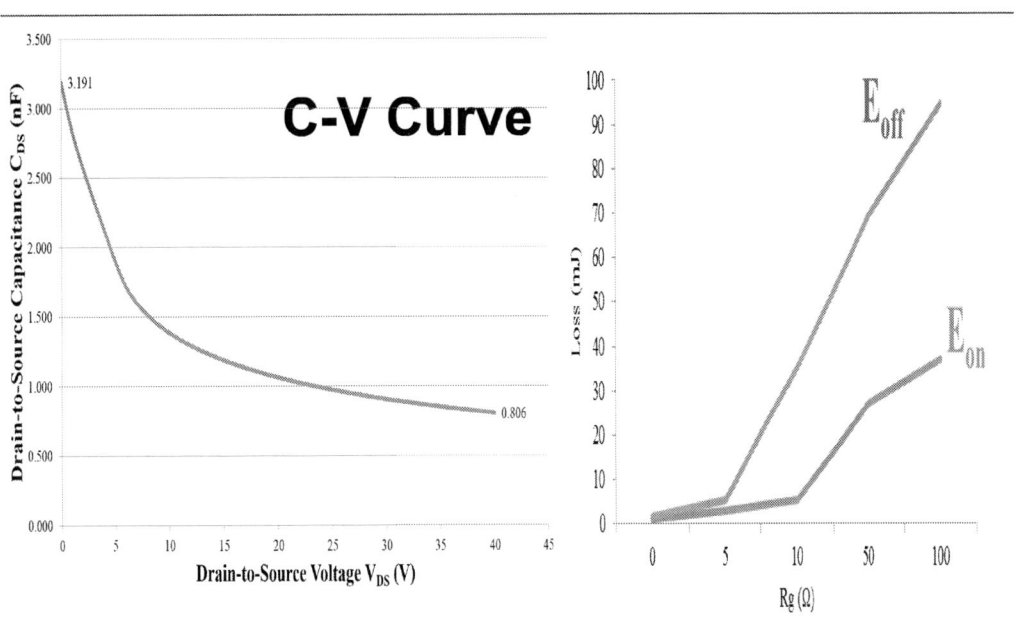

 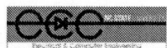

1200V / 45A SiC JFET Module Results

$R_g = 5\ \Omega$		$R_g = 0\ \Omega$	
V_{DC}	600V	V_{DC}	600V
R_g	5Ω	R_g	0Ω
T	25˚C	T	25˚C
I_{load}	15A	I_{load}	24A
I_{peak_spike}	72A	I_{peak_spike}	80A
t_{on}	200ns	t_{on}	140ns
t_{off}	225ns	t_{off}	125ns
$\frac{di}{dt}_{on}$	0.72A/ns	$\frac{di}{dt}_{on}$	0.80A/ns
$\frac{di}{dt}_{off}$	0.50A/ns	$\frac{di}{dt}_{off}$	0.75A/ns
$\frac{dv}{dt}_{on}$	3.00V/ns	$\frac{dv}{dt}_{on}$	4.29V/ns
$\frac{dv}{dt}_{off}$	2.67V/ns	$\frac{dv}{dt}_{off}$	4.80V/ns
E_{on}	2.90mJ	E_{on}	1.30mJ
E_{off}	5.50mJ	E_{off}	1.90mJ

Performance Evaluation and Converter Design with 15kV / 20A SiC IGBT Module and 10kV / 10A SiC MOSFET Module

15 kV, 20 A SiC IGBT Co-pack Modules

 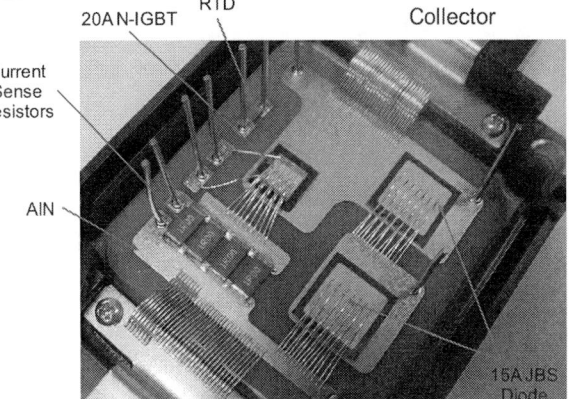

The module includes:

- 15 kV, 20 A SiC IGBT
- 20 kV (10*2), 10 A SiC JBS Diodes
- Current sense resistors
- Thermistor

10kV SiC MOSFET Co-pack Modules

Single 10kV SiC MOSFET Module

978-1-4799-5494-0/14 $31.00 © 2014 IEEE   311

SSPS System based on 10 kV, 120 A Silicon Carbide Half-Bridge Module

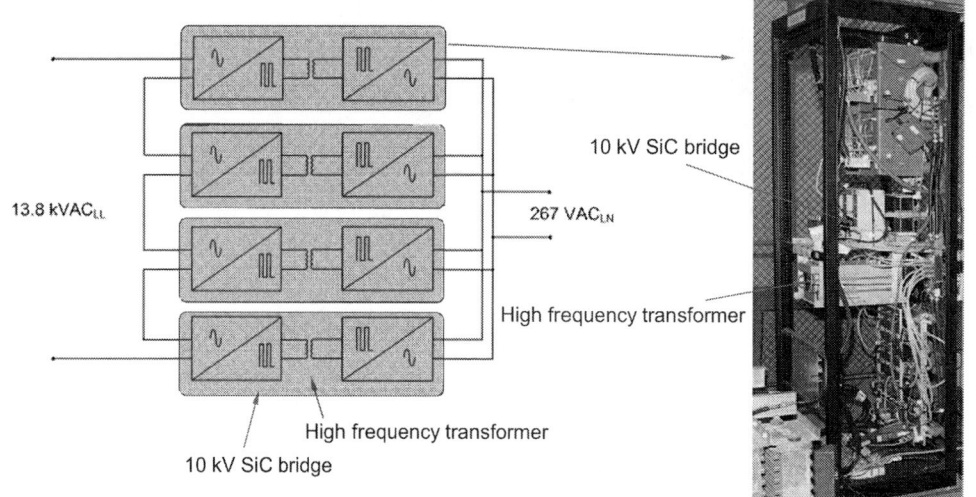

13.8 kVAC_LL

267 VAC_LN

10 kV SiC bridge

High frequency transformer

High frequency transformer

10 kV SiC bridge

source: Design, Development and Testing of a 1 MVA, 13.8 kV/ 465 V solid-state transformer with 10 kV silicon carbide switches - Developed under the DARPA/ ONR High Power Electronics Program IEEE ECCE 2011 SS4 - Ravi Raju, Robert Steigerwald, Michael Schutten, Mark Dame, Jeff Nasadoski, GE Global Research; David Grider, Mrinal Das Cree Inc.; Scott Leslie, Powerex; W. Reass, LANL; Tony Chalitas, IAP Research, Inc; Terry Ericsen, Sharon Beermann-Curtin, Office of Naval Research

---

SSPS System

Si IGBT assembly, 10 kV, 160 amps
- **Conduction drop > 12 V**
- **Switching time > 3 □s**

- **Demonstrated at rated power, 13.8 kV/265 V**
- **Efficiency at full load ~97%**
- **1/3rd weight of conventional transformer**
- **Clean 20 kHz waveforms**
- **Balanced sharing of voltages/ currents**
- **Met 60 Hz waveform THD requirements**

SiC Module, 10 kV, 120 amps (Cree,Powerex)
- Conduction drop < 6 V
- Switching time < 100 ns

Switching Characterization of 10 kV/10 A SiC MOSFET

10 kV, 10 A SiC MOSFET Characteristics

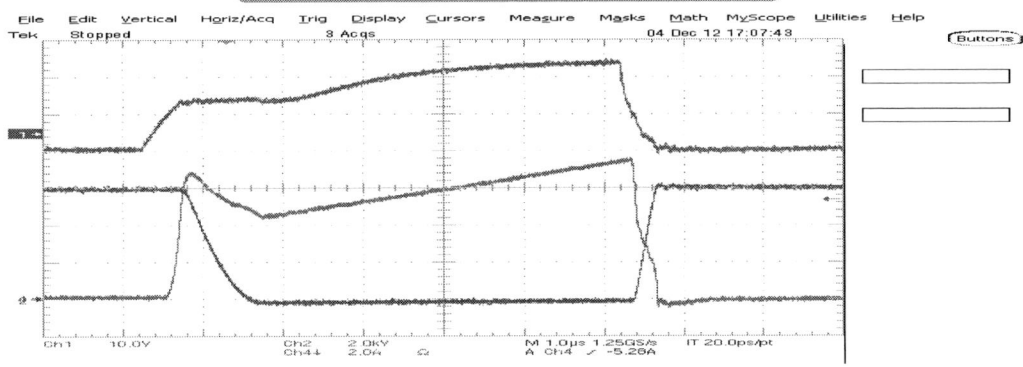

V_{GE} (10 V/div); V_{CE} (2 kV/div); Vdc = 6kV;
I_C (5 A/div); Time scale: 1000 ns/div

Note: For fair comparison, 12 kV, 10 A SiC N-IGBTs are used in this analysis.

Presented at IPEC 2014

Comparison of 10 kV/10 A SiC MOSFET, 12 kV/10 A, 2 µm and 5 µm SiC IGBTs

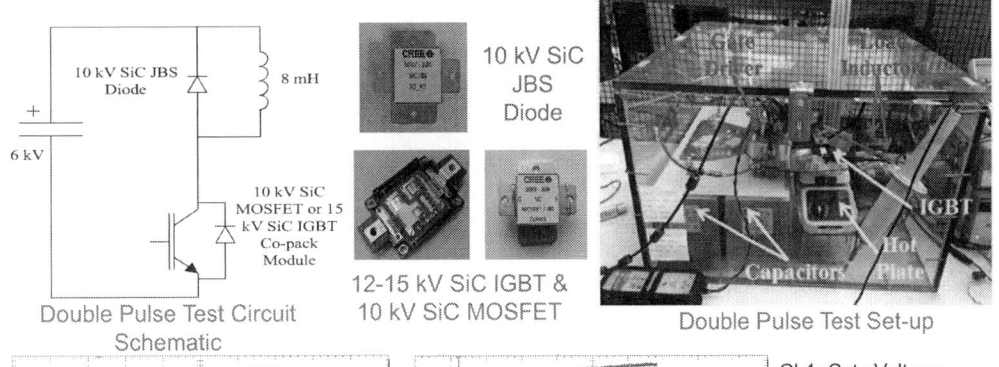

10 kV SiC JBS Diode

8 mH

6 kV

10 kV SiC JBS Diode

10 kV SiC MOSFET or 15 kV SiC IGBT Co-pack Module

Double Pulse Test Circuit Schematic

12-15 kV SiC IGBT & 10 kV SiC MOSFET

Double Pulse Test Set-up

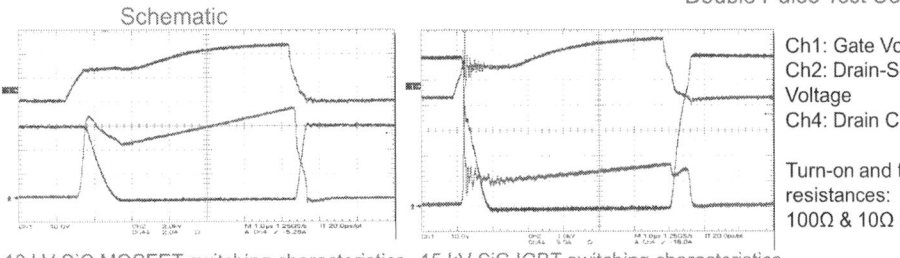

10 kV SiC MOSFET switching characteristics   15 kV SiC IGBT switching characteristics

Ch1: Gate Voltage
Ch2: Drain-Source Voltage
Ch4: Drain Current

Turn-on and turn-off gate resistances:
100Ω & 10Ω respectively

Presented at IPEC 2014

Comparison of 10 kV/10 A SiC MOSFET, 12 kV/10 A, 2 µm and 5 µm SiC IGBTs

Parameter	10 kV, 10 A MOSFET	12 kV, 10 A, 2 µm IGBT	12 kV, 10 A, 5 µm IGBT
Turn-on loss at 6 kV, 4 A	15.1 mJ	10.7 mJ	10.7 mJ
Turn-off loss at 6 kV, 8 A	1.9 mJ	11.1 mJ	4.4 mJ
Turn-on current spike magnitude	7.5 A	24 A	12.5 A
$R_{G(ON)}$	100 Ω	100 Ω	100 Ω
$R_{G(OFF)}$	10 Ω	10 Ω	10 Ω
Forward drop at 10 A	4.1 V	4.4 V	5.3 V

- The MOSFET and 5 µm IGBT turn-on loss can be reduced by using lower $R_{G(ON)}$ to generate same dv/dt as 2 µm IGBT, for fair comparison.
- MOSFET has lowest turn-off loss due to majority carrier physics.
- At 10 A MOSFET is comparable in conduction loss; For higher currents, IGBTs are efficient.

10kV SiC MOSFET 3-phase 2-level Inverter

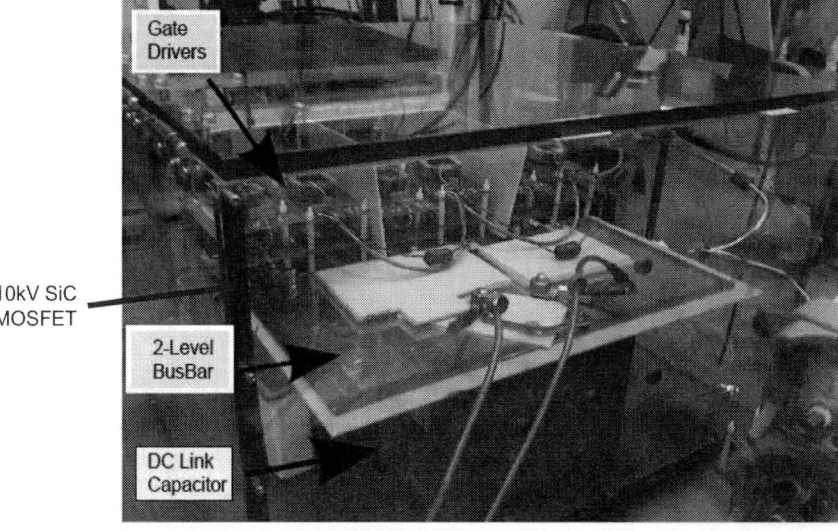

2-level 3-phase Inverter built using 10kV SiC MOSFET

10kV SiC MOSFET 3-phase 2-level Inverter – Experimental Results

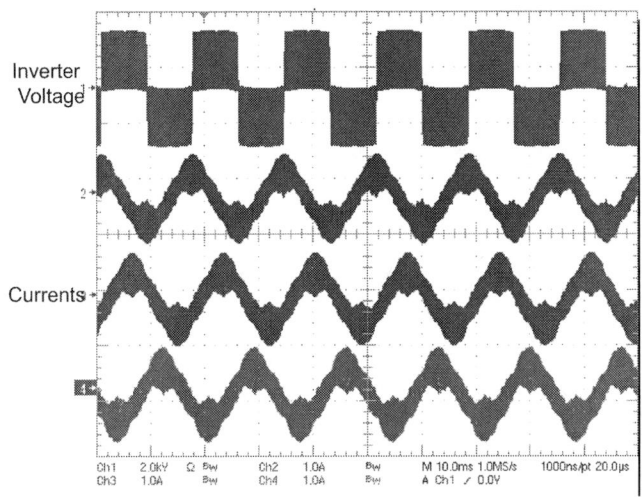

Test Conditions:
- DC link voltage 2kV
- RL load: R=1.4kOhm, L=20 mH
- Switching frequency fsw = 10kHz
- Fundamental Frequency = 60Hz

3-phase 2-level Inverter Waveforms

Transformerless Intelligent Power Substation (TIPS)

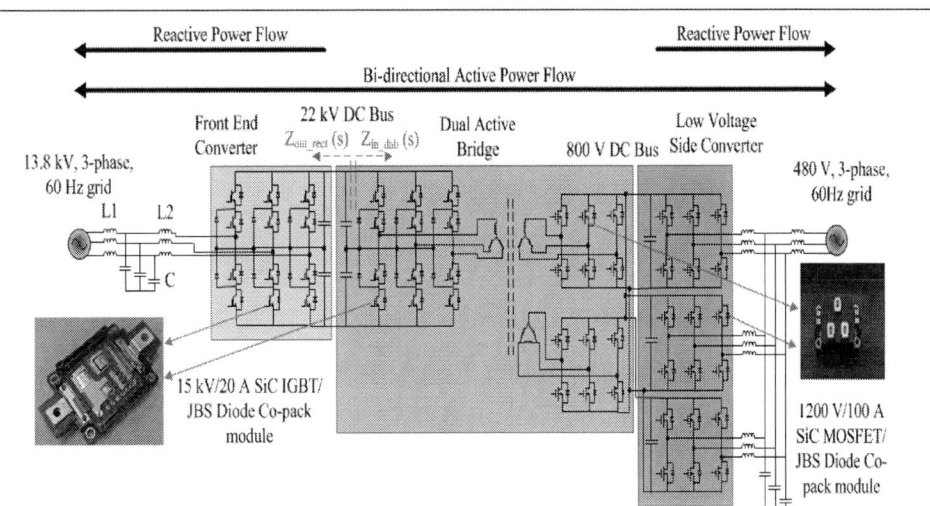

- Three-Phase SiC Devices based Solid State alternative to conventional line frequency transformer for interconnecting 13.8 kV distribution grid with 480 V utility grid.
- Smaller and Light Weight High Frequency Transformer operating at 10 kHz used for Isolation.
- Advantages – Better Power Quality, Controllability, Bi-directional Power Flow, VAR Compensation, Small Size/Light Weight, lower Cooling Requirement, Integration of Renewable Energy Sources/Storage Elements.

978-1-4799-5494-0/14 $31.00 © 2014 IEEE

Gen-I Si-IGBT 6.5kV based Solid State Transformer (SST): Topology and Prototype

Specifications:

- Input: **7.2kV**

- Output: **240Vac/120Vac; 400Vdc**

- Power rating: **20kVA**

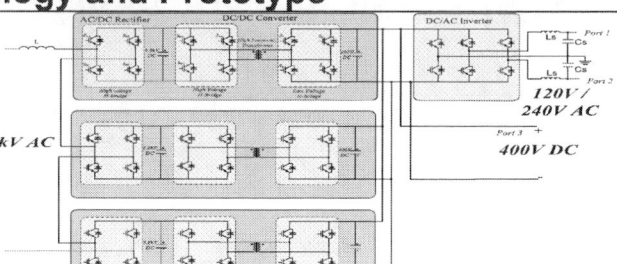

Named to MIT Technology Review's 2011 list of the world's 10 most important emerging technologies

---

Challenges: HV - HF Magnetics and Transformer design

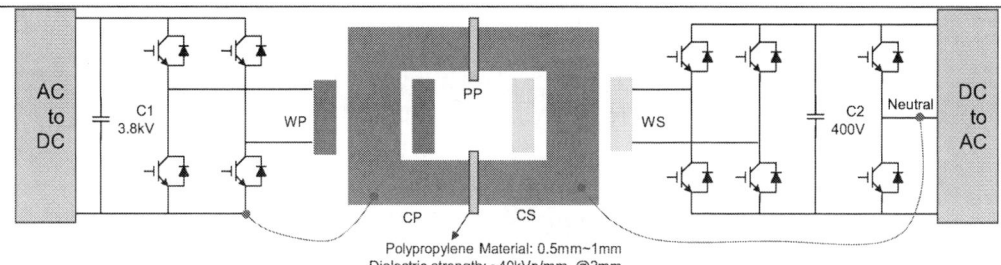

Polypropylene Material: 0.5mm~1mm
Dielectric strength: ~40kVp/mm, @2mm

Specifications and design parameters:

- Frequency: 3 kHz
- High voltage DC-link: 3.8kV
- Low voltage DC-link: 400V
- Power rating: 20/3=6.7kVA
- Turns ratio: 9.5
- Insulation: 15kV
- Number of primary turns:190
- Number of secondary turns:20
- Magnetizing Inductance: 235mH
- Leakage Inductance: 36mH

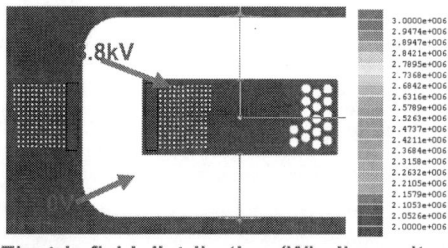

Electric field distribution (Winding voltage is evenly distributed to each wire in order)

Challenges: HV - HF Magnetics and Transformer design

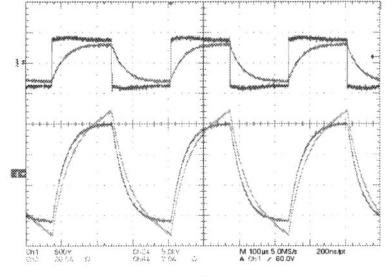

Ch1=Vac_low side;
Ch2=Vac_high side;
Ch3= Iac_low side;
Ch4=Iac_High side,

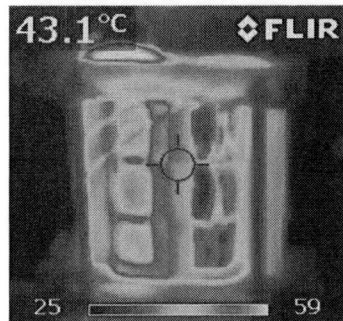

Test Waveforms   **Insulation Capability**

	3.9KW	7.0KW
CORE LOSS (W)	80.6	80.6
HV WINDING LOSS (W)	16.4	62.1
WINDING LOSS (W)	28.7	85.4
TOTAL (W)	126	228
EFFICIENCY	96.9%	96.8%

VOLTAGE (KV)	LEAKAGE CURRENT (nA)
10	3.6
15	5.8
20	10.5
25	16

Efficiency

2-level & 3-level 3-Phase 15kV SiC IGBT Inverter

Gate Drivers

DC link Capacitor

 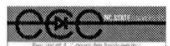

Three Level NPC (Neutral Point Clamped) Pole

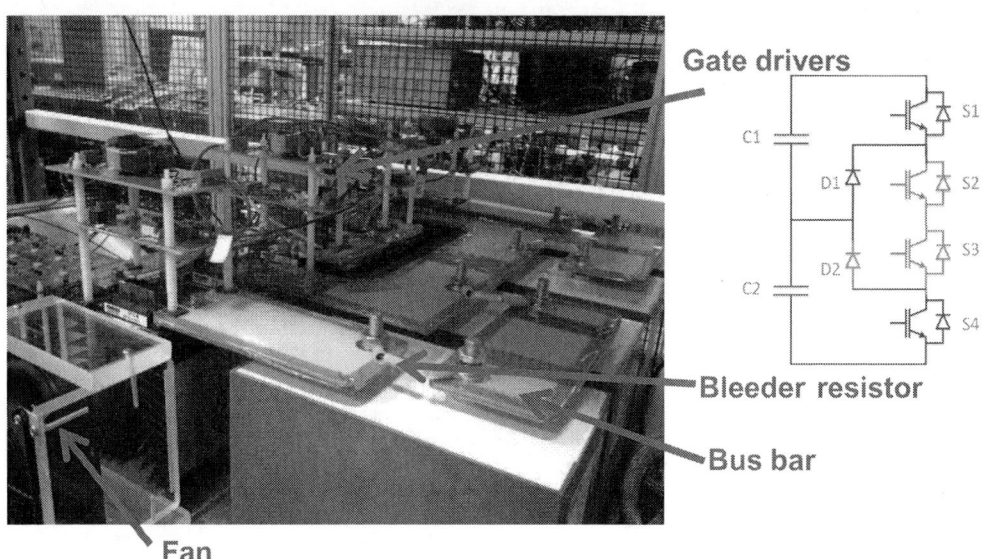

Gate drivers

Bleeder resistor

Bus bar

Fan

3-phase 13.8kV AFE Rectifier Hardware

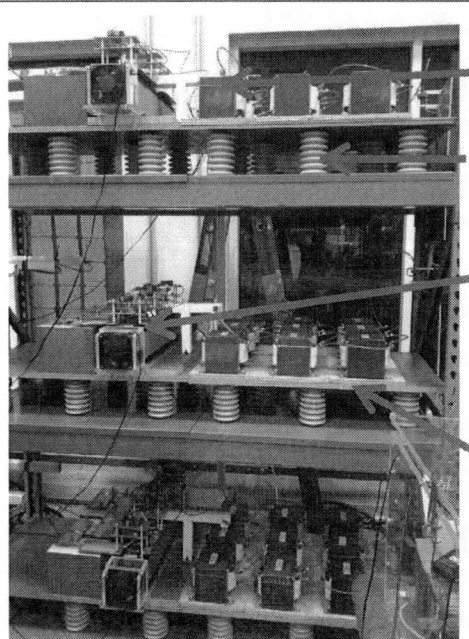

Filter inductors

Porcelain stand offs

Converter leg

Fiber glass Insulation sheet

- Three legs of FEC arranged in three different levels of the cabinet

- The filter inductors per phase is 180 mH

- Porcelain stand offs are used with fiber glass insulation sheets to maintain the isolation from the cabinet

22kV / 11kV – 800V DAB DC-DC converter at 10kHz

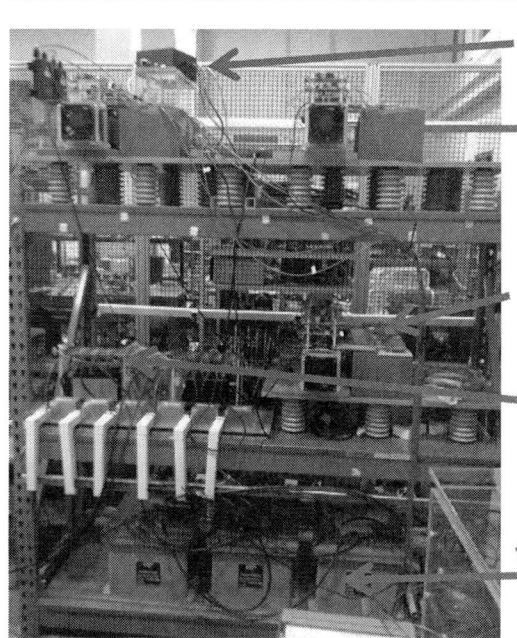

HV pole 1

HV pole 2

HV pole 3

LV converters

DAB transformers

1200 V SiC MOSFET Based Low Voltage Side Converter

Single Phase High Frequency Transformer

Front End Converter 3-Level NPC Pole Testing at 10kV

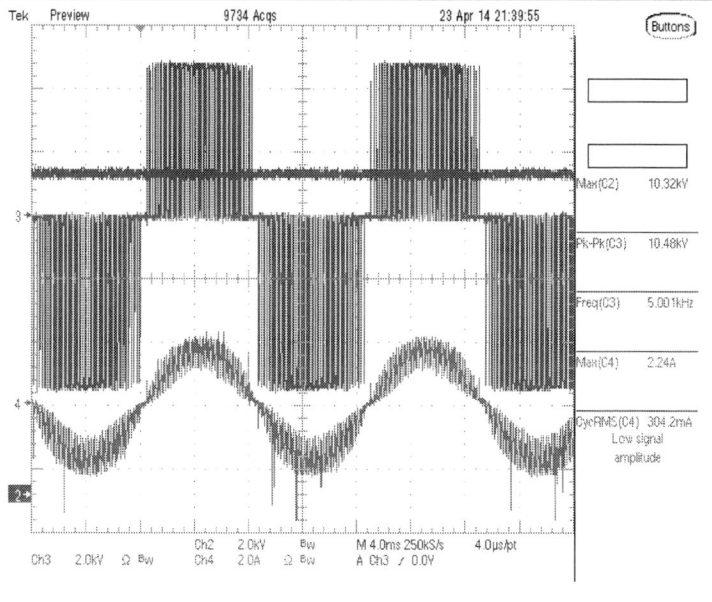

- FEC Y-Phase Pole tests

- Testing using Boost Converter at the input

- 10 kV, 5 kW, 5 kHz SPWM

- 15kV / 20 A IGBT modules

Front End Converter 3-Level NPC Pole Testing at 10kV

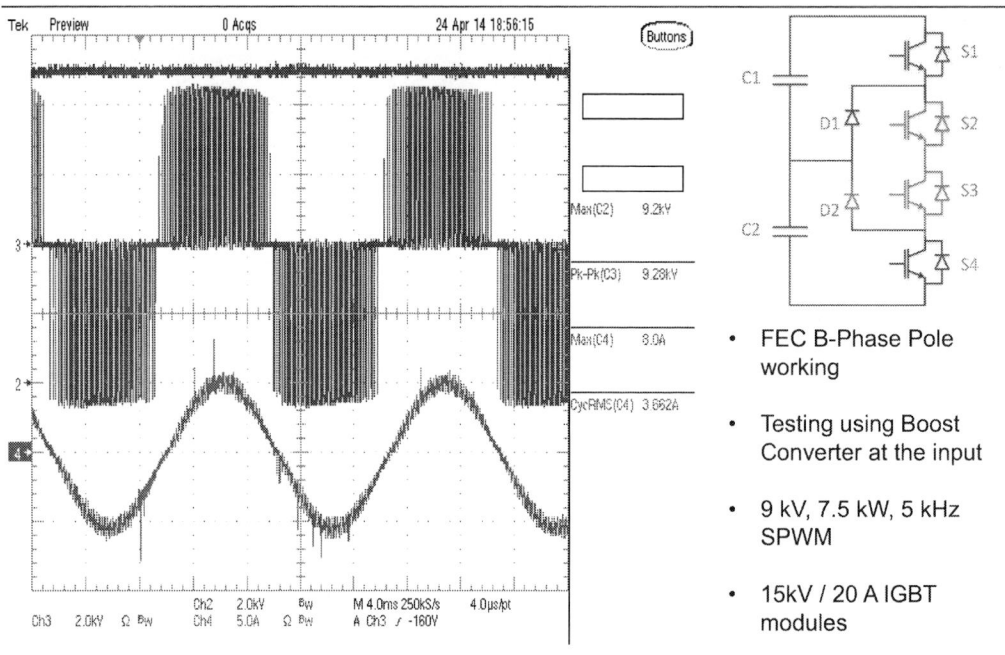

- FEC B-Phase Pole working

- Testing using Boost Converter at the input

- 9 kV, 7.5 kW, 5 kHz SPWM

- 15kV / 20 A IGBT modules

Dual Active Bridge Converter 3-Level NPC Pole Testing at 10kV

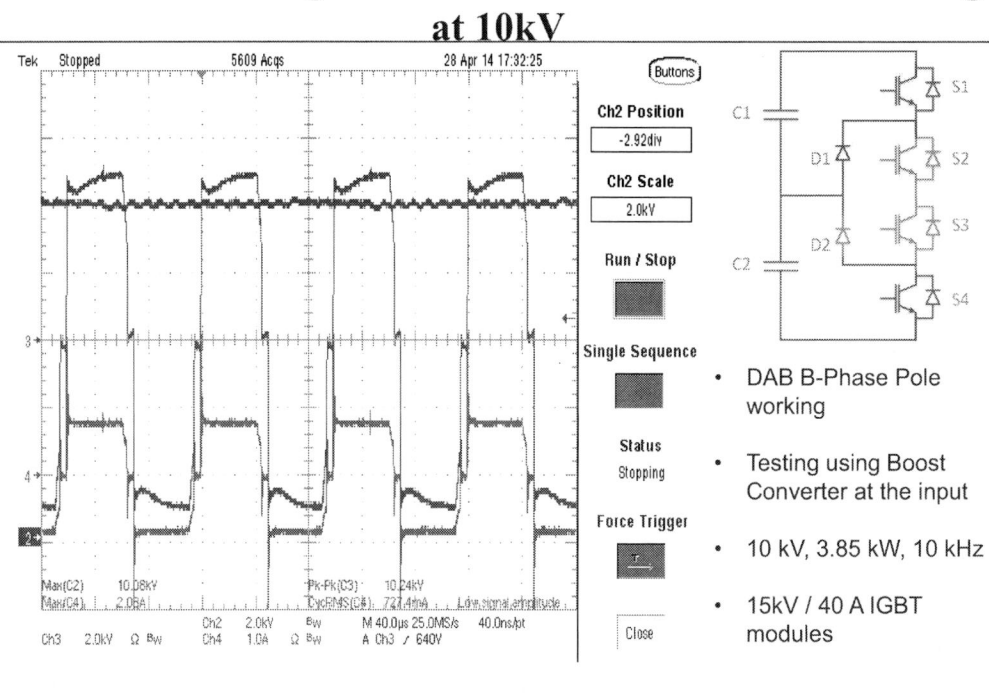

- DAB B-Phase Pole working

- Testing using Boost Converter at the input

- 10 kV, 3.85 kW, 10 kHz

- 15kV / 40 A IGBT modules

Dual Active Bridge Converter 3-Level NPC Pole Testing at 10kV

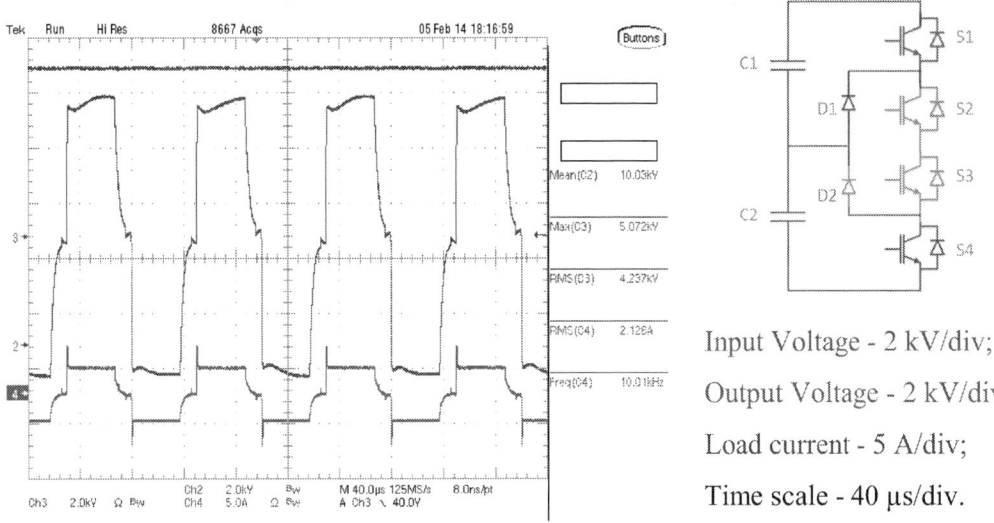

Input Voltage - 2 kV/div;

Output Voltage - 2 kV/div;

Load current - 5 A/div;

Time scale - 40 µs/div.

- The 3-level converter pole is tested up to 10 kV, 10 kHz with 9 kW Resistive loading.
- This serves as validation up to 27 kW with three-phase converter (3-poles in parallel)

Bidirectional DC-DC Converter Setup

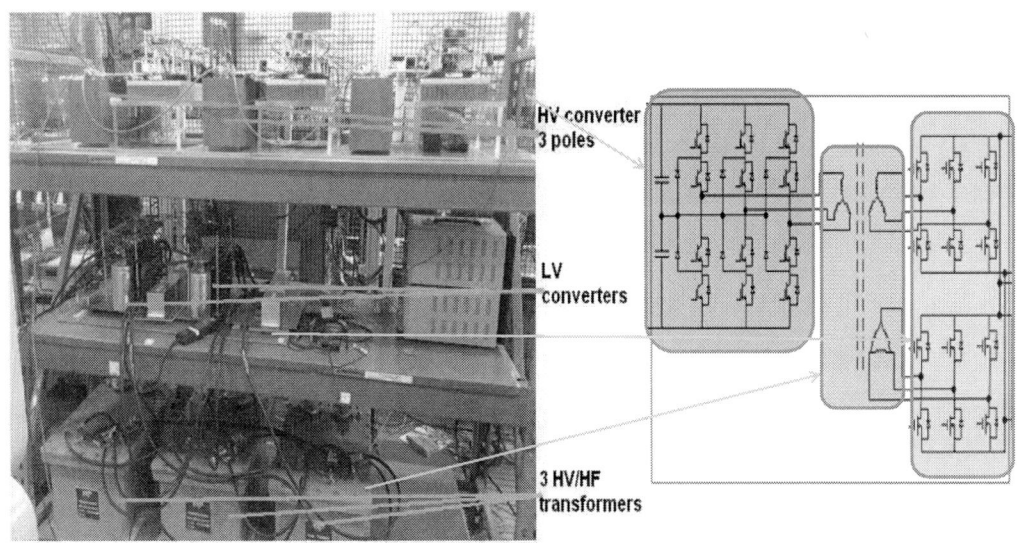

978-1-4799-5494-0/14 $31.00 © 2014 IEEE

TIPS Integration
FEC and DAB Closed Loop Testing using 15kV/40 A IGBT Devices

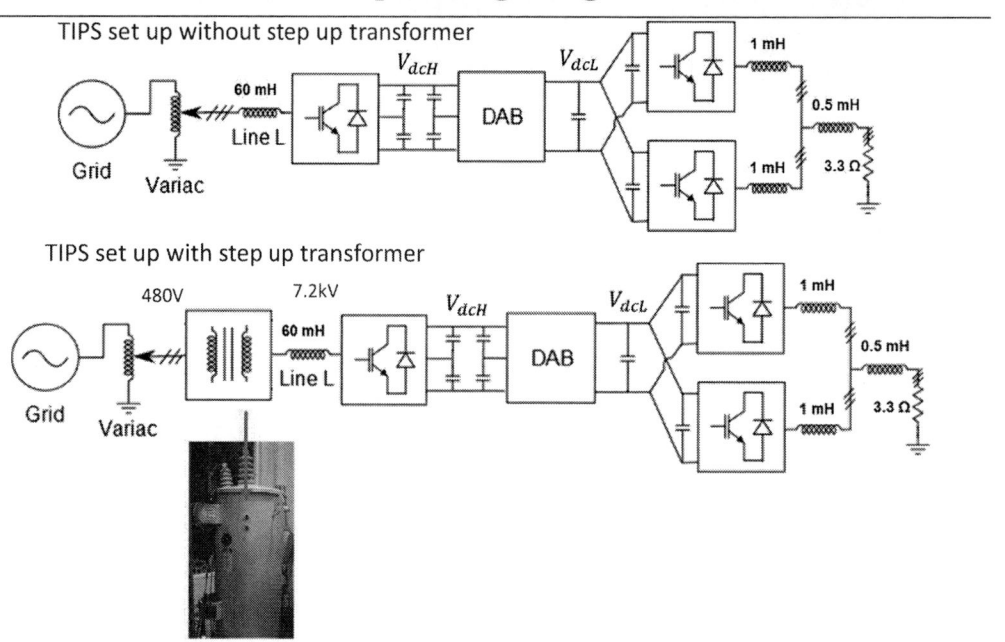

Integration of TIPS Converters
Operation at 4.2kV dc and 2.2kW load

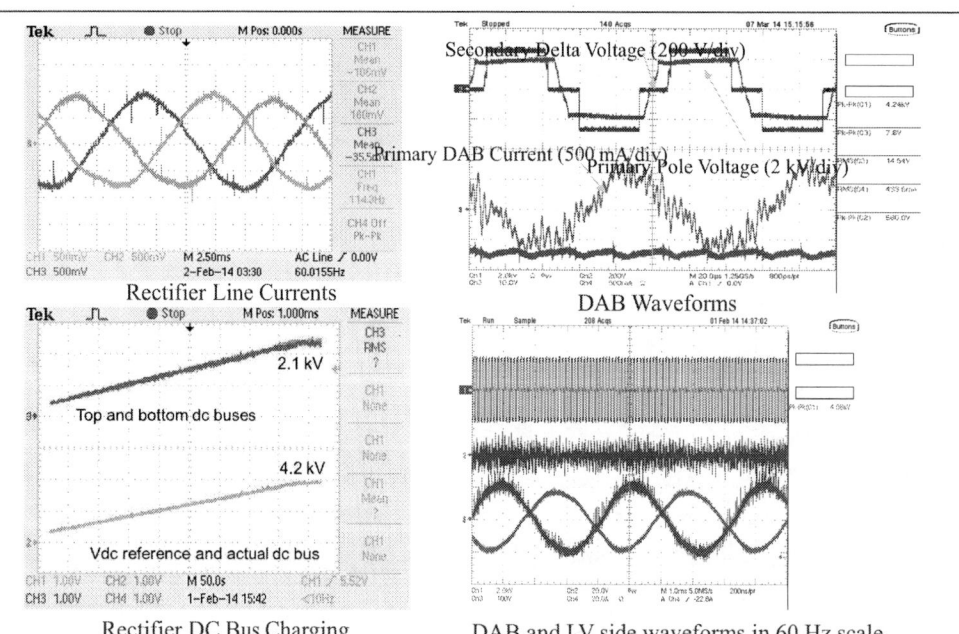

978-1-4799-5494-0/14 $31.00 © 2014 IEEE   322

3-Phase DAB Results

DAB operation at 3 kV primary side DC link with 5 kW
load
HVDC: 3kV
LVDC : 210V

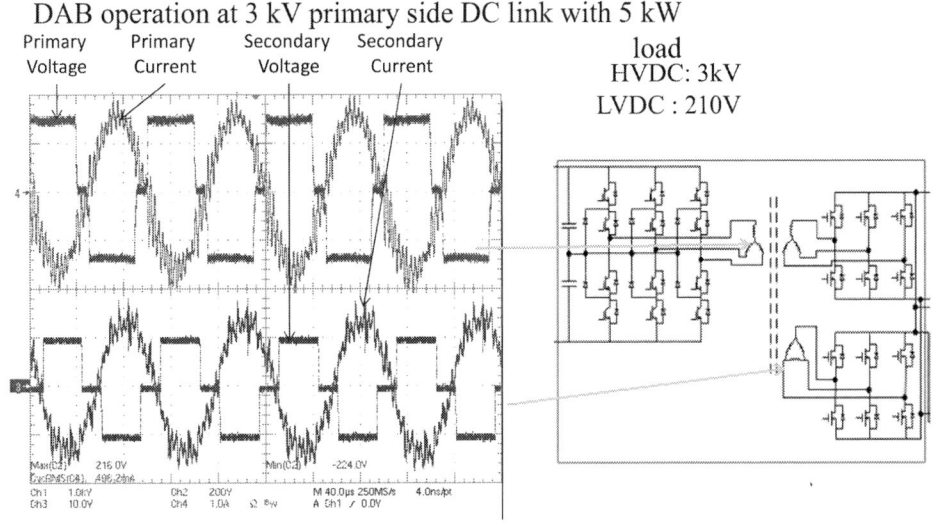

- Near sinusoidal current in DAB due to 3-level 3-phase topology and control advantages

- LC ringing in both primary and secondary currents due to transformer parasitic capacitors

Characteristics of 15 kV, 20 A SiC IGBT

Turn-off transition at 11 kV, 5 A, 25°C

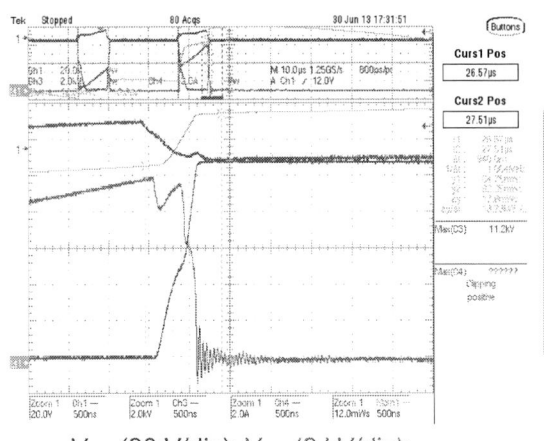

V_{GE} (20 V/div); V_{CE} (2 kV/div);
I_C (5 A/div); E_{OFF} (15 mJ/div);
Time scale: 1000 ns/div

- $R_{G(OFF)}$ = 10 Ω.
- Voltage has two different slopes. (punch-through design)
- Current bump; tail ringing.
- Total duration of the transition = 650 ns.
- Energy loss = 17.6 mJ

978-1-4799-5494-0/14 $31.00 © 2014 IEEE   323

Characteristics of 15 kV, 20 A SiC IGBT

Turn-on transition at 11 kV, 10 A, 25°C

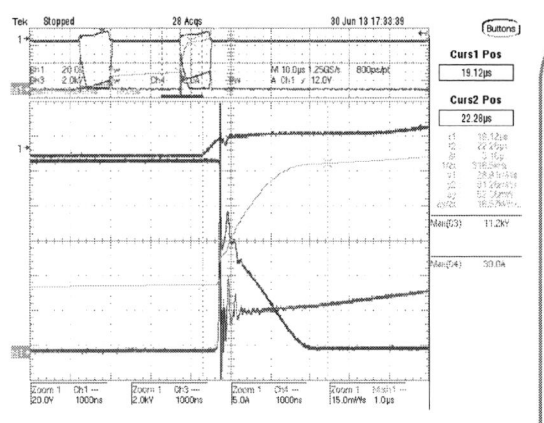

- $R_{G(ON)}$ = 200 Ω.
- Voltage has two different slopes. (very high dv/dt at the beginning of the transition)
- Current spike; followed by ringing.
- Total duration of the transition = 2.2 μs.
- Energy loss = 52.4 mJ.
- High $R_{G(ON)}$ is used to limit the initial dv/dt.

V_{GE} (20 V/div); V_{CE} (2 kV/div); I_C (2 A/div); E_{OFF} (12 mJ/div); Time scale: 500 ns/div

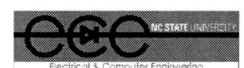

Gate Driver for 15 kV SiC IGBT

HV Gate Driver
Challenges: Materials, dv/dt, dielectric losses

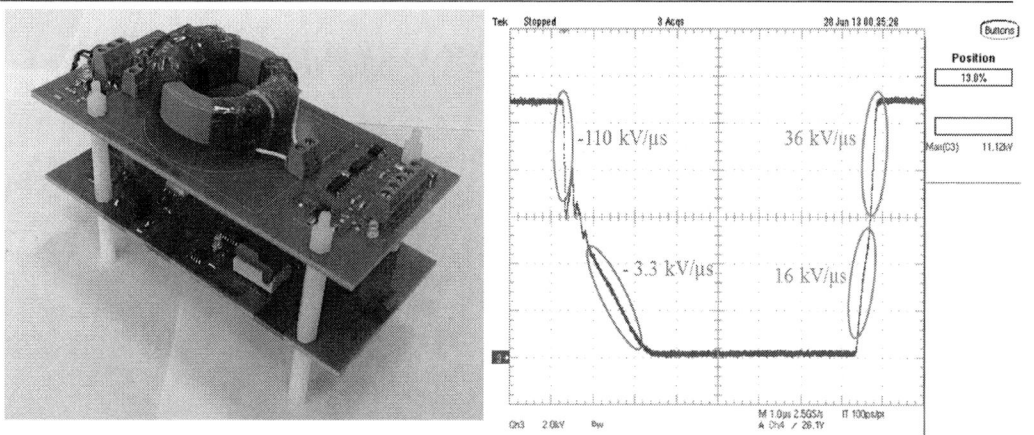

High Isolation Gate Driver

11 kV, high dv/dt switching
voltage of the IGBT

Device maximum Turn ON and Turn OFF dv/dt
- Turn ON dv/dt = 100.6 kV/µs, Turn OFF dv/dt = 28.29 kV/µs

The Gate driver has been exposed to 100 kV/µs at 11 kV

Gate Driver Isolation Transformer

- Isolation transformer inter-winding coupling capacitance (a crucial element for high dv/dt immunity) w.r.t frequency.

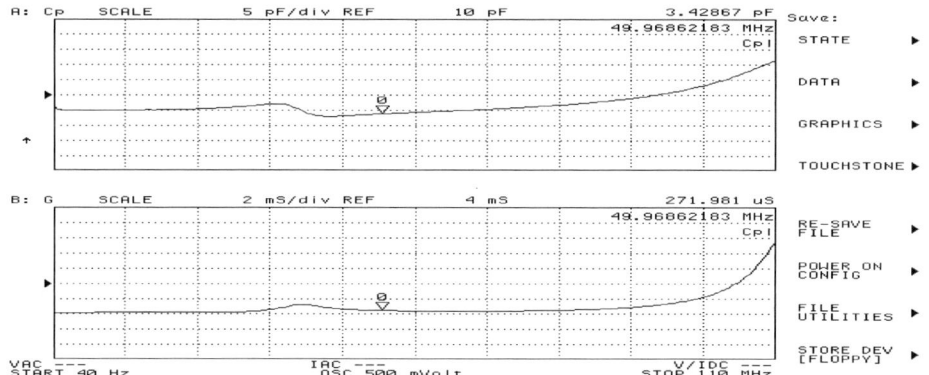

- Measured up to 110 MHz using Agilent Impedance Analyzer.
- 3.4 pF at 50 MHz; 13 pF at 100 MHz.

978-1-4799-5494-0/14 $31.00 © 2014 IEEE

High MegaWatt MV Drives

MV drives issues and needs

- **60-70% of the MV motors are not driven by drives – why?**
- **Estimate yearly energy savings**
- **Capital cost requirement – payback period**
- **Space important ? – match footprint of soft-starter ?**
- **MV motors are used typically in critical processes – steel mills, cement kilns, air-handling, compressors, pumps**
- **Reliability is key; downtime is allowed with concept of "modular replacement" possible by semi-skilled people**

- **MV Motor energy efficiency – what does higher NEMA energy efficiency standards mean – lower losses -> lower damping – need for higher current bandwidth control required?**
  - **Can SiC play a real role?**

500 kW – 1 MW Si-IGBT based drive

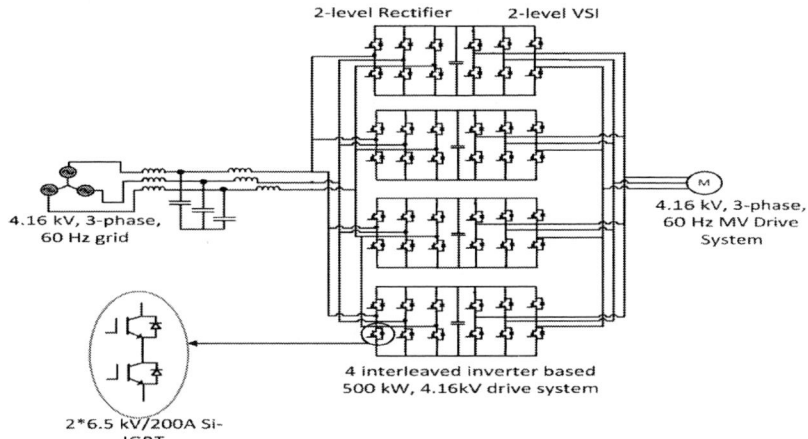

For 4.16 kV, 500 kW system, line rms current = 69A, peak current of each phase=98 A. Peak of line to line voltage is 5.88kV. Hence, two 6.5kV/200A Silicon IGBT devices are required to be in series to block forward voltage.

Si-IGBTs cannot be switched at high frequency while conducting high current. Hence, around 4 interleaved inverters are required to switch at around 1.5 kHz. Each inverter block is switched at 300-400Hz.

500 kW – 1MW SiC MOSFET based drive

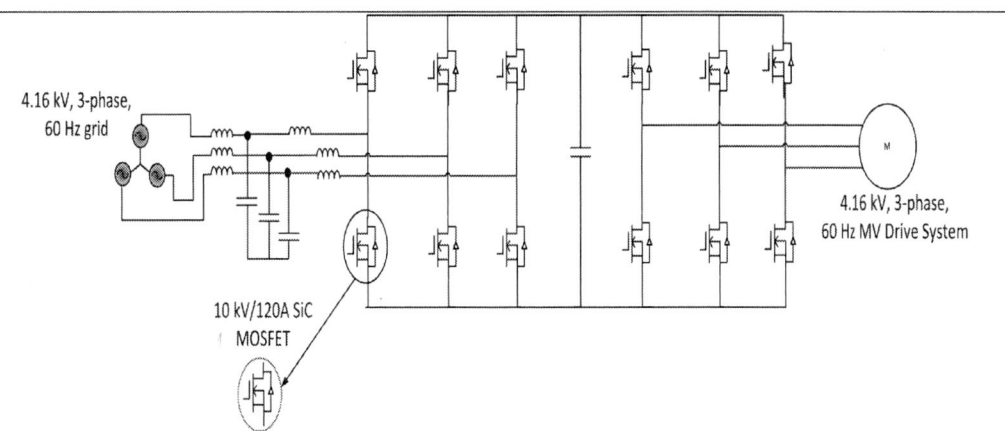

SiC MOSFET kV rating? 10kV, 12kV, 15kV

For the same 4.16 kV, 500 kW drive system, using 10 kV/120A SiC-Mosfet, it is possible to have a 2-level topology. The SiC devices can be switched at 5 kHz, for 69A rms (98A peak) current, and a single device can withstand the forward blocking voltage of Vdc = 6kV.

978-1-4799-5494-0/14 $31.00 © 2014 IEEE

FACTS : Convertible Static Compensator (CSC)
2 x 100 MVA at Marcy Substation – NYPA

Marcy 345kV SC is 18,000 MVA

There are 11 possible configurations

- STATCOM
- SSSC
- UPFC
- IPFC

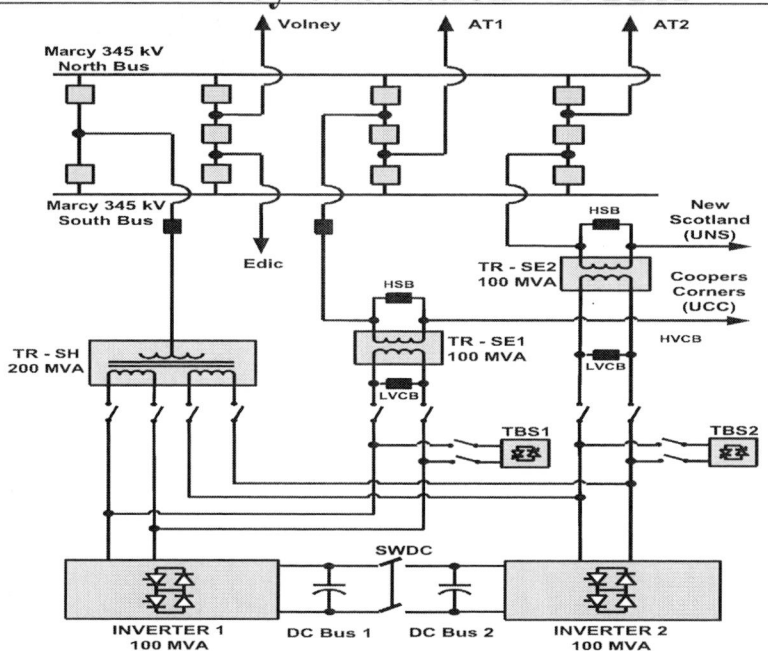

Convertible Static Compensator (CSC) Inverter Hall
2 x 100 MVA at Marcy Substation – NYPA

Main Components of 3-Level NPC Inverter Pole

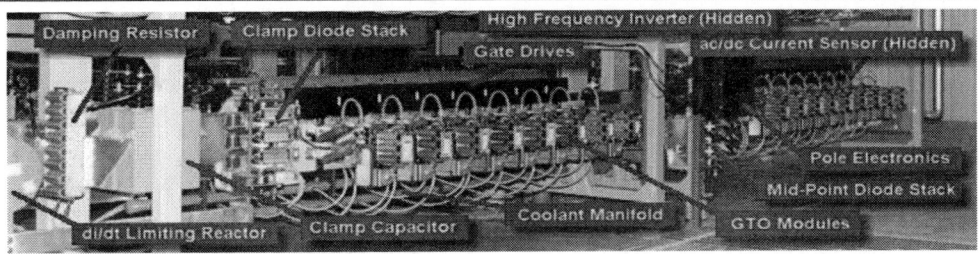

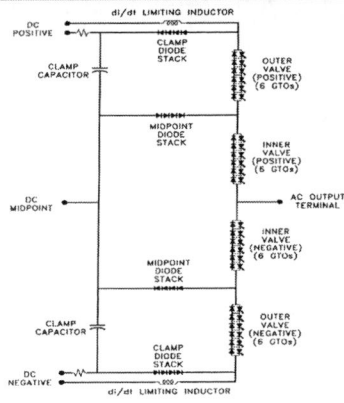

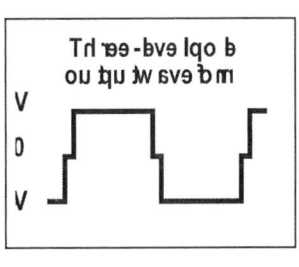

IGCT based 3-level NPC inverter pole

3 series IGCTs (4.5kV / 4kA) per valve – can be replaced by HV SiC devices

978-1-4799-5494-0/14 $31.00 © 2014 IEEE   329

H-bridge test – two IGCT based 3-level NPC poles

3 series IGCTs (4.5kV / 4kA) per valve – can be replaced by HV SiC devices

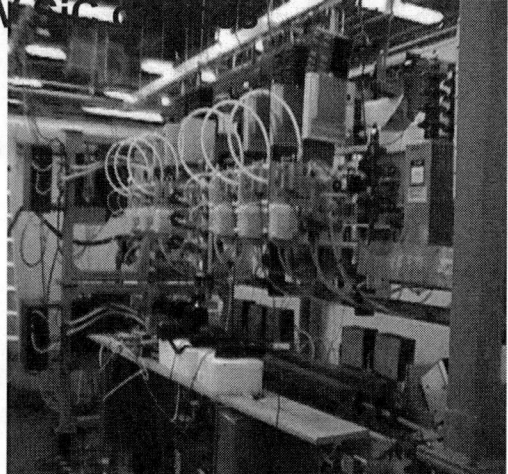

Acknowledgements

- This work is supported by US Government through the DOE ARPA-E program under contract no. DE-AR0000110.

- This work made use of FREEDM ERC shared facilities supported by National Science Foundation under award no. EEC-0812121.

Thank You!!!

Questions

**Acknowledgements:
FREEDM Systems Center
ARPA-E and DOE
Dept. of ECE, NC State University**

 ELECTRIC POWER
RESEARCH INSTITUTE

WBG Devices Enabled MV Power Converters for Utility Applications – Opportunities and Challenges:
Power Electronics – Electric Utility Needs

Dr. Ram Adapa, Technical Leader, EPRI

radapa@epri.com

Tutorial Session at the 2nd IEEE Workshop on Wide Bandgap Power Devices and Applications

Knoxville, TN

October 13, 2014

SMART Grid – Many Definitions – But All Pointing to:
Sensors…Two Way Communication…Intelligence…Response

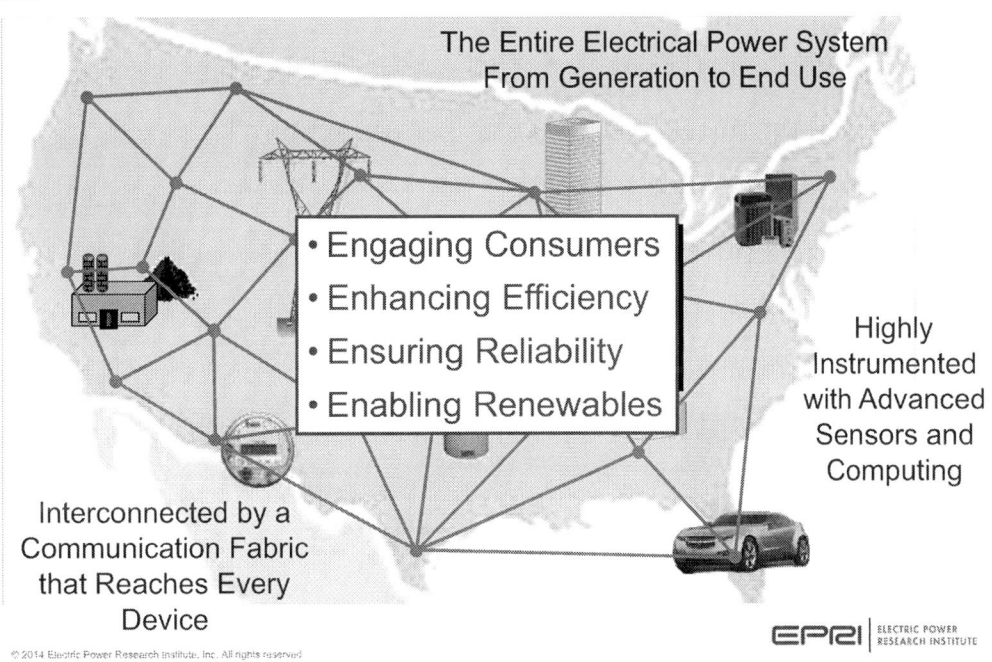

The Entire Electrical Power System From Generation to End Use

- Engaging Consumers
- Enhancing Efficiency
- Ensuring Reliability
- Enabling Renewables

Highly Instrumented with Advanced Sensors and Computing

Interconnected by a Communication Fabric that Reaches Every Device

© 2014 Electric Power Research Institute, Inc. All rights reserved

EPRI | ELECTRIC POWER RESEARCH INSTITUTE

Power Electronics – A Key Technology for SMART Grid

- SMART Grid   =   SMART Generation +
                   SMART Transmission +
                   SMART Distribution +
                   SMART End Use
- SMART Generation
  - Gen Control – e.g. Inverters for Wind and Solar
- SMART Transmission
  - Transmission Control – e.g. HVDC & FACTS
- SMART Distribution
  - Distribution Control – e.g. Custom Power Devices
- SMART End Use
  - Industrial, Commercial, & Residential – e.g. PV & EV

© 2014 Electric Power Research Institute, Inc. All rights reserved.

EPRI | ELECTRIC POWER RESEARCH INSTITUTE

The Power Electronics Revolution – Opportunities

- Reduce energy use

- Enhance the functionality of the power system

- Enable the integration of renewables

- Facilitate the creation of a low-carbon energy future

© 2014 Electric Power Research Institute, Inc. All rights reserved.

EPRI | ELECTRIC POWER RESEARCH INSTITUTE

Power Electronics – Gateway to the Electromagnetic Spectrum

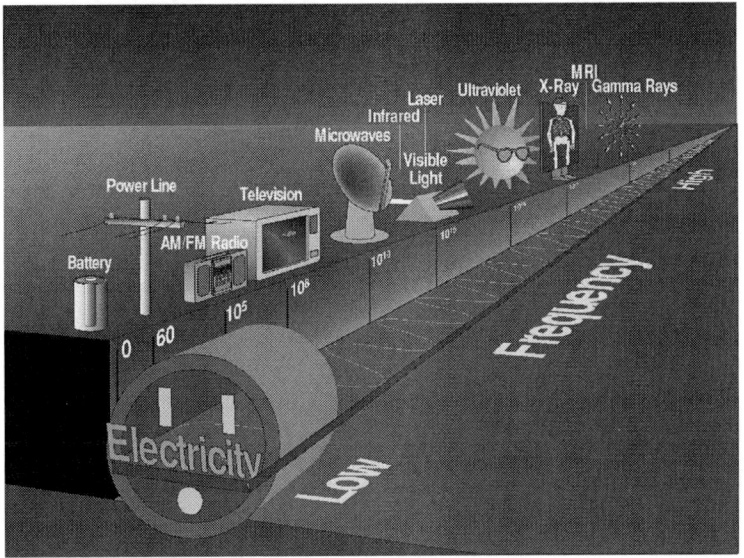

How Do Power Electronics Work?

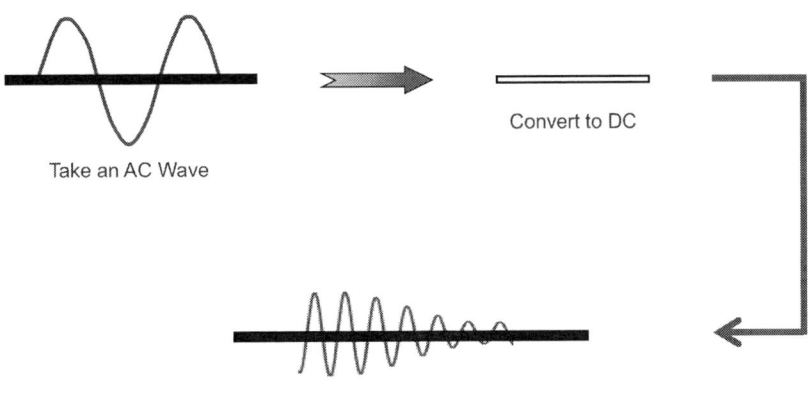

Take an AC Wave

Convert to DC

Convert to AC Wave of Varying
Frequency and Amplitude

Quantifying Energy Losses Along KCPL Electricity Value Chain

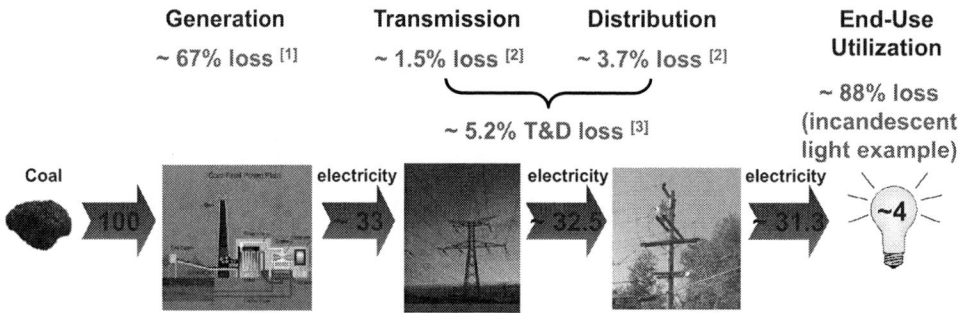

Generation	Transmission	Distribution	End-Use Utilization
~ 67% loss [1]	~ 1.5% loss [2]	~ 3.7% loss [2]	~ 88% loss (incandescent light example)

~ 5.2% T&D loss [3]

[1] Based on weighted average efficiency of KCPL coal fleet, derived from 2006 data
[2] Source: KCPL ("Building the Delivery System of the Future", November 2005)
[3] Compares favorably to national average estimates of 6 - 8 %

© 2014 Electric Power Research Institute, Inc. All rights reserved.

EPRI | ELECTRIC POWER RESEARCH INSTITUTE

Using Power Electronics to Improve Energy Efficiency

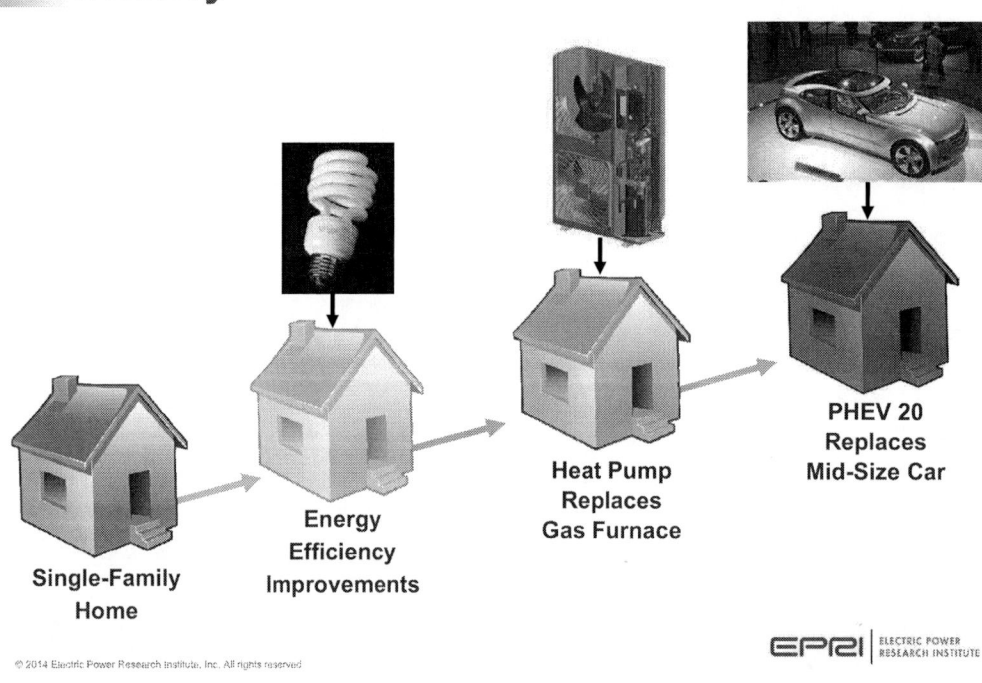

Single-Family Home

Energy Efficiency Improvements

Heat Pump Replaces Gas Furnace

PHEV 20 Replaces Mid-Size Car

© 2014 Electric Power Research Institute, Inc. All rights reserved.

EPRI | ELECTRIC POWER RESEARCH INSTITUTE

Using Power Electronics to Improve Energy Efficiency

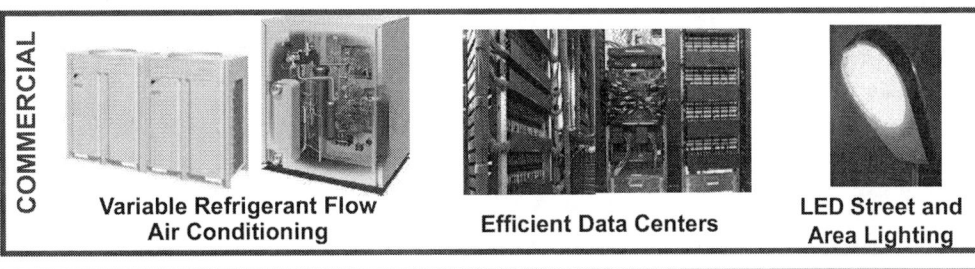

COMMERCIAL

Variable Refrigerant Flow Air Conditioning

Efficient Data Centers

LED Street and Area Lighting

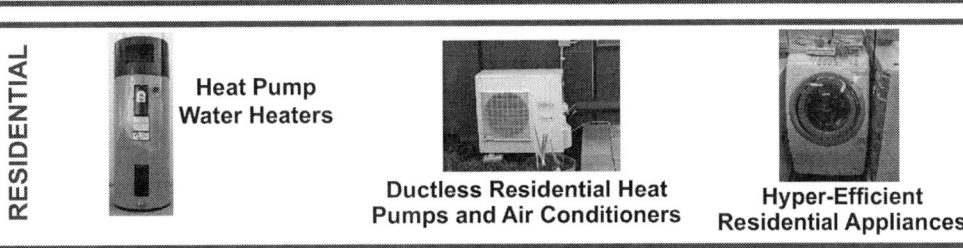

RESIDENTIAL

Heat Pump Water Heaters

Ductless Residential Heat Pumps and Air Conditioners

Hyper-Efficient Residential Appliances

© 2014 Electric Power Research Institute, Inc. All rights reserved.

EPRI | ELECTRIC POWER RESEARCH INSTITUTE

Example: Power Electronics in Appliances

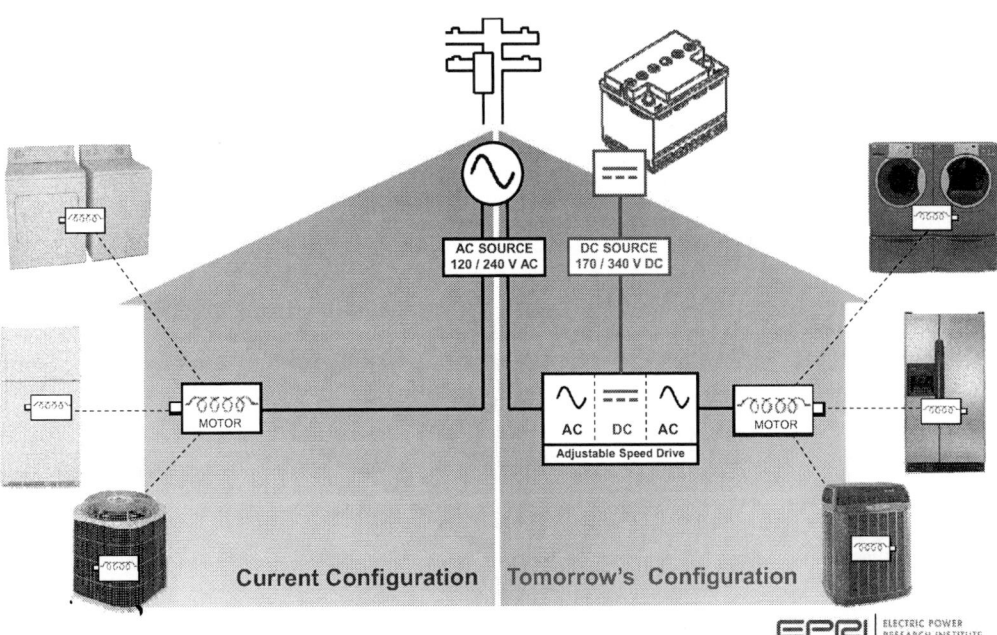

AC SOURCE 120 / 240 V AC

DC SOURCE 170 / 340 V DC

MOTOR

AC | DC | AC
Adjustable Speed Drive

MOTOR

Current Configuration   Tomorrow's Configuration

© 2014 Electric Power Research Institute, Inc. All rights reserved.

EPRI | ELECTRIC POWER RESEARCH INSTITUTE

978-1-4799-5494-0/14 $31.00 © 2014 IEEE   335

Power Electronics can Reduce Electricity Consumption
<u>Technical Potential</u> Electricity Savings

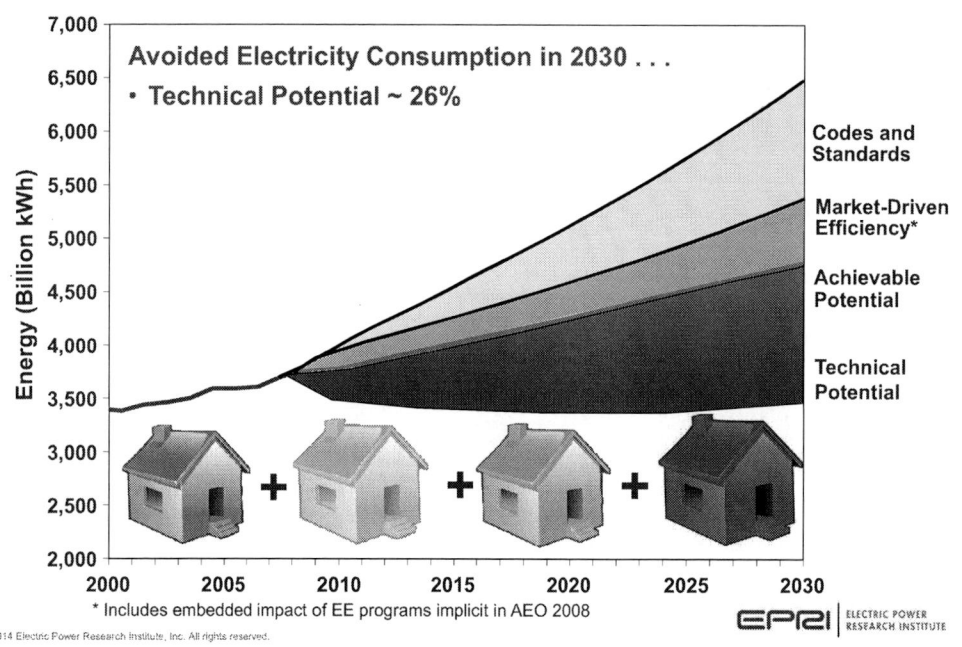

Avoided Electricity Consumption in 2030 . . .
• Technical Potential ~ 26%

* Includes embedded impact of EE programs implicit in AEO 2008

© 2014 Electric Power Research Institute, Inc. All rights reserved.

Technical Potential

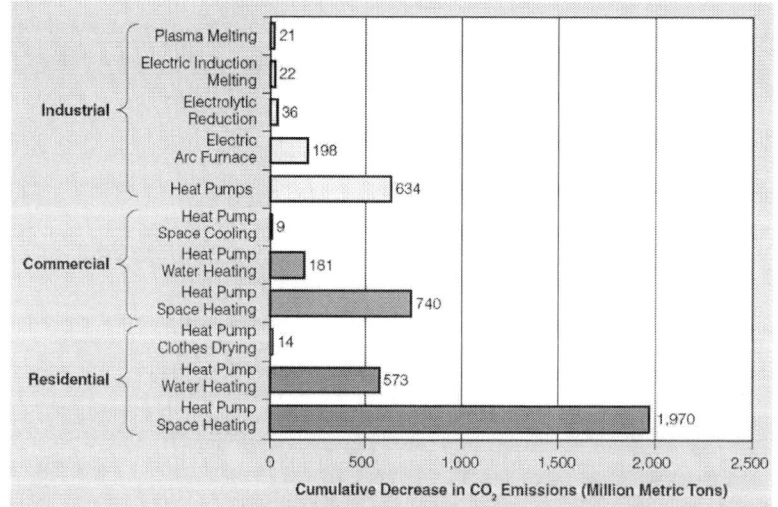

Cumulative Decrease in Energy-Related CO_2 Emissions Between 2009 and 2030 by Sector and Efficient Electric End-Use Technology

© 2014 Electric Power Research Institute, Inc. All rights reserved

Power Electronics: Close the Gap Between Wind Resources & Population Density

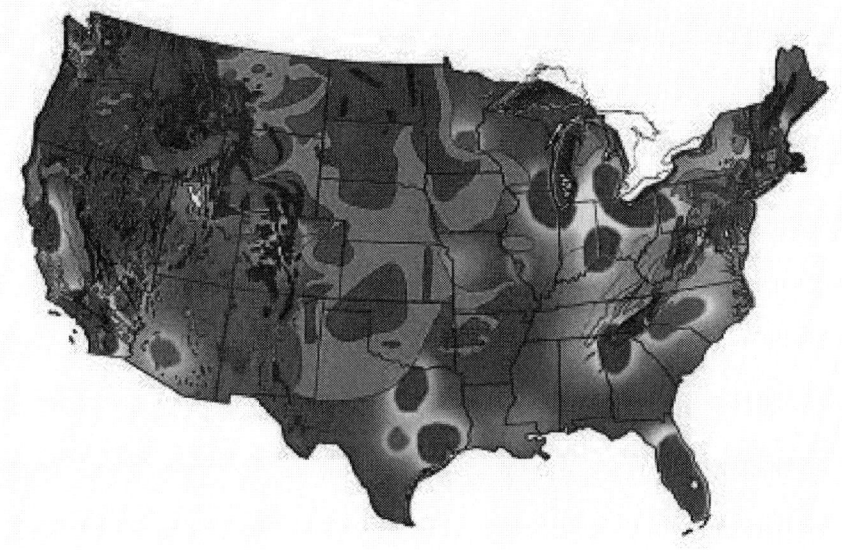

© 2014 Electric Power Research Institute, Inc. All rights reserved.

EPRI | ELECTRIC POWER RESEARCH INSTITUTE

Power Electronics Enable Renewable by Providing Dance Partners

California Energy Policy Targets for Renewable Energy Penetration:

20% by 2010 & 33% by 2020

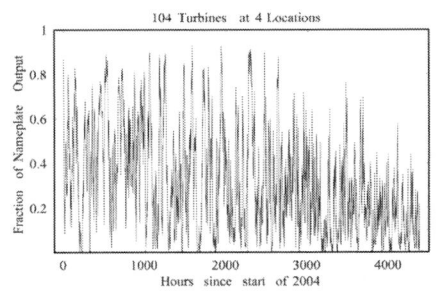

Wind Intermittency

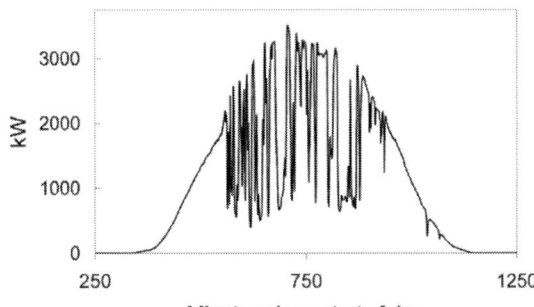

Solar Intermittency

© 2014 Electric Power Research Institute, Inc. All rights reserved.

EPRI | ELECTRIC POWER RESEARCH INSTITUTE

Examples: Dance Partners for Renewab

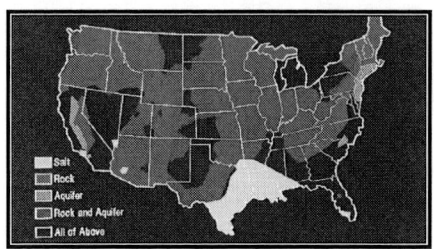

Compressed Air Potential

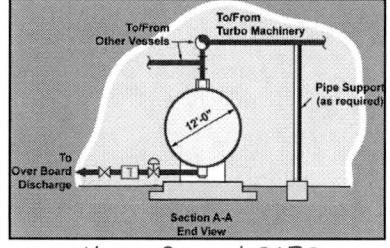

Above Ground CAES

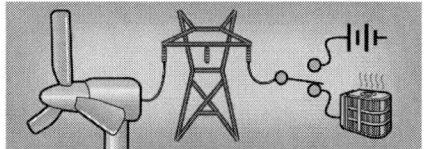

Fast-Reacting Demand Response

© 2014 Electric Power Research Institute, Inc. All rights reserved.

EPRI | ELECTRIC POWER RESEARCH INSTITUTE

Power Electronics in Building-Level Local Energy Networks

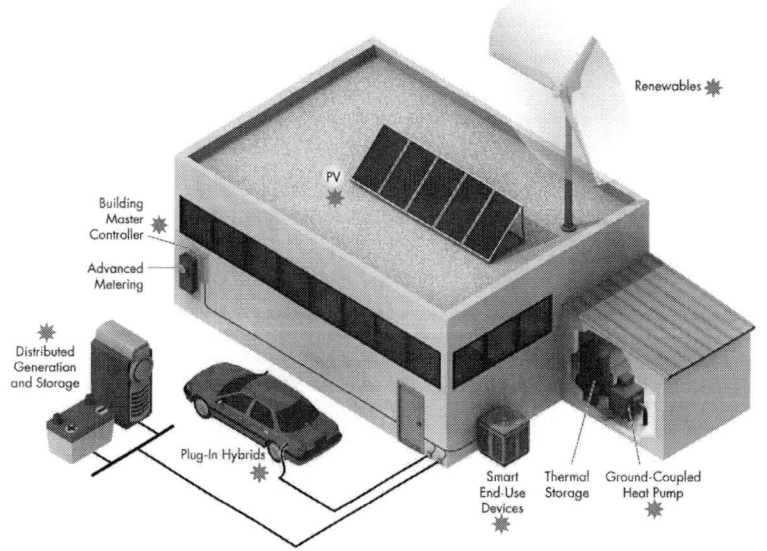

✷ Power Electronics

© 2014 Electric Power Research Institute, Inc. All rights reserved.

EPRI | ELECTRIC POWER RESEARCH INSTITUTE

Power Electronics in Campus-Level Local Energy Networks

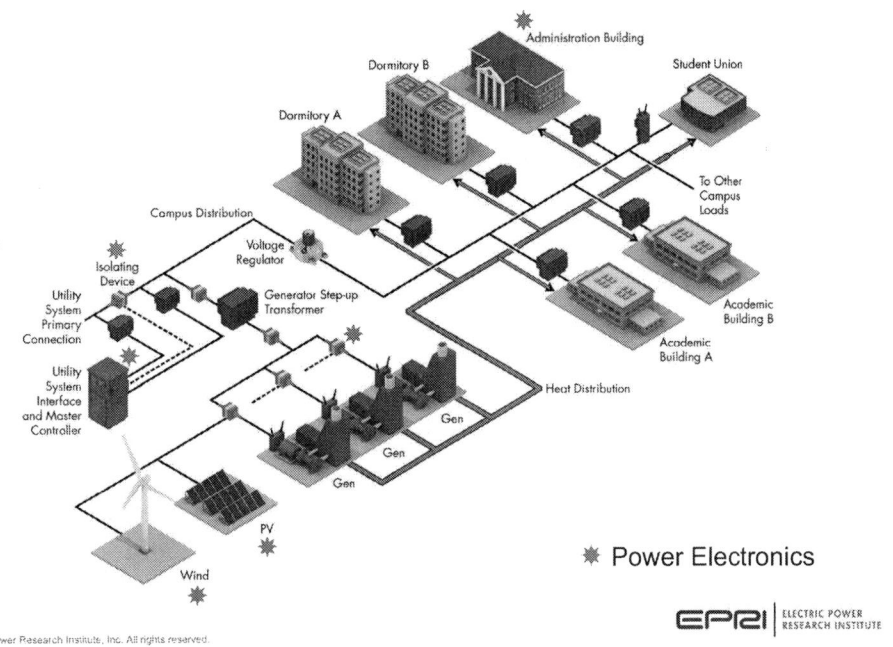

Power Electronics in Bulk Power Systems

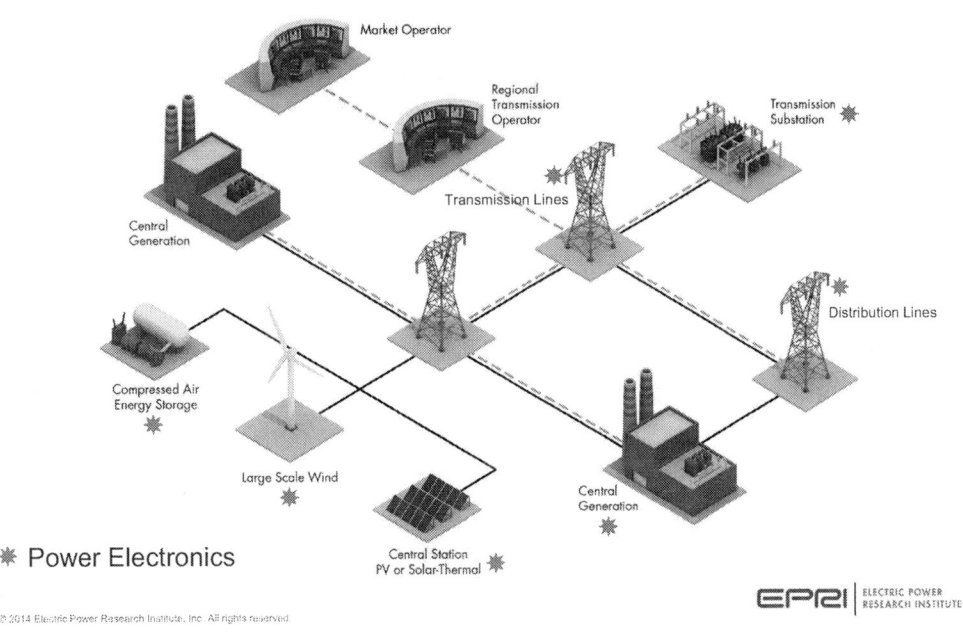

The Power Electronics Revolution – Opportunities

Future Power System Equipment could be more power electronics based such as:

- – Transformers (Solid State Transformer)

- – Fault Current Limiters (Solid State Fault Current Limiter)

- – Circuit Breakers / Switches

Needed: Thinking out of the Box

© 2014 Electric Power Research Institute, Inc. All rights reserved.

EPRI | ELECTRIC POWER RESEARCH INSTITUTE

Power Electronics will Enable the Intelligent Universal Transformer

Core Technologies Needed

New State-of-the-Art Power Electronic Topology

New High-Voltage, Low-Current Power Semiconductor Device

Interoperable with Open Communication Architecture

Needed: Reliable Low Voltage & High Voltage (High Power) Solid State Transformers

All Solid-State Replacement for Distribution Transformers

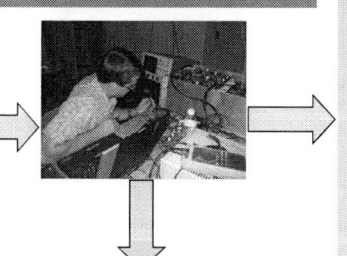

Product Spin-offs

Emergency EHV / Recovery transformer replacement (substations)

Other power electronic applications

Functions & Value

Traditional voltage stepping, plus...

New service options, such as DC

Real-time voltage regulation, sag correction, system monitoring, and other operating benefits

Other benefits: standardization, size, weight, oil elimination

Cornerstone device for advanced distribution automation (ADA)

© 2014 Electric Power Research Institute, Inc. All rights reserved.

EPRI | ELECTRIC POWER RESEARCH INSTITUTE

Solid State Fault Current Limiter

Needed: Distribution & Transmission Level Fault Current Limiters

- 15kV, 1200 Amp Distribution Class SSCL
- Current passes through SGTO modules in normal operation
- Drive modules off when detect fault event
- 2kV, 1000A turn off per block. Stack in series / parallel for increased voltage / current.
- Tune inductors for let-through current

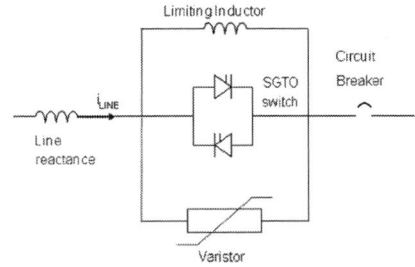

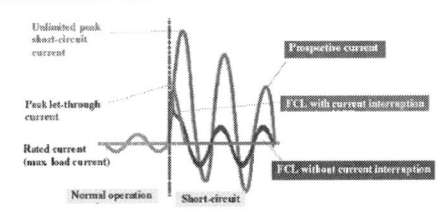

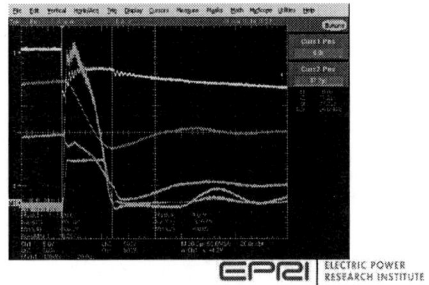

© 2014 Electric Power Research Institute, Inc. All rights reserved.

EPRI | ELECTRIC POWER RESEARCH INSTITUTE

IEEE Spectrum – January 2011 issue
Top 11 Technologies of the Decade

1. SMART Phones
2. Social Networking
3. Voice Over IP
4. LED Lighting
5. Multicore CPUs
6. Cloud Computing
7. Drone Aircraft
8. Planetary Rovers
9. *FACTS*
10. Digital Photography
11. Class-D Audio

FACTS Applications are increasing - NanoMarkets Survey - Global FACTS installations from US $330 M this year to $775 M in 2017

© 2014 Electric Power Research Institute, Inc. All rights reserved.

EPRI | ELECTRIC POWER RESEARCH INSTITUTE

FACTS (Flexible AC Transmission Systems) – Expanding Laws of Physics

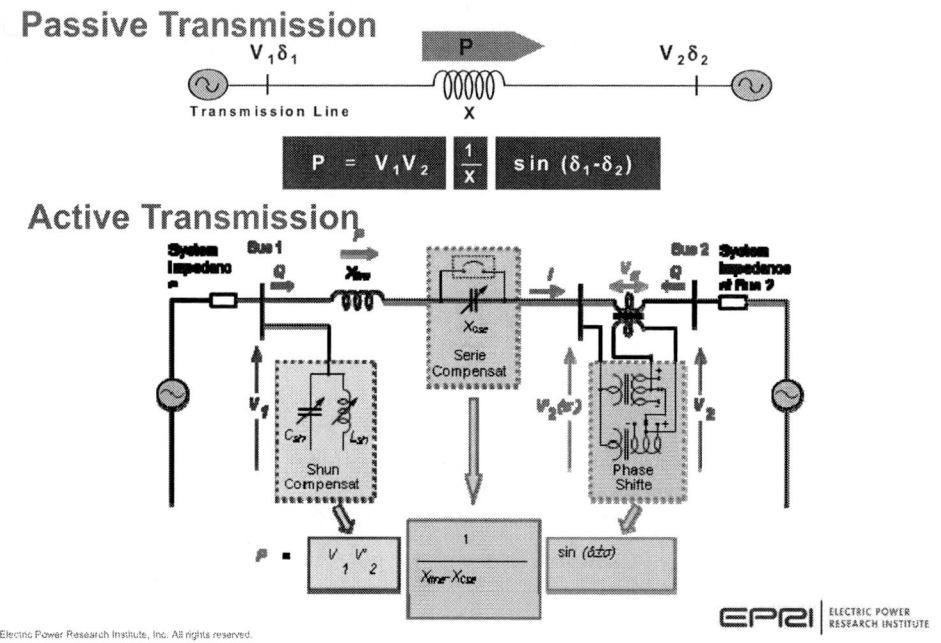

Passive Transmission

$$P = V_1 V_2 \; \frac{1}{X} \; \sin(\delta_1 - \delta_2)$$

Active Transmission

FACTS Hardware

Traditional Technologies

- Thyristor controlled reactors
- Thyristor switched reactors
- Thyristor switched capacitors
- Static Var Compensators (SVC)

"New" Technologies

- Thyristor Controlled Series Compensation (TCSC)
- STATic synchronous COMpensator (STATCOM)
- Static Synchronous Series Compensator (SSSC)
- Unified Power Flow Controller (UPFC)
- Interphase Power Flow Controller (IPFC)

NanoMarkets Survey – Global FACTS installations from U.S. $330M in 2011 to $775 M in 2017

Transmission Applications of VSC

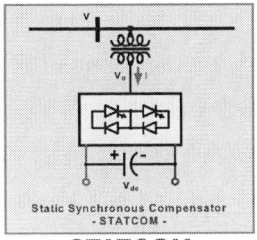
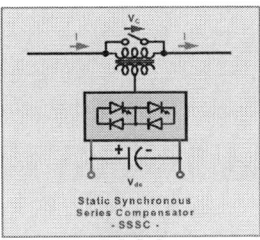
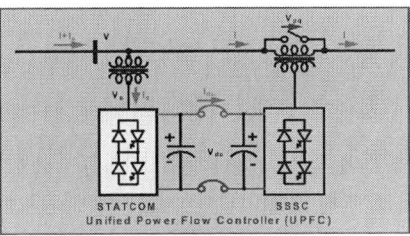

STATCOM
Voltage Control

SSSC
Line Impedance Control

UPFC
Voltage, Line Impedance
& Phase Angle Control

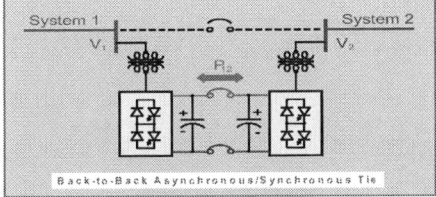

Back-to-Back
Voltage & Power
Transfer Control

IPFC
Interline Power
Exchange

© 2014 Electric Power Research Institute, Inc. All rights reserved.

EPRI Sponsored FACTS Installations

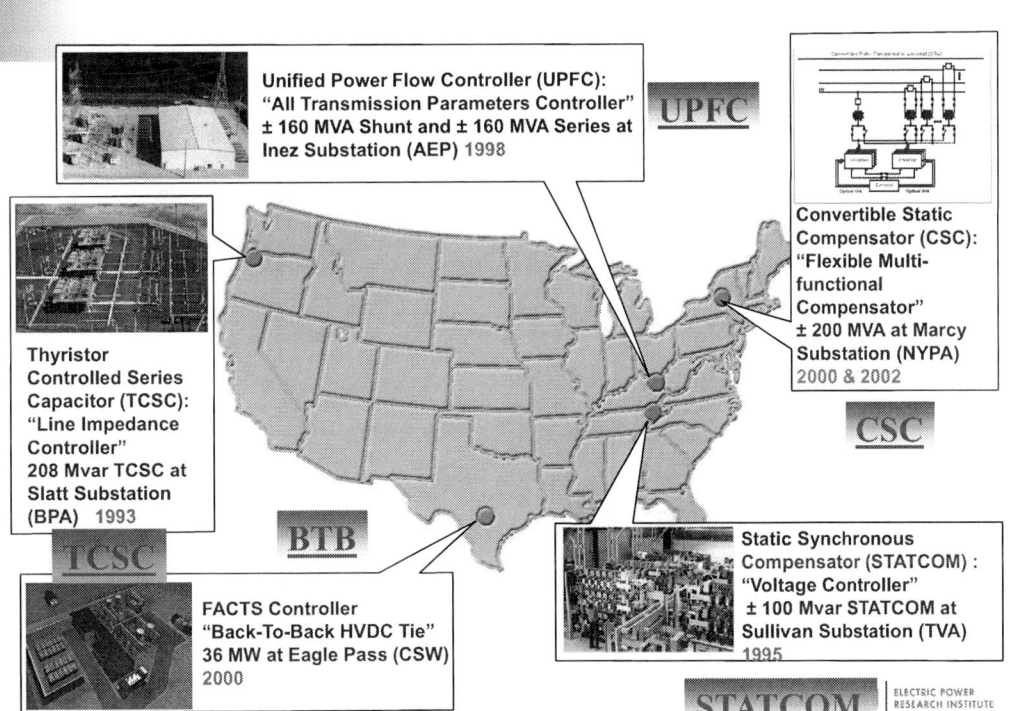

Unified Power Flow Controller (UPFC):
"All Transmission Parameters Controller"
± 160 MVA Shunt and ± 160 MVA Series at
Inez Substation (AEP) 1998

Convertible Static
Compensator (CSC):
"Flexible Multi-
functional
Compensator"
± 200 MVA at Marcy
Substation (NYPA)
2000 & 2002

Thyristor
Controlled Series
Capacitor (TCSC):
"Line Impedance
Controller"
208 Mvar TCSC at
Slatt Substation
(BPA) 1993

FACTS Controller
"Back-To-Back HVDC Tie"
36 MW at Eagle Pass (CSW)
2000

Static Synchronous
Compensator (STATCOM) :
"Voltage Controller"
± 100 Mvar STATCOM at
Sullivan Substation (TVA)
1995

© 2014 Electric Power Research Institute, Inc. All rights reserved.

978-1-4799-5494-0/14 $31.00 © 2014 IEEE

FACTS System Studies:
FACTS can increase power flows

EXAMPLES:

Transmission line in the Southwestern US:

- **Boost power flow from 300 MW to 400 MW (+33%)**

Ties between Southern US and Florida:

- **Boost power flow from 3400 MW to 4100 MW (+21%)**

Ties between upstate New York to New York City:

- **Boost power flow from 2600 MW to 3200 MW (+23%)**

© 2014 Electric Power Research Institute, Inc. All rights reserved.

Basic HVDC Transmission

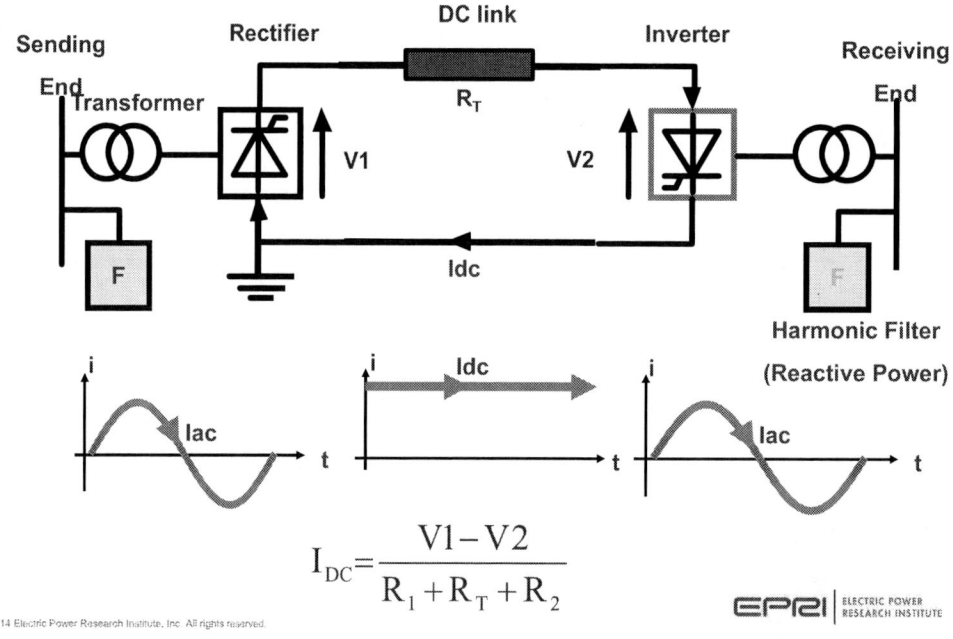

$$I_{DC} = \frac{V1 - V2}{R_1 + R_T + R_2}$$

© 2014 Electric Power Research Institute, Inc. All rights reserved.

HVDC Scheme Types

- Back-to-Back
  - frequency changing
  - asynchronous connection

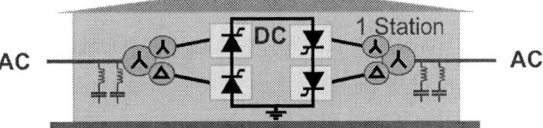

- Point-to-Point Overhead Line
  - bulk transmission
  - overland

- Point-to-Point Submarine Cable
  - bulk transmission
  - underwater or underground

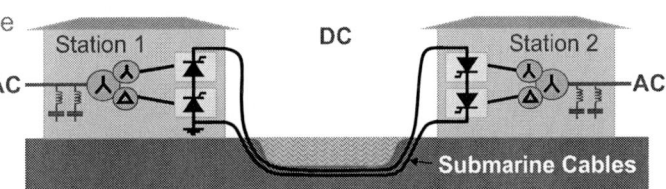

HVDC Converter Technology: LCC Versus VSC

Line Commutated Converter
(or Current Source Converter)

- Thyristor based
- Switches on-off one time per cycle

Voltage Source Converter

- IGBT Based

(Insulated Gate Bipolar Transistor)

- Switches on-off many times per cycle

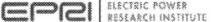

VSC : Recent new Topology (MMC)

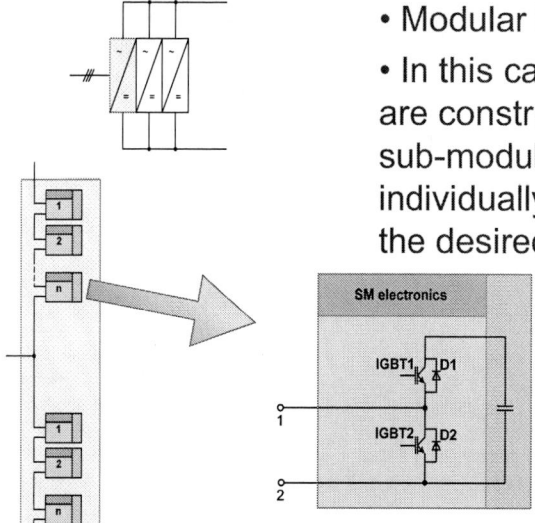

- Modular Multilevel Converter
- In this case the converter arms are constructed from identical sub-modules that are individually controlled to obtain the desired ac voltage.

Half-Chain Links shown here.

Full-Chain Links can be used to reduce fault currents on DC side

© 2014 Electric Power Research Institute, Inc. All rights reserved.

EPRI | ELECTRIC POWER RESEARCH INSTITUTE

VSC

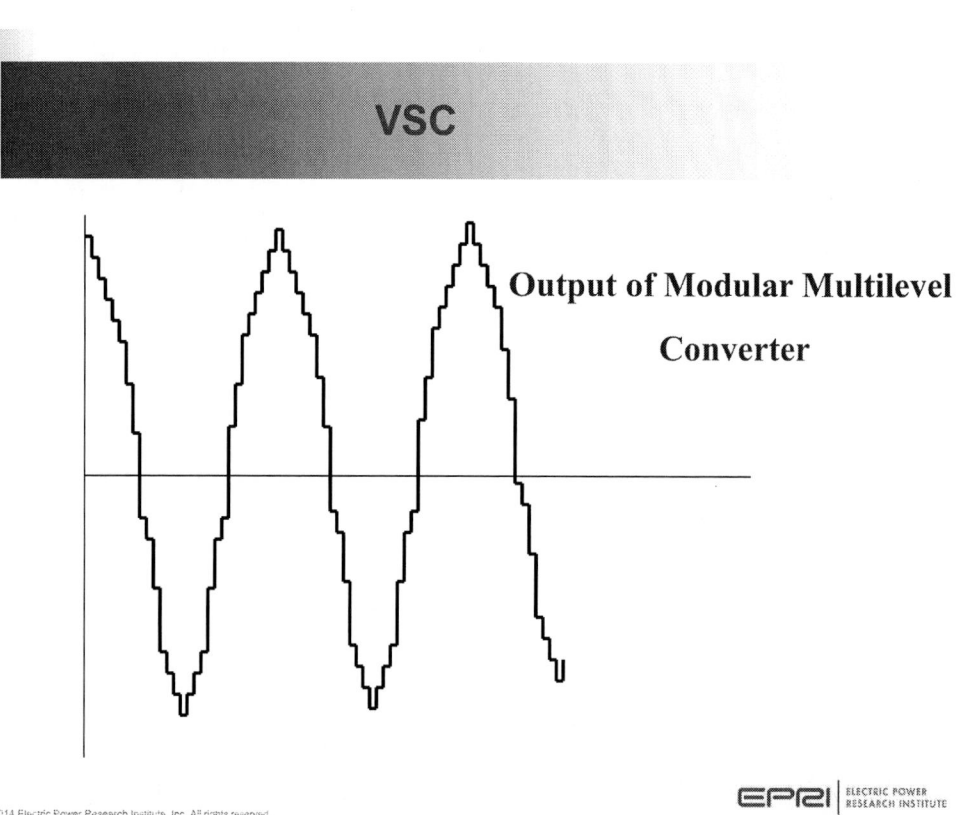

Output of Modular Multilevel Converter

© 2014 Electric Power Research Institute, Inc. All rights reserved.

EPRI | ELECTRIC POWER RESEARCH INSTITUTE

Converter Technology: LCC vs. VSC

Function	LCC	VSC
Semi-Conductor Device	Thyristors currently 6 inch, 8.5 kV and 5000 Amps. No controlled turn off capability	IGBTs with anti-parallel free wheeling diode, with controlled turn-off capability. Current rating 4.5 to 6 kV and turn off current of 1200 Amps.
DC transmission voltage	Up to +/- 800 kV bipolar operation. 1000 kV under consideration in China	Up to +/- 320 kV currently limited by HVDC cable if extruded XLPE cable is used. Up to +/- 350 kV with Overhead line, can go higher
DC power	Currently in the range of 6000 MW per bipolar system	Currently in the range of 600 to 1000 MW per pole
Reactive Power requirements	Consumes reactive power up to 60% of its rating	Does not consume any reactive power and each terminal can independently control its reactive power.
Filtering	Requires large filter banks	Requires moderate size filter banks or no filters at all.
Black start	Limited application	Capable of black start and feeding passive loads

© 2014 Electric Power Research Institute, Inc. All rights reserved.

EPRI | ELECTRIC POWER RESEARCH INSTITUTE

Trans Bay VSC DC Cable

- **Trans Bay Cable Project**

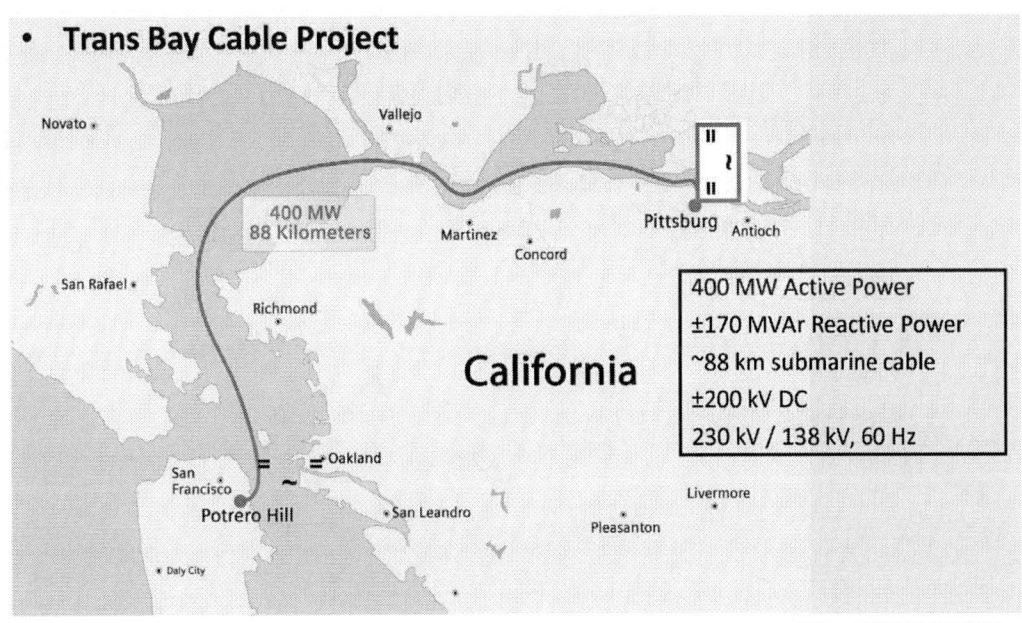

400 MW Active Power
±170 MVAr Reactive Power
~88 km submarine cable
±200 kV DC
230 kV / 138 kV, 60 Hz

© 2014 Electric Power Research Institute, Inc. All rights reserved.

HVDC IN NORTH AMERICA

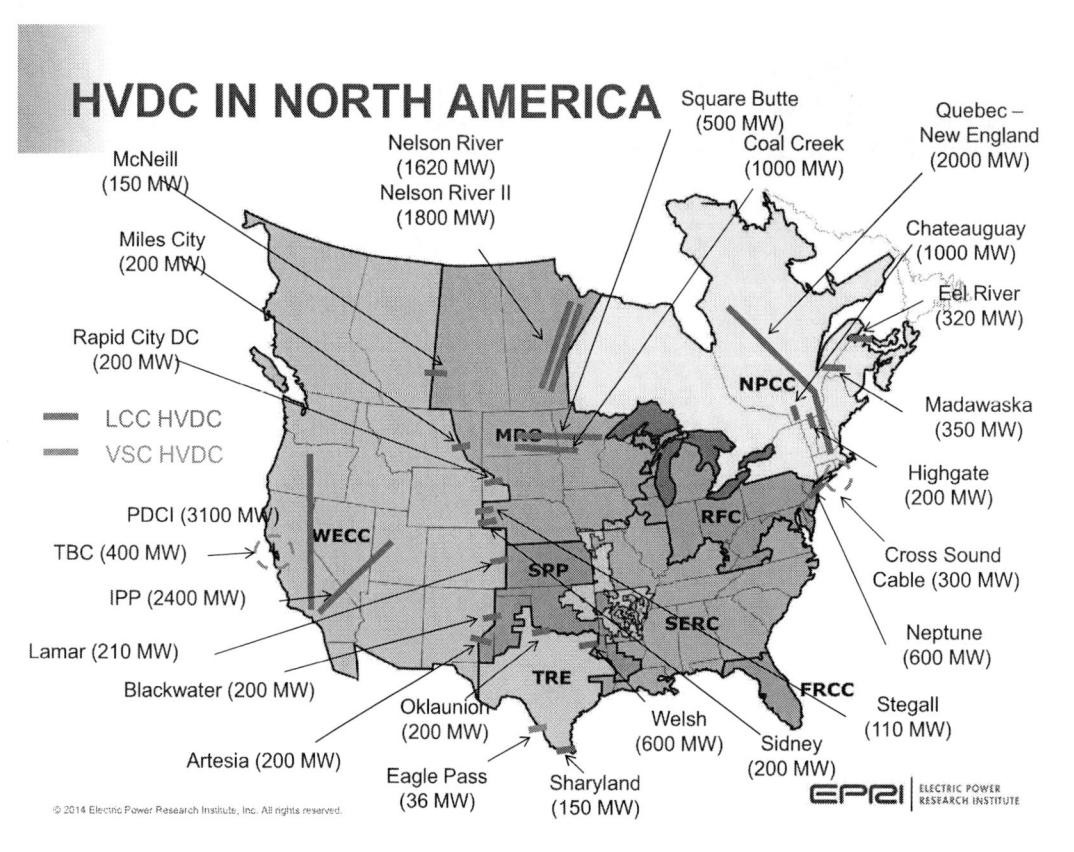

McNeill (150 MW)
Miles City (200 MW)
Nelson River (1620 MW)
Nelson River II (1800 MW)
Square Butte (500 MW)
Coal Creek (1000 MW)
Quebec – New England (2000 MW)
Chateauguay (1000 MW)
Eel River (320 MW)
Rapid City DC (200 MW)
Madawaska (350 MW)
Highgate (200 MW)
LCC HVDC
VSC HVDC
PDCI (3100 MW)
TBC (400 MW)
IPP (2400 MW)
Lamar (210 MW)
Blackwater (200 MW)
Artesia (200 MW)
Oklaunion (200 MW)
Eagle Pass (36 MW)
Sharyland (150 MW)
Welsh (600 MW)
Sidney (200 MW)
Cross Sound Cable (300 MW)
Neptune (600 MW)
Stegall (110 MW)
WECC
MRO
RFC
SPP
SERC
TRE
FRCC
NPCC

© 2014 Electric Power Research Institute, Inc. All rights reserved.

DC Grids – The Future of DC Transmission

- DC Grids for Offshore Wind
- Considered more in Europe than in other countries
- Need to resolve many issues
  - Power & Voltage control
  - DC circuit breakers
  - Standard DC voltages
  - Communication needs

- CIGRE/IEEE WGs

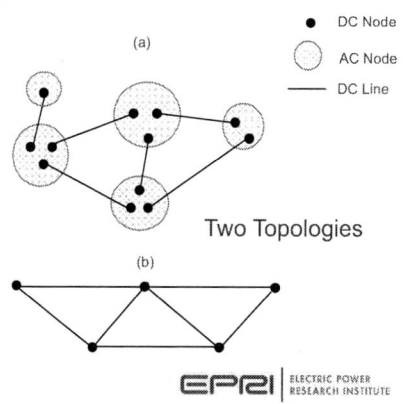

- DC Node
- AC Node
- DC Line

(a)

Two Topologies

(b)

© 2014 Electric Power Research Institute, Inc. All rights reserved.

DC Grid Configurations:
Offshore Development – Point to Point System

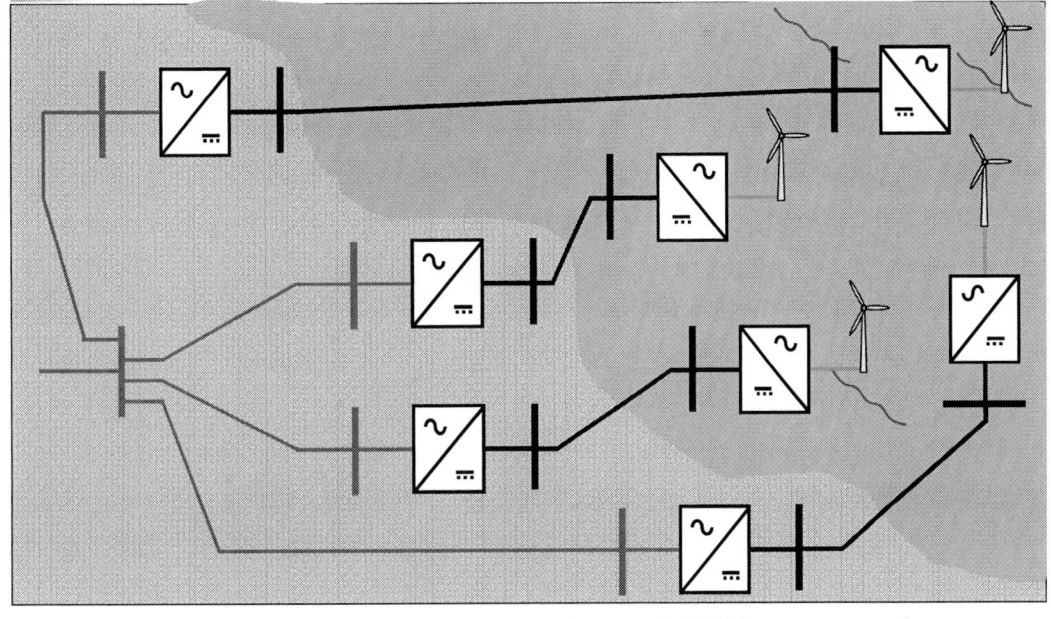

Source: ALSTOM

DC Grid Configurations: Offshore Grid System

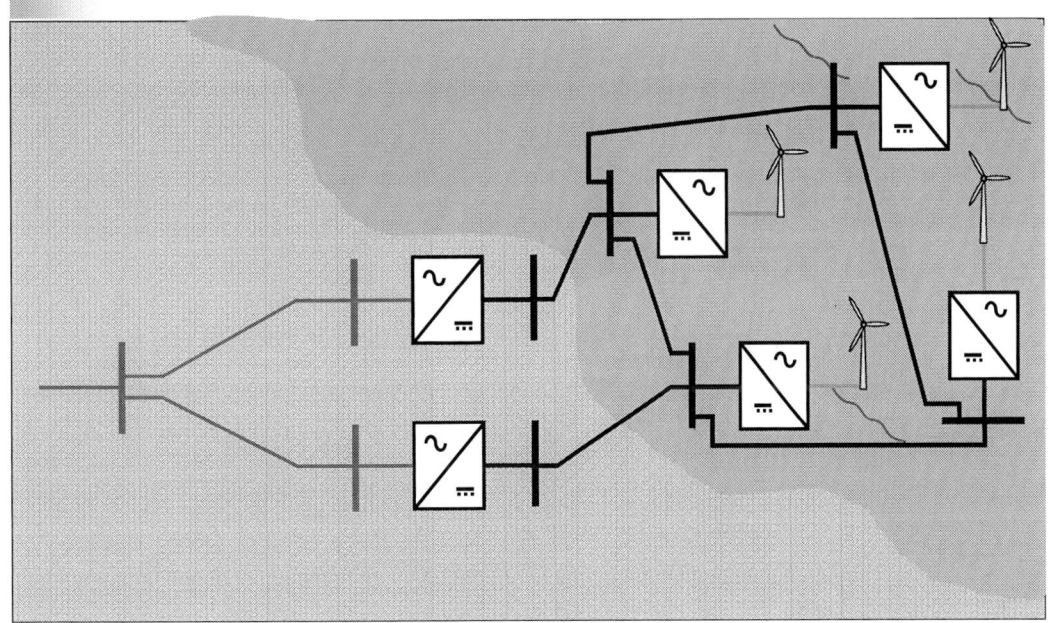

Source: ALSTOM

Overlay DC Grid Gives Access to Renewable Sources within Europe

- Interconnection of remote renewable energy sources
- Overcoming "bottlenecks" in the existing AC grids
- Low loss (HVDC) transmission systems
- Controllable power flows over a wide area
- Avoidance of synchronisation over a wide area
- Less environmental impact than AC reinforcement

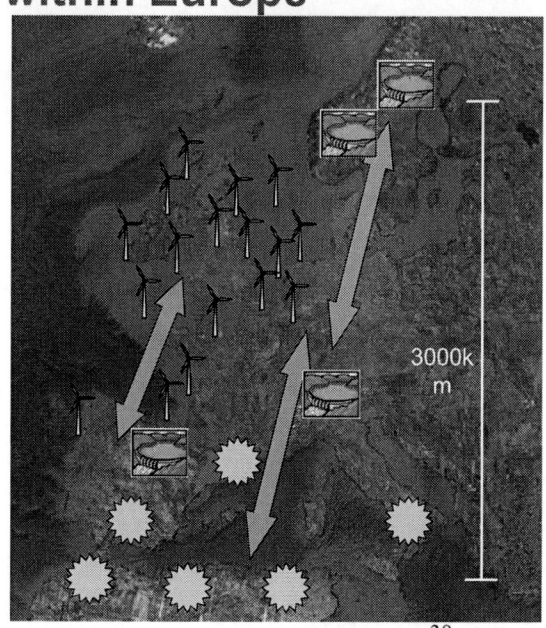

© 2014 Electric Power Research Institute, Inc. All rights reserved.

DC Grid Standardisation Activities

- Cigrè started five further DC grid working groups;

  - B4-56: Guidelines for the preparation of "connection agreements" or "Grid Codes" for HVDC grids

  - B4-57: Guide for the development of models for HVDC converters in a HVDC grid

  - B4-58: Devices for load flow control and methodologies for direct voltage control in a meshed HVDC Grid

  - B4-59: Protection of Multi-terminal HVDC Grids

  - B4-60: Designing HVDC Grids for Optimal Reliability and Availability performance

© 2014 Electric Power Research Institute, Inc. All rights reserved.

A Sample of European Proposals

 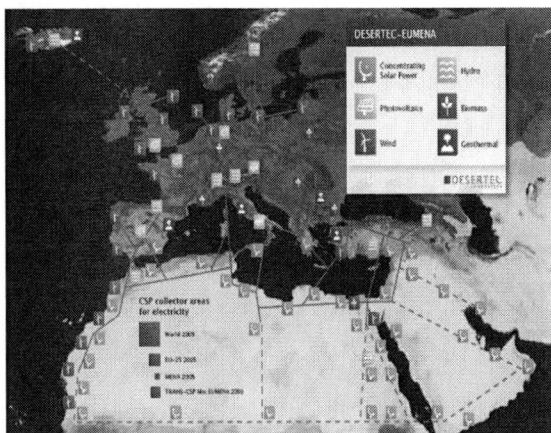

G. Asplund, B. Jacobson, B. Berggren, K. Lindén "Continental Overlay HVDC-Grid", Cigré conference, B4-109, Paris, 2010

Atlantic Wind Connection

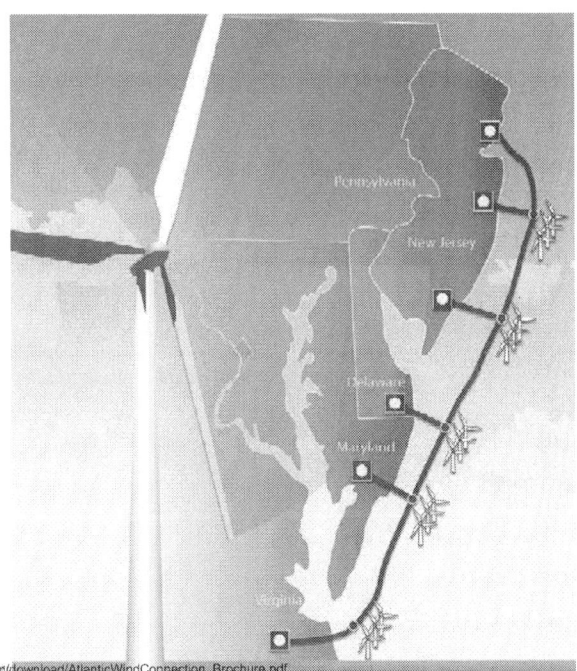

http://atlanticwindconnection.com/download/AtlanticWindConnection_Brochure.pdf

Atlantic Wind Connection Project

(see: www.atlanticwindconnection.com/ferc/2010-12-filing/Petition_for_Declaratory_Order.pdf)

What
- A sub-sea HVDC backbone transmission system

Where
- Extending from northern New Jersey to southern Virginia.

Who
- Google
- Marubeni
- Good Earth
- Elia

Why
- Serve as an efficient collector of ac power from offshore wind farms
- Relieve transmission congestion on the eastern ac grid
- Improve regional system reliability.

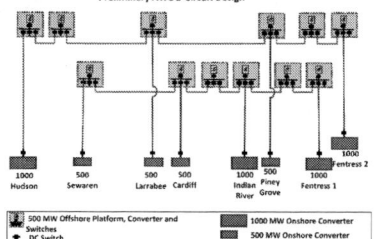

Preliminary AWC 2-Circuit Design

DC Breakers

When closed the DC breaker must have very low losses
- optimum solution mechanical switch

AC

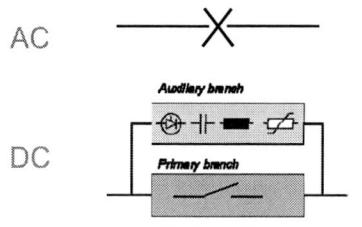

Main switch

DC

Unlike an AC breaker the DC current never experiences a current zero. Hence, to interrupt the DC current the DC breaker must drive the load current to zero.

Modular hybrid solution to drive current to zero

- critical component is the mechanical switch as it has to operate **VERY** fast to minimise the peak current to be interrupted by the auxiliary branch

HVDC Circuit Breaker Developments

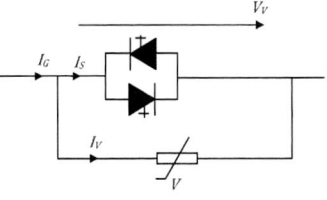

Solid State Circuit Breaker

- Many ideas are explored
- Fast growing area
- Numerous R&D projects
- Minimize size, cost, & interruption time

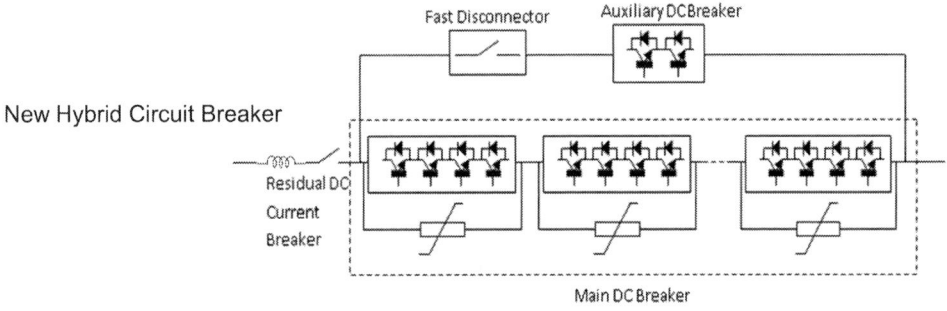

© 2014 Electric Power Research Institute, Inc. All rights reserved.

EPRI | ELECTRIC POWER RESEARCH INSTITUTE

New HVDC Circuit Breaker Developments

ABB develops world's first circuit breaker for HVDC

November 7, 2012
By PennEnergy Editorial Staff

ABB (NYSE: ABB), the leading power and automation technology group, has announced a breakthrough in the ability to interrupt direct current, solving a 100-year-old electrical engineering puzzle and paving the way for a more efficient and reliable electricity supply system.

After years of research, ABB has developed the world's first circuit breaker for high voltage direct current (HVDC). It combines very fast mechanics with power electronics, and will be capable of 'interrupting' power flows equivalent to the output of a large power station within 5milliseconds- that is thirty times faster than the blink of a human eye.

The breakthrough removes a 100-year-old barrier to the development of DC transmission grids, which will enable the efficient integration and exchange of renewable energy. DC grids will also improve grid reliability and enhance the capability of existing AC (alternating current) networks. ABB is in discussions with power utilities to identify pilot projects for the new development.

ABB has written a new chapter in the history of electrical engineering," said Joe Hogan, CEO of ABB. "This historical breakthrough will make it possible to build the grid of the future. Overlay DC grids will be able to interconnect countries and continents, balance loads and reinforce the existing AC transmission networks. "

The Hybrid HVDC breaker development has been a flagship research project for ABB, which invests over $1 billion annually in R&D activities. The breadth of ABB's portfolio and unique combination of in-house manufacturing capability for power semiconductors, converters and high voltage cables (key components of HVDC systems) were distinct advantages in the new development.

978-1-4799-5494-0/14 $31.00 © 2014 IEEE

Current State of HVDC versus HVAC

- Many Existing HVDC systems are old (30 - 50 years old)
  - Life extension is taking place
- Highest DC Voltage is UHVDC at +/- 800 kV in China & India
  - South Africa & Brazil are also considering
  - For long distances over 3000 km
  - For Bulk Power Transfer (3000 to 6000 MW)
- UHVDC of +/- 1000 to 1100 kV is planned in Asia for up to 8000 MW - China & India
- VSC HVDC is increasing (+/- 320 kV up to 1000 MW)

- Max AC Voltage in North America is 765 kV (EHVAC)
- UHVAC (1000 kV to 1200 kV) is considered in China (highest in the world)

Future Trends in HVDC

- For transfers of above 6,000 MW over 4,000 km, the optimum voltage rises to 1,000–1,200 kV.
  - Technological developments in LCC converter stations seem to be ready to handle these voltages.

- HVDC and HVAC overlays for regional interconnections

- Segmenting AC grids with DC back-to-backs for improved reliability

- Growth of VSC DC applications – more dc cable projects

- DC Grids for renewable integration

Power Electronics
(EPRI Technology Innovation Program Overview)

The Challenge

- Increased penetration of variable generation and the growth in electric vehicles are driving industry needs to apply power electronic control to increase reliability and efficiency.
- Innovative power electronic technologies are needed to realize **SMART GRID** for real time monitoring and control

If Successful

- Technologies under investigation in this program have the potential to increase the existing asset utilization by >50% and/or reduce the equipment failures and thus increasing revenues or saving millions of dollars to the utilities

Timeline & Requirements

- A 7-10 year sustained TI effort is needed to enhance wide-band-gap based power electronics development and their application to the electric power industry
- Technologies are being transferred to base and supplemental programs for near-term deployment, beginning immediately

R&D Objective

- Fund innovative, game changing, higher risk ideas and technologies with high potential for performance and financial improvements to members

R&D Approach

- Leverage funding working with others such as DOE to develop wide-band-gap materials and their applications to electricity chain – generation, transmission, distribution, & end use
- Develop and demonstrate new concepts / approaches for smart transmission & distribution using power electronics to solve the present industry issues such as fault current limiting, renewable integration, and electric vehicle charging

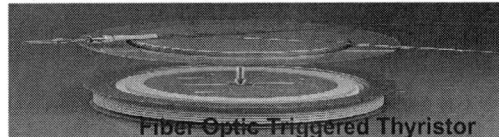

Fiber Optic Triggered Thyristor

© 2014 Electric Power Research Institute, Inc. All rights reserved.

Power Electronics
(Major R&D Gaps)

Power Electronics (PE) Devices

- Today's utility applications use PE devices based on silicon which are reliable and less costly compared to earlier mercury arc valves
- There is a need for high voltage and high power PE devices based on wide-band-gap materials such as SiC and GaN which can further reduce losses and increase system efficiency & reliability

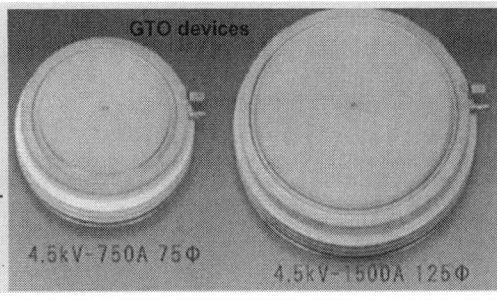

Power Electronics (PE) Applications

- Smart grid needs not only smart meters but also smart PE controllers on the electric grid.
- Fundamental breakthroughs in PE applications are needed to develop solid state counterparts of most of the power equipment such as transformers, breakers, and fault current limiters so that new demands like renewable integration and electric vehicle charging are reliably met.

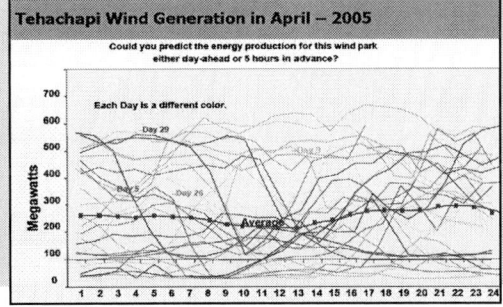

© 2014 Electric Power Research Institute, Inc. All rights reserved.

Power Electronics: Common Denominators & Industry Value

Common Denominators for PE Applications for Electric Utilities, Navy, Army, & others:

- Power Electronic Devices
- Novel Converter Topologies
  - e.g. Modular Multilevel VSC Converters

Needed: Collaboration and leveraging of R&D funding among all Players - EPRI, DOE ARPA-E, Navy, Army, National Labs, Industries (PE, Auto, & others), Universities, & others

Value

- Advanced power electronics technologies can enable a fully functional and controllable power system resulting in a *potential value in excess of $1 trillion over next 20 years by:*

  - Reducing power system losses and increasing efficiency and reliability and thus reduceivng energy cost.
  - Making power grid more controllable and allowing more power throughput
  - Allowing increased penetration of renewables and electric vehicle charging stations
  - Increasing the use of power electronics for the entire electric infrastructure - all the way from generation to the load centers.

© 2014 Electric Power Research Institute, Inc. All rights reserved.

Advanced Power Electronics

Major fundamental research is required for advances in PE materials

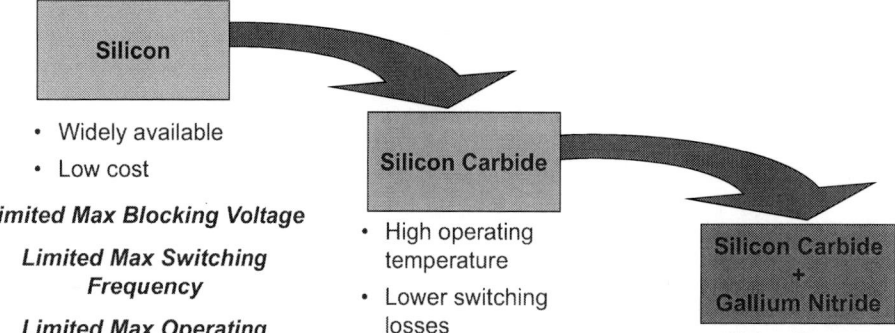

Silicon
- Widely available
- Low cost

Limited Max Blocking Voltage

Limited Max Switching Frequency

Limited Max Operating Temperature

Silicon Carbide
- High operating temperature
- Lower switching losses

Silicon Carbide + Gallium Nitride
- Higher Temperature
- Higher Voltage
- Optical switching

New power electronics materials enable newer applications and benefits

© 2014 Electric Power Research Institute, Inc. All rights reserved.

Needed: Wide Bandgap Power Electronics Devices

Choices:
- Silicon Carbide (SiC)
- Gallium Nitride (GaN)
- AIN (Aluminum Nitride)
- Diamond

Only feasible options before 2030

Issues:

Defects in the material

Lower voltage and current levels

Less experience & Lower reliability

Higher Cost of material

Benefits:
- High Operating Temperature
- High Current Density
- High Blocking Voltage
- Lower Specific Resistance

Needed:

High Voltage & High Current (High Power) & Reliable Devices

© 2014 Electric Power Research Institute, Inc. All rights reserved.

EPRI | ELECTRIC POWER RESEARCH INSTITUTE

Technical Innovations – Power Electronics

Needed: Roadmap and R&D funding to achieve these higher ratings for Wide Bandgap Devices

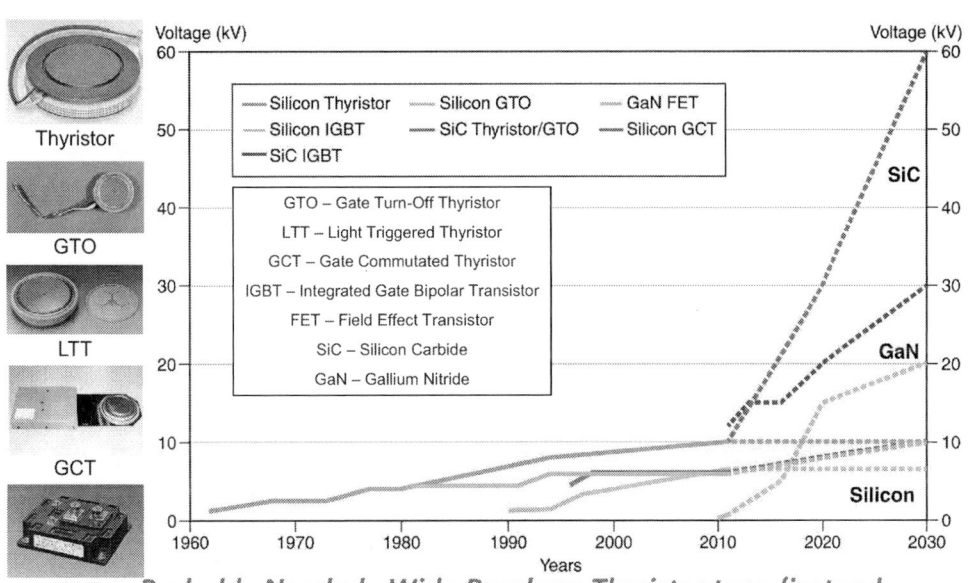

Probably Needed : Wide Bandgap Thyristor type (instead of transistor type) devices for High Power applications

© 2014 Electric Power Research Institute, Inc. All rights reserved.

EPRI | ELECTRIC POWER RESEARCH INSTITUTE

In Conclusion

Power Electronics WBG devices is key for:
- Future SMART Electric Grid (Intelligent Grid / Integrated Grid)
- Integration of renewables such as wind and solar with interconnections to the main grid
- Reducing system losses and increasing efficiency
- Reducing carbon footprint
- Enhancing quality of life

Future Needs:
- *High Power Devices*
- *New Converter Topologies*
- *New PE Applications*

© 2014 Electric Power Research Institute, Inc. All rights reserved.

EPRI | ELECTRIC POWER RESEARCH INSTITUTE

Others Are Interested Too!

© 2014 Electric Power Research Institute, Inc. All rights reserved.

EPRI | ELECTRIC POWER RESEARCH INSTITUTE

Thank you for your attention

Together…Shaping the Future of Electricity

© 2014 Electric Power Research Institute, Inc. All rights reserved.

EPRI | ELECTRIC POWER RESEARCH INSTITUTE

WiPDA 2014 Tutorial

Packaging Technologies to Exploit the Attributes of WBG Power Electronics

Zhenxian Liang

R&D Staff Member
Power Electronics and Electric Machinery Research Group
OAK RIDGE NATIONAL LABORATORY

2360 Cherahala Boulevard
Knoxville, Tennessee 37932

TEL: (865) 946-1467
FAX: (865) 946-1262
EMAIL: liangz@ornl.gov

WiPDA 2014 Tutorial

➢Welcome to the "WBG Packaging" Seminar;

➢The intent of this seminar is to introduce the integration approaches of electrical, thermal and thermo-mechanical designs in packaging technology for WBG (specially SiC) power electronics modules. It is aimed at providing the fundamental and specific knowledge for you to design, manufacture, use, apply, or specify power electronics modules well.

Outline

❑ **Overview of Power Electronics Packaging**

   ➢ *Module Based Power Electronics System*

   ➢ *Evaluation of Power Electronics Packaging*

❑ **Fundamentals of Power Module Packaging**

   ➢ *Structure, Materials and Processing Technologies*

   ➢ *Performance Characteristics*

❑ **Advanced Packaging of WBG Power Modules**

   ➢ *Attributes of the WBG Power Semiconductors*

   ➢ *Integrated SiC Power Modules (IPMs)*

   ➢ *Integrated Cooling of SiC Power Module*

   ➢ *High Temperature Packaging*

   ➢ *Planar-Bond-All : Multi-functional Integration*

❑ **Summary**

Module based Power Electronics System

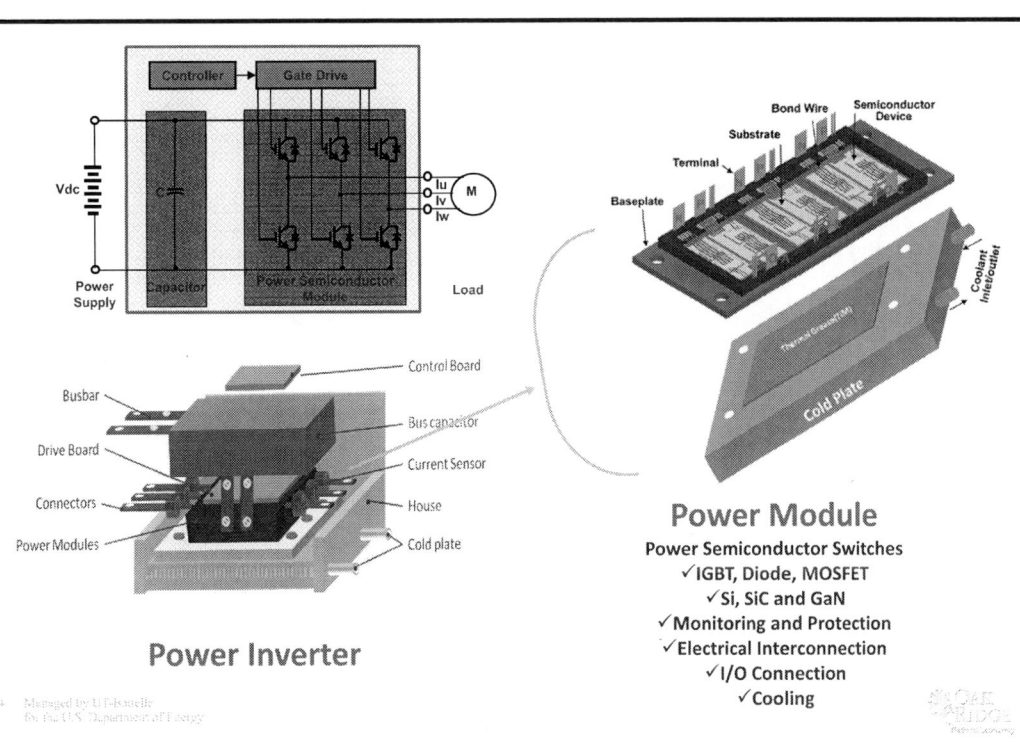

Power Inverter

Power Module
Power Semiconductor Switches
✓IGBT, Diode, MOSFET
✓Si, SiC and GaN
✓Monitoring and Protection
✓Electrical Interconnection
✓I/O Connection
✓Cooling

Power Semiconductor Switches

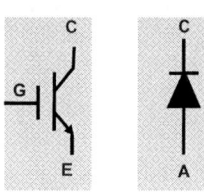

Electric Symbol

Semiconductor Die (Chip)

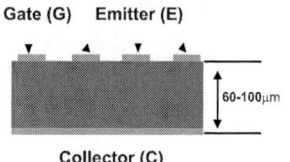

Cross-sectional View
(Schematic)

Die in Wafer

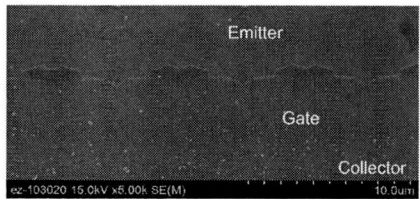

Cross-sectional View (SEM)

Electrical Operation of Power Switch

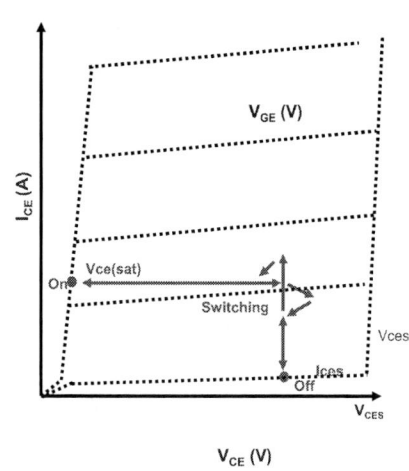

Conduction

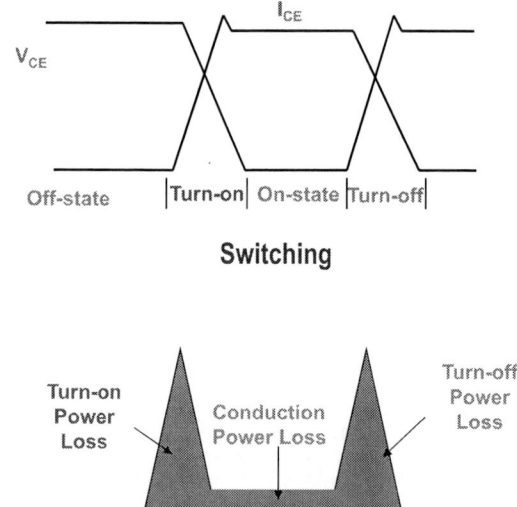

Switching

Electric Power Loss

Temperature Characteristics of Power Semiconductor

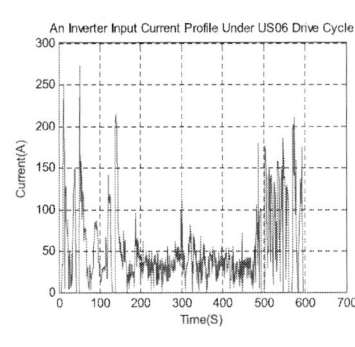

An Inverter Input Current Profile Under US06 Drive Cycle

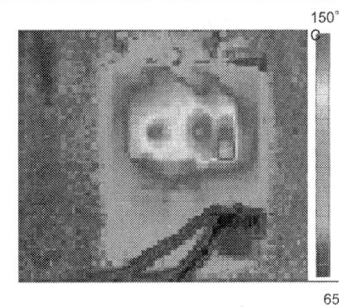

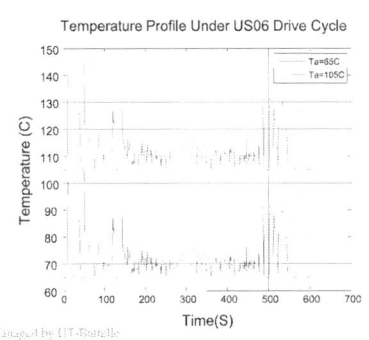

Temperature Profile Under US06 Drive Cycle

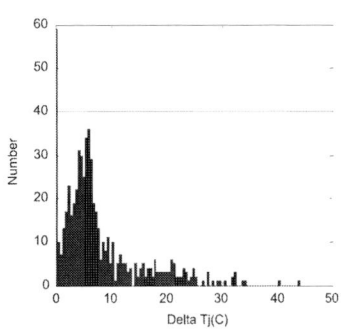

Temperature Dependence and Limits of Power Switch

Serial No. Electrical Parameter	Formula Used	Parameter Value @25C	Parameter Value @125C
1. Electron Mobility	$\mu_n = 1500(300/T_j)^{2.5}$	1500cm²/V-sec	739.92cm²/V-sec
2. Hole Mobility	$\mu_p = 450(300/T_j)^{2.5}$	450cm²/V-sec	221.98cm²/V-sec
3. Electron Mobility near surface	$\mu_{ns} = 400(300/T_j)^{2.5}$	400cm²/V-sec	197.31cm²/V-sec
4. Electron Diffusion Coefficient	$Dn(T_j) = \mu_n kT/q = 2.0187 \times 10^5 (T_j)^{-1.5}$	38.85cm2/sec	25.4cm2/sec
5. Hole Diffusion Coefficient	$Dp(T_j) = \mu p kT/q = 6.0561 \times 10^4 (T_j)^{-1.5}$	11.66cm2/sec	7.62cm2/sec
6. Ambipolar Diffusion Coefficient	$Da(T_j) = 2Dn(T_j)Dp(T_j)/(Dn(T_j)+Dp(T_j))$ $= 9.3171 \times 10^4 (T_j)^{-1.5}$	17.93, 11.72cm2/sec	
7. Ambipolar (high-level) lifetime	$\tau_{HL}(Tj) = \tau_{HL0}(T_j/T_0)^{1.5} = 1.9245 \times 10^{-4}(T_j)^{1.5}$ τ_{HL0} (0.9 μsec)		1.375μsec
8. Threshold Voltage	$V_{TH}(T_j) = V(T_0) - 0.009(T_j - T_0)$	4V	3.1V
9. Ambipolar Diffusion Length	$L_a(T_j) = \sqrt{(Da(Tj)\, \tau HL(T_j)} = La$		
10. Current Gain	$\alpha_T = 1/\cosh(W/La(T_j))$		
12. Intrinsic carrier concentration	$n_i = 1.042 \times 10^{15}(T_j)^2 \exp(-6884/T_j)$	$1.015 \times 10^{10} \times cm^{-3}$	$5.08 \times 10^{12} \times cm^{-3}$

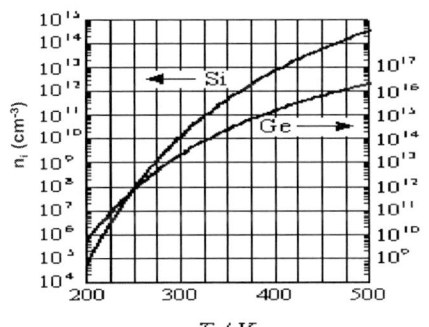

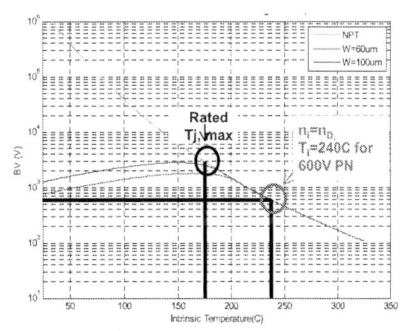

978-1-4799-5494-0/14 $31.00 © 2014 IEEE   363

Temperature Dependence and Limits of Power Switch

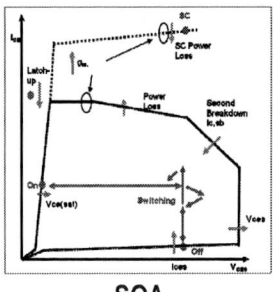

SOA

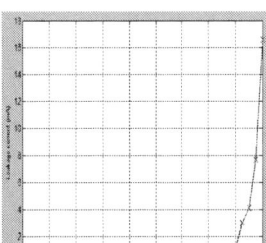

Leakage Current

Electric Power Loss

Power Semiconductor Failure:

HCl, TDDB, Thermal Runaway, Dynamic Avalanche, Latch-up, Short Circuit, Electro-migration, etc.

$$Nf = \alpha \cdot (\Delta Tj)^{-\beta} \cdot \exp(Ea/Tj)$$

Reliability in Device

Power Module Packaging: Multifunction Integration

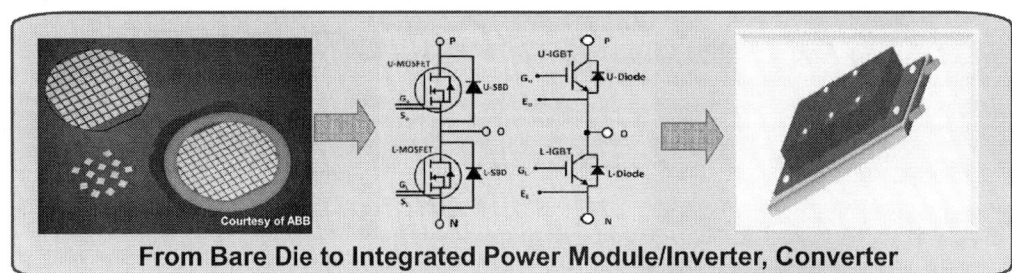

From Bare Die to Integrated Power Module/Inverter, Converter

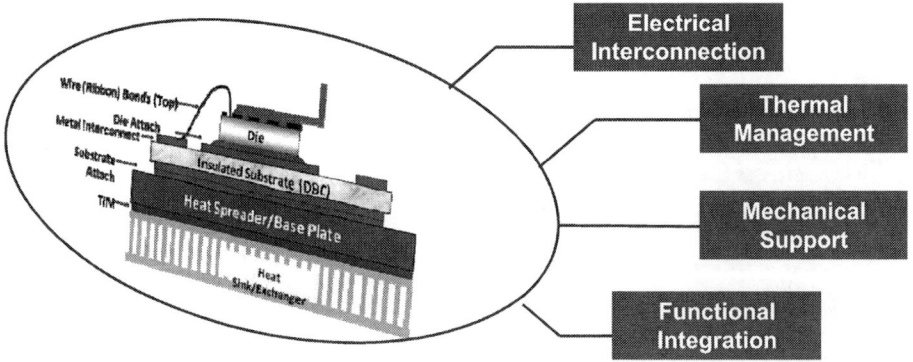

Electrical Interconnection

Thermal Management

Mechanical Support

Functional Integration

Outline

❑ **Overview of Power Electronics Packaging**
  ➢ *Module Based Power Electronics System*
  ➢ ***Evaluation of Power Electronics Packaging***

❑ Fundamentals of Power Module Packaging
  ➢ *Structure, Materials and Processing Technologies*
  ➢ *Performance Characteristics*

❑ Advanced Packaging of WBG Power Modules
  ➢ *Attributes of the WBG Power Semiconductors*
  ➢ *Integrated SiC Power Modules (IPMs)*
  ➢ *Integrated Cooling of SiC Power Module*
  ➢ *High Temperature Packaging*
  ➢ *Planar-Bond-All : Multi-functional Integration*

❑ Summary

Evaluation of Power Module

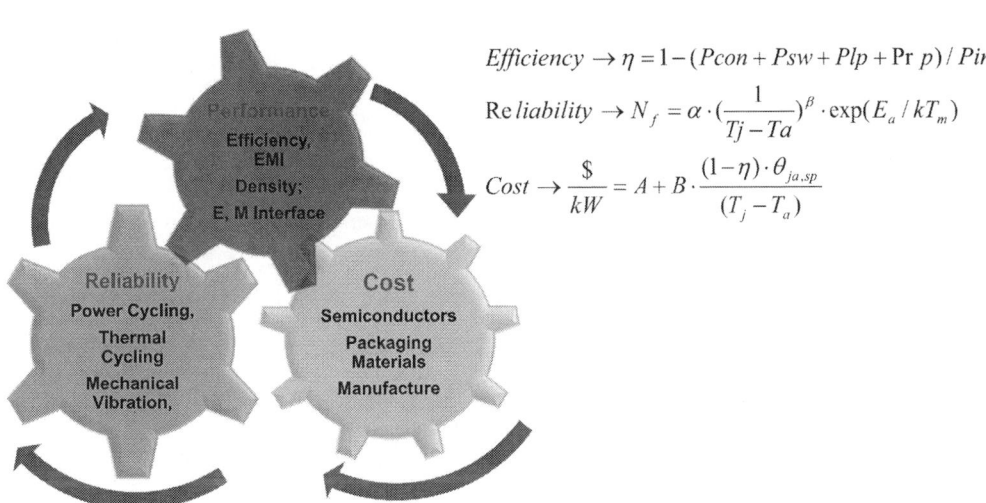

$$Efficiency \rightarrow \eta = 1 - (Pcon + Psw + Plp + \Pr p)/Pin$$

$$\mathrm{Re}\,liability \rightarrow N_f = \alpha \cdot (\frac{1}{Tj - Ta})^{\beta} \cdot \exp(E_a / kT_m)$$

$$Cost \rightarrow \frac{\$}{kW} = A + B \cdot \frac{(1-\eta) \cdot \theta_{ja,sp}}{(T_j - T_a)}$$

Technical Metrics of Power Module Packaging

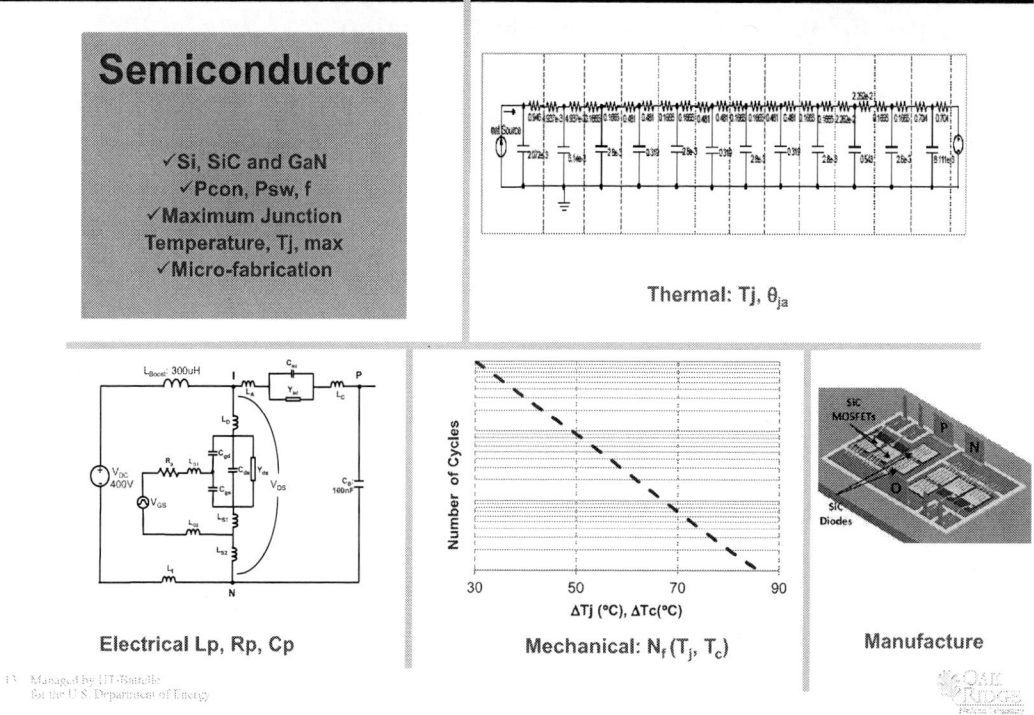

Semiconductor

- Si, SiC and GaN
- Pcon, Psw, f
- Maximum Junction Temperature, Tj, max
- Micro-fabrication

Thermal: Tj, θ_{ja}

Electrical Lp, Rp, Cp

Mechanical: $N_f(T_j, T_c)$

Manufacture

Comprehensive Design of Power Module Packaging

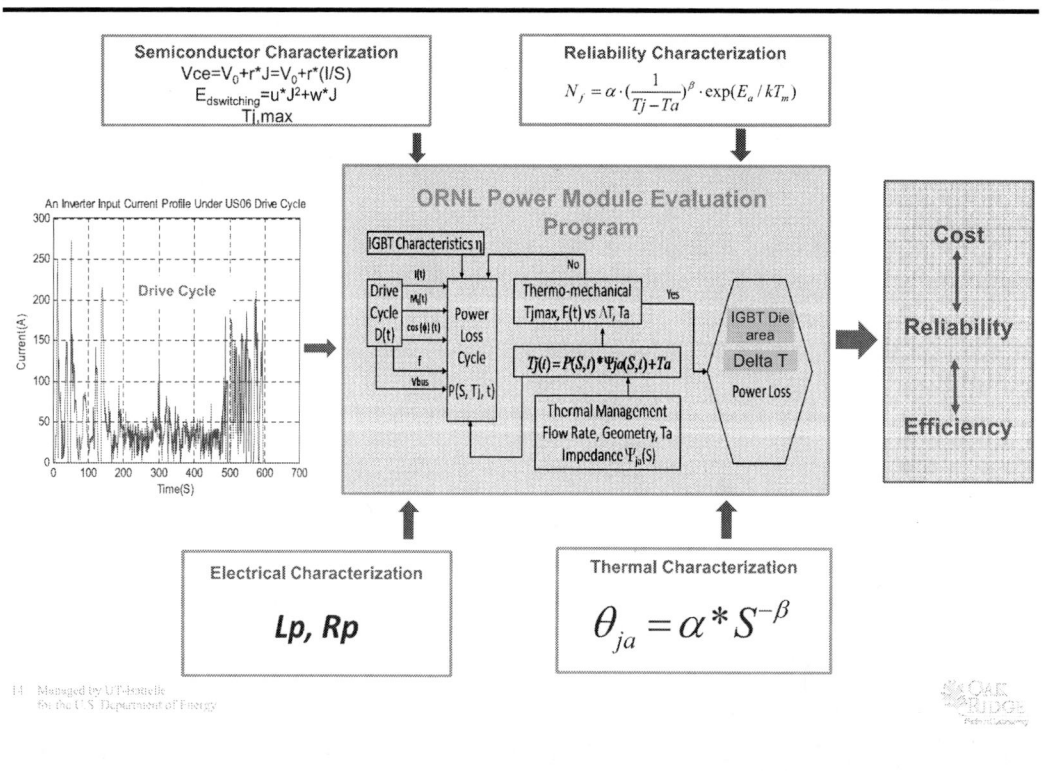

Semiconductor Characterization
$$Vce = V_0 + r*J = V_0 + r*(I/S)$$
$$E_{dswitching} = u*J^2 + w*J$$
$$Tj, max$$

Reliability Characterization
$$N_f = \alpha \cdot (\frac{1}{Tj - Ta})^\beta \cdot \exp(E_a / kT_m)$$

Electrical Characterization
$$Lp, Rp$$

Thermal Characterization
$$\theta_{ja} = \alpha * S^{-\beta}$$

Power Electronics Packaging: Integrated Multidisciplinary Approach

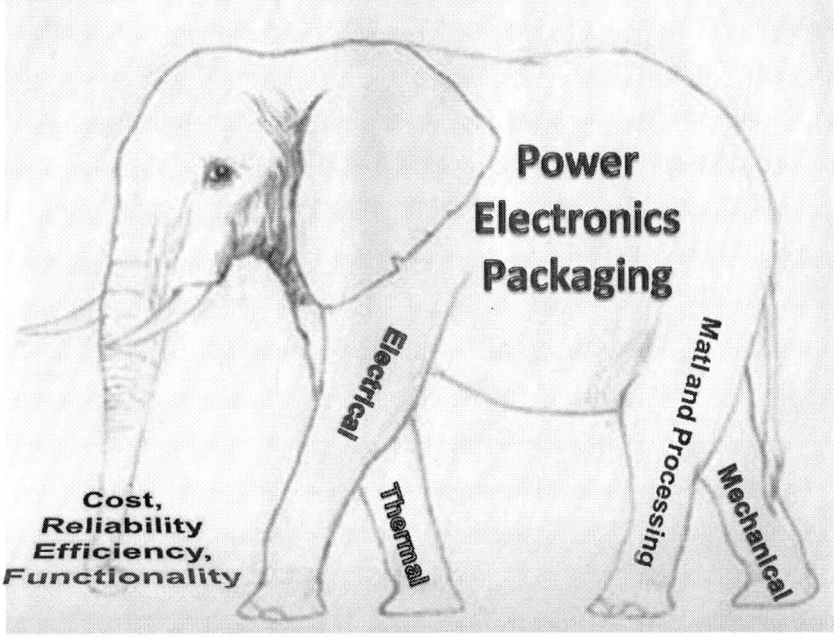

Outline

❑ Overview of Power Electronics Packaging
  ➢ Module Based Power Electronics System
  ➢ Evaluation of Power Electronics Packaging

❑ **Fundamentals of Power Module Packaging**
  ➢ *Structure, Materials and Processing Technologies*
  ➢ Performance Characteristics

❑ Advanced Packaging of WBG Power Modules
  ➢ Attributes of the WBG Power Semiconductors
  ➢ Integrated SiC Power Modules (IPMs)
  ➢ Integrated Cooling of SiC Power Module
  ➢ High Temperature Packaging
  ➢ Planar-Bond-All : Multi-functional Integration

❑ Summary

Power Module Packaging: Structure

 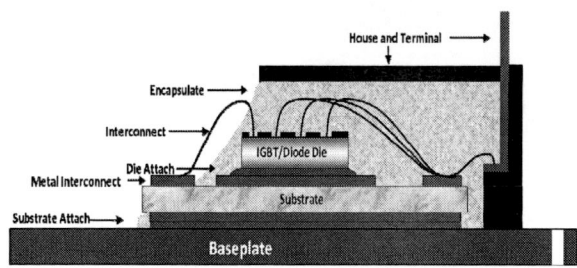

Part	Function	Material	Process	Remark
Base plate	Mechanical support, heat spreading	Cu, AlSiC	Cast	Premade
Substrate Attach	Mechanical joint	Solders	Reflow	Foil or paste
Substrate	Electrical conduction, insulation	DBC, DBA, SiN AlO, ALN	Direct metal bond	Premade
Die attach	Electrical connection, mechanical joint	solders	Reflow	Foil or paste
Interconnect	Electrical connection	Al or Cu, Gold	Wire bonding	In wire or ribbon
Encapsulate	Electrical insulation, mechanical protection	Silicone Gel	Dispense/cure	Chemical resins
House and Terminal	Mecahnical Protection, Electrical connect	Cu, Plastic	Mold	Premade

Power Module Packaging: Manufacture Process

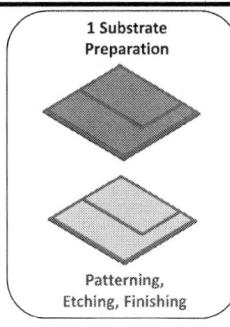

1 Substrate Preparation

Patterning, Etching, Finishing

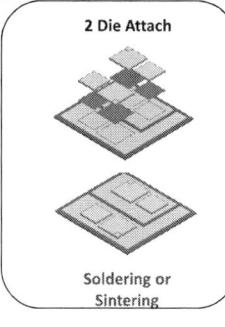

2 Die Attach

Soldering or Sintering

3 Substrate Attach

Soldering

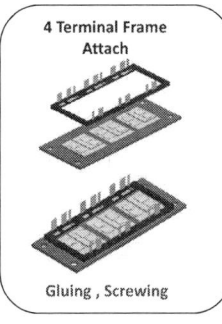

4 Terminal Frame Attach

Gluing , Screwing

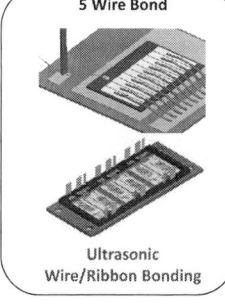

5 Wire Bond

Ultrasonic Wire/Ribbon Bonding

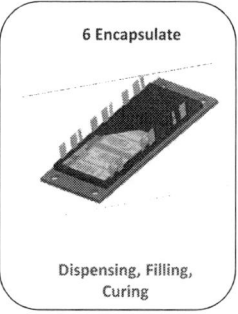

6 Encapsulate

Dispensing, Filling, Curing

Power Module Packaging: Soldering

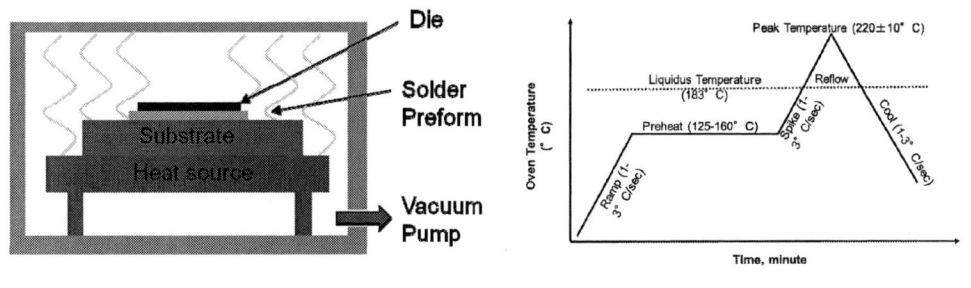

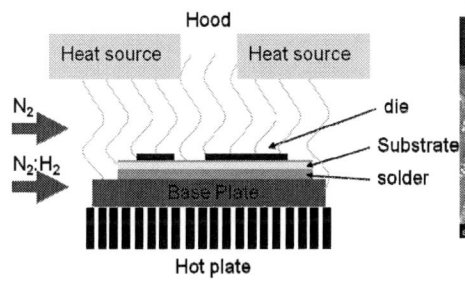

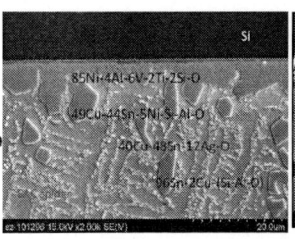 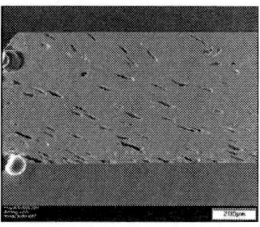

- Solder Metallurgy, phase diagram; Intermetallic formations;
- Surface finishing of parts;
- Voids formation; Operation Temperatures; Environmental and Etc.

Power Module Packaging: Wire Bonding

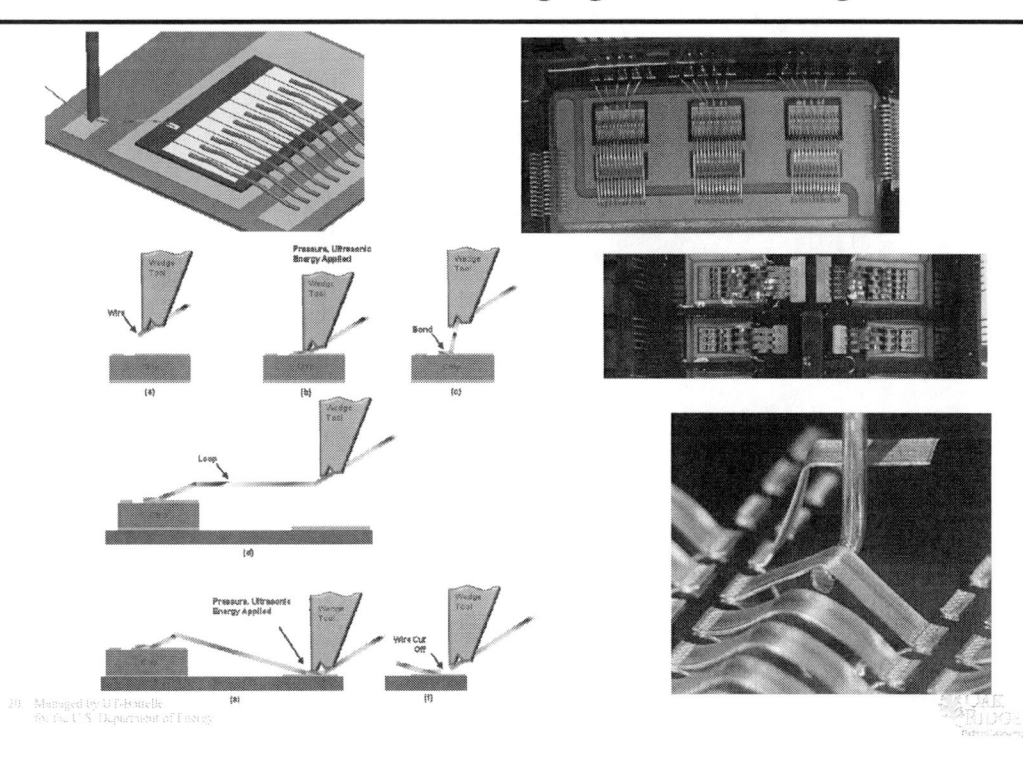

Power Module Packaging: Encapsulate

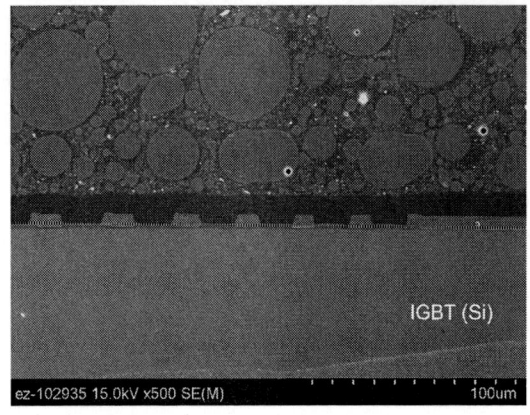

ez-102935 15.0kV x500 SE(M)   IGBT (Si)   100um

Key Issues

- Chemistry of polymers;
- Electrical properties;

- Adhesion and mechanical strength;
- Moisture resistance;
- Operation temperatures;
- Environmental and Etc.

Outline

❏ Overview of Power Electronics Packaging
  ➢ *Module Based Power Electronics System*
  ➢ *Evaluation of Power Electronics Packaging*

❏ **Fundamentals of Power Module Packaging**
  ➢ *Structure, Materials and Processing Technologies*
  ➢ ***Performance Characteristics***

❏ Advanced Packaging of WBG Power Modules
  ➢ *Attributes of the WBG Power Semiconductors*
  ➢ *Integrated SiC Power Modules (IPMs)*
  ➢ *Integrated Cooling of SiC Power Module*
  ➢ *High Temperature Packaging*
  ➢ *Planar-Bond-All : Multi-functional Integration*

❏ Summary

Performance Characteristics: Electrical Model

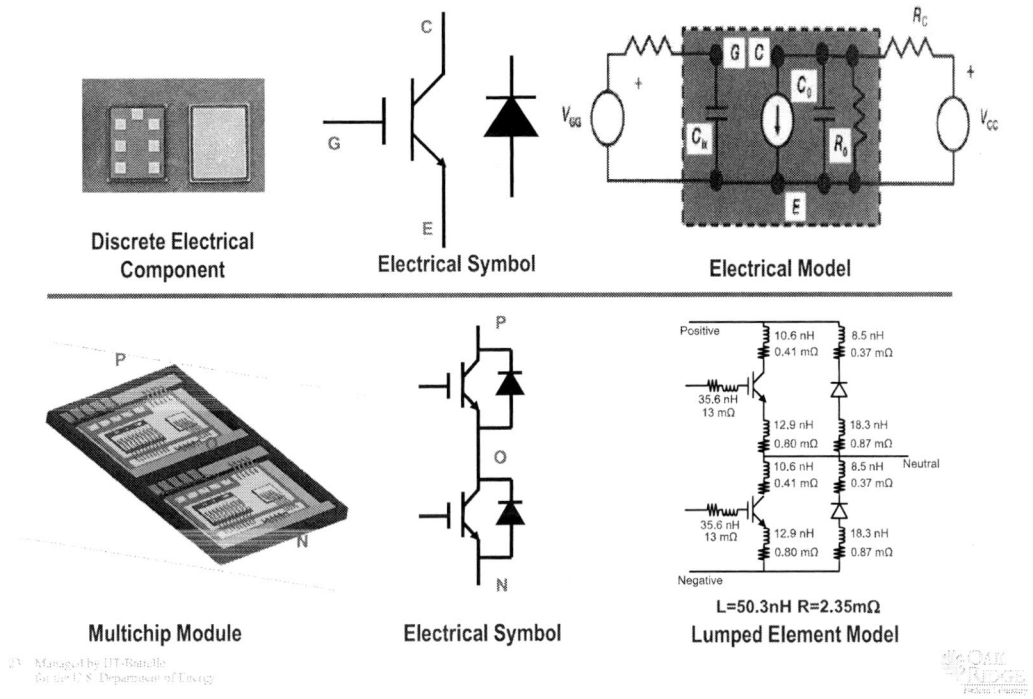

Discrete Electrical Component

Electrical Symbol

Electrical Model

Multichip Module

Electrical Symbol

L=50.3nH R=2.35mΩ
Lumped Element Model

Electrical Characteristics: Experimental Measurement

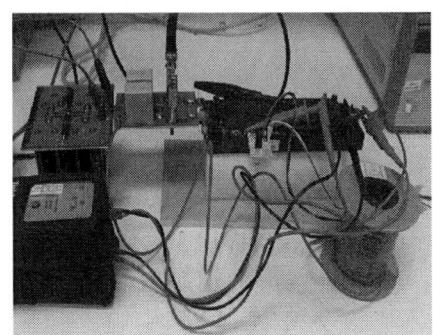

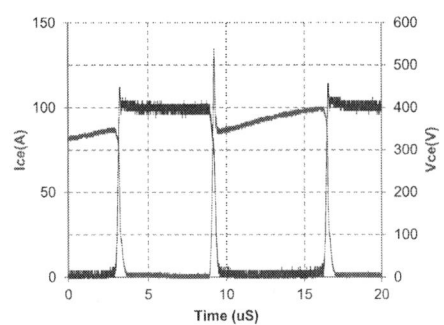

IGBT Switching Curve

Key Issues

- Optimization in design and fabrication of busbar system;
- Identification of module interconnection parasitic parameters;
- Main current control.

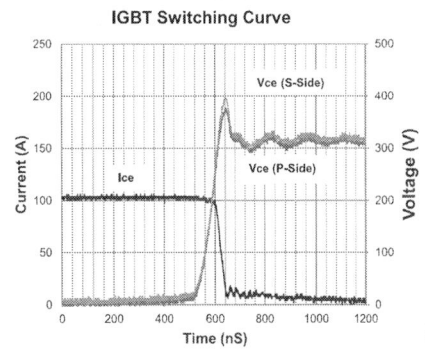

978-1-4799-5494-0/14 $31.00 © 2014 IEEE   371

Effects of Parasitic Parameters on Electrical Performance

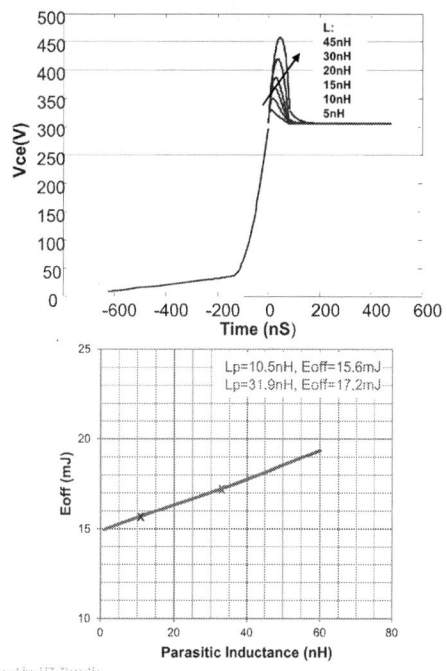

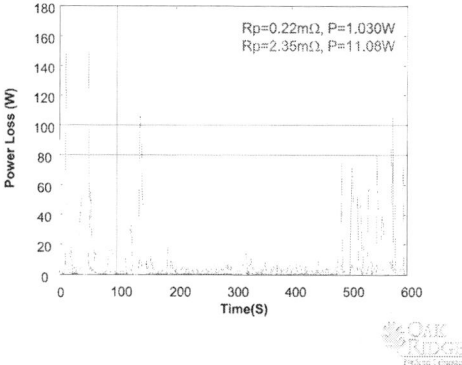

- Current profile with a standard drive cycle;
- Power loss by parasitic inductance;
- Power Loss by parasitic resistance.

Electrical Characterization: Continuous Switching

Key Issues

- Power losses from semiconductors and packaging parasitic components;
- Identification of the effects of parasitic parameters;
- Identical switches and thermal management.

Performance Characteristics: Thermal Model

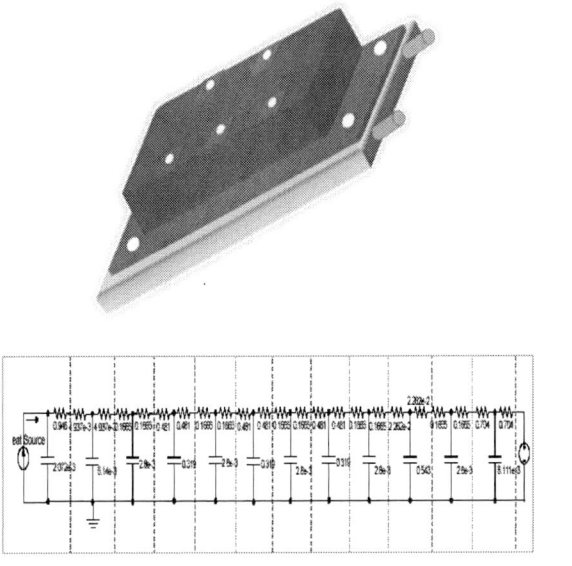

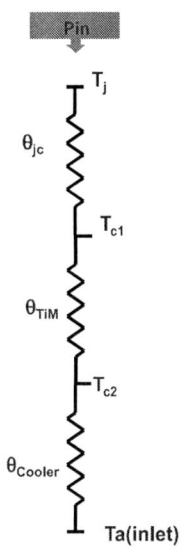

Thermal Characterization: Simulation and Experimental

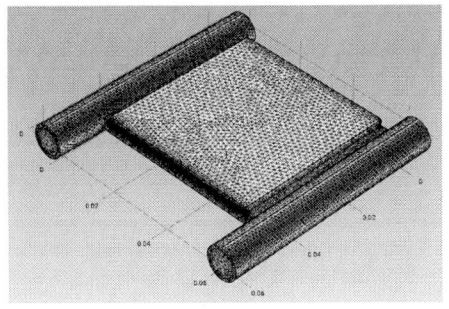

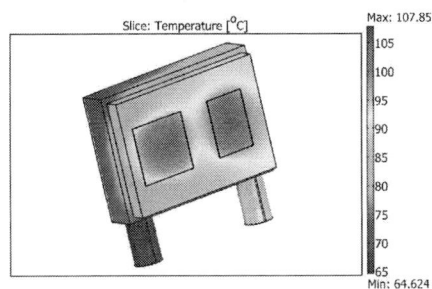

IR Image

Thermal Characterization: Sensor on Chip and Self Sensing

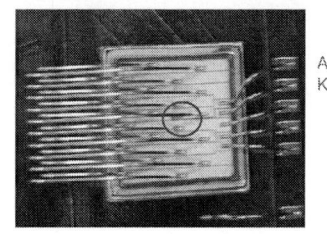

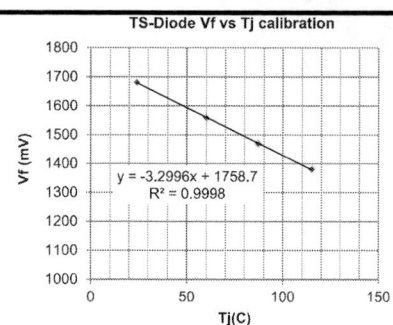

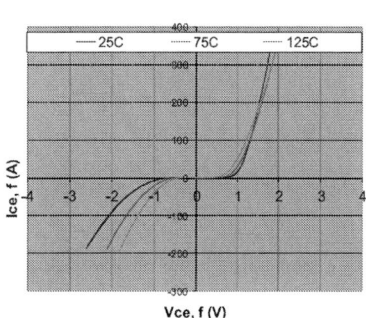

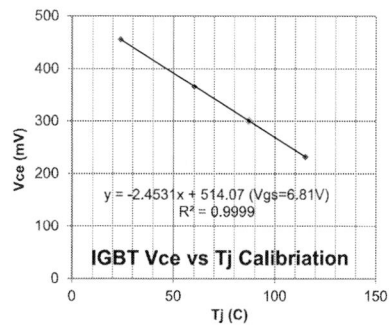

Thermal Characterization: IGBT Cooling Down Curve

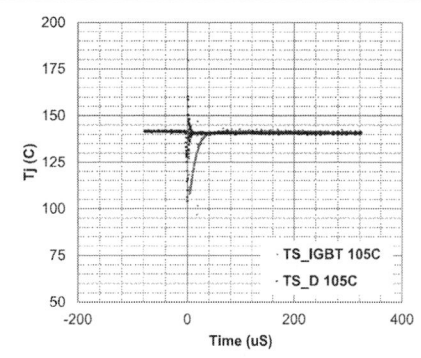

Comparison of Sensor on Chip and Self Sensing

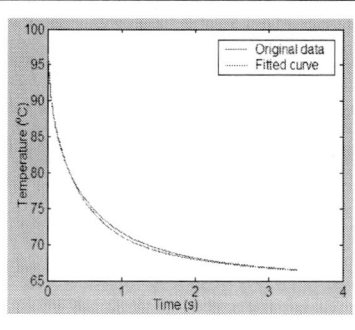

$$y = 3.72\,e^{-861.9\,x} + 7.199\,e^{-38.18\,x} + 3.434\,e^{-5.288\,x}$$
$$+ 15.74\,e^{-2.309\,x} + 7.339\,e^{-0.4991\,x} + 65.12$$

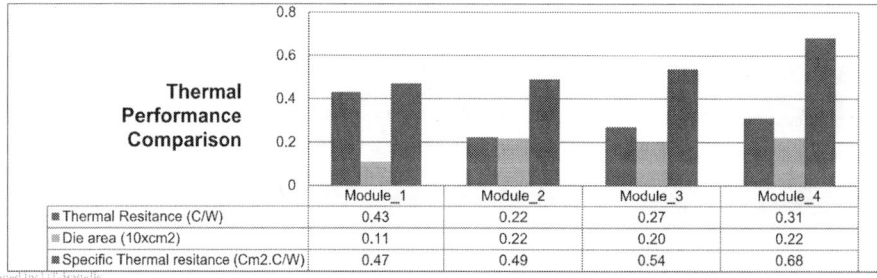

Thermal Performance Comparison	Module_1	Module_2	Module_3	Module_4
■ Thermal Resistance (C/W)	0.43	0.22	0.27	0.31
■ Die area (10xcm2)	0.11	0.22	0.20	0.22
■ Specific Thermal resistance (Cm2.C/W)	0.47	0.49	0.54	0.68

Performance Characteristics: Thermo-mechanical Fatigue

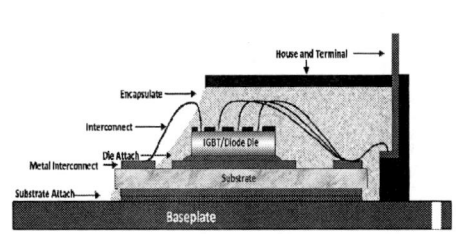

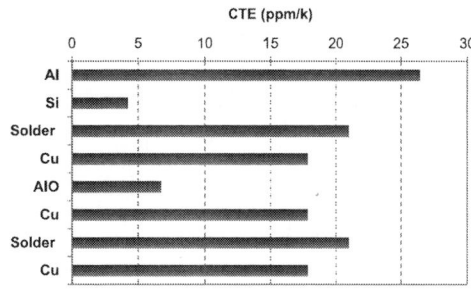

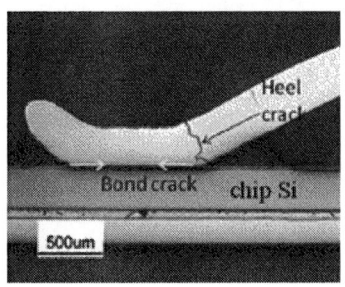

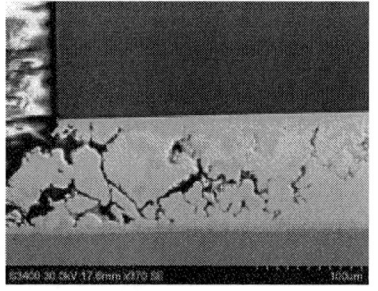

Thermal Characteristics: Thermal/Power Cycling

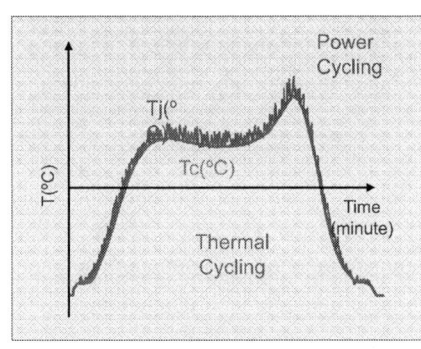

Zoom-in of IGBT temperature profile

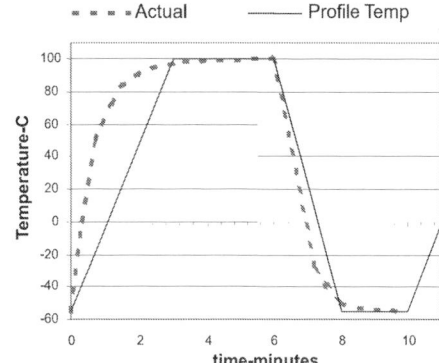

Voids and Extending in Solder Layer during Thermal cycling

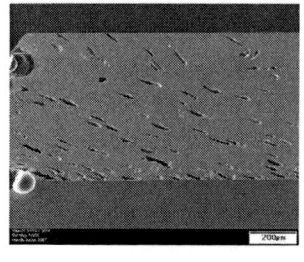

SEM

X-Ray

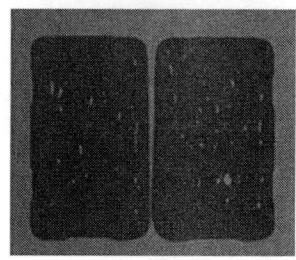

SAM

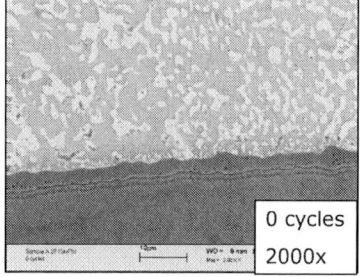

0 cycles
2000x

60 cycles
2000x

Microstructure of PbSn Solder

Dimos Kaisis, Ph D dissertation, Virginia Tech

Power Cycling and Wire Bond Failure

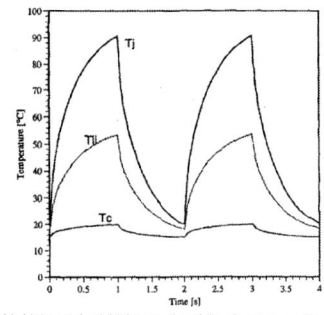

M, Held, etal, 1997 International Conference on Power Electronics and Drive Systems,

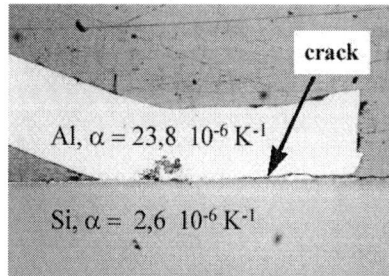

crack

Al, α = 23,8 10^{-6} K^{-1}

Si, α = 2,6 10^{-6} K^{-1}

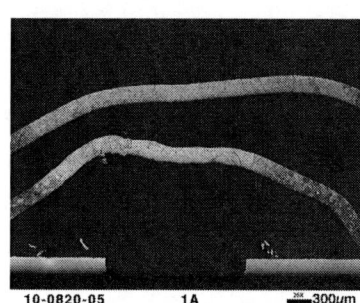

http://www.ornl.gov/sci/propulsionmaterials/pdfs/FY10_Qtr3.pdf

@www.emeraldinsight.com

Outline

❑ Overview of Power Electronics Packaging
> Module Based Power Electronics System
> Evaluation of Power Electronics Packaging

❑ Fundamentals of Power Module Packaging
> Structure, Materials and Processing Technologies
> Performance Characteristics

❑ **Advanced Packaging of WBG Power Modules**
> ***Attributes of the WBG Power Semiconductors***
> Integrated SiC Power Modules (IPMs)
> Integrated Cooling of SiC Power Module
> High Temperature Packaging
> Planar-Bond-All : Multi-functional Integration

❑ Summary

WBG Power Semiconductor Attributes

Property	Si	GaAs	SiC	GaN	Diamond
Bandgap, Eg (ev)	1.12	1.43	3.26	3.45	5.45
Breakdown Electric Field E_c (kV/cm)	300	400	2,200	2,000	10,000
Intrinsic Carrier Concentration n_i (cm^{-3})	9.65E9	1.8E6	1.6E-6	1E-7	1E-27
Electron Mobility μ_n (cm^2/V•s)	1,500	8,500	500-1,000	1,250	2,200
Hole Mobility μ_p (cm^2/V•s)	600	400	100-115	850	850
Dielectric Constant ε_r	11.9	13.1	10.1	9	5.5
Thermal Conductivity κ (W/cm•K)	1.5	0.46	4.9	1.3	22
Saturated Electron Drift Velocity υ_{sat} (10^7 cm/s)	1	1	2	2.2	2.7

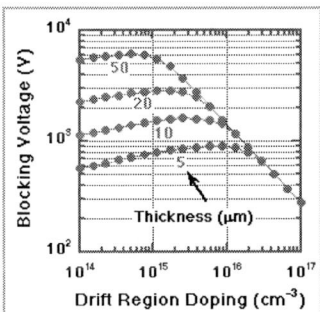

High Voltage

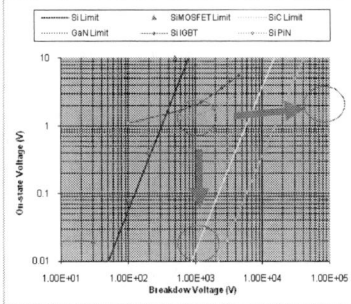

Low Conduction Loss

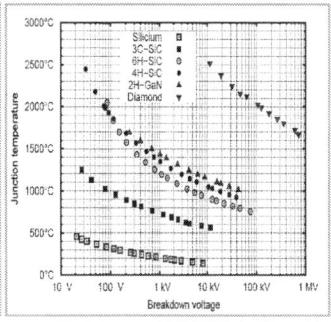

High Intrinsic Temperature

WBG Power Semiconductor Attributes

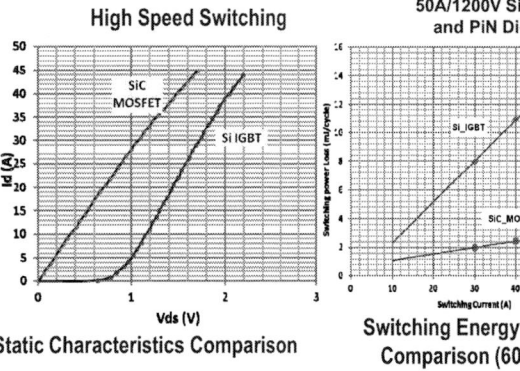

High Speed Switching

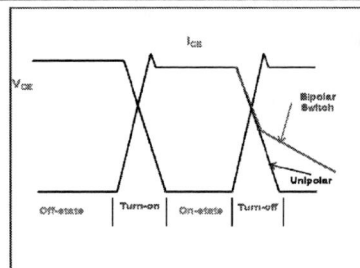

50A/1200V SiC
MOSFET and SBD

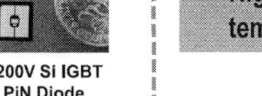

50A/1200V Si IGBT
and PiN Diode

Static Characteristics Comparison

Switching Energy Loss
Comparison (600V)

- High current density, low loss
- High frequency switching
- High operation temperature

✓ High power conversion efficiency
✓ High power density (per volume, weight)
✓ Low cost (less semiconductor, less cooling)

Industry SiC Power Module Samples

CREE 1200V/100A
Phase-leg Module

Fuji 1200V/120A Phase-leg

Infineon 1200V/30A JFET
Phase-leg

Powerex 100A/1200V Phase
leg

Rohm 1200V/120A
Phase-leg

Mitsubishi Full SiC DIPPFC
600V/20Arms

978-1-4799-5494-0/14 $31.00 © 2014 IEEE   378

Packaging and Performance Limitations

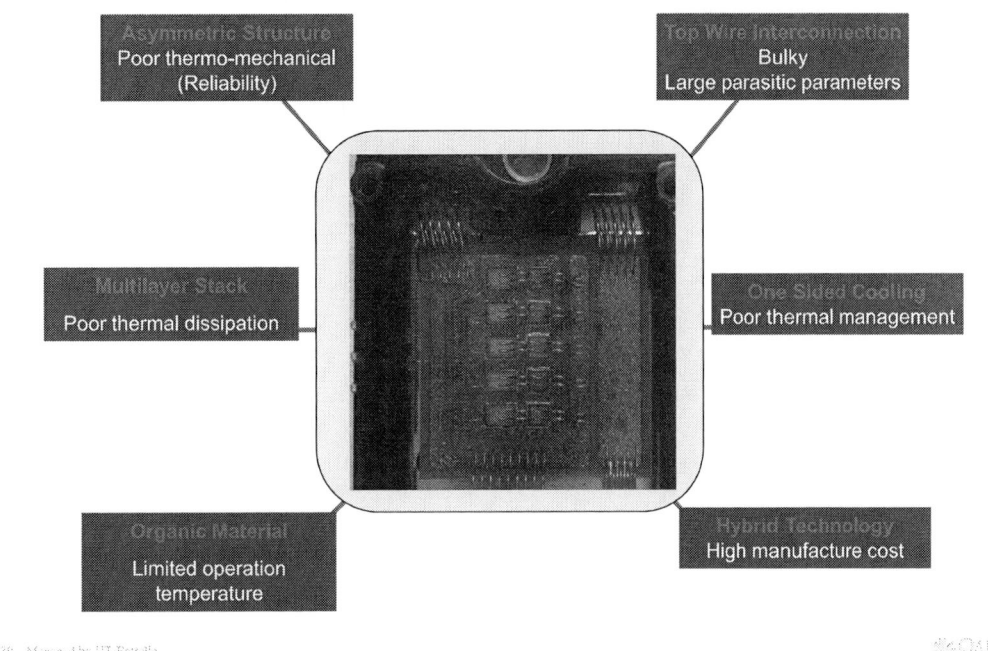

Asymmetric Structure
Poor thermo-mechanical (Reliability)

Top Wire Interconnection
Bulky
Large parasitic parameters

Multilayer Stack
Poor thermal dissipation

One Sided Cooling
Poor thermal management

Organic Material
Limited operation temperature

Hybrid Technology
High manufacture cost

Advance WBG Power Electronics Through Packaging

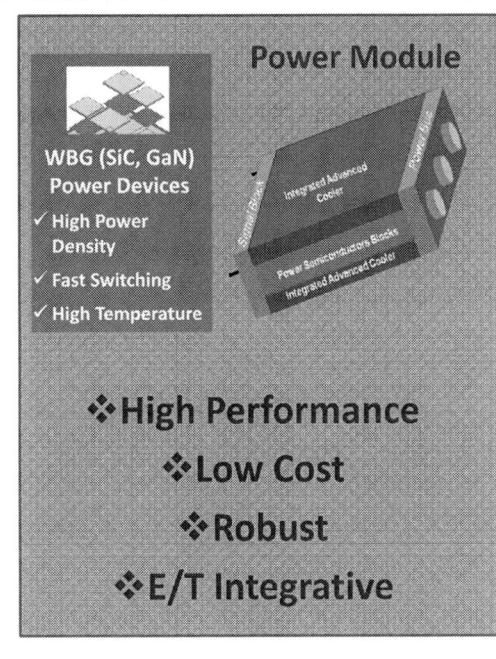

Power Module

WBG (SiC, GaN) Power Devices
✓ High Power Density
✓ Fast Switching
✓ High Temperature

❖ **High Performance**
❖ **Low Cost**
❖ **Robust**
❖ **E/T Integrative**

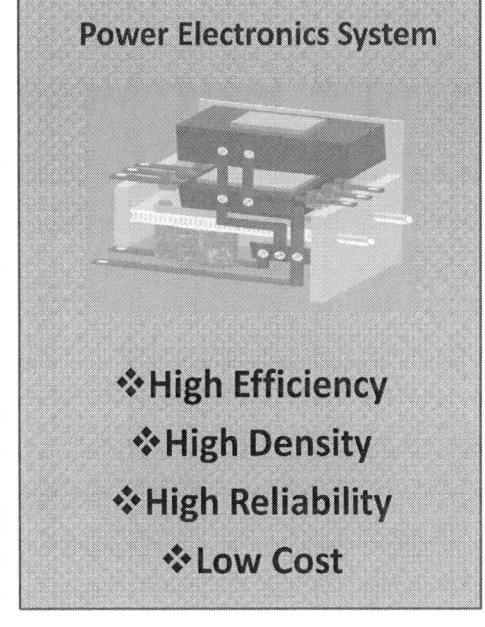

Power Electronics System

❖ **High Efficiency**
❖ **High Density**
❖ **High Reliability**
❖ **Low Cost**

978-1-4799-5494-0/14 $31.00 © 2014 IEEE

Power Electronics Packaging Advances

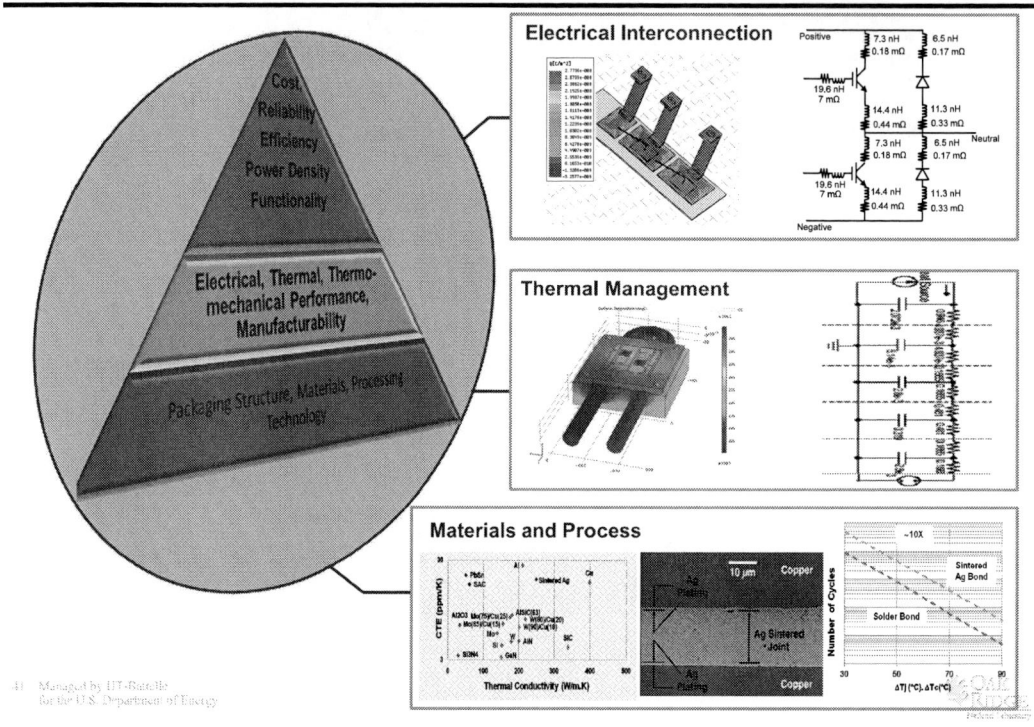

Outline

❑ Overview of Power Electronics Packaging

> Module Based Power Electronics System

> Evaluation of Power Electronics Packaging

❑ Fundamentals of Power Module Packaging

> Structure, Materials and Processing Technologies

> Performance Characteristics

❑ **Advanced Packaging of WBG Power Modules**

> Attributes of the WBG Power Semiconductors

> *Integrated SiC Power Modules (IPMs)*

> Integrated Cooling of SiC Power Module

> High Temperature Packaging

> Planar-Bond-All : Multi-functional Integration

❑ Summary

Optimize Electrical Interconnection Through Integration

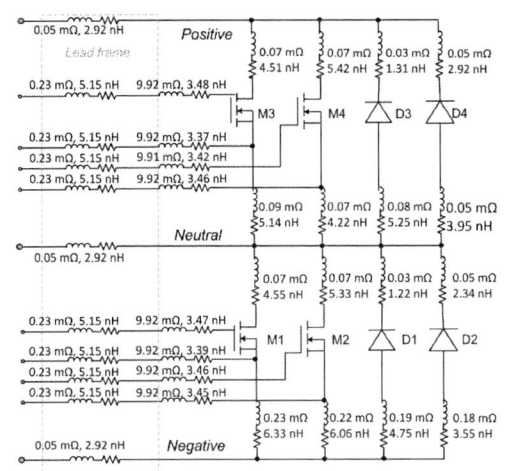

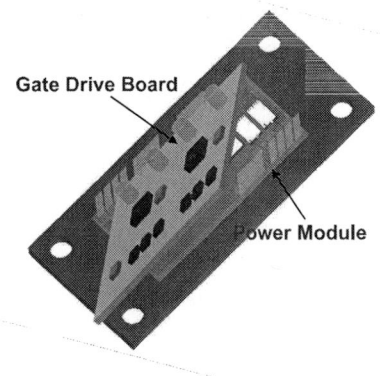

SiC Integrated Power Module (IPM)

Integrated SiC Power Module Packaging

UT/ORNL Si SOI Gate Driver Chip-on-Board

Power Supply Board

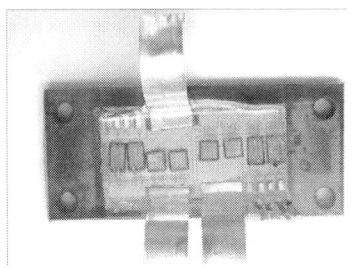

SiC Phase-leg Module

SiC Integrated Power Module (IPM)

978-1-4799-5494-0/14 $31.00 © 2014 IEEE   381

Operation of the IPM in a Converter

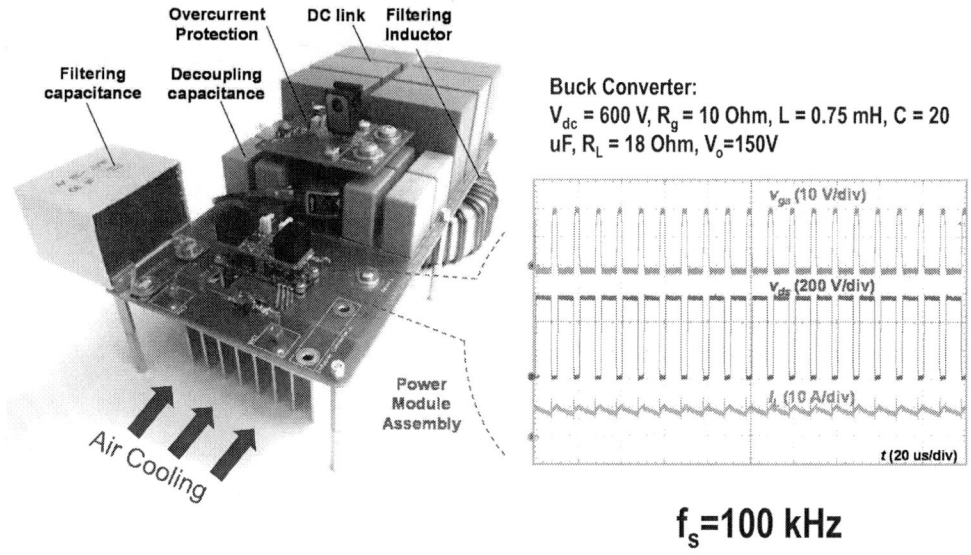

Buck Converter:
V_{dc} = 600 V, R_g = 10 Ohm, L = 0.75 mH, C = 20 uF, R_L = 18 Ohm, V_o=150V

f_s=100 kHz

Evaluated SiC IPMs in an Inverter

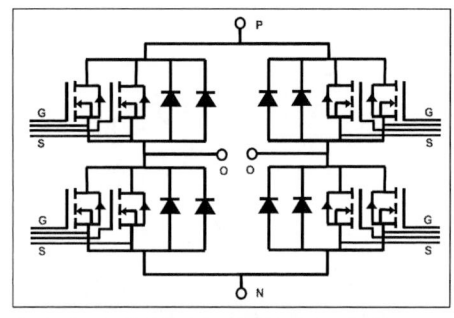

Two 100A/1200V SiC Power Modules in a HF converter (liquid cooled)

SiC Integrated Power Module Performance

Frequency (kHz)	23	23.5	48	81
Sample	Efficiency (%)			
ORNL SiC Phase-leg Module (100A/1200V)	97.84		98.37	95.45
Commercial Phase-leg Module (120A/1200V)		97.67	98.77	96.55

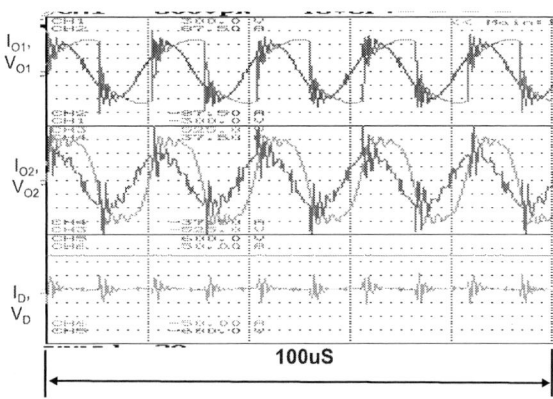

Waveforms at 48KHz, 3.3kW Switching

Outline

❑ Overview of Power Electronics Packaging

  ➢ Module Based Power Electronics System

  ➢ Evaluation of Power Electronics Packaging

❑ Fundamentals of Power Module Packaging

  ➢ Structure, Materials and Processing Technologies

  ➢ Performance Characteristics

❑ **Advanced Packaging of WBG Power Modules**

  ➢ Attributes of the WBG Power Semiconductors

  ➢ Integrated SiC Power Modules (IPMs)

  ➢ **Integrated Cooling of SiC Power Module**

  ➢ High Temperature Packaging

  ➢ Planar-Bond-All : Multi-functional Integration

❑ Summary

Thermal Resistance In Power Module Assembly

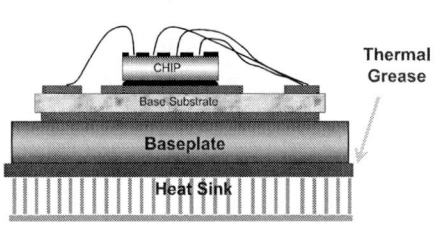

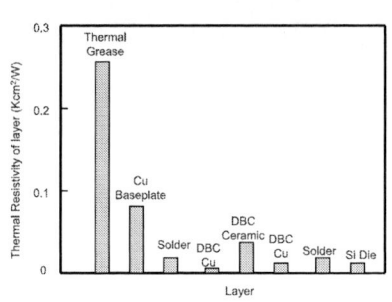

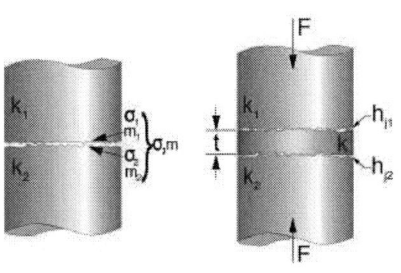

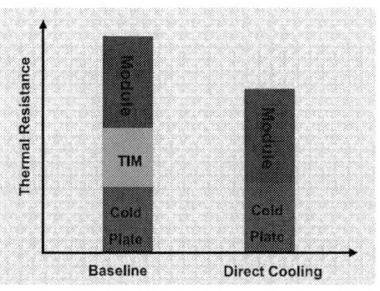

Integrated Cooling Packaging

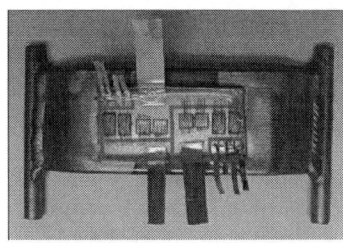

100A/1200V Phase-leg Module

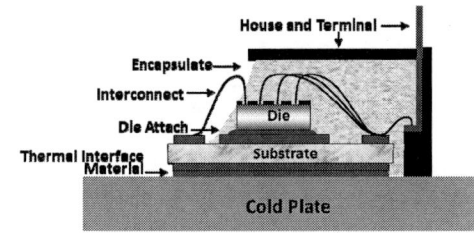

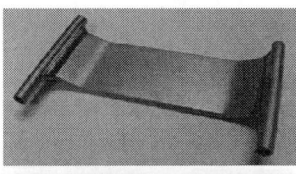

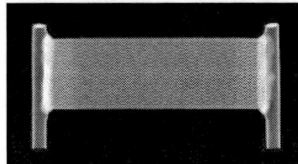

Flat Tube Cold Base Plate

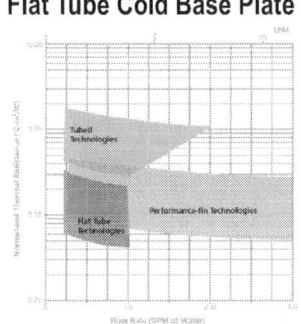

SiC Packaging Thermal Evaluation and Comparison

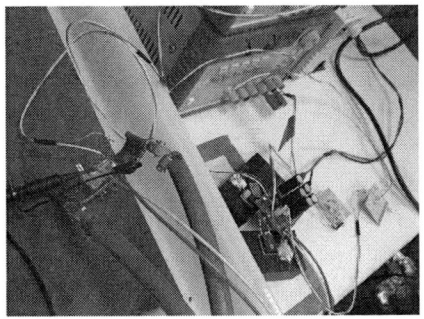

Specific Thermal Resistance Test Setup

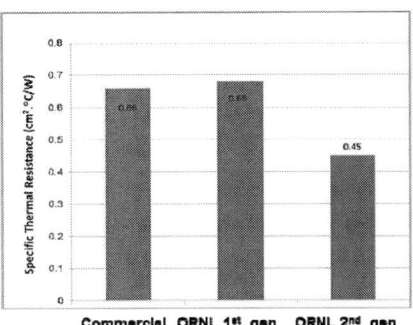

Specific Thermal Resistance Comparison

Power Module Evaluation and Comparison

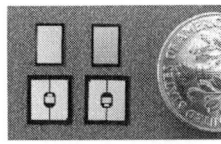

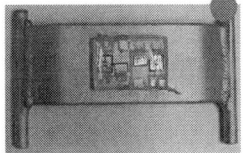

Si Modules

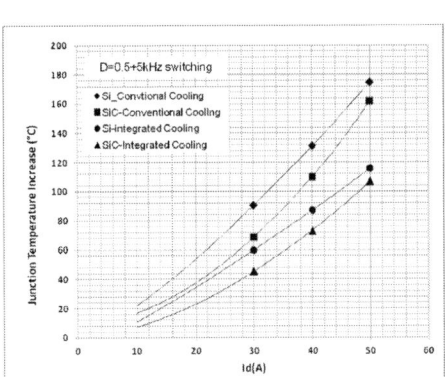

Junction Temperature vs Current for Different Packages

Current Density Allowed at ΔTj=100°C for a Typical Operation

SiC Modules

Item	Si_Con. Cooling	SiC_Con. Cooling	Si_Integ. Cooling	SiC_Integ. Cooling
Current Density J_d (A/cm²)	65.35	144.97	97.57	184.98

978-1-4799-5494-0/14 $31.00 © 2014 IEEE   385

Outline

❏ Overview of Power Electronics Packaging

  ➢ *Module Based Power Electronics System*

  ➢ *Evaluation of Power Electronics Packaging*

❏ Fundamentals of Power Module Packaging

  ➢ *Structure, Materials and Processing Technologies*

  ➢ *Performance Characteristics*

❏ **Advanced Packaging of WBG Power Modules**

  ➢ *Attributes of the WBG Power Semiconductors*

  ➢ *Integrated SiC Power Modules (IPMs)*

  ➢ *Integrated Cooling of SiC Power Module*

  ➢ **High Temperature Packaging**

  ➢ *Planar-Bond-All : Multi-functional Integration*

❏ Summary

Operation Temperature of Power Electronics

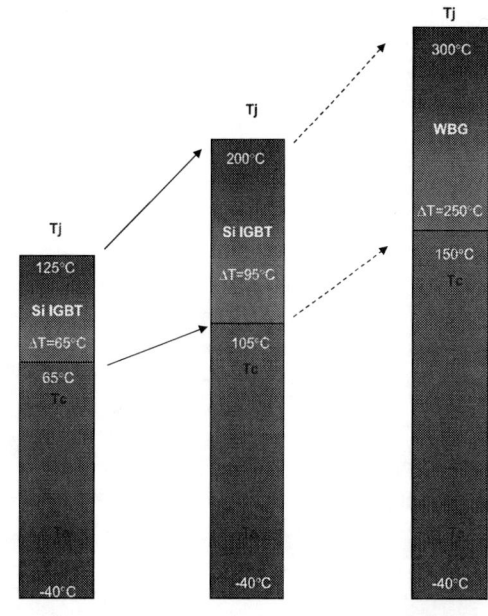

Economy

$$\frac{\$}{kW} \propto \frac{S_{Die\,Area}}{P} = \frac{(1-\eta)\cdot\theta_{ja,sp}}{(T_j - T_a)}$$

HT Applications

petroleum and geothermal

commercial and military aircraft

automotive vehicle

satellite, spacecraft, and space station,

Temperature Dependence of SiC Device: Conduction

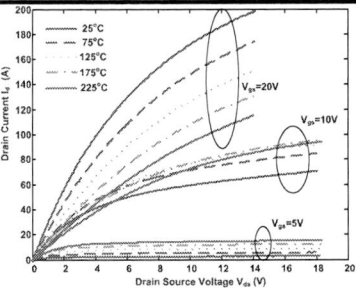

I-V Characteristics of SiC MOSFET

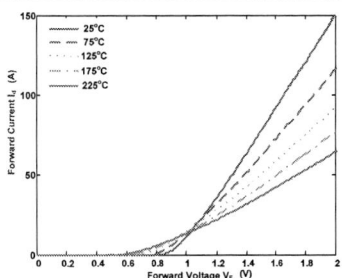

I-V Characteristics of SiC JBS Diode

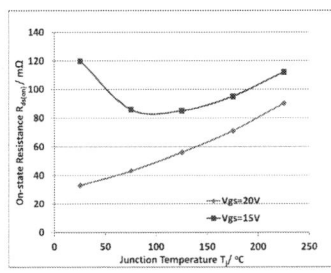

Rds,on vs Temperature of SiC MOSFET

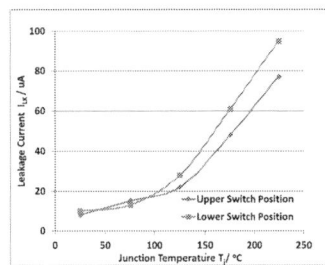

I_{dsr} vs Temperature of SiC MOSFET

Temperature Dependence of SiC Device: Switching

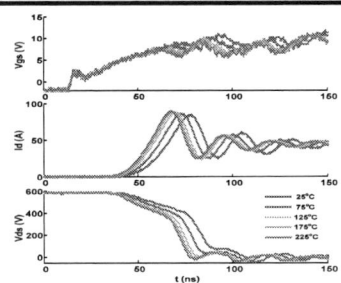

Turn-on vs Temperature of SiC MOSFET

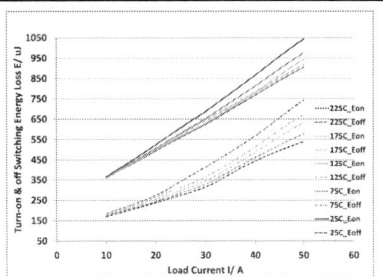

Switching Losses Energy vs Current of SiC MOSFET

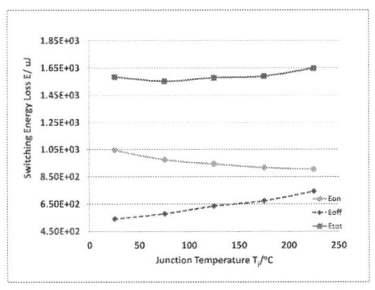

Switching Loss Energy vs Temperature of SiC MOSFET

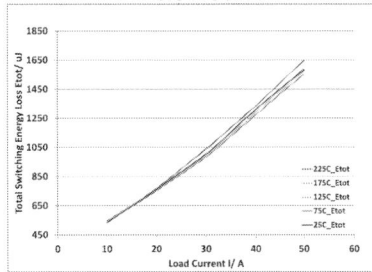

Total Switching Loss Energy vs Current of SiC MOSFET

Reliability of Package

$$N_f = \alpha \cdot \left(\frac{1}{Tj - Ta}\right)^{\beta} \cdot \exp(E_a / kT_m)$$

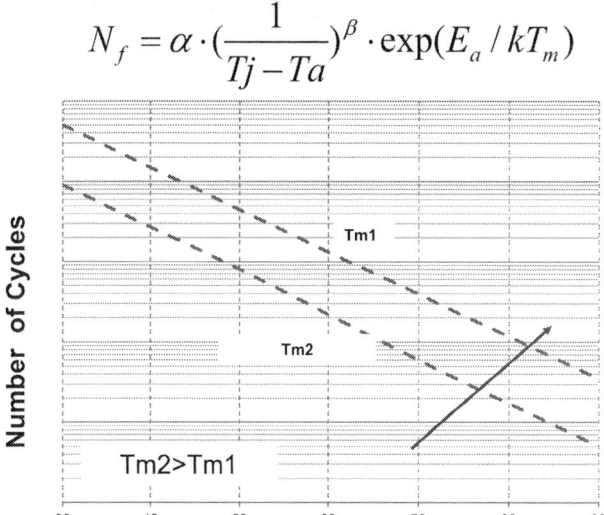

High Temperature Packaging: Material Properties

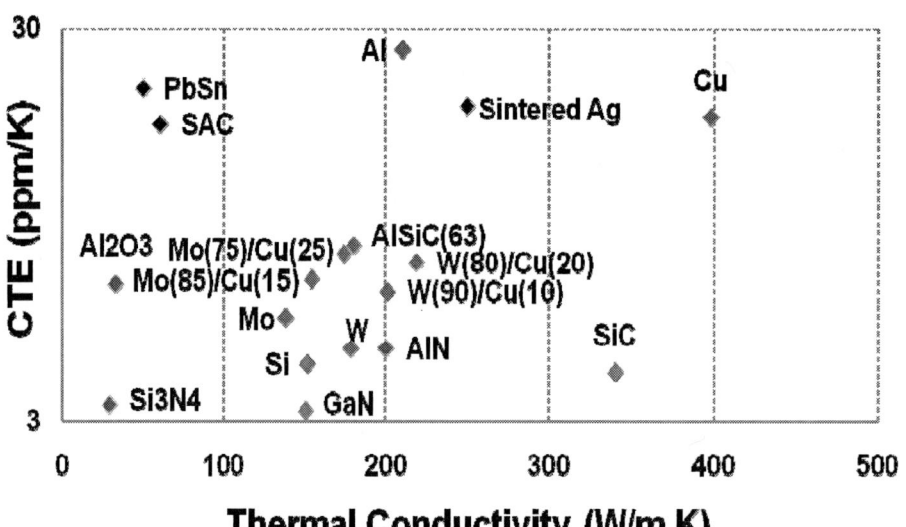

DBC Reliability

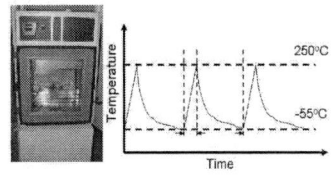

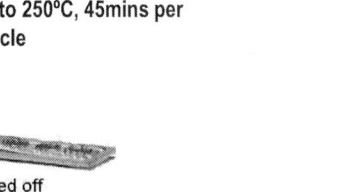

Cycle profile: -55ºC to 250ºC, 45mins per cycle

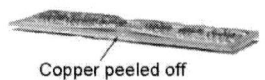

Copper peeled off

DBC failed after 20 cycles

Puqi Nig, Ph D Dissertation, Virginia Tech

Die Attach from Soldering to Ag Sintering

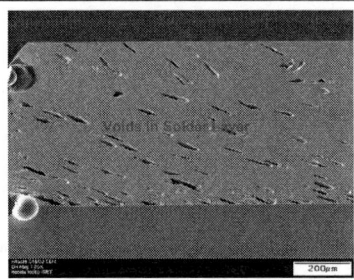

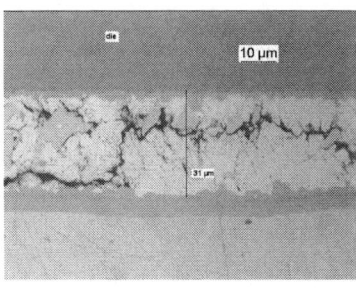

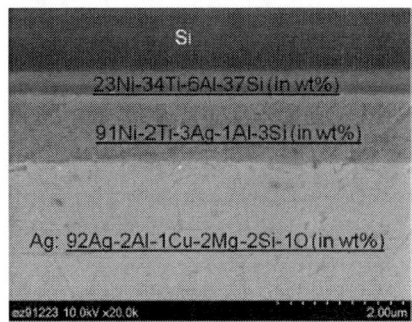

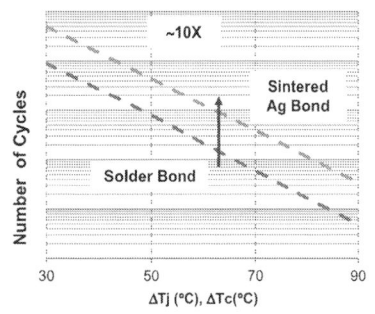

Packaging Process: Ag Sintering

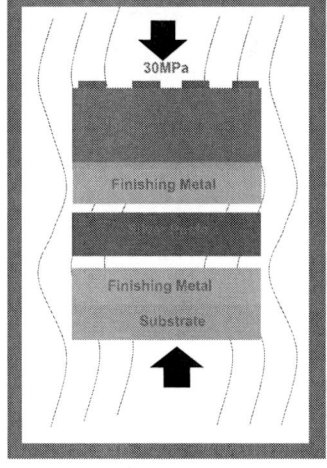

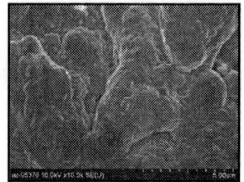
Ag on Cu

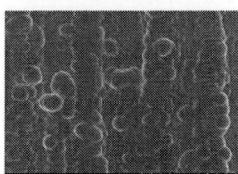
Pd on Al

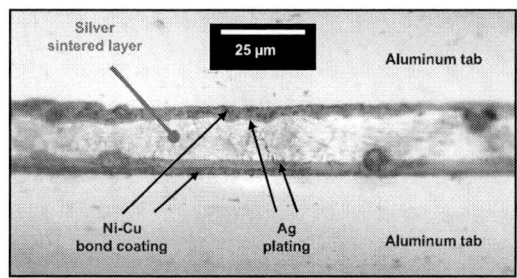

Transient Liquid Phase Bond

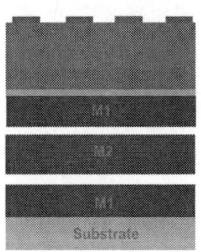

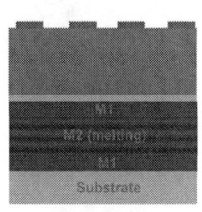
M2 Melting and Diffusion into M1

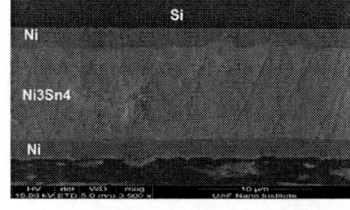
Sang Won Yoon, etal, APEC 2012, P. 478.

Intermetallic M1xM2y Formation (Solidification)

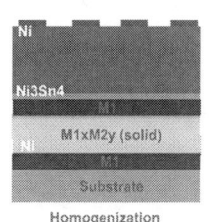
Homogenization

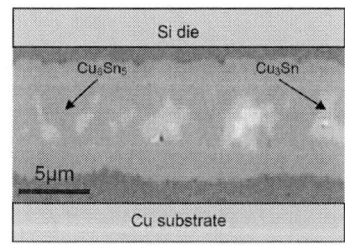
Karsten Guth, etal, PCIM 2010

978-1-4799-5494-0/14 $31.00 © 2014 IEEE

High Temperature *Embedded_Power* SiC Module

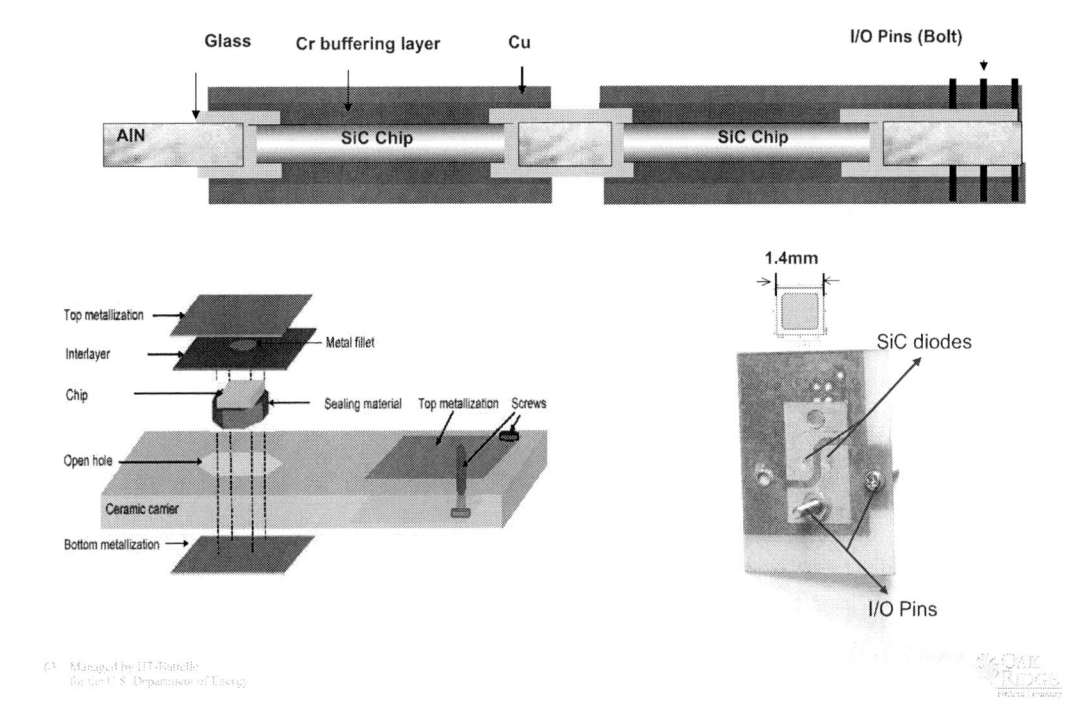

Complex Metallization and Thermal Performance

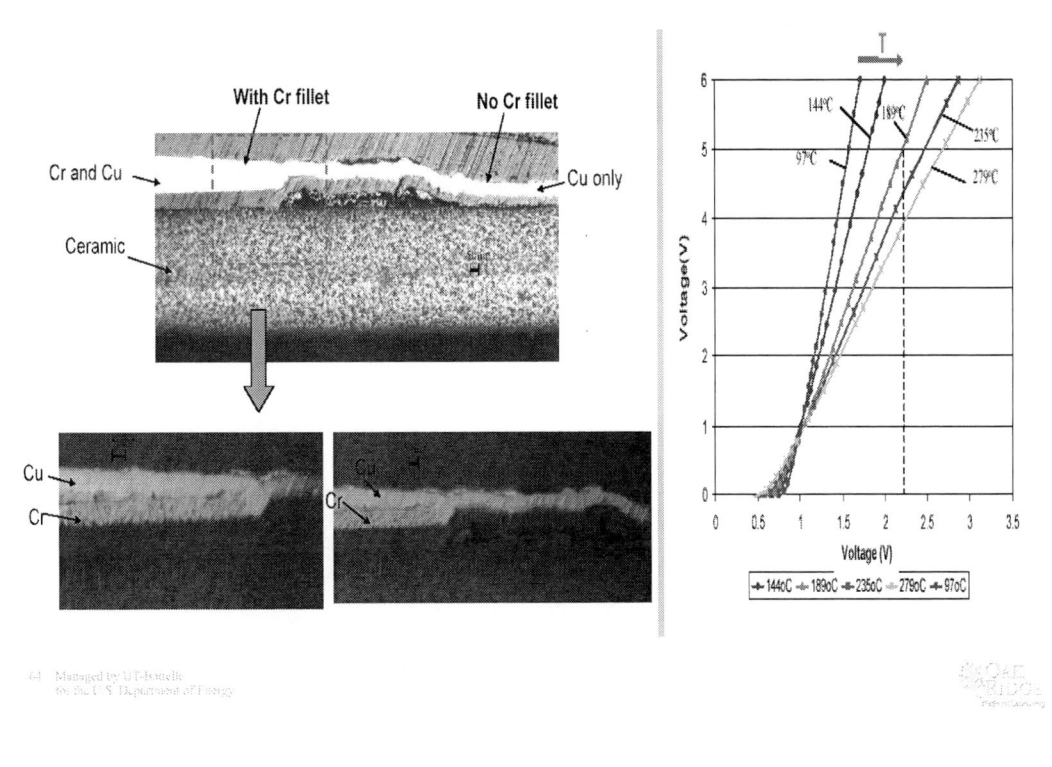

Outline

❑ Overview of Power Electronics Packaging

 ➤ Module Based Power Electronics System

 ➤ Evaluation of Power Electronics Packaging

❑ Fundamentals of Power Module Packaging

 ➤ Structure, Materials and Processing Technologies

 ➤ Performance Characteristics

❑ **Advanced Packaging of WBG Power Modules**

 ➤ Attributes of the WBG Power Semiconductors

 ➤ Integrated SiC Power Modules (IPMs)

 ➤ Integrated Cooling of SiC Power Module

 ➤ High Temperature Packaging

 ➤ **Planar-Bond-All : Multi-functional Integration**

❑ Summary

ORNL Planar-Bond-All Power Module Packaging

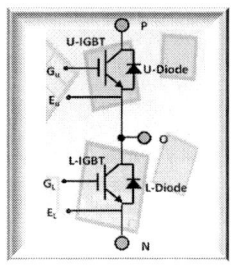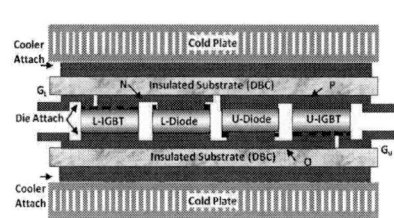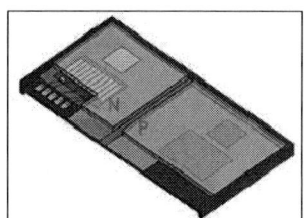

❖ *3-D, Planar Power Interconnection*

❖ *Integrated, Double Sided Cooling*

❖ *Symmetrically Mechanical Structure*

❖ *Simplified Manufacture*

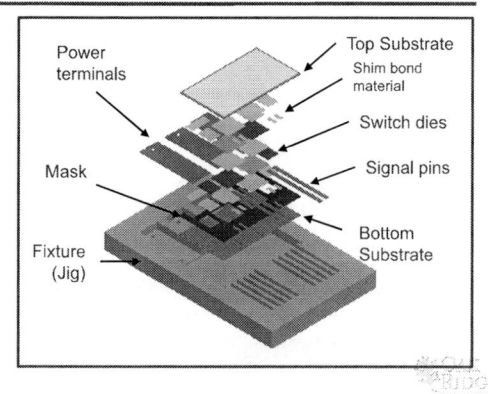

978-1-4799-5494-0/14 $31.00 © 2014 IEEE

Solderable Front Metallization

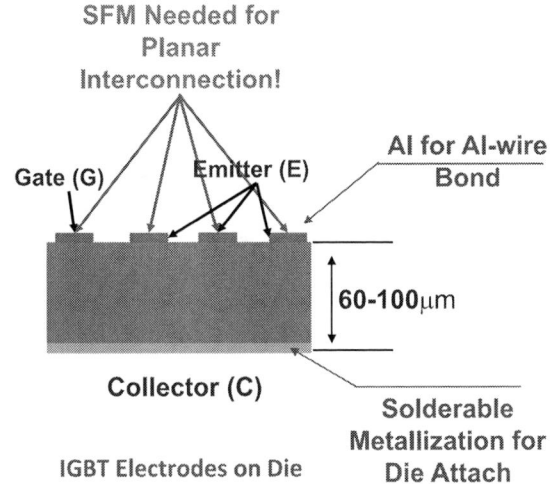

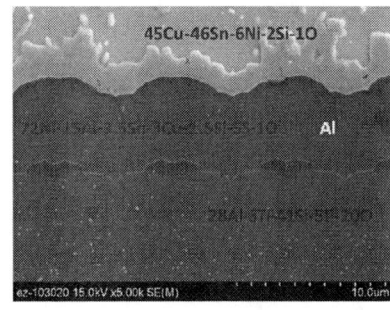

IGBT Electrodes on Die

Microstructure View of a SFM IGBT Package

Prototype: Planar-Bond-All Si Power Module

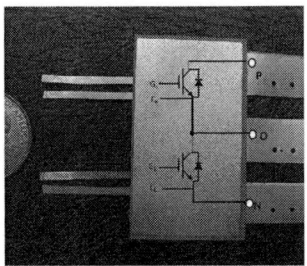

Planar Bond Power Module Stage

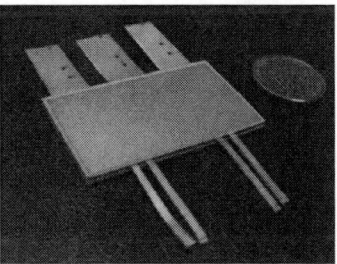

Planar Bond Power Module Stage

Single Sided Cooled Power Module

Double Sided Cooled Power Module

978-1-4799-5494-0/14 $31.00 © 2014 IEEE   393

Electrical Characteristics Improvements

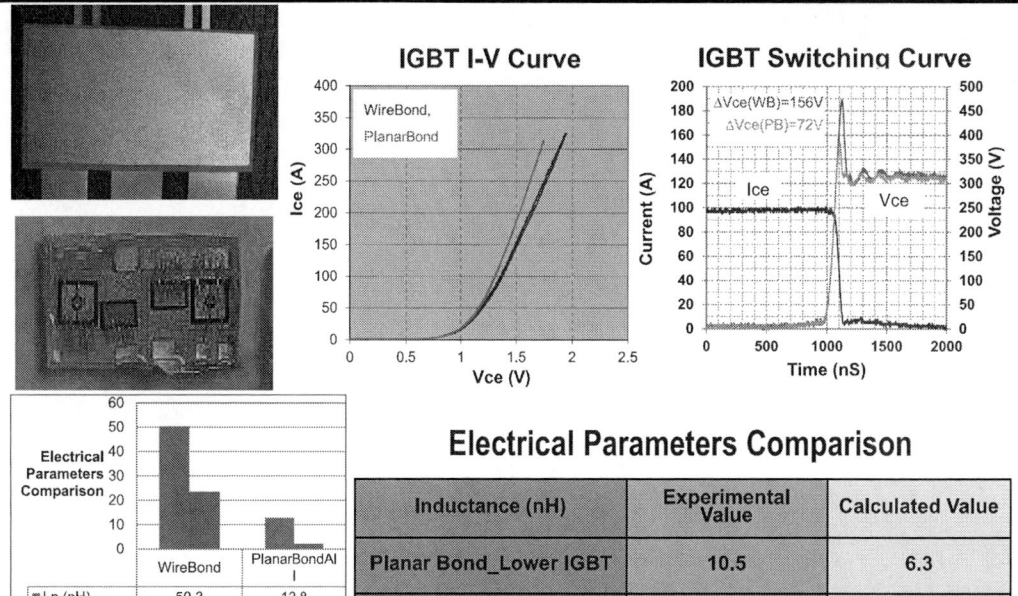

IGBT I-V Curve

WireBond, PlanarBond

IGBT Switching Curve

ΔVce(WB)=156V
ΔVce(PB)=72V
Ice
Vce

Electrical Parameters Comparison

Electrical Parameters Comparison	WireBond	PlanarBondAll
Lp (nH)	50.3	12.8
Rp (0.1xmOhmic)	23.5	2.2

Inductance (nH)	Experimental Value	Calculated Value
Planar Bond_Lower IGBT	10.5	6.3
Wire Bond-Lower IGBT	31.9	23.5

Thermal Parameters: Experiment and Simulation

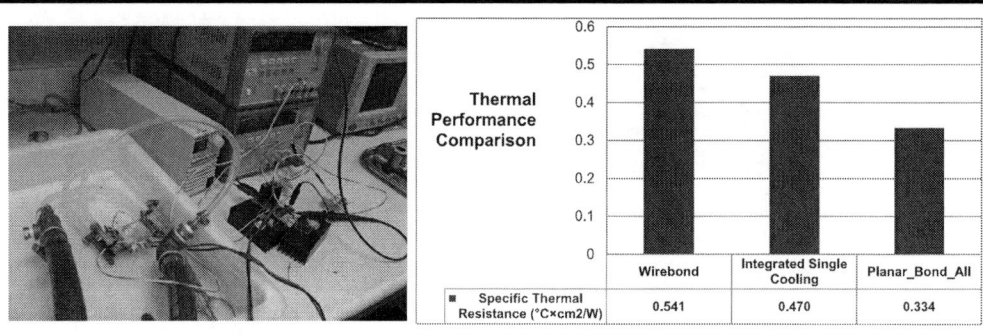

Thermal Performance Comparison	Wirebond	Integrated Single Cooling	Planar_Bond_All
Specific Thermal Resistance (°C×cm2/W)	0.541	0.470	0.334

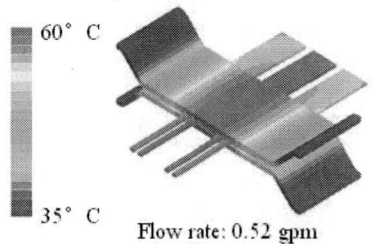

60° C
35° C
Flow rate: 0.52 gpm
Pressure drop: 22 psi

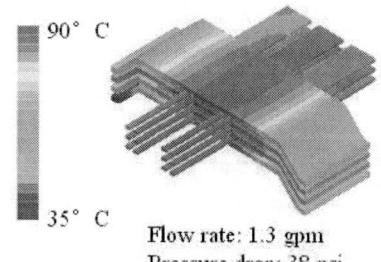

90° C
35° C
Flow rate: 1.3 gpm
Pressure drop: 38 psi

ORNL Planar-Bond-All SiC Power Module Packaging

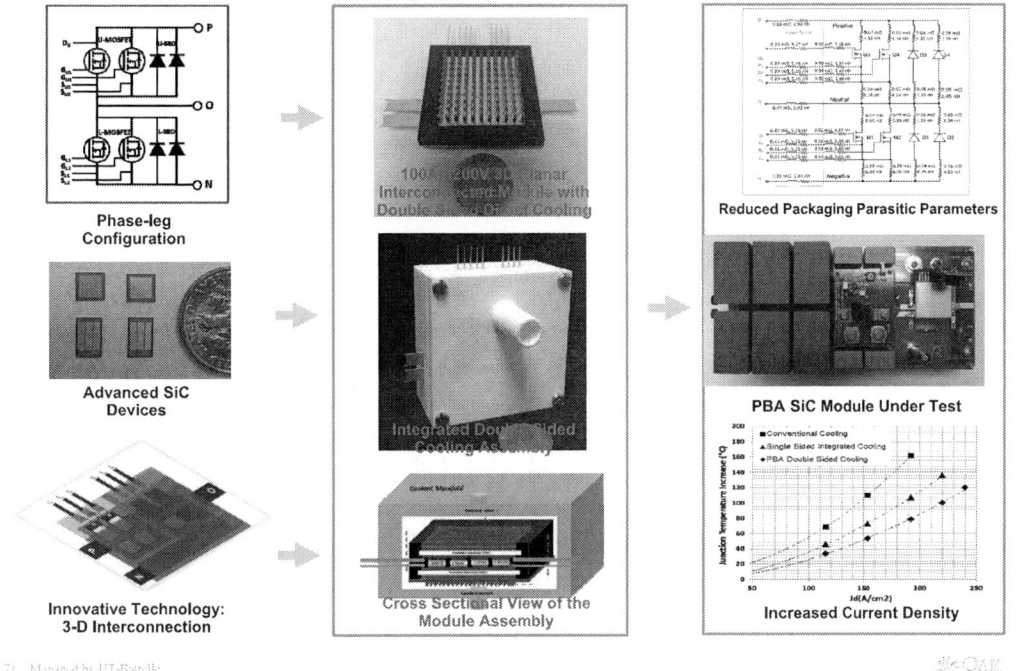

Summary

- ➢ Power electronics packaging is critical to exploit the superior attributes of the wide bandgap (WBG) power semiconductor devices;

- ➢ Highly comprehensive system requirements need advancements on all aspects of packaging technology: electrical, thermal, thermo-mechanical performance;

- ➢ Integration of thermal management, high temperature, high reliability for cost-effective WBG power modules is an promising approach;

- ➢ Advancing materials, innovative cooling mechanism, modern fabrication processing are a few among the key technologies.

ACKNOWLEDGMENT

This seminar contains citations of the research achievements by author's colleagues, and knowledge from literatures, which are highly appreciated.

The WBG Research sponsored by the Advanced Power Electronics and Electric Motors Program, DOE Office of Vehicle Technologies, under contract DE-AC05-00OR22725 with UT-Battelle, LLC.

Thanks and Questions

Zhenxian Liang
TEL: (865) 946-1467
FAX: (865) 946-1262
EMAIL: liangz@ornl.gov

AUTHOR INDEX

A

Abell, William

125 W Multiphase GaN/Si Hybrid Point of Load Converter for Improved High Load Efficiency

A 12 to 1 V Five Phase Interleaving GaN POL Converter for High Current Low Voltage Applications

Agarwal, Anant

Optically-Switched High-Voltage Bipolar SiC Device

Alexandrov, Petre

6.5 kV Normally-off JFETs Technology Status

Allen, Scott

Reliability and Stability of SiC power MOSFETs and Next-Generation SiC MOSFETs

Alsolami, Mohammed

A Full-Bridge Current-Source Isolated DC/DC Converter with Reduced Number of Switches and Voltage Stresses for PV Applications

Andarawis, Emad

Silicon Carbide Transient Voltage Suppressor for Next Generation Lightning Protection

Anderson, Travis

Degradation Mechanisms of AlGaN/GaN HEMTs on Sapphire, Si, and SiC Substrates under Proton Irradiation

Process Optimization of Multicycle Rapid Thermal Annealing of Mgimplanted GaN

Ayers, Curt

A 10-kW SiC Inverter with A Novel Printed Metal Power Module With Integrated Cooling Using Additive Manufacturing

B

Bai, Hua

High-Frequency Wireless Charging System Study Based on Normally-off GaN HEMTs

Bhalla, Anup

6.5 kV Normally-off JFETs Technology Status

Blalock, Benjamin

Understanding the Limitations and Impact Factors of Wide Band-gap Devices' High Switching-Speed Capability in Voltage Source Converter

Wide Band Gap Power Devices Based High Efficiency Power Converters for Data Center Application

Bolotnikov, Alexander

Silicon Carbide Transient Voltage Suppressor for Next Generation Lightning Protection

Boroyevich, Dushan

10 kV, 120 A SiC MOSFET Modules for a Power Electronics Building Block (PEBB)

Britton, Charles

High-Temperature SiC Power Module with Integrated SiC Gate Drivers for Future High-Density Power Electronics Applications

An Integrated Gate Driver in 4H-SiC for Power Converter Applications

Brockman, Jared

Reflected Wave Phenomenon in Motor Drive Systems Using Wide Bandgap Devices

Burgos, Rolando

10 kV, 120 A SiC MOSFET Modules for a Power Electronics Building Block (PEBB)

Burk, Al

Reliability and Stability of SiC power MOSFETs and Next-Generation SiC MOSFETs

C

Campbell, Steven

A 10-kW SiC Inverter with A Novel Printed Metal Power Module With Integrated Cooling Using Additive Manufacturing

Casady, Jeff

Reliability and Stability of SiC power MOSFETs and Next-Generation SiC MOSFETs

Cheng, Lin

Optically-Switched High-Voltage Bipolar SiC Device

Reliability and Stability of SiC power MOSFETs and Next-Generation SiC MOSFETs

Chinthavali, Madhu

A 10-kW SiC Inverter with A Novel Printed Metal Power Module With Integrated Cooling Using Additive Manufacturing

Cole, Zach

High-Temperature SiC Power Module with Integrated SiC Gate Drivers for Future High-Density Power Electronics Applications

Costinett, Daniel

Understanding the Limitations and Impact Factors of Wide Band-gap Devices' High Switching-Speed Capability in Voltage Source Converter

Wide Band Gap Power Devices Based High Efficiency Power Converters for Data Center Application

Cui, Yutian

Wide Band Gap Power Devices Based High Efficiency Power Converters for Data Center Application

Cvetkovic, Igor

10 kV, 120 A SiC MOSFET Modules for a Power Electronics Building Block (PEBB)

D

Dean, Robert

125 W Multiphase GaN/Si Hybrid Point of Load Converter for Improved High Load Efficiency

A 12 to 1 V Five Phase Interleaving GaN POL Converter for High Current Low Voltage Applications

Dhar, Sarit

The Influence of SiC/SiO_2 Interface Morphology on the Electrical Characteristics of SiC MOS Structure

DiMarino, Christina

10 kV, 120 A SiC MOSFET Modules for a Power Electronics Building Block (PEBB)

E

Eddy, Charles

Process Optimization of Multicycle Rapid Thermal Annealing of Mgimplanted GaN

Erdman, William

4500 Volt Si/SiC Hybrid Module Qualification for Modern MegaWatt Scale Wind Energy Inverters

Ericson, Nance

High-Temperature SiC Power Module with Integrated SiC Gate Drivers for Future High-Density Power Electronics Applications

An Integrated Gate Driver in 4H-SiC for Power Converter Applications

Ezell Dianne

An Integrated Gate Driver in 4H-SiC for Power Converter Applications

Emad, Andarawis

Silicon Carbide Transient Voltage Suppressor for Next Generation Lightning Protection

F

Feigelson, Boris

Process Optimization of Multicycle Rapid Thermal Annealing of Mgimplanted GaN

Feldman, Leonard

The Influence of SiC/SiO$_2$ Interface Morphology on the Electrical Characteristics of SiC MOS Structure

Francis, Matt

An Integrated Gate Driver in 4H-SiC for Power Converter Applications

High-Temperature SiC Power Module with Integrated SiC Gate Drivers for Future High-Density Power Electronics Applications

Frank Shane

High-Temperature SiC Power Module with Integrated SiC Gate Drivers for Future High-Density Power Electronics Applications

An Integrated Gate Driver in 4H-SiC for Power Converter Applications

Fu, Lixing

A Full-Bridge Current-Source Isolated DC/DC Converter with Reduced Number of Switches and Voltage Stresses for PV Applications

Discussions on the Semiconductor-based Galvanic Isolation

Reflected Wave Phenomenon in Motor Drive Systems Using Wide Bandgap Devices

The Development of a High-Voltage Power Device Evaluation Platform

Fu, Yongsheng

High-Frequency Wireless Charging System Study Based on Normally-off GaN HEMTs

Fursin, Leonid

6.5 kV Normally-off JFETs Technology Status

Furuhashi, Masayuki

Enhanced Oxidation of SiC Substrates using La_2O_3 Capped Annealing and a Proposal for Uniform LaSiON Gate Dielectric Formation

Passivation of SiO_2/SiC Interface with La_2O_3 Capped Oxidation

G

Gajewski, Don

Reliability and Stability of SiC power MOSFETs and Next-Generation SiC MOSFETs

Glover, Michael

An Integrated Gate Driver in 4H-SiC for Power Converter Applications

High-Temperature SiC Power Module with Integrated SiC Gate Drivers for Future High-Density Power Electronics Applications

Greenlee, Jordan

Degradation Mechanisms of AlGaN/GaN HEMTs on Sapphire, Si, and SiC Substrates under Proton Irradiation

Process Optimization of Multicycle Rapid Thermal Annealing of Mgimplanted GaN

Grider, David

Reliability and Stability of SiC power MOSFETs and Next-Generation SiC MOSFETs

4500 Volt Si/SiC Hybrid Module Qualification for Modern MegaWatt Scale Wind Energy Inverters

Guerrero, Jose

Advances in Reliability and Operation Space of High-voltage GaN Power Devices Grown on Si Substrates

Guo Ben

Wide Band Gap Power Devices Based High Efficiency Power Converters for Data Center Application 30

Guo, Feng

A Full-Bridge Current-Source Isolated DC/DC Converter with Reduced Number of Switches and Voltage Stresses for PV Applications

Guo, Suxuan

An Isolated Bi-directional Soft-Switched High-Frequency-AC Link DC-AC Converter Using SiC MOSFETs

H

Han, Di

Understanding the Influence of Dead-time on GaN Based Synchronous Boost Converter

Harris, Daniel

125 W Multiphase GaN/Si Hybrid Point of Load Converter for Improved High Load Efficiency

A 12 to 1 V Five Phase Interleaving GaN POL Converter for High Current Low Voltage Applications

Hite, Jennifer

Degradation Mechanisms of AlGaN/GaN HEMTs on Sapphire, Si, and SiC Substrates under Proton Irradiation

Process Optimization of Multicycle Rapid Thermal Annealing of Mgimplanted GaN

Hobart, Karl

Degradation Mechanisms of AlGaN/GaN HEMTs on Sapphire, Si, and SiC Substrates under Proton Irradiation

Process Optimization of Multicycle Rapid Thermal Annealing of Mgimplanted GaN

Honea, Jim

Investigation of Drive Circuits for GaN HEMTs in Leaded Packages

Hostetler, John

6.5 kV Normally-off JFETs Technology Status

Hu, Boxue

Reflected Wave Phenomenon in Motor Drive Systems Using Wide Bandgap Devices

Huang, Alex

An Isolated Bi-directional Soft-Switched High-Frequency-AC Link DC-AC Converter Using SiC MOSFETs

Design and Fabrication of High Current AlGaN/GaN HFET for Gen III Solid State Transformers

Huang, Qingyun

An Isolated Bi-directional Soft-Switched High-Frequency-AC Link DC-AC Converter Using SiC MOSFETs

Hull, Brett

Reliability and Stability of SiC power MOSFETs and Next-Generation SiC MOSFETs

I

Iwai, Hiroshi

Dependence of Ti/C Ratio on Ohmic Contact with TiC Electrode in AlGaN/GaN Structure

Enhanced Oxidation of SiC Substrates using La_2O_3 Capped Annealing and a Proposal for Uniform LaSiON Gate Dielectric Formation

Passivation of SiO_2/SiC Interface with La_2O_3 Capped Oxidation

J

Janke, Devon

An Integrated Gate Driver in 4H-SiC for Power Converter Applications

Jenkins, Luke

125 W Multiphase GaN/Si Hybrid Point of Load Converter for Improved High Load Efficiency

A 12 to 1 V Five Phase Interleaving GaN POL Converter for High Current Low Voltage Applications

Ji, In-Hwan

Design and Fabrication of High Current AlGaN/GaN HFET for Gen III Solid State Transformers

Jiao, Chunkun

The Influence of SiC/SiO$_2$ Interface Morphology on the Electrical Characteristics of SiC MOS Structure

Jones, Edward

Application-Based Review of GaN HFETs

K

Kakushima, Kuniyuki

Dependence of Ti/C Ratio on Ohmic Contact with TiC Electrode in AlGaN/GaN Structure

Enhanced Oxidation of SiC Substrates using La$_2$O$_3$ Capped Annealing and a Proposal for Uniform LaSiON Gate Dielectric Formation

Passivation of SiO$_2$/SiC Interface with La$_2$O$_3$ Capped Oxidation

Kashyap, Avinash S.

Silicon Carbide Transient Voltage Suppressor for Next Generation Lightning Protection

Kataoka, Yoshinori

Dependence of Ti/C Ratio on Ohmic Contact with TiC Electrode in AlGaN/GaN Structure

Kataota, K.

Passivation of SiO$_2$/SiC Interface with La$_2$O$_3$ Capped Oxidation

Kawanago, Takamasa

Enhanced Oxidation of SiC Substrates using La$_2$O$_3$ Capped Annealing and a Proposal for Uniform LaSiON Gate Dielectric Formation

Passivation of SiO$_2$/SiC Interface with La$_2$O$_3$ Capped Oxidation

Ke, Haotao

Design and Fabrication of High Current AlGaN/GaN HFET for Gen III Solid State Transformers

Koehler, Andrew

Degradation Mechanisms of AlGaN/GaN HEMTs on Sapphire, Si, and SiC Substrates under Proton Irradiation

Kub, Francis

Process Optimization of Multicycle Rapid Thermal Annealing of Mgimplanted GaN

Kub, Fritz

High-Temperature SiC Power Module with Integrated SiC Gate Drivers for Future High-Density Power Electronics Applications

L

Lamichhane, Ranjan

An Integrated Gate Driver in 4H-SiC for Power Converter Applications

High-Temperature SiC Power Module with Integrated SiC Gate Drivers for Future High-Density Power Electronics Applications

Lautner, Jennifer

Impact of Current Measurement of Switching Characterization of GaN Transistors

Lee, Bongmook

Design and Fabrication of High Current AlGaN/GaN HFET for Gen III Solid State Transformers

Effect of Post Deposition Annealing for High Mobility 4H-SiC MOSFET Utilizing Lanthanum Silicate and Atomic Layer Deposited SiO_2

Lei, Yiming

Enhanced Oxidation of SiC Substrates using La_2O_3 Capped Annealing and a Proposal for Uniform LaSiON Gate Dielectric Formation

Passivation of SiO_2/SiC Interface with La_2O_3 Capped Oxidation

Leng, Mingzhi

Discussions on the Semiconductor-based Galvanic Isolation

Li, He

A Full-Bridge Current-Source Isolated DC/DC Converter with Reduced Number of Switches and Voltage Stresses for PV Applications

Li, Hui

Investigation of Drive Circuits for GaN HEMTs in Leaded Packages

Li, Xueqing

6.5 kV Normally-off JFETs Technology Status

Liang, Zhenxian

Development of Packaging Technologies for Advanced SiC Power Modules

Liu, Gang

The Influence of SiC/SiO_2 Interface Morphology on the Electrical Characteristics of SiC MOS Structure

Liu, Li

The Influence of SiC/SiO_2 Interface Morphology on the Electrical Characteristics of SiC MOS Structure

Long, Yu

Wide Band Gap Power Devices Based High Efficiency Power Converters for Data Center Application

Lostetter, Alex

High-Temperature SiC Power Module with Integrated SiC Gate Drivers for Future High-Density Power Electronics Applications

Lu, Xintong

The Development of a High-Voltage Power Device Evaluation Platform

M

Majab, Alireza

Optically-Switched High-Voltage Bipolar SiC Device

Mantooth, Alan

An Integrated Gate Driver in 4H-SiC for Power Converter Applications

High-Temperature SiC Power Module with Integrated SiC Gate Drivers for Future High-Density Power Electronics Applications

Marlino, Laura

High-Temperature SiC Power Module with Integrated SiC Gate Drivers for Future High-Density Power Electronics Applications

An Integrated Gate Driver in 4H-SiC for Power Converter Applications

Martin, Daniel

High-Temperature SiC Power Module with Integrated SiC Gate Drivers for Future High-Density Power Electronics Applications

Mastro, Micheal

Degradation Mechanisms of AlGaN/GaN HEMTs on Sapphire, Si, and SiC Substrates under Proton Irradiation

Mazumber, Supid

Optically-Switched High-Voltage Bipolar SiC Device

McKay, J.

Advances in Reliability and Operation Space of High-voltage GaN Power Devices Grown on Si Substrates

McMahon, James

Silicon Carbide Transient Voltage Suppressor for Next Generation Lightning Protection

McNutt, Ty

High-Temperature SiC Power Module with Integrated SiC Gate Drivers for Future High-Density Power Electronics Applications

An Integrated Gate Driver in 4H-SiC for Power Converter Applications

Misra, Veena

Design and Fabrication of High Current AlGaN/GaN HFET for Gen III Solid State Transformers

Effect of Post Deposition Annealing for High Mobility 4H-SiC MOSFET Utilizing Lanthanum Silicate and Atomic Layer Deposited SiO_2

Miura, Naruhisa

Enhanced Oxidation of SiC Substrates using La_2O_3 Capped Annealing and a Proposal for Uniform LaSiON Gate Dielectric Formation

Passivation of SiO_2/SiC Interface with La_2O_3 Capped Oxidation

Munekiyo, Shu

Enhanced Oxidation of SiC Substrates using La_2O_3 Capped Annealing and a Proposal for Uniform LaSiON Gate Dielectric Formation

Passivation of SiO_2/SiC Interface with La_2O_3 Capped Oxidation

N

Natori, Kenji

Dependence of Ti/C Ratio on Ohmic Contact with TiC Electrode in AlGaN/GaN Structure

Passivation of SiO_2/SiC Interface with La_2O_3 Capped Oxidation

Nishiyama, A.

Passivation of SiO_2/SiC Interface with La_2O_3 Capped Oxidation

O

O'Loughlin, Michael

Reliability and Stability of SiC power MOSFETs and Next-Generation SiC MOSFETs

Okamoto, Mari

Dependence of Ti/C Ratio on Ohmic Contact with TiC Electrode in AlGaN/GaN Structure

Ozpineci, Burak

A 10-kW SiC Inverter with A Novel Printed Metal Power Module With Integrated Cooling Using Additive Manufacturing

Application-Based Review of GaN HFETs

P

Pala, Vipindas

Reliability and Stability of SiC power MOSFETs and Next-Generation SiC MOSFETs

Palmour, John

Reliability and Stability of SiC power MOSFETs and Next-Generation SiC MOSFETs

Passmore, Brandon

High-Temperature SiC Power Module with Integrated SiC Gate Drivers for Future High-Density Power Electronics Applications

An Integrated Gate Driver in 4H-SiC for Power Converter Applications

Piepenbreier, Bernhard

Impact of Current Measurement of Switching Characterization of GaN Transistors

Porter, Matthew

Degradation Mechanisms of AlGaN/GaN HEMTs on Sapphire, Si, and SiC Substrates under Proton Irradiation

R

Radhakrishnan, Rahul

Temperature Dependence Design of Silicon Carbide Schottky Diodes

Rhea, Benjamin

125 W Multiphase GaN/Si Hybrid Point of Load Converter for Improved High Load Efficiency

A 12 to 1 V Five Phase Interleaving GaN POL Converter for High Current Low Voltage Applications

Richmond, Jim

Reliability and Stability of SiC power MOSFETs and Next-Generation SiC MOSFETs

Ryu, Sei-Hyung

An Integrated Gate Driver in 4H-SiC for Power Converter Applications

Reliability and Stability of SiC power MOSFETs and Next-Generation SiC MOSFETs

S

Saito, Wataru

Dependence of Ti/C Ratio on Ohmic Contact with TiC Electrode in AlGaN/GaN Structure

Sandvik, Peter

Silicon Carbide Transient Voltage Suppressor for Next Generation Lightning Protection

Sarlioglu, Bulent

Understanding the Influence of Dead-time on GaN Based Synchronous Boost Converter

Schuderer, Jürgen

Packaging SiC Power Semiconductors - Challenges, Technologies and Strategies

Scott, Mark

Reflected Wave Phenomenon in Motor Drive Systems Using Wide Bandgap Devices

Scozzie, Charles

Optically-Switched High-Voltage Bipolar SiC Device

Shen, Zhiyu

10 kV, 120 A SiC MOSFET Modules for a Power Electronics Building Block (PEBB)

Shepherd, Paul

An Integrated Gate Driver in 4H-SiC for Power Converter Applications

High-Temperature SiC Power Module with Integrated SiC Gate Drivers for Future High-Density Power Electronics Applications

Shi, Lei

High-Frequency Wireless Charging System Study Based on Normally-off GaN HEMTs

Shi, Yuxiang

Investigation of Drive Circuits for GaN HEMTs in Leaded Packages

Smith, Kurt

Advances in Reliability and Operation Space of High-voltage GaN Power Devices Grown on Si Substrates

Specht, Petra

Degradation Mechanisms of AlGaN/GaN HEMTs on Sapphire, Si, and SiC Substrates under Proton Irradiation

Sugii, Nobuyuki

Dependence of Ti/C Ratio on Ohmic Contact with TiC Electrode in AlGaN/GaN Structure

Passivation of SiO_2/SiC Interface with La_2O_3 Capped Oxidation

T

Tadjer, Marko

Degradation Mechanisms of AlGaN/GaN HEMTs on Sapphire, Si, and SiC Substrates under Proton Irradiation

Process Optimization of Multicycle Rapid Thermal Annealing of Mgimplanted GaN

Tolbert, Leon

Understanding the Limitations and Impact Factors of Wide Band-gap Devices' High Switching-Speed Capability in Voltage Source Converter

Development of Packaging Technologies for Advanced SiC Power Modules

Wide Band Gap Power Devices Based High Efficiency Power Converters for Data Center Application

Traub, Felix

Packaging SiC Power Semiconductors - Challenges, Technologies and Strategies

Tsutsui, Kazuo

Dependence of Ti/C Ratio on Ohmic Contact with TiC Electrode in AlGaN/GaN Structure

Enhanced Oxidation of SiC Substrates using La_2O_3 Capped Annealing and a Proposal for Uniform LaSiON Gate Dielectric Formation

Passivation of SiO_2/SiC Interface with La_2O_3 Capped Oxidation

V

VanBrunt, Edward

4500 Volt Si/SiC Hybrid Module Qualification for Modern MegaWatt Scale Wind Energy Inverters

Reliability and Stability of SiC power MOSFETs and Next-Generation SiC MOSFETs

Vemulapati, Umamaheswara

Packaging SiC Power Semiconductors - Challenges, Technologies and Strategies

W

Wakabayashi, Hitoshi

Dependence of Ti/C Ratio on Ohmic Contact with TiC Electrode in AlGaN/GaN Structure

Enhanced Oxidation of SiC Substrates using La_2O_3 Capped Annealing and a Proposal for Uniform LaSiON Gate Dielectric Formation

Passivation of SiO_2/SiC Interface with La_2O_3 Capped Oxidation

Wang, Fred

Application-Based Review of GaN HFETs

Development of Packaging Technologies for Advanced SiC Power Modules

Understanding the Limitations and Impact Factors of Wide Band-gap Devices' High Switching-Speed Capability in Voltage Source Converter

Wide Band Gap Power Devices Based High Efficiency Power Converters for Data Center Application

Wang, Jin

A Full-Bridge Current-Source Isolated DC/DC Converter with Reduced Number of Switches and Voltage Stresses for PV Applications

Discussions on the Semiconductor-based Galvanic Isolation

Reflected Wave Phenomenon in Motor Drive Systems Using Wide Bandgap Devices

The Development of a High-Voltage Power Device Evaluation Platform

Wang, Mengqi

An Isolated Bi-directional Soft-Switched High-Frequency-AC Link DC-AC Converter Using SiC MOSFETs

Wang, Sizhen

Design and Fabrication of High Current AlGaN/GaN HFET for Gen III Solid State Transformers

Wang, Zhan

Investigation of Drive Circuits for GaN HEMTs in Leaded Packages

Weatherford, Todd

Degradation Mechanisms of AlGaN/GaN HEMTs on Sapphire, Si, and SiC Substrates under Proton Irradiation

Weaver, Bradley

Degradation Mechanisms of AlGaN/GaN HEMTs on Sapphire, Si, and SiC Substrates under Proton Irradiation

Werner, Frank

125 W Multiphase GaN/Si Hybrid Point of Load Converter for Improved High Load Efficiency

A 12 to 1 V Five Phase Interleaving GaN POL Converter for High Current Low Voltage Applications

Whitaker, Bret

High-Temperature SiC Power Module with Integrated SiC Gate Drivers for Future High-Density Power Electronics Applications

An Integrated Gate Driver in 4H-SiC for Power Converter Applications

Wiles, Randy

A 10-kW SiC Inverter with A Novel Printed Metal Power Module With Integrated Cooling Using Additive Manufacturing

Wilson, Christopher

125 W Multiphase GaN/Si Hybrid Point of Load Converter for Improved High Load Efficiency

A 12 to 1 V Five Phase Interleaving GaN POL Converter for High Current Low Voltage Applications

Witt, Tony

Temperature Dependence Design of Silicon Carbide Schottky Diodes

Woodin, Richard

Temperature Dependence Design of Silicon Carbide Schottky Diodes

Wu, Yifeng

Advances in Reliability and Operation Space of High-voltage GaN Power Devices Grown on Si Substrates

X

Xu, Fan

Wide Band Gap Power Devices Based High Efficiency Power Converters for Data Center Application

Xu, Longya

Reflected Wave Phenomenon in Motor Drive Systems Using Wide Bandgap Devices

Xu, Yi

The Influence of SiC/SiO$_2$ Interface Morphology on the Electrical Characteristics of SiC MOS Structure

Y

Yamakawa,S.

Passivation of SiO$_2$/SiC Interface with La$_2$O$_3$ Capped Oxidation

Yang, Xiangyu

Effect of Post Deposition Annealing for High Mobility 4H-SiC MOSFET Utilizing Lanthanum Silicate and Atomic Layer Deposited SiO$_2$

Yu, Wensong

An Isolated Bi-directional Soft-Switched High-Frequency-AC Link DC-AC Converter Using SiC MOSFETs

Z

Zamora, Rachid Darbali

Reflected Wave Phenomenon in Motor Drive Systems Using Wide Bandgap Devices

Zhang, Jie

A Full-Bridge Current-Source Isolated DC/DC Converter with Reduced Number of Switches and Voltage Stresses for PV Applications

Zhang, Jon

Reliability and Stability of SiC power MOSFETs and Next-Generation SiC MOSFETs

Zhang, Weimin

Wide Band Gap Power Devices Based High Efficiency Power Converters for Data Center Application

Zhang, Xuan

A Full-Bridge Current-Source Isolated DC/DC Converter with Reduced Number of Switches and Voltage Stresses for PV Applications

Discussions on the Semiconductor-based Galvanic Isolation

The Development of a High-Voltage Power Device Evaluation Platform

Zhang, Zheyu

Understanding the Limitations and Impact Factors of Wide Band-gap Devices' High Switching-Speed Capability in Voltage Source Converter

CURRAN ASSOCIATES INC.
proceedings
.com

9781479954940